MECHANICAL PROPERTIES OF CERAMICS AND COMPOSITES

MATERIALS ENGINEERING

1. Modern Ceramic Engineering: Properties, Processing, and Use in Design: Second Edition, Revised and Expanded, *David W. Richerson*
2. Introduction to Engineering Materials: Behavior, Properties, and Selection, *G. T. Murray*
3. Rapidly Solidified Alloys: Processes • Structures • Applications, *edited by Howard H. Liebermann*
4. Fiber and Whisker Reinforced Ceramics for Structural Applications, *David Belitskus*
5. Thermal Analysis of Materials, *Robert F. Speyer*
6. Friction and Wear of Ceramics, *edited by Said Jahanmir*
7. Mechanical Properties of Metallic Composites, *edited by Shojiro Ochiai*
8. Chemical Processing of Ceramics, *edited by Burtrand I. Lee and Edward J. A. Pope*
9. Handbook of Advanced Materials Testing, *edited by Nicholas P. Cheremisinoff and Paul N. Cheremisinoff*
10. Ceramic Processing and Sintering, *M. N. Rahaman*
11. Composites Engineering Handbook, *edited by P. K. Mallick*
12. Porosity of Ceramics, *Roy W. Rice*
13. Intermetallic and Ceramic Coatings, *edited by Narendra B. Dahotre and T. S. Sudarshan*
14. Adhesion Promotion Techniques: Technological Applications, *edited by K. L. Mittal and A. Pizzi*
15. Impurities in Engineering Materials: Impact, Reliability, and Control, *edited by Clyde L. Briant*
16. Ferroelectric Devices, *Kenji Uchino*
17. Mechanical Properties of Ceramics and Composites: Grain and Particle Effects, *Roy W. Rice*

Additional Volumes in Preparation

MECHANICAL PROPERTIES OF CERAMICS AND COMPOSITES

Grain and Particle Effects

Roy W. Rice

Consultant
Alexandria, Virginia

MARCEL DEKKER, INC. NEW YORK • BASEL

ISBN: 0-8247-8874-5

This book is printed on acid-free paper.

Headquarters
Marcel Dekker, Inc.
270 Madison Avenue, New York, NY 10016
tel: 212-696-9000; fax: 212-685-4540

Eastern Hemisphere Distribution
Marcel Dekker AG
Hutgasse 4, Postfach 812, CH-4001 Basel, Switzerland
tel: 41-61-261-8482; fax: 41-61-261-8896

World Wide Web
http://www.dekker.com

The publisher offers discounts on this book when ordered in bulk quantities. For more information, write to Special Sales/Professional Marketing at the headquarters address above.

Current printing (last digit)
10 9 8 7 6 5 4 3 2 1

PRINTED IN THE UNITED STATES OF AMERICA

This book is dedicated to my colleagues at the Naval Research Laboratory (NRL) in Washington, D.C. Although I also benefited from the many interactions with other colleagues in my positions at the Boeing Company and at W. R. Grace before and after my 17 years at NRL, respectively, much of my work that forms a basis for this book took place at NRL. Colleagues with whom I interacted extensively on various topics in this book are listed in alphabetical order: Paul Becher, Barry Bender, Bill Coblenz, Steve Freiman, Dave Lewis, Bill McDonough, Karl McKinney, Jack Mecholsky, Bob Pohanka, Dick Spann, Basil Walker, and Carl Wu. I would also like to thank numerous support personnel, including many students on various temporary assignments.

Preface

This book is intended to provide both a summary and a comprehensive view of the dependence of ceramic mechanical properties on grain and particle sizes and other parameters (e.g., shape and spatial orientation and distribution). It complements another book of mine, *Porosity of Ceramics* (also published by Marcel Dekker, Inc.), on effects of porosity. Together, both books reflect my interest in microstructural dependence of properties, especially mechanical properties on which I have spent a major portion of my technical career. The books were written to fill a critical need, given the importance of understanding microstructural dependence of properties as the critical link between fabrication and performance of ceramics and the diversity, frequent incompleteness, and confusion in the literature and lack of any comprehensive treatment. This book is intended for a variety of interests and backgrounds ranging from active research scientists and engineers studying or developing ceramics to those producing, recommending, or using ceramics, especially where mechanical properties are critical (some observations on other related properties are given both directly and indirectly). The intent is not only to summarize pertinent models and review data but also to identify strengths and weaknesses in each (e.g., serious problems with many fracture toughness values and their pertinence to normal strengths of conventional and composite ceramics, and steps to improve understanding).

The comprehensive aspect of this book is provided by the scope of materials, properties, and test conditions covered. Properties extensively addressed include crack propagation, fracture energy and toughness, tensile strength and thermal shock, compressive srength, hardness, erosion, and wear. Materials evaluated range from traditional to high-tech "monolithic" ceramics—traditional ceramic composites such as porcelains and crystallized glasses, and whisker, platelet, and particulate ceramic composites. The latter are emphasized since there is more comprehensive data on them, and they encompass some materials in production, including ZrO_2 toughened materials. Test conditions considered include effects of environment (especially of water), low to high temperatures (mostly in flexure) along with limited tensile and compressive fatigue, ballistic testing, substantial hardness, compression, wear, and erosion. The emphasis is on lower and moderate temperatures where brittle fracture dominates, but high-temperature creep and stress rupture are outlined. The comprehensive character of the book is reflected by both the scope outlined above and the magnitude and depth of the data presented. It is intended to provide an extensive view of the practical aspects of the microstructural dependence (i.e., engineering to achieve realistic balances of properties). The underlying physics and models are also addressed to aid in engineering applications as well as to provide a basic understanding.

Finally, note important aspects of the presentation in this book. A key reason for preparing it was that much valuable, earlier information is widely neglected and many results and evaluations are presented as though they are of broad use and applicability when in fact they are often, at best, of very narrow or questionable utility or validity. Also, there are cases in which data is inconsistent and the applicability of some widely investigated phenomena are very uncertain. A key example is application of much fracture toughness—and especially wake, R-curve effects—to most strength behavior. For both this reason and for completeness, chapters are presented in a review, rather than a textbook, fashion. The book is organized by property and microstructural parameters with substantial cross-correlation of behavior, but each chapter is meant to stand alone. This is accomplished by cross-correlations within a given chapter, as well as between chapters for completeness, self consistency, and emphasis. Also, overall behavior trends and needs are summarized in the final section of each chapter and in the final chapter, along with processing/engineering aspects.

ACKNOWLEDGMENT

Several people have contributed in a variety of ways to the development of this book. Dr. Melvin Leitheiser of 3M Company indirectly planted the idea of this and a companion book in my mind. Dr. Dave Lewis and especially my daughter and son-in-law, Colleen and Philip Montgomery, aided in establishing my computer capability. Several have assisted by reading drafts of chapters (numbers shown below in parentheses), providing comments and sometimes additional references, as follows: Dr. J.-L. Chermant, LERMAT, ISMRA (1); Dr. S. Jahanmir, NIST (5, 7, 10, 11); Dr. J. Lankford, Southwest Research Institute (5, 7, 10, 11); Dr. Dave Lewis, U.S. Naval Research Laboratory (1-12); Dr. J. Mecholsky,

University of Florida (8); Dr. Robert Ruh, Air Force Materials Laboratory (1, 2); Mr. J. Swab, U.S. Army (2); and Dr. J. Varner, Alfred University (4). I thank Dr. Wu, and especially Dr. Lewis, for help with some figures, as well as Mr. Dan Wilhide for extensive assistance. Finally, I thank Drs. Steve Freidman and Sheldon Wiedreharn, and Mr. George Quinn of NIST for making me a visiting scientist there, hence giving me library access.

Roy W. Rice

Contents

Symbols and Abbreviations

In covering such a diverse array of properties it is not possible to avoid some overlap of mathematical symbol designation. However, widely used symbols are defined as follows, and they, and especially any alternative uses, are given in the text in conjunction with their use.

SYMBOLS

b	exponential parameter for porosity dependence (e^{-bP})
c	flaw size (typically the radius or depth)
D	particle size (diameter)
E	Young's modulus
F	force, e.g., on an indentor or abrasive particle
G	grain size (diameter)
H	hardness (subscripts for type, e.g. V for Vickers)
k	bulk modulus
K	fracture toughness (K_{IC} critical stress intensity)
m	Weibull modulus
P	volume fraction porosity
α	thermal expansion coefficient
γ	fracture energy
ΔT_c	quench temperature difference for serious strength loss
ε	strain
λ	mean free path

ν	Poisson's ratio
σ	strength (subscripts T and C tensile and compressive)
ϕ	volume fraction second phase

ABBREVIATIONS

CNB	chevron notch beam (toughness test)
CVD	chemical vapor deposition
CT	compact tension (toughness test)
DCB	double cantilever beam (toughness test)
DPOD	diamond pin on disk scratch hardness/wear test
DT	double torsion (toughness test)
EA	elastic anisotropy
F	fractography (toughness test)
HIP	hot isostatic press
IF	indentation fracture (toughness test)
NB	notch beam (toughness test(s), also SENB)
PSZ	partially stabilized zirconia
PZT	lead zirconate titanate
RSSC	reaction-sintered silicon carbide
RSSN	reaction-sintered silicon nitride
SCG	slow crack growth
TEA	thermal expansion anisotropy
TZP	Tetragonal zirconia polycrystal
W	wear
WOF	work of fracture

1

Introduction to Grain and Particle Effects on Ceramic and Ceramic Composite Properties

I. GRAIN AND PARTICLE PARAMETER DEFINITIONS AND TREATMENT OUTLINE

A. Grain and Particle Parameter Definitions

This chapter introduces the role of grain and particle parameters in determining properties of ceramics and ceramic composites. First we define them, and then we outline their effects on properties, especially important mechanical effects addressed in detail in this book. Next, grain and particle variations that occur and make their characterization challenging are discussed, followed by an outline of measuring these parameters and the properties they impact, focusing on issues that can aid in better evaluating and understanding effects of these parameters.

After this introductory chapter, this book first addresses the effects of the size and other parameters of grains on properties of nominally dense, single phase, i.e. "monolithic," ceramic bodies. Then it similarly addresses effects of both the matrix grains and especially of the dispersed particles (or platelets, whiskers, or fibers) on mechanical properties of both natural and designed ceramic composites. In both cases, the focus is on properties at moderate temperatures, but some behavior at higher temperatures is addressed.

Before addressing definitions it is important to note that there is a diversity of microstructures ranging from ideal to complex ones, with the importance and occurrence of the latter often varying with the material system or fabrication technology considered. Such diversity makes generating a clear and self-consistent set of definitions to cover all bodies complex and uncertain. The first of two points resolving this issue for this book is that the great majority of microstructures considered can be reasonably described by classical or ideal microstructures or limited modifications of these. Second, while increased sophistication in defining and measuring microstructures is important, the more immediate need is to better document key microstructural dependences of mechanical properties considered in order better to define or refine the mechanisms involved. These tasks are the focus of this book, which is limited to existing microstructural characterization. However, some of the complexities of real microstructures and limitations or challenges in meeting measurement needs for even typical microstructures are discussed along with some information on improved definitions and measurements.

Consider now the definition of grain and particle, since they are fundamental to this book and to understanding the behavior of most ceramics and ceramic composites. The terms are related, partially overlapping and basic to describing most, but not all, solid bodies, though they are not always adequate, at least by themselves, for some amorphous or complex bodies. Grain refers to the primary microstructural unit in polycrystalline bodies or in other polycrystalline entities, e.g. fibers or some powder particles, as well as the basic crystalline units in partially crystalline bodies. Grains are typically single-phase single crystals, whose identity is delineated by the difference in crystal orientation or structure from abutting grains or amorphous material, and the resultant grain boundaries formed between them to accommodate these differences. Thus an ideal polycrystalline body consists of one phase whose grain structure describes its microstructure.

Particle is a broader term, referring in general to a discrete solid entity of either glassy, i.e. noncrystalline, structure, or of single-crystal or polycrystalline character, or some combination of these, i.e. a particle may consist of differing compositions of one or more compositionally or structurally different phases as well as some possible porosity. The term particle is commonly used in two somewhat more restricted senses in the literature and in this book. The first is as the solid entities of a powder, i.e. as used to make many single phase or composite bodies, where the particle size, compositional, structural, and morphological character are selected or controlled to give the desired body consistent with the fabrication and related processing parameters. The second, more restricted, use of the term particle is to identify a discrete, typically minority, microstructural unit in a body containing two or more phases differing in composition, structure, or both. While such particles may be of one or more impurities, they more commonly are an intentionally added phase in a body, i.e. in a matrix of different composition than of the particles. The matrix is typically the larger volume fraction constituent, as well as the continuous phase (though again more complex bodies can occur). The matrix and the dispersed particles can both be crystalline, or one can be amorphous, but usually not both (unless they are immiscible, which again leads to microstructures not totally described by the above concepts of grains and particles). In the case of a crystalline matrix its grain parameters are still often important in the behavior of the composite along with the particle parameters discussed below. An amorphous matrix derived by sintering may retain some of the original particle character of its powder origin, most commonly when the particles are delineated by substantial residual porosity. Thus the ideal or classical composite consists of a continuous amorphous single-crystal, or polycrystalline matrix with a disperion of solid particles of a distinct second phase, usually mainly or exclusively of single-crystal or polycrystalline character.

The above ideal microstructures can be broadened to encompass many other bodies of interest via two modifications. The first is to include some porosity of either intergranular or intragranular locations, and often varying combinations of both, e.g. depending on grain size. (Porosity can be considered a [nonsolid] second phase. Modeling its effects by using composite models in which properties of the "pore phase" are set to zero is commonly used, but this approach, while often used, can also be very misleading [1].) Note that while real microstructures often have some porosity, which typically plays a role, often an important one in physical properties, their effects are treated extensively elsewhere [1]. In this book the focus is on microstructural effects with no or limited porosity, the latter commonly corrected for the effects of the porosity [1]. The second and broader modification is the introduction of a solid phase in addition to, or with no, pore phase in one or more of the following fashions: (1) along part, much, or all of the matrix grain, particle–matrix, or both boundaries or (2) within the grains, particles, or both, e.g. from some phase separation process such as precipitation.

Finally, a note on ceramic composites, which were originally the result of empirically derived bodies based on processing natural raw materials, with compositions based on the raw materials and processing available and the resultant behavior. Porcelains and whitewares, which are important examples of composites of silicate glasses and oxide crystalline phases still in broad use today, are treated to some extent in this book. Another related, more recent and often complex family of ceramic composites are those derived by controlled crystallization of glasses of compositions selected for their processing-property opportunities. These are also treated to some extent in this book. In more recent years, many composites of designed character, primarily of a crystalline matrix with significant dispersions of primarily single crystal particles, have become of interest. These composites of either or both oxide or nonoxide phases are more extensively covered in this book following evaluation of the grain dependence of properties of nominally single-phase, i.e. monolithic, ceramics.

B. Grain and Particle Parameters

The most fundamental parameter of grains and particles is their composition, hence generally of the body (i.e. in the absence of significant reaction or impurity phases). For composites, both the compositions (and associated crystal or amorphous structures) and the volume fractions of each phase are needed to define the composition. Beyond body composition, the key microstructural parameters of grains or particles are their amounts, sizes, shapes, orientations, and the numerical and spatial distributions of these. These parameters, while having some relation to composition, especially in composites, are extensively dependent on fabrication/processing parameters. This combined composition-fabrication/processing dependence of microstructural parameters, which determines resultant body properties, is addressed in this book via its focus on microstructural effects.

Size naturally refers to the physical dimensions of grains and particles. While this can be complicated in definition, measurement, or both, e.g. due to effects of shape and orientation, it is most easily and commonly defined when the shape is approximately spherical or regular polyhedral. In this simple, common, useful, but not universal, case, the grain or particle size is typically simply and logically defined by an average grain or particle diameter (G or D respectively). However, even in such simple cases there are important questions of which average is most pertinent to the properties or phenomena of interest, as discussed in Section IV. Further, the statistical variation in size, e.g. the width of the size distribution, is also of importance, as may be also the spatial distribution of grains of different sizes, especially the clustering of larger grains.

While grain and particle size are commonly the most important microstructural parameters beyond the body composition (and the amount and character of porosity, if present), grain and particle shape and orientation are also often important individually, as well as through frequent interrelations among them and other parameters. Shape clearly refers to the three-dimensional shape of grains and particles, which becomes important primarily when it deviates considerably from an equiaxed shape, i.e. a spherical or regular polyhedral shape (typically with > six sides). While shape is a basic factor for all grains and particles, it is particularly important as it also relates to their crystal structure for single crystal grains or particles. Thus while larger grains or particles resulting from substantial growth are often equiaxed, they can also frequently be tabular or acicular in shape, i.e. respectively either more like platelets or rods/needles. Composites of ceramic platelets or whiskers in ceramic matrices, which have been the focus of substantial research, a few being in commercial production, are important examples of more extreme shaped particles. Grain and particle shapes impact both the designation and the measurement of the size of such grains or particles and are themselves difficult to measure in situ since conversion of the typical two-dimensional measurement to a three-dimensional shape requires assumptions about the grain or particle shape.

Grain or particle shapes may also significantly affect their orientation, both locally and globally in the bodies in which they occur; both again in turn affect size measurements and body properties. Property effects occur most extensively when grain or particle shapes reflect aspects of the crystallographic character, which is common. In such cases, global orientation of such shaped grains or particles imparts a corresponding degree of anisotropy of properties in proportion to the crystalline anisotropy and the degree and volume fraction of orientated grains or particles. This encompasses all properties of noncubic materials, as well as various properties of cubic materials such as elastic moduli, fracture (i.e. cleavage, fracture energy, and toughness), and strengths, since cubic materials, while being isotropic in some properties such as dielectric constant and thermal expansion, are generally anisotropic in other properties such as those noted. Local grain or particle orientation effects can impact properties via effects on local crack generation (i.e. microcracking), propagation, or both. It can also result in grain clusters, i.e. colonies, that act collectively as a larger grain or particle, as discussed below and later.

Composites where the particles are single crystals, i.e. like grains in a polycrystalline matrix, are most common and most extensively treated in this book, including those where the dispersed particles are of more extreme morphological shapes, i.e. platelets or whiskers. When such latter composites are discussed, they will typically be specifically identified as platelet or whisker composites, and composites with more equiaxed particles will be referred to as particulate composites. However, the term particle or particulate composite will sometimes be used in the genetic sense irrespective of particle shape/morphology. Grain or particle orientation refers to the spatial orientation of either their physical shape or especially of their crystal structure, which are commonly related.

Measurement adequately to reflect size, shape, and orientation and their interrelation and spatial distributions is a large, imperfectly met challenge that is often inadequately considered. Thus the location, shape, and orientation, and sometimes the size, of pores are often related to grain, and especially particle, shape and orientation. For example, pores in platelet composites commonly remain at the platelet–matrix interface [2], and such pores often are larger and typically somewhat platelet in shape. Similar effects have been indicated in fiber composites (Chap. 8, Sec. III.E, Ref. 1) and are likely in whisker composites. Particle, and especially grain, growth can be significantly enhanced by impurities, with resultant larger grains or particles being equiaxed or often tabular or acicular in shape. Some of these effects are illustrated in the next section, and later in the book. Resulting effects are further frequently complicated by the need to define and address their correlation with other microstructural factors such as other phases or pores, e.g. grain boundary or interfacial ones. Issues of such measurements and effects are discussed further in this chapter and elsewhere in this book.

The grain and particle parameter addressed most extensively is their size, which is generally of greatest importance and is widely addressed in the literature. However, the shape and orientation parameters, as well as the statistical and spatial distributions of these parameters and of sizes of grains and particles, are also treated to the extent feasible, since these parameters can also be important. Though interaction and distributions of these parameters are often neglected, causing considerable variation in the literature, they often play a role in determining many important physical properties of ceramic composites and monolithic ceramics.

II. RELATIVE EFFECTS AND INTERACTIONS OF MICROSTRUCTURAL PARAMETERS ON PROPERTIES

In focusing on grain and particle parameters it is important first to recognize the relative roles and interactions of microstructural features on ceramic properties, especially porosity, which is a dominant factor, as recently comprehensively reviewed [1]. In making ceramic (and other) bodies, processes used can result in either substantial or limited porosity, depending on both the fabrication process and the parameters selected. While a variety of factors impact the choice of fabrication method, e.g. the size, shape, and cost of components to be made, the amount and character of the porosity sought or tolerated in the component is also an important factor. Often, one is dealing with either of two, extreme, cases. In one case a desired, ideally a designed, pore structure is sought for favorable attributes needed from the amount and character of the porosity balanced against limitations of other pertinent properties such as stiffness, strength, conductivity, etc. imposed by the porosity. In the other case one seeks to minimize porosity to approach, or achieve, high levels of properties limited by porosity as a function of cost and performance. In polycrystalline materials grain parameters, and for composites particle parameters, play important, often similar, roles in many properties, depending on porosity content.

Typically the most significant role of grain and particle parameters on properties occurs where low porosity is sought for high levels of properties. This arises because some key ceramic properties, such as strength and fracture, hardness, wear, and erosion behavior, are significantly impacted by grain and particle parameters, of which size is often most important. Thus in order to achieve high levels of important properties in a selected material, porosity must first be minimized, since this commonly reduces properties substantially, e.g. by 1 to 2 orders of magnitude at intermediate and higher porosities. However, beyond increases from reduced porosity, some important properties can be further increased, e.g. by 50 to a few hundred percent, by obtaining desired grain or particle sizes, often as small as feasible. Achieving this entails trade-offs in terms of amounts and character of residual porosity, since at lower levels of porosity, further heating to reduce porosity commonly leads to grain growth and attendant sweeping of grain boundaries past previously intergranular pores. Resultant intragranular pores are often less detrimental to properties, and may even possibly counter some reductions of properties due to grain growth. While this approach of accepting some residual, especially intragranular, porosity is ultimately limiting in properties, it is often a factor in production of bulk ceramics. At the other extreme of product size, i.e. production of ceramic fibers, achieving very fine grains (typically << 1 μm) with ~ 0% porosity is essential to achieving strengths up to an order of magnitude higher than in bulk bodies. Achieving such benefits of fine grain sizes with ~ 0% porosity in bulk bodies has also been sought. However, this can entail serious problems of residual impurities in some materials, particularly in attempting to achieve nano-scale grain or particle sizes for effects expected from extapolations of conventional grain or particle size dependences, or for possible novel behavior at such fine grain or particle sizes. Such problems, which have not been addressed in the nanomaterial literature, are addressed in this book.

In cases where porosity is needed for functions, grain and particle parameters often still play a role in determining properties, but such effects are usually secondary to those of the pore structure and are typically at fine grain and particle sizes. Obtaining such fine sizes often entails retaining substantial porosity due to low processing temperatures used to limit grain or particle growth, and in the common case of sintering, starting with fine particles. However, since some growth of grains, particles, or both occurs, often inversely to the level of porosity, grain size effects must also be considered, especially as porosity decreases, as is extensively addressed elsewhere [1].

A qualitative overview of the relative impacts of pores, grains, and particles on various properties, mainly at moderate temperatures (the primary focus of this book), is summarized in Table 1.1, by indicating whether these features have a primary (P) or a secondary (S) effect on that property. Primary effects are intrinsic to the presence of the pore, grain, or particle structure in the body, but vary with compositions and their amounts, sizes, shapes, and orientations. Secondary effects, which can occur alone or in addition to primary effects arise from combinations of composition, size, shape, and orientation via local and especially global grain (or particle) orientation and hence anisotropy or microcracking, the latter occurring only above a critical grain or particle size for a given body composition. While many primary effects are very substantial, e.g. most porosity effects, and many secondary effects are of lesser impact, this is not always so. Thus while particles in a composite have intrinsic effects on all properties shown, these are often modest, since particle property impacts typically

TABLE 1.1 Summary of Effects of the Three Primary Microstructural Elements on Ceramic Properties[a]

Properties	Pores[b]	Grains[c]	Particles[d]
Elastic	P	S (microcracks, orientation)	P
Fracture toughness	P	P	P
Tensile strength	P	P	P
Compressive strength	P	P	P
Hardness	P	P	P
Wear/erosion	P	P	P
Density	P	S (microcracks)	P
Thermal expansion	S[e]	S (microcracks, orientation)	P
Conductivity	P	S (microcracks, orientation, boundary phases)	P
Electrical breakdown	P	P, S (microcracks, boundary pores & phases)	P
Dielectric constant	P	S (orientation)	P
Optical scattering	P	S (microcracks, pores, boundary impurities)	P

[a–d] P = a primary, i.e. an inherent, dependence on the presence of pores, grains, or particles and their parameters. S = a secondary dependence, which occurs only for some body compositions and some aspects and ranges of microstructural parameters. Secondary effects arise in addition to or instead of primary effects, but arise only for some compositions and often only some ranges of grain and particle parameters and associated microcracking, and gram boundary impurities or phases, or pores (primary or common ones noted for different properties). Particles dispersed in a matrix, i.e. for a composite, have primary effects on all properties, commonly proportional to the property difference of the particle and matrix phases, i.e. varying across a substantial range, but often less than from pore or grain parameters. Significant secondary effects often occur and can be ≥ than primary effects.

[e] Thermal expansion only depends on porosity via its effects on other factors, primarily microcracking.

scale with the difference in that property for the particles and the matrix and with the volume fraction of particles.

The summary of Table 1.1 can be put in better perspective by recalling basic aspects of the isotropic versus anisotropic behavior of crystalline materials. Thus noncubic materials often have substantial variation of properties with crystal direction, i.e. are anisotropic. On the other hand cubic materials are intrinsically isotropic for several important properties such as thermal expansion, electrical and thermal conductivities, dielectric constant, and refractive index. However, cubic materials are intrinsically anisotropic in some properties, particularly elastic moduli, fracture energy and toughness, tensile and compressive strength, and electrical breakdown.

Grain parameters, especially size, play a primary role in most mechanical properties as well as a number of secondary effects that occur due to other factors. Most of this dependence is similar in both cubic and noncubic materials, but there can be some differences in noncubic materials due to secondary effects, especially microcracking, which can play an important role in fracture toughness. Key exceptions to this grain size dependence of mechanical properties are elastic properties, which have no intrinsic dependence on grain size, only a secondary dependence to the extent that it impacts microcracking (i.e. typically above a certain, material dependent, grain size). Elastic properties also have a secondary to substantial dependence on overall, i.e. global, preferred grain orientation in all noncubic materials, and essentially all cubic ones due to their anisotropy as noted above. Grain size has no primary effect on thermal expansion of cubic or noncubic materials. It has limited or no effect on thermal conductivity in noncubic materials at, and especially above, room temperature, but it can have increasing, generally modest, effects at lower temperatures in noncubic materials. Electrical conductivity generally follows similar trends but can have greater grain size dependence in noncubic materials, since it covers a much broader range of values as a function of material composition and crystal phase character than thermal conductivity. Optical scattering also has a primary, but generally limited, dependence on grain size, as well as some significant secondary dependence in noncubic materials. However, these intrinsic effects are commonly much smaller than the effects of microcracking (in noncubic or composite materials), or pores or boundary impurities in either cubic or noncubic materials. Dielectric constant has some crystalline anisitropy in noncubic materials, hence some dependence on global, i.e. preferred, grain orientation, but no intrinsic grain size dependence. Electrical breakdown, which is intrinsicaly dependent on crystal orientation in both cubic, and especially noncubic, materials, is quite sensitive to secondary effects correlating with grain size, i.e. microcracking, intergranular pores, and phases.

Composite properties have a primary, as well as a secondary, dependence on the particle character. However, these dependences, especially the primary ones, are generally dependent, in decreasing order of importance, first on the differences between the matrix and particle properties (hence their compositions), second on the volume fraction of the particulate phase, and third on the particle size, shape, and orientation. Particle size is mainly a factor in the primary dependence of mechanical, other than elastic, properties, where such effects generally parallel those of grain size in nominally single-phase materials. Particle size effects on nonmechanical properties occur mainly in conjunction with the often substantial effects of particle volume fraction and shape and orientation via their effects on contiguity of the particulate phase. These effects are generally greater for thermal and particularly electrical conductivity, as well as electrical breakdown, especially where the second phase has substantially higher conductivity or lower electrical breakdown than the matrix. Particle size effects are important in the often substantial secondary effects on properties via associated microcracking.

Boundary, i.e. interfacial, pores and second phases between grains and particles or both are often present and can play significant roles in body properties via either or both of two avenues. The first is their often interactive effects impacting grain or particle sizes, shapes, and orientations and thus on properties dependent on these parameters. The second is the direct effect of such interfacial phases on properties by modifying the dependence of properties on unaltered grain or particle sizes, shapes, and orientations, e.g. in the extreme by dominating properties. Electrical breakdown across a range of temperatures and mechanical failure at higher temperatures are examples of the latter.

III. RANGE, CHARACTER, AND CHALLENGES OF GRAIN AND PARTICLE PARAMETERS IN MONOLITHIC AND COMPOSITE CERAMICS

A. General Issues and Challenges

Before proceeding to outline the methods of characterizing grain and particle structures, and associated measurements to understand the dependence properties of ceramic or composite ceramic bodies on them, it is important to recognize the challenges to such characterization and understanding. Basic challenges in normal characterization of grain and particle parameters are often significantly compounded primarily by two factors, which are also keys to the successful engineering of properties via fabrication-processing control of microstructure. The first is the variety of microstructures that can occur both locally and globally. The second is the particular sensitivity of some properties to local rather than general microstructure, especially larger grains, particles, or clusters of them along with their shape and orientation. However, the location of such microstructural heterogeneities in the body and their association with defects such as pores and cracks are also important. Properties particularly affected by such microstructural heterogeneities are those involving macrofracture, i.e. in approximate order of decreasing impact: tensile strength, electrical breakdown, fracture toughness, and compressive strength. Other properties sensitive to local microstructure, with or without local fracture, are hardness, wear, and erosion.

An important complication in determining grain or particle sizes occurs when the grains or particles are not nominally equiaxed. In such cases not only do both the shape and the orientation need to be characterized but also their often substantial impact on both the definition and the measurement of grain or particle size is a challenge. While such interactions pose a challenge for determining the average grain size, they can be of greater concern in describing the numerical and spatial distributions of grain sizes. In composites, another challenge is describing the spatial distribution of the particles in the matrix. Interactions of size, shape, and orientation and the challenges of describing numerical, i.e. statistical, and spatial distributions of these have important parallels in describing some important variations of grain parameters.

Another issue is boundary, i.e. interfacial, phases, such as those that commonly result from impurities or residual densification additives. These often present measuring challenges, e.g. their presence and extent at low concentrations. At higher concentrations, both their uniformity and the contiguity of the grains or particles that such phases partially surround can be issues. The following section addresses grain and especially particle parameters in two-phase or multiphase ceramics, especially in designed composites, but natural composites such as whitewares and crystallized glasses are also addressed.

B. Grain Variations, Mainly Size in Nominally Single-Phase Ceramics

Significant challenges arise from grain variations within a given part as well as between different ones, which arise from both the fabrication method and its associated processing parameters, as well as from heterogeneities, e.g. impurities, in the body. A primary variation is commonly grain size, but interrelated factors of shape and orientation can also be important, with characterization to reflect more than one grain population, i.e. bi- or multimodal grain size distributions being a particular challenge. More extreme cases of bimodal grain distributions, which are those often of greater impact, are typically readily identifiable (e.g. Fig. 1.1). In such cases, the primary problem is commonly that of determining the average and variation (e.g. standard deviation) for each grain population. Accurately determining the volume fraction of each distribution and its spatial distribution are also challenging.

A common and important aspect of grain size variation in polycrystalline bodies is the heterogeneous distribution of larger grains having little or no relation to the normal grain size distribution(s). These larger grains, which are typically the result of local exaggerated grain growth, can be an isolated or dominant large grain (Fig. 1.2), or clusters of two to a few larger grains (Fig. 1.3), and occasionally clusters of several larger grains (Fig. 1.4) that can be of a wide variety of character. The grains in such clusters often have a common cause for their occurrence, in which case they have similar character. Of particular impact is the occurrence of much larger individual grains or of clusters of larger grains, which are of sufficient size to be all, or most, of fracture origins. Though this appears to require that they be associated with another defect such as a crack or pore to initiate fracture, this coincidence is common, but not universal. The impact of such isolated larger grains, or clusters of larger grains on fracture, scales inversely with the square root of their size, which can be an order of magnitude or more

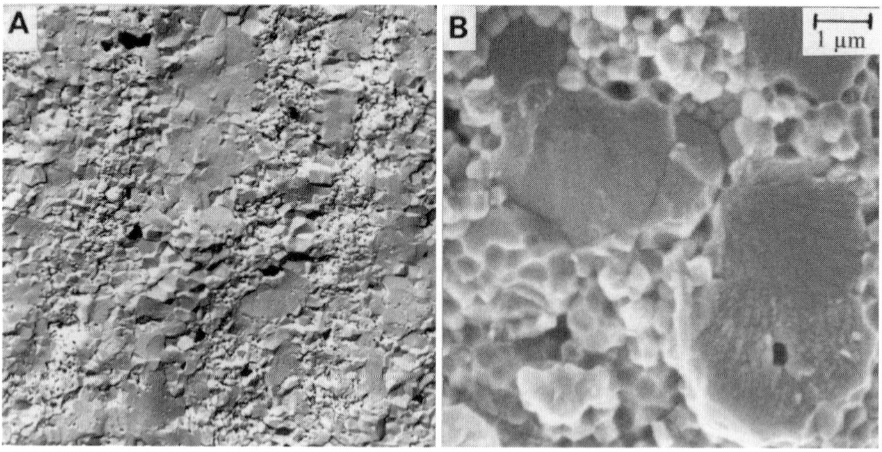

FIGURE 1.1 Clear, more extreme examples of bimodal grain structures seen on room-temperature fractures. (A) Pure, dense Al_2O_3 (TEM of fracture surface replica) showing two grain populations occurring as approximately interleaved (~ vertical) laminations that are nominally perpendicular to the hot pressing direction. (B) Sintered ZrO_2–8 wt% Y_2O_3 (SEM of fracture surface) showing a fine grain matrix with approximately random distribution of much larger grains singly or as clusters of two or three larger grains. Note in both cases that the smaller grains exhibit predominately intergranular and the larger ones predominately transgranular fracture.

larger than the average of the surrounding grains (e.g. Figures 1.2–1.4). The impact of such larger grains or grain clusters can be significant, especially on tensile strength, even if their occurrence is sparse, which makes determining their character and whether they were the actual source of failure challenging, especially if they are not specifically identified at fracture origins.

While the occurrence and character of larger isolated grains or grain clusters is quite variable, there are certain materials and processing factors that favor or exacerbate their occurrence and character. The material aspects entail the basic composition itself and local concentrations of additives or impurities, or sometimes deficiencies of them, especially additives. Consider first the basic material composition, where a common factor is crystal structure. Large grains or grain clusters are most common in some noncubic materials, e.g. CVD ZnSe [3, 4] (Fig. 1.3C), more extensively in hot pressed or sintered B_4C [4–6] (Figures 1.2D, 1.3D) and Si_3N_4 [7–10] and especially in SiC [5, 6, 11], Al_2O_3 [3–5, 12–18], and beta aluminas [5, 19–21]. In Si_3N_4 they are commonly rod-shaped β grains often associated with local concentrations of oxide densification additives [8–10]. In SiC they are commonly long, narrow alpha platelets nucleated

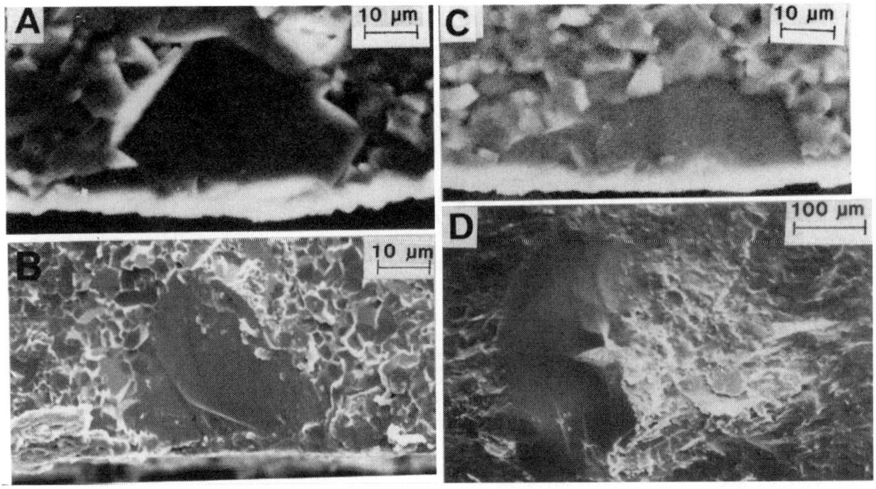

FIGURE 1.2 Examples of fracture origins from a single or dominate large grain seen on room temperature fractures. (A) and (B) Commercial sintered 96% alumina fracture origins from isolated larger grains at the flexure surface respectively at 22°C (376 MPa) and -196°C (593 MPa) after Gruver et al. [14]. (C) Origin from large grain at the flexure surface in pure hot pressed Al_2O_3 (396 MPa, giving K_{IC} ~ 2.7 MPa·$m^{1/2}$). (D) Origin from a large grain internal from the flexure surface of a hot pressed B_4C (<350 MPa at the grain). Smaller lighter area at the lower left side, and the larger one on the right of the large grain, are areas of some porosity and cracking. Note varying degrees of truncation of larger grains at the surface from machining.

in bodies of β grains during sintering, typically at temperatures of ≥ 2000°C, but they also occur as clusters in CVD SiC at much lower temperatures, as discussed below. Their lengths are often > 10 times their widths, with probable, but poorly characterized, thicknesses a fraction of their widths [11]. Larger isolated grains in beta alumina bodies are (usually thin) platelets, which are typical of such bodies and are attributed to exaggerated growth of grains of the same or modified beta aluminas [19–21], e.g. due to fluctuations in sodium content. Large grain character in Al_2O_3 is more varied, ranging from equiaxed to platelet or rod shapes. However, noticeable or significant exaggerated grain structures have apparently not been observed in some noncubic materials such as MgF_2 and mullite, which commonly have finer grain structures. The frequency, extent, and character of larger isolated grains or heir clusters is generally less in cubic materials.

An important compositional impact on the occurrence and character of

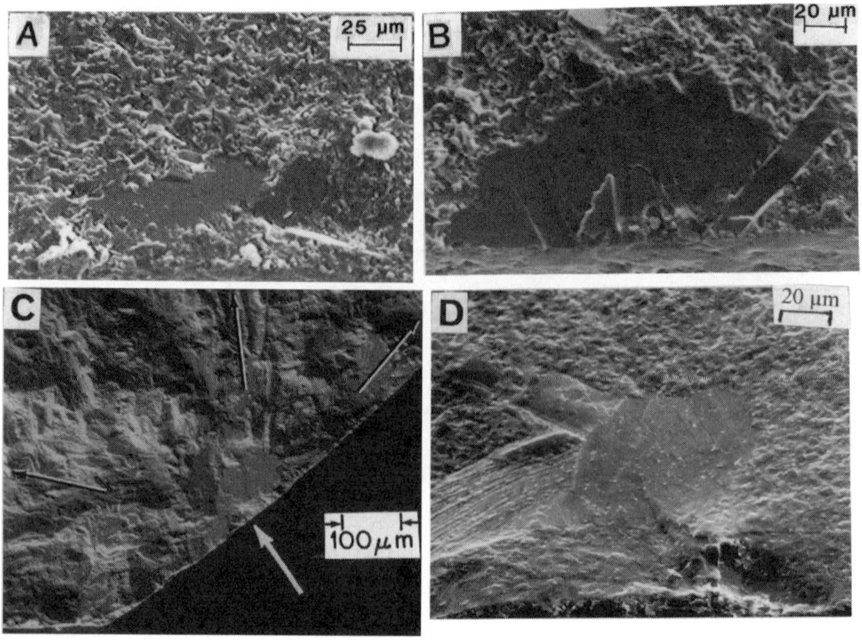

Figure 1.3 Examples of a few larger grains at ceramic fracture origins in flexure at 22°C. (A) Pure hot pressed Al_2O_3 origin (483 MPa) via intergranular fracture from a large platelet grain in the plane of fracture, bottom center of photo, and two similar intermediate sized grains to the right, all three just inside the tensile surface and at progressively higher angles to the fracture surface. The third (right) grain is a white streak in this view, since it is seen nearly on edge, showing the thin platelet nature of many such grains in Al_2O_3. (B) Pure hot pressed Al_2O_3 origin from the tensile surface (405 MPa) from a large intergranularly fractured platelet grain in the plane of fracture, bottom center of micrograph, showing impressions of two or three intermediate sized tabular grains from the mating fracture surface and a transgranularly fractured grain (right). (C) CVD ZnSe flexure origin from a large grain in the plane of fracture at the specimen corner with an adjacent intermediate sized grain (both transgranularly fractured). (D) Hot pressed B_4C origin (376 MPa) from a machining flaw (bottom center) and three to five larger grains above and to the left. Note light laminar striations on some of the grains indicating twinning and its interaction with the fracture.

larger grain structures in ceramics is the effects of additives, (local) impurities, or both. Again, a number of noncubic materials, especially Al_2O_3, show a greater extent and diversity of such effects, apparently reflecting significant anisotropy in crystal (grain) growth characteristics. Thus thin platelets are common in bodies of high purity Al_2O_3 hot pressed without MgO additions ([3–5, 12–16], for

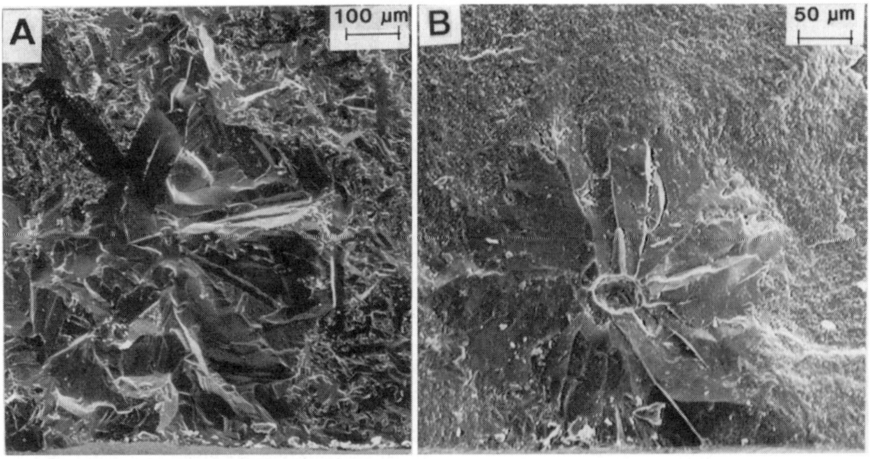

FIGURE 1.4 Examples of flexure fracture origins from clusters of larger grains in alumina bodies in flexure at 22°C. (A) Pure hot pressed Al_2O_3 origin (443 MPa) from a cluster of tabular grains just inside the tensile surface, with mostly transgranular fracture and mixed radial and random orientations. (B) Experimental commercial sintered higher purity alumina origin (266 MPa) inside the tensile surface from a cluster of tabular grains radiating from a central pore.

grain growth control, which can eliminate them if it is homogeneously distributed; other grain variations can occur with heterogeneous MgO distribution [22]). Such platelet grains, which are apparently due to small amounts of Na-based impurities (whose volatilization may be inhibited in hot pressing) triggering formation and growth of beta alumina grains in the absence of MgO or other grain growth inhibition, are often primarily indicated by fracture occurring intergranularly between them and the adjoining finer grain structure (Fig. 1.3A). Thicker platelet or rod-shaped grains, as indicated by their transgranular fracture (Fig. 1.3B), and some nominally equiaxed (Fig. 1.2A) from similar or other sources, are also common. Many larger grains arise from impurities, especially when there is some radial growth pattern (Fig. 1.4), where the central void may be due to a volatilized impurity, e.g. sodium. In some cases larger grains associated with pores may not be immediately around the pore surface but somewhat into the surrounding solid, e.g. indicating diffusion of the species causing excess grain growth reaching the concentration range favoring such growth [22].

Much of the above occurrence of elongated grains or grain clusters reflects directional growth, e.g. from a local impurity source. However, more homogeneous impurity contents or excess additives commonly result in more equiaxed and homogeneous larger grain clusters This is commonly the case for the more limited occurrence of larger grains in cubic materials, e.g. MgO

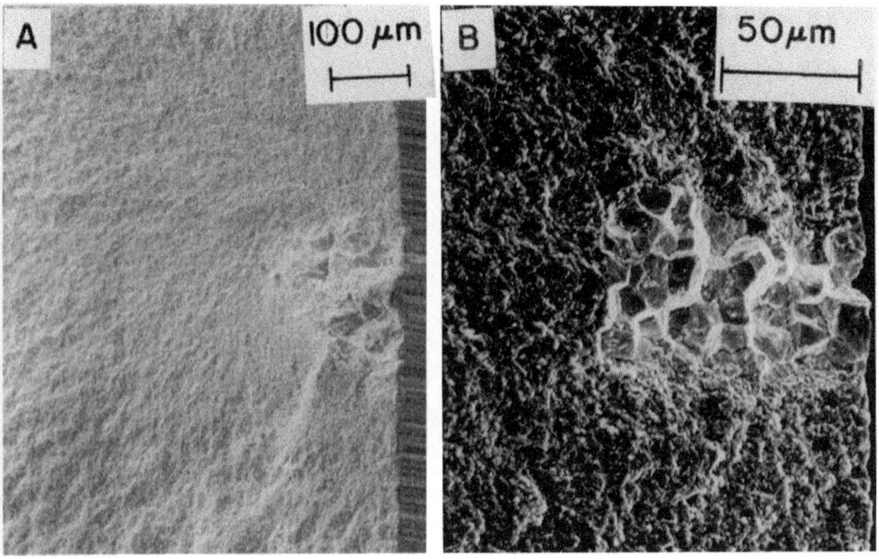

FIGURE 1.5 Examples of clusters of larger equiaxed grains at flexure fracture origins in Si_3N_4 at 22°C. (A) Commercial hot pressed material (HS-110), origin (432 MPa) from a larger grain cluster associated with primarily excess Fe and secondarily Mn impurities. (B) Experimental hot pressed specimen origin (520 MPa) from a region of excess ZrO_2 densification aid. (From Ref. 22. Published with the permission of Plenum Press.)

[4, 22]. However, similar effects also occur in noncubic materials, e.g. Si_3N_4 [22] (Fig. 1.5) and Al_2O_3(22).

The above common occurrence of variable and often significant roles that larger grains can play in mechanical failure raises important measurement issues. The arbitrary use of the size of the largest grains or particles for correlating with strength, as proposed by some, fails to recognize critical aspects of grains and particles as fracture origins, namely sufficient sizes and associated defects such as pores or cracks, which are discussed extensively in this book More fundamentally, effects of larger grains, as well as practical factors, argue against continuing the traditional dominance of obtaining grain and particle parameters from polished surfaces.

While microstructural measurements from polished surfaces can be of some use, their preparation and examination is far more time-consuming and costly than that of fracture surfaces, especially from property tests whose results are to be related to their microstructure. Fracture surfaces commonly reveal the microstructure, without etching, more clearly than polished surfaces, even with good etching (e.g. Fig. 1.6). More fundamentally, polished surfaces fail to provide adequate information of two types. First, even with ideal polishing and etching there is essentially zero probability of such surfaces showing the extremes of grain structure that are often fracture origins. Such limitations are exacerbated when thin tabular or acicular grains are involved, since their

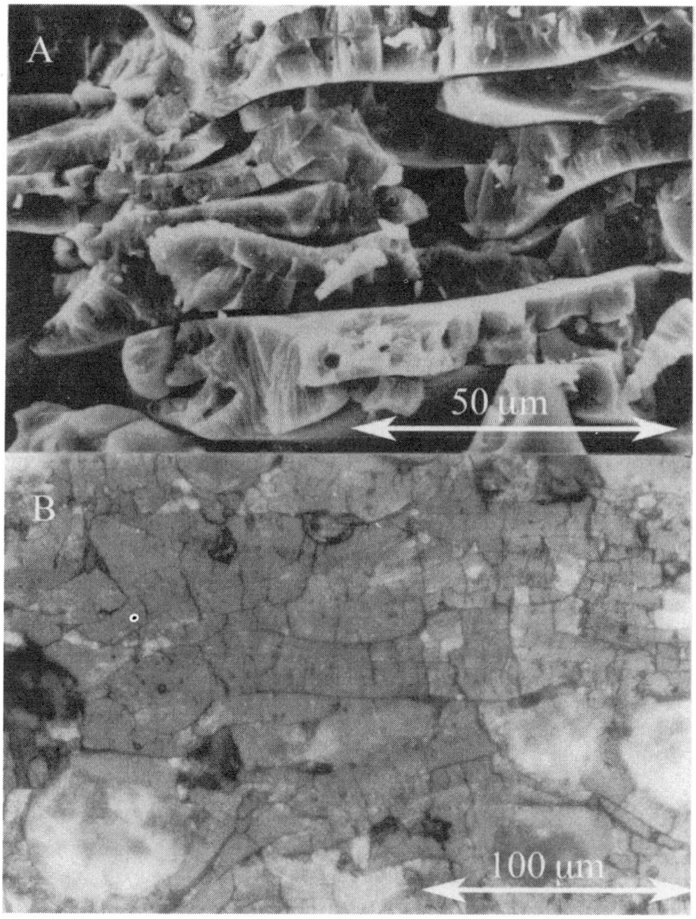

FIGURE 1.6 Comparison of essentially identical cross-sectional microstructures of the same Al_2O_3–wt% TiO_2 plasma sprayed coating prepared by (A) fracture and (B) conventional polishing and etching. Note much greater definition of the microstructure, especially the grain structure on the fracture surfaces, e.g. indicating a fine columnar grain structure within the individual splatted molten droplets. (From Ref. 23. Photos courtesy of Murakami et al.)

observation and impact on grain size is dependent on their orientation relative to the surface of examination, with very low probability of a random polished surface giving results similar to that from fractures, especially fracture origins. The second and even greater deficiency of examining polished versus fracture surfaces for necessary grain structure information is that other key information is only available on the fracture surface. This includes whether the fracture origin was from an atypical area of grain structure of sufficient size and with necessary associated other defects (e.g. pores of cracks) to be the cause of failure, or whether failure was from the primary flaw population with little or no effect of grain structure variations. It also includes other important information such as the fracture mode, and especially for intergranular fracture, information on grain boundary character, e.g. phases and porosity.

C. Grain Variations, Mainly Shape and Orientation in Nominally Single-Phase Ceramics

The above outline of the occurrence of larger isolated grains or grain clusters has necessarily touched on grain shape and orientation, primarily in radial clusters. However, there are material and fabrication-processing factors that further affect either or both of these factors in a systematic fashion. Thus columnar, i.e. longer, typically rod-shaped, grains characteristically form normal to a solidification front with grain diameters and lengths typically inversely related to the speed of the solidification front and the density of nucleation sites for new grains. Such columnar grain structures occur in fusion cast refractories, especially near the mold surface where the solidification front is more uniform. The more rapid cooling, especially near the mold surface, commonly results in smaller grains there, though they are usually still substantially larger than grains in typical sintered bodies. More rapid cooling results in smaller size columnar grains, e.g. Fig. 1.7A. An extreme of the size/solidification speed is melt, e.g. plasma, sprayed ceramic coatings (Fig. 1.6B) [23]. Thus unless there are complications of other phases or formation of an intermediate amorphous phase, the molten droplets splatting onto the deposition surface directionally solidify normal to the surface, typically as fine columnar grains. A large-scale manifestation of such columnar structures and the common preferred crystal orientation in the columnar grains that typically accompanies it is the slow, controlled directional solidification of large polycrystalline ingots of cubic zirconia for the jewelry trade (Fig. 1.7B) [24]. Such grains can reach dimensions of 10 cm or more in large commercial ingots.

Vapor deposition processing, e.g. chemical vapor or physical vapor deposition (CVD or PVD), of coatings, or in the case of CVD also of bulk bodies, commonly results in various clustered grain and columnar structures that occur together or separately. The occurrence and character of these depends on deposi-

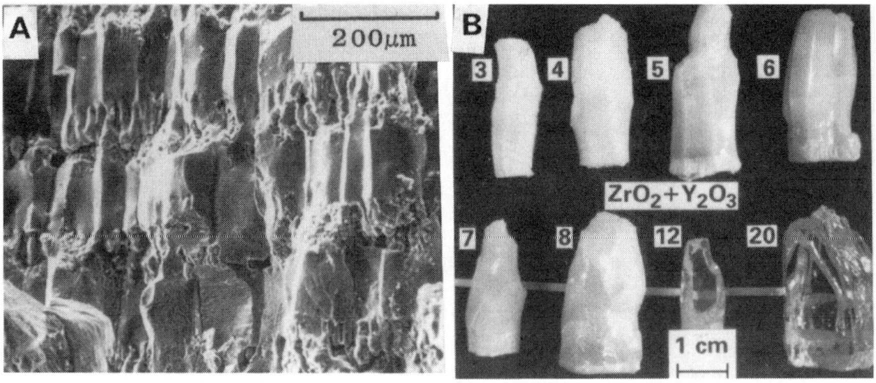

FIGURE 1.7 Examples of columnar grains from directionally solidified melts of partially and fully stabilized zirconia bodies. (A) ZrO_2 + 5 wt% Y_2O_3 sequentially solidified as thin layers on a solid surface (columnar grains form over the whole thickness of the layer with their length essentially being the layer thickness). (B) Examples of individual readily extractable grains (since the grains commonly are not bonded to one another, especially at such large grain sizes) from directionally solidified skull melts of differing w/o contents of Y_2O_3 (shown by each grain). (From Ref. 24)

tion rates (e.g. on pressure and temperature) and the phases involved [4, 25]. At high deposition rates, larger columnar grains result that are highly aligned axially (often with random radial crystal orientations), similar to those in directional solidification. (In such cases the ends of the columnar grains on the free surface of the deposit typically reflect single crystal faceting and are thus often referred to as faceted deposits or surfaces, e.g. in TiN [26].) Such columnar grains may have substantial aspect ratios, ranging from ~1 to several-fold.

In the deposition range below that for producing larger aligned columnar grains, deposition often occurs by successively nucleating colonies of grains of similar orientation, especially axially with the growth direction. The colonies of grains of related orientation from a common nucleus, typically with higher axial than lateral growth rates, result in a knobby, bumpy, i.e. botryoidal or kidney-like free surface of the deposition (Fig. 1.8). Such colonies of grains with preferred axial orientations can occur with varying grain shapes, including blocky grains, but they commonly occur with bundles (colonies) of long, narrow columnar grains (Fig. 1.9). The colonies broaden out as growth occurs but then become constrained in their lateral growth by increasing competition with other colonies and in their axial growth by nucleation of new colonies, in all but very thin depositions.

Several factors should be noted about grain colonies. First, as a result of

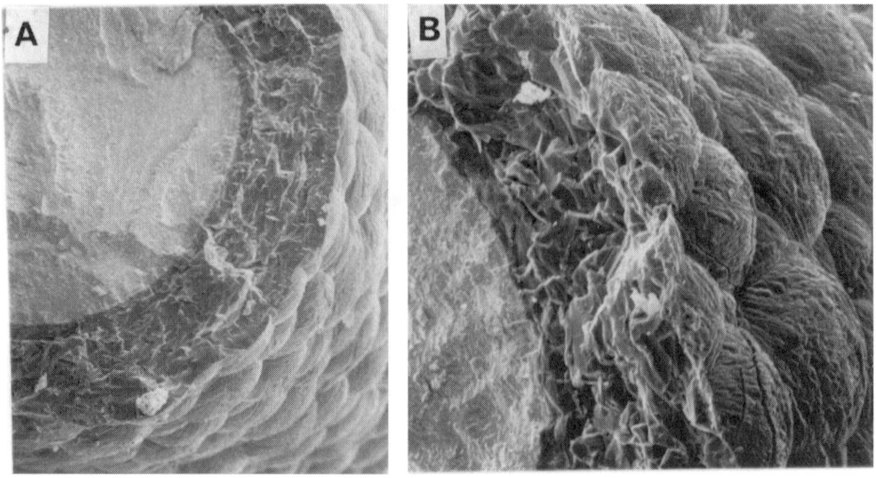

Figure 1.8 Example of the knobby, botryoidal surface of many deposited materials, in this case of a Si_3N_4 coating. (A) and (B) respectively lower and higher magnification SEM photos of deposition surface and cross section.

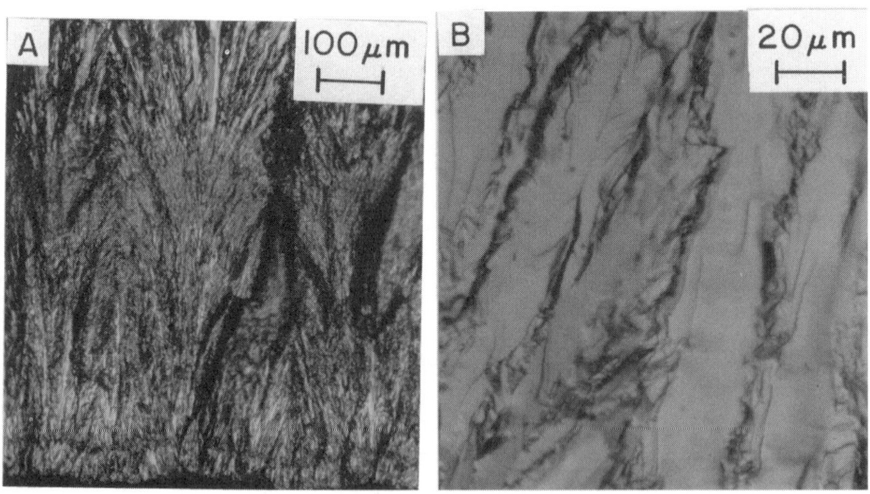

Figure 1.9 Example of fine, higher aspect ratio axially aligned grains on room temperature fracture of CVD SiC. (A) Lower magnification SEM showing the fan or sheaf cross section aligned with the axial growth direction (upwards). (B) Higher magnification showing the columnar character of the individual (transgranularly fractured) SiC grains.

the directional growth they have varying preferred orientations, mainly or exclusively in the axial direction, giving a preferred orientation to the resultant coating or body. This preferred axial orientation within each colony causes it to behave in some fashion as a pseudo larger, oriented grain or particle. Such grain-like behavior commonly includes impacting fracture propagation (Fig. 1.10), and possibly initiation, due to the preferred orientation, as well as possible weaker bonding to adjacent colonies because of orientation-property differences

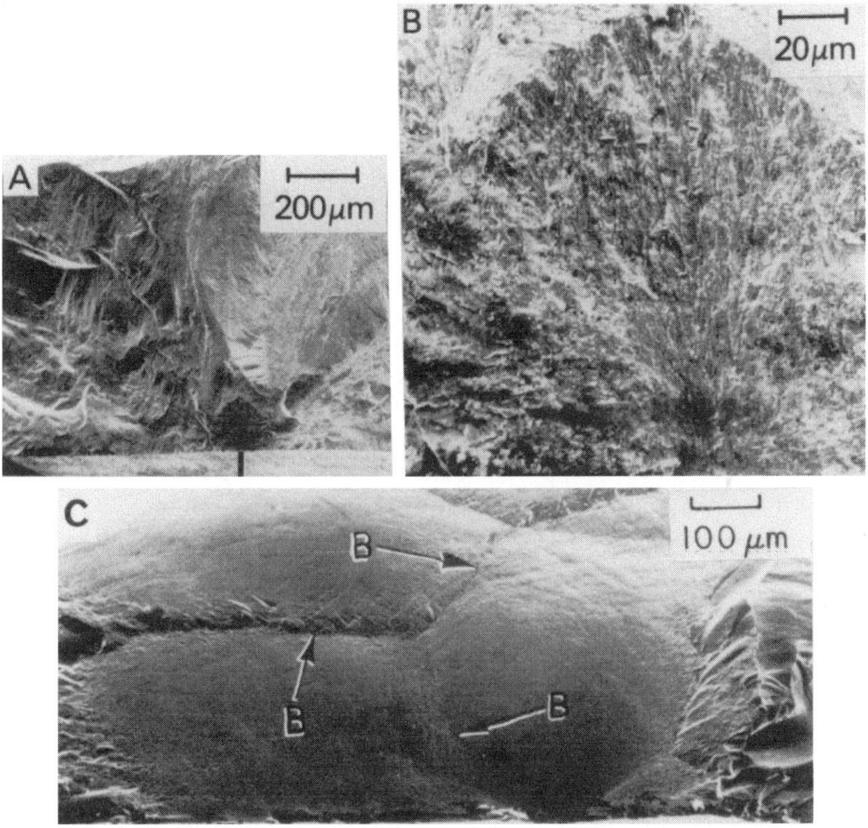

FIGURE 1.10 Examples of the effects of grain colonies in CVD SiC on fracture at 22°C. (A) and (B) Lower and higher magnification SEMs of fracture initiation (583 MPa) from a colony for stressing normal to the CVD growth direction. (C) Fracture surface from stressing parallel with the growth direction, hence fracture nominally parallel with the plane of deposition. The three major colony boundaries are indicated by B. (After Rice [4], published with the permission of the ASTM.)

and possible accumulation of impurities or other second phases and pores along colony boundaries. [Pores can more readily form near the base or along the shaft of the colony (bundle of grains) due to slower lateral versus axial growth of the colonies and shadowing of these lower regions due to greater lateral growth of the upper portion of the colony.] Again, the termination of the colonies on the completed deposition surface (i.e. the side opposite from the initiation of the deposition) is the characteristic boytrioidal (i.e. kidneylike) structure. While its occurrence and specifics vary with the material and deposition process, it is a very common aspect of all deposition processes. It is well known in the deposition of graphite, BN, SiC, and Si_3N_4, since these materials are common candidates for such deposition, and also is common in many other materials, e.g. TiN [26]. In the case of graphite it is common to cause some gas phase nucleation of particles, which on settling to the deposition surface act as nuclei for colonies, so their number is increased and their size reduced, hence mechanical properties improved. The boytrioidal structure and the related growth of grain colonies is also common, but often on a much larger colony scale in a number of mineral deposits, e.g. of important Fe and Cu ores [27].

Turning to other aspects of grain orientation and shape, consider the interrelation of these via particle shape effects in various fabrication methods. Consolidation of powder particles into green bodies results in some preferred particle orientation whenever some of the particles have measurable shape deviations from the equiaxed, i.e. spherical or regular polyhedral. Deviations in terms of particle surface geometry, especially larger flat surfaces, e.g. reflecting crystallographic faces formed in the growth or fracture, can play a role in orientation. However, particle aspect ratio, i.e. of the axial length to the lateral dimensions, is a key parameter. These particle effects, which can be additive, are a necessary condition for orientation, but they become sufficient only in conjunction with varying aspects of different forming methods and their parameters. Thus in pressing operations the aspect ratio of the green part to be formed is a key factor, e.g. lower ratios in die (cold) and hot pressing result in increasing alignment normal to the pressing directions, but ~ random orientation in the plane of pressing. Variations of alignment near the die walls and axial gradients can occur, with the latter often being greater in single versus double acting pressing. Tape casting and lamination typically increase the planer alignment, though the extent and uniformity of this may decrease with increasing tape thickness and decreasing ratio of particle dimensions to tape thickness. This results from shear being an important factor in particle alignment, and shear is highest at the doctor blade surface and grades inward. Similarly, extrusion can result in substantial orientation, which will be greatest at the surface and grade inward, the extent of gradation increasing with the extrudate thickness and decreasing with the ratio of the particle size to the extrudate dimensions. Much more complex orientation effects can occur in injection-molded bodies where particle orientation can occur in the stream f material injected into the mold, but then twisted, turned, and deformed in complex ways that depend on the character of the mold shape and its inlet ports.

Grain and particle parameters are also frequently intertwined with other microstructural parameters, especially porosity and impurities. Thus the location, shape, and orientation, and sometimes the size, of pores are related to grain, and especially particle, shape and orientation. For example, as noted earlier, pores in platelet composites commonly remain at the platelet–matrix interface [2], with such pores often being larger and typically somewhat platelet in shape. Similar effects have been indicated in fiber composites [1] and are likely in whisker composites. The relationship of particle and grain parameters being related to impurities results primarily from impurities often being the source of larger grains. Further, such larger grains are often tabular or acicular in shape, which may have implications for orientation and porosity interactions. Some of these effects are illustrated in the next section, and later in the book.

D. Grain and Particle Variations in Ceramic Composites

Grain and particle structures are both important and interrelated in natural and designed composites, and as with nominally single-phase ceramics, they depend substantially on the fabrication method and parameters. Consider first conventional powder-based fabrication methods, where the volume fraction of added particles, whiskers, fibers, or platelets and their sizes relative to the matrix grain size are key parameters. Typically the matrix grains inhibit the growth of the particles and vice versa. The relative degree of inhibition of the added phase on the matrix phase increases as the volume fraction of the added phase increases and its particle size decreases. This mutual inhibition of growth of the matrix and dispersed phases greatly reduces or eliminates the extremes of exaggerated grain growth in either phase and hence the complication of exaggerated grain (or particle) growth noted earlier for nominally single-phase ceramics. Thus just a few volume percent of fine, homogeneously distributed particles of an insoluble second phase can be quite effective in controlling grain growth, e.g. as clearly demonstrated in Al_2O_3 without MgO but with fine Mo or W [28–30] or ZrO_2 particles [31]. As discussed later, this inhibition of matrix grain growth can play an important role in the improvements of strengths of some ceramics and ceramic composites.

A partial exception to the mutual inhibition of growth of grains and particles in ceramic composites can occur in making such composites by in situ reaction of powder ingredients. Thus, Cameron et al. [32] showed that larger grains of various phases, e.g. Al_2O_3 (Fig. 1.11) and graphite, in such composites, though not as extreme in size as often found in nominally single-phase ceramics, were

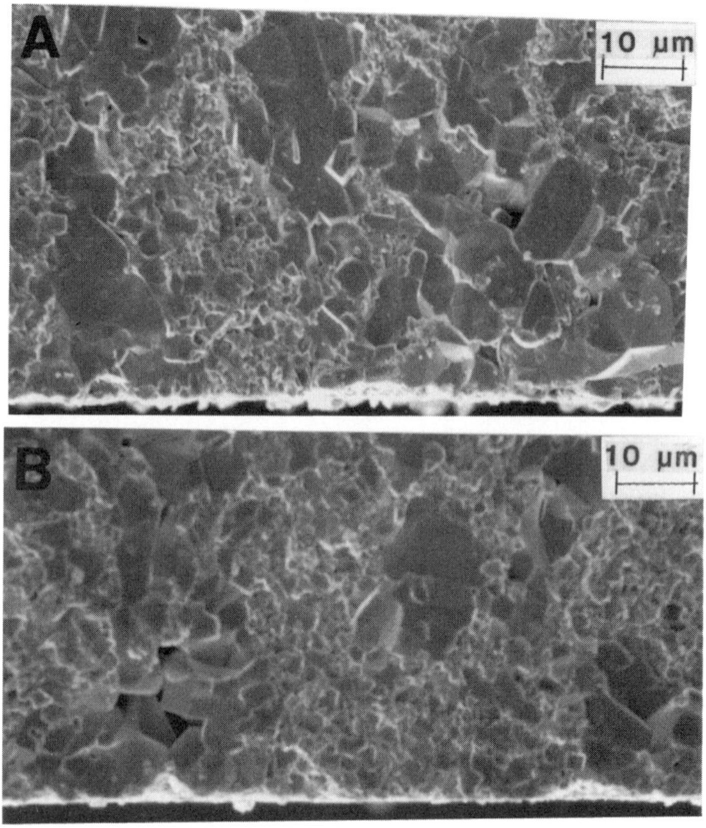

FIGURE 1.11 Example of more extreme microstructural heterogeneity in an Al_2O_3–27 vol% TiB_2 composite made by reactive processing. The larger, primarily or exclusively Al_2O_3, grains are attributed to melting and resulting "agglomeration" of the Al during its reaction with the TiO_2 and B_2O_3 during hot pressing. (From Ref. 32.)

frequently at fracture origins. At least some of these larger grains were attributed to transient formation of a liquid of a precursor, intermediate, or product phase, thus circumventing normal solid-state mutual inhibition of growth of the phases. The sizes of these grains is commonly large enough so that clusters of them are a factor in the failure of such specimens, but due to inhibiting effects in the composite it is still substantially less extreme than large grains that often occur in nominally single-phase ceramics.

In the above processes, the forming methods and their parameters have a major influence on the orientation of the dispersed phase, as well as some on the matrix grains as in nominally single-phase ceramics. All of this depends substantially on the volume fractions, sizes, and especially the shapes of the particles to form the matrix grains and the dispersed phase, and the relation of their shapes to their crystal structure. Use of whiskers, fibers, and platelets for composites are common and has great effects on their orientation in fabricating composites with them. The aspect ratio of the body and the degree of shear resulting from forming operations are key parameters. Thus lower body aspect ratios in die and hot pressing result in increasing alignment normal to the pressing directions, but—random orientation in the plane of pressing. Variations of alignment near the die walls and axial gradients can occur, with the latter often being greater in double versus single-action pressing. Tape casting and lamination typically increase the planar alignment. While characterization of the size, shape, orientation, and spatial distribution of the dispersed phase, as well as of the matrix grain structure, is important, so can be that of the residual pore structure. This arises first since dispersed phases inhibit densification, generally increasingly in the order of particles, whiskers (or short fibers), and platelets. Second, pores are commonly associated with the dispersed phase, the size and character of which impacts the size, shape, and orientation of the pores. Thus platelets in particular commonly have larger laminar or lenticular pores at the platelet–matrix interface.

Processing ceramic composite bodies from the melt is a large and complex subject beyond the scope of this chapter, since it entails many variations and complexities such as varying degrees of liquid solution or immiscibility, and subsolidus phase separation processes (e.g. Refs. 33–35). The latter include varying precipitation within grains and along grain boundaries, which all depend on kinetics, interfacial energies, and their interactions with thermal aspects of the solidification. The latter include the directionality and uniformity of the solidification, the degree of columnar grains formed, and especially their extent of preferred crystallographic orientation. However, beyond the above broad comments, two key points should be noted. First, some of the effects of the multiphase character are used to control the microstructure of the cast bodies, i.e. similar to the effects of multiphase compositions on grain and particle structures in sintered composites.

Second, the interaction of solidification, grain structures, and pore formation should be noted. Pores form in solidifying bodies due to extrinsic and intrinsic effects, the former arise mainly from two sources, as has been recently summarized [1]. An extrinsic source is the release of gases adsorbed on the surfaces of the particles to be melted, e.g. such a pervasive problem that most melt-grown crystals such as sapphire are made from previously melted material. The other extrinsic source is the exsolution of gases upon solidification that were dissolved in the melt. The intrinsic source of porosity in bodies solidified from the melt arises from the typical reduction of volume on solidification. Many ceramics have volume reductions of 5–10%, as is common for most metals, but they can be much higher for some ceramic materials, e.g. to > 30%, with alumina compositions commonly having ~ 20% [36].

Introduction of porosity, especially intrinsically generated porosity, can be eliminated by controlled directional solidification wherein the final solidification occurs at a free surface. Such directional solidification typically results in columnar grains of preferred orientation. In bodies at or near eutectic compositions, proper control of the solidification front can result in aligned columnar grains or of single crystals with axially aligned rod or lamellar second phase structures [37–39]. This is a large and extensive topic that is briefly noted here, since such solidified eutectic structures have been of interest for their mechanical behavior (Chap. 8, Sec. V. E, Chap. 9, Sec. III.F, and Chap. II, Sec. IV.E).

Another aspect of melt-processed materials is the crystallization of liquid or amorphous materials, e.g. of glasses. While there are various aspects to this, three are of particular note based on the source and scope of crystallization. The first is homogeneous nucleation throughout the body, which occurs only in a minority of cases. It provides more uniform microstructures, which can be complex at higher levels of crystallization, where crystallites begin to impinge on one another, which occurs sooner as the aspect ratio of the crystallites increases. The second, and more common, case is nucleation from free surfaces, either of a bulk body of the crystallizing glass, or particles of it, e.g. in the latter case prior to or during their consolidation into a dense body. In either case gradients of crystallization commonly result.

Nucleation of a group of crystals from a common point (often a particle of an added nucleation agent) commonly results in the formation of colonies of rod or needle grains radiating from the common nucleation point (Fig. 1.12). The colony grain structure noted in vapor deposited materials in the previous section (Figures 1.7–1.9) is an important subset of a broader occurrence involving the same basic underlying nucleation and growth of grains that is referred to as spherulitic crystallization. Such crystallization extensively occurs from a variety of liquids, including salt solutions, polymers, and melts, e.g. of some fusion cast refractories [34, 35], and especially for glasses (e.g. Refs. 40 and 41). Thus such crystallization occurs in various ceramic and related materials such as cementitious materials and in inorganic, i.e. many silicate-based, glasses [42, 43] of particular pertinence here. There are two aspects of such crystallization that significantly impact the nature of the resultant colony structure, namely the directionality of the nucleation and the subsequent growth of the colony grains and the mutual impingement of growing colonies. Colonies originating from a nucleus small in comparison to the resultant colony size and growing in an isotropic medium may vary from a slightly dog-bone-shaped cross section to a fully circular cross section, depending on parameters that are characteristics primarily of the material and secondarily of the growth conditions. Fully spherical colonies reflect central nucleation and radial growth (Fig. 1.12).

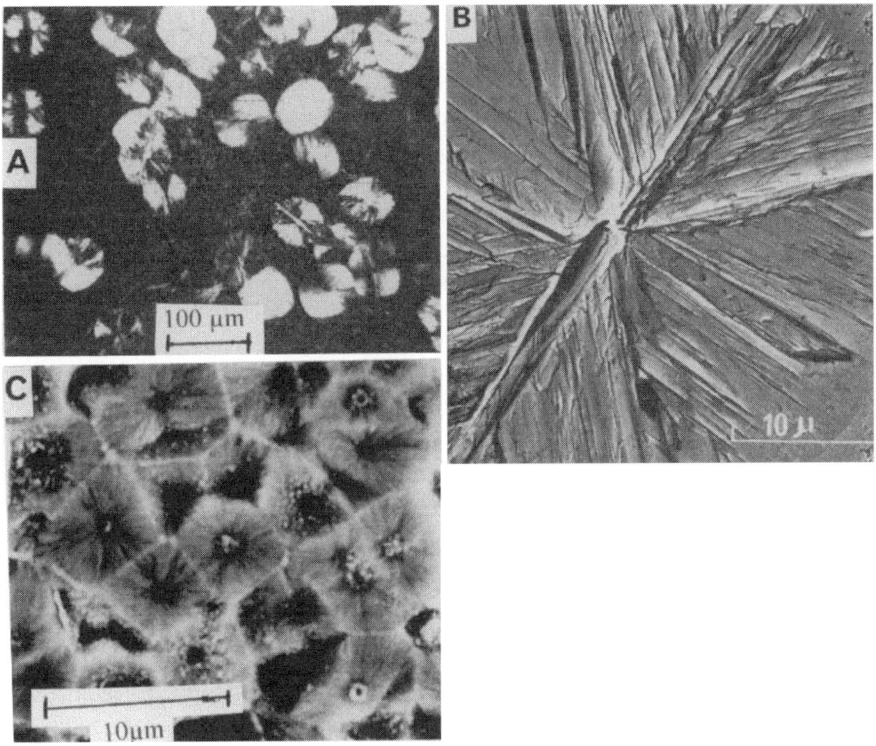

FIGURE 1.12 Examples of spherulitic crystallization in silicate glasses. (A) Lower magnification of an optical thin section showing an earlier stage of such crystallization in a LiO_2–SiO_2 glass. (B) Replica electron micrograph of cross section of a more mature spherulite. (C) Fully impinged spherulites in crystallized $3BaO$–$5SiO_2$ glass. (From Refs. 42 and 43. Photos courtesy of Dr. S. Freiman of NIST Published with the permission of the Journal of the American Ceramic Society.)

Impingement of growing colonies on one another constrains and changes their resulting shapes, e.g. from spheres to polyhedrons (Fig. 1.12C) similar to grain shapes, e.g. tetrakaidecahedrons. Both directional nucleation, e.g. from larger particles (as in many cementitious materials), or in conjunction with growth fronts (as in the above-discussed vapor deposition processes), and resultant constraint of the growth directons significantly alter the geometry, as shown for CVD materials. However, other complications often occur, e.g. in crystallizing glasses. Important examples are differing separate or overlapping stages of crystallization, whether spherulitic or not, that start, finish, or both the crystallization of the glass. Thus for example prior spherulitic crystallization of $3BaO$-$5SiO_2$

glass is subsequently destroyed by subsequent crystallization into larger lath grains at higher temperatures [43].

Individual grains in such spherulites typically have a preferred orientation in the radial direction but random orientation about their radial axis. Thus one of the characterization challenges that such spherulite structures pose is the contrast of radial orientation in each spherulite versus the global orientation of the body. The latter is typically random, since the orientation within each spherulite is radially symmetric, and there is rarely any coordination of the orientation of different spherulites. However, besides complicating the microstructure, these spherulites can play a role in some key, especially mechanical, properties since they often act as large grains, i.e. as noted earlier for spherulitic clusters of larger grains in sintered bodies (Fig. 1.3) and grain colonies in deposited, e.g. CVD bodies (Figures 1.8–1.10).

IV. GRAIN AND PARTICLE CHARACTERIZATION AND PARAMETERS

A. Overview

Thorough microstructural characterization to address most, especially all, of the challenges and uncertainties noted in previous sections is a large task that is often beyond the scope and funding allowable for such evaluations. Detailed description and discussion of all, or even the most important, characterization methods and how to make them more effective is a task beyond the scope of this book. The reader is referred to a few references on general characterization (Refs. 44–47, e.g. more current sources as recommended by academic colleagues). The primary problem, however, is that much of the pertinent microstructural characterization in the literature is marginal or inadequate. This arises in part from limitations of many techniques as well as their often being incompletely, inadequately, or inappropriately used, as well as test methods, parameters, or results often being inadequately described. Thus an overview of the process and discussion of some key needs is presented.

The basic approach to the microstructural characterization needed to resolve important issues effectively from both technical and practical standpoints generally consists of two aspects. First and most fundamental is to draw upon information on the material, fabrication-processing, and microstructural trends, preferably from both the literature, e.g. as outlined earlier, and the experience of the investigator and colleagues as a guide for probable needs. The second aspect is a multiple stage, often iterative, characterization of samples, starting with a screening stage of the body being investigated.. Though blanket application of routine evaluations, e.g. basic stereology, can be valuable, selective characterization in a staged fashion based on initial screening and subsequent information is often more effective.

The initial screening stage should identify the basic microstructural characteristics pertinent to the behavior of concern, as well as indications of the presence or absence of complications such as anisotropies and variations, e.g. gradations and heterogeneities. Such screening should typically entail a range of samples and tests (e.g. as discussed in the next section) aimed at indicating what microstructural factors may be pertinent, variable, and in need of more definition. It should commonly include examination of representative fracture surfaces for both time and cost effectiveness, as well as fracture information, especially for mechanical, but also frequently for nonmechanical, property evaluations. Subsequent characterization stages should be guided by results from previous stages and remaining needs to clarify or confirm the microstructure-material behavior of concern.

The first of three general guides for such evaluations is remembering that the need for, and the effectiveness of, microstructural characterization can depend on the nature of the properties measured and the specifics of the measurements conducted. Thus one microstructural value or technique will not meet all needs as discussed below, e.g. to reflect different effects or extents of variations of the grain size distribution, shapes, and orientations that have differing effects on different properties. Second, qualitative or semiquantitative characterization should always be given in the absence of, as well as often with, detailed quantitative characterization, e.g. illustrative microstructural photos (with scales and comments) can be valuable. This is also true of descriptions of the fabrication, process parameters, and resulting samples. Examples of more detailed, but still incomplete, grain structure descriptions are those of Ting et al. [48] and especially McNamee and Morrell [49]. Third, it is usually valuable to compare different related measurements, e.g. this is often of more value than one more detailed measurement, since all have their limitations.

B. Grain and Particle Size Measurements

Turning to actual measurements of grain and particle size, two aspects of this are detection of the grain and particle structure and then its measurement, which together can entail several stages. Most grain and particle structures are detected by various microscopies, typically in the order of decreasing simplicity and increasing magnification ability and cost: optical (OM), scanning electron (SEM), and transmission electron microscopy (TEM, occasionally still of replicas, but mainly of thin sections). TEM for very fine (submicron) grain or particle structures overlaps with analysis of x-ray and neutron diffraction data to yield grain and particle size information. OM and SEM can be done directly on either polished or fracture surfaces, and OM can also be done on optical thin sections. While both polished and fracture surfaces may require etching to reveal the grain and particle structures, this is primarily so for polished surfaces, which are far more time-consuming and costly to prepare. Thin sections are also more time consuming and costly to prepare but typically readily define size and shape aspects of the microstructure and can often give substantial orientation and some crystallographic information. Fracture surfaces also have the advantage of revealing the grain structure of fracture, which can be critical for revealing effects of isolated larger grains or grain clusters, especially of thin platelet grains. Replicas of fracture are sometimes used in OM and SEM to increase the amount of reflected light or contrast definition of the microstructure.

Microscopies typically give photo or screen images from which measurements can be made by various, especially stereological, techniques. A comprehensive discussion of these techniques is a large and complex topic beyond the scope of this book, since there are variations of, limitations of, and complications to many of these techniques. The reader is referred to other sources on the subject [50–63]. The goal here is instead to note basic techniques and limitations and suggest basic and practical approaches to provide guidance, and to stimulate further development of the techniques.

At the lower, basic end of a potential hierarchy of needs is a simple but important nominal average grain or particle size value. An approximate value can often be obtained by measuring a few representative grains or particles, and a more accurate value from the commonly use linear intercept and related techniques. These give $G=\alpha l$, where l is the average intercept length for grains along a random sample line and α = a constant (commonly ~ 1.5) to account for the fact that neither the sampling lines nor the plane on which they are taken cut grains at their true diameters. However, both theory and experiment show α values ranging from <1 to >2, due to only partially understood dependencies on grain shape and size distributions [64, 65] and possibly on whether the surface is polished or fractured [66, 67] and the degree of inter- vs. trans-granular fracture. (See Refs. 68 and 69 for other limited microstructural evaluation from fracture surfaces, and Ref. 70 for characterization of intergranular and transgranular fracture.) It is thus important to give measurement specifics, including what value is used to obtain the "true" grain size, since there is no single conversion value. Many investigators simply use α = 1 but often do not state this nor give enough other information.

Next, consider several aspects of linear intercept and related measurements starting with the reliability and repeatability of determining an average G value based on detailed comparative round robin tests [71, 72]. These showed about a 10% scatter among 25 international laboratories on an ideal computer-generated grain structure, based on counting at least 100 linear (or circular) intersections (and is estimated to be only cut in about half by going to 1000 intersections). Scatter increased to ~ 25% for a "nice" (96% sintered alumina) microstructure (equiaxed grains of relatively uniform size with clearly marked grain boundaries). The increase was attributed to factors such as differences in polishing and etching techniques and their effectiveness. Analyses of a more complicated but not uncommon (sintered 99% alumina) microstructure resulted in an ~ doubled scatter of results (and somewhat higher scatter still for similar measurements of the limited porosity). Thus measurements of grain sizes are likely to vary by 25–50% or more, and uncertainties in conversions to a "true three-dimensional size" can double or triple this variation or uncertainty.

Consider now four issues of linear intercept (and related circle and grid) measurements. These are, in order of decreasing development and increasing seriousness, (1) that it is limited in giving values reflecting the range of grain size, e.g. the standard deviation (since analysis is based on a single-size, equiaxed grain), (2) that these are limitations in handling nonequiaxed grains or particles, e.g. of tabular or rod-shaped grains or particles, (3) there being no precise way to relate an average G or D value with an individual, e.g., maximum value, i.e. G_m or D_m (e.g. there is no way accurately to relate the measurement of a single grain diameter on a sample surface to random grain chords or linear intercepts), and (4) a single G or D value may often not be sufficient, i.e. there can be a need for different grain or particle size values to reflect differing impacts of the grain or particle structure itself, or of its variation in size, shape, or orientation. Though progress has been made allowing estimates of the grain size distribution, it still assumes uniform, e.g. spherical or tetrakaidecahedral, grains [50–56, 64–66]. Exact accounting for a mixture of grain sizes can be made by computations for mixtures of a few groups of different size grains, with each group consisting of uniform size spherical grains [50]. Methods for a broader range of grain sizes have been presented [73].

Progress has also been made on conversion of measurements of nonequiaxed grains or particles, e.g. of tabular or rod-shaped grains or particles and their shape factors on a plane (typically polished) surface to their true three-dimensional character. An exact relation has been derived and validated assuming identical cylindrical grains as an extreme of elongated grains [74]. More recently, using similar idealizing assumptions based on spheroids of uniform size and shape [75], a shape factor, R, has been recommended, defined as

$$R = (A_T)^{-1}\Sigma_i A_i f_i \qquad (1)$$

where A_i is the area on the sampling plane of the individual grain, f the corresponding grain aspect ratio, and A_T the total area of all grains on the surface being evaluated. Handling of bi- or multimodal distributions of such nonequiaxed grains is recommended via a rule of mixtures based on the areal weighting of each population of elongated grains or particles. However, as noted below for analysis of fracture behavior, specific dimensions are still important. Further, while the above-outlined procedures are of significant help, they involve uncertainties, which can be significantly compounded by factors such as shape, size, and orientation distributions in a single grain structure.

Consider now the comparison of average values of grain or particle sizes in a given body when there are grains or particles present representing other size (or shape) populations. There are at least three cases of concern. The first is bi- or multimodal populations (e.g. Fig. 1.1). This presents the least problem, especially when the differing populations are clear and present in sufficient quantities so that normal techniques can be applied to each distinct population. The second, and particularly serious, case is when there are only a limited number of substantially larger (or differently shaped, or both) clustered or individual grains, especially as fracture origins (e.g. Figures 1.2–1.5). Accurate comparison of an average size from linear intercept and related measurements with the size of one or a few larger grains or particles on a fracture surface is not possible, though it can be addressed by the use of two-dimensional sizes as discussed below. A similar problem with essentially the same solution is that commonly found in bodies with some, and especially most, or all large grains. The problem arises since the large grains at the surface, though commonly substantially truncated by machining (e.g. Figures 1.2 and 1.3C) are still often the fracture origin, but the body average grain size does not reflect their, often substantially, reduced dimension.

However, even with a normal monomodal population there are issues of which average size is appropriate. While a linear average size, i.e. weighted by the first power of G or D for each grain or particle size respectively, is commonly used, since it is the simplest to obtain, it is often not the most physically meaningful [5]. Such a linear average gives a high weighting to small gains and a low weighting to larger ones [76], which is opposite to important trends for some key mechanical properties. Thus an average based on the volume of the grains or particles, hence weighted by G^3 and thus substantially by the presence of larger grains or particles, may often be more appropriate for some property comparisons. Examples of this are where mass distribution or volume absorption (e.g. of radiation), or diffusion in composite, or more commonly in single-phase, bodies are important. More pertinent to this book is where properties are related to the surface areas of the grains or particles. Thus where diffusion, conduction, or fracture along grain boundary surfaces is pertinent, an average weighted by G^2, which gives more, but not extreme, emphasis to larger grains or particles may be appropriate. Similar, and of somewhat greater interest in this book, are cases where properties or behavior depends on the cross-sectional area of grains or particles. Key examples are transgranular fracture in crack propagation tests and especially tensile and compressive failures, and hardness, wear, and erosion resistance evaluations. Electrical and thermal conduction, especially in composites, may often fall into this latter category. For example, weighting based on grain area, i.e. $G = [\Sigma_i G_i^3][\Sigma_i G_i^2]^{-1}$, as opposed to grain diameter, i.e. $G = (1/n)\,[\Sigma_i G_i]$, increases the impact of larger grains on the average (e.g., in measuring diameters of 30 grains on a commercial lamp envelope Al_2O_3, the area-weighted and normal G_a were respectively 51 and 29 μm) [3, 5]. An alternant, direct weighting method for obtaining a "composite" grain size was suggested by Goyette et al. [77], who noted that it was important in correlating the microstructure of various commercial alumina bodies and their response to single-point diamond machining.

The other aspect of which *G* or *D* value to use is whether it should be a two- or a three- dimensional value [5]. Some properties and behavior are better correlated with a three-dimensional grain or particle size, e.g. elastic properties (of composites) and electrical and thermal conductivities. However, since fracture is an area-dependent process, it is more realistic to make measurements of grain or particle dimensions actually exposed on the fracture surfaces. Similarly, the area intercepted by larger grains on wear or erosion surfaces is more important than the three-dimensional sizes. Converting a linear intercept measurement to an average surface grain diameter, G_s, might be done using $G_s \sim (\{1 + \alpha\}/2)\, l$ (i.e. assuming that half of the correction, α - 1, is due to the randomness of the sampling plane cutting the grains and half to the randomness of the linear intercept itself), but it is uncertain in both the form and the actual α value. Overall, it appears better to measure actual grain or particle diameters exposed on the fracture surface, e.g. selected by using random lines as in the linear intercept method. Such measurements would be directly related to measuring individual, e.g. the largest, grains on fracture. Having actual grain or particle diameters on a fracture surface allows the calculation of an average grain size (*G*) or particle size based on various weightings. Such two-dimensional values should be more pertinent to bulk fracture properties as well as wear and erosion phenomena.

Two further points should be noted. First, for fracture from elongated grains or particles, use of a size reflecting their area on the fracture surface is typically a good approximation [78], but more accurate calculations require the actual dimensions. Some used the maximum grain or particle dimension, i.e. length [11], but fracture mechanics uses the smaller dimension of an elongated, e.g. elliptical, flaw as *C* (the larger dimension impacts the flaw geometry parameter), so this is inappropriate (as is the smallest grain dimension by itself, as previously suggested [41]); an intermediate grain or particle size value is appropriate. However, the aspect ratio (and orientation effects) may be important, e.g. as indicated by Hasselman's [79] modeling of effects of elastic anisotropy of grains on mechanical properties. Second, it is important that the different, i.e. two- and three-dimensional grain or particle size values be relatable, which requires substantial further analytical and experimental evaluation.

C. Spatial Distribution and Orientation Measurements

Besides grain or particle size and shape, the spatial distributions of these parameters can also be important, especially if fracture origins were not identified. Systematic spatial variations of sizes and shapes, e.g. between the surface and the interior, e.g. from loss of additives near the surface or machining truncating large surface grains (Figures 1.2 and 1.3C), are easier to handle. Handling random or irregular distributions of larger grains or particles (e.g. due to variations in initial particles, additives or impurities, or porosity) indicates the need for statistical methods, especially to determine spacings of larger grains, particles, or their clusters. Stoyan and Schnabel [80] used a pair correlation approach to address this problem, characterizing the frequency of interpoint distances (e.g. between grain vertices or centers—the latter was preferred). They showed a higher correlation of strength for nearly dense Al_2O_3 bodies than with G_a (~ 9 to 15 μm) itself. Modern stereological tools make such characterization more practical (and potentially applicable to pores and pore–grain or pore–particle associations), but again fractography is the most assured method of addressing this.

Overall, i.e. global, preferred crystallographic orientation of grains or particles clearly occurs in varying degrees as a function of forming methods and processing and material parameters, as is discussed in Chaps. 2–12. Such orientation can clearly affect properties, especially mechanical ones, in a desirable or undesirable fashion, and is often a major issue in understanding the microstructural dependence of properties, especially when its presence is not accounted for. When associated with grain or particle geometry, especially elongation, substantial orientation information can be obtained from stereological measurements. However, x-ray diffraction techniques, which range from qualitative Laue patterns, to comparisons of intensities of various x-ray lines, to complete pole figures, are typically more versatile, effective, and widely used. Pole figures are the most accurate, comprehensive, costly, and time-consuming but have been greatly aided by modern computer aided characterization, and possibly by newer, additional methods of determining grain orientations [81–83]. Also of potential importance is local grain or particle orientation, e.g. of individual elongated or platelet grains or particles, or clusters of them, especially at fracture origins. Such information can be important to determine if microcracking from thermal expansion anisotropy was a factor in fracture. In the past such information was difficult to obtain. However, modern microscopic analytical techniques are providing increasing capabilities in this area [84].

Thus, in summary, at least two different but related *G* or *D* values may often be needed, one based on cross-sectional, especially fracture, area for fracture, as well as probably wear and erosion, and one on three-dimensional size for elastic and conductive properties. Both values should be relatable, which is a challenge due to varying size, shape, and orientation not necessarily being independent of one another and their interrelations probably varying for different properties. Further, the impact of these grain parameters on properties can depend significantly on the spatial distributions of each of these variables. Thus the size of isolated larger platelet grains is likely to have limited effect on some properties such as electrical or thermal conductivities, provided they are not associated with other important microstructural complications such as accumulations of second phases or microcracking. However, even in the absence of the latter complications, they can have varying effects on fracture properties ranging from erosive particle impacts, through crack propagation such as for fracture toughness, to tensile failure. Effects of platelet grains on erosion depend on both the extent of their being in particle impact zones and their effects on erosion via local fracture in these zones, both of which depend separately and collectively on their volume fraction and their size, shape, and (local and global) orientation. However, once the critical size, shape, and orientation range for local fracture is reached for the given erosive environment, then the volume fraction dominates. Similarly, volume fraction of platelet grains is important in fracture toughness, but their orientation, shape, size, and spatial distribution, and the scale of these relative to the scale of crack propagation, can all be factors that may impact typical fracture toughness values in a different fashion than erosion. Typically more extreme is tensile failure, which is determined more by the occurrence of individual or clustered platelets near or above the critical flaw size, oriented at or near normal to the tensile axis, and in combination with another defect such as a pore or a crack (e.g. from machining the surface, or mismatch stresses). Thus it is important to recognize that grain and particle sizes and related characterization for correlation with strength typically represent a fair amount of uncertainty, e.g. a factor of 2 or more, and can thus be a factor in variations between different studies along with uncertainties and differences in the properties themselves. Poor coordination of property and microstructural measurements can exacerbate such problems.

V. COORDINATION OF PROPERTY AND MICROSTRUCTURAL MEASUREMENTS

The microstructural dependence of fracture properties, especially tensile strength, is a good example of the importance of judicious interrelation of test and microstructural evaluations. Thus, as shown extensively later, larger isolated or clustered grains or particles commonly determine the strength of a given specimen. Clearly, both the grain or particle size and the strength values used to relate strength and microstructure need to be self-consistent with each other. However, strength is very commonly based on the outer, i.e. maximum, fiber stress in flexural failure (σ_m, e.g. as used by all investigators coordinating strengths with maximum grain size). Such a maximum strength is, however, basically inconsistent with use of a maximum grain size (G_m) and is often more consistent with use of an average grain size (G_a), since the latter has a moderate to very high probability of being associated with σ_m, while G_m has a moderate to very high probability of being associated with $< \sigma_m$ (indicating lower larger G slopes for those using G_m and σ_m, as is discussed later). These probabilities and the errors involved in using G_a or G_m depend on both the size and spatial distribution of grains and the stressed volume and surface, along with the extent or absence of stress gradients. Smaller volumes under high stresses more likely reflect less deviation from the G_a, i.e. the use of three- versus four-point flexure, as well as smaller specimen cross sections and corresponding shorter spans. Round flexure rods have the smallest stressed volume but a larger surface area from which surface-related flaws can be activated at variable stress. Progressing to true tensile tests (e.g. from three- to four-point bending, to hoop tension then uniaxial tension) as well as increasing specimen sizes in each test gives grater emphasis to failure from microstructural extremes and to the role of associated defects (mainly pores and cracks). Of equal or greater importance is the effect of temperature. A recent review [85] and Chaps. 6, 7, and 11 show that temperature changes of a few hundred degrees Celsius can shift the microstructural dependence of behavior in different fashions for different materials and thus can be a valuable tool in defining the mechanisms controlling behavior.

Fractography, besides being important for microstructural characterization, is also the first and most fundamental of three approaches to properly correlate fracture, especially strength, and grain or particle parameter values. This, if successful, allows both the actual G and the location of fracture initiation to be determined. With the latter the failure stress (if $<\sigma_m$) can be corrected for stress gradients into the sample depth. (Correction for off-center failures due to gradients along the sample length, e.g. for three-point flexure, is a separate operation from fractography.) However, as noted above, even with fractography there can still be considerable uncertainty; hence the need for other approaches. The second approach is to use the various microstructure measurements and analysis discussed in the previous section, especially in conjunction with specimen stress–volume and surface area relations noted earlier. Thus smaller specimens and stressed volumes reflect less G variation, so G_a is more reasonable, while larger specimens and more uniform stressing emphasizes effects of microstructural extremes. The third approach is to use the known property-microstructure behavior as a guide. This and the other approaches are best when done in combination with one another, e.g. for specimens known to have a range of G, the statistical fit of its σ with other data, especially for more homogeneous grain structures, at the pertinent G values can be used as a guide for the placement (or rejection) of a data point probably also aided by fractography. Lack of such combinations and comparison has been a serious shortcoming of many earlier studies, including those using G_m.

The need for fractography for other mechanical tests has also been demonstrated in crack propagation–fracture energy and toughness tests, e.g. again showing the impact of microstructural heterogeneities such as larger grains [86]. Thus tests with different crack sizes and extents of propagation should be of value, especially with fractographic examination. Similarly, evaluation of wear and erosion as a function of impacting particle sizes, velocities, and materials can be important, again especially when coupled with microstructural (i.e. often local fractographic) examination of various wear or impact sites. Again tests as a function of even limited temperature increases can be very valuable.

Finally, the importance of demonstrating isotropy of properties, instead of assuming it without substantial reason, is critical because of the frequent occurrence of some anisotropy in bodies commonly assumed to be isotropic. Such evaluations should include measurements of other properties, which are of broader importance than just the issue of isotropy. Thus elastic property measurements are important to correlate not only with other mechanical properties but also with non-mechanical ones, e.g. electrical and thermal conductivities, as well as of the latter with other mechanical properties, especially in composites. An important aspect of such intercomparison of properties is not only the specific property values but also their distribution. Thus the Weibull modulus of failure from mechanical testing is similar, but not identical, to that for failure from dielectric breakdown [87, 88], reflecting the similarities and differences of the sources of such fracture and breakdown [1]. While both failures are impacted by locally higher porosity, mechanical failure is determined more by a compact area of more, larger, or both pores, usually intergranular, often close or connected, ones, acting as much or all of the failure causing flaw. Thus the cross-sectional area of the pores parallel with the stress is not a key factor, while their area normal to the stress is, along with their close spacing. Electrical breakdown also is fostered by accumulations of pores, especially intergranular ones (often associated with larger grains, and probably boundary phases). However, pores that are most serious in electrical breakdown are those in an (often discontinuous) chain forming a failure path with the least solid material through the body. The closeness or connection of the pores and their cross-sectional area normal to the breakdown is not as critical, while their net area along the resultant breakdown path is. Another factor is that most mechanical tests are in flexure, and hence in a stress gradient, while most electrical breakown tests are in a uniform electrical field analogous to uniaxial tensile testing. Recognition and use of such differences in the details of the microstructural effects of varying tests and properties is thus a very valuable, but seldom used, tool to separate out specifics of the microstructural dependences of properties.

VI. SUMMARY AND CONCLUSION

Grain and particle parameters, especially but not exclusively size, play an important role in many, especially mechanical, properties. Though often less in magnitude than effects of porosity, grain and particle shapes and orientations and especially sizes can commonly vary important properties by up to a few to several fold. Since increased properties from control of grain or particle parameters are typically over and above those obtained by minimizing porosity, they are essential factors in obtaining high property levels. While their impact is greatest through primary effects, secondary effects can also be important but have received much less attention.

Microstructures range from fairly simple to very complex, offering both challenges for characterization and opportunities for achieving desired properties. However, accurate characterization of even simple microstructures can be complex and uncertain in meeting needs. For example, at least two grain and particle sizes are needed, one based on area and one on volume (i.e. respectively two- and three-dimensional) sizes for respectively fracture and wear and erosion behavior versus elastic and conductive properties. Interrelation of these sizes is also needed, but many current measurements reflect mainly a single size value with substantial variation and uncertainty, e.g. by factors of the order of 50 to a few hundred percent. This allows useful comparison of different studies of the same property, but these variations must be considered as a source of difference along with property variations due to differences in specimens and the tests used.

There are a diversity of tools for improved microstructural measurements that can be of considerable aid if more extensively applied, but these are not the complete answer. Some alternate or additional measurements are needed, e.g. to compare maximum with average grain or particle sizes, with the latter reflecting two- or three-dimensional, i.e. surface or volume effect, values. However, the most immediate and important needs are in first perspective on properties and second in their measurement. Thus given the diversity of microstructures, materials, and resultant properties, improved information, perspective, and understanding is needed in three areas. The primary need is for a more balanced perspective, in particular recognition of two key factors: (1) many toughness results are of uncertain or limited pertinence to much strength behavior, and (2) much of the grain and particle dependence of strengths derives from their impact on the size of flaws introduced that control failure, as extensively shown in this book. The other two needs are for better and refined observations and improved documentation via more and better microstructural characterization and more data on a broader range of bodies and especially microstructures.

REFERENCES

1. RW Rice. Porosity of Ceramics. Marcel Dekker, New York, 1998.
2. IK Cherian, MD Lehigh, I Nettleship, WM Kriven. Stereological observations of platelet-reinforced mullite- and zirconia-matrix composites. J Am Cer Soc 79(12):3275–3281, 1996.
3. RW Rice. Microstructure Dependence of Mechanical Behavior of Ceramics, Treatise Mat Sci Tech Properties and Microstructure, 11 (R. C. McCrone, ed.). Academic Press, NY, pp. 199–381, 1977.
4. RW Rice. Ceramic Fracture Features, Observations, Mechanisms, and Uses, Fractography of Ceramic and Metal Failure (J. J. Mecholsky, Jr. and S. R. Powell eds.). ASTM STP 827. Philadelphia, PA, pp. 5–103, 1984.
5. RW Rice. Review ceramic tensile strength-grain size relations: grain sizes, slopes, and branch intersections. J Mat Sci 32:1673–1692, 1997.
6. RW Rice. Machining Flaws and the Strength-Grain Size Behavior of Ceramics. The Science of Ceramic Machining and Surface Finishing II (B. J. Hockey and R. W. Rice, eds.). NBS Special Pub. 562. US Govt. Printing Office, Washington, DC, pp. 429–454, 1979.
7. RW Rice. Microstructural dependence of fracture energy and toughness of ceramics and ceramic composites versus that of their tensile strengths at 22°C. J Mat Sci 31:4503–4519, 1996.

8. C-W Li, J Yamanis. Super-tough silicon nitride with *R*-curve behavior. Cer Eng Sci Proc 10(7–8):632–645, 1989.
9. JA Salem, SR Choi, MR Freedman, MG Jenkins. Mechanical behavior and failure phenomenon of an In Situ toughened silicon nitride. J Mat Sci 27:4421–4428, 1992.
10. N Hirosaki, Y Akimune, M Mitomo. Effect of grain growth of β-silicon nitride on strength, weibull modules, and fracture toughness. J Am Cer Soc 76(7):1892–1894, 1993.
11. S Prochazka, RJ Charles. Strength of boron-doped hot-pressed silicon carbide. Am Cer Soc Bull 52(12):885–891, 1973.
12. RW Rice. Fabrication and Characterization of Hot Pressed Al_2O_3. NRL Report 7111, 1970.
13. RW Rice. Strength/grain-size effects in ceramics. Proc Brit Cer Soc 20, 205–213, 1972.
14. RM Gruver, WA Sotter, HP Kirchner. Fractography of Ceramics. Dept. of the Navy Report, Contract N00019-73-C-0356, 1974.
15. RW Rice. Fracture Identification of Strength-Controlling Flaws and Microstructure. Fracture Mech Cer (R. C. Bradt, D. P. H. Hasselman, and F. F. Lange, eds.) 1:323–343, 1974.
16. RE Tressler, RA Langensiepen, RC Bradt. Surface-finish effects on strength-vs-grain-size relations in polycrystalline Al_2O_3. J Am Cer Soc 57(5):226–227, 1974.
17. HP Kirchner. Strengthening of Ceramics: Treatments, Tests, and Design Applications. Marcel Dekker, New York, 1979.
18. RW Rice. Specimen size-tensile strength relations for a hot-pressed alumina and lead zirconate/titanate, Am Cer Soc Bull 66(5):794–98, 1987.
19. TL Francis, FE Phelps, G MacZura. Sintered sodium beta alumina ceramics. Am Cer Soc Bull 50(7):615–619, 1971.
20. JN Lingscheit, GJ Tennenhouse, TJ Whalen. Compositions and properties of conductive ceramics for the Na-S battery. Am Cer Soc Bull 58(5):536–538, 1979.
21. AV Virkar, RS Gordon. Fracture properties of polycrystalline lithia-stabilized β″-alumina. J Am Cer Soc 60(1–2):58–61, 1977.
22. RW Rice. Processing Induced Sources of Mechanical Failure in Ceramics. Processing of Crystalline Ceramics (H. Palmour III, R.F. Davis, and T.M. Hare, eds.). Plenum Press, New York, pp. 303–319, 1978.
23. K Murakami, U Osaka, S Sampath. Private communication regarding plasma sprayed alumina-13 w/o TiO_2. Centre for Thermal Spray Research, SUNY, Stoney Brook, NY, 1998.
24. RP Ingel, RW Rice, D Lewis. Room-temperature strength and fracture of ZrO_2-Y_2O_3 single crystals. J Am Cer Soc 65(7):C-108–109, 1982.
25. RF Bunshah. Handbook of Deposition Technology for Films and Coatings: Science, Technology, and Applications. Noyes, Park Ridge, NJ, 1994.
26. JP Dekker, PJ Van der Put, HJ Veringa, J Schoonman. Particle-precipitation-aided chemical vapor deposition of titanium nitride. J Am Cer Soc 80(3):629–636, 1997.
27. EH Kraus, WF Hunt, LS Ramsdell. Mineralogy, An Introduction to the Study of Minerals and Crystals. McGraw-Hill, New York, 1959.
28. CO McHugh, TJ Whalen, J Humenik, Jr. Dispersion-strengthened aluminum oxide. J Am Cer Soc 49(9):486–491, 1966.

29. DT Rankin, JJ Stiglich, DR Petrak, R Ruh. Hot-pressing and mechanical properties of Al_2O_3 with an Mo-dispersed phase. J Am Cer Soc 54(6), 277–281, 1971.
30. P Hing. Spatial distribution of tungsten on the physical properties of Al_2O_3-W cermets. Sci Cer 12:87–94, 1984.
31. S Hori, R Kurita, M Yoshimura, S Somiya. Influence of Small ZrO_2 Additions on the Microstructure and Mechanical Properties of Al_2O_3. Sciences and Technology of Zirconia III, Advances in Ceramics, Volume 24A (S Somiya, N Yamamoto, H Yanagida, eds.). American Ceramics Society, Westerville, OH, 1988, pp. 423–429.
32. CP Cameron, JH Enloe, LE Dolhert, RW Rice. A Comparison of reaction vs conventionally hot-pressed ceramic composites. Cer Eng Sci Proc 11(9–10):1190–1202, 1990.
33. AM Alper. Inter-Relationship of Phase Equilibria, Microstructure and Properties in Fusion-Cast Ceramics. Science of Ceramics, 3 (G. H. Stewart, ed.). Academic Press, New York, 1967, pp. 335–369.
34. RN McNally, GH Beall. Crystallization of fusion cast ceramics and glass-ceramics. J Mat Sci 14:2596–2604, 1979.
35. P Bardhan, RN McNally. Review: fusion-casting and crystallization of high temperature materials. J Mat Sci 15:2409–2427, 1980.
36. JJ Rasmussen. Surface tension, density, and volume change on melting of Al_2O_3 systems, Cr_2O_3, and Sm_2O_3. J Am Cer Soc 55(6):326, 1972.
37. DJ Rowcliffe, WJ Warren, AG Elliot, WS Rothwell. The growth of oriented ceramic eutectics. J Mat Sci 4:902–907, 1969.
38. DJ Viechnicki. Single crystal growth and eutectic solidification of oxide ceramics. Mat Sci Monographs 26:281–303, 1985.
39. VS Stubican, RC Bradt, FL Kennard, WJ Minlord, CC Sorrel. Ceramic Eutectic Composites, Tailoring Multiphase and Composite Ceramics (RE Tressler, GL Messing, CG Patano, RE Newnham, eds.). Plenum Press, New York, 1986, pp. 103–114.
40. HD Keith, FJ Padden, Jr. A phenomenological theory of spherulitic crystallization. J Appl Phys 34(8):2409–2420, 1963.
41. HW Morse, JDH Donnay. Optics and structure of three-dimensional spherulites. Am Minerologie 21(7):391–426, 1936.
42. SW Freiman, LL Hench. Kinetics of crystallization in Li_2O-SiO_2 glasses. J Am Cer Soc 51(7):382–87, 1968.
43. SW Freiman, GY Onada, AG Pincus. Controlled spherulitic crystallization of 3BaO-5SiO_2 glass. J. Am Cer Soc 55(7):354–359, 1972.
44. RE Whan, coord. Metals Handbook, Vol. 10. 9th ed ASM, Metals Park, OH, 1983.
45. JI Goldstein, DE Newbury, P Echlin, DC Joy, AP Romig, Jr., CE Lyman, C Friori, E Liffhin. Scanning Electron Microscopy and X-Ray Microanalysis. Plenum Press, New York, 1992.
46. T Rochow, P. Tucker. Introduction to Microscopy by Means of Light, Electrons, X-Rays, or Acoustics. 2nd ed. Plenum Press, New York, 1994.
47. JB Wachtman. Characterization of Materials. Butterworth-Heinemann, Stoneham, MA, 1993.
48. J-M Ting, RY Lin, Y-H Ko. Effect of powder characteristics on microstructure and strength of sintered alumina. Am Cer Soc Bul 70(7):1167–1172, 1991.
49. M McNamee, R Morrell. Textural effects in the microstructure of a 95% alumina ceramic and their relationship to strength. Sci Cer 12:629–634, 1984.

50. RL Fullman. Measurement of particle sizes in opaque bodies. Trans. AIME. J Metals: 447–452, 4, 1953.
51. RT De Hoff, FN Rhines. Quantitative Microscopy. McGraw-Hill, New York, 1968.
52. EE Underwood, AR Colcord, RC Waugh. Quantitative Relationships for Random Microstructures, in Ceramic Microstructures; Their Analysis, Significance, and Production (R. M. Fulrath and J. A. Pask, eds.). John Wiley, New York, 1968, pp 25–52.
53. EE Underwood. Quantitative Stereology. Addison-Wesley, New York, 1970.
54. J Serra. Image Analysis and Mathematical Morphology. Academic Press, New York, 1982.
55. M Coster, JL Chermant. Précis d' Analyse d' Images. Editions du CNRS, Paris, 1985; 2d ed. Presses du CNRS, Paris, 1989.
56. KJ Kurzydlowski, B Ralph. The Quantitative Description of the Microstructure of Materials. CRC Press, Boca Raton, FL, 1995.
57. M Coster, JL Chermant, M Prodhomme. Quantification of Ceramic Microstructures. Q-MAT 97, Proc. Int. Conf. on Quantitative Description of Materials Microstructure. Warsaw, Pologne, 4/16–19, 1997, pp. 115–124.
58. J-L Chermant, J Serra. Automatic image analysis today. Microsc Microanal Microstruct 7:279–288, 1996.
59. J-L Chermant, M Coster. Granulometry and granulomorphy by image analysis. Acta Stereol 10(1):7–23, 1991.
60. S Michelland, B Schiborr, M Coster, J-L Chermant. Size distribution of granular materials from unthreshold images. J Microscopy 156(3):303–11, 1989.
61. J-L Chermant, M Coster, G Gougeon. Shape Analysis in $\boldsymbol{R}^2$ Space using Mathematical Morphology. J Microscopy 145(2):143–157, 1987.
62. WK Pratt. Digital Image Processing. John Wiley, New York, 1991.
63. M Coster, G Gauthier, S Mathis, J-L Chermant. Roughness and $IR^2 \times IR$ function analysis: applications in material science. Microsc Microanal Microstruct 7:533–539, 1996.
64. MI Mendelson. Average grain size in polycrystalline ceramics. J Am Cer Soc, 52(8):443–446, 1969.
65. NA Haroun. Grain size statistics. J Mat Sci 16:2257–2262, 1981.
66. ED Case, JR Smyth, V Monthei. Grain size determinations. J Am Cer Soc C-24, 1981.
67. PH Crayton. Microstructural Analysis from Fractures Surfaces. Surfaces and Interfaces of Glass and Ceramics, Mat. Sci. Res. 7 (V. D. Frechette, W. C. LaCourse, and V. L. Burdick, eds.). Plenum Press, New York, 1974, pp. 427–437.
68. M Coster. Morphological Tools for Analysis of Non Planar Surfaces. 8th Int. Congr. For Stereol., Irvine CA, USA, 8/27–30, 1991. Acta Sterol. 11:639–650, 1992.
69. S. Michelland-Abbe, M Coster, J-L Chermant, BL Mordeke. Quantitative fractography on microcrystalline steels by automatic image analysis. J Comp Ass Micr 3:23–32, 1991.
70. W-J Yang, CT Yu, AS Kobayashi. SEM quantification of transgranular vs. intergranular fracture. J. Am. Cer. Soc. 74 (2):290–295, 1991.
71. L Dortmans, R Morrell, G de With. CEN-VAMAS Round Robin on Grain Size Measurement for Advanced Technical Ceramics—Final Report. Nat. Physics Lab., Tedington, Middlesex VAMAS Report No. 12, 12/1992.
72. LJMG Dortmans, R Morrell, G de With. Round robin on grain size measurement for advanced technical ceramics. J Eur Cer Soc 12:205–213, 1993.

73. RG Cortes, AO Sepulveda, WO Busch. On the determination of grain-size distributions from intercept distributions. J Mat Sci 20:2997–3002, 1985.
74. T Kinoshita, T Wakabayashi, Hiroshi, H Kubo. An equation which relates the real three-dimensional aspect ratio to the apparent cross-sectional aspect ratio. J Jap Cer Soc Intl Ed 99(9):793–796, 1991.
75. VJ Laraia, IL Rus, AH Heuer. Microstructural shape factors: relation of random planar sections to three-dimensional microstructures. J Am Cer Soc 76(6):1532–1536, 1995.
76. EM Chamot, CW Mason. Principles and use of microscopes and accessories, physical methods for the study of chemical problems. Handbook Chem Microscopy 1, 1938.
77. F Goyette, TJ Kim, PJ Gielisse. Effect of grain size on grinding high density aluminum oxide. Bul Am Cer Soc 56(11):1018, 1977.
78. GK Bansal. Effect of flaw shape on strengths of ceramics. J Am Cer Soc 59(1–2):87–88, 1976.
79. DPH Hasselman. Single Crystal Elastic Anisotropy and the Mechanical Behavior of Polycrystalline Brittle Refractory Materials. Anisotropy in Single-Crystal Refractory Compounds 2 (FW Vahldiek and SA Mersol, eds.). Plenum Press, New York, 1968, pp. 247–265.
80. D Stoyan, H-D Schnabel. Description of relations between spatial variability of microstructure and mechanical strength of alumina ceramics. Cer Intl 16:11–18, 1990.
81. F Mehran, KA Muller, WJ Fitzpatrick, W Berlinger, MS Fung. Characterization of particle orientations in ceramics by electron paramagnetic resonance. J Am Cer Soc 64(10):C-129–130, 1981.
82. LA Boatner, JL Boldu, MM Abraham. Characterization of textured ceramics by electron paramagnetic resonance spectroscopy: I. concepts and theory. J Am Cer Soc 73(8):2333–2344, 1990.
83. JL Boldu, LA Boatner, MM Abraham. Characterization of textured ceramics by electron paramagnetic resonance spectroscopy: II. formation and properties of textured MgO. J Am Cer Soc 73(8):2345–2359, 1990.
84. J Glass, JR Michael, MJ Readey, SI Wright, DP Field. Characterization of Microstructure and Crack Propagation in Alumina Using Orientation Imaging Microscopy (OIM). Sandia Report, Sandia Natl. Lab., Albuquerque, NM, SAND 96-1019•UC-404, 12/1996.
85. RW Rice. Review effects of environment and temperature on ceramic tensile strength–grain size relations. J Mat Sci 32:3071–3087, 1997.
86. RW Rice, SW Freiman, JJ Mecholsky, Jr. The dependence of strength-controlling fracture energy on the flaw-size to grain-size ratio. J Am Cer Soc 63(3–4):129–136, 1980.
87. A Kishimoto, K Endo, Y Nakamura, N Motohira, K Sugi. Effect of high-voltage screening on strength distribution for titanium dioxide ceramics. J Am Cer Soc 78(8):2248–2250, 1995.
88. A Kishimoto, K Endo, N Motohira, Y Nakamura, H Yanagida, M Miyayama. Strength distribution of titania ceramics after high-voltage screening. J Mat Sci 31:3419–3425, 1996.

2

Grain Dependence of Microcracking, Crack Propagation, and Fracture Toughness at ~ 22°C

I. INTRODUCTION

This chapter addresses grain effects on crack propagation, fracture toughness, and related phenomena in nominally dense, single phase, i.e. monolithic, ceramics. Grain parameters include grain size, shape, and orientation (the latter reflecting effects of intrinsic property changes as a function of grain, crystal, orientation), as well as grain boundary phases. Grain size effects, though often neglected, have received a fair amount of investigation, and grain shape and orientation substantially less (mostly qualitative for shape). Studies, which have focused on tests designed to provide controlled crack propagation well prior to, and not necessarily culminating in, catastrophic propagation under uniaxial tensile stress, are addressed along with slow crack growth (SCG) due to environmental effects, microcracking, crack branching and bridging, and related formation of crack wake zones and *R*-curve effects. These lead up to, or are aspects of, the basic fracture mechanics parameters of fracture toughness (K_{IC}) and related fracture energy (γ) via

$$K_{IC} \sim (2E\gamma)^{1/2} \tag{2.1}$$

where the ~ sign reflects the fact that for plane stress the term in the square root should be multiplied by $1-\nu^2$ (Poisson's ratio)^2 but is exact for plane strain conditions. Attention is also given to crack size effects and fracture mode, i.e. the extent of trans- versus intergranular fracture.

A fundamental complication in addressing fracture toughness is that different tests often give similar results for some materials and microstructures, but different results, often significantly so, for other materials and microstructures for the same test parameters. While some factors causing these differences, such as crack size and microstructure, are recognized, much understanding is still needed, e.g. of interrelations between crack size and geometry, microcracking, crack branching, slow crack growth, and crack bridging. Another complication is the frequent inadequate material, especially microstructural, characterization, a concern it is hoped this book will aid in rectifying. However, since the full range of test and materials parameters has not been evaluated, understood, or recognized, compounding confusion and uncertainty, it is necessary to note some aspects of the different tests to indicate some probable contributions to these complexities. A further complication is that while there are many common and self-consistent trends in crack propagation tests considered in this chapter, there are differences, e.g. in fracture mode, and especially differences between grain dependences of fracture toughness and tensile strength, which are partly addressed here and more extensively in Chaps. 3, 6, 8, 9, 11, and 12.

Despite complications outlined above, much understanding has been established or can be obtained by integrating observations of fracture mode, SCG, microcracking, toughness, and strength. Thus fracture is mainly intergranular for finer grains, becoming transgranular for most grain sizes and materials; but microcracking, which can be environmentally, and is clearly *G*, dependent, and SCG can extend intergranular fracture to larger *G*. The significant maxima seen in many fracture energy and toughness tests as a function of *G*, originally attributed to microcracking, are now generally attributed to wake bridging or *R*-curve effects. However, microcracking can play a role in intergranular fracture, crack branching, and crack wake bridging or *R*-curve effects, as well as some other toughness and mechanical property behavior. Fracture toughness–tensile strength differences are due to basic differences in propagation of their respectively large versus small cracks, e.g. their scale to that of the microstructures.

II. TEST AND MECHANISMS BACKGROUND

A. Crack Propagation–Fracture Toughness Test Methods, Factors, and Differences

A variety of crack propagation tests exist for slow crack growth and fracture energy (γ) and toughness (K_{IC}), which form an extensive subject themselves. Here the focus is primarily on grain effects and secondarily on related parameters, e.g. whether the assumed crack character and scale is pertinent to behavior to be evaluated. The reader is referred to other sources for the mechanics and evaluations of these methods [1–10]. These methods can be approximately ranked by the typical crack sizes they entail and these in turn broken into subgroups based on both test specimen nature and partly on the extent of stable crack propagation. Going from larger to smaller potential crack sizes, major tests employ (1) plates, i.e. double cantilever (DCB), double torsion (DT), and compact tension (CT), (2) beams the same as or very similar to those for flexural strength, i.e. notch (NB), chevron notch (CNB), indent (I, or indent fracture, IF), and (3) fractography (F, i.e. calculated from the strength and subsequently observed failure initiating flaw size, location, and geometry). This ranking by crack size, though only approximate, since it also depends on specimen sizes and the extent of crack propagation, is important, since it approximately correlates with toughness values (Table 3 of Ref. 11) and is a factor in different microstructural dependences of toughness and strength. Recent terminology distinguishing "long" and "short" cracks is a step in the right direction, but it is incomplete, because this refers to only one, neglecting the other, dimension of crack propagation; "large" and "small" are more accurate descriptions for crack size effects.

The diversity of tests is greater since there are different versions of them, especially of DCB and NB tests; no test is dominant, since each has advantages and disadvantages. Thus plate-based tests use the most material and indentation (I) tests the least, but the latter as well as DCB and CT tests are generally not applicable for differing, especially high, temperature tests. DCB and DT tests typically provide the greatest extent of constant stress intensity propagation and the lowest achievable controlled crack velocities for SCG tests. DT, which is useful at high temperatures, uses torsion loading with a resultant long curved crack front (the direction of which depends on the direction of torsion loading) that may limit in situ observations of crack-microstructure interactions, while DCB and CT tests, along with I tests, are most amenable to such observations. NB tests using the same or similar test beams as for flexure strength are readily used at high temperatures but are less amenable to in situ crack-microstructural observations and may present serious uncertainties in the nature of the crack character involved. Sufficiently fine notches for forming a sharp crack at the notch root has been identified in finer grain alumina bodies [2, 12] but not other bodies, e.g. there is still some uncertainty even in fine grain Si_3N_4 [13]. Less investigated is whether cracks from the notch are of the assumed slit nature on which the NB test is based. Fractography has shown significant deviations in NB toughnesses values due to failure from ~ half penny rather than the assumed slit cracks, despite being done in materials that should be ideal for such tests (i.e. ZrO_2 single crystals [14] and especially a silicate glass [9]). Residual stresses from indents, which depend on indent load and type (mainly Vickers versus Knoop), and whether the indents are ground off (which can raise issues of the nature of the resulting crack) can be an issue for indentation tests.

Recognizing the above issues, variations, and uncertainties, it is essential to gain further understanding of crack propagation and its microstructural dependence, in which case differences between the various tests can become a tool to enhance this understanding. The first of three needs is intercomparison of several tests on the same bodies, instead of only one test (often on materials of limited or no microstructural characterization). Second is much broader comparison of the microstructural dependence of toughness and tensile strength, i.e. measuring both, not just toughness. Third is more evaluation of the resultant fracture surfaces, especially where there are test variations for which a few studies have noted significant test and microstructural [15] deviations.

Observations of grain structure dependence of crack bridging in the crack wake zone (i.e. behind the crack front) are important but require more attention [16]. Thus they have typically been made close to the intersection of a free machined surface with large cracks arbitrarily introduced into the samples that were propagated at limited, mostly unknown, velocities and then arrested for observation. This is in contrast to cracks for normal tensile failure from the most serious defect, propagated at an accelerating, rather than an arresting, velocity, mostly into the sample bulk, not primarily along the specimen surface. Better quantification of the bridging occurrences is needed, e.g. their frequency-density as a function of microstructure, surface finish, and crack velocity and size. Except for one study using microradiography [6] (which qualitatively indicates some similar crack complexities in the interior of samples as on the surface), and some optical observations, many bridging observations are made in situ in a SEM precluding environmental effects, which may be a factor. Bridging observations, though invoked in biaxial behavior, have been made mostly or exclusively under uniaxial tensile stress and have been applied without consideration of crack geometry, despite indications that crack geometry may be a factor [16]. Thus increased toughness due to wake effects may scale with wake area, its rate of increase, or both, possibly normalized, all of which depend on basic crack shape (Fig. 2.1). Similarly, crack tip dependent processes, e.g. microcracking, crack branching (and possibly some bridging) may depend on the periphery (L) of the crack tip, its rate of increase (dL/dt), or both, again normalized by $1/L$. Such normalized values again differ, e.g. are 0 for expected cracks in DCB and NB, but is $(1/C)(dL/dt)$ for some other ideal cracks, e.g. half penny cracks [17]. Recent analysis by Gilbert et al. [18] indicated that bridging is independent of crack size, if bridging scales wth crack area, but neglects possible threshold crack area/grain size or crack periphery/grain size effects on the onset and growth of bridging.

B. Slow Crack Growth

Griffith's theory of brittle fracture [19] (originally for silicate glasses) assumed crack propagation started at a critical stress and rapidly accelerated to catastrophic failure. However, it is now known that slow crack growth (SCG) occurs in a variety of ceramics [9, 20] and other materials such as intermetallic alu-

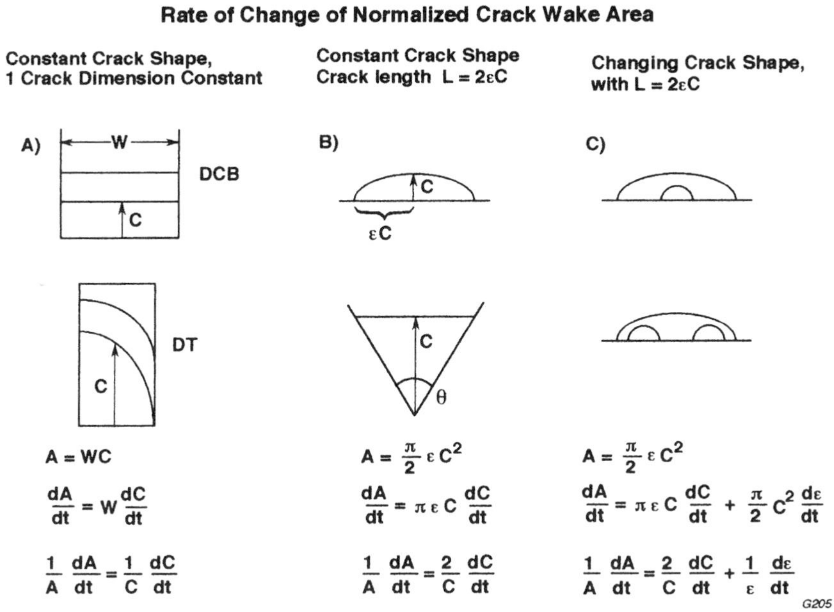

FIGURE 2.1 Idealized crack geometries for crack propagation and strength tests and their rate of generating normalized crack wake area per unit crack area (and of crack periphery, see text). Note (1) differences in wake area effects and (2) $(\pi/2)\epsilon$ is replaced by tan $\theta/2$ for a triangular crack, but the net expression is the same as for elliptical cracks of constant shape. (After Rice [17], published with the permission of The Journal of the American Ceramic Society.)

minides of Ni and Fe [21]. Such SCG, which is also referred to by various terms such as stress corrosion, static fatigue, or delayed failure, occurs prior to, or in the absence of, catastrophic failure. Such growth results in reduced strengths from those for the original crack sizes, but the resultant tensile strength (σ) is generally consistent with those predicted by Griffith's equation when the increase of crack-flaw size c is accounted for [9, 21–22] i.e.

$$\sigma = YK_{IC}(c)^{-1/2} \tag{2.2}$$

where Y = a geometrical factor to account for flaw shape, orientation, and location.

While there are still uncertainties about SCG, some of which concern its microstructural dependence as discussed later for grain and particle effects, and elsewhere for porosity effects (Chap. 4 of Ref. 5), the overall mechanisms and nature of the process are understood, at least for species that effect crack propagation at and near room temperature [10, 20], the focus of this chapter. (More limited data on SCG due to active species at higher temperatures, e.g. of molten aluminum, are discussed in Chap. 6.) Thus SCG is an important mechanism of crack propagation, much of it at orders of magnitude lower crack velocities than those associated with initiation of catastrophic failure (Fig. 2.2). SCG occurs as a result of weakening and failure of stressed atomic bonds at crack tips due to interactions with active environmental species at lower stresses than in the absence of such species. While water is one of the most active and prevalent species for SCG in many materials, there are other species such as NH_4 and hydrazine that can be active with some of the same, as well as different, materials susceptible to effects of H_2O [20]. Details of chemical interactions at the crack tip may vary, since crack growth entails some balance between crack tip sharpening and blunting via local corrosion. Bulk chemical corrosion may also lower strengths, but

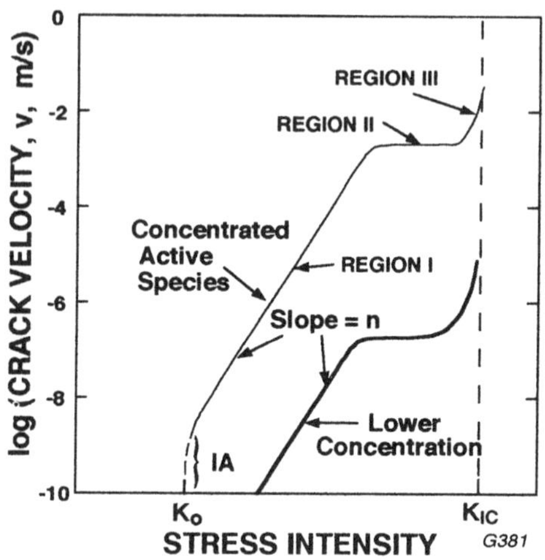

FIGURE 2.2 Schematic of log crack velocity versus stress intensity for environmentally driven slow crack growth (SCG). Upper and lower curves represent respectively maximum and more dilute concentration of the active environmental species, often a liquid, such as H_2O. The three main stages or regions of SCG, starting from the lowest crack velocity, are controlled by (I) the reaction rate of the active species and crack tip bonds, (II) diffusion of the active species down the crack to its tip, and (III) effects of the active species dielectric properties on electrostatic interaction of bonds at the crack tip. Eq. (2.3) is used for stage I where the crack spends its greatest time (above a possible limiting stress intensity, K_0, below which no environmental effect is observed, indicated mainly in soda lime glasses). Though experimental variations occur, the slope (n) is assumed to be constant, characterizing the process with curves for different concentrations, being differentiated mainly or exclusively by A [Eq. (2.3)].

due to other mechanisms, e.g. corrosion often occurs when SCG does not and vice versa.

SCG measurements, e.g. by DCB and DT tests, are typically characterized by region I crack velocity (v) versus stress intensity (K) portion of the crack growth (Fig. 2.2), per the most common expression for this:

$$\mathrm{v} = AK^n \tag{2.3}$$

where A = a constant of proportionality and n = the slope, since stage I growth normally entails the longest time of the growth process. Attention is commonly focused on the exponent n as a characterization of SCG, since the parameter A changes with concentration of the active species at the crack tip, but n is generally assumed independent of concentration and reflects, and is experimentally found to be approximately so. Two other tests, dynamic fatigue (DF) and delayed failure, measure respectively strength as a function of stressing rate and the time to tensile (typically flexure) failure as a function of constant loads applied to various specimen sets. Slower loading rates and longer times of loading (at lower loads) allow more slow crack growth and hence lower strengths, and calculation of n values. Indentation induced flaws are used, especially in DF tests, to give a more uniform and identifiable starting flaw population. Details of possible effects of crack sizes (e.g. including greater possible effects of residual surface and indent stresses for smaller cracks), microstructural variations, and test factors such as inadvertently using some data points from stage IA or II in calculating n values have received limited attention but are possible factors in the scatter of n values. SCG and fast fracture are shown to be typically via respectively intergranular and transgranular fracture in polycrystals [23, 24] (Figures 2.5, 2.6), and grain size dependence of n is indicated (Fig. 2.8).

C. Occurrence and Character of Microcracking

Microcracking occurs due to microstructural mismatch strains, most commonly from differential expansion (e.g. on cooling from processing) between (1) adjacent grains of noncubic phases having inherent thermal expansion anisotropy (TEA) as a function of crystal direction or (2) particles and a surrounding matrix of different composition and thermal expansions [25–34]. Phase transformations, e.g. the tetragonal to monoclinic transformation of ZrO_2, can also cause similar microcracking, as can other sources of strain differences between grains as well as grain clusters or colonies. Strain differences due to elastic anisotropy (EA, amplifying applied stresses in cubic and noncubic materials, as well as TEA stresses in the latter) or elastic differences between particles and a matrix can also contribute to such microcracking [25].

Regardless of the source of microcracking, it generally has a basic microstructural dependence in that it does not occur until a critical size, G_S, which depends on local properties [25–34] per expressions of the following nature:

$$G_S \sim 9\gamma[E(\Delta\in)^2]^{-1} \tag{2.4}$$

where γ = the local fracture energy pertinent to the microcracking (e.g. grain boundary or single crystal values depending on whether the microcracks are respectively inter- or transgranular), E = Young's modulus, and $\Delta\in$ = the mismatch strain. Strains derived from differences in thermal expansion ($\Delta\alpha$) between two grains are $\Delta\in = (\Delta\alpha)(\Delta T)$, where ΔT = the temperature difference between the onset of stress buildup (i.e. where stress relief is no longer possible, often ~ 1200°C) and the temperature where the microcracks are formed (often > 22°C). As grain sizes increase beyond G_S for a specific material, the number and size of microcracks increases until the process saturates, e.g. due to reduced body stiffness.

Different derivations of Eq. 2.4 (mostly from two-dimensional models) give its functional form with similar numerical values to 9, as does the one simple but reasonable three-dimensional model [29]. The latter equates a volume source of strain energy (e.g. about 1.5 to 2 times the grain or particle diameter) with a surface sink for that energy due to fracture around or through most or all of the grain. More detailed analysis shows somewhat more severe microcracking in three-versus two-dimensional models and differing effects of the specific character of the grain junctions [34]. Both grain shape and the location and extent of microcracking affect the process. While intergranular cracks about the size of the grains causing them are assumed and observed, transgranular cracks can frequently occur, especially at larger G, as is noted later.

Equation (2.4) does not directly address possible environmental effects on microcracking but is probably consistent with them, e.g. by reducing the γ value. This is important, since there is evidence that microcracking can depend on environmental effects, as is shown by microcracking in various HfO_2 bodies not commencing until moisture was present, and taking a few days to saturate [35, 36]. Such effects (also seen in ceramic composites, Chap. 8) are probably variable due to differing (often limited) degrees of connectivity of microcracks to provide paths into the bulk of the body, but connected porosity may provide more access for environmental species even at limited porosity levels.

Microcrack observations on polished or fracture surfaces are uncertain because of effects of the free surface on the occurrence and character of microcracks intersecting such surfaces. Neutron scattering [37], and more commonly following the progressive increase of microcracking via elastic moduli and electrical or thermal conductivity measurements (and the hysteresis of these changes with temperature cycling), can provide valuable microcrack characterization (Ref. 5, Chaps. 4 and 10), as can effects on fracture toughness and acoustic emission, which are discussed in subsequent sections. Tensile (and also compressive) stresses applied to a sample before or during testing can also be important in microcracking and whether it occurs throughout the body, locally in conjunction with propagation of a macrocrack, or combinations of these. Modeling indicates that microcracking in conjunction with macrocracking commences when $G \sim 0.4G_S$ [32, 33].

The relevance of microcracking to this book arises from its occurrence depending on G per Eq. (2.4) and the resultant G dependence of properties. Thus microcracks introduce some grain size dependence to thermal expansion (normally independent of grain size, Figure 7.10B of Ref. 5) [38] and to electrical and thermal conductivities (also often independent of grain size, especially above room temperature), as well as on mechanical properties. Effects of microcracking on fracture energy and toughness are a focus of this chapter, while tensile and compressive strengths, hardness, wear, and erosion, are discussed (to the extent that data exists) respectively in the next and subsequent chapters. Microcracking may also have broader applicability as a mechanism of crack propagation, e.g. a qualitative mechanism to explain fracture mode changing from mainly intergranular at finer G to mainly transgranular at larger grain sizes (Ref. 5, Fig. 2.3) [39, 40]. A quantitative two-dimensional model of fracture mode has also been proposed by Tatami et al. [41], which provides some guidance. But further development is needed.

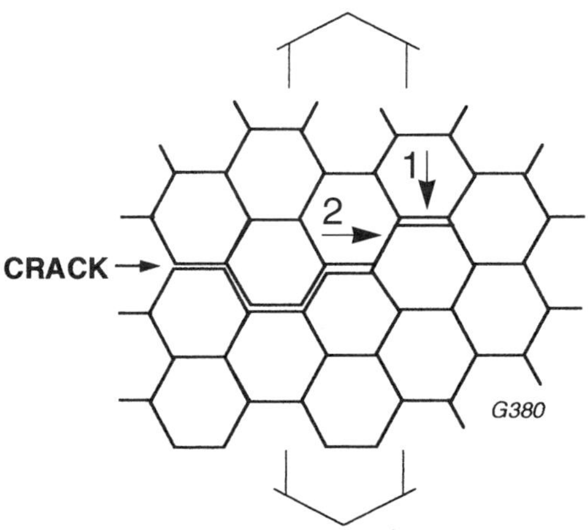

FIGURE 2.3 Schematic of a possible microcrack mechanism of intergranular fracture. Crack tip stress concentrations may cause fracturing of grain boundary facets at and closely ahead of the crack (see 1 above), which are then linked to the main crack via cracking of a previously uncracked boundary (see 2 above). This favors intergranular fracture at finer grain sizes, since the stress concentrations are higher over boundary facets closer to the crack tip, while larger grains have lower stresses on boundary facets because of their inherent greater distance from the crack tip. (After Rice [39], published with the permission of the ASTM.)

D. Microcracking and Wake Bridging Effects on Crack Propagation and Toughness

A few models have been developed to explain the often substantial grain size dependence of fracture energy and toughness found with large crack tests of some noncubic materials (Fig. 2.16) which were first attributed to microcracking. Fu and Evans [33] presented a model (later updated [42]) that at least semiquantitatively explained such noncubic fracture toughness–grain size data. Rice and Freiman [43] derived the following equation for microcracking effects on fracture energy (γ_m) in noncubic materials based on competition between increased energy requirements from microcrack formation ahead of the main crack versus reduced energy requirements for crack advance by linking with microcracks as:

$$\gamma_m = M(\Delta\epsilon)[(9E\,\gamma_b G)^{1/2} - \Delta\epsilon\, EG] \tag{2.5}$$

where M = a proportionality factor, e.g. ~ 3, $\Delta\epsilon$ = the mean strain mismatch between grains, γ_b = the grain boundary fracture energy (assuming intergranular microcracking), and E = the local average (often the body) Young's modulus. This model gives the maximum increase in fracture energy (hence also fracture toughness) occurring at $G = (1/4)G_S$, independent of M. This and other microcrack models based on more rigorous mechanics [44–47] were developed based on microcracks forming at or somewhat ahead of the main crack. A model proposing that microcracks occurred mainly in two lobes extending somewhat ahead of the main crack, one well above the main crack plane and one well below it [44] (Fig. 8.2A) has not been supported by observations [48–50] and has been abandoned. However, the cessation of most study of microcracking, especially at and ahead of the main crack tip due to the focus on R-curve and related wake effects, as discussed later, is premature.

The shift to a focus on crack wake or R-curve effects was driven by observations of variable toughness increases with increasing crack propagation (then plateauing after propagation of up to 5–10 mm in typical toughness tests), the essence of R-curve effects. These occur with transformation toughening or with crack bridging in a number of, especially, larger grain noncubic, ceramics as the extent, typically the length, of the macrocrack increases. Wake effects refer to phenomena found in the region behind the tip of a propagating crack (Fig. 8.2B, typically substantially larger than natural flaws), i.e. in the region through which the crack has propagated. That R-curve effects are due to wake effects was clearly shown by the disappearance of increased toughness when the wake region was removed [51], as is discussed later. Bridging refers to portions of, complete, or multiple, grains left connecting the two sides of a crack in regions through which it has propagated (Fig. 2.4). However, as noted earlier, there are critical gaps in the data and tests, in particular the onset and extent of such effects at smaller crack sizes, especially of natural failure causing flaws and sur-

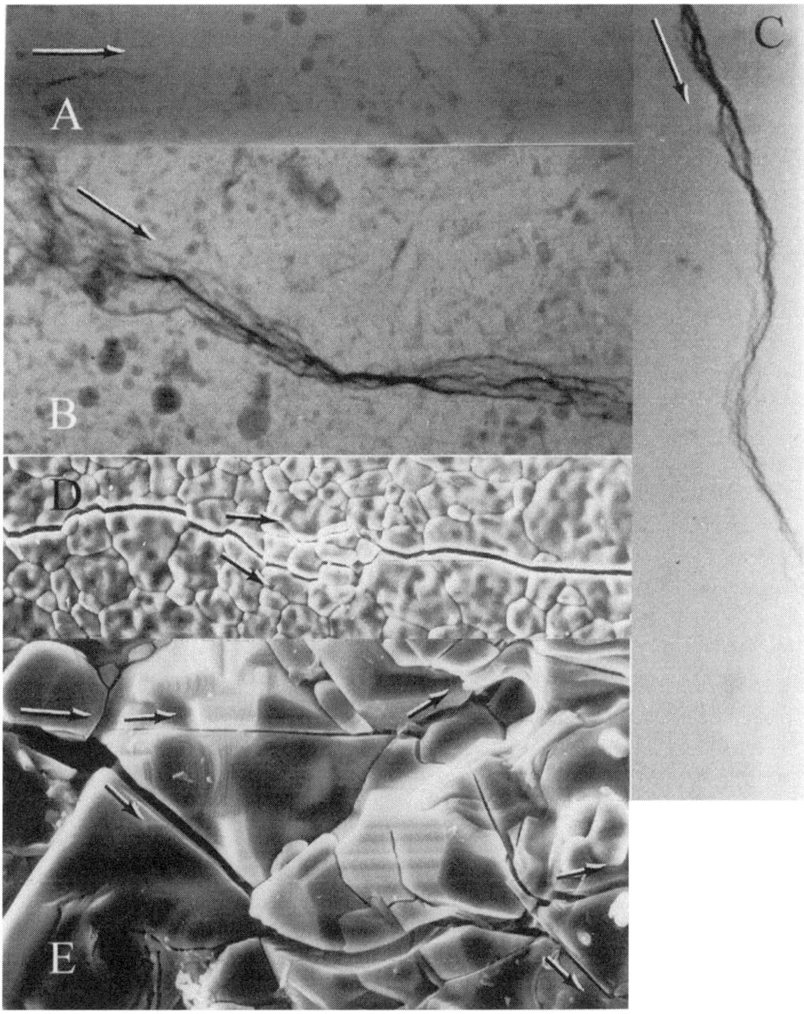

FIGURE 2.4 Examples of crack bridging and related complexity in the wake region, (A) and (B) Microradiographs of fused alumina refractory ($G \sim 50$–$2000\ \mu m$) showing preexisting microcracks, large ~ spherical pores, and crack tortuosity. (C) Microradiograph of dense MgO (mixed $G \sim 25$–$50\ \mu m$) showing greater crack tortuosity attributed to heterogeneous grain structure (corroborated by SEM examination). (D) In situ SEM test of dense alumina ($G \sim 30$–$50\ \mu m$) with an as-fired surface; note the multigrain nature of the isolated crack branch-bridge (~ photo center). (E) In situ SEM test of dense alumina (Lucalox) with an as-fired surface showing a fairly isolated multigrain bridge having substantial cracking (After Wu et al. [6], published with the permission of the ASTM.)

face and environmental limitations of most observations, e.g. in situ in a SEM [6, 49, 50, 52–55] and along cracks from indents. Recent alumina testing by Tandon and Faber [56] showed a substantial strain rate dependence of toughness-type tests, indicating that the increased toughness can have substantial dependence on environmental effect, i.e. SCG of microcracks, as is discussed further later.

Several investigators have developed various crack bridging models, often focused more on mechanics of the process and less on specific microstructural effects, but some have explicitly addressed effects of grain or particle sizes and volume fractions of these involved, e.g. Refs. 57–59. Most treat bridging elements as uniform in character and spatial distribution, often require iterative solutions, and depend on adjustable parameters often obtained by data fitting which is often not precise. Typically, crack tip shielding is assumed, with some models noting that the process starts with microcrack formation, but that frictional pullout of bridges is an important factor as shown experimentally, and by simpler order of magnitude models (Sec. III.G).

Before reviewing grain structure dependence in detail it is again important to note that fracture toughness and strength behavior must be correlated, as is done elsewhere [5, 11, 60] and here in this book, since there are often major inconsistencies in the grain size dependence of these two properties. Similar discrepancies, which will be shown for ceramic composites and have been previously shown for effects of porosity [5], are critical in designing and using ceramics where fracture is an important concern. They are a reminder of the importance of measuring both strength and toughness (preferably by more than one test, especially for toughness), instead of a single measurement, as has unfortunately become common, e.g. via indentation, assuming, but not showing, that strengths will show similar trends with composition, microstructure, etc. Grain shape effects have been considered a limited amount (mainly in conjunction with bridging), as have grain orientation effects (commonly associated with grain elongation). Global effects of grain orientation on fracture toughness should generally parallel those of E, since, per Eq. (2.1), K_{IC} and E as well as E and γ are closely related [61], as long as fracture is a simple elastic process and there is no flaw anisotropy, which occurs much more than is commonly assumed.

III. GRAIN DEPENDENCE OF CRACK PROPAGATION

A. Grain Size Dependence of Fracture Mode

A detailed review by Rice [62] showed that the overall fracture mode of ceramics, the average trend over a complete fracture surface, is for finer grain fracture to be mainly, often exclusively, intergranular, then transition to mainly, often exclusively, transgranular fracture as grain size increases for a broad range of stress conditions. The more limited data for nonoxide ceramics is consistent with the more extensive results for oxide ceramics (Fig. 2.5). The transition between inter- and transgranular fracture at finer grain sizes is often completed by the time grain sizes have reached 1 to several microns, but it may not be completed till larger grain sizes. This transition is seen not only between different grain size bodies but also as a function of the grain distribution on a given fracture where its grain size range straddles much or all of the range of the transition found for bodies of different average grain sizes (Fig. 1.1).

The first two of three factors besides higher temperatures that can shift the transition to transgranular fracture to larger grain sizes are grain boundary phases and porosity These are probably factors in intergranular fracture often not beginning till $G \sim 1$ or so microns, since obtaining finer grain sizes typically entails lower processing temperatures that may leave residual boundary impurities and porosity that enhance intergranular fracture. Porosity effects are important when there is substantial boundary coverage by pores on the scale of, or smaller than, the grains [5]. Boundary phases, important examples of which are those from oxide densification aids for AlN, Si_3N_4, and SiC, commonly play an important role in extending intergranular fracture by themselves and in conjunction

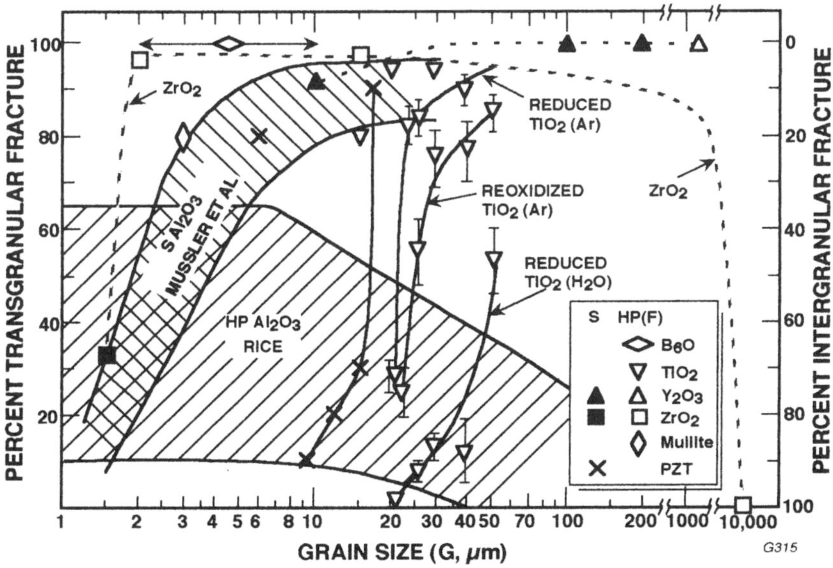

FIGURE 2.5 Trans- and intergranular fracture of oxide ceramics versus gram size. Note that most materials have mainly transgranular fracture, except at finer grain sizes, and, at least for some materials, at large (Al_2O_3 here) and very large grain sizes (ZrO_2 here). Very large *G* for Y_2O_3 and ZrO_2 are from fused ingots. (After Rice [62], published with the permission of the American Ceramic Society.)

with the third source of shifting the transition to larger *G,* namely SCG. Thus while the transition between inter- and transgranular fracture is typically shifted to larger grain sizes with SCG in the absence of boundary phases, such phases can cause SCG, shift the transition to larger grain sizes, or both, as is discussed in the next section.

A mechanism of intergranular fracture has been proposed based on microcracking of grain boundary facets at, and just ahead of, the macrocrack tip (Fig. 2.3), which then connects with them as a mechanism of its advance. Such microcracking should be more extensive at finer grain sizes due to higher net stress concentrations across smaller grain facets at and close to the crack tip, and less as grain size increases. TEA stresses, which occur only in noncubic materials, are independent of, but can add to the, applied stresses, extending the range of microcracking. EA, which occurs in both cubic and noncubic materials, locally alters the applied and TEA stresses, so its effects on microcracking clearly drop off significantly as a function of distance from the crack tip. The two-dimensional model of Tatami et al. [41] also provides an explanation for the inter–transgranular fracture transition but does not account for TEA and EA, which are probably basic reasons why the predictions are inconsistent with some increased intergranular fracture trends at intermediate and larger grains (discussed below).

Partial to complete transition back to intergranular fracture occurs at larger, often very large, grain sizes in at least some ceramics. Though there is often substantial variation, there is a common trend for some increase in intergranular fracture of alumina as grain size increases at larger grain sizes (Fig. 2.5). This contrary trend in the normal *G* range versus other common ceramics is also indicated in TiB_2 [62], suggesting that the substantial TEA in both these materials is a factor. This is supported by the observations of transgranular fracture being the dominant mode at $G \sim 2$ μm, diminishing to a minority mode by ~ 20 μm in less anisotropic $MgTi_2O_5$ [63], this transition being at even finer grain sizes in more anisotropic Fe_2TiO_5 [26, 27]. Thus the decrease in transgranular fracture with increasing grain size in noncubic materials is attributed to increasing effects of TEA on fracture. Such attribution to TEA effects is tentative, since the database is limited, and there are some potentially inconsistent observations; B_4C, with nearly identical TEA as Al_2O_3 [64] retains transgranular fracture to at least $G \sim 200$ μm. However, this may be associated with the extensive twinning that is commonly indicated in B_4C, especially in larger grains.

Transitioning back to primarily intergranular fracture is also indicated in cubic materials, but at much larger than normal grain sizes [62]. Thus polycrystalline ingots from skull melting of CaO, MgO, and ZrO_2 (Figure 1.7) show progressively increasing intergranular separation in the order listed as the grain diameters increase typically to 1 to several centimeters. Elastic anisotropies of these materials, respectively ~ 0, 2+, and 6–10% at room temperature, all increase as temperature increases [65], i.e. are higher where intergranular cracking is likely to initiate. (Note that at ~ 8% EA the maximum-to-minimum E ratio is 2.) This correlation of intergranular fracture at very large grain sizes to EA is consistent with and supports similar TEA effects, but further evaluation is necessary. Varying combinations of TEA and EA effects is probably an important factor in variations in fracture mode, e.g. between TiB_2 and Al_2O_3 (and other related behavior).

B. Grain Dependence of Slow Crack Growth and Other Environmental Effects

While only a few key studies of slow crack growth directly address its grain dependence, these and other data clearly show that grains frequently can have substantial effect on slow crack growth. Consider first a comparison of single crystal and polycrystalline data on the same material, with a key case being MgO. Shockey and Groves's [66] DCB tests showed that there was no SCG, i.e. water vapor had no effect, on {100} fracture (cleavage) of MgO crystals; in fact, tests in liquid water increased toughness by ~ 30%. They correlated increased toughness with increased fracture surface roughness [66, 67] but did not associate the increased roughness with enhanced dislocation activity. Freiman and McKinney [68] also found no SCG in DCB tests of MgO crystals for {100} or {110} cleavage. In contrast to no SCG in MgO crystals, Rhodes et al. [69] showed SCG occurred in dense polycrystalline MgO, but by intergranular fracture, which appeared to be associated with grain boundary impurities (in bodies with G ~ 26–46 μm). Little or no SCG occurred in the highest purity (~99.98%), largest grain body, but SCG clearly occurred in bodies with higher impurity content (to ~ 0.4%), much of which was at grain boundaries, e.g. single or mixed oxides of Ca, Na, and Si (typical of most high-purity MgO powders), as well as residual Li and F in bodies made with LiF additions.

Grain boundaries allowing the occurrence of SCG via even very limited levels of grain boundary phases in materials that have no intrinsic susceptabiliy to environmentally driven SCG were also shown in a study of Si_3N_4 [70] and a review of fracture mode in ceramics [62]. These showed nonoxides commonly made with oxide additions leaving a few percent residual oxide-based grain boundary phases, e.g. SiC (+Al_2O_3), AlN (+CaO or Y_2O_3), and Si_3N_4 (+MgO, Al_2O_3, Y_2O_3, etc.) showed SCG due to H_2O via predominant to exclusive intergranular fracture. On the other hand, bodies made without such additives, e.g. CVD SiC or Si_3N_4, and RSSN, or SiC made with B+C or B_4C additions, showed no SCG. Further, tests of Si_3N_4 bodies with oxide additions also often showed a crack size dependence of SCG, i.e. bodies with oxide additives commonly showed delayed failure, hence SCG, with normal strength controlling flaws, but not in larger scale crack, e.g. DCB or DT, tests. In some cases, large cracks showed no SCG, and in other cases they showed evidence of some initial, limited, erratic SCG, which then ceased. This significant difference as a function of crack size was attributed to incomplete second-phase coverage of boundaries to provide a sufficiently continuous boundary phase path for SCG with larger cracks. However, many such bodies should have sufficient contiguity of boundaries with adequate second-phase content for SCG along the much smaller crack fronts to allow sufficient crack growth from natural flaws in many delayed failure tests. Thus SCG due to H_2O could occur with small flaws because of the limited number of grain boundaries needed with sufficient second-phase content balanced by the significant number of small flaws, while with much larger cracks, regions of insufficient boundary oxide phase were frequently encountered either initially or with limited, erratic propagation of the large crack. This evaluation implies that strength losses in delayed failure of such nonoxides made with oxide additions should be limited when cracks grow to sizes where sufficient contiguity of boundary phases no longer occurs. Nonoxides made without oxide additions, but having some oxide contamination, may also show some, probably variable, SCG, due to limited or irregular distribution of the oxide contamination, e.g. as indicated by limited, variable SCG in materials such as TiB_2 and B_4C.

While intergranular fracture via SCG may often occur due to boundary phases that are susceptible to environmental SCG, such SCG fracture due to environment is also observed in materials that exhibit SCG in single crystal form, e.g. Al_2O_3 and MgF_2 [23, 24, 62, 71]. Thus while boundary phases can play an important role in intergranular SCG, this may reflect other environmental and microstructural effects. For example, some intergranular fracture may reflect varying environmental effects of the range of crystal orientations required to accommodate transgranular fracture in polycrystalline bodies, especially randomly oriented ones; i.e. some intergranular SCG may simply reflect grains unfavorably oriented for transgranular SCG.

An important aspect of intergranular SCG is that following catastrophic fracture is often transgranular, i.e. SCG often shifts the transition to mainly or exclusively transgranular fracture (Fig. 2.5) to larger G only over the SCG area. This inter- to transgranular fracture transition with the change from SCG to fast fracture often provides definition of the SCG region (Fig. 2.6). Particularly clear occurrences of such inter- to transgranular transitions have been observed in Al_2O_3, MgF_2, and ZnSe [23, 24, 71]. However, at least one important deviation from this, i.e. the reverse trend, has been observed by Beauchamp and Monroe [52, 53], namely SCG via more transgranular fracture and fast fracture via more intergranular fracture in a larger grain (~ 35 μm, with less Ca at the grain boundaries) versus a finer grain (~ 24 μm) Mn Zn ferrite (see also the note at the end of this chapter). They also observed that crack velocity and the stress intensity for crack growth initiation were dependent on prior history, as have a few others.

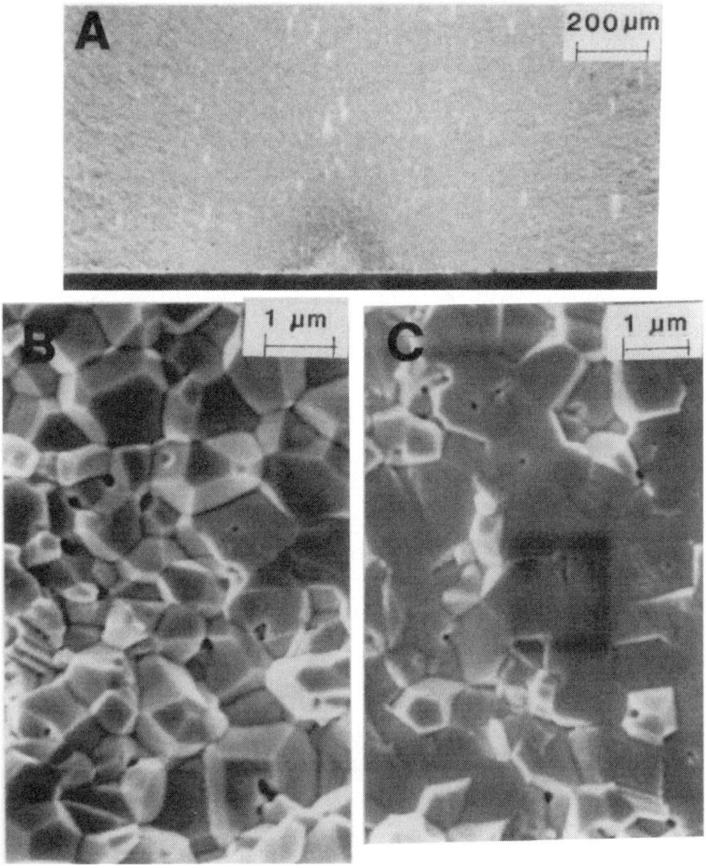

FIGURE 2.6 Example of the change from intergranular SCG to fast transgranular fracture in hot pressed alumina (~ 2% porosity, strength 290 MPA). (A) Lower magnification SEM showing fracture origin (bottom center). (B) and (C) Higher magnification SEMs of respectively intergranular fracture due to SCG and transgranular fracture adjoining this and for the rest of the fracture (After Rice [71], published with the permission of the Journal of the American Ceramic Society.)

Tests by Rice and Wu [72, 73] comparing the strengths of specimens in air at 22°C versus in liquid nitrogen (-196°C, eliminating SCG with only a modest change in strength due to a 2–4% increase in *E* with decreased temperature) showed SCG in various materials. Single crystals of both stoichiometric $MgAl_2O_4$ (for {100} or {110} fracture) and cubic ZrO_2 (+20 w/o Y_2O_3)

both exhibited SCG, via respectively ~ 30 ± 10% and ~ 50% increased strengths in liquid N_2, while ZrO_2 crystals partially stabilized with 5 w/o Y_2O_3 only increased strength 9%. However, DCB tests by Wu et al. [74, 75] of SCG of stoichiometric $MgAl_2O_4$ crystals oriented for {110} <110> crack propagation instead showed a zigzag {100} fracture at lower crack velocities that transitioned to the planned {110} <110> fracture at higher crack velocities (Fig. 2.7). Similarly, DCB tests of other orientations, e.g. where (100) and (110) or (111) planes were parallel to the specimen axes, but at different angles to the specimen surfaces, always resulted in crack propagation on (100) planes, despite this requiring 30% or more fracture area to be generated. However, flexure tests of bars oriented for fracture on (100) or (110) planes did so from machining flaws formed on these planes (including the typical strength anisotropy as a function of machining direction relative to the stress axis). Thus despite a strong preference for (100) fracture with large, low-velocity cracks in DCB tests, flexure tests showed no significant preference of (100) over (110) planes for formation of machining flaws and subsequent failure. Crack propagation studies of quartz crystals by Ball and Payne [76] showed the opposite type of crack propagation variation on some fracture planes, namely an increasing amplitude of zigzag crack propagation at high crack veocities, due to mist and hackle formation as a precursor to crack branching in quartz. Thus the spinel results show that important changes in

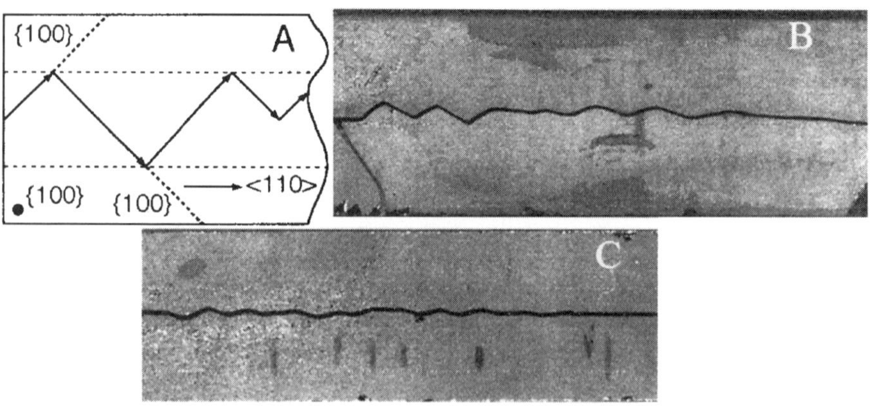

FIGURE 2.7 Change from {100} zigzag to planar {110} crack propagation in stoichiometric $MgAl_2O_4$ crystals oriented for {110}<110> fracture in DCB SCG tests. (A) Schematic of low-velocity fracture. (B) and (C) Lower magnification SEMs of fractures with lower velocity (e.g. 2–50 m/s) crack propagation at the left and higher velocity (including catastrophic) propagation to the right. (After Wu et al. [74], published with the permission of the Journal of Materials Science.)

crack propagation, and thus presumably in fracture toughness, can occur as crack velocity changes.

The above tests at -196 versus 22°C of polycrystalline cubic ZrO_2 (6.5 mol-% Y_2O_3 bodies, $G \sim 20$ μm) also showed that SCG with the same or very similar strength increases as the cubic stabilized crystal specimens. Polycrystalline bodies of stoichiometric $MgAl_2O_4$ with the same surface finishing as the above $MgAl_2O_4$ crystals, showed decreasing strength increases in liquid nitrogen versus air at 22°C. Bodies with $G \sim 100$ μm showed the same or similar strength increases as the single crystals did, i.e. ~ 35%. On the other hand, bodies with $G \sim 20$, 8, and 3 μm gave respectively decreasing strength increases of 31%, 27%, and 16%, i.e. the latter ~ 1/2 the single crystal and large grain increases, indicating reduced SCG as grain size decreases (but still with predominant transgranular fracture) [72]. This G dependence is similar to that of alumina (Figure 8A). Polycrystalline bodies of CeO_2, Y_2O_3, and TiO_2 (with 1 of 4 ~1/4% oxide additions) having respective grain sizes of ~ 20, ~ 100, and 5–10 μm showed respective strength increases of ~ 12%, ~ 25%, and 40–75% [72]. All exhibited predominant or exclusive transgranular fracture at the fracture origin (and elsewhere), implying that SCG occurs in single crystals of these materials.

Similar strength tests of nonoxide materials in air (22°C) and liquid N_2 corroborated and extended results, e.g. showing that very limited (e.g. 5%) or no strength increase in liquid N_2 occurred, indicating little or no SCG in ZrB, B_4C, TiC, ZrC, and Si_3N_4(CVD) bodies made without additives [73]. This was corroborated by the predominate or exclusive transgranular fracture of grains (typically 20–50 μm) at the fracture origins and over the total fracture. Bodies such as TiB_2, SiC, and Si_3N_4 (commonly, especially the latter two, made with oxide additions) exhibited some to substantial strength increase, hence SCG, with mixed to substantial intergranular fracture at origins, and often across the fracture. Similarly, comparative tests of a silicate glass, and finer grain bodies of Al_2O_3, and MgF_2 were consistent with other tests, i.e. substantial strength increases in liquid N_2 and mainly or exclusively intergranular fracture in the SCG region of polycrystalline fractures.

Two specific evaluations using v–K or dynamic fatigue (DF) tests showed a grain size dependence of SCG in Al_2O_3. Gessing and Bradt [77] showed from their own studies as well as from a literature survey that the crack growth exponent, n [Eq. (2.3)] increased with decreasing alumina grain size (Fig. 2.8A), with this trend being fitted by

$$n \sim 10 + 225G^{-1} \quad \text{from } K\text{–}v \text{ data} \quad \text{and}$$
$$n \sim 20 + 114G^{-1} \quad \text{from DF data} \tag{2.6}$$

Becher and Ferber [78] subsequently corroborated this v–K data trend for Al_2O_3 using some of the same and some of their own DCB data (Figure 8A) for high purity Al_2O_3 [79] e.g. giving n values of ≥ 110, 107, and 38 respectively for G =

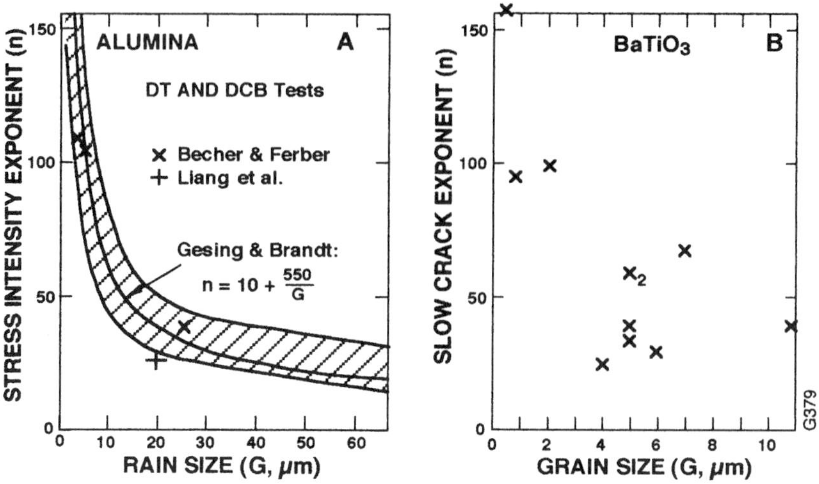

FIGURE 2.8 Grain size dependence of the exponential crack growth parameter *n*[Eq. (2.3)] from difference *v–K* studies of: (A) alumina of Gessing and Bradt [77] and Becher and Ferber [78, 79], and (B) $BaTiO_3$ data from de With's compilation [86].

4, 5, and 25 μm, corroborating the above trend. However, $n = 22$ for a body with bimodal grain size (averages ~7 and 25 μm) indicates complexities of the *G* dependence. Liang et al. [80] obtained $n \sim 26$ and 78 for two aluminas with $G \sim 20$ and 2 μm, consistent with the above data. Despite higher scatter, *n* values of 26–38 for various alumina bodies ($G \sim 3$–8 μm) with various SiO_2 contents, while somewhat lower, are reasonably consistent with the trends of these more extensive studies [81–83]. Data of Ferber and Brown (*n:* 36–42, *G*: not given) [84] are also reasonably consistent with the above trends but clearly showed significant effects of variations in the water due to additives (as do other studies). However, data of Byrne et al. [85] for various commercial aluminas gave some results inconsistent with the above trend of *n* to increase with decreasing grain size (and again showed significant effects of pH). Some of these differences are probably due to varying degrees of chemical attack, e.g. dissolution of some of the boundary phase in some bodies (which is related to composition-pH effects discussed by them). Such effects of dissolution are probably consistent with differing effects of crack sharpening versus crack blunting in stress corrosion.

Three other data sets indicate decreasing *n* values as *G* increases. De With's compilation of data on $BaTiO_3$ for multilayer capacitors [86], though scattered (probably reflecting differing compositions), also shows a decrease in *n* values as grain size increases (Fig. 2.8B). Hecht et al.'s [87] data for various toughened zirconia bodies ($G \sim 0.3$–45 μm), though widely scattered, again probably reflecting composition differences, also suggested a possible significant increase in n values at finer grain sizes. Becher and Ferber [79] obtained n values for two TiB_2 bodies ($G = 5$ and 11 μm, hot pressed with 10 wt% Ni, resulting in ~ 1.3–1.5 wt% NiB) respectively of 150 and 62, which again indicated a strong increase of n with decreasing grain size. Thus while there can be significant effects of compositional details of the bodies and the corrosion medium, there is substantial indication that n increases substantially with decreasing grain size, which is corroborated by results for strengths of $MgAl_2O_4$ in liquid nitrogen and in air at 22°C noted earlier. Therefore, much more attention to effects of grain size on SCG is needed than has been given. Considerably more SCG data exists, but often with only one, often unspecified, grain size, e.g. Yamade et al. [88] demonstrated SCG in dense sintered mullite ($K_{IC} \sim 2$ MPa·m$^{1/2}$) due to H_2O.

Gessing and Bradt [77] modeled the G dependence of n based on microcracking at the crack tip due to TEA stresses, i.e. in noncubic materials. Becher and Ferber [78] also presented a model, again based on TEA stresses in noncubic materials, but considering the superposition of TEA and applied stresses at finer grain sizes, and crack shielding from wake effects at larger grain sizes. However, the first and most fundamental of three points is that the focus on grain facet fracture due to TEA stresses neglects the frequent significant transgranular fracture of many of the materials, particularly at intermediate grain sizes, e.g. of aluminas at 5–20 μm, Figures 2.5 and 2.6. Second, the focus on TEA stresses and associated microcracking of grain boundary facets implies effects only in noncubic materials of sufficiently high (but unspecified) TEA, i.e. either neglecting, or assuming no, effects in cubic materials (as did much work on crack bridging and shielding). However, though limited and not directly giving n values, the data for $MgAl_2O_4$ implies a similar grain size effect in cubic materials (where bridging has also been found). Third, the form of Gessing and Bradts' model [Eq. (2.6)] would be consistent with a crack pinning-bowing model (Figure 8.4), in this case crack pinning due to points of greater difficulty of SCG. Pinning points could be grain boundary facets with low tensile or especially substantial compressive TEA stresses between them, or simply grains unfavorably oriented for SCG around or through them. Such a model would be at least partially focused on impediments to propagation of the macrocrack rather than just on microcracks in advance of it and would be consistent with both inter- and transgranular fracture and grain size effects in not only all noncubic materials but also cubic ones.

The above indications of a G dependence of the slow crack growth parameter n are but one of many examples of the limited data directly showing substantial effects of G, hence putting much data not giving reasonable G characterization in question. Both the probability of effects of G on n and the uncertainties and complexities this introduces with inadequate documentation of results and microstructure is shown by work of Singh et al. [89]. They made differing measurements on two sintered β″ - alumina, both of which had mixtures of finer and larger (often more tabular) grains, which gave substantially different *n* values for the two bodies. For example, they obtained *n* values of 76 and 26 for the two respective bodies via dynamic fatigue tests, again indicating an important effect of grain structure like that noted earlier. However, there were two important variations indicating further complexities of test-grain structure interactions. First, they obtained different *n* values, e.g. 26 and 16 respectively, for the same larger *G* body via dynamic fatigue, depending on whether they used a diametrally loaded ring or flexure testing. Second, and more serious, they obtained an opposite trend, i.e. $n = 64$ and 96 respectively in the finer and coarser *G* bodies by DCB/DT stress relaxations tests. Both discussion and modeling indicated that these variations reflected combined effects of the crack-grain size and the microstructural variation in the bodies.

Consider briefly grain size dependence of other environmental, especially corrosion, effects on mechanical properties, which is a large and complex subject with limited attention to microstructural effects. Some TZP bodies partially stabilized with Y_2O_3 are subject to serious degradation, beginning with attack along grain boundaries and resultant microcracking from transformation of tetragonal to monoclinic ZrO_2, which lowers crack resistance and strength, and in the extreme causes crumbling to a granular or powder character due to effects of water (Fig. 2.9). Degradation increases with the activity of the water (to an upper limit of ~300–500°C), decreasing Y_2O_3 content, and increasing grain size (with the latter two generally being inversely related) [90]. More recent data corroborates this general trend but shows a more complex variation with *G* than a simple monotonic change for a two m/o Y_2O_3 TZP over the limited *G* range (~0.5–1 μm) investigated [91]. Swab [92] has clearly outlined the composition–grain size variations of such degradation.

More generally, ceramics, while being more chemically inert, can be subject to a variety of other corrosive attacks at modest to high temperatures that reduce strengths at all subsequent testing temperatures. Such attack is commonly most severe along grain boundaries, which introduces a grain size dependence from intrinsic and extrinsic sources such as TEA stresses and frequent greater accumulation of more corrosion susceptible species at grain boundaries as *G* increases. Some probable *G* dependence can also occur in lower, e.g. room, temperature corrosion of materials susceptible to H_2O attack, e.g. CaO and MgO, where expansion of hydroxide products in pores or cracks causes failure as observed in CaO single- and polycrystals (e.g. over periods of weeks to months) [93] and in polycrystalline MgO (e.g. over years) [94, 95]. A *G* dependence of strength after corrosion is expected from the *G* dependence of tensile strength (Sec. 2.2).

FIGURE 2.9 Example of gross degradation of a commercial Y_2O_3 TZP ball in 100 psi steam exposure for 30 hours at 200°C. (Photo courtesy of Dr. T. Quadir.)

C. Grain Size and Other Dependence of Microcracking

Equations of the form of (2.2) with the same or similar numerical value give reasonable estimates of grain sizes for the onset of spontaneous microcracking due to microstructural stresses from phase transformation or TEA, as a function of material (mostly local) properties (Fig. 2.10), as is shown by the studies of Cleveland and Bradt [26, 27], Hunter et al. [35], Rice and Pohanka [29], and Yamai and Ota [96, 97]. Some of these evaluations have explicitly or implicitly used as a first approximation the γ/E ratio being ~ constant, leaving $G_s \propto (\Delta\epsilon)^{-2}$, and for TEA-driven microcracking that $G_s \propto (\Delta\alpha)^{-2}$, e.g. per a recent compilation [96], since ΔT does not vary widely for most ceramics. Over a broad range such approximations give reasonable estimates, but these estimates can often be improved by using known or estimated values of E, γ, and ΔT and adjustments of these. Note that Telle and Petzow's designation of spontaneous microcracking at $G \sim 6$ μm [98] is not consistent with the strength behavior they presented (Figure 3.25), other TiB_2 data, or the general trends of Fig. 2.10, but may be consistent with decreasing toughness (Fig. 2.16) due to possible stress-induced microcracking.

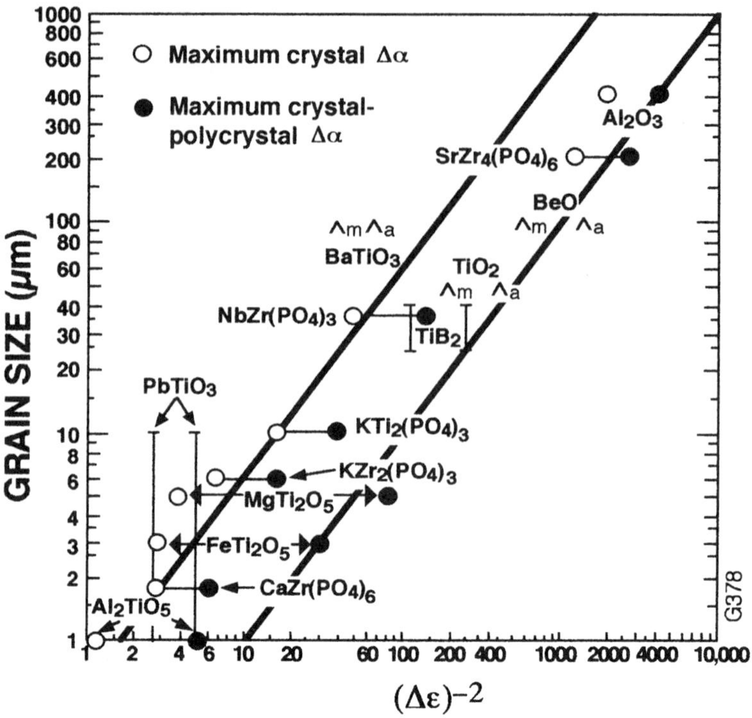

FIGURE 2.10 Log–log plot of the grain size for the onset of spontaneous microcracking as a function of $(\Delta\epsilon)^{-2}$ for transformation ($BaTiO_3$ and $PbTiO_3$) and TEA derived strains. Note the general, but imperfect, correlation over a broad range and the use of the maximum minus either the minimum or the average grain (crystal) strain mismatch, with the latter indicating better correlation, as is discussed in the text. (Data from Rice and Pohanka [29], and from Yamai and Ota [96, 97], using respectively ΔT values of 1200 and 1000°C [though the latter may be somewhat high].)

The first of two related sets of uncertainties in values used in Eq. (2.4) is its derivation, which, as noted earlier, entails either a two-dimensional approximation with more precise mechanical analysis or in one case a more appropriate three-dimensional model but with less precise mechanics analysis for idealized grains. Neglecting the ideal grain character still leaves uncertainty in the equation, e.g. in the numerical factor of 9, and adds to the second set of uncertainties of property values to use. Differing orientations of adjacent grains in quite anisotropic materials may vary the appropriate local E value, and varying grain shape may change the numerical factor. Greater uncertainty arises for γ, since this depends on the nature of the microcrack, e.g. trans- or more commonly intergranular fracture. The latter presents the greatest uncertainty, since γ for grain boundary fracture, commonly approximated as ~ 1/2 of the lowest grain cleavage energy, is quite dependent on the character of specific boundaries involved and the resultant microcrack, e g. the number and spatial relation of grain boundary facets it encompasses. As much or more uncertainty arises for $\Delta\in$, since it is commonly quite dependent on the specific crystallographic orientations of the grains involved, whether the strain mismatch arises from TEA or from phase transformations. For TEA, the ΔT involved is often also uncertain since the temperature where stresses start to build up (commonly assumed to be ~1200°C, as used for all TEA-based calculations in Figure 10 except for the phosphates) depends on material and cooling rates, and the temperatures where the microcracks occur is often not known. Another uncertainty in $\Delta\in$ is associated with statistically significant microcracking. The maximum strain mismatch, e.g. due to the maximum–minimum single crystal thermal expansion msmatch, is often used, but typically pertains to only ~1 grain facet of a limited number of grains. Limited microcracking probably commences when microcrack conditions are met with such a maximum grain (crystal) strain mismatch, but more should occur when conditions are met for a strain mismatch of the maximum minus the average grain (crystal) strain. Fig. 2.10 indicates better correlation with the latter, which can vary $(\Delta\in)^{-2}$ by a few, to nearly 10, fold, as well as a deviation at larger grain sizes and related lower anisotropy; but more evaluation is needed.

Equal or greater uncertainty is associated with reported or implied G_s values due in part to uncertainty in the measured grain size values, which as noted earlier can be a factor of 2 (and involve the issue two- versus three- dimensional G values, the latter appearing more pertinent). Greater uncertainty often accompanies direct microscopic observations of microcracks on, or close to, free (mainly fracture) surfaces. Values from property measurements, e.g. expansion, elastic moduli (or damping), and thermal or electrical conductivities, and their temperature dependences and resultant hysteresis, are generally more valuable and offer important additional information (Chap. 3, Sec. III.D, Chap. 10, Sec. II.E, and Figure 9.3, Refs. 5,26–35,97,99). In particular, plotting of the temperature difference between the forming or opening of microcracks and their closing versus the inverse square root of G, as shown by Case et al. [30], is particularly effective. However, such information should be accompanied by data on grain size and shape, their distributions in the body, and the degree and character of preferred grain orientation. Of two bodies with the same average G, the one with the wider grain size and shape distributions, or both, is likely to have more microcracks, provided there is random orientation.

A more comprehensive, but still incomplete, characterization of microcracks and their effects is Tomaszewski's [100] study of dense sintered Al_2O_3 (Fig. 2.11)

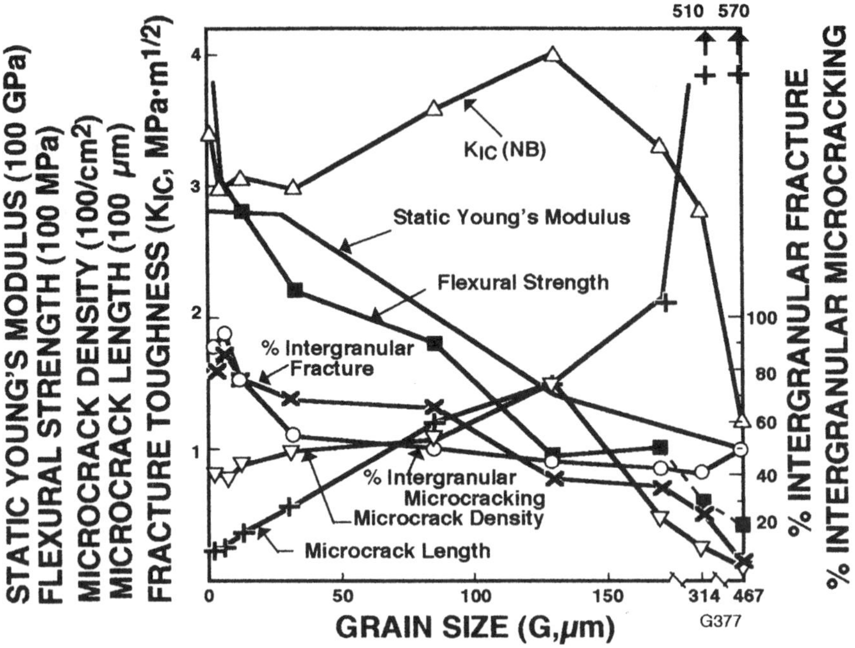

FIGURE 2.11 Summary of Tomaszewski's characterization of microcracks and resultant properties in a dense sintered alumina as a function of grain size [100]. Note general decreases, except for maxima in K_{IC} and microcrack density at $G \sim 130$ μm, and increasing microcrack length as G increases.

showing that while microcracks in finer G bodies were ~5 times the grain size, this reduced to ~1.2 at G ~470 μm. (These results for stress generated microcracks in alumina are in contrast to the limited data of those from spontaneous microcracking.) Qualitative data indicates that spontaneously formed microcracks are more consistently similar in size to the grains. While the question of the effects of as-fired surfaces on such microcracking were not addressed, he showed that the general fracture mode decreased from ~ 85% intergranualar fracture at the finest G to ~5% at the largest G (i.e. generally consistent with Fig. 2.5). However, the ~95% intergranular microcracking at the finest G initially decreased faster, then less rapidly, than the general fracture mode. The net effect was substantially more intergranular microcracking than for general fracture (i.e. ~50 versus 5% respectively), but still with substantial transgranular microcracking over most of the G range. He also showed that microcrack densities reached a maximum at G ~ 130 μm. Calculations from his data indicate that the microcrack area per unit sample area reaches a maximum at this or larger G, e.g. ~300 μm.

Quantitative data of Kirchner and Gruver [101] for TiO_2 shows microcrack size to average ~1.5 to 2 times *G,* over the *G* range ~5 to 150 μm; but much more documentation is needed.

Two sets of potentially related observations are important. First, as noted earlier, most observations of the cyclic change of microcrack dependent properties, while exhibiting hysteresis, are typically repeatable with multiple cycling. However, while such behavior was shown in evaluations of most HfO_2 bodies, it ceased to be so in larger grain (~17 μm) bodies, which showed serious progressive degradation of elastic moduli with thermal cycling [35, 36]. This clearly implied that microcracks no longer closed, or healed at such larger HfO_2 grain sizes, since such closure and healing are basic to the observed recovery of the normal temperature dependence of properties in the absence of microcracking. The second observation is that while microcracking is commonly intergranular, some transgranular microcracking is observed, especially as grain size in a particular material increases sufficiently as found in TiO_2 [29,101], and clearly shown in $2ZrO_2 \cdot P_2O_5$ bodies [95], e.g. at *G* to ~200 μm in the former and ~10 to 20 μm in the latter as well as in Al_2O_3 [99] (Fig. 2.11). Such transgranular microcracking at larger (probably material dependent) grain sizes suggests an explanation for the progressive degradation of larger grain microcracking bodies on thermal cycling. While intergranular fracture results in well-defined fracture surfaces with potential for mating and possible healing due to sintering (especially with some boundary phase), transgranular fracture often generates more complex fracture surfaces that may be more difficult to close and heal. The occurrence of some transgranular microcracking at larger grain sizes may be consistent with general fracture mode trends, though probably shifted due to the TEA and other causes of the microcracking.

Another important aspect of microcracking is its occurrence due to the combination of microstructural stresses from phase transformation or TEA with applied stresses, either prior to or during measurement of crack propagation behavior. Limited investigations indicate that microcracks can be generated, enlarged, or both in some bodies and tests, e.g. *E* measured in flexure strength testing decreasing as *G* increased while dynamic measurements were independent of *G* at ~393 GPa [99]. In more anisotropic graphite, Brocklehurst's review [102] cited decreases in elastic moduli due to prior application of either a uniaxial tensile or compressive stress that varied with the microstructure and stress. The moduli decreases increased nonlinearly, and sometimes irregularly, as the stress level increased, e.g. reaching *E* reductions of > 30%, but could be fully or partially recovered by subsequent treatments, especially annealing. Greater decreases may occur from compressive stressing (possibly in part because of the greater stress levels achievable in compression) than with tensile stressing. Other differences occur, e.g. density and thermal expansion reduced from tensile, and increased from compressive, loading, and properties such as electrical conductivity are also affected, but strengths are often not significantly reduced. Major sources of these complications in graphite are effects of grain and pore parameters, e.g. microcracks originating and branching from, or terminating on, pores have been reported [102, 103].

Consider now the case of introducing, or generating, more microcracks in conjunction with propagation of a macrocrack. While the originally predicted occurrence of microcrack formation in two lobes ahead of the crack tip significantly above and below the macrocrack plane [44] is contrary to experimental results, and attention has now been focused on wake region bridging, the issue of some microcrack formation at or closely ahead of the macrocrack is not settled. Ultrasonic probing around a stressed macrocrack by Swanson [49, 50] in both granite and alumina samples showed microcracking being confined mainly in or near the wake region of the macrocrack, mainly or exclusively associated with bridging. However, while Hoagland et al.'s [48] results (Figure 4.5 of Ref. 5) showed much wake microcracking, they also clearly showed considerable microcracking ahead of and around the macrocrack, much less in spatial extent and location than first proposed, but more than indicated by Swanson. The extent to which the sandstone porosity was a factor in this broader spatial distribution of microcracks is unknown, but Hoagland et al.'s and other results [102–105] showed that even quite limited porosity can play a significant role in some microcracking in conjunction with a macrocrack, indicating complex behavior that is not fully understood. Additionally, even where most toughening occurs via wake bridging, it appears that microcracks may initiate at or slightly ahead of the macrocrack tip as precursors for bridges, so bridging evidence is also probable evidence for microcracking. Acoustic emission, which is probably more sensitive but is not widely used, is an indication of this (Sec. III.G). There may also be contributions of other mechanisms such as crack deflection, branching, and tortuosity (see Sec. IV.B of Chap. 8), depending on material and microstructural parameters as discussed for K_{IC}–G dependence of materials such as $Al_2Ti_2O_5$ with extensive microcracking [104]. Lawn [106] theoretically considered crack tip microcracking, concluding that there was a limited opportunity for microcracking ahead of the crack tip, but noted that this might be expanded by second phase particles. This may imply some effects of porosity, though he did not address this or extreme anisotropy, e.g. of graphite.

The first of five additional observations is Peck et al.'s [107] qualitative assessment of crack propagation and fracture energy of rocks in their and other studies. While they showed that porosity (and weak interface bonding) gave lower fracture energies, as expected, higher fracture energies and greater distances of crack propagation for steady state fracture occurred when there was a preexisting network of interconnecting microcracks (or when rock texture provided multiple, incipient fracture surfaces); but the highest fracture energies were found in crystalline rocks without interconnected microcracks. Second, significant maxima of fracture toughness versus grain size (Figures 2.16 and 2.17), initially attributed to microcracking effects from TEA or from crystal structure transformation (e.g. $BaTiO_3$), have since been implied to be due to bridging. Third, tests of bodies in which the microstructural stresses arise from phase transformations that occur close to room temperature, e.g. $BaTiO_3$, offer opportunities to verify the role of microstructural stresses by testing above and below the transformation temperature. The fourth and related point is that in piezoelectric and ferroelectric materials, electric fields can increase the grain boundary stresses and microcracking (Sec. III.I). Fifth, as noted above and later, acoustic emission can be an important tool for detecting and evaluating microcracking.

D. Grain Size Dependence of Fracture Toughness of Cubic Ceramics

Though many studies of fracture toughness have been made with little or no grain size characterization, there is considerable data on the grain size dependence of fracture toughness. While there are uncertainties of test parameters such as crack size effects, important data trends are seen from four previous reviews [11, 60, 108, 109], especially the two most recent. This section addresses toughness for cubic materials, particularly overall trends, and notes variations and some of the known or indicated test differences and limitations. The following section treats single-crystal toughnesses and the transition between them and polycrystalline values for both cubic and noncubic ceramics. Section F addresses the *G* dependence of noncubic ceramics, Sec. G, crack wake effects such as bridging (which occur in cubic materials, and more often in noncubic ones) and Sec. H, effects of preferred grain orientation, again mainly in noncubic ceramics.

The grain size dependence of fracture energy and toughness of cubic materials is generally simpler and apparently less prone to wide variations in comparison to noncubic materials, but it is still subject to variations. The overall trend for cubic ceramics is for limited dependence on grain size over the *G* ranges generally covered (e.g. a few to ~100 μm, Figs. 2.12–2.14), but some, typically modest, deviations from this are common and possibly universal. Thus most MgO studies do not indicate a *G* dependence (Fig. 2.12), but this may reflect the more limited *G* ranges, and numbers of different grain sizes covered; an earlier compilation of data from 1 to 200 μm [110] indicated a possible limited maximum of fracture energy at intermediate *G* (e.g. 10–40 μm). (The low value at 1 μm reflects in part residual additive effects, since fine-grain MgO as hot pressed with 1–2 wt% LiF had ~1/2 the fracture energy as of material with similar grain sizes annealed to remove residual LiF or MgO hot pressed without LiF; both of the latter show substantially more transgranular fracture than material as-hot pressed with LiF [110, 121]). MgO with limited porosity, e.g. < 1 to ~4% porosity, showed K_{IC} increasing ~3-fold from values for dense bodies, with much of

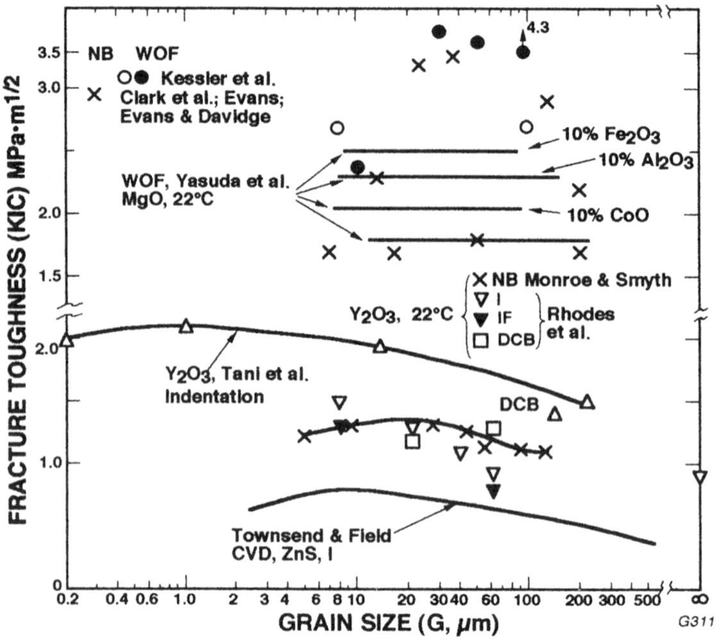

FIGURE 2.12 Fracture toughness (K_{IC}) versus average grain size (G) at 22°C for various investigations of MgO (upper portion), Y_2O_3 (middle), and ZnS (lower portion). Note (1) no clear G dependence of K_{IC} for MgO with ~ 0% porosity (P), but for MgO with <0.1 to >1% porosity (solid circles), K_{IC} clearly increased above P = 0 levels with increasing G as P decreases, shifting to a more intragranular porosity [110, 111]; (2) some increases in toughness levels of MgO for 10% additions of different (partly to fully soluble) oxides in MgO [112–115]; (3) similar maxima trends, but different values for them for two of the three Y_2O_3 studies [116–118]; (4) the lower value (from indentation tests) for single crystals of Y_2O_3 [115]; values for {100} fracture of MgO crystals are typically ~ 1 MPa·m$^{1/2}$ [118]; and (5) a clear maximum for the CVD ZnS [120]. (After Rice [11], published with the permission of the Journal of Materials Science.)

this limited porosity transitioning from inter- to intragranular locations as G increased from ~10 to 100 μm, which may reflect possible general effects either of pore location or of slip associated with pores [11], consistent with dislocation activity accompanying MgO fracture at modest temperatures (Sec. III.B). Similarly, at first glance, no grain size dependence is shown for the fracture energy of dense FeO being ~ 7.5 J/m^2 over the G range ~10–90 μm [122], but there is a probable 4–8% decrease as G increases. Two data points of Evans and Davidge [123] for UO_2 with G ~8 and ~ 30 μm had fracture energies of respectively ~8 ±1

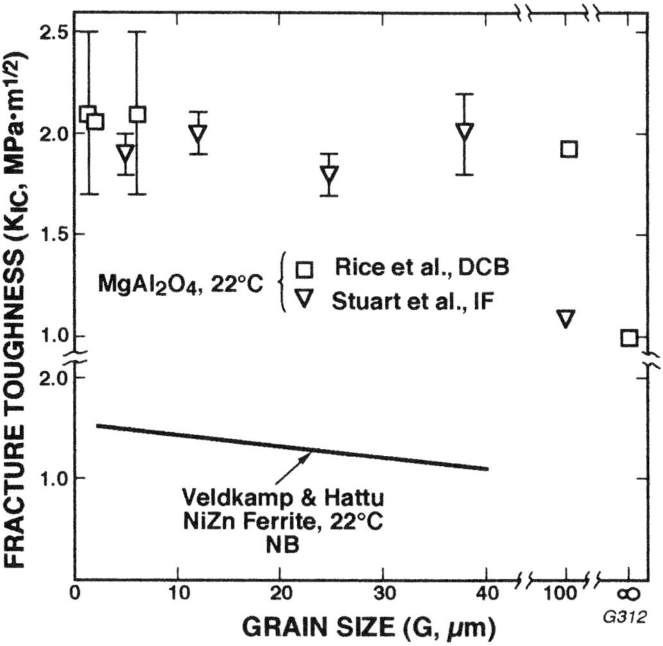

FIGURE 2.13 Fracture toughness versus average grain size at 22°C for various indicated investigators for spinel and a ferrite. Upper portion: $MgAl_2O_4$ data of Stuart et al. [124] and Rice et al. [109], all made without additives, except for a low value at G = 100 μm for a body made with LiF additions. Note (1) the good agreement of the two studies showing no clear intrinsic dependence on grain size and (2) the value for fracture on either {100} or {110} single crystal planes $G = \infty$. Lower portion: a soft (cubic) NiZn ferrite [125] showing some decrease in toughness as G increases. (After Rice [11], published with the permission of the Journal of Materials Science.)

and ~ 6 ±1 J/m^2. Though some of this decrease may reflect differences in the limited porosity in the two bodies, some of it may be intrinsic.

Several data sets for cubic materials, commonly with no measurable porosity and no densification aids, showed modest, but probably statistically significant, toughness maxima at intermediate grain sizes. Thus all data sets for Y_2O_3 [116–118] (though at substantially different grain sizes of ~1 and ~20 μm) showed higher toughness at intermediate G as did CVD ZnS [120] at ~1 μm, and various $BaTiO_3$ bodies indicate ~ twice the finer and larger grain fracture energy of ~2.4 J/m2 at ~40 μm [126–128] at 150°C where they are cubic. (A more definitive maximum is shown in the noncubic state at this same G and corroborated by strength trends [129]. Data over a more limited G range (~ 2–40 μm) for a NiZn ferrite showed a decrease as G increased [125]. A significantly lower value for large G $MgAl_2O_4$ made with LiF addition (which is apparently more difficult to remove from $MgAl_2O_4$ than MgO) is reflected in the ~100% intergranular fracture versus the ~100% transgranular fracture of material without LiF additions [11, 110]). Equally or more definitive is Niihara's data for CVD β SiC giving a maximum at ~1–2 μm [130], and Kodama and Miyoshi's data for HIPed polycarbosilane-derived ultrafine grain powder showed a pronounced maximum at G ~0.6 μm [131] (Fig. 2.14). An earlier compilation of SiC data [109], though complicated by reflecting more noncubic, α, SiC (discussed below), also indicated a possible decrease at larger grain sizes.

Thus, in summary, cubic, in contrast to noncubic, materials generally show no G dependence of fracture energy and toughness, or modest decreases at finer, or larger grain sizes, or both. These trends are indicated despite the variations in test methods, especially between different investigators (e.g. for indentation techniques). Extreme results reported for the fine grain SiC from HIPed polycarbosilane powder are unexplained. More broadly, the mechanism(s) for the com-

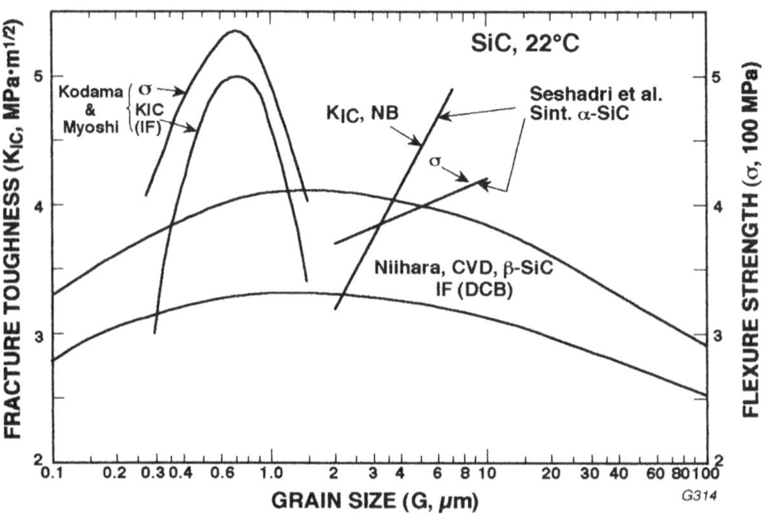

FIGURE 2.14 Fracture toughness (K_{IC}) (and some corresponding flexure strength) versus average grain size (G) at 22°C for α and β SiC from a previous survey [60]. The data of Kodama and Miyoshi [131] is for polycarbosilane derived SiC, Niihara [130] for CVD SiC, and Seshardi et al. [132] for sintered alpha SiC. (After Rice [11], published with the permission of the Journal of Materials Science.)

monly indicated K_{IC} decrease at finer G are uncertain, but extrinsic effects common to preparing many fine-grain bodies may be a factor in some, possibly many, of these results (Chap. 3, Sec. IV.A). Though also uncertain, the first of two possible, not mutually exclusive, intrinsic mechanisms for decreasing toughness at larger G is elastic anisotropy, which has similar effects to TEA in noncubic materials [65]. Unfortunately EA data is often either uncertain or nonexistent, so quantitatively testing is limited. The second intrinsic reason for decreased toughness at larger grain sizes is the transition to single crystal or grain boundary toughness, as is discussed in the next section.

E. Single Crystal/Grain Boundary-Polycrystalline Fracture Toughness Transition

This section addresses single crystal and grain boundary toughnesses for cubic, then noncubic, materials and their relation to typical polycrystalline values. Single crystal fracture toughnesses are important to understanding mechanical failure of not only single crystal but also polycrystalline bodies, e.g. for transgranular grain microcracking. Generally toughness of polycrystalline bodies with normal, e.g. machining, flaw populations must transition to those of the corresponding single crystal or grain boundary values as G increases to larger sizes, as is noted above and discussed below and in Chap. 3. Further, polycrystalline toughnesses for any grain size with increasing preferred grain orientation often begins to approach that of the corresponding single crystal orientation, provided bridging and other polycrystalline toughening mechanisms are not significant. Grain boundary fracture toughnesses are pertinent to the broader case of intergranular microcracking and are expected to be the toughness values that large grain polycrystalline bodies will approach where intergranular failure is dominant (Chap. 3, Sec. I.B).

Single crystal fracture toughnesses of greatest pertinence are those for lower toughness cleavage or preferred fracture planes, since these control most failure of single crystals and are dominant in most transgranular fracture; the latter usually with varying degrees of mixed mode failure for the one or more lower toughness crystal planes involved. Such single crystal values are invariably substantially below those of polycrystalline bodies of the same material and crystal structure and hence provide a general check on the probable validity of limited data for single- or polycrystal values, especially where one or both sets of data are limited. Table 2.1 summarizes much single crystal data and corresponding typical, usually finer grain, polycrystalline values. Again, note that single crystal values approaching or exceeding lower polycrystalline values, or lower polycrystalline values falling below single crystal values for preferred fracture surfaces, should raise suspicion about the validity of one or both values. Thus, for example, values for YAG single crystals seem high, e.g. in view

Table 2.1 Comparison of Ceramic Single Crystal and Polycrystalline Fracture Toughnesses[a]

Material	Typical Sxl fracture	Sxl K_{IC} ($MPa \cdot m^{1/2}$)	Pxl K_{IC} ($MPa \cdot m^{1/2}$)	Pxl fracture[b]
(A) Cubic Ceramics				
MgO	{100} cleavage	1	2	Mainly TGF
ThO_2	{111} cleavage	0.6–1	1.1	Mainly TGF
Y_2O_3	{111} cleavage	<0.9	1.3	Mainly TGF
ZrO_2	Possibly {110} cleavage[c]	1.1–2.4	2.6	Mainly TGF
$MgAl_2O_4$	{100} and {110} cleavage	1	2	Mainly TGF
$Y_3Al_5O_{12}$	{111} cleavage	2.2	1.7–3.1	Mainly TGF?
TiC	{100} cleavage	1.2	3–4	Mainly TGF
CaF_2	{111} cleavage	0.4	1.1	Mainly TGF
ZnSe	{110}	0.3	0.9	Mainly TGF
Diamond	{111} cleavage	2.9–4.1	5.3–13[d]	Mainly TGF
(B) Noncubic Ceramics				
Al_2O_3	Rhomb. & other nonbasal fract.	1.5–2	3.5–4	Mixed IGF & TGF
β-Al_2O_3[e]	Basal, (0001)	0.16	2.7–3.3	Mainly TGF
TiO_2	{110}	0.8	2.5	Mainly TGF
SiC	Probably (1010)or (1120)[f]	2	3–4	Mixed IGF & TGF
Si_3N_4[g]	?	~1–1.5	4	Mainly TGF
Calcite	$\{10\bar{1}1\}$	0.23	1	Mainly IGF?

[a] Sxl = single crystal, Pxl = polycrystalline, K_{IC} = fracture toughness (polycrystalline fracture toughness generally in the absence of significant microcracking, wake, etc. effects)
[b] Pxl fracture: typical or representative fracture mode, with TGF = transgranular fracture and IGF = intergranular fracture.
[c] While {111} cleavage may be expected due to the fluorite structure, cleavage is often on {110} but may be variable, depending on stabilization, composition, and stoichiometry, e.g. as indicated by indent crack complexities as reported by Pajares et al. [134].
[d] Higher polycrystalline toughness is for metal bonded, sintered diamond [142].
[e] In such extreme cases of a single very low toughness plane, other fracture planes are more important, fracture normal to the basal plane gives K_{IC} of 2 $MPa \cdot m^{1/2}$ [144].
[f] Respectively first- and second-order prismatic cleavage (per private communication with P. T. B. Shaffer).
[g] CVD Si_3N_4, i.e. without densification aids and resultant grain boundary phases.
Source: Refs. 4, 11, 39, 40, 60, 76, 124, 133–150. See also Chap. 6, Figs. 1, 5, and 6.

of Pardavi-Horváth's [151] values of two other garnets (GGG and CaGeGG) respectively of 1.2 and 0.8 $MPa \cdot m^{1/2}$ for (111) fracture.

There is typically substantial anisotropy of fracture toughness in single crystals of various crystal structures. However, except for materials of extreme crystal anisotropy, i.e. noncubic materials with very platy structures such as micas, beta-aluminas, graphite, and hexagonal BN, there is no clear difference in the anisotropy of single crystal toughnesses between cubic and noncubic materials. Many crystals, whether of cubic or noncubic structures, have one significantly preferred fracture surface, usually a cleavage plane, with {100} cleavage in many NaCl structure materials being an example, e.g. in CaO and MgO, while showing very limited cleavage on {110} planes, does cleave on {110} planes only under conditions that are not fully understood. This is not surprising in view of the elastic anisotropy of crystals and the relation of elastic moduli and toughness. (Thus the comment by some that cubic ZrO_2 has unusually anisotropic toughness is partly true as well as misleading, since like its elastic moduli, it is fairly anisotropic.) A good test for the primacy of a cleavage or fracture surface is that cracks propagating on other planes will often branch onto the primary, i.e. the lowest toughness, plane, but the reverse crack branching, i.e. from the primary to other planes, occurs only occasionally, if at all. Stoichiometric $MgAl_2O_4$ is unusual in having two cleavage planes of very similar toughnesses [73–76, 124] (Table 2.1). (MnZn ferrite approaches this similarity; see the note at the end of this chapter.) Table 2.2 lists a few other crystals with two or three sets of cleavage planes with similar toughnesses. See also Chap. 6, Sec. IV for additional single crystal toughness data.

Turning to the transition from single- to polycrystal fracture toughnesses, this is seen as a natural consequence of grain sizes increasing to approach and subsequently become larger than the strength controlling flaw. Thus fracture (energy and) toughness must decrease toward single crystal values for transgranular fracture origins given the commonly significantly lower single crystal fracture toughness for preferred single crystals fracture planes, e.g. by factors of 2–3 (Table 2.1, Fig. 2.15). Some mapping of such polycrystalline–single crystal transitions has been by DCB tests of bodies of increasing *G* and especially from fractographic evaluations of larger grain, transgranular, polycrystalline fracture origins [15, 70] (Fig. 2.15). For example, lower than expected DCB toughness

TABLE 2.2 Fracture Toughnesses for Different Preferred Fracture/Cleavage Planes of Some Crystals with More Than One Such Plane[a]

Crystal	Structure	Fracture/cleavage plane- K_{IC} (MPa·m$^{1/2}$)	
Si	Cubic[a]	(111) 0.8	(110), (100) 0.9–0.95
SiO_2	Quartz	{11$\bar{2}$0} 0.85–0.95	{10$\bar{1}$1} 0.85–1.0
$KalSi_3O_8$[b]	Triclinic	{110} 0.31	{001}, {010} 0.39
$Be_3Al_2(SiO_3)_6$[c]	Hexagonal	{10$\bar{1}$0} 0.2	{11$\bar{2}$0} 0.25

[a] Diamond structure.
[b] Microline feldspar.
[c] beryl, which with green coloration, e.g. from Cr doping, is the precious gem emerald.
Source: Refs. 149, 151–155.

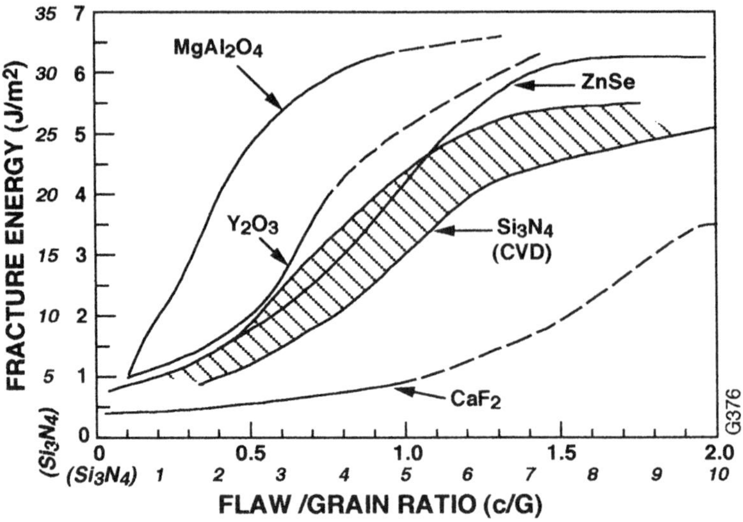

FIGURE 2.15 Data on the single crystal to polycrystalline fracture toughnesses transition as a function of flaw to grain size ratio [15, 62, 70]. Note different scales for CVD Si_3N_4.

values were found to be associated with larger grains at the crack tip [15], showing the utility of fractographic examination of toughness test specimens, not just strength specimens or components. Indentation $MgAl_2O_4$ data of Sakai et al. [156] is consistent with such a transition, as is single crystal data of Chen and colleagues [152, 153] for two polycrystalline bodies, but with flaw sizes slightly smaller than *G*. More recent evaluation of crack tip stresses [157] may be an alternative approach to such evaluation. The level of polycrystalline fracture toughnesses versus corresponding lower single crystal values must reflect the range of single crystal values, especially the lower ones, but is impacted by two factors. The first is the multiplicity of the sets of planes with low toughnesses. A single, e.g. noncubic basal, fracture plane having only one member, while limiting polycrystalline toughness, must have less effect in limiting polycrystalline toughnesses since it allows more effect of other, higher toughness, crystal fracture planes. Increased multiplicity of lower toughness fracture planes, e.g. {100} cleavage in a cubic material such as CaO or MgO which reflects three planes probably limits polycrystalline toughness [15, 62]. Second is the resultant mixed mode combinations of different crystal planes making up transgranular polycrystalline fracture, which is probably a major factor in the greater polycrystalline versus single crystal fracture toughness.

A similar transition must occur from polycrystalline to grain boundary toughness values for intergranular fracture origins. Such boundary values should typically be lower than the lowest single crystal values, often substantially so, especially for the extreme case of fracture of a single grain boundary facet, as represented in fracture of bicrystals. It is commonly estimated that fracture surface energies of grain boundaries relative to crystal fracture, e.g. cleavage, energies would parallel those of surface energies, where boundary energies are often ~ 1/2 those of lower crystal surface energies for a given material. Such estimated trends are reasonably supported by the limited measurements. A classic study is the DCB fracture energy measurements of Class and Machlin [158] of KCl bicrystals grown from the melt with grain boundaries parallel with (100) surfaces, but with controlled twist angles between the two "grains." Their results showed fracture energies decreasing rapidly as the twist angle increased from zero to ~ 0.08 J/m^2 at ~ 5° twist, then bottoming out at 0.04–0.05 J/m^2 at twist angles of 15–45°, in contrast to (100) values of ~ 0.11 J/m^2 (independent of angle of propagation on the (100) plane). This reflects an average boundary fracture energy ~ 1/2 that of the cleavage energy as suggested by surface energy differences and translates to a decrease to a minimum boundary toughness versus that for cleavage of respectively 0.06 and 0.09 MPa·m$^{1/2}$.

More recently, Tatami et al. [159] reported similar NB measurements of sapphire bicrystals made by pressure sintering crystals with a common *c*-axis, i.e. <1000>, normal to the boundary, which was parallel with the (0001) plane, and typically contained 0–10% porosity. (Such purely twist boundaries avoid significant boundary stresses from thermal expansion anisotropy, as shown by Mar and Scott [160] for sapphire bicrystals with twist angles about other axes.) Fracture energies at 0° twist varied from ~ 6 to ~ 15 J/m^2 (for toughnesses of ~ 2.7–4.3 MPa·m$^{1/2}$) with higher values consistent with values for basal fracture (Fig. 6.1) and lower values with increasing residual porosity. The overall trend for fracture energies with increasing twist angle was for substantial decrease, with a few higher spikes of fracture energy at boundary angles where adjacent "grains" had coincident lattices, but the maximum of these spikes also decreased as the twist angle increased. Minimum fracture energies of ≤ 1 J/m^2 occurred over most twist angles over the range of ~ 25–45°, which corresponds to a fracture toughness of ~ 1 MPa·m$^{1/2}$. Such decreases were greater than expected from surface energies, e.g. by a factor of 2 or more.

F. Grain Size Dependence of Fracture Toughness of Noncubic Ceramics

The previous two sections have addressed the *G* dependence of toughness in cubic ceramics, then single crystal toughnesses, their anisotropy, and relation to typical polycrystalline values. This section addresses the *G* dependence of toughness of noncubic polycrystalline ceramics in the order: Al_2O_3 (the most studied ceramic), BeO, and TiO_2; some mixed oxides and Nb_2O_5 (some with phase transformations or extensive microcracking); TiB_2,B_4C, and Si_3N_4 (including some effects of grain elongation and boundary phases), and then TZP bodies with phase transformation and possible microcracking. A few more comments on the transition to single crystal and especially grain boundary values are made at the end. Again the following two sections address crack wake and *R*-curve effects and effects of preferred grain orientation, both particularly for noncubic ceramics.

Overall, the fracture energy and toughness of noncubic ceramics often show more pronounced grain size dependences as well as more variation with different materials, microstructures, and measurements (Fig. 2.16). While there are variations in the relative magnitude of the values with general agreement in the trends with grain size, there can also be significant differences in the trends

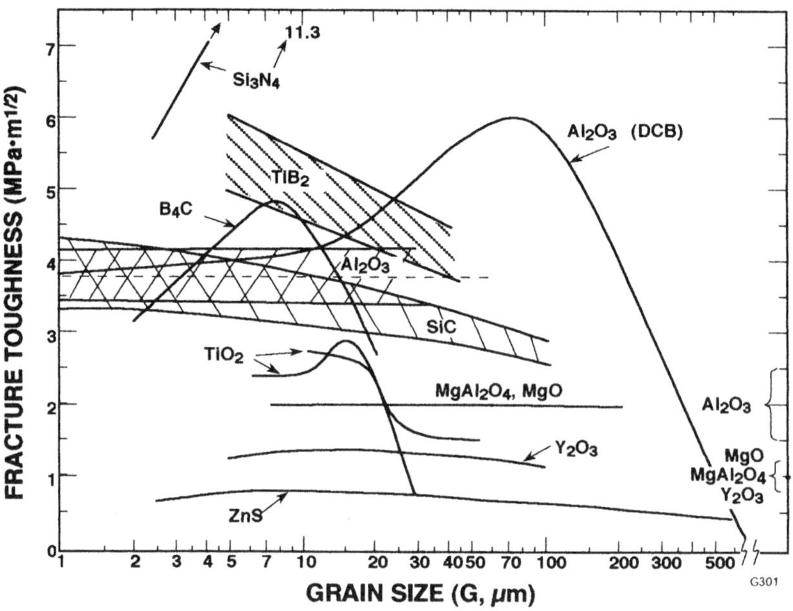

FIGURE 2.16 Summary of the fracture toughness versus grain size for various noncubic ceramics at 22°C by various tests (along with cubic material data from Figure 2.11). Note that while these are common trends shown by one or two techniques (mainly WOF or DCB) for some materials and several techniques for Al_2O_3, some investigators have obtained significantly different results, mainly for Al_2O_3 (the most extensively tested ceramic), as is discussed in the text, along with differences among these material results. Note representative single crystal values for some of the materials shown along the lower right scale.

with *G*. These differences reflect incompletely documented and understood changes in mechanisms and their dependence on test effects. Thus per Fig. 2.16 and previous reviews [11, 60, 108] DCB and DT, and sometimes NB, tests show pronounced Al_2O_3 toughness maxima, e.g. at *G* between 50 and 150 μm, which as noted earlier correlates with stress-induced microcracking (Figs. 2.11 and 2.16). Such trends appear consistent with a transition from mainly or exclusively stress-induced microcracking at lower *G* and spontaneous microcracking at larger *G*. However, other studies, especially with NB tests, show toughness constant or decreasing slightly as *G* increases over the typical *G* range of ~ 3–30 μm, with the more limited NB (and IF) data at *G* ~ 40–50 μm showing further, often more pronounced, toughness decrease. Some NB tests also show marked increases below *G* ~ 10 μm (which may reflect inadequate notch-crack effects), while other NB data shows toughness independent of *G* or slightly decreasing as *G* increases to the limits of much testing at ~ 30 μm. Note that the overall NBAl_2O_3 data shows the typical toughness for it of ~ 3–4 MPa·m$^{1/2}$. Also, even tests designed to give toughness values without bridging done in Al_2O_3 (e.g. *G* 1–16 μm) give K_0 (i.e. "crack tip") values of 1.8–2.7 and values controlling strength failurenearer 3 MPa·m$^{1/2}$ [157], but there are various uncertainties in such measurements. Further, much more limited data for BeO (Fig. 2.17) indicates a similar trend as for DCB Al_2O_3 data, consistent with the two materials having very similar Young's moduli and TEA (hence microcracking).

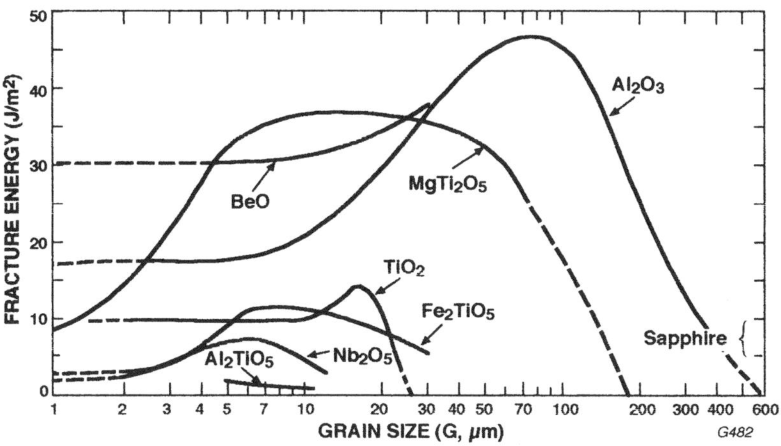

FIGURE 2.17 Fracture energy versus grain size for several noncubic oxides, including DCB data for Al_2O_3 and TiO_2 data in Figure 2.16 and for three titanates and Nb_2O_5 with higher TEA and thus more microcracking at finer *G*. (After Rice [60] published with the permission of the Journal of Materials Science.)

More limited TiO_2 DCB and WOF data (Figs. 2.16, and 2.17) shows some differences, but reasonable overall consistency. More recent data [60, 161, 162] also showed marked toughness decreases between G = 20–30 μm from higher values (e.g. 6+ down to 3–4 MPa·$m^{1/2}$), and probable effects of stoichiometry. The TiO_2 data thus also appears to reflect effects of microcracking, e.g. probably much of it stress induced, but also spontaneous cracking at the lower toughnesses at larger *G* (Fig. 2.10).

Consider now noncubic ceramics of more extreme TEA, e.g. titanates of MgO, Fe_2O_3, and Al_2O_3 along with Nb_2O_5, which have extensive microcracking at modest to fine *G* (Fig. 2.10). These also generally show, often pronounced, maxima of fracture energy and toughness versus *G* (Fig. 2.17), with overall trends generally consistent with those expected from Eq. (2.5), and the previously outlined microcracking model [43]. However, comparison of two groups of materials: (1) Al_2O_3 and TiO_2 (as well as presumably BeO and $BaTiO_3$), and (2) $MgTi_2O_5$, Fe_2TiO_5, and Nb_2O_5 (and presumably also Al_2TiO_5) shows that fracture energy peaks of group 1 are narrower and occur at finer *G,* typically substantially so, than predicted for spontaneous fracture per Eq. (2.4) than those of group 2; the peaks of group 2 occur at *G* values at or beyond those predicted for spontaneous fracture. The former is consistent with the microcracking model, the latter is not. Second, the fracture mode of group 1 typically involves some to substantial transgranular fracture, while that of group 2 entails mainly intergranular fracture, e.g. for *G* above half the peak toughness value. Third, and particularly important, while the tensile (flexural) strengths of both groups decrease with increasing grain size (Chap. 3), those of group 2 do so at a faster initial rate as compared to group 1, as well as most ceramics of cubic or noncubic crystal structure (Figures 3.1 and 3.23). Fourth, dynamic *E* values are independent of *G* for group 1 but decrease with increasing *G* for group 2, and for the latter, along with tensile strength, increase with temperature [164] Hamano et al. [105] showed that as microcracking increased as *G* increased, the work of fracture more than doubled as *E,* flexure strength, and crack velocity decreased substantially. They attributed these effects to crack deflection and blunting due to the microcracks.

Noncubic (i.e. piezoelectric) $BaTiO_3$ shows a more definite and pronounced fracture energy maximum, 2–3 times that found at the same *G* as cubic $BaTiO_3$ (above the Curie temperature, 126–129). More limited data of Freiman et al. [163] on PZT as-fired or poled showed toughness increased from ~ 1.3 to ~ 1.55 MPa·$m^{1/2}$ as *G* increased from ~ 2 to ~ 14 μm and was ~ 1.7 MPa·$m^{1/2}$ at *G* ~ 100 μm. A toughness maximum was assumed between the finer and large *G* values, i.e. similar to $BaTiO_3$, but much broader. Their data for pressure depoled samples indicated a similar trend but shifted upward to ~ 2.2 to 2.5 MPa·$m^{1/2}$ at the lower and large *G* values respectively, again consistent with a maximum at intermediated *G.* At least some of these toughness changes in the noncubic state are related to stresses from the transformation (and possible microcracking Fig. 2.10) and from poling (Sec. III.I).

Consider next nonoxides. The fairly definitive toughness decrease for TiB_2 (Fig. 2.16) [60] as *G* increases from ~ 5 to 25 μm may be in part due to stress-induced microcracking, consistent with strength changes, prompting Telle and Petzow [98] to suggest, probably incorrectly as noted in Sec. III.C, spontaneous microcracking. The general trend for K_{IC} of TiB_2 to decrease above $G \sim 5$ μm and a maximum for B_4C at $G \sim 8$ μm is generally consistent with Eq. 2.4 and possibly also with the microcracking model for fracture energy and toughness versus *G*. Thus Skaar and Croft [165] give the maximum single crystal $\Delta\alpha$ of $\sim 3 \times 10^{-6}$ C^{-1}, and the maximum single crystal–polycrystal $\Delta\alpha$ of $\sim 2 \times 10^{-6}$ C^{-1}, giving grain sizes for microcracking of the order of those in Fig. 2.4.

Turning to hot pressed B_4C, DCB data of Rice [60] is generally consistent with more extensive and definitive NB data of Korneev et al. [166] and agrees well with DCB data of Niihara et al. [167] on CVD material. All of these showed toughness rising from 2.5–4 $MPa{\cdot}m^{-1/2}$ at G ~ 3 μm through a maximum of ~ 5.5 $MPa{\cdot}m^{-1/2}$ at $G \sim 10$ μm, then decreasing, to ~ 2.5 $MPa{\cdot}m^{-1/2}$ at $G \sim 30$ μm. This trend is consistent with data of Schwetz et al. [168] for sintered-HIPed B_4C + 2, 3, 4, or 5 w/o carbon, which rose, rapidly initially, then at a diminishing rate, from ~ 2.5 to ~ 4 $MPa{\cdot}m^{1/2}$ as *G* increased from 1–2 to 10 μm. The only noticeable effect of the excess carbon was generally to give somewhat finer *G*. Thus, except for one data point of Rice at $G \sim 80$ μm and ~ 4 $MPa{\cdot}m^{1/2}$, the data sets are generally consistent. The occurrence and scale of the B_4C toughness maxima is also consistent with it having nearly identical TEA and Young's modulus as Al_2O_3 [60], thus also suggesting a microcracking effect. This would suggest that the *G* value for its K_{IC} maximum would be similar to that of Al_2O_3. However, it clearly deviates significantly from this expectation and hence from Eq. 2.4 and the microcracking model for the grain size dependence of fracture energy and toughness, being about an order of magnitude too low. Two other deviations are the essentially 100% transgranular fracture and extensive twinning on the fracture surfaces of larger B_4C grains in contrast to mixed to intergranular fracture mode of Al_2O_3 with little or no obvious twinning. Twinning in B_4C may provide a mechanism of microstructural stress relief or redistributin to avoid microcracking at large *G*, but it leaves uncertain the reason for the toughness maxima at $G \sim$ 10 μm, where transgranular fracture still dominates, but twinning appears to be much less.

The substantial fracture energy and toughness data for various Si_3N_4 bodies is complicated by limited characterization, as well as microstructural complexities similar to and beyond those for most other ceramics. Complexities include the α versus β phase contents and residual additive-boundary phases and contents, a limited range of grain diameters achievable (commonly < 10 μm, but often with substantial grain elongation, e.g. aspect ratios to ~ 10), often with considerable heterogeneity of grain structure, along with varying degrees of grain orientation. However, overall trends are indicated by individual and some collections of tests as follows, with effects of bridging and grain shape and orientation in Secs. III.G and H.

First, the absence or presence, amount, and character of residual boundary phases play a key role, since, while varying with the additive chemistry, fracture energy and toughness generally increase with oxide additive content, as shown for a variety of types and amounts of additive contents by Rice et al. [169], and more recently in a narrower study by Choi et al. [170]. Extrapolation of such trends to zero additive levels gives $K_{IC} \sim 4$ MPa·m$^{1/2}$, i.e. below levels with oxide additives, which is consistent with extrapolation of fracture toughness data of RSSN (i.e without additives) to $P = 0$ [169]. It is also consistent with data for Si_3N_4 bodies: (1) hot pressed with nonoxide additives ($BeSiN_2$) by Palm and Greskovitch [171], giving no significant residual boundary phases, (2) hot pressed or HIPed at high pressures without additives of Shimada et al. [172] and Tanaka et al. [173], and (3) from CVD Si_3N_4, as shown by Rice et al. [169] and Niihara [130, 174]. The latter CVD bodies ($G \sim 1$, 4, and 10 μm) give K_{IC} respectively of ~ 5, 4, and 3.5 MPa·m$^{1/2}$ (indentation), while the former CVD body with $G \sim 100$ μm also gives $K_{IC} \sim 4$ MPa·m$^{1/2}$. Bodies made without additives commonly have G of the order of 1 μm, as did those made with $BeSiN_2$. Thus these results show little or no G dependence of K_{IC}, or it is decreasing some with increasing G (e.g. from Niihara's data), either of which is in contrast to results for bodies made with oxide additives (discussed below). Another important aspect of the above bodies is their predominately transgranular fracture mode (and in those tested, no SCG, Sec. III.B) across the complete range of G values, even with substantial to complete β phase content and grain elongation. This is in ontrast to substantial intergranular fracture in Si_3N_4 bodies made with oxide additives. The frequent absence of significant, if any, effects of β phase content in RSSN [175] is attributed to their smaller grain size, less elongation, and especially common transgranular fracture.

The toughness of Si_3N_4 bodies made with oxide additives, while generally increasing with additive content, also generally increases with grain size and grain elongation, the latter typically arising from α to β transformation with increased temperature exposure (mainly in processing). Studies show K_{IC} increasing over part or all of the limited practical G range, e.g. Tani et al. [176], Matsuhiro and Takahashi [177], and especially Kawashima et al. [178]. The latter data (Fig. 2.16) implies a probable K_{IC} maximum at G of the order of 10+ μm based on the high K_{IC} level, as well as on strengths (Chap. 3), but such a maximum is at an order of magnitude or more lower G than expected from the TEA of Si_3N_4, i.e. for α Si_3N_4, $\alpha_a = 3.7$, and $\alpha_c = 3.8 \times 10^{-6}$ C^{-1} and for β Si_3N_4 (i.e. the larger, elongated grains in most sintered bodies) $\alpha_a = 3.3$ and $\alpha_c = 3.8 \times 10^{-6}$ °C^{-1} [179] are both too small for microcracking in the grain size of such a K_{IC} maximum. On the other hand, Si_2ON_2 [179] and most oxide boundary phases would have expansion differences that could give microcracking closer to the indicated G range for the probable K_{IC} maximum. Though specific grain sizes were not given, Himsolt et al. [180] showed a toughness maximum of ~ 8 MPa·m$^{1/2}$ at similar G values, correlating with nearly complete α to β conversion and increased grain growth (e.g. due to less inhibition by diminishing α content). This evaluation is consistent with that of others as summarized in the review of Pyzik and Carroll [181] showing projected toughness mxima at G ~ 14 and ~ 20 μm respectively for Y_2O_3-MgO and Y_2O_3-Al_2O_3 additives. Such additive effects are consistent with Peterson and Tien [182] correlating increased toughness and related crack tortuosity (and bridging) with increasing thermal expansion of the oxynitride boundary phase. Correlation of K_{IC} with G is complicated by the extent of grain elongation, the volume fraction of such grains, and their orientation, but reasonable overall understanding and corroboration of the expected trends has been established. Sajgalik et al. [183] showed toughness increasing as the volume fraction of grains with larger aspect ratios increases (e.g. 4), and Ohji et al. [184] (Sec. III.E) showed substantially higher toughness (and bridging) with crack propagation normal to aligned tabular β grains, and much less for crack propagation parallel with such elongated grains.

Fine-grain tetragonal ZrO_2, i.e. TZP, bodies also show high maximum toughnesses at fine G (Fig. 2.18) with transformation toughening a major factor. Thus Wang et al. [185] showed a definite (NB) toughness maximum at G ~ 1.2 μm for compositions with 2, 2.5, or 3 m/o Y_2O_3, and a similar result, but more differentiation of compositions and some differences in details of the G dependence, by indent methods (and overall similarity, but some variations in the G dependence of strength). Indent toughness of Cottom and Mayo [186] for 3 m/o Y_2O_3 (~ 4 MPa·m$^{1/2}$ at G ~ 0.25 μm and ~ 8 MPa·m$^{1/2}$ at G ~ 1.4 μm), while not inconsistent with the G dependence of Wang et al., are substantially lower in values, which are more consistent with indent values of Swain [187], which suggest a maximum at larger G. Data of Ruiz and Readey [188] on 2 m/o Y_2O_3 using the average of their four different calculations of indent toughness imply a possible toughness maximum at > 5 μm, but one set of calculations was reasonably consistent with a maximum at G ~ 1–1.5 μm (respectively solid and dashed lines in Fig. 2.18). The three data points of Theunissen et al. [189] are in the same range, except shifted to finer G values than those of Ruiz and Readey. Indent results of Duran et al. [190] for two compositions of ~ 3 m/o Y_2O_3 (with some Er_2O_3) showed modest toughness maxima of ~ 5 and nearly 8 MPa·m$^{1/2}$ at G ~ 0.3 μm, which is clearly both different from, and similar to, data of Masaki [191] for 2, 2.5, 3, and 4 m/o Y_2O_3 for G = 0.2–0.6 μm showing no G dependence (but toughness of 15+ and ~ 5 MPa·m$^{1/2}$ for respectively 2 and 2.5–4 m/o Y_2O_3), consistent with some to substantially greater toughness with 2 m/o Y_2O_3. Combined indent data of Duh et al. [192] for 1 m/o $YO_{1.5}$ + 11 mol% CeO_2 and SENB data of

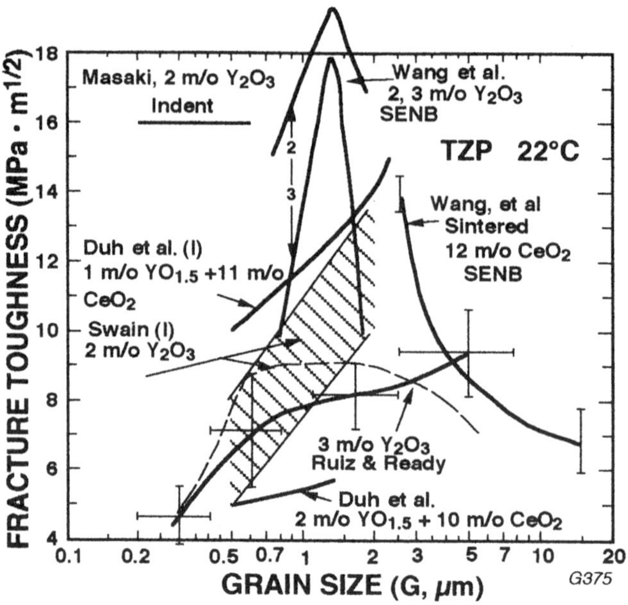

Figure 2.18 Indent and SENB fracture toughness versus grain size for various TZP bodies, investigators, and tests [185–194]. Note clear or implied, pronounced toughness maxima, some similar and different, and a possible maximum for one of four indent calculations of Ruiz and Readey [188], respectively dashed and solid lines.

Wang et al. [193] for 12 m/o CeO_2 indicate a pronounced maximum at G ~. 2–2.5 μm. While there are uncertainties in such combination, later indent results of Duh and Wan [194] for bodies with 5.2 m/o CeO_2 and 2 m/o Y_2O_3 indicated a fairly broad toughness maximum of ~ 19 MPa·m$^{1/2}$ at *G* ~ 1.5 μm. Thus, while there are considerable variation and differences, there is substantial evidence for often pronounced toughness maxima as a function of *G*. The indicated *G* values are approximately consistent with Eq. (2.4), and possibly (2.5), but other related effects of the transformation process and its effects need to be considered, e.g. as reviewed by Becher et al. [195].

Note the range of K_{IC} values for some noncubic single- and polycrystals (Fig. 2.16, Table 2.1), mostly showing ratios of polycrystalline to single crystal K_{IC} values of 2–3, as for cubic materials. This is not surprising, since cubic materials often have similar, sometimes more, anisotropy of elastic moduli [65] and cleavage and fracture as noncubic materials. Thus, except where TEA and other microstructural stresses are dominant, noncubic materials should show similar

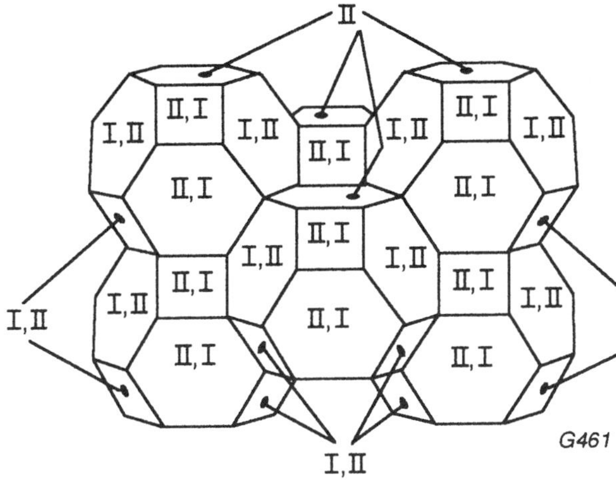

FIGURE 2.19 Schematic of local fracture modes (I, II, or III) of grain boundary facets for local microcracking (or crack propagation). (After Rice [39], published with the permission of the ASTM.)

transitions between polycrystalline and single crystal or grain boundary fracture energies or toughnesses as cubic materials do, e.g. as indicated for Si_3N_4 (Fig. 2.15). This is also indicated by indent tests of Liang et al. [80] of larger Al_2O_3 grains showing toughness decreasing from ~ 2–4 MPa·m$^{1/2}$ (G ~ 50 μm) and 1.5–2.5 MPa·m$^{1/2}$ (G ~ 100 μm) (respectively for 5 and 10 N loads). Such a transition has also been outlined for CVD Si_3N_4 (Fig. 2.15), which has limited anisotropy and all transgranular fracture. However, it should again be noted that the combination of TEA and increasing G of many noncubic ceramics should result in increased intergranular fracture, which would shift the transition to grain boundary rather than single crystal values. While such grain boundary values will often be lower, they will often probably be more complex due to the local mixed mode fracture entailed when more than one grain boundary facet is involved (Fig. 2.19).

G. Grain Dependence of Crack Wake Bridging and Its Relation to Fracture Toughness

Crack bridging in the wake region is an established phenomenon based on direct observation in this region and is the source of increased toughness in R-curve effects. The latter was clearly shown by Knehans and Steinbrech [51] who first showed R-curve effects in NB tests of alumina. Then they extended the original sawn notch, removing the wake region generated from the initial crack propagation from the original notch. Upon retesting the specimen with the wake region from the original test removed, toughness returned to the initial level at the beginning of the first test but again increased in the same fashion as in the original test with subsequent crack propagation and new wake development.

Crack wake bridging observations are substantial, especially in Al_2O_3 and Si_3N_4 bodies, but have not been extensively reviewed. However, a review by Sakai and Bradt [196] is summarized and followed by a broader review, generally corroborating and extending key points in their review. These include bridging and resultant *R*-curve effects widely occurring and generally being more pronounced with increasing crystalline (grain) anisotropy. Thus *R*-curve effects are substantial in materials such as graphite (and especially in fiber reinforced ceramic composites), but results are probably not unique to a given material and microstructure. Further, while Weibull modulus may increase, the highest toughness does not necessarily yield the highest strength (Chap. 3), and there probably are crack geometry effects (e.g. along the lines depicted in Fig. 2.1).

Earlier observations of Wu et al. [6] using DCB tests in an SEM as well as microradiographic and some optical observations of crack character in ceramics were made before wake bridging was recognized. Though their purpose was to demonstrate that the then common idealization of cracks in polycrystalline bodies as simple ~ planar cracks as commonly found in glass was often not true, their results also clearly showed what is now recognized as crack wake bridging in both similar and different bodies (Fig. 2.4) from those subsequently investigated. Thus they showed via in situ SEM examination of Al_2O_3 of larger than normally investigated grain size (e.g. 30–50 μm) limited occurrence of bridges, which often consisted of several grains, and that some bridging in such bodies resulted in considerable local microcracking within bridges. Whether these reflect effects of larger grain sizes, as-fired, rather than the normal polished, surfaces, or both is not known, but they are one of many indicators of the need for wider study of surface and grain size effects on bridging. They also conducted microradiography of a commercial fused-cast alumina refractory sample (with machined surfaces, an average *G* ~ 200 μm, with grains ~ 50 to 2000 μm, and preexisting microcracks a few hundred μm in size). These microcracks, especially larger ones or clusters of them, interacted with the propagating macrocrack, e.g. each deflecting to join with each other (but not with isolated spherical pores, Fig. 2.4).

Other representative and important alumina observations for finer grain bodies with machined surfaces are outlined as follows. Swanson et al. [49, 50] showed a bridging zone length of ~ 100 grains in a body with *G* ~ 20 μm. Rodel et al. [55] also showed multigrain bridges, and their fracturing, i.e. microcracking along grain boundaries (*G* ~ 11 μm). Vekins et al. [197] using DT tests of a commercial 96% alumina, some heat treated to extend the starting *G* from ~ 4 to ~ 10 μm in several steps, showed progressively faster rises in toughnesses plateauing at higher values as grain size increased, distinguishing bodies of < 1 μm difference in G. They showed increasing reloading hysteresis as G increased in the R-curve region, and the ultimate fracture energy rose slightly faster than linearly with increasing G, while the length of the bridging zone behind the crack decreased nonlinearly with G from ~ 11 to ~ 6 mm, i.e. respectively ~ 3000 and ~ 600 grain diameters. Predominate transgranular fracture was observed, sometimes forming bridges a fraction of a grain in width and a substantial fraction to over a grain in length, i.e. similar to yttria and the ferrite noted above. However, the most significant bridging was attributed to intergranular fracture forming bridges of complete, often larger, grains or grain clusters. These were verified by locating the bridging grain on one half of the resultant fracture surface and the mating hole for it on the other fracture half along with wear markings from the pullout of the grain [197]. These observations are directly contrary to later claims that fractography can provide virtually no information on bridging [198]. Vekins et al. found no evidence of microcracking ahead of the crack or in its wake, other than that involved in forming the bridges. They concluded that, rather than bending of elastic ligaments, work expended in rotating and pulling bridged grains out of the mating fracture surface against friction was the main source of toughening, which they estimated entailed ~ 10% of the grains in their case, but could be less if larger grains dominate the bridging process. They had no explanation for the large decrease, by ~ 2, in the bridging zone length as grain size approximately doubled. Steinbrech et al. [199] showed a similar trend for greater toughness increases (but starting from lower initial values) for a 2 versus a 10 μm body. They also concluded that frictional effects were probably an important factor, that microcracks in the wake zone were mainly associated with debris, and that R-curves were not unique for a given body but depend on test parameters and microstructural variations. Thus crack wake bridging, extensively observed in Al_2O_3, may be an alternate explanation to microcracking for toughness maxima versus G (Fig. 2.16), but bridging-R-curve effects have been observed only over part of the G range (mostly G ~ 1 to 10–20 μm) and may involve some microcracking and thus correlate with microcracking models.

Bridging and R-curve effects clearly occur in many Si_3N_4 bodies but are quite dependent on microstructural parameters, apparently occurring only in bodies with sufficient grain size, elongation, and amount and character of grain boundary phase. Such effects have not been observed in bodies without oxide additives due to resultant transgranular fracture (and no slow crack growth, Secs. III.A and B) in the absence of resultant grain boundary phases. Even with oxide additives, such effects are observed not to occur, or to occur only in very modest amounts, in finer grain, equiaxed bodies, e.g. as reported by Mizuno and Okuda in their evaluation of round robin K_{IC} tests [13]. All observations of significant bridging in Si_3N_4 are with substantial amounts of oxide additives, larger, especially elongated (β) grains with substantial intergranular fracture, e.g. as shown by Li et al. [200], Peterson and Tien [182], Sajgalik et al. [183], and Okada and Hiroshi [201], with the latter also showing the expected accompaniment of environment-driven SCG (Sec. III.B). However, it is also generally observed that effects of grain elongation become limited due to increasing fracture of more elongated grains. It is widely accepted that bridging is the dominant or exclusive source of the substantially higher K_{IC} values, e.g. above 4–5, to values of ~ 10+ MPa·m$^{1/2}$. However, other observations indicate that bridging is not the whole story. Thus Hirosaki et al. [202] concluded that crack bridging by elongated beta grains was the main source of their high toughness (10.3 MPa·m$^{1/2}$, 600 MPa strength and Weibull modulus m of 25) and in their lowest toughness (8.5 MPa·m$^{1/2}$, 690 MPa strength and $m = 53$) bodies. However, they attributed toughness in their intermediate toughness bodies (8.8 MPa·m$^{1/2}$, 515 MPa strength and $m = 19$) to microcracking from excessively large grains. (Also, Salem et al. [203] reported that large grain fracture initiation limited strengths in their Si_3N_4.)

Turning to limited data for a very anisotropic material, graphite, Sakai et al. [204, 205] reported both microcracking and bridging in the crack wake region of an isotropic graphite (coke grain size ~ 15 μm, 15% porosity, mean pore dia. ~ 5 μm) using CT and in situ SEM tests. The latter were reported to show microcracks in both the wake region and ahead of the main crack within ± 60° of the crack plane at about 60% of the failure load that were about 1–20 μm long, with some of them joining and growing to > 100 μm long. These microcracks underwent partial closure around (often quite elongated, multigrain) bridges in the wake region. That microcracks formed considerably ahead of the main crack was indicated by reductions in crack resistance as the main crack tip approached but was still some distance from the specimen edge. They corroborated that the crack resistance/toughness effects arose primarily from the wake region via renotching à la Knehans and Steinbrech [51]. These observations are supported by those on microcracking in graphites (102, 103, and Secs. 3.3.4, 9.3.1D. of Ref. 5).

Graphite results, while more extreme in some aspects, are similar overall to other ceramic bridging results, adding to the correlation of bridging, microcracking, and TEA. Microcracks as a factor in forming bridges are logical, as noted earlier, but their observed formation at or ahead of the main crack tip is still inconsistent with the earlier, abandoned, concept of microcrack zones ahead of, but well above and below, the main crack being the primary source of toughening. However, the extent of the microcrack occurrence, e.g. in terms of their size and distance ahead of the main crack in graphite, appear more extreme, raising the question of the relative roles of the extreme anisotropy and of pores in graphites that are not adequately understood. Despite the common assumption and observations that pores and microcracks have independent effects on behavior, pores and microcracks have been observed to interact not only in graphites but also in a porous sandstone (46, and Figure 4.5 of Ref. 5) and some ceramic composites (Chap. 8. Sec. V). The high graphite EA may also play a role as indicated by the following results for cubic materials.

While much of the attention on wake effects from bridging focused on noncubic Al_2O_3 and Si_3N_4 with TEA as a cause of their increased toughness, both earlier and subsequent results clearly show that bridging also occurs in cubic materials. Thus Wu et al. [6] showed in dense MgO bodies (Fig. 2.4) by microradiography that (1) crack twisting, tortuosity, and branching-bridging occurred with $G \sim 50$ and ~ 200 μm, with the latter showing more extreme deviations, and via both microradiography and in situ SEM observations that (2) heterogeneous grain sizes resulted in more complex-tortious crack character, e.g. due to multiple microcracks formed in larger grain clusters. They also showed by in situ SEM observations that dense, large (~ 100 μm) grain Y_2O_3 samples with as-fired surfaces had long, narrow bridging ligaments formed from the essentially universal transgranular fracture (i.e. cracks apparently avoiding grain boundaries) that were sometimes a grain or more long, but a fraction of a grain thick.

Beauchamp and Monroe showed that bridging occurred in the crack wake zone of Mn Zn ferrites undergoing slow crack growth in DCB [52] or indent-flexure tests [53]. DCB tests showed crack wake bridges similar to those in the Y_2O_3 noted above, reflecting the predominant SCG transgranular fracture (and more intergranular fast fracture) in their ferrites. This atypical reversal of the SCG to fast fracture mode transition (Sec. III.B) was associated with a general increase of intergranular fracture as the crack stress intensity increased, and more intergranular fracture for any given condition in the finer grain (~ 24 μm by DCB tests) ferrite with more Ca at grain boundaries than for the larger grain (~35 μm) body. Their indentation tests showed that cracks with numerous bridges were formed with indentation under oil, while indentation under water produced cracks with essentially no bridges, i.e. again different from most other observations. (See the note at the end of this chapter for corroborating data.)

White and colleagues investigated crack bridging by DCB tests [206, 207] in two dense, transparent $MgAl_2O_4$ bodies made with LiF additions, with bimodal grain distributions of (1) 10–40 and 50–100 μm ranges (average $G \sim 35$ μm) and (2) 50–200 (average ~ 150 μm) and 200–500 μm ranges. The smaller grain body had higher toughness over the ranges of crack propagation (1 to ~ 10 mm), i.e. starting at ~ 1.7 and going to ~ 3.4 $MPa{\cdot}m^{1/2}$, versus ~ 1.4 to 2.9 $MPa{\cdot}m^{1/2}$, with grains > 80–100 μm generally fracturing transgranularly and the rest intergranularly. This toughness behavior was repeated when specimens of either grain size had the original wake area removed by sawing a fresh notch à la Knehans and Steinbrech [51]. In contrast to tests of finer grain aluminas, the *R*-curve increase in toughnesses were ~ linear and had not saturated in their tests since full development would have required larger specimens because of the larger grain sizes. They did not comment on the larger grain body starting from, and remaining at, lower *K* values than the finer grain body, which is opposite to grain size effects found with finer grain alumina bodies (discussed below). However, this may reflect part of the transition from polycrystalline to single crystal or grain boundary toughnesses. In a subsequent evaluation they concluded that models for bridging effects in alumina did not fit their $MgAl_2O_4$ results well, implying that the assumed grain interlocking mechanism was less effective for the $MgAl_2O_4$ [208].

Noncubic (i.e. α) SiC behaves quite similar to cubic (β) SiC (e.g. Figures 2.14 and 2.16) provided there is no significant grain boundary phase to promote grain elongation, intergranular fracture, and frequently associated crack deflection, branching, and bridging. Lee et al. [209] sintered β (cubic) powders with either Al_2O_3 + Y_2O_3 or B + C additions yielding various admixtures of α (noncubic) SiC, giving mostly elongated, platelet grains. *R*-curve effects and related crack wake bridging occurred in bodies made with oxide additives having equiaxed grains or tabular grains (e.g. aspect ratios of ~ 3) and toughnesses for crack extensions of 0.1 to 1 mm respectively of 5.5–6 and 5–7 MPa·m$^{1/2}$, all with mixed trans- and intergranular fracture. In contrast, the body sintered with B +C additions, with similar grain sizes and shapes as those made with oxide additions, had low fracture toughness ~ 3 MPa·m$^{1/2}$, independent of crack extension, i.e. no *R*-curve effect. These bodies had essentially complete transgranular fracture (consistent with earlier discussion of the effects of such B + C additives giving transgranular fracture and no slow crack growth, Sec. III B) and hence no significant grain bridging. Gilbert et al. [18] similarly showed that commercially sintered α-SiC (with B + C and equiaxed grains, *G* ~ 5 μm, and ~ 3% porosity) had low toughness (IF, ~ 2.5 MPa·m$^{1/2}$), no *R*-curve behavior (i.e. was independent of crack size) and transgranular fracture. However, SiC hot pressed to similar *G*, but with Al, B, and C additions yielding elongated grains and an amorphous grain boundary phase ~ 1 nm thick with predominantly intergranular fracture, showed significant *R*-curve effects. IF tests with smaller craks extrapolated to a zero crack size toughness the same as the α-SiC, and increased to ~ 9 MPa·m$^{1/2}$ at crack extensions of > 400 μm, while large crack, CT tests started at nearly 6 MPa·m$^{1/2}$ at zero crack length, rising to an ~ plateau of > 9 MPa·m$^{1/2}$ at crack extensions of 600 μm.

It is clear from the above that crack wake bridging plays an important role in large crack fracture toughness for many noncubic, and some cubic, bodies. However, there are critical issues of the nature and mechanisms of such crack-dependent toughness and its role in controlling strength. With regard to the former, issues of starting crack character, size, and its distance and conditions of propagation have been noted, including multiple cracks from notches, crack branching, or both [210] and the extent to which they are precluded by side grooves in tests (e.g. DCB and DT). As noted earlier, comparison of Wu et al.'s

[6] observations on specimens with as-fired surfaces versus other studies, with machined surfaces, indicates substantially less bridging on as-fired surfaces. Also, Ewart and Suresh [211] noted that crack propagation from notches in machined alumina NB specimens subjected to cyclical compressive loading showed ~ 50% greater extent of crack growth from the notch at and near the machined surfaces as opposed to the center of the specimen.

Much more attention is needed on environmental and grain boundary phase effects, since these play a possibly critical role that has been only partly addressed as shown earlier (and by Choi and Horibe [212] and more recently [213] and is probably significant in limited attempts to determine *R*-curve effects at nearly normal strength controlling flaw sizes [214]. Recent studies of Becher and colleagues provide more insight into effects of grain boundary phases in Si_3N_4 bodies, and a potentially valuable experimental tool of evaluating behavior of a specific boundary by observing the stopping of a crack from a nearby indent at the boundary and the occurrence and extent of deflection of the crack along the boundary as a function of the angle between the crack and the boundary [215, 216].

The above issues are partly related to the limited observations of some crack velocity decreases with bridging, microcracking, or both [52, 213], environmental [217] and strain rate effects [56] and their making resultant toughness being nonunique, not only between different tests supposedly measuring the same property but even for the same material and test, which raises large issues. Another, probably related, issue is the role of microcracking not only in generating bridges but also as a competitive or accompanying mechanism [e.g. 218]. These issues are interrelated to those of the size, extent, and nature of propagation of large cracks versus those factors for flaws causing normal strength failure, which is particularly critical. While bridging observations are valuable for understanding large crack behavior, e.g. in serious thermal shock or impact damage, they often have, at best, limited effect on normal strengths unless sufficient bridging occurs at finer grain sizes. All of these issues are factors in the nonuniqueness of *R*-curve results, e.g. Cox [219] reported NB testing that showed "the profound influence of the extrinsic factors of specimen shape and load distribution on the propagation of bridged cracks when the bridging zone length is comparable to any dimensions of the crack or specimen."

All the above issues are factors in the second and larger issue of the relation of these large crack effects to normal strengths controlled by small cracks. Thus comparison of bridging and crack deflection [220] showed that deflection is not significant at large crack sizes, but comparison at smaller crack sizes was not considered, and possible mechanisms operative at smaller crack sizes have been widely neglected. That these are critical issues is shown by the fact that there are extensive and basic inconsistencies between much toughness data, primarily that with larger crack, e.g. *R*-curve, effects and normal strength data as extensively discussed further below and especially in Chap. 3. The key to these critical differences is the difference in the microstructural dependence of toughness and strength, especially the grain size dependence in monolithic ceramics (Chaps. 3 and 6) and the particle size dependence in ceramic composites(Chaps. 8, 9, and 11).

H. Grain Shape and Orientation Dependence of Crack Propagation

Grain shape, especially elongation, often plays an important role in crack propagation behavior via effects on microcracking, crack deflection and roughness, and especially crack wake bridging. Such effects are most pronounced with intergranular fracture, often due to weaker grain boundaries from residual additive phases, e.g. oxides in Si_3N_4 and SiC, and LiF in MgO and especially $MgAl_2O_4$, as was discussed earlier. Thus when such boundary phases are not present, elongated grains are generally ineffective in increasing toughness due to their common transgranular fracture. Larger, equiaxed or elongated grains, or grain clusters, can also be strength limiting by acting as failure sources in conjunction with pores, or cracks from machining, handling, or TEA or other microstructural stresses, e.g. Figs. 1.7B, and 3.35. Grain boundary phases or impurities to enhance intergranular fracture may exacerbate such weakening, as is discussed in Chap. 3, Sec. III.G.

Koyama et al. [221] directly observed effects of progressive, generally uniform, increases in grain size (~ 1–50 μm) and elongation in dense Al_2O_3, separately controlled respectively by sintering temperature and limited, increasing CaO and SiO_2 additive (e.g. to 0.25 m/o) levels to transition from equiaxed to blocky, platelet grains. They showed both increased *R*-curve and resultant (SENB) toughness with both the size of equiaxed grains (from ~ 3.5 to 4.5 $MPa{\cdot}m^{1/2}$) and the length and thickness of the platelet grains (from ~ 4 to 6.6 $MPa{\cdot}m^{1/2}$). While all bodies had predominant to exclusive intergranular fracture, the platelet grains, though showing some transgranular fracture, had progressively wider crack deflections correlating with their greater crack resistance. However, flexure strengths varied inversely with toughness, following conventional grain dependence (Sect. III.A of Chap. 3). Some similar effects have been reported in alumina bodies doped with Na_2O + 4 MgO [222] or SiO_2 + $LaAl_{11}O_{18}$ [223], producing composites with platy second-phase grains (particles), hence discussed in Chap. 8, Sec. V.D.

Much broader effects of grain elongation occur when it is coupled with grain orientation locally, and especially globally (the latter often is in part a result of starting grain shape, as is discussed below, while the former commonly results from in situ elongated grain development, e.g. as in most bridging effects). Global grain orientation commonly occurs since, for a given shape-crystallographic particle morphology, it tends to increase in the approximate order:

die pressing, tape casting, and extrusion of green bodies. Orientation of particles in fields (e.g. magnetic) or in various hot working and deposition processes (including a number of rocks and minerals) often produces substantial preferred orientation, as is discussed further later. Key examples of limited studies of effects of preferred grain orientation on crack propagation behavior of ceramics are summarized below and in Table 2.3. While many cubic ceramics have similar anisotropy of mechanical properties to those of noncubic ceramics, the latter are anisotropic in other properties, that aid in giving noncubic bodies more preferred orientation. Thus the primary examples of effects of preferred grain orientation are in noncubic ceramics.

Fryxell, Chandler and colleagues [224, 225] sintered BeO made by either green body extrusion or isopressing of powders designated as either AOX (from $Be(OH)_2$) or UOX (from $BeSO_4$), often with annealing after sintering to increase grain size. While their results were obtained before fracture toughness measurements had been established, they are still quite useful, since they observed that extrusion aligned the considerable population of acicular UOX particles (*c*-axis parallel with particle lengths) to give a substantial *c*-axis texture parallel with the extrusion axis. This basal texture normal to the extrusion axis, and its increases with grain growth, was observed to give x-ray intensities of preferred peaks of as much as 100 times that of a random body. However, no preferred orientation was found in either extruded or isopressed bodies from equiaxed AOX powders, or from isopressed UOX powders, i.e. such bodies were isotropic. The axial thermal expansion of extruded rods from AOX powder was independent of grain size at the normal value, while that of extruded UOX rods decreased as grain size (and preferred orientation) increased, consistent with a lower expansion in the *c* direction, saturating at $G \sim 100$ μm with ~ 6% decrease. Extrapolation of axial Young's and shear moduli of oriented and unoriented bodies to zero porosity gave values respectively of 416 and 389 GPa, i.e. an increase of ~ 7% in the oriented material. While exact flexural strength comparisons are complicated by differing grain sizes and less data for a given grain size, limited or no difference was indicated at 22°C between oriented and unoriented bodies over most of the grain size range 5–50 μm, but oriented bodies with larger grains, $G \sim 80$ and 100 μm, were ~ 30% stronger than bodies without oriented grains. As test temperatures increased, strengths of finer grain (–20 μm) bodies did not decrease significantly till 1000, and especially 1200°C (possibly first increasing a few percent, peaking at 500–800°C). However, strengths of larger grain bodies increased as test temperature increased, especially in the 500–800°C range, commonly equaling or exceeding those of the finer grain bodies at 1200°C, with oriented bodies typically maintaining their strength advantages.

Virkar and Gordon [146] found considerable preferred orientation of basal planes in the pressing plane of hot pressed beta alumina bodies (1–2% porosity). Average toughness values at 22°C were of the order of 20% lower for

Table 2.3 Summary of Grain Orientation Effects on Mechanical Properties[a]

Material	Orientation method[b]	$G\|/G\perp$ (G in μm)[c]	$E\|/E\perp$ (E in GPa)[d]	$K_{IC}\|/K_{IC}\perp$ (K_{IC} in $MPa \cdot m^{1/2}$)[e]	$\sigma\|/\sigma\perp$ (σ in MPa)[f]	Investigator
Al_2O_3	GB Ex.	5.2/4.2 ~ 1.2	392/376 ~ 1.04	2.5/4.1 ~ 0.61	—	Salem et al. [226]
$BaO \cdot 6Fe_2O_3$	Mag. field	—	317/154 ~ 2.1	2.8/0.96 ~ 2.9	—	Iwasa et al. [227]
$SrO \cdot \sim 5Fe_2O_3$	Mag. field	1.5/2.5 ~ 0.6	169/179 ~ 0.94	2.5/1.7 ~ 1.5	265/162 ~ 1.6	de With and Hattu [228]
Jadite	Natural	40/20 ~ 2	—	2.6–3.1/3.6[g]	—/107[g]	
Hornblend	Natural	40/5 ~ 8	—	3.7–3.9/3.5[g]	—/93[g]	Wu et al. [229]
Hornblend	Natural	100/1 ~ 100	—	7.2–8.2/16.6[h]	40/288[h]	

[a] For measurements with stress parallel (|) and perpendicular (⊥) with the axial orientation, all at ~ 22°C.
[b] GB Ex = green body extrusion. Mag. field refers to die pressing with a magnetic field applied along the axis of the die and resultant part.
[c] G = grain size.
[d] E = Young's modulus.
[e] K_{IC} = fracture toughness (values for Al_2O_3 are extrapolated to zero crack extension in R-curve tests, as discussed in the text).
[f] σ = tensile (flexure) strength.
[g] Transgranular fracture.
[h] Intergranular fracture. Note some variation in K_{IC} values for jade from tests on two orthogonal planes parallel to the fiber axis indicating effects of variable fiber orientation.

crack propagation normal versus parallel to the basal texture, i.e. respectively 2.7–3.0 versus 3.5–3.7 MPa·m$^{1/2}$, with the higher values of the preceding ranges being for large grain (200–300 μm) and lower values for finer grain (2–10 μm) bodies, with both orientations showing nominally 100% transgranular fracture. Values for an isopressed and sintered body (~ 3% porosity, G ~ 100 μm) with no significant preferred orientation were ~ 3.2 MPa·m$^{1/2}$, i.e. in between those for the two orientations of the hot pressed bodies. The hot pressed toughness anisotropy is opposite of what would be expected from Hitchcock and De-Jonghes' [144] single crystal results (Table 2.1). However, the flexure strength anisotropy for the same crack propagation orientations was nearly 40%, but opposite to that for toughness, i.e. lower strengths for crack propagation parallel with the basal texture, while the limited strengths of the isopressed and sintered bars were approximately in between the two sets of hot pressed values, as for toughness. While compositional differences between these polycrystalline and crystal specimens (Table 2.1) may be a factor, Virkar and Gordon's suggestion that the toughness for crack propagation parallel with the basal texture was higher due to possible multiple basal cracking, e.g. due to imperfect alignment, is a reasonable possibility. The opposite and greater anisotropy in strength from toughness indicates that large scale toughness values are not fully consistent with strengths. Alternatively or additionally the strength anisotropy may reflect oriented pores, larger grains, or clusters of them acting as fracture origins of lower strenth for stressing normal (and fracture parallel) to the basal texture, as also indicated in other studies below.

Salem et al. [226] showed a c-axis texture along the green body extrusion axis of alumina (i.e. a basal texture normal to it), versus isotropic isopressed bodies. The latter gave isotropic Young's modulus (E, 368 GPa) and K_{IC} (~ 3.7 versus 4.9 and 3.6 MPa·m$^{1/2}$ for the extruded bodies, the latter being the plateau values from R-curve tests, and hence different from values in Table 2.3). The isotropic E value was slightly below that for the lowest value for oriented bodies and the isotropic K_{IC} at the low end of the range of oriented values, probably reflecting differences in the amount and especially character of porosity, as is discussed below. All three crack propagation orientations in extruded specimens showed some R-curve effect. Fractographic examination of samples showed clear differences in overall appearance of fractures perpendicular versus parallel with the extrusion axis. Differences reflect in part definite effects of pore elongation in the extrusion direction along with heterogeneous distribution of grains of varying size that complicate normal effects of just the grain orientation.

Two related sintered, hard (hexa-)ferrites oriented using a magnetic field axial with the die ram both gave a c-axis texture parallel with the field-pressing direction, i.e. a basal plane texture normal to the pressing direction (Table 2.3). Resultant sintered grains were somewhat elongated normal to the c-axis and typically failed intergranularly for fracture parallel with the basal texture and transgranularly for fracture normal to the basal texture at 22°C. Iwasa et al. [227] also reported that samples pressed without an applied magnetic field were isotropic and showed parallel decreases of Young's moduli in the isotropic body and perpendicular and parallel to the axial texture in the oriented samples as temperature increased to 900°C (the limit of testing). K_{IC} values similarly decreased for the isotropic body and for fracture parallel to the basal (i.e. perpendicular to the *c*-axis) texture, but no decrease in toughness occurred for crack propagation normal to the basal (i.e. parallel with the *c*-axis) texture till 1000°C. (From 1000 to 1200°C, the limits of testing, all toughnesses decreased drastically.)

De With and Hattu [228] showed lower *E,* but substantial toughness and strength, anisotropy in similar, but less orientated, related hard ferrite at 22°C. They also showed an average decrease in both toughness and strength from "dry" versus "wet" tests averaging ~ 20%, i.e. with no clear differentiation as a function orientation of the stress relative to the texture. However, they showed that there was a marked effect of orientation on the crack growth exponent, *n,* of Eq. (2.3), i.e. values of 39 and 320 respectively for fracture perpendicular and parallel with the *c*-axis texture. Byrne et al. [85] noted some anisotropy in SCG *n* values, i.e. respectively 160 and 200 for crack propagation parallel and normal to the green body extrusion axis for fired porcelain samples, which presumably introduced some preferred orientation of the SiO_2 and other particulate phases in the glass matrix.

The first of two other sets of tests further showing effects of preferred orientation on fracture toughness and related properties of rock materials is the study by Wu et al. [229] of toughness and flexural strength of jade, a natural gem material known for its toughness, attributed to its common fibrous structure. Actually jadite and hornblend, two naturally occurring minerals that are common sources of "jade" with very similar structure and common [110] cleavage with varying fibrous grain structures, were studied [Table 2.3]. DCB toughness results for propagation parallel versus perpendicular to the fiber orientation showed modest to no significant anisotropy with lower grain aspect ratios, and especially larger grain (fiber) diameters and associated transgranular fiber (grain) fracture (Fig. 2.20). Toughness values for crack propagation parallel with the fiber axis, but on orthogonal planes to each other, differed some, i.e. similar to variations found in the extruded alumina (Table 2.3). Hoagland and Embury [45, Fig. 4 of Ref. 5] also showed significant anisotropy, i.e. fracture toughness as much as 2–5 times higher for crack propagation normal to the bedding planes versus parallel to the bedding planes and differences of 1.2–2.2-fold for the two planes normal to each other and the bedding planes, as well as differences in both magnitude and direction between the limestone and sandstone used.

Pertinent to earlier bridging effects is the work of Ohji et al. [184]. They incorporated 2 v/o of fine ~ 1.3 μm dia. β Si_3N_4 rod grains (i.e. whiskers with as-

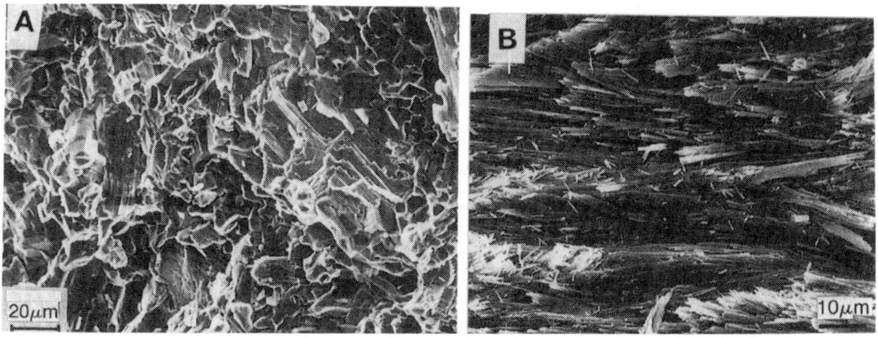

FIGURE 2.20 Examples of jade fractures and microstructures. Note the coarser, blocky, transgranularly fractured, grain structure with no clear preferred orientation in jadite (A) and the clear horizontal orientation of fine, very elongated grains of a fine intergranularly fractured, grain hornblend body (B). (After Wu et al. [229], published with the permission of the Journal of Materials Science.)

pect ratios of ~ 4) into an α Si_3N_4 matrix by tape casting to orient the elongated β grains in the casting direction and stacking of cast sheets to maintain the orientation between the sheets. This resulted in highly oriented elongated grains giving very anisotropic toughness, i.e. ~ 7 and 11 MPa·m$^{1/2}$ for crack propagation respectively parallel and perpendicular to the oriented grains. However, neither crack propagation direction showed much *R*-curve effect. This was attributed to the close spacing (e.g. ~ 5 μm) of the elongated grains causing near or complete saturation of *R*-curve effects at much smaller crack sizes–propagation distances than normally required in microstructures with lower densities and larger spacing of the elongated grains for bridging. Comparative tests of a similar Si_3N_4 with randomly oriented rod grains showed substantial *R*-curve effects starting at toughness for cracks parallel with the elongated grains and ending at levels for crack propagation normal to the elongated grains in bodies with the highly oriented, elongated grains. They reported almost twice the strength (~ 1100 MPa) and Weibull moduli (~ 46) in bars with elongated grains aligned normal to the crack propagation of failure versus bars with randomly oriented elongated grains.

Several studies have addressed anisotropy in hot pressed and especially hot forged S_3N_4, e.g. an earlier study of Lange [230] showed that flexure bars oriented for fracture parallel to hot pressing direction had strengths ~ 35% higher than bars oriented for fracture perpendicular to that direction. This was attributed to greater difficulty of fracture around the elongated grains (due to the grain boundary phase) resulting from the transformation of the starting α powder that had a preferred orientation normal to the hot pressing direction. DCB fracture energy/toughness tests partially supported this, showing an anisotropy of 18% in fracture energy (thus probably less for toughness), i.e. accounting for about half or less of the strength difference (while he subsequently corroborated increased toughness with increased β grain content and elongation [231]). Weston [232] corroborated the type of texture found by Lange, as well as similar strength anisotropy (26–38% higher for fracture parallel versus perpendicular to the hot pressing direction) of hot pressed S_3N_4. However, he also did a fractographic study, showing flaws initiating fracture parallel to the hot pressing direction averaged ~ 2.7 times the size of flaws initiating fracture normal to the hot pressing direction, which leads to a strength anisotropy of ~ 65%, i.e. nearly twice that observed. However, this difference was generally accounted for by the anisotropy in K_{IC} calculated from his fractographic results, which was ~ 23%, but opposite that of Lange, i.e. higher for fracture normal versus parallel to the hot pressing direction. He also reported that fracture parallel to the hot pressing direction was predominately intergranular, while that for fracture perpendicular to the hot pressing direction was mixed inter- and transgranular. Ito et al. [233] also showed anisotropy in hot pressed Si->3N_4, i.e. for fracture parallel versus perpendicular to the hot pressing axis the strengths were ~ 20% and the Weibull modulus ~ 40% higher. They correlated this with larger average size and broader spatial distribution (i.e. depth from the tensile surface) of fracture origins in the pressing plane.

More recently Lee and Bowman [234] corroborated the *c*-axis texture normal to the hot pressing direction and showed strong increases in this by press forging (e.g. to 6 times random and hence also of sintered bodies they examined). Indentation tests of K_{IC} for fracture parallel versus normal to the pressing axis varied from 23–46% higher for hot pressed and 43–106% higher for press forged bodies, i.e. higher values for fracture normal versus parallel with the *c*-axis texture (and intermediate K_{IC} values for fractures at 45° in between these two tests). These three more comprehensive examples of S_3N_4 studies clearly showed common textures and resultant significant anisotropy of strength and fracture toughness that occur with hot densification or forming, but clearly some important differences, some of which are addressed in Chap. 3, Sec. V.D. It is clear that fractography, which has not been adequately used, is an important tool and that flaw character, especially orientation and size, is probably an important factor in the strength anisotropy, with probable contributions from toughness anisotropy. While less extreme and more dependent on fabrication parameters, preferred orientation also results from other forming operations, e.g. die pressing, as is shown by Goto et al. [235].

Recently Kim et al. [236] showed that hot pressing of 80 w/o β-SiC with 12 w/o Al_2O_3 and 8 w/o Y_2O_3 resulted in substantial {111} orientation parallel with the pressing plane and that annealing converted the body to mostly α-SiC with substantial (004) orientation parallel with the pressing plane. I toughness for crack propagation normal and parallel with the pressing plane were respectively 5.7 and 4.4 MPa·m$^{1/2}$.

Graphite materials frequently have varying, often substantial, degrees of local or global orientation, or both, depending on a variety of factors, especially their fabrication. However, evaluating grain orientation effects on properties is often challenging due to the generally substantial porosity present, and especially since varying amounts of the aniosotropy of properties also occur due to varying degrees of orientation of anisotropically shaped porosity, which is a function of both fabrication and microstructure. However, CVD graphite (also referred to as PG, pyrolytic graphite), while generally being complicated some by varying colony structures and their impact on local and global orientation and mechanical behavior, has no significant porosity and thus offers a clear indication of the anisotropy of graphite properties due solely to grain orientation. (CVD graphite results should also be indicative of such trends for CVD BN, since both have very similar properties and anisotropy as well as CVD structures. However, bodies of the two materials made by consolidation of particles may differ considerably, since liquid phase densification aids are typically used for BN and not for graphite, which can result in important differences in the behavior of the two materials processed from particulates.)

Sakai et al. [237] reported substantial anisotropy of toughness as well as Young's modulus and flexure strength in CVD graphite, which has a high degree of preferred orientation of the basal plane parallel with the deposition plane (Table 2.4). Their results on the latter properties were generally consistent with other data in the literature for Young's modulus and flexure strength given expected variations in CVD bodies, especially colony structure and hence orientation and other microstructural factors. Their toughness results were from a mix of DCB, CT, and NB testing, since crack propagation was not always well behaved, especially in the high toughness orientation, which was for both the crack propagation plane and the direction normal to the plane of deposition and the basal texture. The high and variable toughness for this orientation of 1.4 to 7.5 MPa·m$^{1/2}$ was associated with variable but generally extensive delamination that occurred parallel with the basal texture normal to the crack. Such delamination occurred at distances from the notch and with spacings of the order of mm, i.e. on a large scale. The lowest toughness of 0.53 MPa·m$^{1/2}$ was for both the crack propagation plane and the direction parallel with the basal texture. The third orientation, i.e. with the crack propagation plane and direction being respectively perpendicular and parallel with the basal texture, gave a toughness of 0.93 MPa·m$^{1/2}$, which is between the other two values, but much closer to the lowest value. Their strength and Young's modulus values roughly scale with each other, as do the two lowest toughness values, but their high toughness value does not, again showing high toughness behavior with large cracks with limited or no correlation with smaller crack behavior as for normal strength, as is discussed extensively in the next chapter.

TABLE 2.4 Effects of Preferred Grain Orientation on Mechanical Properties of CVD Graphite

Property	Orientation[a]		
	ca	ab	bc
E(GPa)	5.5	28	20
σ(MPa)	9.6	190	158
K_{IC}(MPA·m$^{1/2}$)	0.53	0.93	1.4–7.5

[a] Designations of Sakai et al. [237] referring respectively to crack propagation planes and directions relative to the basal plane texture: parallel, parallel; perpendicular, parallel; and perpendicular, perpendicular.
Source: From Sakai et al. Ref. 237.

I. Other Factors and Evaluations

There are several factors that impact on, or have correlation with, effects of grain stresses that should be considered; effects of electrical fields in piezoelectric and ferroelectric materials are important examples. McHenry and Koepke [238] showed in DT tests of poled PZT that both the K and the n value [Eq. (2.3)] decreased substantially with increased electric fields perpendicular to the crack, with somewhat less effect of ac versus dc fields. They also showed that similar field application to poled samples deflected propagating cracks. Kim et al. [239] showed that a clear transition from predominately intergranular fracture below ~ 10 μm to predominately transgranular fracture by 18–20 μm in sintered PZT (P ~ 3%) was progressively shifted to larger grain sizes after poling with progressively higher poling fields. Thus a body with G ~ 18 μm and ~ 76% transgranular fracture unpoled had ~ 26% transgranular fracture after poling with 2 kV/mm and ~ 2% after poling at 3 kV/mm. Chung et al. [240] showed that microcracks began forming at grain boundaries and increased in number and size in sintered $BaTiO_3$ (G 36–53 μm), and in a tetragonal, and a rhombohedral, PZT (G 10–17 μm) as grain size and poling field (1 or 3 kV/mm) increased. However, while the cracks started at the grain boundaries, they typically propagated into the grains. The above and following cracking associated with poling and other applied electric fields is consistent with expected stress concentrations of such fields at crack tips [127, 238].

Earlier studies of Halloran and Skaar [241] on Navy Type III Pt-8 sonar rings showed that all four apex cracks from Vicker's indents (500 gm, 4.9 N) were equal in thermally depoled samples whether the cracks were parallel or perpendicular to the original (axial) poling field (giving K_{IC} ~ 0.7 MPa·m$^{1/2}$, Fig. 2.21A). However, they showed that stresses from poling resulted in significant anisotropy of the indent cracks due to some shortening of cracks parallel and especially significant elongation perpendicular to the poling direction (Fig. 2.21 B), which they attributed respectively to compressive stresses normal, and tensile stresses parallel, to the poling direction. They calculated that these stresses would increase hoop, and decrease axial, tensile strengths by ~ 60% and 40% respectively if they were linearly additive to the applied stress. Subsequent work [242] showed poling stresses increased hoop tensile strengths of poled rings, but by only ~ 10%, not 60%, and that the Weibull modulus for hoop strengths of poled and depoled rings were respectively ~ 3.6 and 7.6. The lower increased hoop strength and Weibull modulus for poled rings must reflect other complexities, e.g. effects of pores.

Zhang and Raj [243] corroborated the above anisotropy of cracks from Vickers indents relative to the poling direction in poled samples (3.5 kV/mm) of a hot pressed tetragonal PZT. Indent (3.5 N) toughness of unpoled material was isotropic but decreased from ~ 1.9 at $G = 0.5$ μm to ~ 1.4 MPa·m$^{1/2}$ at $G = 3$ μm, then was constant at the latter value to the limits of their grain sizes (15 μm). Toughness at $G = 0.5$ and 1.5 μm were respectively 2.8/0.8 and 2.2/0.7 MPa·m$^{1/2}$, where the first and second values are respectively for crack propagation parallel and perpendicular to the poling direction, i.e a ratio of >3.

Lynch [244] has recently demonstrated similar effects of poling and has shown effects of applied fields using Vickers indents in a ferroelectric composition of PLZT (G ~ 5 μm). He showed that indent cracks in unpoled material were isotropic but that cracks normal to a subsequently applied electric field elongated above a threshold field. Such elongated cracks also occurred with indentation under an electrical field (again above a threshold field) and in poled

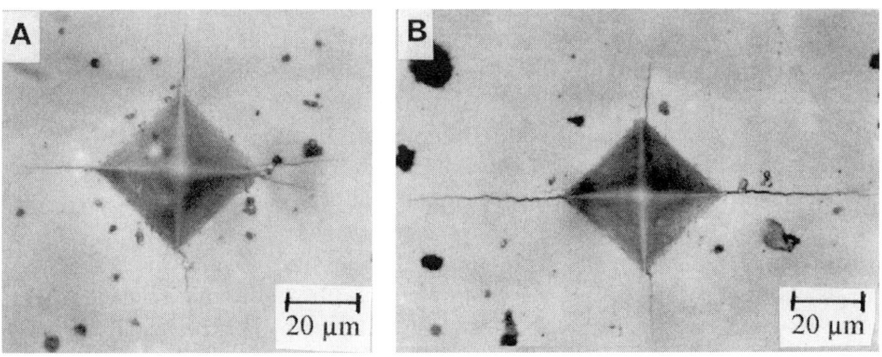

FIGURE 2.21 Photomicrographs of Vickers indents (4.9 N) on surfaces of PZT: (A) depoled and (B) poled (in vertical direction). (Photos courtesy of Dr. J. Hallorin, University of Michigan.)

samples. Microcracking was also observed near indents made in poled samples with high applied fields. The fracture mode for these effects, which were attributed to grain mismatch stresses from the fields, was mainly intergranular. A similar relaxor composition always had isotropic indent crack patterns, consistent with little or no mismatch stresses. A model was presented for toughness accounting for effects of residual strain energy released, showing that finer *G* bodies would have higher toughness normal to the field direction and less spontaneous cracking during polarization switching.

Wan and Bowman [245] recently showed that poling of PZT also leads to elastic anisotropy consistent with that of toughness. They further showed that progressive thermal depoling progressively reduces the anisotropy, returning to isotropic properties on complete depoling, and that these changes are consistent with changes in domain character revealed by x-ray diffraction.

The above static applications of electric fields and mechanical stresses raises the question of the grain dependence of microcracking and macrocrack propagation when either or both are applied in a cyclic fashion. Though direct studies of grain dependence are lacking, the previous and following results indicate that such dependence probably occurs. Jiang and Cross report that electrical fatigue can be a serious life limiting factor for ferroelectric and piezoelectric materials such as PZT and PLZT, where porosity [246] and surface contamination [247] can be factors. White and colleagues found some similarities, as well as complexities relative to the above results in delayed fatigue and electrical cycling tests of a PZT (-8, $G \sim 3$ μm) with and without indent flaws at temperatures of < 86 and > 150°C. TEM showed microcracking at both lower and higher temperature cyclic loading [248]. At the higher temperatures crack extension was observed with no evidence of environmental effects or reduction of toughness and microcracking occurring in small clusters throughout the specimen. Little crack extension was seen at the lower temperature, where microcracking was confined to a region of ~ 600 μm radius from the indentation. Subsequent cycling of such specimens at > 100°C resulted in immediate failure from undefined sources. Further studies [249] showed that high microcrack densities formed in mechanically cycled samples and in samples electrically cycled at < 80°C, with the microcracks originating from second-phase particles consisting of Pb, Ti, and Fe at the triple points. Electrically cycled samples allowed to heat to 180°C showed much lower crack densities, but were depolarized, while samples simply heated in a furnace to 180°C were not. Higher densities of microcracking found in mechanically cycled samples were related to the higher mechanical stresss. Note that other mechanical properties can be affected by electric fields, e.g. hardness of both single and polycrystalline $BaTiO_3$ [250] (Chap. 4, Sec. II.E) and other complications occur (see the note at the end of this chapter).

The developing field of mechanical fatigue of ceramics typically uses fracture toughness-type specimens and resultant v–*K* curves and the related Eq.

(2.3), (2.6) for evaluation. Such testing has focused on the use of large cracks, so grain bridging in the crack wake region is a major factor in fatigue crack propagation, which implies substantial grain dependence (Sec. III.E). While much attention has been focused on the mechanics of testing and behavior, some data on grain dependence does exist. Thus Suresh's review [251] reports that ceramics, like other materials, typically follow the Paris relation, i.e.

$$\frac{dc}{dN} = (\Delta K)^m \tag{2.6}$$

where c = crack length (for large cracks), N = number of cycles, K = stress intensity on the crack, and m = a constant for a given material, e.g. $m \sim 14$ for ceramics and 2–4 for metals. He also showed that increased grain size in WC-Co bodies reduces fatigue crack growth with predominately intergranular fracture along the Co binder (as in crack growth under static loading), but transgranular fracture increased as G increased. He also summarized work showing overaged (i.e. microcracked and lower toughness) PSZ had substantially faster crack growth than the same material of more optimum aging and higher toughness, as did Dauskardt [252]. The latter also reported modeling and limited experimental results for sintered Al_2O_3 ($G \sim 8$ and 13 μm) and Si_3N_4 with elongated beta grains showing reduced fatigue crack growth as toughness increased, consistent with bridging expectations. Hoffman et al. [253], however, showed that varying grain size of PSZ did not significantly affect fatigue crack growth. Healy et al. [254] reported "long" and "short" (indent) crack tests respectively in a pure sintered alumina ($P \sim 0.02$, $G \sim 10$ μm) and a commercial 90% alumina ($G \sim 4$ μm), the former, but not the latter, following a Paris law. They attributed the failure of the short (indent) crack tests to follow a Paris law to differences of short crack behavior, e.g. dependence on local microstructure, and also noted differences in grain size effects on fatigue crack growth and (rotary) fatigue strength. They also observed that the "long" crack fatigue behavior of the two aluminas was similar, being dominated by crack wake bridging. However, in situ fatigue tests in a SEM resulted in mixed transgranular fracture with discontinuous propagation and frequent crack arrest, versus continuous propaation and entirely intergranular fracture in air, implying environmental effect on bridging, consistent with earlier results (Sec. III.B).

Kishimoto et al. [255] evaluated cyclic tensile fatigue behavior of two dense, pure aluminas, $G = 1$ and 19 μm, using compact tension specimens in testing at room temperature. They showed that while crack propagation was accelerated by cyclic loading, i.e. gives higher crack propagation rates than static loading, the stress intensity for the onset of cyclic crack propagation is greater in the larger grain body, which also had a subsequent lower rate of crack increase.

Thus the crack propagation rate for the fine-grain body increased as crack length increased, while that for the coarser grain body was independent of crack length. They attributed these differences to the significant grain bridging and related *R*-curve effects in the larger grain body, which is consistent with more intergranular fracture in the larger grain body and more of this fracture mode in cyclic versus fast fracture.

Besides use of electric fields and cyclic loading to alter and hence probe effects of grain stresses and associated micro- or macrocracking, there are other important tools that can be of value but have not been widely used. Thus measuring elastic moduli and damping as a function of temperature cycling can be a valuable indicator of micro and other cracking [35, 63], e.g. from serious thermal shock [256]. Similarly, while measurement of thermal expansion during thermal cycling is valuable for determining aspects of microcracking, careful measurements of specimen length or volume can give useful estimates of microcrack content [99]. Further, measurements of precision elastic limit, the stress beyond which a particular body will exceed its original dimensions upon release of the stress, while essential for some high-precision applications such as telescope mirrors, can be valuable indicators of microcrack formation (see Note at end of chapter).

Finally, the first of two powerful, but highly neglected, tools is acoustic emission. This is useful not only for identifying microcracking, e.g. in Al_2TiO_5 [257, 258], but also potentially for learning more about various aspects of crack propagation, including bridging and fatigue. Both recent results [49, 259–261] showed promise, and advances in electronics increase opportunity. Thus Kishi et al. [262], using acoustic emission and fractography in crack propagation (including *R*-curve) studies of two aluminas ($G \sim 5$ and 20 μm), showed formation of 15–20 μm microcracks at the crack tip and subsequently some ~ 100 μm cracks. They reported that in the finer *G* body smaller microcracks formed intergranularly, e.g. from pores, and larger ones from the coalescence of smaller ones, while in the larger *G* body microcracks were mainly transgranular with the size determined by the grain size. Sklarczyk [263] showed microcracking and subsequent microcrack coalescence ahead of the crack tip, substantial, e.g. ~ 20% of events, occurring in the wake area, so one or both of these probably indicate greater sensitivity for such events. The second tool is fractography, which while used more than acoustic emission is still widely underutilized. Thus, for example, while fractography has been used to identify bridging mechanisms, its use has been limited, and even claimed (incorrectly) to be ineffective. Besides broader use, more quantitative and sophisticated use of fracture mirror and fracture origin–fracture mechanics evaluation is needed [39, 40, 60]. A key need is to use these tools in addition to broader measurement of different properties on the same body under a broader range of conditions, as shown by the limitations of so much testing of a limited range of material, microstructural, and environmental conditions, usually for one physical property.

While the application of fractography to fracture of bulk materials is important, as discussed elsewhere in this chapter and book, its use for evaluating properties of ceramic fibers allows obtaining fiber property data not obtainable by other methods. Thus Sawyer et al. [264, 265] obtained a fracture toughness of ~ 2 MPa·m$^{1/2}$ for polymer-derived SiC-based ceramic fibers with nanoscale grains (Figure 3.13) by measuring fracture mirror sizes versus failure stress. This result was also consistent with indicated flaw sizes or a few microns to a fraction of a micron and is consistent with values obtained for variations of fiber composition [167]. Similarly, Vega-Boddio et al. [266] reported a fracture energy of ~ 4.6 J/m^2 from the strengths of CVD B filaments and fracture initiating flaw sizes and character; this gives a similar toughness, i.e ~ 2.2 MPa·m$^{1/2}$ [267].

IV. DISCUSSION, SUMMARY, AND CONCLUSIONS

The transition from intergranular to transgranular fracture at finer grain sizes (e.g. from a fraction of a micron to ~ 10 or fewer microns) is widely found for most ceramics (Fig. 2.5, and often on a given specimen whose grain size range covers much of the fracture transition mode range). However, this transition is often shifted to larger grain sizes by boundary phases, SCG under static and dynamic loading, superposition of other stresses such as from electric fields, and by higher temperatures. Much less data indicates partial to complete transition back to intergranular fracture at larger to very large (e.g. multi cm scale) grain sizes, with a probable inverse correlation between the grain size range and the degree of TEA, EA, or both. Because of these factors influencing fracture mode and its role in many mechanical properties, failure to determine the fracture mode, preferably as part of a larger fracture surface evaluation, is a serious limitation to establishing a sound self-consistency to mechanistic interpretations of test results. However, there are also at least two cases where SCG is transgranular and fast fracture intergranular (Sec. III.B and the note at the end of this chapter), showing the need for further characterization and study.

Microcracking from mismatch strains between grains is clearly established, with grain sizes and parameters for this approximately predicted by Eq. (2.4) (Fig. 2.10), but there is a clear need to refine the equation, e.g. its numerical constant and parameters that are used in it as a function of material and microcrack character. There are limited, but clear, indications of (1) a significant dependence of microcracking on environmental effects, i.e. SCG, (2) changing microcrack/grain ratios as G changes, (3) closure and healing of microcracks diminishing, then disappearing as grain size increases, at least in HfO_2 (at G ~ 17 μm), and (4) some transgranular microcracking occurring in larger grains of at least some bodies. The latter two observations, while typically neglected or assumed not to occur, are probably of fairly broad occurrence. Observation 4 indicates a fracture mode transition similar to the conventional one but shifted to larger G by boundary stresses, probably occurring in larger grains with less mismatch strain with adjacent grains (hence favoring microcracking at or nearer room temperature where intergranular fracture is less favored). Observations 3 and 4 are probably related, since transgranular microcracks would appear to be less amenable to closure and healing than intergranular microcracks.

Grain structure effects on SCG are clearly shown by differences between single- and polycrystals, e.g. in the high directionality of SCG in many single crystals (Fig. 2.7) versus polycrystals, and especially by the occurrence of intergranular SCG in polycrystals (Fig. 2.6), including materials having no SCG in single crystal form. While some of the latter differences may be an intrinsic result of grain boundary character, it commonly is an extrinsic result of grain boundary phases as found from effects of oxide additives for sintering nonoxide materials such as AIN, Si_3N_4, and SiC. Additional grain structure effects on SCG are indicated by effects of grain orientation. A direct effect of grain size is indicated by reduced SCG, i.e. higher n values [Eq. (2.3)], in some finer grain bodies of both cubic ($MgAl_2O_4$) and noncubic (Al_2O_3, and TiB_2) bodies (Fig. 2.8).

Effects of grain structure on fracture energy and toughness are clearly shown by differences between polycrystals and the lower fracture toughness surfaces of single crystals, e.g. by a factor of 2–3 (Table 1) (Figs. 2.12, 2.13, 2.16, and 2.17) and the transition between these as a function of the crack-to-grain size ratio (Fig. 2.15). In extreme cases of one very low toughness plane (e.g. the basal plane of beta alumina), comparison to primarily or only the lower fracture toughness crystal values is generally appropriate, as is indicated by the dominance of such planes in the transgranular fracture of most polycrystals. This is also indicated by a possible trend for a more rapid single crystal to polycrystal toughness transition as the multiplicity of lower toughness crystal surfaces increases. Grain size dependence of fracture energy and toughness can be significant depending on test parameters, especially with larger crack sizes and extents of propagation, as well as material and microstructure character. Significant maxima are commonly seen as a function of G for noncubic materials and lesser ones in cubic materials, but some of the latter can be substantial. The significant increases in toughness, yielding maxima over a sufficient G range in noncubic materials, first attributed to microcracking ahead of the macrocrack tip, is now attributed to crack bridging in the macrocrack wake, i.e. behind its tip. However, experimental evidence shows that some microcracking occurs closely ahead of, in or near the plane of, and around, and possibly somewhat behind, the macrocrack tip. Thus microcracking probably occurs, though not in the zone configurations previously proposed, and it may be a precursor to bridging, independent of bridging or both. Crack branching, though mostly neglected, probably because many tests limit or preclude it, may be another factor in, or in addition to, bridging.

Bridging clearly occurs in the wakes of large cracks in crack propagation tests in a variety of materials and microstructures and is an important factor in R-

curve effects and resultant higher toughnesses observed with such cracks and tests. While bridging by fragments of individual grains, or clusters of grains as a single bridge occurs, it appears most common and effective for individual grains via intergranular fracture, which is enhanced by grain boundary phases, e.g. in $MgAl_2O_4$, Si_3N_4, and SiC. It is enhanced by elongated or platelet grains, again mainly with intergranular fracture, again aided by grain boundary phases as in Si_3N_4 and SiC. However, bridging has not been documented over a broad G range for any material, e.g. only to ~ 15 μm in Al_2O_3.

Besides uncertainties of its interaction with other mechanisms such as microcracking, crack deflection and branching, there are other basic issues. Bridging was originally attributed to TEA effects due to earlier studies being mainly or exclusively on noncubic materials such as Al_2O_3. However, bridging has now been shown to occur in some bodies of cubic materials (again mainly or exclusively with grain boundary phases, e.g. in $MgAl_2O_4$, some ferrites, and β-SiC, with little apparent difference from α-SiC, which has limited anisotropy). Thus TEA is probably a basic factor in noncubic materials, while EA is probably a factor in cubic materials, with the differences in these two mechanisms probably being important in the differences between the two types of material, e.g. of $MgAl_2O_4$ and Al_2O_3.

However, more fundamental to the issue of the applicability of bridging and R-curve effects to strengths are issues of observations of these phenomena as noted in Sec. II.A, as well as of direct comparison with strength results. Thus, as noted earlier, arbitrarily introduced large cracks propagated at low velocities and often arrested, with observations mainly along arrested crack–surface intersections (neglecting indicated basic differences on machined versus as-fired surfaces, Fig. 2.4D and E) are issues. That higher crack velocities may destroy bridges is indicated by the loss of zigzag crack propagation in Fig. 2.7, and other crack velocity effects are discussed in Chaps. 6 and 12. Direct comparison of the microstructural dependence of tensile strengths and fracture toughness–R-curve effects in Chaps. 3, 6, 8, 9, 11, and 12 shows they are often inconsistent, including opposite dependences. Also, the lack of any detailed tensile fracture observations supporting the applicability of most R-curve-bridging effects in such failure will be noted, and limited fractography data questioning the applicability of such effects to normal strength behavior of most ceramics will be presented.

Both complexity and the potential for further insight is indicated by interactions between the various crack phenomena as shown by limited but self-consistent tests. These include limited demonstration of the environmental dependence of microcracking, bridging, related strain rate dependence of fracture toughness, and resultant common intergranular fracture (e.g. via EA, Fig. 2.3). These interrelations would also be consistent with the role of grain boundary phases in bridging and intergranular fracture and SCG of materials such as AlN, Si_3N_4, and SiC, as well as on microcracking as a factor in bridging. These interrelations are also probably a major factor in the nonuniqueness of bridging results, which raise key questions for strength and life predictions. Thus much more work is needed in evaluating these and other interactions, which requires more comprehensive tests and characterization to evaluate self-consistency of results. This should entail a broader range of test parameters, e.g. environment, specimen configurations and sizes, surface finishes, loading conditions (e.g. biaxial), with different materials and microstructures especially grain structures. Broader evaluation of fractography (e.g. fracture mode and character), acoustic emission, and crack velocity, as well as properties such as *E,* damping, and strength is needed. A key aspect of broader evaluation should be evaluation of the toughness–tensile strength relationship, with fractography to determine fracture origins as a key tool.

The primary need is for better perspective, particularly that the crack sizes used in crack propagation–toughness tests are pertinent to the flaw sizes controlling strength, i.e., give similar microstructural effects of strength controlling flaws, e.g. typically from machining. While much work has assumed that large crack test results predict small flaw effects, it has been stressed in this and subsequent chapters that this is not necessarily so; in particular it is only true if the results with test cracks are applicable to strength controlling flaws. This is usually true if similar actual cracks/flaws are used, as in fractographic determinations of crack propagation and toughness behavior, or if the scale of both the test cracks and strength flaws relative to the strength controlling microstructure give similar statistical samplings on both, e.g of grains or other resultant phenomena such as crack deflection, branching, or bridging. This is typically, at least approximately, the case for TZP bodies due to their finer *G* and is often similarly true for many Si_3N_4 bodies. The second factor of microstructural impact on flaws controlling strength, especially from machining, is particularly important, since it has been widely neglected, i.e. the focus on understanding strength dependence on microstructure has been sought primarily via its effects on toughness. However, as is shown extensively in Chaps. 3 and 8, body microstructure has an important and often dominant effect on the microstructural dependence of strength, via effects on strength controlling flaws introduced (e.g. as noted in the anisotropy of oriented hot pressed Si_3N_4), a key factor often lost on a frequently singular focus on toughness.

Preferred orientation of grains, especially elongated ones, though widely neglected, occurs to varying, often significant, extent depending on material and especially fabrication. While much of the resultant effects, such as anisotropy in toughness, are significant, their effects may be greater with intergranular fracture, which is often a function of grain boundary phases. Further, anisotropy of some properties such as toughness and especially strength, while in part determined by properties along appropriate single crystal axes, may often be complicated by other polycrystalline factors such as preferred orientation of anisotropically shaped pores. Fractography is thus a critical tool in resolving these, especially strength, effects, as it is in so many other cases.

NOTE

After completing this chapter, a few additional and important references were obtained, whose results pertinent to this chapter are briefly summarized. Donners [268] conducted a more comprehensive study of fracture of MnZn ferrites. He corroborated the opposite change in fracture mode, i.e. from more transgranular for SCG to more intergranular for fast fracture reported in such ferrites reported by Beauchamp and Monroe [52, 53] instead of the usual reverse of this, and cites evidence that this is related to effects of stoichiometry. He also showed that (1) fracture mode could thus vary with location on the fracture surface as a function of distance from the specimen surface due to reduced H_2O diffusion, (2) fracture toughness decreased from ~ 2 to 1.3 MPa·m$^{1/2}$ as the relative humidity increased from 0 – 10 to 100%, (3) NH_3 had the greatest effect on SCG, H_2S, the next, with NO and CO having a similar effect as H_2O, and (4) large, exaggerated grains (e.g. >> 100 μm due to Al_2O_3 from contact with refractories) were preferred fracture origins. A threshold for SCG was also reported, as was fracture initiation for areas of exaggerated grain growth. He also cited date of Tanaka et al. [269] showing fracture on both (100) and (110) crystal planes with respectively ~ 1–1.3 and 0.9–1 MPa·m$^{1/2}$, consistent with a preference for (110) fracture.

Swab et al. [270] have shown that fracture initiation in dense, transparent AlON with large *G* (~ 150–200 μm) was from one or a few transgranularly fractured grains followed by mainly intergranular fracture indicating another case of the reversal of the typical SCG to fast fracture mode transition. They also showed that isolated large grains in dense, transparent $MgAl_2O_4$ with a bimodal grain distribution (G - 5–20 and ~ 200 μm) were frequent fracture origins, apparently with transgranular fracture, with more intergranular fracture of the surrounding finer grains.

Further testing of electric field effects on fracture of piezoelectric ceramics shows wider and conflicting results. Thus, while some, e.g. Park and Sun [271] showed that positive fields normal to a crack aid propagation while negative ones retard propagation, and questioned the use of stress intensity factors in such cases, Fu and Zhang [272] found crack propagation enhanced by either field polarity, e.g. apparently consistent with effects of ac fields [237]. However, these and other differences, and reports of microcracking and fatigue, respectively implying and showing nonreversible effects, indicate that this is a broader and more complex area of research that requires much more comprehensive study and analysis than it has generally received so far.

Finally, further note the very high dimensional stability of materials, which consists of two aspects: (1) the precision elastic limit (PEL) commonly defined as the stress to produce a detectable positive residual strain, e.g. 1 ppm, upon removal of a temporarily applied stress and (2) dimensional stability (DS) with a stress applied for extended times, often expressed as a percentage of the PEL, i.e. a "creep" resistance. While the PEL definition is based on stability to 1 part in 10^{-6} some (e.g. optical and gyroscope) applications, may require higher stability, e.g. 1 part in 10^{-7}–10^{-8}. PEL values for metals are a fraction of their yield stress, e.g. < 1/2, while the more limited values for ceramics may be their ultimate, usually true tensile, strength, but can be less [273–275]. Thus, three measurements showed the PEL of sintered BeO varying from 70–81% of their true tensile strengths (of 105–145 MPa) and one measurement in compression gave a PEL of 82% of the ultimate (1.7 GPa) [273]. The DS of the three tensile tests were 50–80% of their tensile PEL, while the one compressive test was 6–12% of the compressive PEL, i.e. ~ twice the stress level for compression versus tension. Other tests of fused SiO_2 and a low expansion, highly crystallized glass (Cer-Vit®) showed elastic behavior to a strain sensitivity of 5×10^{-8} to stresses of 34 MPa (which was true for specimens with etched or unetched surfaces, with some specimens failing at such a stress) [275]. Such tests may be valuable for detecting effects such as SCG (though the similarity of etched and unetched results with glassy materials may question this) and microcracking, e.g. as may occur in BeO (in tension and possibly compression).

REFERENCES

1. AG Evans. Fracture Mechanics Determinations. Fracture Mechanics of Ceramics 1 (RC Bradt, DPH Hasselman, FF Lange, eds.). Plenum Press, New York, 1974, pp. 17–48.
2. RF Pabst, Determination of K_{IC}-Factors with Diamond-Saw-Cuts in Ceramic Materials. Fracture Mechanics of Ceramics 2. Plenum Press, New York, 1974, pp. 555–565.
3. RF Pabst, K Kromp, G Popp. Fracture toughness—measurement and interpretation. Proc Brit Cer Soc, No. 32 (R Davidge, ed.), pp. 89–103, 1982.
4. RW Rice. Fractographic Determination of K_{IC} and Effects of Microstructural Stresses in Ceramics. Ceramic Transactions, 17:Fractography of Glasses and Ceramics II (VD Frechette and JR Varner, eds.). Am. Cer. Soc., Westerville, OH, 1991, pp. 509–545.
5. RW Rice. Porosity of Ceramics. Marcel Dekker, New York, 1998.
6. CCm Wu, RW Rice, PF Becher. The Character of Cracks in Fracture Toughness Measurements of Ceramics. Fracture Mechanics Methods for Cements, Rocks, and Ceramics (SW Freiman and ER Fuller, Jr., eds.). Am. Soc. for Testing and Materials, STP 745, 1981, pp. 127–140.
7. CCm Wu, SW Freiman, RW Rice, JJ Mecholsky. Microstructural aspects of crack propagation in ceramics. J Mat Sci 13:2659–2670, 1978.
8. RW Rice. Test-Microstructural Dependence of Fracture Energy Measurements in

Ceramics. Fracture Mechanics Methods for Cements, Rocks, and Ceramics (SW Freiman and ER Fuller, Jr., eds.). Am. Soc. for Testing and Materials, STP 745, 1981, pp. 96–117.
9. KR McKinney, RW Rice. Specimen Size Effects in Fracture Toughness Testing of Heterogenous Ceramics by the Notch Beam Method. STP 745, 1981, pp. 118–126.
10. SW Freiman. Stress-Corrosion Cracking of Glasses and Ceramics. In:Stress-Corrosion Cracking—Materials Performance and Evaluation (RH Jones, ed.). ASM, Materials Park, OH, 1992, pp. 337–344.
11. RW Rice. Grain size and porosity dependance of fracture energy and toughness of ceramics at 22°C. J Mat Sci 31:1969–1983, 1996.
12. T Nishida, Y Hanaki, G Pezzotti. Effect of notch root radius on the fracture toughness of a fine-grained alumina. J Am Cer Soc 77(2):606–608, 1994.
13. M Mizuno, H Okuda. VAMS round robin on fracture toughness of silicon nitride. J Am Cer Soc 78(7):1793–1801, 1995.
14. RP Ingel. US Naval Res. Lab., Washington, DC, private communication.
15. RW Rice, SW Freiman, JJ Mecholsky, Jr. The dependence of strength-controlling fracture energy on the flaw-size to grain-size ratio. J Am Cer Soc 63(3–4):129–136, 1980.
16. RW Rice. Comment on role of grain size in the strength and *R*-curve properties of alumina. J Am Cer Soc 76(7):1898–1899, 1993.
17. RW Rice. Crack-shape-wake-area effects on ceramic fracture toughness and strength. J Am Cer Soc 77(9):2479–2480, 1994.
18. CJ Gilbert, JJ Cao, LC De Jonghe, RO Ritchie. Crack-growth resistance-curve behavior in silicon carbide:small versus long cracks. J Am Cer Soc 80(9):2253–2261, 1997.
19. AA Griffith. The Phenomena of Rupture and Flow in Solids. Phil. Trans. Roy. Soc. London A221:163–198, 1920.
20. TA Michalske, BC Bunker, SW Freiman. Stress corrosion of ionic and mixed ionic/covalent solids. J Am Cer Soc 69(10):721–724, 1986.
21. EP George, M Yamaguchi, KS Kumar, CT Liu. Ordered intermetallics, Ann Rev Mater Sci 24:409–451, 1994.
22. JJ Mecholsky, RW Rice, SW Freiman. Prediction of fracture energy and flaw size in glasses from measurement of mirror size. J Am Cer Soc 57(10):440–443, 1974.
23. JJ Mecholsky, Jr. Intergranular slow crack growth in MgF_2. J Am Cer Soc 64(9):563–566, 1981.
24. JJ Mecholsky, SW Freiman. Fractographic Analysis of Delayed Failure in Ceramics. Fractography and Materials Science (LN Gilbertson and RD Zipp, Eds.). Am. Soc. for Testing and Materials, STP 733, 1981, pp. 246–258.
25. V Tvergaard, JW Hutchinson. Microcracking in ceramics induced by thermal expansion or elastic anisotropy. J Am Cer Soc 71(3):157–166, 1988.
26. JJ Cleveland. The Critical Gain Size for Microcracking in the Pseudo-Brookite Structure. M.Sci. thesis, Pennsylvania State University, March, 1977.
27. JJ Cleveland, RC Bradt. Grain size/microcrack relations for pseudobrookite oxides. J Am Cer Soc. 61(11–12):478–481, 1978.
28. AG Evans. Microfracture from thermal expansion anisotropy—I. single phase systems. Acta Met 26:2 1845–1853, 1978.

29. RW Rice, RC Pohanka. Grain-size dependence of spontaneous cracking in ceramics. J Am Cer Soc 62(11–12):559–563, 1979.
30. ED Case, JR Smyth, O Hunter. Grain-size dependence of microcrack initiation in brittle materials. J Mat Sci 15:149–153, 1980.
31. RW Davidge. Cracking at grain boundaries in polycrystalline brittle materials. Acta Metall 29:1695–1702, 1981.
32. YM Ito, M Rosenblatt, LY Cheng, FF Lange, AG Evans. Cracking in particulate composites due to thermalmechanical stress. Int J Fract 17:483–491, 1981.
33. Y Fu, AG Evans. Microfracture zone formation in single phase polycrystals. Acta Metall 30:1619–1625, 1982.
34. F Ghahremani, JW Hitchinson, V Tveraard. Three-dimensional effects in microcrack nucleation in brittle polycrystalline materials. J Am Cer Soc 73(6):1548–1554, 1990.
35. O Hunter, Jr, RW Scheidecker, S Tojo. Characterization of metastable tetragonal hafnia. Ceramurgica Intl 5(4):137-, 1979.
36. SL Dole, O Hunter, Jr., DJ Bray. Microcracking of monoclinic HfO_2. J Am Cer Soc 61(11–12):486–490, 1978.
37. ED Case, CJ Glinka. Characterization of microcracks in $YCrO_3$ using small-angle neutron scattering and elasticity measurements. J Mat Sci 19:2962–2968, 1984.
38. FJ Parker, RW Rice. Correlation between grain size and thermal expansion for aluminum titanate materials. J Am Cer Soc 72(12):2364–2366, 1989.
39. RW Rice. Ceramic Fracture Features, Observations, Mechanisms, and Uses. Fractography of Ceramic and Metal Failures ASTM STP 827 (JJ Mecholsky, Jr., and SR Powell, Jr., eds.), 1984, pp. 5–103.
40. RW Rice. Perspective on Fractography. Advances in Ceramics 22, Fractography of Glasses and Ceramics (JR Varner and VD Frechette eds.). Am. Cer. Soc., Westerville, OH, 1988, pp. 3–56.
41. J Tatami, K Yasuda, Y Matsuo, S Kimura. Stochastic analysis on crack propagation path of polycrystalline ceramics based on the difference between the released energies in crack propagation. J Mat Sci 32:2341–2346, 1992.
42. AG Evans. Perspective on the development of high-toughness ceramics. J Am Cer Soc 73(2):187–206, 1990.
43. RW Rice, SW Freiman. Grain-size dependence of fracture energy in ceramics:I, a model for noncubic materials. J Am Cer Soc 64(6):350–354, 1981.
44. AG Evans, KT Faber. Toughening of ceramics by circumferential microcracking. J Am Cer Soc 64(7):394–398, 1981.
45. RG Hoagland, JD Embury. A treatment of inelastic deformation around a crack tip due to microcracking. J Am Cer Soc. 63(7–8):404–410, 1980.
46. H Cai, B Moran KT Faber. Analysis of a microcrack prototype and its implications for microcrack toughening. J Am Cer Soc 70(11):849–854, 1987.
47. GD Bowling, KT Faber, RG Hoagland. Computer simulations of *R*-curve behavior in microcracking materials. J Am Cer Soc 74(7):1695–1698, 1991.
48. RG Hoagland, GT Hahn, AR Rosenfeld. Influence of microstructure on fracture propagation in rock. Rock Mech 5:77–106, 1973.
49. PL Swanson. Tensile fracture resistance mechanisms in brittle polycrystals:An ultrasonic and In Situ microscopy investigation. J Geophy Res 92(B8):8015–8036, 1987.

50. PL Swanson CJ Fairbanks, BR Lawn, YW Mai, BJ Hockey. Crack-interface grain bridging as a fracture resistance mechanism in ceramics:I. experimental study of alumina. J Am Cer Soc 70(4):279–289, 1987.
51. R Knehans, R Steinbrech. Memory effect of crack resistance during slow crack growth in notched Al_2O_3 bend specimens. J Mat Sci Lett. 1:327–329, 1982.
52. EK Beauchamp, SL Monroe. Effect of crack-interface bridging on subcritical crack growth in ferrites. J Am Cer Soc 72(7):1179–1184, 1989.
53. EK Beauchamp, SL Monroe. *R*-Curve Behavior in Ferrite Ceramics. Fractography of Glasses and Ceramics II, Ceramic Trans. 17(VD Frechette and JR Varner, eds.). Am Cer. Soc., Westerville, OH, 1991, pp. 201–218.
54. A Quinten, W Arnold. Observation of stable crack growth in Al_2O_3 ceramics using a scanning acoustic microscope. Mat Sci Eng A122:15–19, 1989.
55. J Rodel, JF Kelly, BR Lawn. In situ measurements of bridged crack interfaces in the scanning electron microscope. J Am Cer Soc 73(11):3313–3318, 1990.
56. S Tandon, KT Faber. On loading rate effects in toughening processes. Scripta materialia 34(5):757–762, 1996.
57. Y-W Mai BR Lawn. Crack-interface grain bridging as a fracture resistance mechanism in ceramics:II. theoretical fracture mechanics model. J Am Cer Soc 70(4):289–294, 1987.
58. BR Lawn, NP Padture, LM Braun SJ Bennison. Model for toughness curves in two-phase ceramics:II. microstructural variables. J Am Cer Soc 76(9):2241–2247, 1993.
59. WA Curtin. Toughening by crack bridging in heterogeneous ceramics. J Am Cer Soc 78(5):1313–1323. 1995.
60. RW Rice. Microstructural dependence of fracture energy and toughness of ceramics and ceramic composites versus that of their tensile strengths at 22°C. J Mat Sci 31:4503–4519, 1996.
61. JJ Gilman. Cleavage, Ductility, and Tenacity in Crystals. In:Fracture (BL Averbach, DK Felbeck, GT Hahn, and DA Thomas, eds.). Tech. Press of MIT, Boston, MA, 1959, pp. 193–222.
62. RW Rice. Ceramic Fracture Mode—Intergranular vs. Transgranular Fracture. Ceramic Transactions, 64:Fractography of Glasses and Ceramics III (JR Varner, VD Frechette, and GD Quinn, eds.). Am. Cer. Soc., Westerville, OH, 1996, pp. 1–53.
63. JA Kuszyk, RC Bradt. Influence of grain size on effects of thermal expansion anisotropy in $MgTi_2O_5$. J. Am Cer Soc 56(8):420–423, 1973.
64. RW Rice. Effects of thermal expansion mismatch stresses on the room-temperature fracture of boron carbide. J Am Cer Soc 73(10):3116–3118, 1990.
65. RW Rice. Possible effects of elastic anisotropy on mechanical properties of ceramics. J Mat Sci Lett 13:1261–1266, 1994.
66. DA Shockey, GW Groves. Effect of water on toughness of MgO crystals. J Am Cer Soc 51(6):299–301, 1968.
67. DA Shockey, GW Groves. Origin of water-induced toughening of MgO crystals. J Am Cer Soc 52(2):82–85, 1969.
68. SW Freiman, KR Mc Kinney. Unpublished MgO data.
69. WH Rhodes, RM Cannon, Jr., T Vasilos. Stress-Corrosion Cracking in Polycrystalline MgO. Fracture Mechanics of Ceramics 2 (RC Bradt, DPH, Hasselman, and FF Lange, eds.). Plenum Press, New York, 1974, pp. 708–733.

70. KR McKinney, BA Bender, RW Rice, C Cm. Wu. Slow crack growth in Si_3N_4 at room temperature. J Mat Sci 26:6467–6472, 1991.
71. RW Rice. Porosity effects on machining direction—strength anisotropy and failure mechanisms. J Am Cer Soc 77(8):2232–2236, 1994.
72. RW Rice, CCm Wu. Slow crack growth in $MgAl_2O_4$ single- and poly-crystals. J Mat Sci Lett. 14:723–727, 1995.
73. CCm Wu, RW Rice. Slow crack growth in oxide and nonoxide ceramics. To be published.
74. CCm Wu, KR McKinney, RW Rice. Zig-zag-crack propagation in $MgAl_2O_4$ crystals. J Mat Sci Lett 14:474–477, 1995.
75. RW Rice, AC Wu, KR McKinney. Fracture and fracture toughness of stoichiometric $MgAl_2O_4$ Crystals at Room Temperature. J. Mat. Sci. 31:1353–1360, 1996.
76. A Ball, BW Payne. The tensile fracture of quartz crystals. J Mat Sci 11:731–740, 1976.
77. AJ Gessing, RC Bradt. A Microcracking Model for the Effect of Grain Size on Slow Crack Growth in Polycrystalline Al_2O_3. Fracture Mechanics of Ceramics, 5 (RC Bradt, AG Evans, DPH Hasselman, and FF Lange, eds). Plenum Press, New York, 1983, pp. 569–590.
78. PF Becher, MK Ferber. Grain size dependence of slow-crack growth behavior in noncubic ceramics. Acta Metall 33(7):1217–1221, 1985.
79. PF Becher, MK Ferber.. ORNI. Unpublished Al_2O_3 and TiB_2 SCG data.
80. KM Liang, KF Gu, G Orange, G Fantozzi. Influence of Grain Size on Crack Resistance. Ceramics Today—Tomorrow's Ceramics (P Vincenzini, ed.) Elsvier Science Publishers B.V. 1991, pp. 1509–1518.
81. AG Evans. A method for evaluating the time-dependent failure characteristics of brittle materials and its application to polycrystalline alumina. J Mat Sci 7:1137–1146, 1972.
82. JE Ritter, Jr., JN Humenik. Static and dynamic fatigue of polycrystalline alumina. J Mat Sci 14:626–632, 1979.
83. JN Humenik. Susceptability of an alumina-mullite ceramic to delayed failure. Am Cer Soc Bul 60(4):497–500, 1981.
84. MK Ferber, SD Brown. Subcritical crack growth in dense alumina exposed to physiological media. J Am Cer Soc 63:(7–8):424–429, 1980.
85. WP R Byrne, MJ Hanney, R Morrell. Slow Crack Growth of Oxide Ceramics in Corrosive Environments. Proc. Brit. Cer. Soc., No. 32 (RW Davidge, ed.), 1982, pp. 303–314.
86. G de With. Structural integrity of ceramic multilayer capacitor materials and ceramic multilayer capacitors. J Eur Cer Soc 12:323–336, 1993.
87. NL Hecht, DE McCallum, GA Graves, SD Jang. Environmental effects in toughened ceramics. Cer Eng Sci Proc 8(7–8):892–909, 1987.
88. Y Yamade, Y Kawaguchi, N Takeda, T Kishi. Slow crack growth of mullite ceramic. J Cer Soc Jpn, Intl Ed 99:452-, 1991.
89. JP Signh, AV Virkar, DK Shetty, RS Gordon. Microstructural Effects on the Subcritical Crack Growth in Polycrystalline β''-Aluimina. Fracture Mechanics of Ceramics 8 Microstructure, Methods, Design, and Fracture (R. C. Bradt, A. G. Evans,

D. P. H. Hasselman, and F. F. Lange, eds.). Plenum Press, New York, 1986, pp. 273–284.
90. M Hirano. Inhibition of low temperature degradation of tetragonal zirconia ceramics—a review. Br Cer Trans J 91:139–147, 1992.
91. J -F Li, R Watanabe. Phase transformation in Y_2O_3-partially stabilized ZrO-polycrystals of various grain sizes during low-temperature aging in water. J Am Cer Soc 81(10):2687–2691, 1998.
92. JJ Swab. Low temperature degredation of Y-TZP materials. J Mat Sci 26:6706–6714, 1991.
93. RW Rice. CaO:II. properties. J Am Cer Soc 52(8):428–436, 1969.
94. RW Rice. Unpublished MgO data.
95. KR Janowski, RC Rossi. Mechanical degradation of MgO by water. J Am Cer Soc 51(8):453–454, 1968.
96. I Yamai, T Ota. Grain size-microcracking relation for $NaZr_2(PO_4)_3$ family ceramics. J Am Cer Soc 76(2):487–491, 1993.
97. I Yamai, T Ota. Low-thermal-expansion polycrystalline zirconyl phosphate ceramic:solid-solution and microcracking-related properties. J Am Cer Soc 70(8):585–590, 1987.
98. R Telle, G Petzow. Strengthening and toughening of boride and carbide hard material composites. Mats Sci Eng A105/106:97–104, 1988.
99. Y Ohya, Z Nakagawa. Measurement of crack volume due to thermal expansion anisotropy in aluminum titanate ceramics. J Mat Sci 31:1555–1559, 1996.
100. H Tomaszewski. Influence of microstructure on the thermomechanical properties of alumina ceramics. Cer Intl 18:51–55, 1992.
101. HP Kirchner, RM Gruver. Strength-anisotropy-grain size relations in ceramic oxides'. J Am Cer Soc 53(5):232–236, 1970.
102. JE Brocklehurst. Fracture in Polycrystalline Graphite. Chem. Phy. Carbon, 13 (P. L. Walker and P. A. Thrower, eds). Marcel Dekker, New York, 1977, pp. 145–279.
103. R Stevens. Fracture behavior and electron microscopy of a fine grained graphite. Carbon 9:573–578, 1971.
104. EH Lutz, MV Swain, N Claussen. Thermal shock behavior of duplex ceramics. J Am Cer Soc 74(1):19–24, 1991.
105. K Hamano, Y Ohya, Z Nakagawa. Crack propagation resistance of aluminum titanate ceramics. Int J High Tech Cer 1:129–137, 1985.
106. BR Lawn. Fundamental condition of the existence of microcrack clouds in monophase ceramics. J Eur Cer Soc 7:17–20, 1991.
107. L Peck, CC Barton, RB Gordon. Microstructure and the resistance of rock to tensile fracture. J Geophy Res 90(*B13*):11533–11546, 1985.
108. RW Rice. Microstructural Dependence of Mechanical Behavior of Ceramics. Treatise on Materials Science and Technology 11, Properties and Microstructure (RK MacCrone, ed.). Academic Press, New York, 1977, pp. 191–381.
109. RW Rice, SW Freiman, PF Becher. Grain-size dependence of fracture energy in ceramics:II. experiment. J Am Cer Soc 64(6):345–350, 1981.
110. JB Kessler, JE Ritter, Jr., RW Rice. The Effects of Microstructure on the Fracture Energy of Hot Pressed MgO. Materials Science Research, 7:Surfaces and Inter-

faces of Glass and Ceramics (VD Frechette, WC LaCourse, and VL Burdick eds.). Plenum Press, New York, 1974, pp. 529–544.
111. FJP Clarke, HG Tattersall, G Tappin. Toughness of ceramics and their work of fracture. Proc Brit Cer Soc 6:163–172, 1966.
112. K Yasuda, S -D Kim, Y Kanemichi, Y Matsuo, S Kimura. Influence of grain size on fracture toughness of MgO sintered bodies. J Cer Soc Japan Intl Ed 98:1110–1115, 1990.
113. K Yasuda, Y Tsuru, Y Kanemichi, Y Matsuo, S Kimura. The influence of grain size and secondary phase on the fracture toughness of sintered MgO bodies (part I—addition of Al_2O_3) J Cer Soc Japan Intl Ed 100:790–794, 1992.
114. Y Tsuru, K Yasuda, Y Matsuo, S Kimura, E Yasuda. The influence of grain size and secondary phase on the fracture toughness of sintered MgO bodies (part 2)—addition of Fe_2O_3. J Cer Soc Japan Intl Ed 101:165–169, 1993.
115. K Yasuda, Y Tsuru, Y Matsuo, S Kimura, T Yano. The influence of grain size and secondary phase on the fracture toughness of sintered MgO bodies (part 3—addition of CoO). J Cer Soc Japan Intl Ed 101:648–653, 1993.
116. LD Monroe, JR Smyth. Grain size dependence of fracture energy of Y_2O_3. J Am Cer Soc 61(11–12):538–540, 1978.
117. WH Rhodes, JG Baldoni, GC Wei. The Mechanical Properties of La_2O_3-Doped Y_2O_3. GTE Lab., Final Rept. for ONR Contract N00014-82-C-0452, July 1986.
118. T Tani, Y Miyamato, M Koizumi, M Shimado. Grain size dependence of vickers microhardness and fracture toughness in Al_2O_3 and Y_2O_3 ceramics. Cer Intl 12:33–37, 1986.
119. JJ Mecholsky, J., SW Freiman, RW Rice. Fracture surface analysis of ceramics. J Mat Sci 11:1310–1319, 1976.
120. D Townsend, JE Field. Fracture toughness and hardness of zinc sulphide as a function of grain size. J Mat Sci 25:1347–1352, 1990.
121. WC Johnson, DE Stein, RW Rice. Analysis of grain boundary impurities and fluoride additives in hot-pressed oxides by auger electron spectroscopy. J Am Cer Soc 57(8):342–344, 1974.
122. MI Mendelson, ME Fine. Fracture of Wustite and Wustite-Fe_3O_4-5 v/0 Fe. Fracture Mechanics of Ceramics 2 (RC Bradt, DPH Hasselman, and FF Lange, eds.). Plenum Press, New York, 1974, pp. 527–540.
123. AG Evans, RW Davidge. The strength and fracture of stoichiometric polycrystalline UO_2. J Nuc Mat 33:249–260, 1969.
124. RL Stuart, M Iwasa, RC Bradt. Room-temperature K_{IC} values for single-crystal and polycrystalline $MgAl_2O_4$. J Am Cer Soc 64(2):C-22–23, 1981.
125. JDB Veldkamp, N Hattu. On the fracture toughness of brittle materials. Phillips J Res 34:1–25, 1979.
126. RC Pohanka, SW Frieman, BA Bender. Effect of the phase transformation on the fracture behavior of $BaTiO_3$. J Am Cer Soc 61(2):72–75, 1978.
127. RC Pohanka, SW Freiman, K Okazaki, S Tashiro. Fracture of Piezoelectric Materials. Fracture Mechanics of Ceramics 5 (RC Bradt, AG Evans, DPH Hasselman, and FF Lange, eds). Plenum Press, New York, 1983, pp. 353–364.

128. SW Freiman, RC Pohanka. Review of mechanically related failures of ceramic capacitors and capacitor materials. J Am Cer Soc 72(12):2258–2263, 1989.
129. RC Pohanka, RW Rice, BE Walker, Jr. Effect of internal stress on the strength of $BaTiO_3$. J. Am. Cer. Soc 59(1–2):71–74, 1976.
130. K Niihara. Mechanical properties of chemically vapor deposited nonoxide ceramics. Am Cer Soc Bull 63(9):1160–1165, 1984.
131. H Kodama, T Miyoshi. Study of fracture behavior of very fine-grained silicon carbide. J Am Cer Soc 73(10):3081–3086, 1990.
132. SG Seshardi, M Srinivasan, KY Chia. Microstructure and Mechanical Properties of Pressureless Sintered Alpha-SiC. Cer. Trans., 2, Silicon Carbide '87 (JD Cawley and CE Semler, eds.). Am. Cer. Soc., Westerville, OH, 1989, pp. 215–226.
133. RP Ingel, D Lewis, BA Bender, RW Rice. Temperature dependence of strength and fracture toughness of ZrO_2 single crystals. J Am Cer Soc 65(9):C-150–152, 1982.
134. A Pajares, F Guiberteau, A Dominguez-Rodriguez, AH Heuer. Microhardness and fracture toughness anisotropy in cubic zirconium oxide single crystals. J Am Cer Soc 71(7):C-332–333, 1988; Indentation-Induced Cracks and the Toughness Anisotropy of 9.4 mol %~ Yttria-Stabilized Cubic Zirconia Single Crystals. J Am Cer Soc. 71(7):859–862, 1991.
135. MO Guillou, JL Henshall, RM Hooper, GM Carter. Indentation hardness and fracture in single crystal magnesia, zirconia, and silicon carbide. special ceramics 9, Proc Brit Cer Soc 49:191–202, 1992.
136. K Keller, T Mah, TA Parthasarathy. Processing and mechanical properties of polycrystalline $Y_3Al_5O_{12}$ (yttrium aluminum garnet). Cer Eng Sci Proc 11(7–8):1122–1133, 1990.
137. T Mah, TA Parthasarathy. Effects of temperature, environment, and orientation on the fracture toughness of single-crystal YAG. J Am Cer Soc 80(10):2730–2734, 1997.
138. G de With, JED Parren. Translucent $Y_3Al_5O_{12}$ ceramics:mechanical properties. Solid State Ionics 16:87–94, 1985.
139. G de With. Translucent $Y_3Al_5O_{12}$ Ceramics:Something Old, Something New. High Tech Ceramics (P Vincenzini, ed.). Elsvier Science Publishers, Amsterdam, 1987, pp. 1063–1072.
140. JA Savage, CJH Worst, CSJ Pickels, RS Sussmann, CG Sweeney, MR McClymont, JR Brandon, CN Dodge, and AC Beale. Properties of Free-Standing CVD Diamond Optical Components. Window and Dome Technologies and Materials, 3060. SPIE Proc. (RW Tustison, ed.), pp. 144–159, 4/1997.
141. MD Drory, CF Gardinier, JS Speck. Fracture toughness of chemically vapor-deposited diamond. J Am Cer Soc 74(12):3148–3150, 1991.
142. TP Lin, GA Cooper, M Hood. Measurement of the fracture toughness of polycrystalline diamond Using the double-torsion test. J Mat Sci 29:4750–4756, 1994.
143. M Iwasa, RC Bradt. Fracture Toughness of Single-Crystal Alumina. Advances in Ceramics 10, Structure and Properties of MgO and Al_2O_3 Ceramics (WD Kingery, ed.). Am. Cer. Soc., Columbus, OH, 1984, pp. 767–779.
144. DC Hitchcock, LC DeJonghe. Fracture toughness anisotropy of sodium β-Alumina. J Am Cer Soc 66(11):C-204–205, 1983.

145. AV Virkar, GT Tennenhouse, RS Gordon. Hot-pressing of LiO_2-stabilized β″-alumina. J Am Cer Soc 57(11):508, 1974.
146. AV Virkar, RS Gordon. Fracture properties of polycrystalline LiO_2-stabilized β-alumina. J Am Cer Soc 60(1–2):58–61, 1977.
147. R Stevens. Strength and fracture mechanisms in beta-alumina. J Mat Sci 9:934–940, 1974.
148. YP Gupta, AT Santhanam. On cleavage surface energy of calcite crystals. Acta Metall. 17:419–424, 1969.
149. K Honma, M Yoshinaka, K Hirota, O Yamaguchi. Fabrication of calcite ($CaC0_3$) ceramics with high density. J Mat Sci Lett 17:745–746, 1998.
150. RA Schultz, MC Jensen, RC Bradt. Single crystal cleavage of brittle materials. Int J Fracture 65:291–312, 1994.
151. M Pardavi-Horváth. Microhardness and brittle fracture of garnet single crystals. J Mat Sci 19:1159–1170, 1984.
152. CP Chen, MH Leipold. Fracture toughness of silicon. Am Cer Soc Bull 59(4):469–472, 1980.
153. CP Chen, MH Leipold, Jr., D Helmreich. Fracture of directionally solidified multicrystalline silicon. J. Am Cer Soc 65(4):C-49, 1982.
154. M Iwasa, RC Bradt. Cleavage of natural and synthetic single crystal quartz. Mat Res Bull 22:1241–1248, 1987.
155. OO Adewoye. Anisotropic flow and fracture in beryl [$Be_3Al_2(SiO_3)_6$]. J Mat Sci 21:1161–1163, 1986.
156. M Sakai, RC Bradt, AS Kobayashi. The toughness of polycrystalline $MgAl_2O_4$. J Cer Soc Jpn, Intl Ed 96:510–515, 1988.
157. J Seidel, J Rodel. Measurement of crack tip toughness in alumina as a function of grain size. J Am Cer Soc 80(2):433–438, 1997.
158. WH Class, ES Machlin. Crack propagation method for measuring grain boundary energies in brittle materials. J Am Cer Soc 49(6):306–309, 1966.
159. J Tatami, T Harada, K Yasuda, Y Matsuo. Influence of twist angle on the toughness of (0001) twist boundary of alumina. Cer Eng Sci Proc 19:211–218, 1998.
160. HYB. Mar, WD Scott. Fracture induced in Al_2O_3 bicrystals by anisotropic thermal expansion. J Am Cer Soc 53(10):555–558, 1970.
161. WP Minnear, RC Bradt. Stoichiometry effects on the fracture of TiO_2. J Am Cer Soc 63(9–10):485–490, 1980.
162. K Yasuda, S Ohsawa, Y Matsuo, S. Kimura. Influence of microcracking on fracture toughness of TiO_2 sintered body. J Jap Soc Mat Sci 41(463):482–488, 1992.
163. SW Freiman, L Chuck, JJ Mecholsky, DL Shellman, LJ Storz. Fracture Mechanisms in Lead Zirconate Titanate Ceramics. Fracture Mechanics of Ceramics 8, Microstructure, Methods, Design, and Fracture (R C Bradt, AG Evans, DPH Hasselman, FF Lange, eds.). Plenum Press, New York, 1986, pp. 175–185.
164. Y Ohya, Z-E Nakagawa, K Hamono. Crack healing and bending strength of aluminum titanate ceramics at high temperature. J Am Cer Soc 71(5):C-232–233, 1988.
165. EC Skaar, WC Croft. Thermal expansion of TiB_2. J Am Cer Soc 56(1):45, 1973.

166. AA Korneev, IT Ostapenko, MA Dolshek, AG Mironova, ND Rybal cenko, IA Lyashenko, VP Podtykan. Effect of annealing on the structure and properties of hot-pressed boron carbide. Poroshkovaya Metallurgiya 1 (349):9969–9972, 1972.
167. K Niihara, A Nakahira, T Hirai. The effect of stoichiometry on mechanical properties of boron carbide. J Am Cer Soc 74(1):C-13–14, 1984.
168. KA Schwetz, LS Sigl, L Pfau. Mechanical properties of injection molded B_4C-C ceramics. J Solid State Chem. 133:68–76, 1997.
169. RW Rice, KR McKinney, CCm Wu, SW Freiman, WJ McDonough. Fracture energy of Si_3N_4. J. Mat Sci 20:1392–1406, 1985.
170. H-J Choi, YW Kim, J-G Lee. Effect of amount and composition of additives on the fracture toughness of silicon nitride. J Mat Sci Let 15:375–377, 1996.
171. JA Palm, CD Greskovitch. Thermomechanical properties of hot-pressed $Si_{2.9}Be_{0.1}N_{3.8}O_{0.2}$ ceramic. Am Cer Soc Bull 59(4):447–452, 1980.
172. M Shimada, M Koizumi, A Tanaka, T YYamada. Temperature dependence of K_{IC} for high-pressure hot-pressed Si_3N_4 without additives. J Am Cer Soc 65(4):C-48, 1982.
173. I Tanaka, G Pezzotti, Y Miyamoto, T Okamoto. Fracture toughness of Si_3N_4 and its Si_3N_4 whisker composite without sintering aids. J Mat Sci 26:201–210, 1991.
174. K Niihara, T Hirai. Fractography of chemical vapor-deposited Si_3N_4. J Mat Sci, 13, 2385–2393 (1978).
175. SC Danforth, MH Richman. Strength and fracture toughness of reaction-bonded Si_3N_4. Am Cer Soc Bull 62(4):501–504, 1983.
176. E Tani, S Umebayashi, K Kishi, K Kobatashi, M Nishijima. Effect of size of grains with fiber-Like structure of Si_3N_4 on fracture toughness. J Mat Sci Lett 4:1454–1456, 1985.
177. K Matsuhiro, T Takahashi. The effect of grain size on the toughness of sintered Si_3N_4. Cer Eng Sci Proc 10(7–8):807–816, 1989.
178. T Kawaehima, H Okamoto, H Yamamoto, A Kitamura. Grain size dependence of the fracture toughness of silicon nitride ceramics. J Cer Soc Jap, Intl Ed 99:310–313, 1991.
179. CMB Henderson, D Taylor. Thermal expansion of the nitrides and oxynitride of silicon in relation to their structures. Trans J Brit Cer Soc 74(2):49–53, 1975.
180. G Himsolt, H Knoch, H Huebner, FW Kleinlein. Mechanical properties of hot-pressed silicon nitride with different grain structures. J Am Cer Soc 62(1–2):29–32, 1979.
181. AJ Pyzik, DF Carroll. Technology of self-reinforced silicon nitride. Ann Rev Mater Sci, Annual Reviews Inc. 24:189–214, 1994.
182. IM Peterson, T-Y Tien. Effect of the grain boundary thermal expansion coefficient on the fracture toughness in silicon nitride. J Am Cer Soc 78(9):2345–2352, 1995.
183. P Sajgalik, J Dusza, MJ Hoffmann. Relationship between microstructure, toughening mechanisms, and fracture toughness of reinforced silicon nitride ceramics. J Am Cer Soc 78(10):2619–2624, 1995.
184. T Ohji, K Hirao, S Kanzaki. Fracture resistance behavior of highly anisotropic silicon nitride. J Am Cer Soc 78(11):3125–3128, 1995.
185. J Wang, M Rainforth, R Stevens. The grain size dependence of the mechanical properties in TZP ceramics. Br Cer Trans 88(1):1–6, 1989.

186. BA Cottom, MJ Mayo. Fracture toughness of nanocrystalline ZrO_2-3 mol% Y_2O_3 determined by Vickers indentation. Scripta Mats 34 (5):809–814, 1996.
187. MV Swain. Grain-size dependence of toughness and transformability of 2 mol%-TZP ceramics. J Mat Sci Lett 5:1159–1162, 1986.
188. L Ruiz, MJ Readey. Effect of heat treatment on grain size, phase assemblage, and mechanical properties of 3 mol% Y-TZP. J Am Cer Soc 79(9):2331–2340, 1996.
189. GSAM Theunissen, JS Bouma, AJA Winnubst, AJ Burggraaf. Mechanical properties of ultra-fine grained zirconia ceramics. J Mat Sci 27:4429–4438, 1992.
190. P Duran, P Reico, JR Jurado, C Pascual, F Capel, C Moure. Y(E)-doped tetragonal zirconia polycrystalline solid electrolyte. J Mat Sci 24:708–716, 1989.
191. T Masaki. Mechanical properties of toughened ZrO_2-Y_2O_3 ceramics. J Am Cer Soc 69(8):638–640, 1986.
192. J-G Duh, H-T Dai, B-S Chiou. Sintering, microstructure, hardness, and fracture toughness behavior of Y_2O_3-CeO_2-ZrO_2. J Am Cer Soc 71(10):813–819, 1988.
193. J Wang, XH Zheng, R Stevens. Fabrication and microstructure-mechanical property relationships in Ce-TZPs. J Mat Sci 27:5348–5356, 1992.
194. JG Duh, JU Wan. Developments in highly toughened CeO_2-Y_2O_3-ZrO_2 ceramic system. J Mat Sci 27:6197–6203, 1992.
195. PF Becher, KB Alexander, A Blier, SB Waters, WH Warwick. Influence of ZrO_2 Grain size on the transformation response in the Al_2O_3-ZrO_2 (12 mol% CeO_2) system. J Am Cer Soc 76(3):657–663, 1995.
196. M Sakai, RC Bradt. The crack growth resistance curve of non-phase transforming ceramics. J Cer Soc Jpn, Inten Ed 96(8):801–809, 1988.
197. G Vekins, MF Ashby, PWR Beaumont. *R*-curve behavior of Al_2O_3 ceramics. Acta Met Mat 38(6):151–162, 1990.
198. BR Lawn, LM Braun, SJ Bennision, RF Cook. Reply to comment on role of grain size in the strength and *R*-curve properties of alumina. J Am Cer Soc 76(7):1900–1901, 1993.
199. R Steinbrech, R Khehans, W Schaarrwachter. Increase of crack resistance during slow crack growth in Al_2O_3 bend specimens. J Mat Sci 18:265–270, 1983.
200. C-W Li, DJ Lee, S-C Lui, J Goldacker. Relation Between strength, microstructure, and grain bridging in-situ reinforced silicon nitride. J Am Cer Soc 78(2):449–459, 1995.
201. A Okada, N Hiroshi. Subcritical crack growth in silicon nitride exhibiting *R*-curve. J Am Cer Soc 73(7):2095–2096, 1990.
202. N Hirosaki, Y Akimune, M Mitomo. Effect of grain growth of β-silicon nitride on strength, Weibull modules, and fracture toughness. J Am Cer Soc 76(7):1892–1894, 1993.
203. JA Salem, SR Choi, MR Freedman, MG Jenkins. Mechanical behavior and failure phenomenon of an in situ toughened silicon nitride. J Mat Sci 27:4421–4428, 1992.
204. M Sakai, J-I Yoshimura, Y Goto, M Ingaki. *R*-curve behavior of a polycrystalline graphite:microcracking and grain bridging in the wake region. J Am Cer Soc 71(8):609–616, 1988.
205. M Sakai, H Kurita. Size-effect on the fracture toughness and the *R*-curve of carbon materials. J Am Cer Soc 79(12):3177–3184, 1996.

206. A Ghosh, KW White, MG Jenkins, AS Kobayashi, RC Bradt. Fracture resistance of a transparent magnesium aluminate spinel. J Am Cer Soc 74(7):1624–1630, 1991.
207. KW White, GP Kelkar. Evaluation of the crack bridging mechanism in a $MgAl_2O_4$spinel. J Am Cer Soc 74(70):1732–1734, 1991.
208. JC Hay, KW White. Grain-bridging mechanisms in monolithic alumina and spinel. J Am Cer Soc 76(7):1849–1854, 1993.
209. SK Lee, DK Kim, CH Kim. Flaw-tolerance and *R*-curve behavior of liquid-phase-sintered silicon carbides with different microstructures. J Am Cer Soc 78(1):65–70, 1995.
210. H Hubner, W Jillek. Sub-critical crack extension and crack resistance in polycrystalline alumina J Mat Sci 12:117–125, 1977.
211. L Ewart, S Suresh. Crack propagation in ceramics under cyclic loads. J Mat Sci 22:1173–1192, 1987.
212. G Choi, S Horibe. Static fatigue in ceramic materials:influence of an intergranular glassy phase and fracture toughness. J Mat Sci 28:5931–5936, 1993.
213. RW Steinbrech, R Khehans, W Schaarwachter. Increase of crack resistance during slow crack growth in Al_2O_3 bend specimens. J Mat Sci 18:265–270, 1983.
214. RW Steinbrech, O Schmenkel. Crack-resistance curves of surface cracks in alumina. J Am Cer Soc 71(5):C-271–273, 1988.
215. PF Becher, EY Sun, C-H Hsueh, KB Alexander, S-L Hwang, SB Waters, CG Westmoreland. Debonding of interfaces between beta-silicon nitride whiskers and Si-Al-Y oxynitride glasses. Acta Mater 44(10):3881–3893, 1996.
216. EY Sun, PF Becher, SB Waters, CH Hsueh, KP Plunckett, MJ Hoffmann. Control of Interface Fracture in Silicon Nitride Ceramics:Influence of Different Fare Earth Elements. In:Ceramic Microstructure:Control at the Atomic Level (AP Tomsia, A Glaeser, eds). Plenum Press, New York, 1998, pp. 779–786.
217. A Okada, N Hirosaki, M Yoshimura. Subcritical crack growth in sintered silicon nitride exhibiting a Rising *R*-Curve. J Am Cer Soc 73(7):2095–2096, 1990.
218. YW Kim, M Mitomo, N Hirosaki. *R*-curve behavior of sintered silicon nitride. J Mat Sci 30:4043–4048, 1995.
219. BN Cox. Extrinsic factors in the mechanics of bridged cracks. Acta Metall Mater 39(6):1189–1201, 1991.
220. J Rodel. Interaction between crack deflection and crack bridging. J Eur Cer Soc 10:143–150, 1992.
221. T Koyama, A Nishiyama, K Niihara. Effect of grain morphology and grain size on the mechanical properties of Al_2O_3 ceramics. J Mat Sci 29:3949–3954, 1994.
222. H-D Kim, I-S Lee, S-W Kang, J-W Ko. The formation of $NaMg_2Al_{15}O_{25}$ in an α-Al_2O_3 matrix and its effect on the mechanical properties of alumina. J Mat Sci 29:4119–4124, 1994.
223. M Yasuoka, KME Brito, S Kanzaki. High-strength and high-fracture-toughness ceramics in the $Al_2O_3/LaAl_{11}O_{18}$ systems. J Am Cer Soc 78(7):1853–1856, 1995.
224. BA Chandler, EC Duderstadt, JF White. Fabrication and properties of extruded and Sintered BeO. J Nuc Mat 8(3):329–347, 1963.

225. RE Fryxell, BA Chandler. Creep, strength, expansion, and elastic moduli of sintered BeO as a function of grain size, porosity, and grain orientation. J Am Cer Soc 47(6):283–291, 1964.
226. JA Salem, JL Shannon, Jr., RC Bradt. Crack growth resistance in textured alumina. J Am Cer Soc 72(1):20–27, 1989.
227. M Iwasa, EC Liang, RC Bradt. Fracture of isotropic and textured Ba Hexaferrite. J Am Cer Soc 64(7):390–393, 1981.
228. G de With, N Hattu. Mechanical behavior of Sr-hexa ferrite. Proc Brit Cer Soc, No. 32, Engineering with Ceramics (RW Davidge, ed.), 191–198, 1982.
229. CCm Wu, KR McKinney, RW Rice. Strength and toughness of jade and related natural fibrous materials. J Mat Sci 25:2170–2174, 1990.
230. FF Lange. Relation between strength, fracture energy, and microstructure of hot-pressed Si_3N_4. J Am Cer Soc 56(10):518–522, 1973.
231. FF Lange. Fracture toughness of Si_3N_4 as a function of the initial α-phase content. J Am Cer Soc 62(7–8):428–430, 1979.
232. JE Weston. Origin of strength anisotropi in hot-pressed silicon nitride. J Mat Sci 15:1568–1576, 1980.
233. M Ito, S Sakai, Y Yamauchi, T Ohoji, W. Kanematsu, S Ito. A study on strength anisotropy of the fracture origin of HP-Si_3N_4. J Cer Soc Jpn, Int Ed 97:492–495, 1989.
234. F Lee, KJ Bowman. Texture and anisotropy in silicon Nitride. J Am Cer Soc 75(7):1748–1755, 1992.
235. Y Goto, A Tsuge, K Komeya. Preferred orientation and strength anisotropy of pressureless-sintered silicon nitride. J Eur Cer Soc 6:269–272, 1990.
236. W Kim, Y-W Kim, D Ho Cho. Texture and fracture toughness anisotropy in silicon carbide. J Am Cer Soc 81(6):1669–1672, 1998.
237. M Sakai, RC Bradt, DB Fischbach. Fracture toughness anisotropy of a pyrolytic carbon. J Mat Sci 21:1491–1501, 1986.
238. KD McHenery, BG Koepke. Electric field effects on subcritical crack growth in PZT. Fracture Mechanics of Ceramics 5 (RC Bradt, AG Evans, DPH Hasselman, FF Lange, eds.) Plenum Press, New York, 1983, pp. 337–352.
239. S-B Kim, D-Y Kim, J-J Kim, S-H Cho. Effect of grain size and poling on the fracture mode of lead zirconate titanate ceramics. J Am Cer Soc 73(1):561–563, 1990.
240. H-T Chung, B-C Shin, H-G Kim. Grain-size dependence of electrically induced microcracking in ferroelectric ceramics. J Am Cer Soc 72(2):327–329, 1989.
241. JW Halloran, EC Skaar. Poling stresses in lead zirconate titanate ceramics:part 1:measurement of residual poling stresses. Unpublished manuscript, 1981.
242. EC Skaar, JW Halloran. Poling stresses in lead zirconate titanate ceramics:part 2:the effect of poling stresses on strength. Unpublished manuscript, 1981.
243. Z Zhang, R Rai. The influence of grain size on ferroelastic toughening and piezoelectric behavior of PZT. J Am Cer Soc 78(12):3363–3368, 1995.
244. CS Lynch. Fracture of ferroelectric and relaxor electroceramics:influence of electric field. Acta Material. 46(2):599–608, 1998.
245. S Wan, KJ Bowman. Thermal depoling effects on anisotropy of lead zirconate titanate materials. J Am Cer Soc 81(10):2717–2720, 1998.

246. QY Jiang, LE Cross. Effects of porosity on electric fatigue behavior in PLZT and PZT ferroelectric ceramics. J Mat Sci 28:4536–4543, 1993.
247. Q Jiang, W Cao, LE Cross. Electric fatigue in lead zirconate titanate ceramics. J Am Cer Soc 77(1):211–215, 1994.
248. GS White, AS Raynes, MD Vaudin, SW Freiman. Fracture behavior of cyclically loaded PZT. J Am Cer Soc 77(10):2603–2608, 1994.
249. MD Hill, GS White, CS Hwang. Cyclic damage in lead zirconate titanate. J Am Cer Soc 79(7):1915–1920, 1996.
250. ET Park, JL Routbort, Z Li, P Nash. Anisotropic microhardness in single-crystal and polycrystalline $BaTiO_3$. J Mat Sci 33:669–673, 1998.
251. S Suresh. Fatigue crack growth in brittle materials. J Hard Mat 2(1–2):29–54, 1991.
252. RH Dauskardt. A frictional-wear mechanism for fatigue-crack growth in grain bridging ceramics. Acta Metall Mater 41(9):2756–2781, 1993.
253. MJ Hoffman, YW Mai, RH Dauskardt, J Ager, RO Ritchie. Grain size effects on cyclic fatigue and crack-growth resistance behavior of partially stabilized zirconia. J Mat Sci 30:3291–3299, 1995.
254. J Healy, AJ Bushby, YW Mai, AK Mukhopadhyav. Cyclic fatigue of long and short cracks in alumina. J Mat Sci 32:741–747, 1997.
255. H Kishimoto, A Ueno, S Okawara, H Kawamoto. Crack propagation behavior of polycrystalline alumina under static and cyclic load. J Am Cer Soc 77(5): 1324–1328, 1994.
256. DC Larson, LR Johnson, JW Adams, APS Teotia, LG Hill. Ceramic materials for advanced heat engines technical and economic evaluation. Noyes, Park Ridge, NJ, 1985.
257. RE Wright. Acoustic emission of aluminum titanate. J Am Cer Soc 55(1):54, 1972.
258. Y Ohya, Z-E Nakagawa, K Hamono. Grain-boundary microcracking due to thermal expansion anisotropy in aluminum titanate ceramics. J Am Cer Soc 70(8):C-184–186, 1987.
259. LJ Graham, GA Alers. Microstructural aspects of acoustic emission generation in ceramics. Fracture Mechanics of Ceramics 1 (RC Bradt, DPH Hasselman, FF Lange, eds.). Plenum Press, New York, 1974, pp. 175–188.
260. Z Xiaoli, W Chongmin, Z Hongtu. Fracture toughness and acoustic emission in silicon nitride. J Mat Sci Lett 6:1459–1462, 1987.
261. DA Rega, DK Agrawal, C-Y Huang, HA McKinstry. Microstructure and microcracking behavior of barium zirconium phosphate ($BaZr_4P_6O_{24}$) ceramics. J Mat Sci 27:2406–2412, 1992.
262. T Kishi, S Wakayama, S Kohara. Microfracture Process During Fracture Toughness Testing in Al_2O_3 Ceramics Evaluated by AE Source Characterization. Fracture Mechanics of Ceramics (RC Bradt, AG Evans, DPH Hasselman, FF Lange, eds.). Plenum Press, New York, 1986, pp. 85–100.
263. C Sklarczyk. The acoustic emission analysis of the crack process in alumina. J Eur Cer Soc 9:427–435, 1992.
264. LC Sawyer, M Jamieson, D Brikowski, MI Haider, RT Chen. Strength, structure, and fracture properties of ceramic fibers produced from polymeric precursors. J Am Cer Soc 79(11):798–810, 1987.

265. RW Rice. Fractographic determination of K_{IC} and effects of microstructural stresses in ceramics. fractography of glasses and ceramics, ceramic trans. 17 (JR Varner, VD Frechette eds.). Am. Cer. Soc., Westerville, OH, 1991, pp. 509–545.
266. Vega-Boddio, J Schweitz, O Vingsbo. Surface energy measurements from Griffith cracks in boron Fibers. J Mat Sci 12:1692–1693, 1977.
267. J Lipowitz, JA Rabe, RM Salinger. Ceramic fibers derived from organosilicon polymers. Handbook of Fiber Science and Technology:Vol. III High Technology Fibers, Part C, Chapter 4 (M Lewin, J Preston, eds.). Marcel Dekker, New York, 1993, pp. 207–273.
268. MAH Donners. Fracture of MnZn ferrites. Proefschrift, Technische Universiteit Eindhoven, 1999.
269. K Tanaka, Y Kitahara, Y Ichinose, T Iimura. Fracture analysis of single crystal manganese zinc ferrites using indentation flaws. Acta Metal 32(1):1719–1729, 1984.
270. JJ Swab, JC LaSalvia, GA Gilde, PJ Patel, MJ Motyka. Transparent armor ceramics:AION and spinel. Cer Eng Sci Proc to be pub. in 2000.
271. S Park, C-T Sun. Fracture criteria for fiezoelectric ceramics. J Am Cer Soc 78(6): 1475–1480, 1995.
272. R Fu, T-Y Zhang. Effects of an electric field on the fracture toughness of poled lead zirconate titanate ceramics. To be published in J Am Cer Soc
273. B Rockower, E Hall. Investigation of Physical Properties of Beryllium Oxide as a Material for Integrating Gyros, MIT Instrumentation Lab. Report R-560, 9/1966, presented at the 1966 Fall Meeting. Electronics Div Am Cer Soc.
274. A G Imgram, M E Hoskins, J H Sovick, R E Maringer, F C Holden. Study of Microplastic Properties and Dimensional Stability of Materials. Battelle Mem. Inst. Tech. Report AFML-TR-67-232, Part II, for Air Force contract AF 33(615)-5218, Project 7351, 8/1968.
275. J W Moberly. Tensile Microstrain Properties of Telescope Mirror Materials. Stanford Res. Inst. Final Report for NASA contract NAS1-9982, 6/1971.

3

Grain Dependence of Ceramic Tensile Strengths at ~ 22°C

I. INTRODUCTION

A. Background

It has been known for years that grain effects play an important role in the tensile, hence flexure, strength (σ) of ceramics. Since grain size (G) is the most important and pervasive variable in both improving and understanding tensile strength, various G functions have been used to show strength decreases with increasing G. Some investigators opted for log σ vs. log G plots, e.g. Knudsen [1] but this compresses data at large G (obscuring important σ–G changes there). An inverse G function is preferred, since this allows use of data for all grain sizes, including single crystals ($G = \infty$), though with some large-G data compression. For mechanistic reasons discussed below, plotting σ vs. $G^{-1/2}$ has become standard, but there is still some confusion and uncertainty concerning specific parameters and interpretations. Related issues of using average or maximum G values and branch intersections and slopes, particularly at finer G, whose clarification [2, 3] is addressed later, need broader recognition and use. The volume fraction porosity (P) and the shape and spatial variations of porosity, grains, and other phases can be important, particularly in view of the challenge of obtaining a significant range of specimens in which G is the primary factor determining σ.

Before proceeding to discussing mechanisms and presentation and evaluation of data it is useful to consider briefly the history of investigating and understanding grain size effects on the strength behavior of ceramics. To obtain reasonable data several limitations had to be overcome, often in an iterative fashion, namely effects of second, e.g. glassy, phases in many oxide ceramics such as Al_2O_3, and varying pore and flaw populations, and their interactions, and limitations and variations of grain size, shape, and orientation. Problems of these limitations were compounded by many studies focusing on one particular mechanism or one particular aspect of a variable, e.g. G distribution, often with a limited range of testing and evaluation, and failing to apply some necessary basic tools, especially fractography. An important step was sorting out the main porosity effects [1–6], thus enhancing identification of grain size effects.

Another major help was improvements in raw materials and processing, especially the introduction and increasing use of hot pressing, to give bodies with little or no porosity and second phases with either finer or larger G [7–11], thus better defining both branches (Fig. 3.1). Larger G behavior generally approximated a $G^{-1/2}$ dependence of strength, suggesting grains being the flaws (recognizing that flaw size, c, is measured by a radius and G by a diameter). However, this relation must be limited, since it implies strength extrapolating to 0 at $G = \infty$, neglecting single crystal strengths generally being similar to many, and higher than some, polycrystalline bodies, as is discussed later. Finer grain samples typically showed some, but much less, grain size dependence than larger grain samples, i.e. laying mostly or completely along the finer grain branch (Fig. 3.1), hence projecting to a positive strength intercept at $G = \infty$ (i.e. $G^{-1/2} = 0$), commonly of the order of magnitude as for single crystal specimens with the same surface finish. The combination of such a positive intercept along with increasing observations of slip or twinning on a micro-, or in some cases a macro-, scale suggested possible microplastic crack nucleation, growth, or both. Carniglia's surveys [8, 9] showed that when a sufficient G range was covered, both regimes of behavior were typically observed as a continuous function of G, showing finer G strengths decreasing with increasing G, extrapolated to $\sigma > 0$ at $G = \infty$, i.e. at $G^{-1/2} = 0$. Intermediate to large G samples showed greater σ decrease with increasing G, generally indicating extrapolation to $\sigma \sim 0$ at $G^{-1/2} = 0$.

Carniglia suggested that the failure mechanism for the finer grain branch was microplasticity, hence referring to this as the Petch or Hall–Petch regime, and that for the large G branch it was Griffith flaw failure with the flaw dimensions being ~ those of the grain size. The change between mechanisms was attributed to stresses for failure from flaws being higher than the stresses to activate microplastic failure at fine G, until G became sufficiently large to allow flaw failure at stresses below those needed for microplasticity. This provided not only a possible explanation for the two-branch character of the $\sigma - G^{-1/2}$ plots commonly observed but also for variability in the intersection of the $\sigma - G^{-1/2}$ branches. Thus, as the quality of surface finish improved in specimens lacking other major sources of flaws e.g., pores, the larger G branch would extend to finer G before microplasticity took over. A few investigators, such as Kirchner and Gruver [10] and Rice [3, 11–13], suggested contributions of other intrinsic

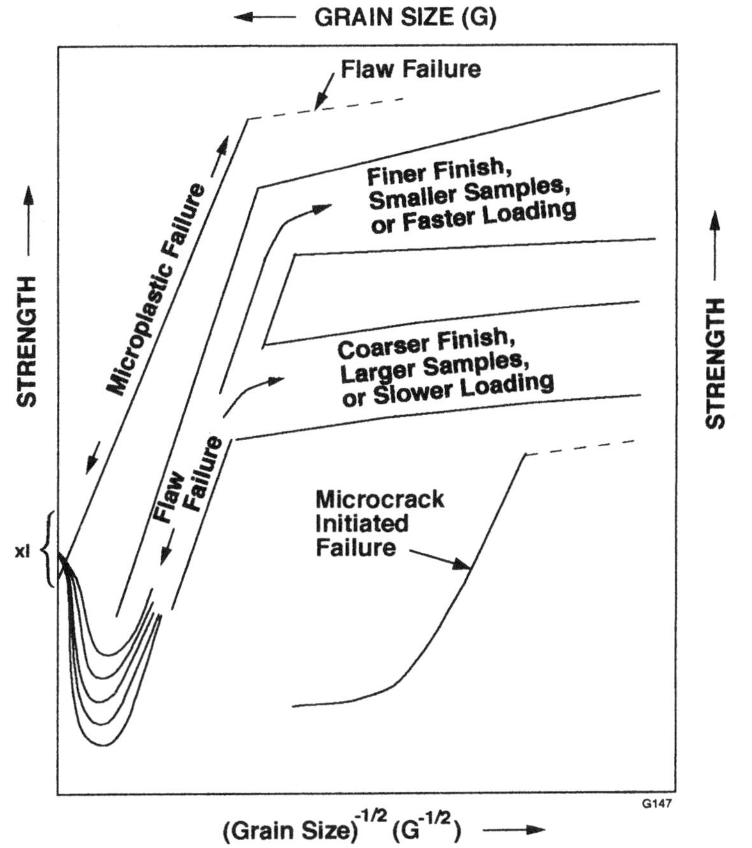

FIGURE 3.1 Schematic of the preexisting flaw model showing the larger *G* branch with strengths decreasing rapidly with increased *G*, falling below single crystal strengths followed by (test and specimen dependent) variable reversal toward single crystal strengths. Finer *G* branches have lower slopes, giving less strength increases with decreasing *G*, and meet the larger *G* branch when $c \sim G/2$. Since the finer *G* branch has flaws > *G*, i.e. not constrained by *G*, it can be more variable, e.g. wider, or may consist of different flaw populations, hence separate branches. Microcrack and microplastic variations shown illustrate similar rates of σ changes, e.g. an initially rapid decrease with the onset of substantial microcracking and then saturation. However, both models are shown at arbitrary strength levels relative to the preexisting flaw model.

factors such as elastic anisotropy (EA) and thermal expansion anisotropy (TEA). Rice, with others, also considered microplasticity (including twinning as a possibility in some materials) but subsequently broadened consideration to include extrinsic factors such as surface finish, preferred orientation, and pore and grain characteristics and heterogeneities (e.g. of larger grains).

As the range of materials, microstructures, and testing broadened, four factors ended consideration of microplasticity as a broad mechanism controlling the G dependence of ceramic tensile strengths. First, the intercept strength levels for finer G branches were typically lower, e.g. by an order of magnitude, than the expected stresses for activating microplasticity in many materials, e.g. Al_2O_3 (one of the key materials in Carniglia's review). Second, while local stresses from EA and TEA can be substantial, they were of uncertain levels and spatial extents to make up the difference between the strength intercepts and the yield stresses. Further, they are only pertinent to some materials and vary substantially from material to material. Third, TEM examination of sharp cracks in single crystal and polycrystalline Al_2O_3 at room temperatures showed no evidence of associated slip [14]. While the possibility of crack-twin relations was not addressed, twinning is pertinent to only a limited and uncertain (more likely noncubic) set of materials. Fourth and most fundamental was subsequent evaluation, with fractography being a key tool, showing that typically both branches result from flaw failure [2, 3, 11–13].

Thus it will be shown that the finer and larger $\sigma - G^{-1/2}$ branches typically arise from changes in the flaw to grain size ratio, with strength levels of both finer and larger G branches being determined by flaw sizes (Fig. 3.1). However, there can be two important variations of the normal flaw controlled $\sigma - G^{-1/2}$ behavior. First, some room temperature failure is determined by microplasticity in a limited number of ceramics and related materials (and becomes somewhat more common and extensive as temperature increases, Chap. 6). However, where this occurs, the large grain branch is the regime of this behavior, not the finer grain branch as proposed by Carniglia. Further, as the stress for microplasticity increases, e.g. as G decreases, there is often a switch to flaw failure when preexisting flaw dimensions are ~ those of the grains, which is identical to the finer grain flaw failure (Fig. 3.1). The other deviation from the now-established normal two-branch flaw controlled strength behavior occurs with substantial microcracking.

Two aspects of $\sigma - G$ behavior should be stressed. First, the above two-branch $\sigma - G^{-1/2}$ behavior is due first to changing c/G ratios and the resultant grain boundary or single crystal polycrystalline toughness transition, and secondarily, at least for machining flaws, to impacts of grain parameters, especially size, and local hardness, elastic moduli, and toughnesses on flaws that are introduced and determine strength. This is contrary to the focus of most mechanical property studies on toughness, as measured by large cracks, as the mechanism of controlling σ and its microstructural dependence. Second, as comparison of this and the previous chapter shows, much of the G dependence of strength is inconsistent with that of large crack toughness, especially at larger G, since much of the larger crack toughening is precluded by the above process.

Beyond the above historical summary of the grain dependence of failure mechanisms, this review first addresses such mechanisms in more detail and then provides an extensive summary of $\sigma - G^{-1/2}$ data, mechanisms, and parameters, updating and extending previous reviews [2, 3, 11, 13]. This starts with materials exhibiting some known or expected microplastic control of strength, often with a transition to flaw failure. Next, $\sigma - G^{-1/2}$ behavior for cubic and then noncubic materials, where strength is generally controlled by preexisting flaws, is presented. In both cases oxide materials are reviewed first. Then pertinent microstructural effects are addressed, including grain shape and orientation as well as moderate levels of impurities and especially additives (bodies having higher levels of second phases are addressed later as composites, Chaps. 8–12). The basic microstructural question of which G to use, specifically the average or the maximum, is also addressed, showing that while tensile failure is a weak link process suggesting use of the maximum G, since it is associated with lower strengths, this is a serious oversimplification. Blind application of this (i.e. without other input, especially fractography) leads to errors and distortions, since grains by themselves are not flaws. At the other extreme, behavior of ceramics with nanoscale grains is discussed, including neglected but often significant effects of second phases commonly left from frequent unusually low processing temperatures used to obtain bodies with very fine grains. Surface finish effects, i.e. of machining flaws and effects versus as-fired surfaces, and test issues such as specimen size and shape along with loading and environmental effects and their interactions, are discussed. Then more detailed evaluation of the $\sigma - G^{-1/2}$ model is given, followed by a comparison of strength-toughness-grain size behavior and discusson of mechanical reliability, primarily via Weibull moduli.

B. Mechanisms, Parameters, and Analysis

Consider now more specific aspects of the $\sigma - G^{-1/2}$ brittle fracture model (Fig. 3.1), whose development and interpretation arose mainly from fractographic study of machining flaws revealing their variation relative to G [2, 3, 11–13]. This showed that, in relatively dense machined ceramics, machining flaws formed in the surface region were the dominant or exclusive source of failure. Such flaw failure in finer G bodies clearly disproved Carniglia's model of microplastic control of the finer G branches. To a first approximation, the depth of such flaws for a given material and machining condition has no dependence on grain size, i.e. did not change significantly in size with G (nor between most ceramics). (However, as discussed later, there is some second-order variation of c

with material and microstructural parameters.) Thus failure of machined finer grain bodies was due to machining flaws whose size, *c*, was > *G*/2 (since flaw size is measured by a radius and grain size by a diameter). Typically two flaw populations are generated in machining, one parallel and the other perpendicular to the direction of the abrasive particle motion, with the depths of both being similar, but the former being substantially elongated and the former being ~ half penny cracks [12, 13, 15–22]. This difference in flaw shape translates to two levels of strength of the finer *G* branches as a function of stress direction relative to the machining direction, again supporting the above model and concepts.

Other flaw sources may have similar or different trends with *G*. An important flaw population that is produced in the body and can change, e.g. shift, $\sigma - G^{-1/2}$ behavior (Fig. 3.1) is that of microcracks. Surface flaws from contact damage or particle impact/erosion may often be similar or ~ proportionate in size to machining flaws, but this needs to be established. Some processing flaws, e.g. some larger pores, may fit this trend, and some flaws may not, e.g. other pores or cracks from thermal stresses (Chap. 6). However, even in such cases, failure to account for flaws causing failure and basing analysis only on toughness can be misleading, e.g. as shown by evaluation of effects of preferred orientation on strength of hot pressed Si_3N_4 (Chap. 2, Sec. III.H).

Another important surface finish condition that often gives similar trends as machining as a function of *G*, at least over the normal *G* range, is failure of samples with as-fired surfaces, e.g. as for some ceramic parts and tests, especially for fibers [2]. This is consistent with Coble's evaluation showing grain boundary grooves from firing (or annealing) acting as sharp flaws whose size is a reasonable fraction of the grain size, e.g. over the range of *G*/15 to ~ *G* [23]. Thus at finer *G* such flaws are < *G*, in part due to the fine *G* and generally less severe grooving from less severe firing conditions, which means that several connected grooves along different boundaries act as the failure causing flaws. As *G* increases, grooving also generally increases, and fewer connected grooves are needed to form a failure causing flaw, reaching a point where failure may occur from a groove along part or all of a single grain, i.e. analogous to the above machining flaw case, as is discussed next.

Typical flaw populations have *c* > *G* at finer *G*, and *c* does not increase substantially as *G* increases, so *c* must become ~, then progressively, < *G*/2 (1/2 arises since *c* is measured as a radius and *G* a diameter) progressing from the finer *G* branch, its intersection with the larger *G* branch when the flaws ~ encompass individual surface grain, then the larger *G* branch where the initial flaw size is < *G*/2. However, while such flaw–grain relations define much of the behavior, there are other issues. The first of three sets of issues are which *G* to use, especially an average, G_a, or the maximum, G_m, as used by some [24–27], and whether there is any *G* dependence in the finer *G* branches, i.e. whether the slope of the finer *G* branch = 0 as claimed by those proposing the use of G_m. It will be shown that the finer G slope of 0, while possibly having pertinence in some limited cases, is not the general case, since there is generally some to substantial decrease in strength as G increases along finer G branches. Second is that while polycrystalline fracture toughness values are appropriate at finer G, a transition to single crystal (Figure 2.15) or grain boundary toughnesses must occur as the intersection of the large and finer G branches is approached or reached [28]. The slope of the larger G branch is later shown to be between polycrystalline and single crystal toughness values. The third issue is contributions of intrinsic effects, such as slow crack growth, EA and TEA, and extrinsic effects, e.g. of secondary phases. Several, or all, of these may affect both branches and their intersection(s), e.g. SCG due to differing effects of grain size and of inter- versus transgranular fracture, but they are also probably important in a less recognizd but clearly established occurrence, namely large G strengths lower than those of the lowest strength crystal orientation with comparable surface finish. Thus as G increases so failure can progressively only occur from fewer and fewer individual grains or grain boundaries, strengths can decrease below those of single crystals with comparable surface finish due to effects of TEA and EA stresses, as well as residual pores or phase, especially at grain boundaries [2, 3, 13]. Such large G effects are likely to depend substantially on specimen size and shape as well as the nature of the stress, e.g. true tension versus flexure, since these will affect the limited statistics of grain structure impacting failure.

While some of the above effects at large G have limited documentation (presented later), there is clear evidence of significant effects of extreme TEA at finer G. Thus an important flaw population that leads to a substantial variation in the σ–$G^{-1/2}$ is for bodies with extensive microcracking, such as that resulting from high TEA, e.g. in $MgTi_2O_5$ and Fe_2TiO_5 (Figs. 3.1, 3.23). Subsequent microcracking results in much faster decreases in strength as G increases above the critical value G_s for microcracking [per Eq. (2.4)], since the effective flaw size is increasing due to more and larger microcracks as G increases, but then the decrease begins to saturate. Such effects are typically paralleled by similar initially rapid and then decreasing reductions in elastic moduli, again due to saturation of the microcracking process.

The above fractographic studies and resultant flaw model development partly overlapped with conceptual arguments based on σ–G data. Thus aspects of this model, i.e. finer G branches due to $c > G/2$, the higher slope large G branch, and higher single crystal strengths than for some large G bodies were schematically indicated by Emrich in his review of crystallized glasses [29]. Similarly, finer and larger G branches both being due to flaw failure with the two branches intersecting when $c \sim G/2$ was proposed by Rhodes and Cannon [30]. Bradt, Tressler, and students [24–26] also subsequently discussed such a model in conjunction with their surface finish studies.

The other basic, but less common, failure mechanism is by microplastic crack nucleation and possible slip assisted crack growth. This also has a distinct *G* dependence, since stress levels for it decrease with increasing slip band length, the maximum of which in a polycrystalline body is normally *G*, since grain boundaries are generally both sources and barriers to dislocations. This inverse relation between slip band length (hence also grain size) and stresses for crack nucleation results because blockage of a slip band at a barrier generates a back stress resisting dislocation motion from the source into the barrier. The effect of the back stress on the dislocation source decreases with increasing source–barrier separation, hence with *G*. The mathematical relation for nucleation of a crack by a slip band extending across a grain being blocked at the opposite grain boundary is the Hall–Petch relation:

$$\sigma_f = \sigma_y + kG^{-1/2} \tag{3.1}$$

where σ_f = the fracture stress (in materials where there is insufficient bulk ductility), σ_y = the yield stress, i.e. the tensile stress to activate the easiest activated slip system in single crystals of comparable composition, crystal perfection, and surface finish, and k a constant. Note (1) both σ_y and k are clearly dependent on the specific material and on the absence or degree of grain orientation relative to the stress axis, and (2) Eq. (3.1) is very similar to the Griffith equation, Eq. (2.2), when the grain and flaw sizes are the same or similar, i.e. in such cases the primary difference is σ_y, which was a major source of confusion, as was discussed earlier. Twin nucleation of cracks or assistance in their growth, though more restricted in the number of materials it is applicable to, is more complex, since twins (though often thin) are three-dimensional defects rather than linear ones as dislocations are. However, twin caused failure is expected to follow the same or similar *G* dependence as Eq. (3.1), since *G* again is typically the upper limit to the scale of twin lengths. Note that microplastic initiated failure can be variable, e.g. due to variable *G* and σ_y, which may increase or decrease σ_f as well as variable grain boundary character. Slip assisted growth of cracks (e.g. from machining) is a transition between slip initiated and preexisting flaw failure, which lowers strengths. While the latter would give some variation similar to flaw initiated failure, microplastic initiated failure should have less extreme variations.

Thus, in summary, the dominant failure mechanism for ceramics is from preexisting flaws, such as from as-fired and especially machined surfaces, giving the σ–$G^{-1/2}$ behavior of Fig. 3.1. Other flaw populations can deviate from this, e.g. flaws from microcracking giving faster initial strength decreases at finer *G* and less at larger *G*. Another fundamental variation is with microplastic initiated or assisted failure, which typically results in cracks in, or along boundaries of, grains. The key distinction between this and typical brittle failure from preexisting flaws from a σ–$G^{-1/2}$ plot is that strengths of the latter at larger *G* commonly extrapolate to σ values well below any realistic σ_y values, e.g. 0, while microplastic failure extrapolates to σ_y at $G= \infty$ ($G^{-1/2} = 0$). Thus it is typical for larger grain bodies having flaw failure at strengths less than those for comparably finished single crystals oriented for the lowest strength. Conversely, data reflecting microplastic caused failure have large grain strengths that extrapolate to σ_y and thus do not fall much below single crystal strengths. Brittle fracture initiated by microplasticity is typically in competition with failure from preexisting flaws, with slip assisted crack growth being a transition between these two mechanisms. Flaw failure is favored by higher stresses for finer G fracture and larger preexisting flaws, e.g. from harsher machining, handling, etc., or increased severity due to second phases or pores. Note that as G decreases, the stress for microplastic failure increases in a similar fashion as for flaw initiated failure increases (e.g. since both are ultimately controlled by cracks). Thus it is both logical and observed that at some finer G, competition between microplastic and flaw initated failure shifts depending on material and surface finish. However, in the absence of significant extrinsic effects, this change should also occur when the size of microplastically introduced crack is ~ $G/2$, i.e. as for the flaw model (Fig. 3.1).

II. CERAMICS AND RELATED MATERIALS WITH KNOWN OR EXPECTED MICROPLASTIC INITIATED FAILURE

Some softer, less refractory ceramics and related materials often exhibit macroscopic yield (Fig. 3.2A). NaCl macroscopic yield stresses of Stokes and Li [11, 31] lie on, or slightly below, while their fracture stresses lie slightly above, that of KCl [32]. CdTe may or may not exhibit yield before fracture [33–35]. Somewhat harder, stronger materials like PbTe [36] and CsI [37], while not showing macroscopic yield, extrapolate to σ values (at $G= \infty$) that are probably ~ their yield stresses. Similarly, ZnSe [38, 39] (large G) extrapolation to σ ~ 15–30 MPa at $G= \infty$ (Fig. 3.2B) may indicate microplastic control of σ (e.g. by twinning). Purity, processing, and testing are all probably factors in σ differences.

Previous data [40–43] for hot pressed and sintered (commercial) $BaTiO_3$ shows two σ–$G^{-1/2}$ branches (Fig. 3.3A), with the large G data for unalloyed (and some alloyed) samples extrapolating to single crystal strengths [41]. This is taken as evidence of microplastic controlled failure given yield shortly preceding fracture of ("butterfly") $BaTiO_3$ crystals and associated substantial birefringence around their fractures (Fig. 3.3B). The higher σ of $BaTiO_3$ hot pressed with 1% LiF + 2% MgO at larger G (hence higher heat treatment for more homogenization) also supports microplastic control via indicated alloying effect [41], as do other indications of microplasticity in $BaTiO_3$ [44, 45]. Statistical analysis of much of the data shows the fine G branch having a slope > 0, both above and below the Curie temperature [42], but with higher strengths (by ~ 50%) above the

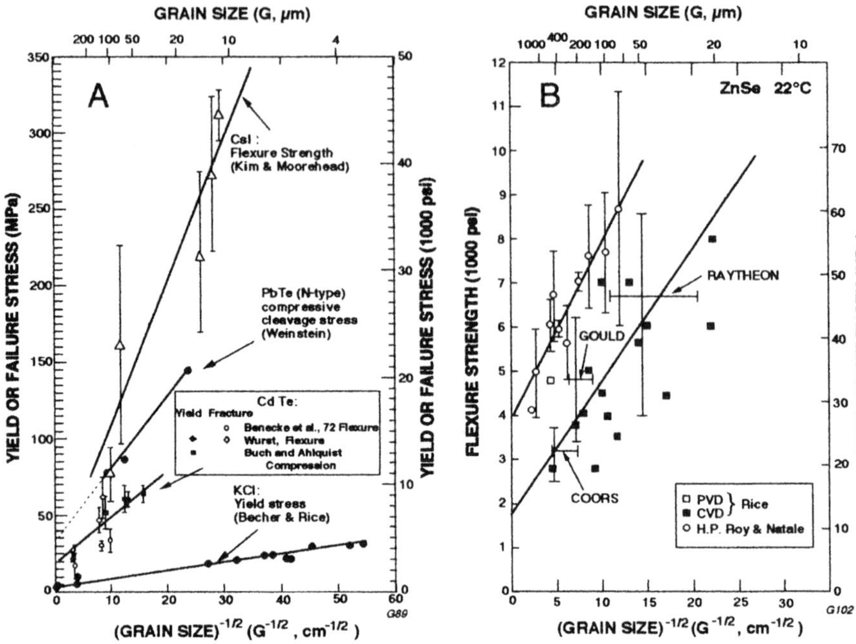

FIGURE 3.2 Yield or failure stress versus $G^{-1/2}$ for materials of known or probable microplastic controlled failure (A) CsI [37], KCl [32], PbTe [36], and CdTe [33]; (B) ZnSe data of Roy and Natale for various hot-pressed (H) and PVD and CVD bodies from a previous survey [38, 39]. Vertical and horizontal bars represent standard deviations.

Curie temperature where it is in a cubic structure with no microstructural stresses.

Strengths of dense CaO, from hot pressing [46] or recrystallizing single crystals (via press forging or hot extruding) [47, 48] all gave single $\sigma - G^{-1/2}$ branches, with different slopes for different bodies (Fig. 3.4). All extrapolate to σ = 20–35 MPa at $G = \infty$, reasonably consistent with macro yield stresses of 40–45 MPa for CaO crystals having some probable solution hardening, suggesting probable microplastic control of strength. This is reinforced by some strength increase for specimens tested with as-sanded (and more frequent internal fracture origins, Fig. 3.4B) versus as-annealed surfaces indicating surface work hardening and tests at ~ 1100 and 1300°C showing limited strength decreases, frequently with macroscopic yield preceding fracture, but maintaining transparency [47, Chap. 6]. It is also indicated by fractography studies of recrystallized CaO single crystal samples (with ~ 100% transgranular fracture, Fig. 3.4B-D) and

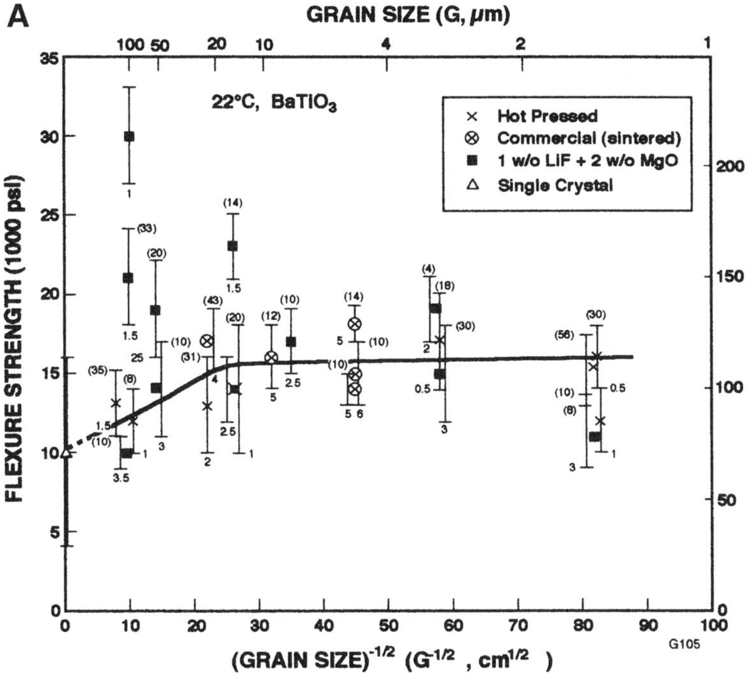

FIGURE 3.3 $BaTiO_3$ data indicating microplastic control of strengths at larger grain size in dense bodies. (A) strength versus G^{-12} for hot-pressed (with or without additions of LiF + MgO as shown), sintered (commercial, with machined surfaces), and single crystal bars. Vertical bars are standard deviations; numbers in parentheses represent the number of tests, and numbers at the bottom of the bars the % porosity. Note data for pure bodies extrapolating to single crystal strengths (which are shortly preceded by yielding). (Data after Rice, Pohanka, and colleagues [40–43].) (B) and (C) Transmitted, polarized light, lower and higher magnification micrographs of failed crystal bend bar showing substantial birefringence of respectively a coarser (B) and fine (C) scale indicative of plastic deformation (the X cut by the fracture in B) was associated with observed yielding preceding failure [41].

probable slip bands at fracture origins [46], similar to (unetched) origins in MgO (Fig. 3.6), including their internal nature indicating work hardening of surface grains from machining [48, 49].

Substantial testing of (transparent) MgO as hot pressed, annealed, or hot extruded (giving a <100> axial texture) showed primarily single $\sigma - G^{-1/2}$ branches with significant slope increases in the respective order listed [48, 50–52] (Fig 3.5). All extrapolate to $\sigma \sim 70$ MPa at $G = \infty$, less than typical single

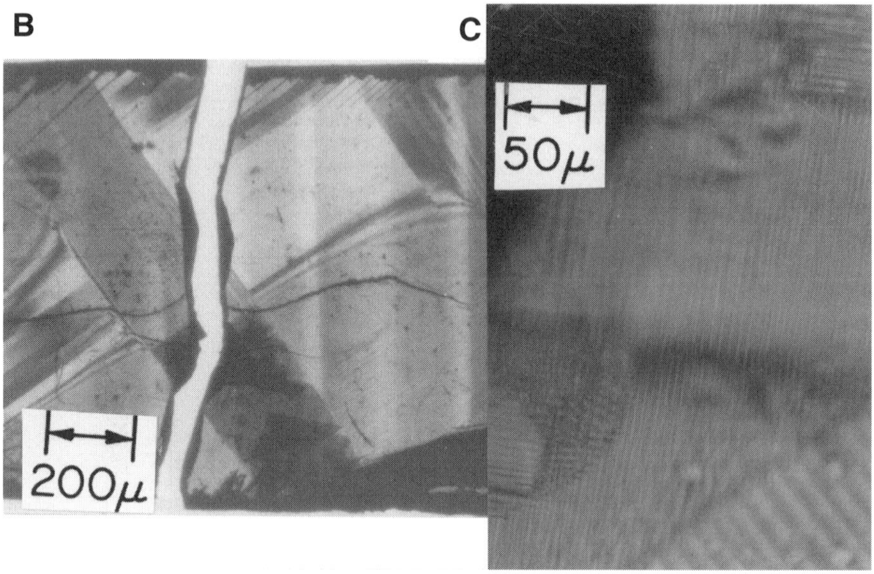

FIGURE 3.3 Continued.

crystal flexure strengths of ~ 150 MPa, but consistent with higher purity crystal yield stresses of ~ 75 MPa [46, 48]. Such indications of microplastic controlled strength at large *G* are reinforced by fractographic observations showing internal failures (attributed to work hardened surface grains) [48–52], and especially substantial identification of slip bands blocked at fracture origins on both unetched and etched fractures (Figs. 6A and B). This data is generally consistent with substantial data for hot pressed and annealed MgO [25, 53–55] and sintered MgO [56, 57], mostly with machined surfaces, which in turn reinforces the above trends and mechanisms. Thus Evans and Davidge's (transparent) hot pressed and annealed MgO with chemically polished versus as-sawn surfaces [55] showed respectively higher and lower σ at larger *G* reflecting respectively microplastic and flaw failure as shown by respective intercepts at ~ the yield stress and 0. Harrison's dense [56], sintered MgO with various machining straddles the hot pressed and annealed curves of Rice with a trend for higher strength with finer surface finish, as well as probable branching to less *G* dependence of σ at finer *G* (e.g., at <~ 10 μm) due to a transition to flaw failure at finer *G*. Spriggs and Vasilos' [53, 54] and Nishida et al.'s [57] sintered MgO data support this, indicating a decreased, but >0, finer *G* branch slope. Such a transition is also indicated by effects of additives [52], and greater reduction of machined versus annealed strengths at finer *G* (5–10 μm). Subsequent studies [15] clearly showed (1) sig-

A

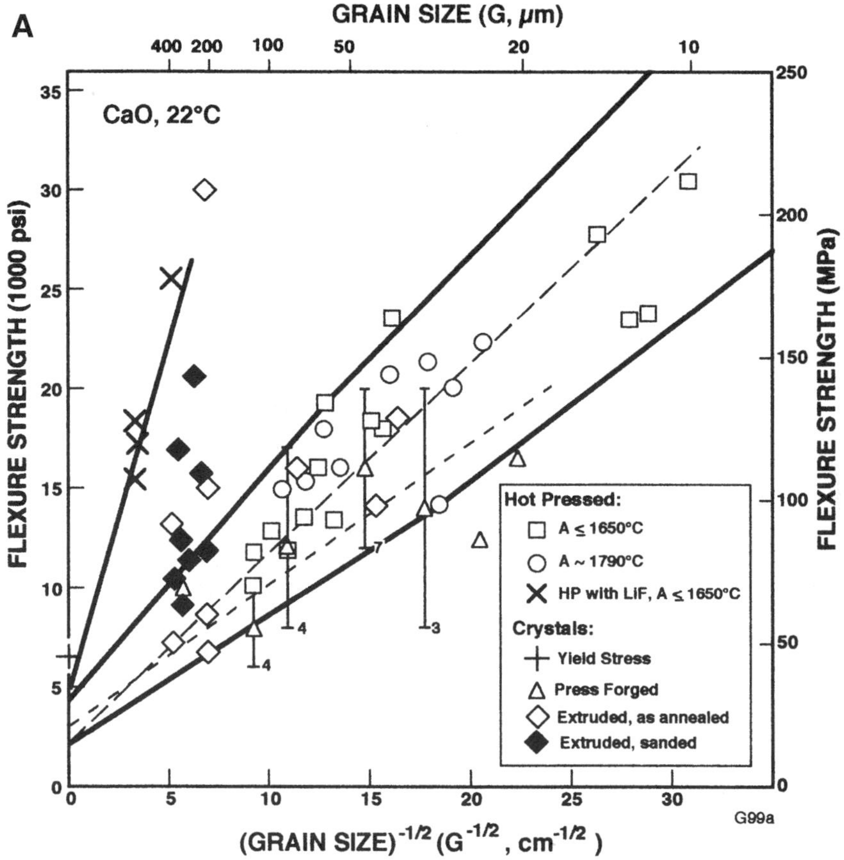

FIGURE 3.4 Strength and fracture of dense CaO indicating microplastic controlled strength at ~ 22°C. (A) Strength versus $G^{-1/2}$ for various bodies hot-pressed [46] (with or without LiF additions) with various annealing as well as of single crystals and bodies recrystallized from these by press forging or hot extrusion [47, 48]. Note intersection of the stress axis at ~ 30–40 MPa in good agreement with the yield stress for CaO crystals, with a higher slope for the extruded material (tested parallel with the <100> axial texture), and similar results for bars tested with as-annealed or annealed and sanded surfaces. (B–D) Fracture origins of dense, recrystallized CaO test bars. In (B) note the internal fracture origin (attributed to surface work hardening) and mist and hackle starting somewhat before the crack reached the grain boundaries. (C) and (D) Fracture origins at or near the surface in surface grains. In (C) note the fracture step (at ~ 45° to the surface) indicating a slip band, i.e. similar to MgO failures (where slip was corroborated by dislocation etching which does not exist for CaO), and the onset of mist and hackle when the crack ~ reached the grain boundary in this smaller G body. In (D), a larger G body; note the onset of mist and hackle ~ 1/2, way through the larger grain of origin.

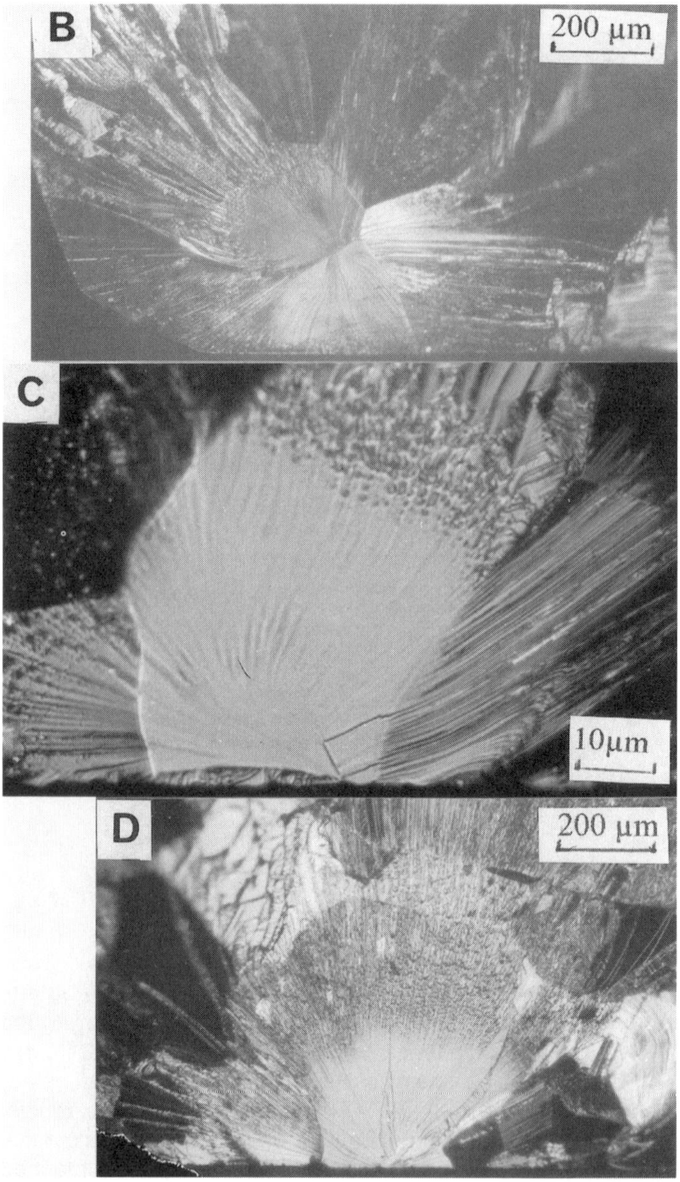

FIGURE 3.4 Continued

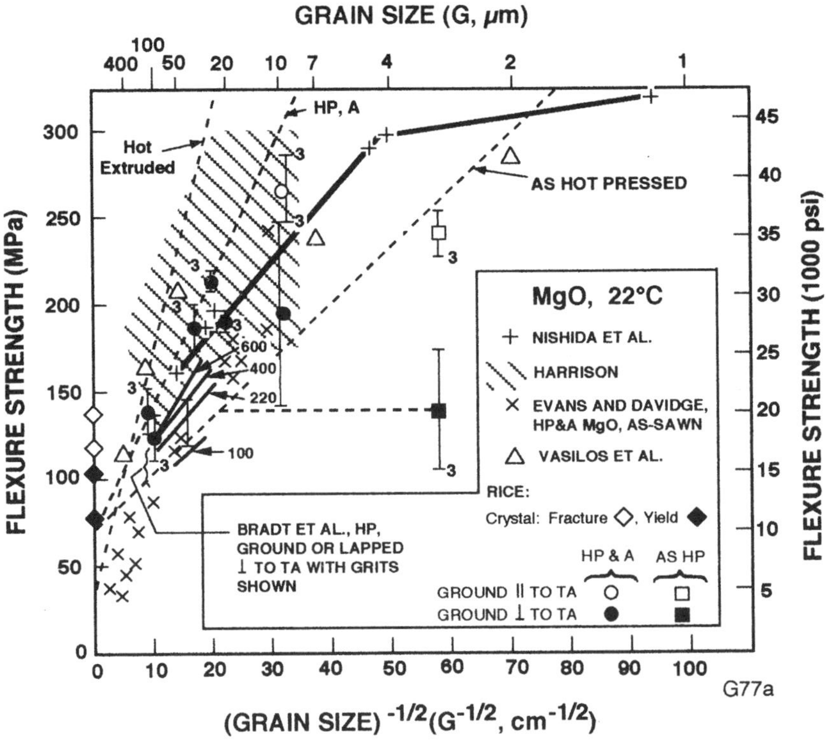

FIGURE 3.5 Strength of dense MgO versus $G^{-1/2}$ from various studies, including substantial earlier data of Rice for as-hot-pressed (AS HP), hot pressed and annealed (HP & A) and hot extruded [48, 50–52] (all with as-sanded surfaces); earlier data of Vasilos and colleagues [53, 54] as well as data for single crystals as-cleaved, machined, or recrystallized [47, 48] or translucent-transparent bodies as hot-pressed (with different grinding directions [13] relative to the tensile axis, TA) or hot-pressed and annealed. Also shown are hot pressed data of Bradt et al. [25] (ground or lapped perpendicular to the tensile axis with grits as shown) and Evans and Davidge [55] for transparent MgO (with as-sawn surfaces); and sintered data of Harrison [56] (with various surface finishes), as well as more recent data of Nishida et al. [57].

nificant σ effects of grinding direction, especially at finer *G* (3–20 μm), and (2) much lower slope (and lower σ) finer *G* branches. Such finer *G* branches were further (1) shown by fractography separating edge and tensile surface failures [17] (i.e. similar to Al_2O_3 studies) and (2) indicated in more recent analysis of MgO data, especially that of MgO with second phases (much of them at grain boundaries) [52]. Similarly Bradt et al.'s [25] tests of hot pressed MgO ground

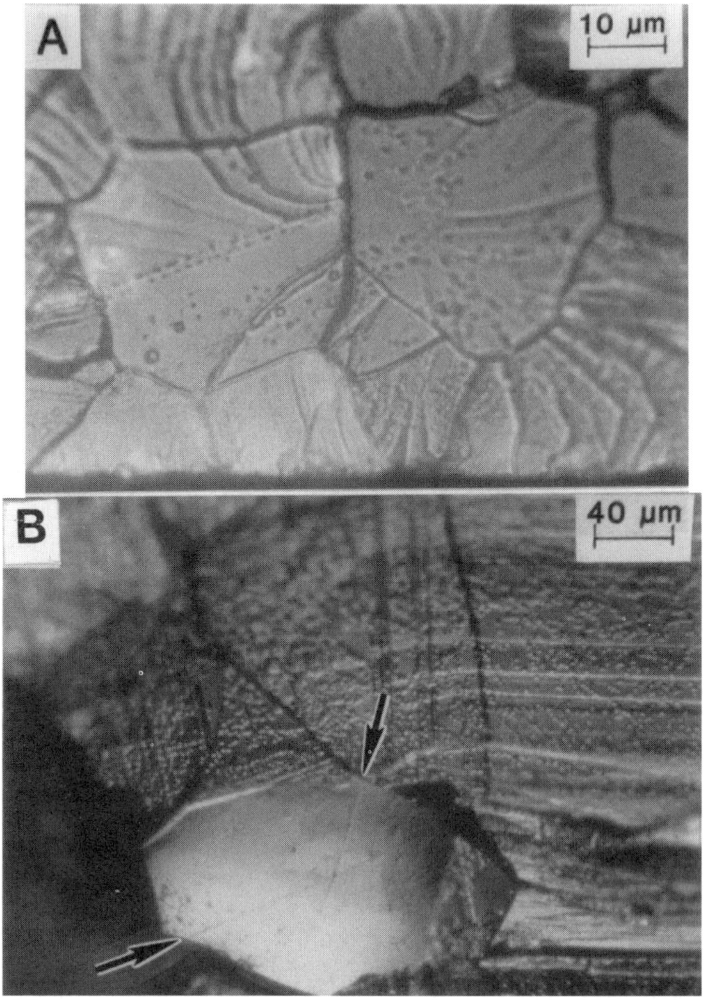

FIGURE 3.6 Fractographs of recrystallized MgO samples showing fracture initiation associated with slip bands blocked at grain boundaries. (A) Fracture origin from two transgranular fractured grains showing etched slip band across the left grain terminating at the origin, which is internal, as was typical for machined surfaces due to work hardening of the surface grains. (B) Less common origin from grain boundary surface, again with associated slip bands (arrows) and internal location. (From Ref. 52, *After Rice* [52] published with the permission of the *Journal of the American Ceramic Society.*)

parallel or perpendicular to the tensile axes with various grit sizes again indicate lower strengths with coarser grits, especially at finer *G*. These all indicate a transition to preexisting flaw control of strength at finer *G* and more competition between failure from such flaws and microplasticity at larger *G*, with harsher surface finish and more residual porosity or other phases shifting the balance in favor of flaws.

Extensive fractography of specimens from hot deformed, recrystallized MgO showed fracture origins from slip bands (Fig. 3.6). Some of these were implied based on expected association with cleavage steps [48, 52]. Etching corroborated such cleavage step–slip band association as well as frequent internal fracture origins being due to work hardening of surface grains due to mechanical finishing [49].

III. CUBIC CERAMICS AND RELATED MATERIALS FAILING FROM PREEXISTING FLAWS

A. Cubic Alkaline Earth Halides and Oxides

While most, if not all, dense alkali halides show microplastic controlled strength (and often some macroplasticity), alkaline earth halides have shown no evidence of this at room temperature [58]. Single crystals and large *G* bodies that have been investigated, mainly CaF_2 and some BaF_2 and SrF_2, have shown three characteristics of failure from preexisting flaws. The first is greater strength sensitivity to surface finishing than expected for microplastic controlled strength, e.g. doubling single crystal and fully dense (press forged) polycrystal strengths [59, 60]. The second is substantial identification of flaws at fracture origins that were consistent with fracture stress and toughness (~ 0.4 $MPa{\cdot}m^{1/2}$) values [61]. Third is some increase in strengths with limited decreases in *G* (e.g. to 100–200 μm) as well as some strengths higher and lower than single crystals at large *G* (~ 1 cm in test bars from fusion cast billets with bar dimensions ~ 1/2 cm). Fractography showed clear cases of single crystal fracture mirrors, i.e. mist and hackle formation within a single grain (Fig. 3.7) [61], showing that failure occurred entirely within such a large grain rather than propagating subcritically into surrounding grains.

Limited data exists for some other common single oxide (machined) ceramics that is not inconsistent with the flaw model or provides some support for it. Thus NiO strength data of Spriggs et al. [62] and of Harrison [63] individually are not inconsistent with the flaw model (but are difficult to compare since they used respectively diametral and flexure tests). However, both showed clear deviations to lower strengths at finer grain sizes achieved, e.g. respectively 0.5–1.5 and < 5 μm by higher pressure hot pressing or hot pressing versus sintering. Thus Spriggs et al.'s bodies hot pressed at higher pressures (70–140 MPa) only had

FIGURE 3.7 Optical fractograph of a higher strength, larger grain, dense CaF_2 from a surface (probably machining) flaw near the test bar edge, photo bottom center, at start of edge rounding. Note grain boundary near right photo edge and the onset of mist and hackle within the grain of origin.

about 2/3 the strength, ~ 240 MPa (P ~ 0, G ~ 1.5 μm) versus samples hot pressed at lower pressures, despite the latter having larger G, e.g. ~ 3-fold. Such lower strengths with finer G as higher hot pressing pressures were used were also shown by Spriggs et al. [64] for other oxide ceramics, Al_2O_3, Cr_2O_3, TiO_2, and MgO, as is discussed in Sec. V.A.

ThO_2 data of Knudsen [65] illustrates a typical problem with much, particularly earlier, data, namely substantial P, especially at finer G. However, correction to P= 0 gives results reasonably consistent with the flaw model, e.g. suggesting a finer G branch of lower but >0 slope transitioning to the larger G branch at G ~ 10–15 μm (Fig. 3.8). Even more limited uniaxial flexure data [28, 67] for dense, transparent sintered Y_2O_3 shows decreasing strength with increasing G, as does biaxial flexure data of Rhodes et al. [68]. Both studies clearly show failure initiation in larger G bodies occurring from machining flaws substantially smaller than the grains, even with the size of surface grains typically being substantially reduced by machining (Fig. 3.9A and B). On the other hand,

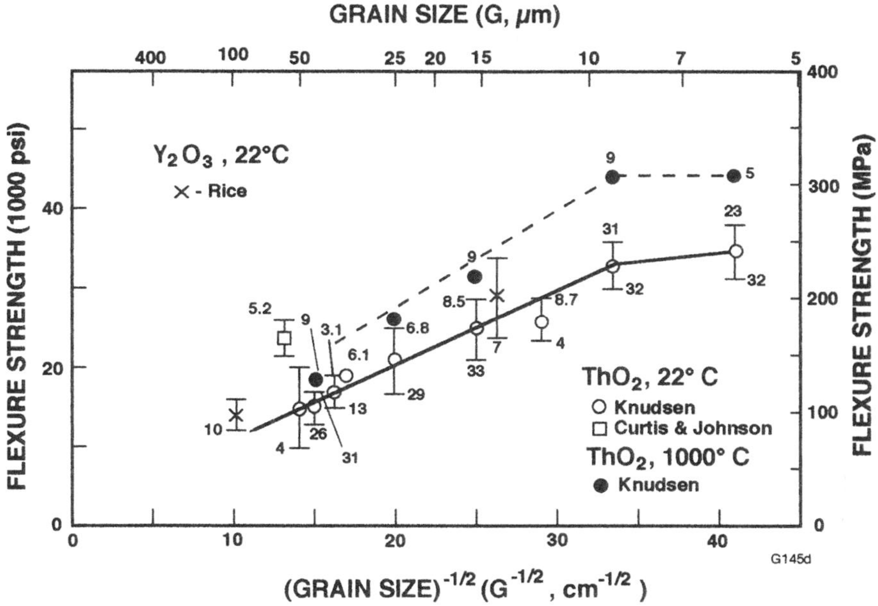

FIGURE 3.8 Strength versus $G^{-1/2}$ for sintered ThO_2 at 22°C of Knudsen corrected for variable porosity (% shown as superscripts) using the exponential relation e^{-bP}, with b ~ 4.4 per his determination [65]. Vertical bars = standard deviations, subscripts = number of tests averaged. Note good consistency between various data points of differing *P*, as well as reasonable agreement with the one data point of Curtis and Johnson [66]. Also included are data on the same ThO_2 at 1000°C [65] and for two grain sizes of dense (transparent)Y_2O_3 at 22°C [28, 67]. Bars are the standard deviations, and subscripts or numbers adjacent to data points the number of tests.

fractography of the finer grain body showed machining flaw fracture origins somewhat larger than *G*, indicating they are from the finer *G* branch, but near its intersection with the larger *G* branch.

Diametral compression strengths of UO_2 by Kennedy and Bandyopadhyay [69] for *G* ~ 2–20 μm showed a finer *G* branch with limited, but >0, slope, and probably transitioning to the larger *G* branch at *G* ~ 10–20 μm, whether strengths are corrected for the limited *P* or not. Flexural strengths from three other studies with *G* ~ 8–50 μm [70–73], again with or without limited corrections for limited *P*, are more consistent with each other than the diametral strength values, being higher than the latter by 2–3-fold. These studies also indicate possible transitions to a larger *G* branch in the range of G ~ 10–20 μm, which would be consistent

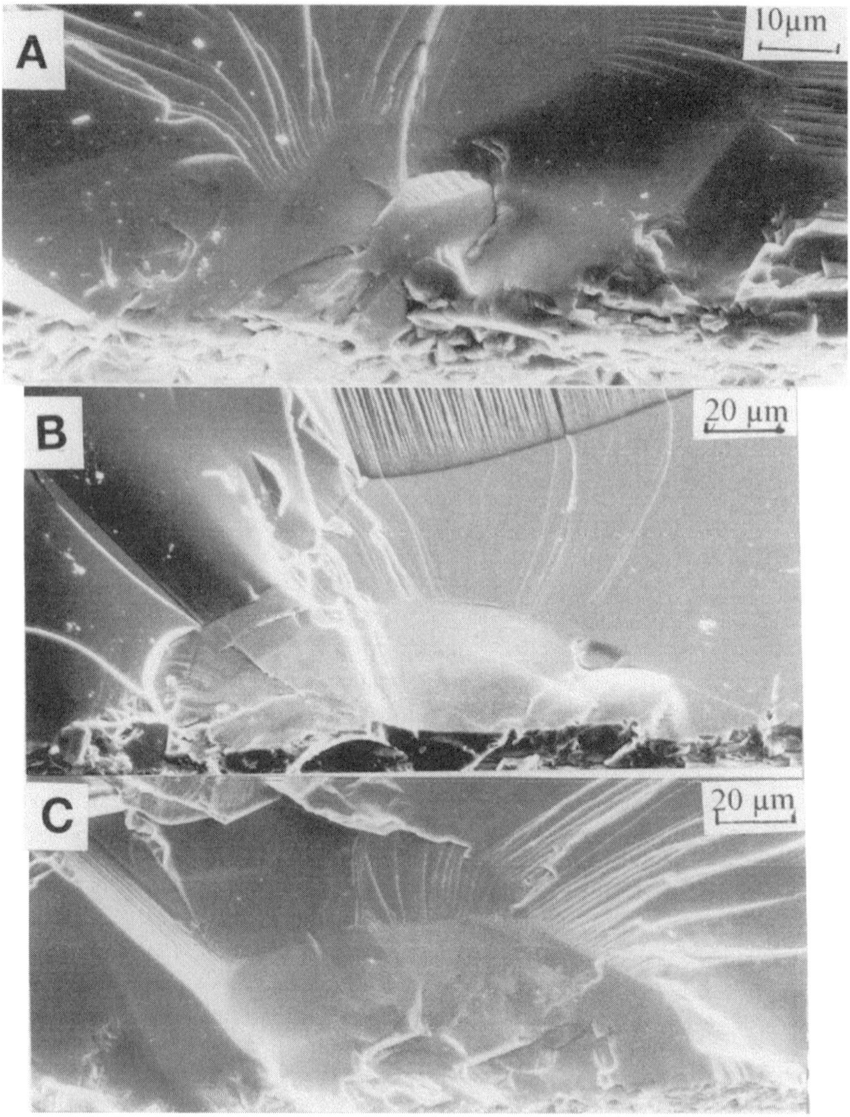

FIGURE 3.9 Sample fractographs of machining flaw origins in dense (transparent) Y_2O_3. (A) Smaller, more irregular flaw from machining parallel to the subsequent tensile axis. Note its location in a large (> 300 μm) grain. (B) Elongated flaw from machining perpendicular to the subsequent tensile axis. Note location near the left boundary of a truncated grain. (C) More uniform flaw in a smaller remnant of a larger grain.

with strength being controlled by toughness values between those for grain boundaries or single crystals and polycrystalline values.

Previous compilations [74, 75] of the limited data for nearly, or fully, cubic ZrO_2 with CaO, MgO, or Y_2O_3 stabilizers showed a substantial decrease in strength with increasing *G* over ~10 μm, consistent with this data being on the larger *G* branch as expected (Fig. 3.10A). More recent limited data of Adams et al. [76] (11 wt% Y_2O_3, *G* ~ 1–50 μm), though generally of lower strengths (in

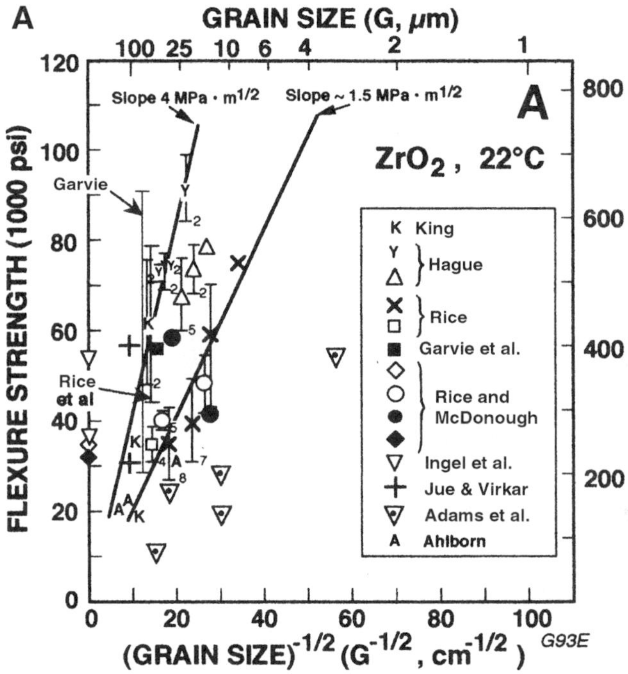

FIGURE 3.10 Strength and fracture of fully stabilized, machined cubic zirconia bodies. (A) Strength–$G^{-1/2}$ data for fully and some partially stabilized bodies (see Fig. 3.21 for TZP data), as well as similarly finished, fully stabilized crystals of the same or similar compositions [80]. (B) Optical fractograph of a fracture origin from a smaller surface grain (surrounded by somewhat larger, but still normal, mostly transgranularly fractured, intermediate size grains) and the fracture mirror (brighter region) in an intermediate to larger grain ZrO_2 + 1lw/o Y_2O_3 body. (C) SEM fractograph of a machining flaw fracture origin in a CZ crystal. (D) SEM of fracture origin (center of photo) in a commercial PSZ (Zircoa 1027 with 2.8 w/o, 8.1 m/o, MgO) from an individual grain boundary facet with common excess tetragonal phase and some pores [82]. (Photos (B) and (D) published with the permission of the ASTM [21].)

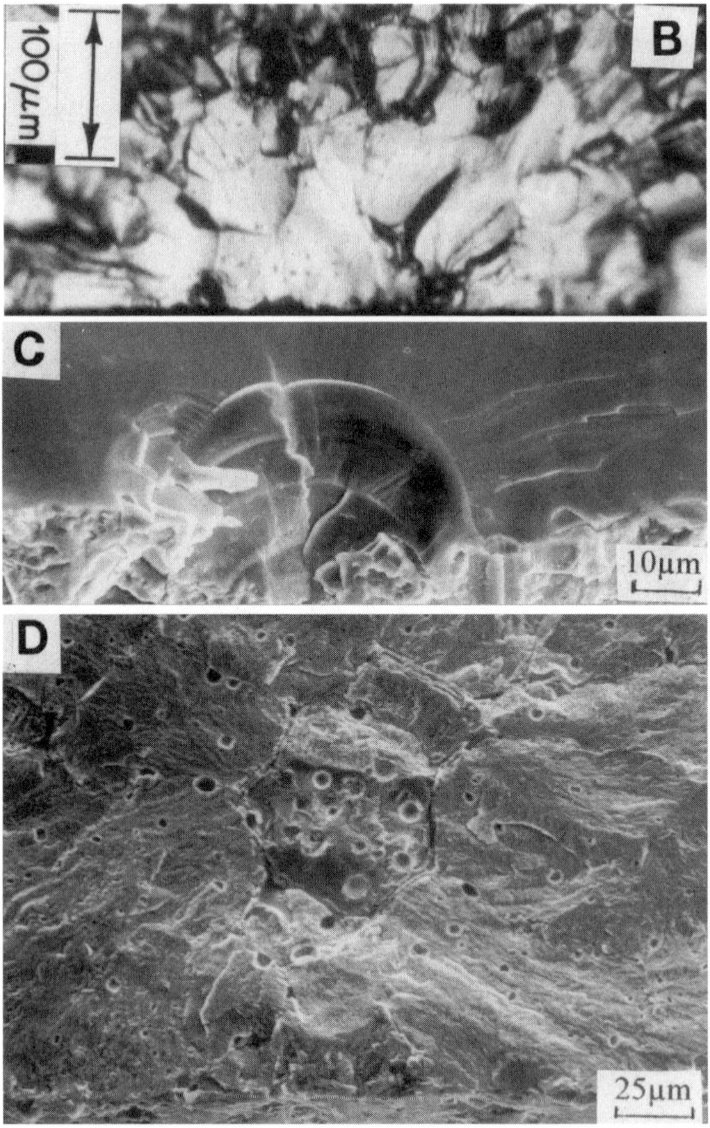

FIGURE 3.10 Continued.

part reflecting no correction for the 1–8%, often heterogeneously distributed porosity, especially at larger *G*), clearly extends into the finer *G* branch, e.g. being consistent with an intersection of the branches at *G* ~ 15 μm. This data and effects of porosity on it are corroborated by data for transparent cubic ZrO_2 (+10 m/o TiO_2 and 7 m/o Y_2O_3) of Ahlbom et al. [77] falling along or slightly above the upper bound of Adams et al.'s data. Data of Hague [78] and King and Fuchs [79], having the highest porosity levels (to ~ 15%, but mostly ≤ 8%), is (possibly over) corrected to zero porosity via the exponential relation e^{-bP}, using $b = 8$, except at the lowest stabilizer contents). Again, note polycrystalline strengths extending substantially below earlier [74] and later, higher quality fully stabilized single crystals of the same or similar compositions (mostly with Y_2O_3) and surface machining [80]. PSZ polycrystalline data (i.e. for cubic grains with tetragonal precipitates) shows similar larger *G* dependence, but with higher strengths, as expected from transformation toughening, and thus steeper slopes [81–83]. (Such PSZ polycrystalline strengths are well below those for PSZ crystals of similar compositions and surface finishes, but are consistent with strengths of TZP bodies, Fig. 3.21). Fractography showed fracture initiation in intermediate *G* Y-CZ from grains or defects on the scale of the typical cubic grains (Fig. 3.10B and C) and fracture mirrors consistent with such grain scale flaws [21]. Grain scale origins from individual grain boundary facets (with some pores) are found in commercial PSZ (Fig. 3.10D), again showing similarities between cubic and PSZ [83].

Extensive $MgAl_2O_4$ data of Bailey and Russell [84, 85] and others in an earlier survey [86] and more recent work [87–92] show two sets of σ–$G^{-1/2}$ branches, with fine *G* branches being determined by various fabrication, e.g. finishing, parameters (Fig. 3.11). Though there is considerable variability, all fine *G* branches show >0 slopes. Most if not all the significant variations of σ with stoichiometery were via *G* [84, 85], e.g. Bailey and Russell's MgO rich $MgAl_2O_4$ showed *G* decreasing ~ 10·fold with corresponding σ increases with increasing excess MgO [85]. The data clearly shows σ (even for highly translucent to transparent specimens) in the large *G* region extending down to ~ 40–70% that of low index orientation of stoichiometric single crystals with similar surface finishes. This is also true for Gentilman's [92] data for fusion-cast transparent ($2Al_2O_3 \cdot 1MgO$) large (2–5mm) *G* specimens tested with grain boundaries perpendicular to the tensile surface. Fig. 3.12 shows one set of the author's data from Fig. 3.11 along with specific fracture origins. Data of Hou and Kriven [93] for $CaZrO_3$ also shows a two-branch behavior. A single data point of Moya et al. [94], is reasonably consistent with their data but probably reflects common differences in preparation, testing, and characterization (especially *G* measurement).

Note that there is generally no significant inconsistency between the *G* dependence of toughness and strength for most cubic materials where there is

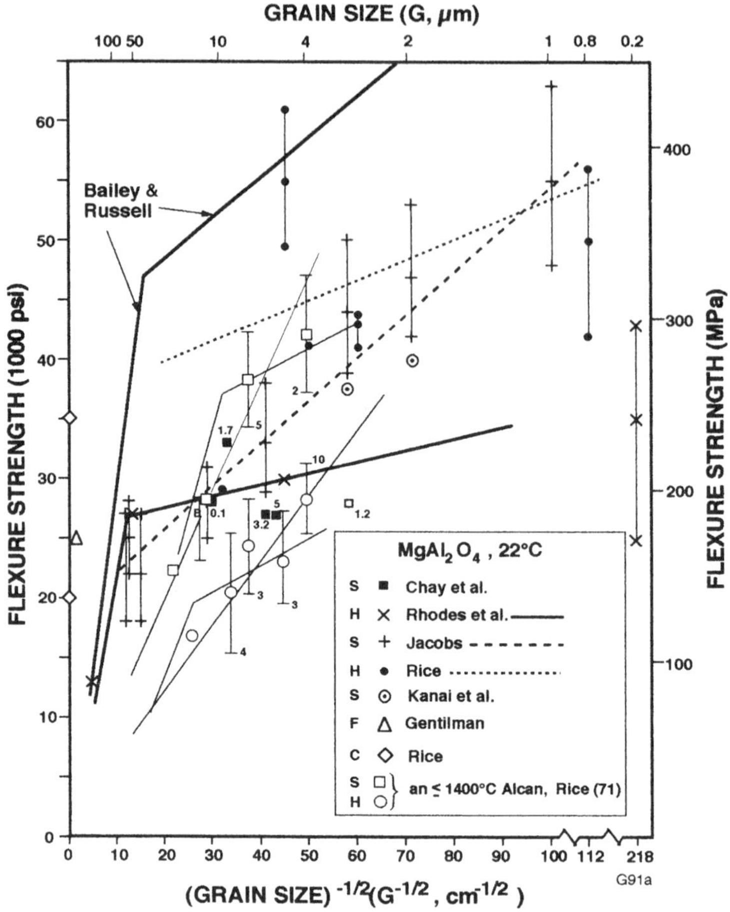

FIGURE 3.11 $MgAl_2O_4$ σ–$G^{-1/2}$ data. The curve marked Bailey and Russell reflects not only their data [84, 85] but also other data of an earlier survey [86]. Subsequent data of Chay et al. [87], Rhodes et al. [88], Jacobs [89], Bakker and Lindsay [90], and Rice [13, 67] are shown corrected to 0 porosity using b=3, 6, and 8 (respectively low, middle, and high points of vertical bars); the limited % porosity of Rice's specimens is shown next to those data points. Data of Kanai et al. [91], Gentleman [92], and Rice are shown as measured with machined surfaces. Vertical bars are the standard deviation, and the numbers underneath them, the number of tests; S=sintering, H=hot-pressing. While there is considerable uncertainty in the specific finer G slopes of all of the data (except that represented by Bailey and Russell), all such slopes are > 0.

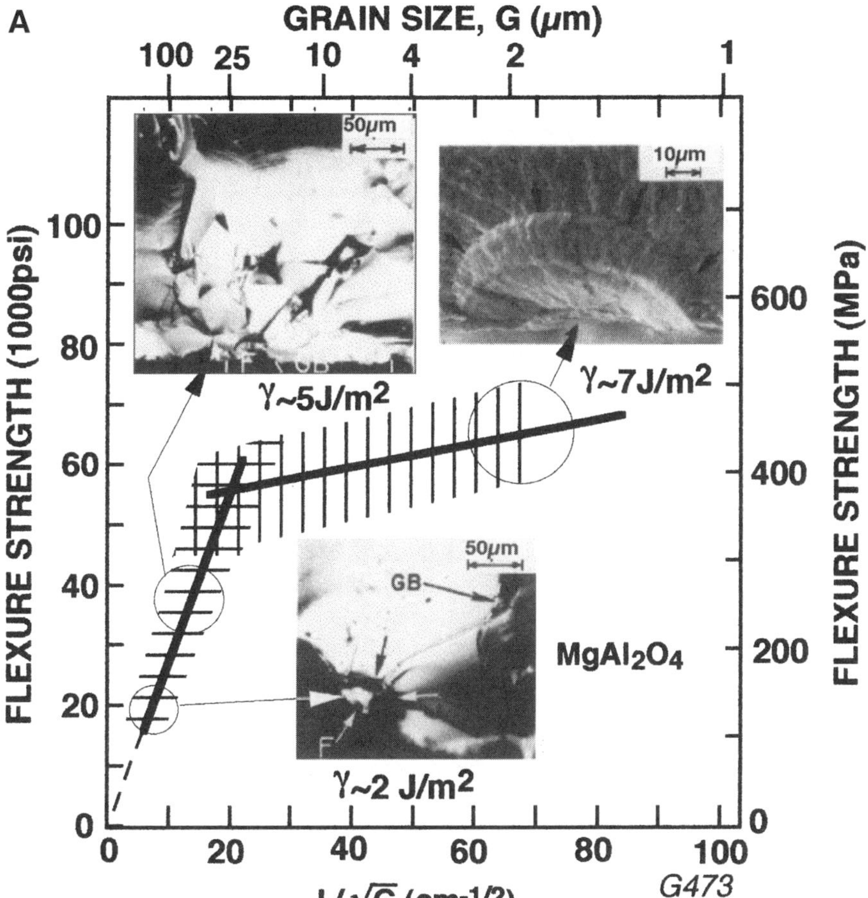

FIGURE 3.12 $MgAl_2O_4$ strength and fracture. (A) σ–$G^{-1/2}$ data for dense sintered and hot pressed $MgAl_2O_4$ from an earlier survey with fractographic inserts showing specific fracture origins in the strength and *G* ranges circled. (After Rice [12], published with the permission of Plenum Press.) (B) Fracture origin from a machining flaw in a stoichiometric $MgAl_2O_4$ crystal machined parallel with the subsequent stress axis normal to (100), on which fracture occurred at ~ 370 MPa. Note the mist and hackle at the left edge of the photo (and in the top left insert of (A)) indicating that the failure crack had become catastrophic well before it exited the large grain.

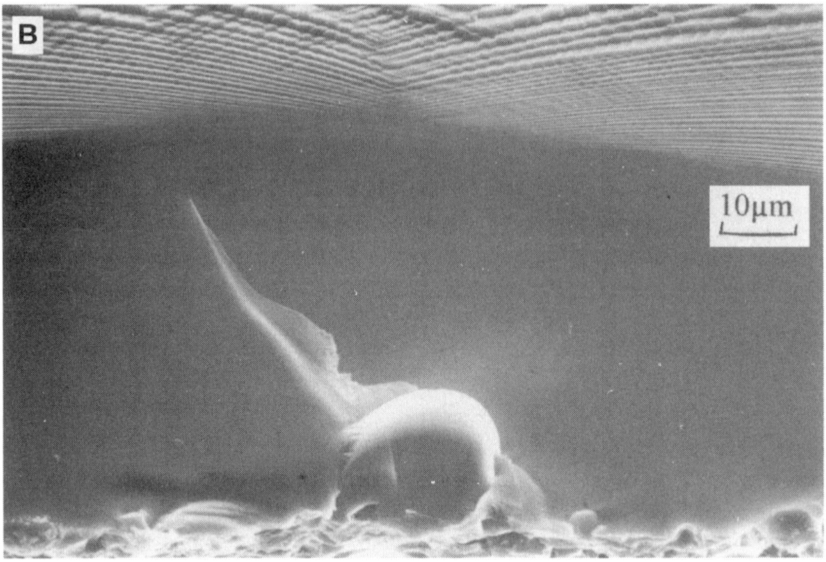

FIGURE 3.12 Continued.

reasonable data, e.g. for Y_2O_3 (Figure 2.12) and $MgAl_2O_4$ (Figure 2.13), which show limited or no toughness dependence on *G*. The limited decrease of strength with increasing *G* in the finer *G* branch is explained by effects on flaw size in Section VI.A and the greater strength decrease at larger *G* by the transition to single crystal or grain boundary toughness control (Chap. 2, Sec. III.E; (Chap. 3) Sec. VII.A). Thus this data is clearly consistent with the two-branch model (Fig. 3.1). Such comments also apply to much of the failure of materials having limited plastic induced fracture, especially those tranisitioning to preexisting flaw failure, e.g. MgO (Figs. 2.12, 3.5). However, SiC, addressed in the following section, shows serious discrepancies between the *G* dependences of toughness (Figure 2.14) and strength (Fig. 3.13) similar to those often found for noncubic materials discussed later.

B. Cubic Carbides and Refractory Metals

The most extensive data for a cubic nonoxide is for SiC (Fig. 3.13). Though some data is for bodies of some to all α-SiC content which is noncubic, it has limited anisotropy and is thus generally consistent with cubic (β) data. Previous surveys [3, 11, 13] of earlier SiC data [95, 96] clearly established part of the large *G* branch and finer *G* branches, based on G_a, generally agreeing, as does that of

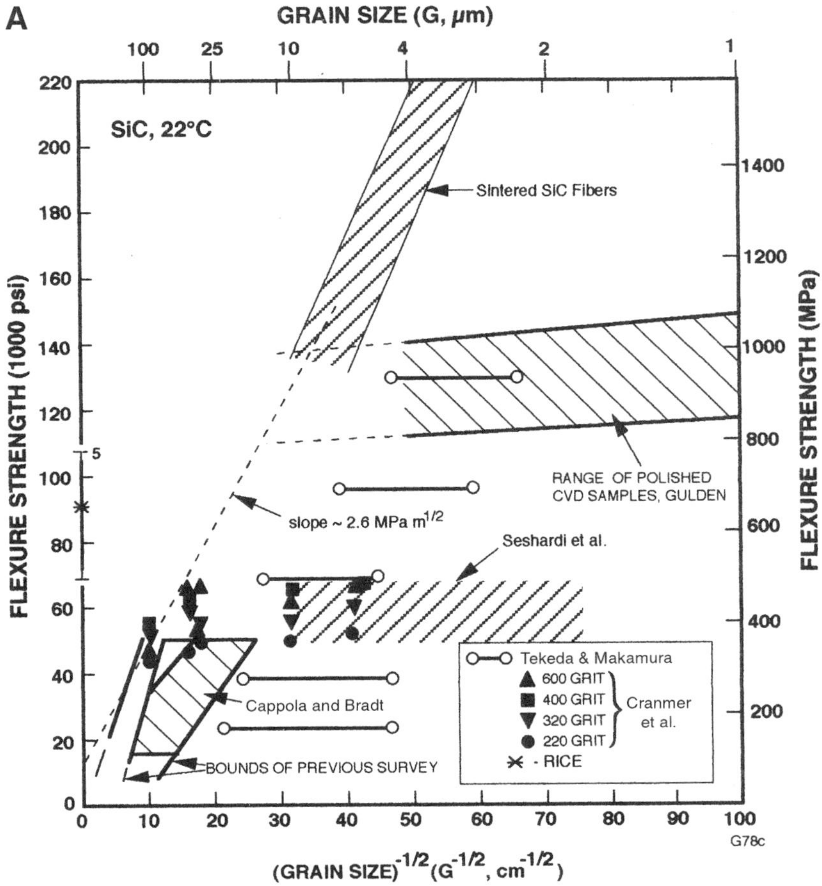

FIGURE 3.13 Strength versus $G^{-1/2}$ for various, mostly β-, SiC bodies at ~ 22°C. (A) Data for mostly dense sintered or hot pressed SiC bodies, including bounds from earlier surveys [3, 11, 13], along with earlier [95–97] and more recent data [98, 99], showing different finer *G* branches and one larger *G* branch consistent with the expected model. This is also supported by data of Larson et al. [100] (not shown for clarity). Again limited strengths for similarly machined small, single crystal bars [101, 102] are above much of the polycrystalline values. Note data for extruded and as-sintered SiC fibers (~ 50 μm dia., 103), which appear to be an extension of the larger *G* branch, consistent with similar $\sigma - G^{-1/2}$ behavior for as-fired surfaces. (B), Data for polymer-derived SiC fibers containing typical [104] and low [105] oxygen levels tested as-fired and as-deposited CVD SiC filaments [106]. Note that the compressed $G^{-1/2}$ scale gives the appearance of high slopes in (B) but they are actually ~ 0.4 and 0.2 MPa·m$^{1/2}$, consistent with being finer *G* branches, as is their fine *G*.

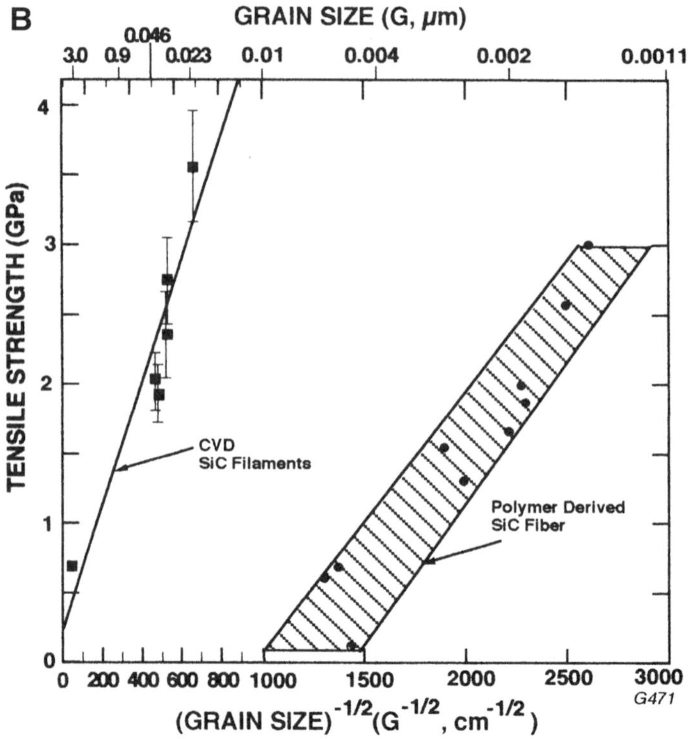

Figure 13 Continued.

Cranmer et al. [26] (using G_m). Prochazka and Charles' data for sintered β bodies [96] also generally agrees with other data when strengths (<140 to ~350 MPa) are plotted versus the effective G (taken as 1/10 the length) of the long tabular (exaggerated) α-grains at fracture origins in the β-SiC matrix, rather than the grain length, as used by the authors. Gulden's [97] CVD SiC data (0.6 ± 0.5% porosity), based on G_a, is also consistent, recognizing the higher strength due to the use of small polished specimens (with some possible preferred orientation), and it supports a positive fine G slope, as does Cranmer et al.'s surface finish study. The latter again show different finer G branches as a function of machining grit size, with less differentiation in the larger G branch. Seshardri et al.'s [98] (sintered α, ~1.6% porosity) SiC data also showed a finer G branch of low slope. Takeda and Nakamura's [99] SiC (sintered to 98–99% theoretical density, mostly with 1–2% of AIN, BN, B, or B_4C) data tends to fall to the right of the other data whether plotted versus G_a or G_m from general microstructure evaluation (measurement specifics not given). This data being to the right of other data probably reflects G

measurement issues, especially the inability to determine a true G_m on other than the fracture surface, so the values given are likely to be $<G_m$ at fracture origins (not determined), especially at medium and larger *G*. Correction for this would likely bring their data into agreement with the other SiC. Larsen et al.'s [100] SiC strength data versus measured G_m (based on fractography), as well as variable G_a values for various SiC materials, is consistent with other SiC data and the model of Fig. 3.1, e.g. similar to their data for Si_3N_4 (Fig. 3.29). This showed that most higher strength failures occurred from undetermined origins (i.e. no dominant features such as large grains in the origin area), thus being more consistent with G_a values and on finer *G* branches (of which there should be more than one, substantial scatter, or both in view of different and varying flaw populations), while lower strengths (e.g. <450 MPa) showed failure from various flaws, tabular grains (e.g. in the GE β SiC), or both. Again, note various polycrystalline strengths extending down to ~ 20% or less of the average strengths of small α single crystal bars (oriented for easy fracture, e.g. on (100) [101, 102] with similar surface machining as the polycrystalline bodies and basic inconsistencies with *K–G* data (Figure 2.14), especially at finer *G*.

While the above bulk SiC data is for machined samples, as is typical for most materials, various SiC fibers provide data on as-fired or as-deposited surfaces (and some possible surface damage), as is the case for alumina-based fibers discussed below (Fig. 3.18). Data of Srinivasan and Venkateswaran [103] for SiC fibers made by green extrusion of SiC powders and sintering clearly falls along an extension of the larger *G* branch for bulk SiC to strengths of ~ 1500 MPa (Fig. 3.13A). This is consistent with its grain size range, the as-fired surfaces, and fractography indicating fracture origins from flaws with $c \sim G$. Data for polymer-derived SiC-based fibers (Fig. 3.13B) also shows a clear *G* dependence [104, 105], extending well into the nanograin regime. Data of Elkind and Barsoum [106] on commercial CVD SiC filaments tested as-received or after heat treated to increase *G,* both by itself and combined with the one much larger *G* specimen [107], shows a distinct *G* dependence. This is corroborated by data of Bhatt and Hull [108, 109] showing similar decreases in strength with heat treatments and associated qualitative changes in *G* and by the microstructural characterization of such fibers by Nutt and Wawner [110]. Differences between these sets of as-fired polymer-derived and as-deposited or as-fired CVD fibers are attributed primarily to the *G* differences and composition, especially significant dilution of the SiC content, primarily in the polymer-derived fibers, as well as to some test differences (e.g. loading rates, gage lengths, etc.). Other microstructural differences, especially elongated grains with an axial preferred [111] alignment with the carbon core of the CVD filaments, also are probably a source of the difference in σ–$G^{1/2}$ behavior between the fibers. However, these fiber results clearly show that such as-fired surfaces follow similar *G* dependences as machined surfaces.

Another SiC data source providing additional insight is very fine G (0.01–0.1 μm), small rods (~ 1.5 mm dia.) of SiC deposited on W wire cores (~ 200 μm dia.). This is a transition between the above SiC bulk and filament data. As-deposited surfaces were very smooth, giving strengths averaging 1400–1800 MPa depending on whether they were respectively tested in four- or three-point flexure, despite an indicated internal stress from W–SiC expansion differences of the order of 700 MPa [111, 112]. Limited fracture origin identification indicated failure from surface flaws from handling damage, indicating such bodies would fall on a fine G branch, as was expected. These results were corroborated by three-point flexure of lightly and more heavily abraded surfaces (respectively with 600 and 320 grit SiC paper in the axial direction), giving average strengths of ~ 1000 and 600 MPa respectively (the latter also obtained for oxidized surfaces). Substantial fracture origin determinations showed that fractures in abraded rods occurred from machining flaws substantially larger than the handling flaws and the G, confirming their placement on finer G branches.

A previous survey [11] of more limited data of Cr_3C_2, HfC, TaC, and ZrC showed larger G behavior like that of other ceramics, as well as reasonable evidence of finer G branches (e.g. TaC, TiC, and ZrC), the latter sufficient to indicate a positive slope. Subsequent data by Miracle and Lipsett [113] for TiC_x (x = 0.66, 0.75, 0.83, or 0.93 with G = 22, 21, 20, and 14 μm respectively) is consistent with the other TiC data; i.e. effects of G dominate over, or are a reflection of, stoichiometry. ZrC data of Bulychev et al. [114] for G ~ 10–24 μm indicates intersection of the finer and larger G branches in the range of G ~ 15–20 μm.

Recently, Savage et al. [115] presented fairly extensive data for CVD diamond showing a clear $G^{-1/2}$ dependence of strength (Fig. 3.14). Though they showed a single line through their data, there is reasonable indication of a change from a finer to a larger G branch, with the intersection of the two branches in the 100–250 μm range and a larger G branch slope of ~ 6 MPa·m$^{1/2}$. Both these values are in reasonable agreement with the model (Table 3.1), especially given uncertainties due to the elongated, oriented grain structure from CVD.

While some refractory metals such as Nb and Ta commonly exhibit macroscopic ductility (hence also commonly following a Petch relation), others such as Cr, Mo, and W commonly have limited or no macroscopic ductility. Thus when there are cracks, pores, impurity particles, etc. of sufficient size and concentration to be failure causing flaws [116], failure is brittle, e.g. as shown by effects of surface finishing on ductility in W [117]. In such cases there may be relations of strength to G via relations of the flaw sizes or of toughness to G, or both. In the absence of such flaws, these commonly brittle materials fail by slip-induced (cleavage) fracture, e.g. down to temperatures of ~ 100° K [118] which is corroborated by their strengths following a Petch relation. However, there are some departures to a lower Petch plot slope, e.g. at G > 0.25 mm in Mo that re-

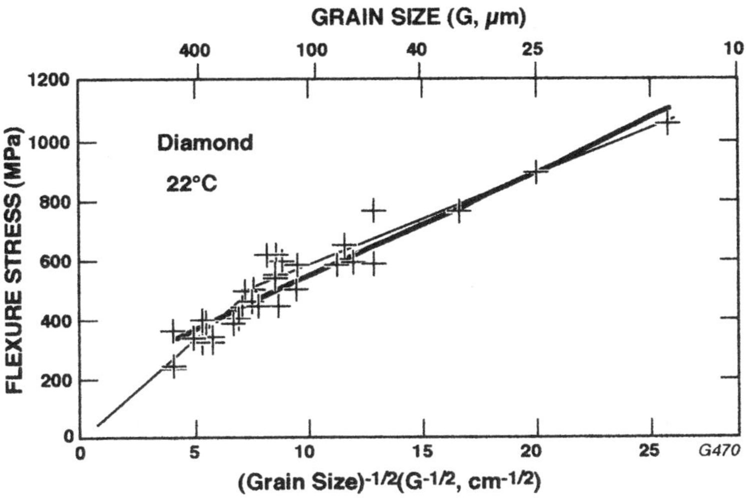

FIGURE 3.14 Flexure strength of CVD diamond at ~ 22°C as a function of the inverse square root of grain size ($G^{-1/2}$). Note that while they drew a single line through all of the data, it is fitted better by two lines consistent with the model for machining flaw–grain interactions. (From Ref. 115, published with the permission of the SPIE.)

sults in intercept stresses higher than those projected from finer *G*, e.g. by ~ 50%. This and other variations may arise from differing brittle-to-ductile transition temperatures at different *G* values, e.g. in W ~ 300° K at fine *G* and < 200° K for large *G* [119] with brittle fracture following the Griffith equation [120]. Such higher transition temperatures in finer *G* bodies would suggest the occurrence of brittle fracture in some finer *G* bodies with slip-induced fracture in larger *G* bodies, i.e. as in Fig. 3.1. However, it is not clear whether this combination as found in ceramics has been clearly demonstrated in brittle metals, though an earlier combination of strengths of a few different W bodies to give a broader *G* range [11] suggests that possibility.

IV. NONCUBIC CERAMICS FAILING FROM PREEXISTING FLAWS

A. Al_2O_3 Machined and As-Fired

Al_2O_3 has the most extensive data base, providing substantial data not only on the basic $\sigma - G^{-1/2}$ and related fractography, but also on effects of different mechanical testing, additives and second phases (especially to inhibit grain growth), and machined and as-fired surfaces. Though much of this data is limited by the grain size range covered and the accuracy and detail of microstructural characterization, it clearly provides extensive demonstration of the two branch behavior of Fig. 3.1, as was shown in previous surveys [2, 8, 9, 11–13] summarized here. Thus the area labeled previous surveys in Fig. 3.15 covers the range of almost all data with only a limited amount of data falling outside of this area and only to a limited extent. Thus earlier data on commercial and similar aluminas, often with some (e.g. 3–8%) porosity, second phase (1–7%, mainly silicate), or both [121–127] falls mostly in the lower half or somewhat below this area but individually or collectively agree with the two-branch model. This includes Alford et al.'s [27] data for specimens machined from somewhat porous pressed discs [2] and that of Evans and Tappin [125] (sintered, diamond ground) Al_2O_3 with 2 σ–$G^{-1/2}$ branches with a reduced but > 0 slope in the fine G region, but somewhat lower σ due to ~ 5% porosity. Correction for effects of porosity and silicate phase (Fig. 3.30) of sintered bodies moves this data further toward the mean of

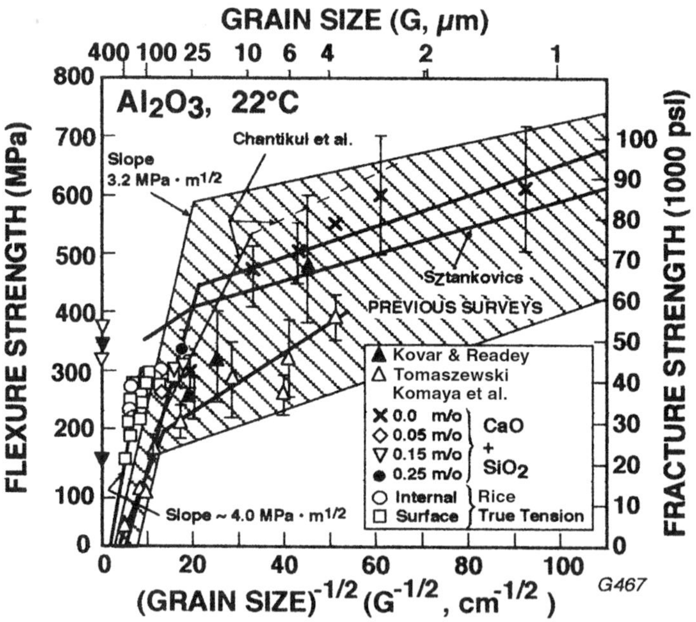

FIGURE 3.15 Strength versus $G^{-1/2}$ for alumina bodies at ~ 22°C. A summary of trends from previous surveys [2, 3, 11, 13] is shown along with more recent results for mainly high-purity, dense sintered bodies [128–136] and results of true tensile testing of large hot pressed specimens [67, 138]. Note data for Verneuil crystals ground parallel or perpendicular to the tensile axis [15] (open and closed symbols respectively).

the data survey and into better agreement with hot pressed data, which fell mostly near the central-to-upper portion of the survey area [13, 53, 128–130]. Subsequent studies of higher purity bodies, mostly sintered close to theoretical density [131–136], also agree individually or collectively with the previous survey, including that of Koyama et al. [135] with small additions of CaO, SiO_2, or both giving elongated Al_2O_3 grains (Fig. 3.17). Two exceptions to these trends are for submicron *G* bodies of Spriggs et al. [64] and Krell and Blank [137] with strengths for (unspecified) machined dense Al_2O_3 of 450–550 MPa (*G* ~ 0.6–0.8 μm) and 582 MPa (*G* ~ 0.4 μm). The lower level of these strengths relative to that expected from extrapolation of data of Fig. 3.15 to finer *G* is attributed to grain boundary contamination left from raw materials and processing used to achieve the finer *G* (Sect. V.A).

Data of Fig. 3.15 also allows comparison of different test techniques, though most data is from various flexure tests, which is probably a factor in the data range. Chantikul et al.'s data [131] in biaxial flexure of small discs tends to fall higher in strength, especially at finer *G*, which may reflect some benefits of avoiding edge failures (and probably also reflects some benefits of polished versus ground surfaces, as is discussed below), but it still falls in the typical range for other studies. True tensile testing of much larger hot pressed Al_2O_3 samples is shifted to larger *G* values due to the large specimen volume tested and the occurrence of isolated single large grains or clusters of a few of them, e.g. reaching dimensions of ≥ 500 μm [67, 138]. While some of these grains or clusters may exceed the flaw size, indicating a somewhat higher than real slope, the slope of the resultant larger *G* branch is still in the range of the other studies of Fig. 3.15 (see also Sec. VII). This true tension data also shows some of the finer *G* branch consistent with the other data. Further, diametral tensile testing of mostly hot pressed bodies [139–141] falls near the lower bound of the survey range in Fig. 3.15 and clearly shows a consistent intersection of larger and finer *G* branches [13]. Thus while different testing may shift the results due to stressing of different volumes of material and thus sampling different ranges of failure causing flaws, the σ–$G^{-1/2}$ behavior shifts accordingly, e.g. in slopes and intersections, but remains consistent with the basic model, which is thus reinforced.

Consider next machining studies, which were an important factor in developing the model of Fig. 3.1. The demonstration that there was significant anisotropy of tensile strength as a function of machining direction relative to the subsequent stress direction in uniaxial tensile or flexure testing [15, 17] was a major factor in indicating the two-branch character of σ–$G^{-1/2}$ behavior. Initial studies showed that machining parallel versus perpendicular to the tensile axis resulted in respectively higher and lower strengths at both finer and large *G* (and also generally in single crystals), thus indicating different finer *G* branches. Subsequent studies of Tressler et al. [24] showed different machining σ effects in the finer, than in the larger, *G* region. Coarser abrasive finishing in the finer *G* region resulted in lower strengths on separate branches, while in the larger G region coarser finishing also resulted in lower strengths, but in one basic larger G branch. The polishing results of Chantikul et al. [131], though also entailing some probable benefit from biaxial testing (i.e. no edge failures), is consistent with higher strengths with finer finishing. [Note that Tresssler et al.'s [24] data for Al_2O_3 hot pressed with MgO (G_a ~ 2 μm) or without MgO (G_a ~ 3, 12, or 18 μm) generally agreed with other data whether plotted vs. G_a or, as they did, vs. G_m (G_m ~ 3.3 and 4.3 μm for the two finest G bodies, but ~ 80 – 100 μm, due to scattered large platy grains found at fracture origins, in their two largest G bodies). However, their data is insufficient to prove their claim that the fine G slope = 0; it is at least as supportive of a positive slope (though complicated by their use of G_m, see Sec. V.B). Also note that (1) Chantikul et al.'s [131] model is not correct, since the σ–$G^{-1/2}$ behavior is basically the same in cubic and noncubic ceramics and thus is not determined by TEA effects as they proposed, and (2) there are important machining effects when the flaw and grain sizes approach each other [142–144] (Fig. 3.33), as is discussed later.

Again an important component of the σ–$G^{-1/2}$ model for understanding failure mechanisms is relating strength and fractography of single crystals of the same material with the same or similar surface finish, i.e. sapphire in this case, focusing on weaker single crystal orientations (which are also typically the ones tested), since these are likely to dominate failure in larger G polycrystals. Rice's [15] data for Verneuil crystals of two different orientations and machining parallel with the tensile axis agrees well with earlier data of Wachtman and Maxwell [145] (σ ~ 350–700 MPa) and of Heuer and Roberts [17, 146] (σ ~ 450 MPa) for unspecified machining (presumably parallel to the tensile axis) of similar sapphire. Again, polycrystalline strengths are frequently below those for single crystals with similar finishing, and machining perpendicular to the tensile axis reduces most strengths [15], often substantially. Mechanical polishing also gave single crystal strengths of ~ 450 MPa, which increased to 600–900 MPa with subsequent annealing [17, 146], consistent with no grain boundary grooving on crystal specimens.

Definitive fractography of polycrystalline Al_2O_3 is more difficult, especially as G increases, but machining flaws have been found in some finer G bodies, where they are > G as expected (Fig. 3.16). In single crystals, definitive machining flaws are frequently found and are of similar size and character to those in similarly finished polycrystalline bodies. For bodies of intervening G, definitive machining fracture origins are generally not found, but probable or certain origins from larger grains or clusters of them occur [3, 11–13, 138] (Figs. 1.2A,B, 1.3A,B, 1.4).

Next consider effects of phases added, usually in limited amounts, for limiting G. Such additions were an important factor in extending the G range, especially with low to 0 P, but raises questions of additive roles in σ – G behavior.

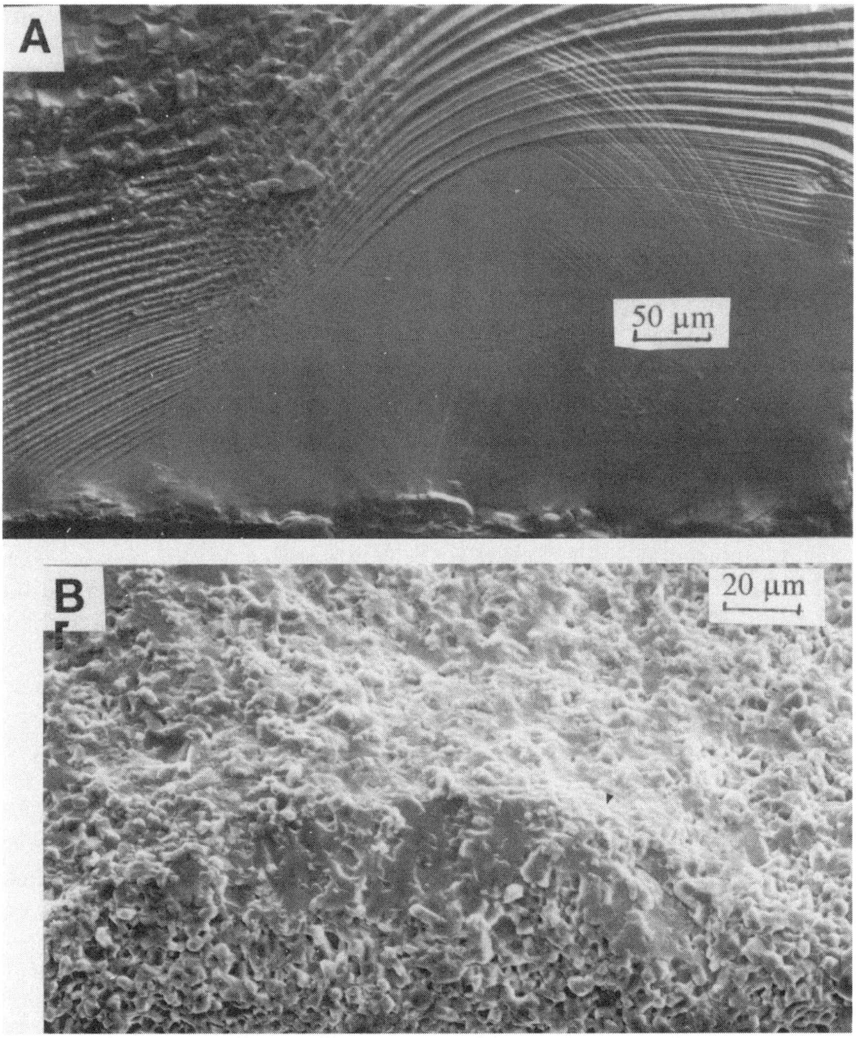

FIGURE 3.16 Examples of machining flaw fracture origins from ~ semielliptical machining flaws in alumina bodies ground perpendicular to the subsequent tensile axis. (A) Flexure tested sapphire bar (~ 300 MPa); flaw: semielliptical feature ~ bottom center of photo. (B) Large tensile tested hot pressed Al_2O_3 round, dumbbell specimen (280 MPa); flaw: large semiellipse across ~ bottom 1/3 of photo. Since these flaws are from different machining conditions, they cannot be directly compared quantitatively, but qualitatively they show overall similarity of flaw sizes given statistical variations of machining and testing, e.g. due to flat versus round machined surfaces and large differences in surface areas and resultant flaw population sampling.

MgO, the most common additive (often at ≤ 1 %), can be very effective in controlling G with resultant strength increases mainly, or exclusively, as a result of the reduced G, as was indicated above (but local, substantial, excesses of it can be quite detrimental [16]. Several other additives also appear to have little or no effect on strength other than increasing it via resultant finer G. Thus McHugh et al. [140] showed that sintered Al_2O_3 strengths increased with additions of fine Mo particles up to ~ 6 v/o due to reductions in G, then remained constant to the limits of their study of 16% addition. Similar improvements were noted in another study using diametral testing [141], and for addition of W or ZrO_2, as is discussed below. While some negative effects of other constituents occur, as is noted later, specimens with up to several percent silicate phase [121–129], 30% AlON [147], and 25% Cr_2O_3 [10, 148], indicate limited, or no, effect of these phases, other than via their effect on G.

Consider now data for specimens with as-fired surfaces, typically from extruded rods. Data of Charles [149] for (lamp envelope quality) Al_2O_3 extruded, as-fired rods (~1.7 mm dia, noting inhomogeneous G only in the next-to-largest G body) clearly shows a two-branch curve with the finer G branch having a significant positive slope (Fig. 3.17). Alford et al.'s [27] extruded (< 1 mm dia., as-fired) rod strengths are much higher, presumably reflecting the small-size) high-quality as-fired finish, and possible (unexamined) preferred orientation (e.g. reflecting the ~ 1.5 aspect ratio of their Al_2O_3 powder particles and the small rod diameter, implying a high reduction ratio). However, their data is at least, if not more, consistent with a positive finer G branch slope rather than their proposed slope of 0. Their one fractograph indicating fracture from a cluster of three larger grains raises issues of the rationale of using the large G in such cases, as is discussed later (Sec. IV.B). Note that their strengths of machined bars (~2.5 mm thick of unreported orientation relative to the pressing direction for the original disks) made from the same Al_2O_3 powder are similar to those of others but tend to lower strengths, especially at finer G in view of their use of G_m (~3 G_a). This lower σ is reasonable in view of the large (e.g. ~ 100 μm) processing defects (e.g. laminar voids), and as-sawn surfaces. Alford et al.'s use of maximum σ (σ_m, i.e. the outer fiber stress) is inconsistent with their use of G_m, since such round rods have such a small σ_m region and thus a low probability of failure occurring at σ_m with significant microstructural heterogeneity. Their proposed 0 fine G slope for their extruded rods (though affected by use of G_m and probable orientation effects) is far less than for similar extruded Al_2O_3 rods. Baily and Barker's [150] and Blakelock et al.'s [151] data for smaller (<1 mm dia.) extruded, sintered Al_2O_3 rods vs. G_a showed positive slopes, which are also supported by these data extrapolating to σ–$G^{-1/2}$ values for Al_2O_3-based fibers (Figs. 3.17, 3.18). Hing's [152] data for as-fired Al_2O_3–W (isopressed) rods (Fig. 3.17) clearly showed a two-branch behavior with > 0 slopes for the fine G branches and agrees with other as-fired or machined Al_2O_3

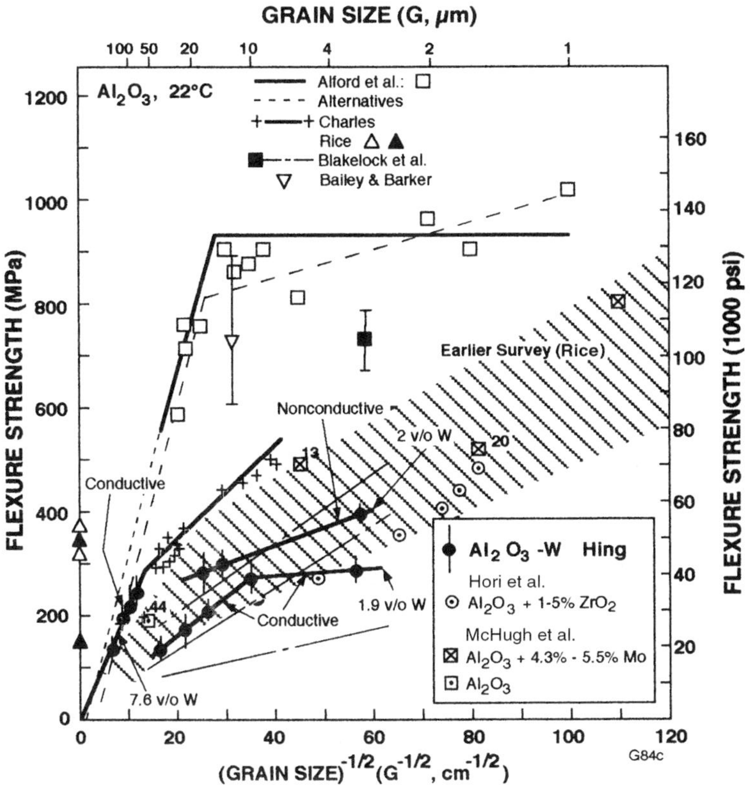

FIGURE 3.17 Strength–$G^{-1/2}$ data for alumina bodies with primarily as-fired surfaces. Nominally pure alumina bodies of Charles [149], extruded ~1.7 mm rods as-fired, and Alford et al. [27], high strength extruded rods (< 1 mm dia.), both tested as-fired. The solid lines for Alford et al.'s data are their original lines; the dashed lines are an equal or more probable alternative. Bailey and Barker's [150] data (vertical bars represent the range of data) for as-fired extruded rods (0.8 mm dia., with 0.25% MgO). The three lines (top to bottom) are for three-point and four-point flexure, then true tension of Blakelock et al.'s rod [151] with ~ 6 to ~ 15% porosity, correction for which would increase strength ~ 15 to ~ 50% (assuming similar correction to that shown by Bailey and Barker). Data of Rice [15] for Verneuil crystals ground parallel or perpendicular to the tensile axis (open and closed symbols respectively). Hing's [152] data (vertical bars = standard deviation) for isopressed, as-fired (~ 9 mm dia.) rods of Al_2O_3 with 1.9 to 7.6 v/o W, and data of Hori et al. [153] for sintered bars with up to 5% ZrO_2; both additions for grain growth inhibition show the same σ–$G^{-1/2}$ behavior as pure alumina, except for Alford et al.'s high strength Al_2O_3. Note designation of whether the bodies were conductive or nonconductive indicating the degree of contiguity of the W phase. (From Ref. 2, [2], published with the permission of the *Journal of Materials Science.*)

data with or without Mo additions. Similarly, data of Hori et al. [153] for sintered Al_2O_3 bars with up to 5 wt% ZrO_2 that were machined and then annealed (at near sintering conditions) showed strengths increasing (but toughness decreasing) due to reduced grain growth. Lange and Hirlinger showed similar results [154].

Data on Al_2O_3-based fibers (Fig. 3.18) [155–168] complements and extends Al_2O_3 data (especially for as-fired surfaces). The typical fine *G* and defects (e.g. pores and cracks) [3] that are the cause of failure in most fibers indicate that all of this data belongs on finer *G* branches, except probably for some of Simp-

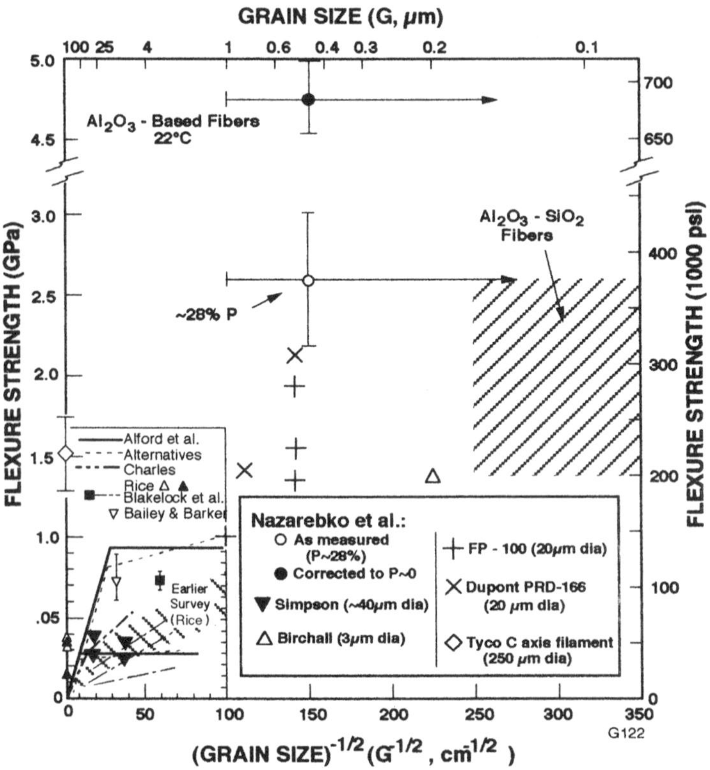

FIGURE 3.18 Al_2O_3-based fiber σ–$G^{-1/2}$, i.e. as-fired, data. Data of Nazarenko et al. [155] is shown as-measured and corrected to $P = 0$. Data of Simpson [156] Hamling [157], Birchall [158], Dupont FP and PRD [159, 160] as well as various Al_2O_3-SiO_2 fibers [159–164, shown as-measured]. Note reduced σ for larger *G* (annealed) Dupont fibers and the σ of commercial single crystal (*c*-axis) filaments [165–168].

son's [3, 156] data. Thus the Dupont fiber data by itself, as well as the other data collectively, show a substantial finer *G* slope. Addition of 20% ZrO_2 maintains finer or more uniform *G,* while SiO_2 additions reduce *G* and give non-α phases and smoother surfaces [159–161], the latter clearly shown to improve strengths. Note that single crystal sapphire filaments have higher strengths [165, 166] than many of the polycrystalline fibers, despite their generally >10-fold larger diameter and frequent failure from residual pores [168], and also fall below strengths of many polycrystalline fibers.

Note two important facts for later discussion. First, there are frequent significant discrepancies between the *G* dependence of toughness of Al_2O_3 (Figs. 2.16, 2.17) for many tests, especially those showing *R*-curve effects, and strength, which is true for much other data for noncubic ceramics. Second, strength data for Al_2O_3 as well as most other noncubic ceramics is consistent with the two branch model (Fig. 3.1).

B. Other Noncubic Oxide Ceramic

Individual BeO studies [169–175] and compilations by Cargniglia [8, 9] and Rice [11, 13] clearly show two σ–$G^{-1/2}$ branches (Fig. 3.19). Again, (1) a few percent porosity present in some samples lowered σ, as did machining perpendicular (e.g., circumferentially for round rods) vs. parallel with their lengths (more so in the finer than in the larger *G* branch), and (2) the primary effect of a nonmiscible second phase (SiC) [175] or a reactive phase (Al_2O_3) [173] in increasing σ was via reduced *G*. The number of specimens (e.g. 50) per test and the demonstration of increasing preferred orientation (from extrusion) as *G* increased indicates that the negative finer *G* slope is real for UOX-derived BeO [169, 170] (Sec. III.N). Again, σ–*G* behavior is inconsistent with γ–(hence also *K*–) *G* behavior (Figure 2.17).

Machined TiO_2 data of Kirchner and Gruver [176] (hot pressed, porosity ≤3%) vs. G_a, Alford et al.'s [27] (lower σ specimens from sintered, die pressed disks) are reasonably consistent regardless of whether G_a or G_m is used for Alford et al.'s data (Fig. 3.20). Alford et al.'s small (<1 mm dia) as-fired extruded rods with expected higher σ have a larger *G* slope < K_{IC}, when it is recognized that $c = G/2$, not G_m as they used [27], and the zero σ intercept at $G = \infty$, finer *G* slopes = 0 are quite uncertain, and not supported by Kirchner and Gruver's data. Additionally, all of the comments for Alford et al.'s extruded Al_2O_3 data apply to their extruded TiO_2 data, since the processing (including a TiO_2 powder aspect ratio of ~ 1.5), testing, characterization, and analysis issues are the same. Again, some large *G* σ's are below those of (Vernuil) crystals [13] (which may be weaker than Czchralsky crystals) with similar surface finishes. Again, γ–*G* and *K*–*G* behavior are inconsistent.

A previous σ–$G^{-1/2}$ data compilation [74] for mainly PSZ ZrO_2, and a more

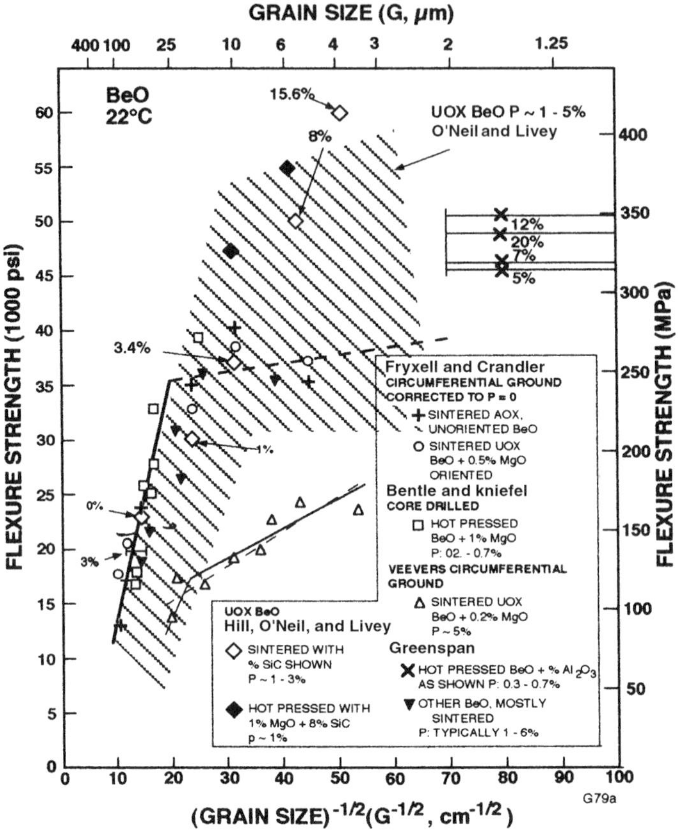

FIGURE 3.19 BeO $\sigma - G^{1/2}$ data of Chandler and colleagues [169, 170], Bentle and Kniefel [171], Veevers [172], Greenspan [173], and O'Neal and Livey [174] for relatively high-purity BeO with various processing and machining. Specimens of Greenspan and Hill et al. [175] with and without substantial levels of respectively Al_2O_3 and SiC additives.

recent survey [75] (especially for TZP bodies) with various types and levels of partial stabilization [177–185], show higher strengths but similar $\sigma - G^{-1/2}$ behavior with transformation toughening from tetragonal ZrO_2 as without (Fig. 3.21). PSZ bodies (some of which are conservatively corrected for porosity, Fig. 3.9A) lie along larger G branches whose slopes generally increase with decreasing stabilization (presumably to an optimum toughness, beyond which their slopes and other behavior should again decrease). Such PSZ strengths commonly extend well below the strengths of similarly machined PSZ crystal specimens of the

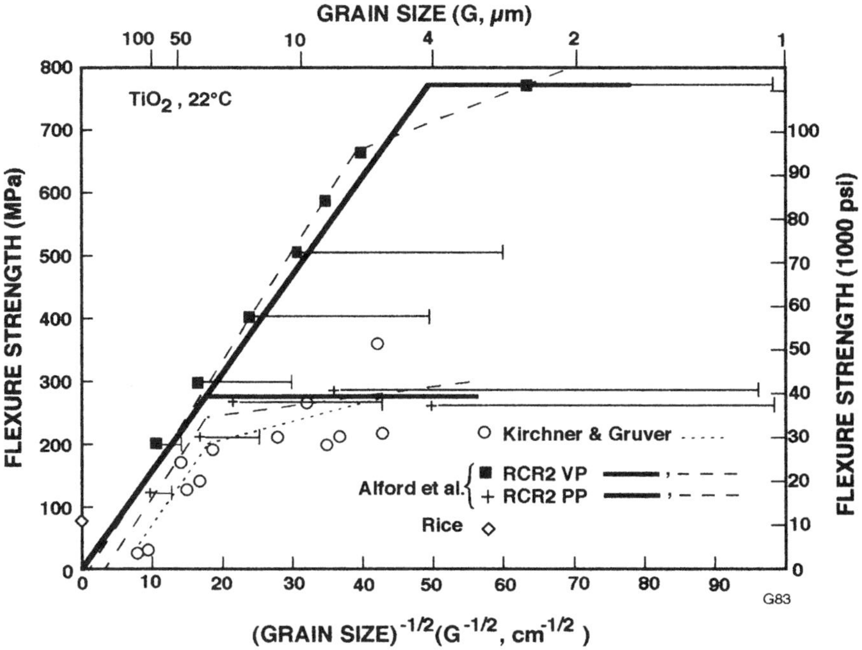

FIGURE 3.20 TiO_2 $\sigma - G^{-1/2}$ data of Kirchner and Gruver [176] (hot-pressed and machined), machined single crystals (Verneuil) of Rice [13] and of Alford et al. [27] for extruded high strength (< 1 mm dia.) rods tested as-fired, and lower strength bars (tested as-sawn from die pressed and sintered disks). Solid lines are those proposed by Alford et al. [27] with equal or more probable dashed alternatives indicated.

same or similar composition and machining (e.g. > 1000 MPa [80], Fig. 3.21). On the other hand, TZP materials all appear to lie along finer *G* branches, again often at lower strength levels than strengths of PSZ crystals of similar finish and composition. Again there are some variations due to residual porosity, e.g. the slope of Wang et al.'s [180] data for ZrO_2 fully stabilized with CeO_2 (12 m/o, i.e., ~16 w/o) would be decreased a limited amount since the limited porosity present (0.8–2.8%) tended to increase as *G* increased, so there would be greater correction of the larger *G* strengths. In either case, this data is approximately an extension to larger grain sizes of the data for similar compositions of Tsukuma and Shimada [177] and of Hecht el al. [182]. While it is possible that Wang et al.'s data reflects the large grain branch for such CeO_2-based compositions, this seems unlikely based on (1) probable limited differences in slope between it and the other similar compositions, and (2) the fine *G* of such an intersection, in view

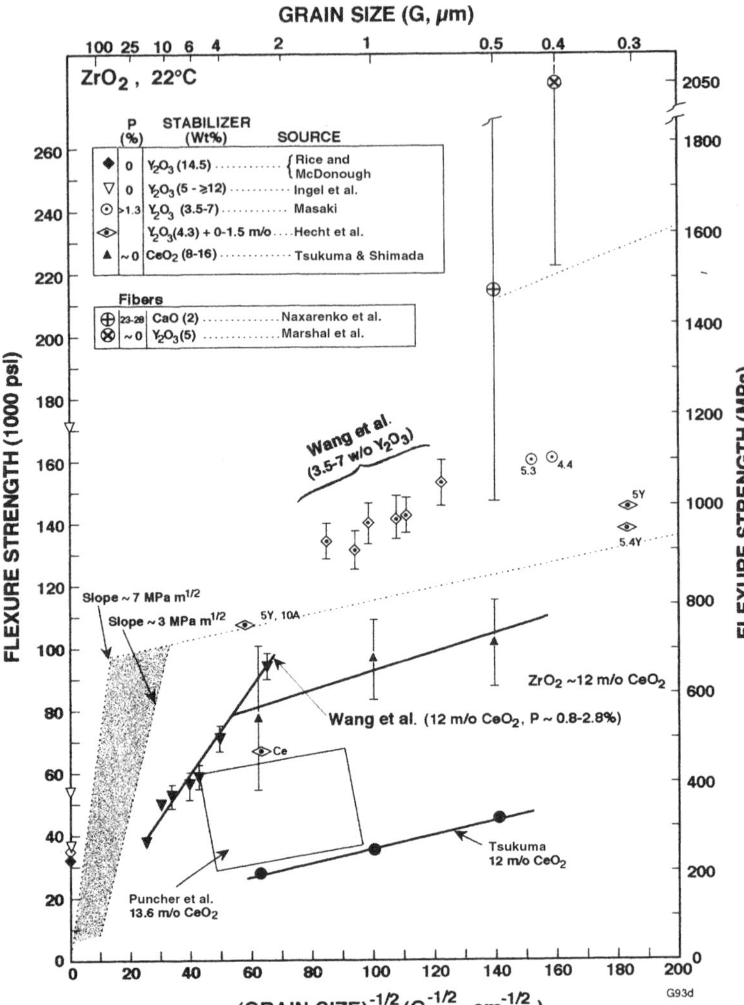

FIGURE 3.21 ZrO_2 σ–$G^{-1/2}$ data for machined bars with various stabilizers. Some PSZ data is shown from a previous survey [74] (Fig. 3.10) along with data for PSZ crystals of Ingel et al. [80]. Most data is for dense TZP bodies of Tsukuma and Shimada [177], Wang et al. [179, 180], Masaki [181], Hecht et al. [182], Tsukuma [183], and Puchner et al. [184]. Stabilizer contents (Y_2O_3 w/o shown next to data points of Masaki). Vertical bars = standard deviations, and numbers at bar bottoms the number of tests. Note (1) strengths of ~ 1 GPa for Y-TZP of Kihara et al. [185] with finer *G* (~ 0.5 μm) from modest additions of Al_2O_3 (to 1.5 m/o) agree well with other data, (2) TZP data appears mainly or exclusively as finer *G* branches for the larger *G* PSZ data, and (3) the general agreement of the bulk TZP data with that for ZrO_2 fibers [74, 75, 186]. (From Ref. 75, published with the permission of the *Journal of Materials Science*.)

of K_{IC} of such material (5–19 MPa·m$^{1/2}$) predicting intersecton per the model for $c(\sim G/2)$ at ~50 – 400 μm. Note that while the Ce-PSZ materials varied in σ (as expected from variations in composition and processing, hence resultant quality), they all show (1) general consistency between data sets, (2) lower ~ and moderate but definitely positive, finer *G* branch slopes, and (3) more definitive and extensive microplasticity.

The above trends show that toughened ZrO_2 bodies follow very similar trends as normal ceramics, but with strength levels often increased substantially due to transformation toughening. Again, larger *G* branches extend to strengths considerably below those for single crystals of comparable composition and surface finish, and finer *G* branches have lower slopes than the larger *G* branches, but often substantial slopes. Although data to define intersection(s) of the larger and finer *G* branches is limited, there are indications that larger and finer *G* branches do intersect when $c \sim G$ (Table 3.1). Extrapolation of this data gives an intersection with the larger *G* PSZ data at $G \sim 10$μm, which is consistent with the expected intersection at $c \sim G$. Further, other miscellaneous TZP data, e.g., of Rice and McDonough [74], Jue and Virkar [82], Masaki [181], and Hecht el al. [182], would all be consistent with their being on the same or similar finer *G* branches, dependent upon their composition and processing. Note that primarily composition and secondarily processing are important factors in strength, and hence in determining the branches bodies appear on, especially in the finer *G* branch regime. Whether the finer *G* branches, especially for bodies (e.g. Ce-PSZ) showing considerable plasticity, extrapolate, at least approximately, to single crystal yield stresses at $G = \infty$ (e.g. to ~ 400 MPa) is not known. Such stress levels may not be unreasonable for a solid solution type system (versus compressive yield stresses in precipitate PSZ crystals, ~1200 MPa for Mg-PSZ [187] and ~2800 MPa for Y-PSZ [188]). Again note that similar trends for machined and as-fired surfaces, including extrapolation to fiber strength–grain size trends, are also shown by ZrO_2 data. There are again some uncertainties in correction for porosity (especially for Nazarenko et al.'s ZrO_2 fibers [155], since the same or similar correction as for their Al_2O_3 fibers would imply a much greater Young's modulus than for theoretically dense ZrO_2. However, data for finer *G* TZP bodies overlaps with the data for ZrO_2 fibers partially stabilized with CaO or Y_2O_3 [155, 186] (Fig. 3.21).

Collectively, these results show that grain size plays a major role in the mechanical behavior of partially stabilized ZrO_2 materials, in essentially the same fashion as it does in other ceramics. The effect of toughening from transformation associated with partial stabilization appears to manifest itself primarily by effects of increased K_{IC} increasing strengths by either allowing the introduction of smaller (e.g. machining) flaws or more difficult flaw propagation, or both, thus increasing σ over and above that for fully stabilized materials, but maintaining a similar *G* dependence. However, there is again a

TABLE 3.1 Summary of Fracture Mechanics Evaluation of σ–$G^{-1/2}$ Behavior of Ceramics

Material (Fig. no.)	K_{IC}[a] (MPa·m$^{1/2}$)	Larger G[b] Slope (MPa·m$^{1/2}$)	Larger–finer G branch intersection σ (MPa)	G (μm)	c(μm)[c]
(A) Noncubic Ceramics					
Al_2O_3 (Fig. 3.13, survey)	3.5 (1.5–2)	3.2 (2.3)	590	25	22 (7)
Al_2O_3 (Fig. 3.13, survey)	3.5 (1.5–2)	4.0(2.8)	290	120	92(30)
Al_2O_3 (Fig. 3.13, Komaya et al.)	3.5 (1.5–2)	2.7 (1.9)	440	23	40 (13)
Al_2O_3 (Fig. 3.13, Tomaszewski)	3.5 (1.5–2)	2.0 (1.4)	180	70	240 (78)
Al_2O_3 (Fig. 3.14, Alford et al.)	3.5 (1.5–2)	3.2 (2.3)	800	16	12 (4)
Al_2O_3 (Fig. 3.14, Charles)	3.5 (1.5–2)	2.2 (1.5)	290	60	92 (30)
Al_2O_3-W (Figure 314, Hing et al.)	~ 3.5	2.3 (1.6)	350	70	62
BeO (Fig. 3.16, survey)	3.5	1.8 (1.3)	250	25	123
TiO_2 (Fig. 3.17, Alford et al.)	2.5 (0.8)	1.6 (1.1)	680	6	11 (0.9)
TiO_2 (Fig. 3.17, Alford et al.)	2.5 (0.8)	1.6 (1.1)	250	30	63 (6.5)
β Al_2O_3 (Fig. 3.19, Virkar and Gordon)	~ 3 (0.16)	1.9 (1.3)	170	110	196 (0.6)
β Al_2O_3 (Fig. 3.19, Virkar and Gordon)	~ 3 (0.16)	1.9 (1.3)	120	190	394 (1.1)
B_4C (Fig. 3.22, survey)	~ 3.3	1.4–2.3 (1–1.6)	550	15	23
Si_3N_4 Fig. 3.23, Larson et al.)	≥4	2–4 (1.4–2.8)	450	18–110	50
(B) Cubic Ceramics					
ThO_2 (Fig. 3.7, Knudsen)	1.1(0.65)	0.7 (0.5)	230	15	19 (5)
ZrO_2 (+Y_2O_3, Fig. 3.9)				15	
$MgAl_2O_4$ (Fig. 3.11, survey)	2 (1)	1.8 (1.3)	400	25	16 (4)
$MgAl_2O_4$ (Fig. 3.10, Bailey and Russell)	2 (1)	2.1 (1.5)	330	30	23 (6)
SiC(Fig. 3.12, Cranmer et al.)	3.5 (2)	2.6–3.5 (1.8–2.5)	400	50	48 (16)
SiC(Fig. 3.12, CVD, Gulden)	3.5 (2)	2.6–3.5 (1.8–2.5)	950	10	9 (3)
C(CVD diamond, Fig. 3.14, Savage et al.)	5	~ 6+(4.3)	500	100	100–250

[a] Polycrystalline and single crystal values per Table 2.1, the latter in ().
[b] Values in () are the slopes multiplied by 0.71 to convert them to comparable toughness values, since grain size (G) is measured as a diameter and flaw size(c) as a radius.
[c] Values shown are calculated for a surface half penny crack per Eq. (3.3) using either polycrystalline or (where known) single crystal fracture toughnesses, the latter results shown in ( ).

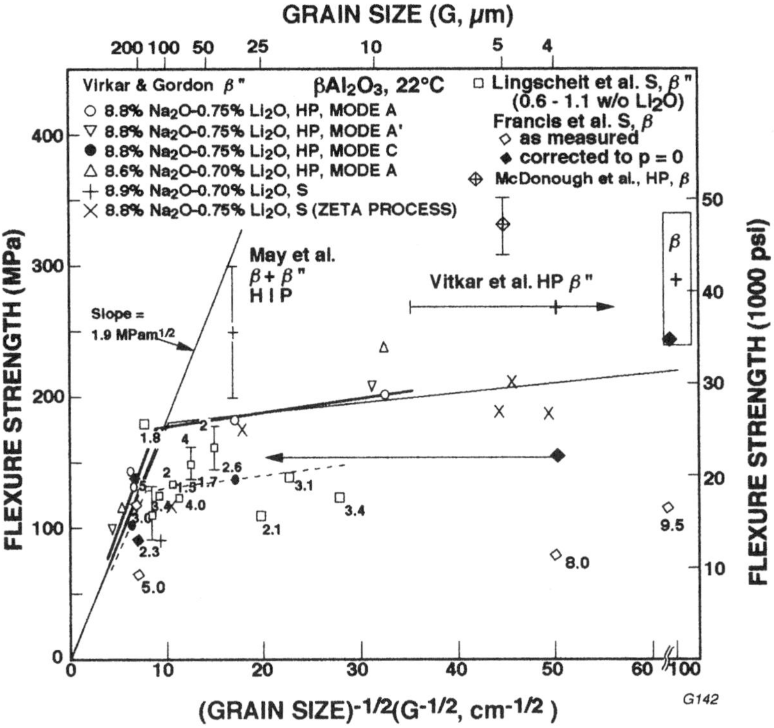

FIGURE 3.22 Strengths of various machined beta aluminas versus $G^{-1/2}$ of Virkar and Gordon [195] (hot pressed or sintered, plotted vs. G_m) and various other studies [196–200] of hot pressed, HIPed, and sintered bodies plotted versus average G, G_a (horizontal bars for some bodies with a wider G range). (From Ref. 2, published with the permission of the *Journal of Materials Science.*)

fundamental difference between the G dependences of strength and toughness (Figure 2.18), i.e. the strength continuously decreasing as G increases, while toughness goes through a, commonly significant, maximum at certain G, e.g. 1–3 μm. Thus there is a need to integrate the $\sigma - G^{-1/2}$ behavior with evaluation of mechanisms for transformation toughening. Some of this may involve increased G increasing the martinsitic start temperature, which reduces K_{IC} [189–194] and hence generally σ, thus also impacting $\sigma - G^{-1/2}$ trends, but such effects are not pertinent to fully stabilized ZrO_2 and are of uncertain applicability to precipitate toughened PSZ. Thus the toughness maximum must reflect testing effects not pertinent to normal strength behavior, though the general increase in toughness from transformation does increase strength in two ways, as is discussed later.

Data for (mostly hot pressed) β''-Al_2O_3 bodies of Virkar and Gordon [195] vs. G_m (G_a values and specifics of the G_m determination were not given) clearly showed the characteristic two-branch $\sigma - G^{-1/2}$ behavior with the larger G slope < K_{IC} (Table 3.1) and the finer G slope probably >0 (Fig. 3.22). The crystallographic orientation from hot pressing was a primary factor in a second probable finer G branch. More limited earlier data of other investigators [196–200] was generally consistent with this data, given effects of some porosity, composition differences, and larger grains from exaggerated grain growth [2]. However, again resultant larger tabular grains were large enough to be fracture origins in some cases and not others. McDonough et al.'s [196] fractography in their uniformly fine grain bodies showed clear fracture origins from typical machining flaws substantially larger than the fine (~5 μm) grains.

Chu et al.'s data for Li_4SiO_4 (G ~ 3–50 μm, P 2–32%, increasing as G decreased) indicated a possible finer–larger G intersection at G ~ 3 – 10 μm using their porosity correction of $e^{-3.8P}$ [201]. The limited data on highly anisotropic

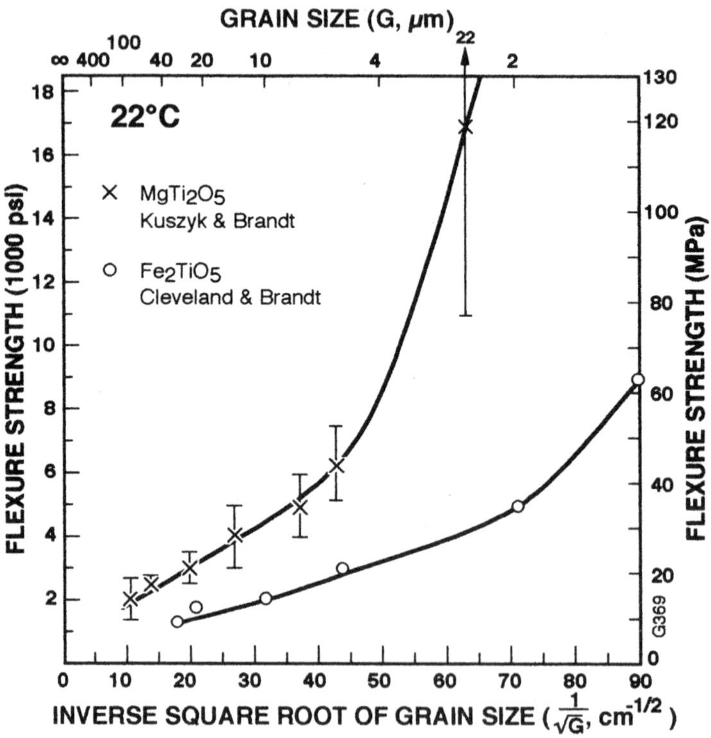

FIGURE 3.23 Strengths of $MgTi_2O_5$ [202] and Fe_2TiO_5 [203] as a function of $G^{-1/2}$.

materials such as $MgTi_2O_5$ [202] and Al_2TiO_4 [203] with extensive microcracking showed a significant change in $\sigma - G^{-1/2}$ behavior, i.e. a much faster decrease in strengths at fine G, then saturating as G increases (Fig. 3.23). Data for these two materials also show nominally identical G dependence of their Young's moduli.

C. Noncubic Nonoxide Ceramics

Data for dense sintered and hot pressed bodies MgF_2 [204 and 205] shows strengths of bodies with average $G \geq \sim 1$ μm substantially decreasing within normal bounds as G increases (Fig. 3.24), with some difference in strengths at a

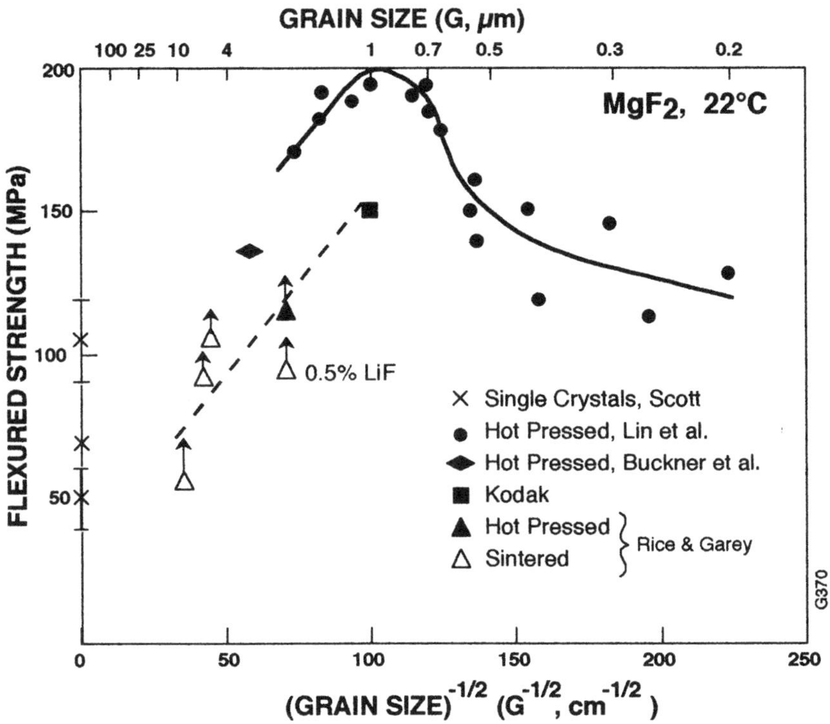

FIGURE 3.24 Strength versus $G^{-1/2}$ for hot pressed and sintered MgF_2 test bars with polished surfaces [204–207], and for three basic orientations of single crystals (error bars are shown for the standard deviations where greater than the symbol height), with no evidence of plasticity observed [204]. Note the decrease in strength with increasing G at $G > 1$ μm and decreasing $G < 1$ μm. Vertical arrows indicate probable corrections for limited porosity.

given G between the different investigators, data of Lin et al. [205] being higher than the others. This probably reflects typical flaw differences due to differences in machining, residual processing defects, and specimen size and other testing effects, which are particularly pertinent to this probably being the finer G branch. This data is consistent with an intersection with the larger G branch being at $\geq G = 10$ μm, which in turn is consistent with typical polishing flaws in dense (transparent) MgF_2 being of the order of ~ 10–20 μm [18]. Note that polycrystalline strengths are below single crystal strengths. Also note the clear decreases in strength as G decreased below ~ 1 μm, attributed to retention of (primarily anion, e.g. OH) impurities [204] from higher pressure hot pressing at lower temperatures, e.g. ~ 210 MPa at 650°C [205] and 240 MPa at 568–713°C [207], both for powders of ~ 0.1 μm particles.

Earlier TiB_2 [11, 208–210] agrees with more recent data of Baumgartner and Steiger [211] as well as that of Becher et al. [212], and Telle and Petzow's [213] evaluation of Watanabe and Kouno's [214] data indicate a clear larger G branch and possible finer G branches at various strength levels (Fig. 3.25). Other limited data is generally consistent with these trends [215–218].

B_4C data from mostly hot pressed bodies in a previous survey [13] showed two branches, with finer G branches showing more effect of grinding parallel or perpendicular to the tensile axis than the larger G branch (i.e. each forming a finer G branch [17]. A further survey [2] (Fig. 3.26) of substantially more data [219–234] corroborated these trends, i.e. generally showing positive finer G slopes, but greater G dependence of the larger G branch, which may or may not extrapolate to $\sigma = 0$ at $G = \infty$. These data cover a considerable composition range, showing no clear, substantial effect of this on σ, other than via G, except for the hot pressed B_4C data of Rybalchenko et al. [225] (~2–5% porosity), which generally falls well below the other data. This may be in part due to lower quality (unspecified) surface finish but appears to reflect processing effects. Thus σ increased at a given G with both hot-pressing temperature and pressure, and for a given pressing pressure σ often even increased with increasing G for $G \leq 5$ μm (for all but one of their bodies). The trend for limited effects of composition in most cases is consistent with evaluations of B_4C hardness data showing that most of the composition effects were via G [231], in contrast to claims of Niihara and colleagues [232, 233].

Fractography has played an important role in better defining the G dependence of B_4C strength, for example some failure from substantially isolated or clustered larger (exaggerated) grains [2, 13]. While this is similar to Al_2O_3, it is also different, since the large grains in B_4C are typically ~ equiaxed, occur less frequently than in Al_2O_3, and have almost, if not universal transgranular fracture. Frequent twins in B_4C, at least in larger grains, may be a factor in these differences, but correction for observed and expected failure from isolated or clustered larger grains (e.g. Figs. 1.20, 1.30) shifts some data points to the larger G branch,

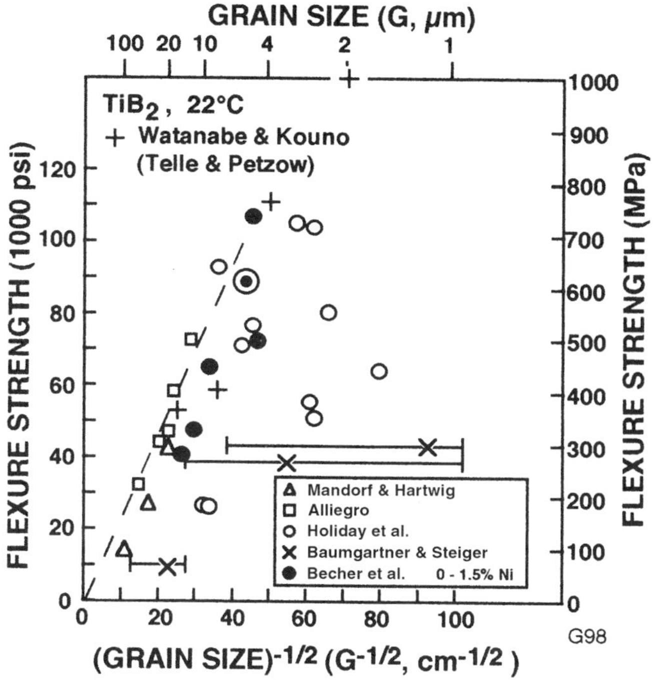

FIGURE 3.25 TiB_2 σ–$G^{-1/2}$ data of Mandorf and Hartwig [208], Alliegro [209], and Holiday et al. [210] from an earlier survey [11] along with more recent data for similar materials by Baumgartner and Steiger [211] as well as data of Becher et al. [212] (for specimens with 0–1.5% Ni additions) and from Telle and Petzow's [213] evaluation of data for nearly dense bodies of TiB_2 with three different additives from Watanabe and Kouno [214]. (From Ref. Rice [2], published with the permission of the British Ceramic Society.)

increasing its slope a limited amount, and removing some more extreme lower strengths from the finer G branch. However, again fractography has also shown that some data points are (incorrectly) moved further to the left than most of the data, and some not far enough by using G_m, indicating respectively that they (1) are not pertinent to σ–G behavior, i.e. reflect a more serious defect and (2) are not large enough to be the flaws by themselves. Further, fractography clearly showed cases of failure initiation not occurring from large grains (Fig. 3.27), again showing that the arbitrary use of G_m even in the vicinity of the fracture origin can lead to serious errors. Both the $\sigma - G^{-1/2}$ data and fractography again show intersection of the finer and larger G branches when $c \sim G$. Fractography also showed (via extensive transgranular fracture) mist and hackle, arguing

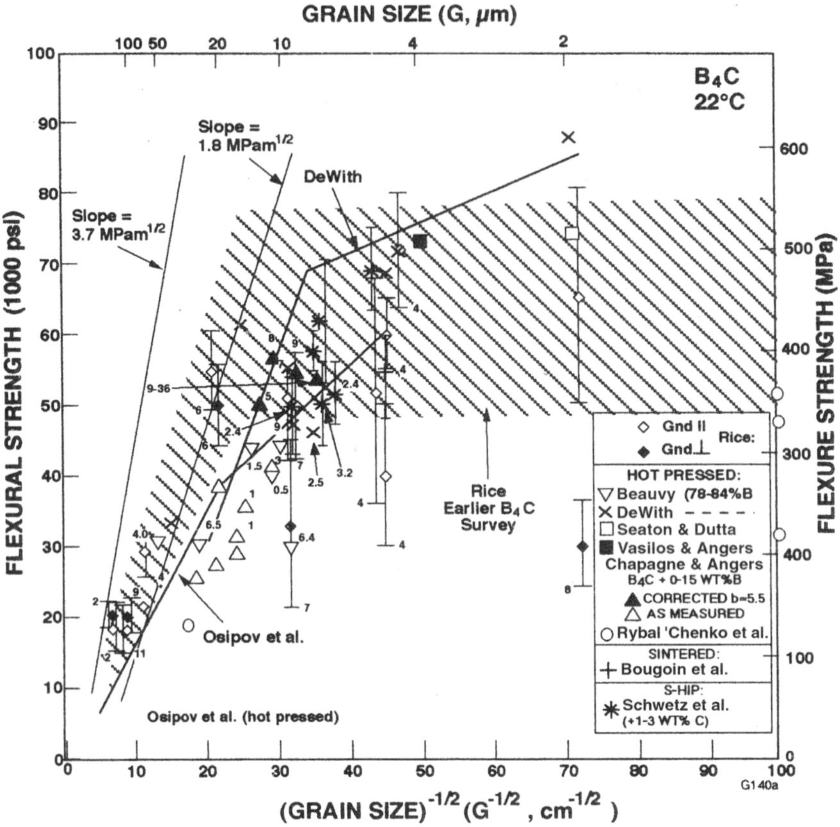

FIGURE 3.26 B_4C σ–$G^{-1/2}$ data (mostly hot pressed) from an earlier survey [13] as well as subsequent studies of Rice [17] of two different bodies ground parallel or perpendicular to the tensile axis are shown along with data of Osipov et al. [219], de With [220] (all hot pressed), Champagne and Angers [221] (with varying excess B) Beauvy [222], Seaton and Dutta [223], Vasilos and Dutta [224], Rybal'chenko et al. [225], Bougoin et al. [226] (sintered), Schwetz and Grellner [227], and Schwetz et al. [228] (sintered-HIPed with varying excess C). Numbers next to points are % porosity where > 1%. Vertical lines are standard deviation, and associated numbers are the number of tests. (From Ref. 2 published with the permission of the *Journal of Materials Science.*)

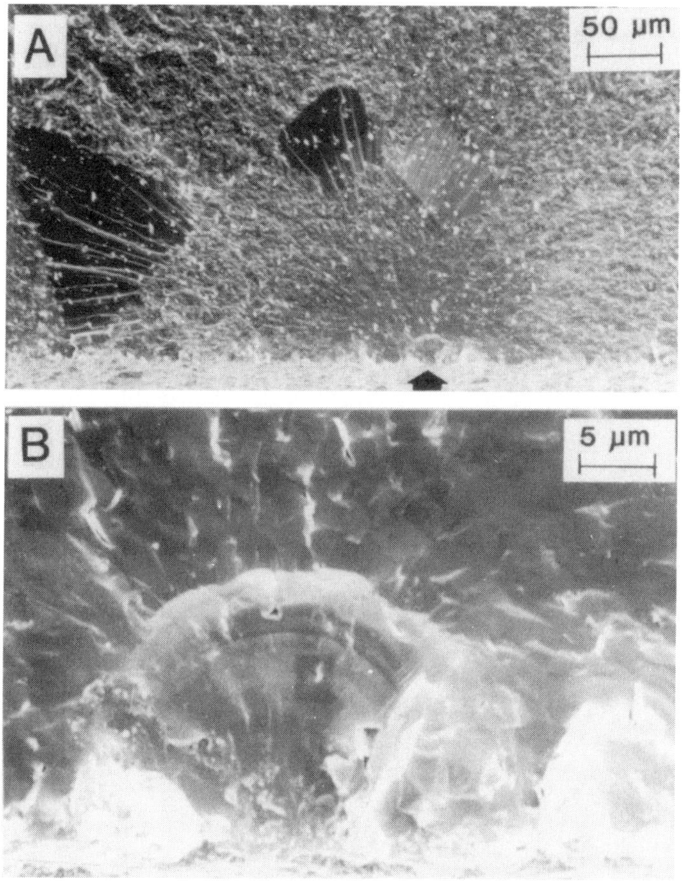

Figure 3.27 Fractographic demonstration of failure not occurring from much larger grains in dense B_4C. (A) Lower magnification SEM showing a nearly half penny origin (arrow) from a machining flaw in the fine grain structure and a higher magnification of it in (B). Note in (A) the two large grains to the left of the origin, the largest one touching the tensile surface, are clearly not the origin (From Ref. [2], published with the permission of the *Journal of Materials Science*.)

against significant subcritical crack growth and bridging, which typically is via mostly intergranular fracture. Finally, note (1) that Rice [234] showed evidence of TEA stresses contributing to tensile failure (Fig. 3.35), and (2) that σ trends with G are clearly inconsistnt with toughness results, typically having significant maxima at $G \sim 10\ \mu m$ [2] (Chap. 2, Sec. III.F).

As is shown in Fig. 3.13, the limited α-SiC data is generally consistent with that for cubic, β-SiC, as is expected from the limited anisotropy of noncubic α-SiC. However, exaggerated growth of α-SiC grains in β-SiC give frequent fracture origins, e.g. as first shown by Prochazka and Charles [96]. A previous survey [11] of more limited data of Cr_3C_2 and WC shows large *G* behavior like that of other ceramics, as well as some evidence of finer *G* branches. Most WC-Co data appears to form fine *G* branches, e.g. for pure (larger *G*) WC data. However, Roebuck's [236] WC-Co data from a novel flexure test concentrating stress in a small test volume showed strengths increasing from 1 to 3 GPa as the origin (pore) size decreased from ~ 100 to ~ 20 μm, which then was constant for failure from larger grains of ~ 6 – 30 μm. This may reflect fracture initiation due to Co deformation, WC-Co miss match strains, or both, again indicating possible changes in mechanisms at small crack sizes.

AlN data of Rafaniello [237] for machined bars from bodies sintered with a few percent of different oxide additions (typically leaving ~ 0.3–2% porosity) clearly showed strengths increasing as G^{-12} increased (Fig. 3.28). More limited data of Rice [15] for a finer *G* sintered AlN (+ Y_2O_3) and a hot pressed body (with CaO addition) is in excellent agreement for machining both parallel and perpendicular to the bar/stress axis, consistent with such machining direction effects. Substantial data of Hiruta el al. [238], for AlN bodies

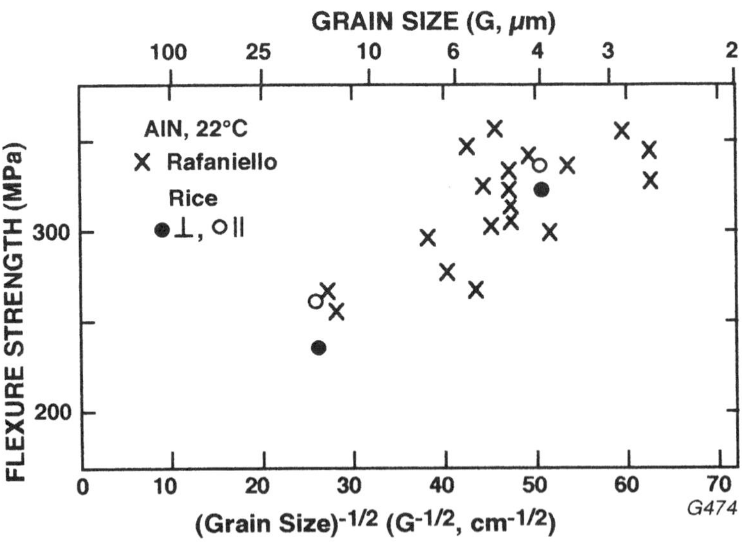

FIGURE 3.28 Flexure strength of dense (< 3% porosity) AlN processed with different additives versus the inverse square root of grain size ($G^{-1/2}$). (From Refs. 15, 237.)

made with various additions (and < 2% porosity) over a similar G range, though somewhat (e.g. ~ 10 – 15%) higher in strength (probably reflecting moderate test-specimen conditions), are also generally consistent with the other data and the G^{-12} dependence.

Though limited in its extent (mainly at intermediate and larger G), a previous compilation of Si_3N_4 data [2] indicates typical large and fine G branches. Plotting of Larsen et al.'s [100] hot pressed and reaction sintered Si_3N_4 (corrected to zero porosity, Fig. 3.29) shows the same trends, i.e. (1) at finer G, higher σ's being from machining flaws rather than microstructural heterogeneities, and lower σ's being from microstructural heterogeneities such as large tabular grains, and (2) the larger G slope (using G_m) is < K_{IC} (≥ 4 MPa·m$^{1/2}$, for P

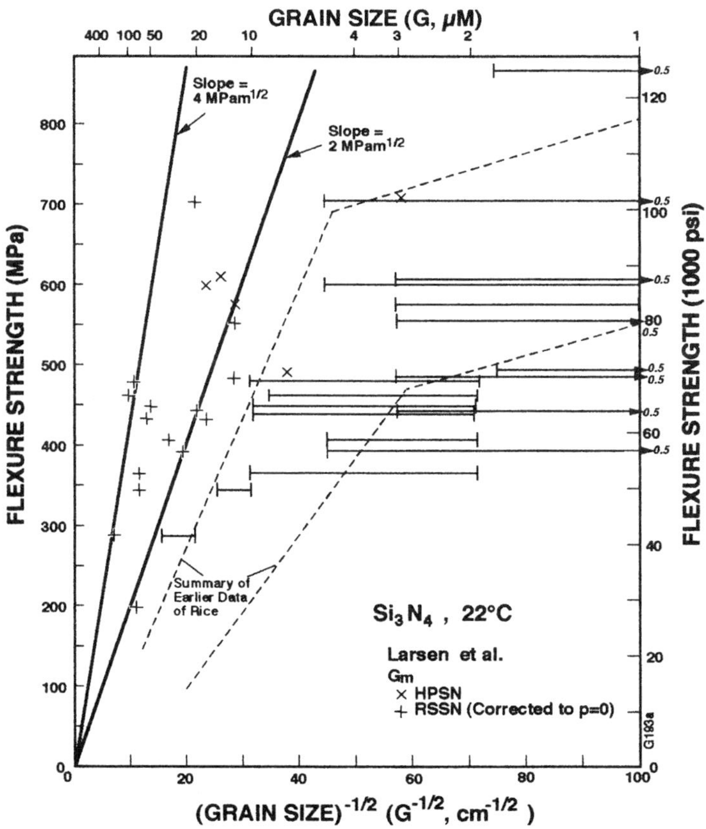

FIGURE 3.29 Strength versus $G^{-1/2}$ for various, mainly sintered or hot pressed, Si_3N_4 bodies. (From Refs. 2, 11, 67, 100.)

= 0). Further, limited fractographic studies showed failure from larger, usually exaggerated grains [100, 239, 240], as has also been found in self-reinforced bodies, often associated with excess additive phase [241–244]. Again note that significant toughness maxima often seen with *R*-curve effects (Chap. 2, Sec. III.F) are inconsistent with $\sigma - G^{-1/2}$ behavior.

Data on graphites and related carbon materials is limited by the extensive complications of the typical substantial porosity and local or global grain orientation and resultant anisotropy of properties reflecting the high crystalline anisotropy, as well as the coupling between these factors. However, limited attempts to sort out the grain size component have consistently shown that strength decreases with increasing *G*. Thus Brocklehurst's review [245] notes such *G* dependence of graphites, citing in particular work of Knibbs [246]. Rice has also presented some analyzed data in a survey [11]. While details of this *G* dependence of strength such as finer and larger *G* branches are uncertain, they are not inconsistent with the basic model (Fig. 3.1).

V. COMPOSITIONAL, MICROSTRUCTURAL, AND SURFACE FINISH EFFECTS

A. Compositional and Nanoscale Grain Effects

A diversity of compositional effects can occur ranging from enhanced to reduced strength, which may involve a *G* dependence or obscure it, e.g. the latter via global effects of body composition on strength through changes in *E*, or more local changes. Effects mainly or exclusively via global changes in properties impacting strength are only outlined here, since they are a very diverse topic and often require large additions to have substantial effect. However, their effects must often be recognized in order to adequately identify and understand σ–G relations. Both local and global compositional effects on micro-, especially grain-, structure are addressed in more detail. An important separation of effects is in part associated with whether the other constituents are as a second phase or in solid solution.

Impurities or additives that end up in solid solution commonly increase strengths of materials in which microplasticity, especially slip, determines their strength, even at modest levels, e.g. CaO [46–48] (Fig. 3.4), MgO [48, 51, 52] (Fig. 3.5), and $BaTiO_3$ [41] (Fig. 3.3). Such increased strength should be independent of *G* if a suitable solid solution is achieved, uniformly raising the σ–$G^{-1/2}$ line. However, a decreased solid solution with reduced temperature–time exposure in obtaining finer *G* would reduce the σ–$G^{-1/2}$ slope. Increased solid solution (and undissolved phase, typically at grain boundaries) would commonly result in transitioning from microplastically controlled strength at larger *G* values. For materials whose strength is controlled by brittle failure from flaws, constituents in solid solution may affect strength by altering properties such as *E*, or EA and especially TEA, but this typically requires substantial contents for significant effects. However, limited study of such possible effects is not definitive (e.g. for Cr_2O_3 in Al_2O_3 [10, 145].

While added constituents as second phases can influence microplastic behavior, e.g. via precipitation toughening, effects of such phases have more extensive effects on failure due to preexisting flaws, and some on failure from microcracks. Again, a general effect is via changes in properties affecting strength; a key example of this is effects on Young's modulus, e.g. of SiO_2 (Fig. 3.30), which can lower *E* substantially. This reduction may or may not depend on the degree of reaction and hence processing conditions that also affect grain structure (though this is often not much of a factor in the Al_2O_3-SiO_2 system, since *E* is not greatly different for mixtures or compounds of these at a given constituent ratio). Another, though probably less common, property effect that is pertinent to the presence of lesser quantities of SiO_2 is surface roughness of as-fired surfaces. Some strength increase of alumina fibers containing or coated with SiO_2 has been attributed to reduced surface roughness [159].

Other diverse compositional effects include effects on microcracking, either increasing this by forming phases with substantial expansion differences with themselves or the matrix, or decreasing this by reducing intergranular stresses, e.g. as some silica glass phases may. The latter often require less second phase and are often particularly dependent on wetting effects, which at elevated temperature can seriously reduce strengths with more extensively coated grain boundaries. However, serious effects can also occur at moderate temperatures, e.g. Wakamatsu et al. [247] recently showed lower strength in Al_2O_3 + V_2O_3 sintered in air due to formation of an $AlVO_4$ grain boundary phase despite some inhibition of grain growth. On the other hand, firing in a reducing atmosphere, which also inhibited grain growth, resulted in solid solution of the V without significant strength degradation, but with a higher surface concentration of V^{4+}.

Another example of variations and complications is with TiO_2 additions to Al_2O_3, which can substantially aid densification [248, 249]. However, net effects can again be dependent on processing and use details. Thus beyond the solubility limit in vacuum sintering, grain growth inhibition occurs [250], but subsequent air annealing can result in preferential diffusion to grain boundaries and the surface with some reduction in strength. Hot pressing with TiO_2 + MgO additions has indicated loss of MgO inhibition of grain growth [251]. High-temperature air sintering with TiO_2 additions results in large grains, substantial intragranular pores, very rough surfaces, and resultant low strengths. However, oxygen partial pressure can have substantial effects on TiO_2 in Al_2O_3 [252]. Thus while extensive second phase precipitation can have various effects, including toughening, as is discussed for ceramic composites (Chaps. 8 and 9), second phase effects can be complex, varying, and direct or indirect.

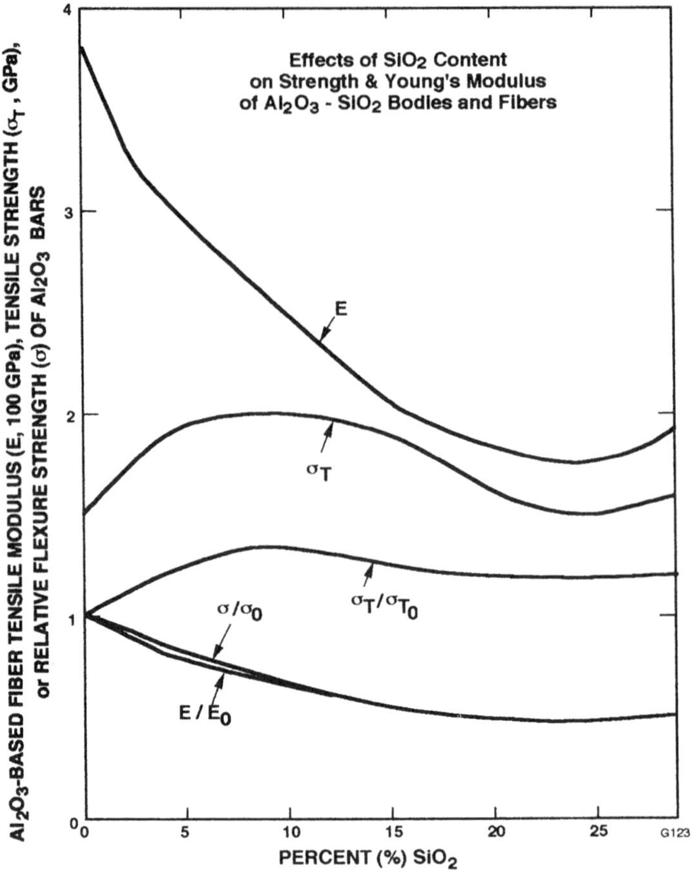

Figure 3.30 Effects of SiO_2 content on Young's modulus and tensile and flexure strengths of alumina bodies based on data for fibers and flexure bars. Such data is only a rough guide, since effects of different processing, wetting and reaction, and microstructure are not fully specified or investigated. Curves were obtained from analysis of Steele et al.'s [122] data for bulk alumina bodies and from data for alumina rich fibers [162] by using the experimental correction of σ for lower density, then subtracting the decreases due specifically to porosity, leaving the reductions due to reduced density from the SiO_2 additions.

Consider next effects of local compositional variations, which can cause microcracks, and especially larger local grains, by themselves, or with microcracks, the local phase itself, or both acting as fracture origins. While occurring in several materials, such origins are commonly observed, especially in larger grains, in Al_2O_3 and in situ toughened Si_3N_4. Thus formation of larger platelet beta alumina grains that act as origins have been noted in Al_2O_3 bodies with alkali [253], e.g. especially Na, impurities (e.g. in Bayer processed Al_2O_3) made without grain growth inhibitors [24] (Figure 1.3A). In Si_3N_4, large rod shaped grains singly or more commonly two or more in combination with excess additive phase are dominant fracture origins as grain sizes and the level of toughness increase [241–244]. While both the number of grains clustered and the extent of excess additive phase, as well as the location of these relative to flexure tensile surfaces, affect the resultant effective flaw size (Fig. 3.31), there has been little evaluation. Such evaluation is of uncertain relevance to basic σ–$G^{-1/2}$ behavior, e.g. for projection of strengths to significantly larger or finer grain sizes, since such combinations often represent a significantly different flaw population. Clues to the validity of such extrapolations are the similarity of the larger grain shapes, phases, associated amount and character of matrix phase, and resultant fracture mode to the body matrix. However, the above questions are relevant for understanding strength behavior of such bodies better to control their microstructure and properties. Such cases of heterogeneous larger grains may also reflect some local preferred orientation due to interactions of possible relations of nuclei or the impurity distribution to fabrication effects. Further, impurities or additives can be interactive with each other, as well as with effects of environment on strength.

Consider now the more general case of more global, hence typically more uniform effects of other constituents on grain structure, which are often sought and used to reduce both average and maximum grain size to increase strength. Thus note that stoichiometry variations in spinel samples [84, 85] (Fig. 3.11) had their primary effects via grain size changes, with possible (but unidentified) secondary effects on physical properties such as Young's modulus. Similar effects are shown with other mechanical properties in subsequent chapters. Other constituents (often nonreactive or nonsoluble ones) are also often added to control grain size, which typically greatly limits exaggerated gram growth and hence also commonly deviations from equiaxed grain shape. Key examples of these for Al_2O_3 are MgO, ZrO_2, Mo, and W (Fig. 3.17). Effects of such additions are commonly primarily or exclusively directly related to the reduction in grain size, increasing strengths, so long as the level of addition is limited. At higher levels of additions such grain growth inhibitors can change effects, e.g. MgO may form larger spinel grains, hence possibly lowering strengths. Both average and maximum alumina strengths increased with increasing Mo additions to ~ 6 vol% and then leveled off (or, especially for the maxima, may decrease slightly) to the limits of investigation of 16 vol% [141]. This limitation on effects of Mo additions was attributed to coalescence of Mo particles limiting further alumina grain size reductions but may also reflect some reduced elastic moduli and possibly bonding. However, while increasing AlON additions to alumina (via reaction processing) from 0 to 40 vol% decreased toughness some, e.g. by 10–20%, it gave intermediate flexural strengths [147] (Fig. 3.17), consistent with the intermediate grain size constant over the range of AlON additions (but better retention of strengths at higher temperatures). On the other hand, increased ZrO_2 additions enhanced-strength and toughness to the extent that transformation toughening occurs and increases with increasing ZrO_2 additions, but at low addition levels (e.g. <6%), the primary benefits come from maintaining fine G [153, 154]. The above constituents, whether additives or impurities, are either metals or compounds with anions giving bond strengths and temperature capabilities similar to those of the matrix material. Significant reductions in strengths at finer grain sizes, attributed to impurities involving anions of much weaker, much less refractory compounds, are discussed below.

While examples of added phases inhibiting grain growth are particularly common in Al_2O_3, as was outlined above, it is demonstrated in other materials, e.g. BeO with SiC [174, 175] or Al_2O_3 [173] additions (Fig. 3.19). Though some nonoxide materials such as Si_3N_4 are less prone in their pure state to substantial grain growth, some important densification aids for nonoxides also significantly limit grain growth, contributing to improved strength via grain size limitations. Thus pure dense WC typically has G values $\geq$ 10 μm, while metal, e.g. Co, additions generally give dense bodies in which G = 1–3 μm, and pure TiB_2 commonly has grains 10 to 20 μm, while densification with Ni reduces G's to 5 to 10 μm (but again with diminishing returns with increasing Ni content) [212]. Certainly in such cermets and other composites there can be other important effects increasing strength over and above that due to G reduction, but the latter can be an important factor.

Consider now effects of impurities characterized by anions that severely limit refractoriness and other properties that can be pervasive and severe in obtaining much finer, especially nanoscale, grain structures in some materials. These impurities and their effects are almost universally neglected, despite there being both reasonable theoretical, and especially experimental, evidence of their occurrence and effects [254, 255]. They are primarily materials such as hydroxides, carbonates, sulfates, etc., e.g. often the sources of common oxide ceramics, or of impurities (e.g. of Ca), and can form on oxide and on some other material surfaces. They are neglected because it is commonly assumed that their normally low decomposition temperatures preclude their presence in a solid body due to densification temperatures. However, this neglects effects in densification that inhibit decomposition, hence allowing retention of limited quantities of these (e.g. from ≪ 1 to a few percent). While there are some chemical interactions (e.g. with densification aids such as LiF) that limit decomposition, a major factor is the pressures commonly used to obtain high densities at low temperatures necessary to limit grain growth and hence retention of finer grains from fine particles. The higher surface areas and lower decomposition temperatures commonly used to obtain finer powders exacerbate the problem by leaving more of these impurities in higher surface area powders to be densified. While such impurities can cause various fabrication problems, they can also be an aid in densification, e.g. the ready direct hot pressing of magnesium hydroxides of bicarbonates to near theoretical density (transparent), high purity (≥ 99%) MgO [256].

There are clear, well documented manifestations of these impurities [254, 255]. A simple direct one is IR transmission of frequently transparent or translucent specimens that commonly clearly show anion species discussed above via their IR absorption bands. More general methods are based on heating, with weight losses, though commonly limited, often being definitive, as is identification of the evolved species, e.g. by mass spectrometry. Particularly demonstrative are clouding, blistering, gross bloating, and in the extreme, crumbling back to powder (e.g. for MgO and Al_2O_3) pressed to ~ full density at room temperature with ~ GPa) [254, 255, 257] due to the large, ~ 10^4 leverage in the expansion of the gaseous species released in solid state thermal decomposition on heating.

Turning to the specific effect of such impurities on σ–*G* relations, it is important to note that such impurities will commonly be mostly or exclusively at grain boundaries, where they will typically have pronounced effects even for very modest quantities, due to their markedly different, usually much lower, properties than the ceramic itself. Thus strengths of MgO, while typically increasing with decreasing *G*, e.g. to *G* ~ 10 μm or less [48, 50–54, 254, 255] start decreasing as *G* decreases further, often falling well below the generally expected strengths at finer *G*, especially with higher pressure processing to obtain finer *G* bodies with decreased grain size. The latter clearly show the presence of such anion species as hydroxides and can have their strengths markedly increased, e.g. doubled, by very slowly annealing such specimens to remove the impurity without serious disruption of the body structure. In such cases, the resultant strengths are those projected from larger grain bodies. Note that the decreased strengths of MgF_2 at finer *G* (Fig. 3.24) were correlated with observed OH absorption bands in the transparent bodies [207]. Similarly, decreased Al_2O_3 strengths at finer *G* from higher pressure fabrication [64] and demonstrated high temperature outgassing of species from hydroxides, carbonates, and sulfates (e.g. from alum precursors) also strongly indicate the same effects [254, 255]. This is also true of similar reductions in strengths at finer *G* as higher pressures are used for processing NiO and Cr_2O_3 in view of their, and their precursor, chemical similarities respectively with those for MgO and Al_2O_3 [64]. A very important corroboration of these effects is the high strengths obtained in fibers of these or similar ceramics. The use of only temperature, not pressure for densification at low temperatures for fine *G*, along with very small cross ections of such fibers, means respectively that limited quantities of such anion impurity species will be in the fiber, and those there are much more likely to diffuse out of the fibers at their modest firing temperatures.

The above effects in a few common ceramic oxides and one related nonoxide, MgF_2, probably reflect only a small fraction of the materials manifesting such anion impurity effects. Thus for example such effects have been indicated in Y_2O_3, manifesting themselves at higher temperatures (due to phosphate precursors) [255]. Also note that residues from the use of LiF additions that can be very beneficial to densification of several oxides, especially MgO and $MgAl_2O_4$, can also be detrimental to resultant strengths unless suitably removed by subsequent annealing, which is much more feasible in MgO than in $MgAl_2O_4$, as was discussed earlier (Figure 2.12).

Higher densification temperatures required for many other refractory ceramics such as borides, carbides, and nitrides indicate that such effects will be greatly reduced or not present in these materials from conventional powder processing. However, other low-temperature processing of nonoxides via preceramic polymers has similarities to processing of oxides from their common precursors and results in similar bulk body (Figure 2.14) [255] to fiber strength differences (Fig. 3.13). Whether such strength differences are related to similar impurity effects, residual stresses, or other effects, e.g. porosity, that would also be more serious in bulk bodies versus fibers is not established.

B. Grain Size Variation Effects, Especially Average Versus Maximum *G*

Even from a purely abstract standpoint, the issue of what *G* value to use for correlation with tensile strengths arises given any significant grain size distribution, strength being controlled by weak links. This issue is heightened by the frequent occurrence of fracture origins from larger grains or clusters of them, especially extremes of these from exaggerated grain growth in some materials such as Al_2O_3 (especially without grain growth inhibitors), beta aluminas, B_4C, SiC, and Si_3N_4 (Figs. 1.2–1.5). However, the use of the maximum grain size (G_m), as proposed by some investigators, is a serious oversimplification whose use leads to serious problems for basic reasons. A fundamental reason from an experimental standpoint is that larger grains or clusters of them are not always fracture origins, even when in a region of high to maximum stress. Thus, though not seriously sought by most investigators (nor for a long time by this author), clear cases of large grains not being fracture origins have been observed (e.g. Fig. 3.27).

A fundamental theoretical reason for larger grains or clusters of them not always being fracture origins, even when in a region of high or maximum stress, is that they may not be sufficiently serious flaws for either or both of two reasons. The first is that larger grains or clusters of them by themselves are not flaws; they must be associated with some other factor, a defect or stress, that causes the combination of it and the larger grain(s) to become a source of failure. Such a factor is typically an associated machining flaw (requiring the larger grains to be sufficiently near the finished surface), some other crack formed prior to testing or during testing in combination with the applied stress (e.g. from expansion differences between adjacent grains or second-phase regions or particles), or pores (e.g. Figure 1.4B). The second reason that larger grains or clusters of them are not always fracture origins is that they may not be of sufficient size. This is a particularly important problem for the finer G branches, which is also often the region where the difference between G_a and G_m has the greatest impact on the $G^{-1/2}$ dependence of strength. Thus, for example, use of a G_m instead of a G_a value can significantly change (often increasing) the slope of the finer G, especially at very fine G (where use of G_m has less relevance), thus distorting the projections of strengths to finer grain sizes. Note that the suggestion of most using G_m that the slope of the finer G branches is zero is generally not supported by their own data and is inconsistent with the great bulk of other data.

The above observations imply a significant statistical variation in the occurrence of larger grains or clusters of them at fracture origins, e.g. due to the statistics of association of larger grains and other defects. These statistical effects are compounded by statistical variations in orientations of easier fracture planes nearly normal to the applied stress. That the statistics of the spatial distribution of larger grains is a factor was shown by Stoyan and Schnabel [259]. Using a pair correlation approach, they showed a higher correlation of the spatial distribution of larger grains with σ for nearly dense Al_2O_3 bodies than with variations of G_a over their limited G range (~9 to 15 μm).

Larger grains or clusters of them frequently being fracture origins raises three broad questions concerning their (1) occurrence in terms of types of materials and body factors, (2) impact on σ–$G^{-1/2}$ relations, and (3) impact on engineering development. With regard to their material occurrence, the preceding sections and a previous survey [2] clearly show larger grains as fracture origins mainly in some noncubic materials such as Al_2O_3, beta aluminas, B_4C, and Si_3N_4, except for SiC where noncubic exaggerated tabular grains of α-SiC commonly form in β-SiC (e.g. above ~ 2000°C). Similar large grain origins have been reported much less in cubic ceramics, since they typically do not exhibit as much exaggerated grain growth common to many noncubic materials. In fact in cubic ceramics cases are seen where some failure initiation occurs from smaller grains, though in such cases smaller grains at the origin commonly abut one or more somewhat larger (bur not unusually large) grains, e.g. Fig. 3.10B.

Fracture initiation arising from larger grains is of practical, e.g. engineering, importance in order to identify their cause(s) so they can be reduced or eliminated to improve strength and reliability. However, such cases may not be relevant to basic σ–$G^{-1/2}$ relations or comparison of grain size dependences of strength and toughness to understand basic factors controlling fracture or projection of strength–grain size data to finer G as an indication of potential strength improvements by reducing grain size. Thus, three cases where larger G strengths may have limited relevance to basic σ–$G^{-1/2}$ behavior [2] are where impurities (1) with a pore (and possibly causing the pore) cause large associated grains (e.g. Figure 1.4B), (2) cause platelet grains (Figure 1.3A), and (3) cause large grains that also have associated cracks (Figure 1.2D), or also cause cracks or reduced properties over the area affected by the impurity.

A fourth, more specialized question is the role of G variations on slip-induced or assisted failure. Theoretically, strength should still be controlled by the largest grain size, since this gives the longest slip source–barrier distance and thus the lowest back stress. However, there are the combined statistical effects of grain orientation on both the resolved shear stress on easily activated slip systems and the stress normal to the resultant initiated crack that may significantly affect which grains become fracture origins. There are also issues of effects of impurities on grain size, e.g. impurities might cause larger grains but increase their yield stress, thus partly or fully counteracting effects of their larger size in favoring their being origins. Experimentally it is observed that slip nucleated fracture does occur from some smaller grains [52], though again there are often one or more somewhat larger abutting gains. In some of these cases it is not clear whether the slip nucleated crack formed in the smaller grain with the slip band or in the abutting grain against which the slip band piled up. Note that slip band length should on average be controlled by the three-dimensional grain size, while the failure causing crack is more related to the grain dimensions on the fracture surface.

Consider now more specific experimental results based mainly on fractography, especially in alumina bodies, with some observations secondarily also from B_4C [2]. As noted earlier, identification of specific fracture origins from larger grains showed plotting strengths versus the observed G_m shifted some data points from the finer to the larger G branches. However, this also left a number of data points still remaining on the finer G branch, indicating that the larger grains or clusters of them were not of sufficient size to be fracture origins by themselves. Larger grain origins frequently are from clusters of larger grains making them even more uncertain in their relation to $\sigma - G^{-1/2}$ relations. Thus their frequently being shifted too far to the left, i.e. past the limiting larger G branch slope for data from failures from a single large or dominating grain, implying that the complete cluster was not the origin, or that the cluster origin was not directly pertinent to $\sigma - G^{-1/2}$ relations, also usually indicated by abnormally low strengths (e.g. Figure 1.4B). However, use of G_m values transferring varying numbers of data points to the larger G branch made limited reductions of data scatter and increases in the larger *G* branch slope, and limited change in the finer *G* branch slopes, e.g. leaving them substantially positive. The number of data points affected by such evaluations varied substantially, from a high of 16 out of 40 data points (i.e. 40%) to much less (e.g. to ~7%) [2, 17]. The extent and character of such G_m corrections increased with the size of the body and fabrication method, e.g. more for larger die pressed disks and less for smaller extruded rods. Similar evaluations of the much more limited data based on grain size averages and ranges (mostly observed on polished surfaces, thus not showing the full extreme of grain size range) ad much more limited impact on scatter and slopes of the larger and finer *G* branches. Thus sintered, machined Al_2O_3 data of Ting et al. (99+%) [133] and McNamee and Morrell (95%) [127] vs. G_m(from microstructural, not fractographic, studies) simply moved some data points leftward in the finer *G* regime or to the larger *G* branch. In either case their data agrees with the other data, as does Alford et al.'s [27] (99+%) machined data vs. G_m(the only values given).

Thus while there is scatter and uncertainty in many details, which can be improved some by more detailed evaluation such as of specific *G* values, there is a clear, very consistent overall pattern (Fig. 3.1). This pattern is observed despite the diversity of materials, test methods, especially characterization. The latter is often overlooked but in fact is probably one of the larger factors in variations, i.e., scatter, between different sets of σ–*G* data. The techniques of measuring *G* are often only partially given or not given at all. Thus while many investigators may state that they used a linear intercept, they frequently do not tell whether they used a factor, and if so what it was, to convert that linear intercept to a "true *G* value." Such values are commonly of the order of 1.5 or more but can vary from <1 to >2, thus commonly giving at least a 50% variation in *G* values and possibly in excess of 100%. It was recommended (Chap. 1, Sec. IV.B) that the average diameters of grains be measured, e.g., along randomly selected lines with the average possibly being weighted based upon grain area considered for σ–*G* relations. It is important that the *G* measurement be compatible with measurement of individual grains, i.e. where they are fracture origins (which generally needs to be corroborated by fractographic analysis). It is also important that the σ and *G* measurements be self-consistent, i.e. use of $G_m$is often not consistent with use of the maximum flexure strength (= outer fiber stress at fracture, Sec. F).

C. Grain Shape Effects

Grain size effects and issues are often intimately related to various other grain factors, of which grain shape is the most immediate one. This affects σ–$G^{-1/2}$ relations directly via effects on *G* measurements (Chap. 1, Sec. IV) as well as possible effects on failure causing flaw sizes and shapes. While, much, if not all, of such grain shape effects can be included in the size, e.g. via equivalent circular area, this needs further evaluation, because of other possible ramifications or correlations. Thus grain shape is often interactive with local and global grain orientation, since this can influence the shape and extent of grain growth. Conversely growth of nonequiaxed grains may impart some, or enhance, local grain orientation. The phase content of the body can be important, e.g. as shown by exaggerated grain growth of α-Al_2O_3 or β-Al_2O_3 grains in α-Al_2O_3, of α-SiC in β-SiC, and of β-Si_3N_4 grains in α-Si_3N_4. Other body constituents can enhance or retard exaggerated grain growth and hence increase or decrease grain elongation effects. Grain shape effects by themselves and body parameters affecting them are discussed next, followed by compositional and fabrication effects on grain shape, then by effects of grain orientation.

Very limited direct assessment of grain shape on tensile strength has been made, with much of the limited data being of some uncertainty due to questions of possible impurity or additive effects, e.g. in AIN and Al_2O_3 bodies. Thus Sakai [260] reported an ~ 1/3 increase in strength (from ~ 310 to ~ 470 MPa) of hot pressed AIN with an increase in oxygen content from ~ 1 to ~ 2.7 wt% (by addition of Al_2O_3). Part of this increased strength was attributed to reduction of AIN grain size (from G_a nearly 10 to ~ 4 μm), but some of this was probably also due to the distinct platelet character of most grains. (Note that while Sakai's AIN without additives had typical equiaxed grains and predominately intergranular fracture, bodies with the platelet grains showed extensive transgranular fracture.) Similarly, Komeya and Inoue [261] showed similar levels and increases of strengths of AIN-based composites with ~ 25 w/o Y_2O_3 additions that resulted in similar elongated, but more fibrous and elongated, AIN grains.

Al_2O_3, data of Koyama et al. [135] with relatively uniform grains having aspect ratios of ~ 2 (e.g. due to limited additives) showed very similar σ–$G^{-1/2}$ behavior to that of bodies with nominally equiaxed grains (Fig. 3.15), though some uncertainty exists due to lack of comparison of G measurement techniques for varying grain shape. While alumina bodies with in situ growth of alumina-based platelet grains really fall into the category of composites (Chaps. 8–12), their results are also reasonably consistent with data for pure Al_2O_3 (Fig. 3.15). Similarly, strengths of the pure Al_2O_3 of Yasuoka et al. [262] of ~ 430 and 660 MPa respectively for G_a ~ 5 and ~ 2 μm are in good agreement with Al_2O_3 data, as are those of bodies with ~ 20 vol% platelet grains of lanthanum aluminate of similar average grain size as the alumina grains (~ 5 μm). There is also agreement of strengths of similar data of Kim et al. [263] on Al_2O_3 with and without additions of Na_2O and MgO to form by in situ reaction during sintering to yield several volume percent long lath, beta-alumina grains of small cross section. Strength increases, commonly ~ 20%, but in the extreme nearly 100%, were attributed to combined effects of reduction of the matrix alumina grain size and formation of more and longer beta-alumina grains. Development of platelet or other exaggerated grains via other constituents is paralleled by some effects of impurities (generally with sufficient solubility or reactivity with the body phase). Note again that alkali, especially Na from Bayer processing of Al_2O_3, results in beta-alumina platelet grains commonly acting as fracture origins (Figure 1.3A), especially in bodies without grain growth inhibitors. Much or all of this data is at least approximately consistent with data for alumina with the same grain size as such larger beta-alumina grains. However, strengths may fall below the σ–$G^{-1/2}$ relation for the pure material because of the platelet size and shape of single, and especially clustered, platelets, acting as fracture origins, possibly aided by additive or impurity effects.

The shape (and orientation) of nonequiaxed grains are important in most properties. It has been recommended that measuring of nonequiaxed grains for σ–$G^{-1/2}$ evaluations can be aided by determining the size of an equivalent equiaxed grain having the same fracture surface area as the actual grain. Generally this would be essentially the same as the area of an equivalent semi- or full-circular flaw if an isolated grain is a possible fracture origin (e.g. for the situation depicted in Fig. 3.31A). This requires recognizing that for single larger grains at fracture origins the smaller grain dimension is the flaw size rather than the larger grain dimension (as incorrectly used in some SiC studies [96]; the larger dimension enters in determining the flaw geometry factor, Section VI.B). However, this is not necessarily pertinent to other nonmechanical properties, and it has limitations for tensile strength and especially other mechanical properties.

Of broader concern for tensile strength are common complications such as grain orientation, and character relative to the matrix grain structure other than size and shape. Other issues are connection with other flaws, e.g. surface flaws (Fig. 3.31), pores (Figure 1.4B), or combinations of these. However, a major issue is grain and local compositional variations or more than one grain with or without local compositional variations acting as fracture origins. While probably fairly common in several materials, this is common in origins from larger grains in situ toughened Si_3N_4. There, large rod or platelet shaped grains singly, or more commonly two or more in combination with excess additive phase, are dominant fracture origins as grain sizes and the level of toughness increases. While both the number of grains in a cluster and the extent of excess additive phase, as well as the location of these relative to flexure tensile surfaces, impact the resultant effective flaw size (Fig. 3.31), there has been little evaluation. Again such evaluation is of uncertain relevance to basic σ–$G^{-1/2}$ behavior, e.g. for projection of strengths to significantly larger or finer grain sizes, since such combinations often represent a significantly different flaw population. However, the above questions are relevant for understanding strength behavior of such bodies to control better their microstructure and properties.

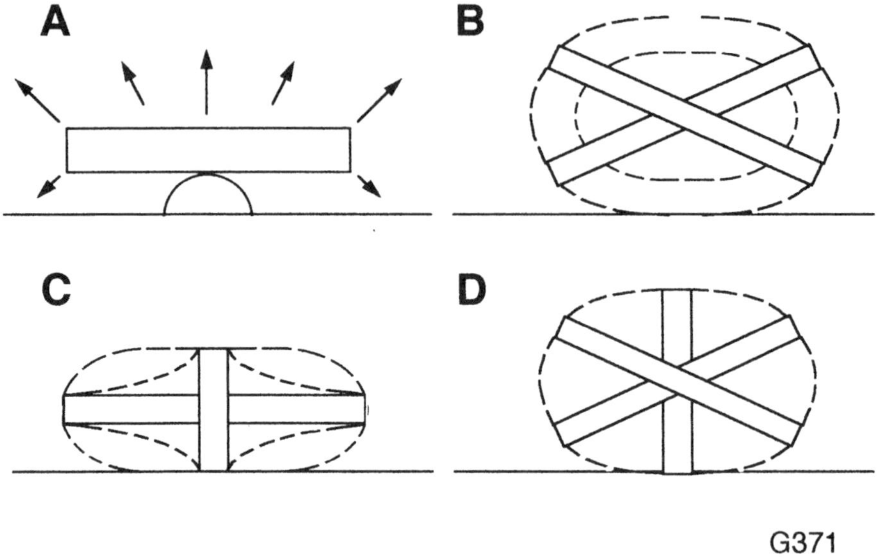

FIGURE 3.31 Schematic of some possible idealized flaws for fracture initiation from one or more larger grains. (A) A single larger tabular or rod grain near the surface with an associated machining flaw. Arrows indicate subsequent crack propagation.(B) and (C) Possible effects of two crossed grains, which clearly increase the effective flaw sizes (dashed lines), with the extent of this probably depending on the extent of associated excess additive phase (i.e. less of these indicated by inner dashed lines). (D) Three grains, increasing probable flaw size (dashed line) some.

D. Grain Orientation Effects

Preferred crystallographic orientation of grains can significantly effect σ. Chandler and colleagues [169, 170] showed higher σ due to preferred orientation in extruded, sintered BeO from UOX-derived powders having a significant fraction of needle shaped BeO grains (Chap. 2, Sec. III.H). The degree of preferred orientation increased with decreasing specimen cross section (i.e. with increasing area reduction in extrusion) and with subsequent grain growth (i.e. ~ 50% orientation at $G = 20$ μm, reaching ~ 80% orientation at $G = 80–100$ μm). Several investigators have shown similar preferred orientation from green body extrusion of Al_2O_3 and resultant increases in σ for fracture perpendicular to the extrusion axis (Chap. 2, Sec. III.H). McNamee and Morrell [127] subsequently corroborated the earlier favorable effects of extrusion on strengths of Al_2O_3 reported by Haney and Morrell [126] and concluded that the preferred crystallographic orientation of the Al_2O_3 grains was the major factor. Salem and Shannon [264] have also recently shown a K_{IC} anisotropy for extruded Al_2O_3, e.g. 2.5 MPa·m$^{1/2}$ for the crack plane, and propagation direction perpendicular to the extrusion direction to 4.1 MPa·m$^{1/2}$ for the crack propagation direction and plane being respectively perpendicular and parallel to the extrusion direction (with an intermediate value for the third basic orientation). Similarly, preferred orientation obtained by hot pressing or hot working (press forging) Al_2O_3 has been shown to effect strengths significantly [11]. Preferred crystallographic orientation of MgO by hot extrusion results in increased strength attributed to reductions in the ease of crack nucleation via blocked slip bands at grain boundaries due to reduced misorientation between the grains [48, 50, 52]. Other studies of preferred orientation corroborate and extend these effects [265–267]. These prevalent effects of orientation raise questions of adequately interpreting data for specimens from materials and processes that may yield (generally unexamined) preferred orientation, e.g. in Al_20_3 studies of Charles [149] and especially in studies of small extruded Al_2O_3 and TiO_2 rods of Alford et al. [27] (Figs. 3.16, 3.20), where effects could be greater because of the small rod size.

Dykins's [268] study of the tensile strength of ice from –27 to –4°C (i.e. from ~90 to ~98% of the melting point) showed marked anisotropy of strength relative to the axis of the freezing direction. Strengths for stressing normal to the freezing direction averaged 39 ± 5% of those for stressing parallel with the freezing direction, with little effect of grain size. Strength, even this close to melting, also commonly followed a $G^{-1/2}$ dependence (Figure 6.13).

Bodies having both preferred grain orientation and pronounced grain elongation often show greater effects on strength, since both often reflect greater crystalline anisotropy. Thus data of Virkar and Gordon [195] for beta-alumina bodies with substantial basal plane texture of the platelet grains in the hot pressing plane showed ~30% higher strength in the finer grain regime for specimens tested for fracture normal, versus parallel, with this basal texture (Figure 3.21, i.e. respectively mode A and A′ versus mode C). Clearly much of this difference is due to the anisotropy of single crystal fracture energy, but it may also reflect some effect of grain elongation. Note that the latter could also affect the branch intersection some, but primarily if the grain elongations were oriented to accommodate machining flaw elongation for tests based on failure from flaws formed parallel with the abrasive particle motion.

More pronounced effects are seen in bodies with substantial uniaxial alignment of rod or needle shaped grains. Thus data of Yon et al. [269] for bodies of SbSI (a piezoelectric material of interest) with highly aligned needle grains showed anisotropies of strengths for four bodies ranging from ~40 to ~170% higher for stress parallel with the C (and elongation) axis of the needle shaped grains as the grain diameter increased from 0.38 to 600 μm and lengths increased from 11 to 6000 μm (Fig. 3.32). Though the much less refractory and much

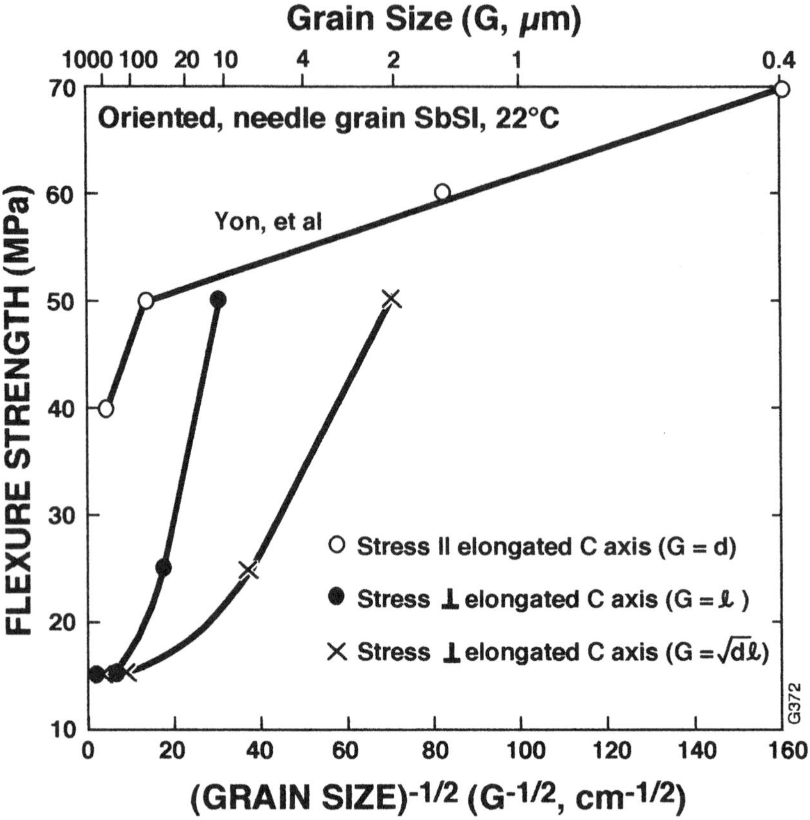

FIG. 3.32 Strength at 22°C versus the inverse square root of the grain size[G] for SbSI bodies with highly oriented needle grains. Strengths plotted using (1) the diameter of elongated grains as *G* for stressing parallel with the axis of elongation with resultant fracture transverse to the grain axis, and (2) either the length (1) of the elongated grains or the equivalent diameter (*d*, assuming an elliptical cross section of the elongated grains) for stressing normal to the grain orientation axis for fracture parallel with the grain elongation. Note that in either of the latter cases the much greater initial rates of strength decrease as the corresponding grain dimension increases. (Data from Yon et al. [269].)

softer nature of this material raise questions of the specific mechanisms, the strengths for stressing parallel with the *c*-axis indicate finer *G* branch flaw failure over most of the *G* range studied with a branch to either larger *G* flaw or microplastic failure at *G* (= needle diameter) of ~55 μm. On the other hand, the much more rapid initial decrease in strengths for stressing normal to the needle axes may indicate microcrack failure, which may be present for two reasons. The first is the much larger needle dimension in the plane of fracture for this test orientation and the high anisotropy of the material with substantial, but imperfect, alignment. Second is that this is an orthorhombic material and thus has unequal thermal expansion (and other properties) not only between the *c* direction and a normal direction but also in the other direction normal to both the *c*-axis and a normal to it. A more extreme case of strength anisotropy due to orientation of elongated grains is that of Ohji et al. [270], who extensively seeded α-Si_3N_4 with fine rod shaped β-Si_3N_4 grains that were oriented by tape casting. Resultant dense sintered bodies had strengths of 1100 MPa for stressing normal to the resultant elongated grains and just over half of this, 650 MPa, for stressing in the normal direction. This strngth anisotropy was despite nearly identical toughness values (~ 11 MPa·m$^{1/2}$) and Young's moduli in the two directions but was similar to that of the Weibull moduli for strengths in these two directions, i.e. 26 and 46 respectively. Much more extreme is the > 7-to-1 ratio of strengths for stressing parallel versus perpendicular to the fibrous grain axis of the fine grain jadeite [271] (Table 2.3). This data, though limited to three strengths for two different jade compositions stressed parallel with (but only one stressed normal to) the fiber axis is consistent with a definite positive slope to the finer *G* branch of the σ–$G^{-1/2}$ relation.

Further note the high strength anisotropy of CVD graphite (Table 2.14), which roughly correlates with the anisotropies of *E* and *K*, but much less so for the orientation giving very high *K* due to delamination normal to the crack. The latter orientation is not the strongest, again showing large crack behavior not reflecting normal strength behavior. More generally note that while the orientation dependence of strengths often at least partly correlates with that of *K* (Sec. III.H), there are a variety of variations, some with opposing toughness and strength trends. Examples of such variations besides CVD graphite are oriented bodies of Al_2O_3, β-Al_2O_3, and Si_3N_4. In all these cases fractography has been necessary to identify possible or definite orientation dependences of failure causing flaws, which together with the anisotropy of *K* explained that of strength.

VI. OTHER FACTORS

A. Flaw Character—Surface Finish Effects

The typically prominent role that surface finish plays in the tensile strength of ceramics is reflected in their resultant σ–$G^{-1/2}$ behavior, with the surface finishing and grain size dependence of tensile strength each providing insight into the role and character of the other. This is especially true for the dominant surface finishing methods of machining that have been most extensively investigated with regard to the range of parameters, materials, and grain sizes and has been supplemented by successful fractography. Thus machining effects, which were the basis for the σ–$G^{-1/2}$ model (Fig. 3.1) for flaw failure, are discussed first and most extensively, but with some discussion of the more limited data on effects for as-deposited or as-fired or annealed and chemically polished surfaces.

Fractography showed that to a first approximation machining flaw sizes causing failure for a given material and machining condition were independent of grain size (including single crystals, i.e. for $G \sim \infty$, $G^{-1/2} = 0$) [13, 15–22, 138, 142–144], e.g. note machining flaws in Figs 3.7, 3.9, 3.10BC, 3.12, and 3.16. A central aspect of this is that the larger and finer *G* branches intersect when the flaw size and grain size are ~ the same (recognizing that the former is measured as a radius and the latter as a diameter, as well as effects of varying flaw and grain factors). The first of two machining factors that corroborate this model is that more severe machining, commonly by use of coarser abrasive grits, reduces strengths across the limited, but useful, finer *G* ranges encompassed [24–26]. Though it has been claimed that such finer *G* branches have no *G* dependence, the data cited for this is at best insufficient, and other data clearly shows some *G* dependence (e.g. Figs. 3.10–3.19), as indicated theoretically below.

The second aspect of machining effects that provides even stronger support for the basic model of Fig. 3.1 is the effect of machining direction on strength. It has been extensively shown in a variety of single- and polycrystal ceramics that tensile testing a machined body as a function of the angle of the stress axis relative to the direction of motion of the abrasive particles in machining plays a major role in resultant strengths [15–22, 142–144]. Thus strengths for samples of the same body machined in a direction parallel with uniaxial stressing are commonly nearly twice the strengths of the same samples tested machined in a direction perpendicular. This has been extensively shown to be due to abrasive motion generating two flaw populations, one of elongated flaws, and the other of ~ half penny cracks respectively parallel and normal to the abrasive motion. Both flaw populations are ~ the same depth, and while there are some other variations in shape, the differences in flaw aspect ratios, commonly of 2 to 4 versus ~1 respectively, is the predominant factor in the resultant strength difference. Thus specimens uniaxially tested in two different directions relative to the abrasive machining direction result in two different finer *G* strength branches. Other limited variations in flaw dimensions are indicated experimentally, consistent with limited theoretical dependence on *E, H,* and *K* per Eq. (3. 2), as is discussed later.

Recent further evaluation or earlier and additional machining direction studies show that such machining direction effects are grain size–dependent, and that this dependence significantly further corroborates the basic model [142–144]. Thus, as summarized in Fig. 3.33, the difference in strength as a function of stress versus machining direction goes to ~ 0 at an ~ common *G* value, designated G_C, but increases as *G* either decreases or increases from this

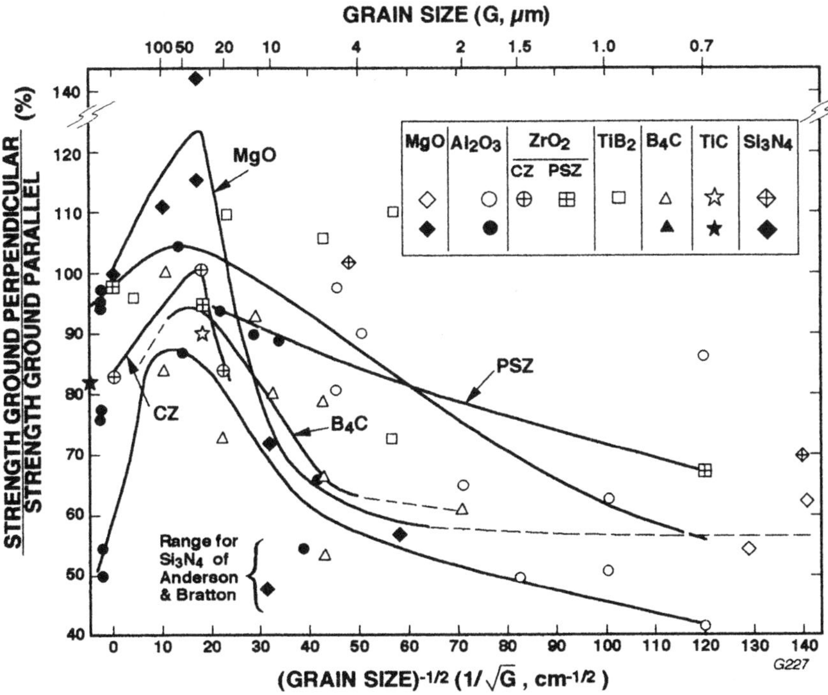

Fig. 3.33 Percent change in σ vs. $G^{-1/2}$ for various materials for machining perpendicular to the tensile axis vs. machining parallel with it. Machining was by diamond grinding except for one set of Al_2O_3 tests sanded with SiC fixed abrasive. All data shows reduction in the strength anisotropy as a function of stress versus machining direction as *G* increases to intermediate *G* values and then increases again consistent with substantial strength anisotropy as a function of machining direction in single crystals. The shift to higher values at intermediate and large *G* as well as single crystal for MgO is attributed to surface work hardening in these ranges [49]. (Published with the permission of the *Journal of the American Ceramic Society* [143].)

G_C value, for a given machining operation. This is completely consistent with, and strongly corroborates, the basic model that as the grain and flaw dimensions approach one another, the grain dimensions begin to constrain, hence become, the flaw dimensions. However, this is more significant for elongated flaws, since their greater length versus depth becomes incompatible sooner with grain dimensions as these approach each other than does the depth of the flaws. (Also, flaw depths are probably a more basic result of the flaw generation process, e.g. as indicated by depths of both flaw populations being similar.) As *G* decreases below

G_C for a given machining operation, flaws become progressively larger than the grain size and thus progressively less constrained in shape by individual grain boundaries. Conversely, for $G > G_C$, flaw dimensions are progressively < the G values, so there is progressively less effect of the boundaries of the single grain in which the flaw formed on the flaw shape.

It is important to note that there are three intrinsic statistical effects impacting the specific relations between flaw and grain dimensions as they approach one another, so the dimensions of the failure causing flaws are only approximately those of the grains at the branch intersections or the strength anisotropy being zero at G_C. First is the variation in grain size, which will vary from specimen to specimen, with greater variations between different bodies, but some between a given set of samples, as well as within specific samples. An important one in larger G machined samples is truncation of surface grains (Figs. 1.2A&C, 3.9). Thus there will be some tendency for branch intersections and G_C values to be toward the larger size of the G range, but other factors such as truncation of larger grains can partially counteract such shifts. Further, orientation of individual surface grains relative to preferred planes for both forming of flaws and their causing subsequent mechanical failure is important. Additionally, favorable orientation of two adjacent surface grains can allow flaw formation partly in both grains instead of just one even though the flaws and grain dimensions are similar. While the above grain orientation affects impact flaw formation, this is also impacted by the second factor, the specific local micromechanics of each abrasive particle–surface interaction, e.g. of the size, shape, and force on the particle forming flaws. The impact of this is clearly indicated by flaws not forming in much larger grains despite their being favorably oriented (as shown by their subsequent cleavage fracture, Fig. 3.27).

There is a general and two specific effects that need to be considered in evaluation of effects on $\sigma - G$ relations. The general one is introduction of surface compressive stresses, which occurs in all materials and obviously does not override effects of machining direction. However, whether such effects change with G and possibly alter the anisotropy of strength due to machining direction with G is unknown. The first of the two specific cases is work hardening of surface grains of materials with easier activated slip such as MgO and CaO. However, this merely shifts the relative machining anisotropy curve (Fig. 3.33) higher. The second specific case is the substantial surface transformation of tetragonal to monoclinic ZrO_2 from machining of TZP bodies with ~ 20% increased strengths attributed to expected surface compression [272]. Thus while the baseline for measuring such strength increases is uncertain, study shows substantial reduction, or complete elimination, of strength anisotropy as a function of machining direction possibly for some TZP and PSZ poly- and especially single crystals [273]. However, there is no evidence that these trends for toughened bodies change $\sigma - G$ relations. In fact, the similar effect from fine G to $G = \infty$ reinforces similar behavior across the complete G range.

An important question is what effects other surface treatments have on $\sigma - G$ relations. Abrasion, wear, and impact (e.g. from tumbling in a mill) are examples of other mechanical effects on surfaces. The nature and size of many impact flaws [16] suggests that they may follow similar trends as for machining flaws. This is also suggested by similar increases in strengths as a result of grit blasting TZP bars as for machining them [8, 274, 275].

Consider now other surface finishing effects, primarily of as-deposited or as-fired or annealed surfaces, which have received only limited investigation. Some single crystal, and especially glass, specimens can commonly be flame or chemically polished to remove surface flaws such as from machining and achieve high strengths, so there is a common impression that as-fired or annealed surfaces yield higher strengths. However, such surface treatments are commonly not applicable to, or much less effective with, polycrystalline samples due to probable intrinsic factors such as TEA and EA, and are clearly limited by a variety extrinsic surface effects. The latter arise from factors such as sintering of dust (e.g. specimen powder, including agglomerates from spray drying) on their surface as well as flashing and other surface and internal variations resulting from imperfect specimen forming and densification, all leading to stress concentrations limiting strengths. Further, machining again commonly introduces some surface compression stress which aids strengths, as directly shown by Hanney and Morrell [126], introducing some, usually moderate, uncertainty in calculating flaw sizes.

For the above reasons, as-deposited, fired, of annealed surfaces often do not give the highest strengths, but even in some, possibly many, of these cases useful $\sigma - G^{-1/2}$ data may be obtained, but probably more scattered. Direct comparison with machined samples is valuable. Thus Hanney and Morrell [126] showed strengths of their as-fired Al_2O_3 bodies generally being less than that for specimens that were diamond ground. (Mechanical polishing resulted in σ roughly intermediate between these two levels.) They also showed that annealing of bars with coarse, but not fine, machining reduced σ. Similarly, Steele et al. [122] showed strengths of ground and polished Al_2O_3 bars being similar to each other, but greater than as-sintered or thermally etched specimens. McMahon [276] showed that Al_2O_3 specimens tested as-fired had lower strength, those ground perpendicular to the specimen and tensile axis intermediate strengths and those ground parallel with the specimens and tensile axis the highest strengths. Lino and Hubner [277] also showed lower as-fired vs. machined strengths of Al_2O_3 (P~0.02, G~4 μm), but an increase of the latter upon annealing (e.g. at 1400–1500°C). They also showed that machining increased strengths of their as-sintered Al_2O_3 bars (P ~ 2%, G ~ 4 μm) by ~100 MPa, i.e. ~50% to ~300 MPa (still low for such bodies), but annealing near the original sintering conditions increased the machined strengths by ~ 100 MPa (Weibull moduli were similar for all three surface conditions at ~ 17). Earlier studies of Rice [17] showed strengths of machined samples of dense Al_2O_3, $MgAl_2O_4$, and ZrO_2 (+ 11.2% Y_2O_3) decreasing upon annealing at 1400, 1600, and 1670°C (in general *G* increased only at 1670°C).

With limited, or no, extrinsic sources of failure, e.g. pores, or surface irregularities, the presence of (intrinsic) grain boundary grooves on as-fired surfaces also limits strengths. Coble [23] has theoretically shown that such grooves act as failure causing flaws. Such grooves are expected to be less severe for finer *G* bodies than for coarser ones due to the relatively lower temperatures, shorter times, or both used in obtaining finer *G*, thus introducing a *G* dependence of such grooves. Coble indicates that the equivalent flaws will generally be related to *G* and vary in the range of *G*/15 to ~*G*. Support for the deleterious effects of grain boundary grooving and its correlation with *G* is provided by the general fiber results as well as Simpson's [156] fiber failures [3] and increasing the σ of FP fibers ~25% by smoothing their surface with a SiO_2 coating [159, 160].

Chemical finishing of polycrystalline materials has received very limited study, mainly because of frequent adverse affects of porosity, impurities, and varying grain orientation, often resulting in rough or irregular, hence weaker, surfaces. Evans and Davidge [72] showed a small (<10%) strength increase for their chemically polished UO_2 (P~ 0.03) over the temperature range compared (to 1200°C), with this increase being greater for their ~25 μm vs. ~8 μm *G* body. Similarly, Gruszka et al.'s [124] chemical finishing of 99.5% Al_2O_3 electronic substrates (G ~ 0.9 μm, P ~ 3%) showed both some σ increases and decreases vs. as-fired or diamond lapped surfaces (both giving σ ~ 400 MPa). They showed an approximate inverse σ-surface roughness correlation, i.e. a maximum σ of 550 MPa from use of molten borax with among the lowest surface roughness (2.9 μin.) similar to that for the as-fired and diamond lapped surfaces (respectively 2.6 and 2.0 μin), with σ ~ 410 MPa. This supports the implication from glass, single crystal, and more specifically fiber, processing that achieving very smooth surfaces of dense, fine grain polycrystalline bodies can give substantially higher σ. Strengths of 2–5 GPa reported for small CVD SiC specimens (P = 0) with very smooth as-deposited surfaces (apparently due in part to the extremely fine G ~ 0.01–0.1 μm) clearly show that such surfaces can yield high strengths [111, 112]. However, abrasion even by only light sanding with SiC abrasive paper readily dropped these very high strengths to normal levels observed for conventional SiC bodies, consistent with the resulting flaw sizes observed (which also indicate lower K_{IC} e.g. by ~ 20%). Such σ decreases with abrasion were greater than those observed for *c*-axis sapphire filaments [167].

An important question is what other flaw sources and populations can give similar σ–$G^{-1/2}$ behavior as for specimens whose strength is primarily determined by surface flaws in machined or as-fired surfaces. Clearly, handling, impact, etc.

of as-fired or deposited surfaces result in flaws that are generally similar to machining flaws, thus expected to give similar $\sigma - G^{-1/2}$ behavior. Two other primary flaw sources are compositional variations (including impurities) and porosity, with heterogeneities of these often being particularly important sources of failure. Impurities and other compositional variations often have their main influence via effects on grain size, especially that local to a serious compositional fluctuation (e.g. Figure 1.5), in which case similar $\sigma - G^{-1/2}$ trends are likely to result (but require identification of effects of such local G changes, typically by fractography). On the other hand, impurity particles of random size, location, or both are likely to be simply a source of scatter for most mechanisms envisioned, e.g. microcracking, thus obscuring or eliminating normal $\sigma - G^{-1/2}$ behavior. The other major source of failure is pores or pore clusters. Rice [6, 144, 273] has shown that machined alumina samples with fine porosity fail from machining flaws similar to those found in dense bodies and thus should show the expected $\sigma - G^{-1/2}$ behavior (at least over the ranges in which G can change without significantly changing the amount or character of the porosity). Recently, Zimmermann et al. [278] showed that large (artificially introduced) isolated, uniform spherical pores that were fracture origins in otherwise dense alumina samples showed strengths decreasing as G increased from 0.8 to 9.2 μm, following the same trend with G as for dense bodies without the large pores, but at about 30% lower strengths for failure from the isolated large pores. Zimmermann and Rödel [279] discussed this trend, indicating that it probably reflected localized cracks, rather than the postulated circumferential cracks around the pores, as also suggested by Rice [6, 280]. This thus indicates that some larger pore failures can give similar $\sigma - G^{-1/2}$ behavior as for dense bodies, but this probably requires that the pore size, shape, spatial distribution, and surrounding microstructure be fairly uniform to limit data scatter, which could obscure or obliterate any consistent trend as a function of G. Another issue that needs to be addressed to assess the extent of such pore-grain failure is demonstrating such pore-grain failure relations over a larger G range, especially into the larger G regime and where spacing of larger pores is variable, which typically occurs and greatly complicates pore induced fracture [6, 280]. The generalization of the two-branch σ–$G^{-1/2}$ behavior proposed by Zimmermann and Rödel [279] is a stimulus for further study but appears limited by the above-noted variations of individual mechanisms, as well as the complications of various combinations of mechanisms, e.g. of pores, second phases, and larger grains.

B. Test Condition Effects on Strength–Grain Size Behavior

Effects of test conditions on resulting σ–$G^{-1/2}$, which entail specimen size and geometry, and test factors, e.g. uniaxial versus biaxial stress, test environment, and cyclic stressing, need to be considered. The primary effects of specimen size, geometry, as well as some aspects of test methodology are due to changes in the amount of materials being at substantial stress levels and hence the availability of different flaw types and sizes to cause failure. As noted earlier, use of flexure specimens of rectangular cross section raises some questions with regard to the appropriateness of using G_m, since the latter usually represents a volume flaw distribution, whereas this test configuration favors surface failure. The use of round versus rectangular cross section flexure rods is even more questionable, since this presents significantly more surface area of varying stress from which failure can occur. Binns and Popper [121] showed σ of round rods (1 cm dia.) averaging 25% higher (ranging from 20% lower, the only case of round rods being weaker than rectangular rods, to as much as 90% higher) than for 1 cm square bars of the same materials for 10 different alumina bodies. Similarly, McNamee and Morrell's [127] 95% Al_2O_3 fabricated by various means (G = 3–7.5 μm, P = 0.035–0.05) showed σ of round rods and rectangular bars made by extrusion higher by 40 and 60% respectively than for bars machined from die pressed, isopressed, or slip cast plates. Further, within an extruded body σ was greater for fracture perpendicular vs. parallel with the extrusion axis. Although all three of their direct comparisons showed as-fired strengths 10% higher than for as-machined surfaces, their machining was with a somewhat coarser [180] grit transverse to the bar lengths, which gives lower σ. Fractographic determination of specific failure origins, if sucessful, is the basic solution to this problem. However, since this is often not done, and is not always practical or feasible, the next best procedure should be to attempt at least an approximate statistical evaluation of the frequency of potential failure sources, e.g. as done by McNamee and Morrell [127]. They showed the occurrence of larger G (> 70 μm) varying from 0.01/mm^2 to 0.45/mm^2 in their samples. Thus three point flexure (preferably with rectangular cross section bars to avoid large surface areas, at variable stress) more closely approximates average G dependence of σ.

That statistical effects continue to small size and high σ is shown by the definitive gauge section dependence of σ in tensile testing fibers [158–165]. This is consistent with the frequently low Weibull moduli in such tests, e.g. 4–6. The frequent correspondence of tensile strengths of fibers and corresponding flexure strengths of small bars (e.g. Figs. 3.17, 3.18, and 3.21) indicate similar volumes under high stress.

An important question is the effects of stress conditions, e.g. biaxial versus uniaxial stress, on σ–$G^{-1/2}$ behavior, both due directly to the differences of the stress states as well as indirectly on other factors such as slow crack growth and microcracking. One of the few tests of these issues is data of Chantikul et al. [131] on biaxial strength as a function of G (Fig. 3.14). Their results cannot be distinguished from uniaxial results, indicating little or no effect of biaxial loading, but more testing of other materials and microstructures is needed. Another stress factor is repeated stressing, i.e. mechanical fatigue. Aspects of such tests were briefly outlined in Chapter 2, Section II.I, but these were focused on large crack behavior with very little attention to *G* effects. One exception to this was Lewis and Rices' [285] tests of Lucalox Al_2O_3, which indicated an intrinsic fatigue mechanism due to microcracking from TEA (or EA), which would thus be *G* dependent.

Consider next effects of environmentally driven slow crack growth, i.e. beyond subcritical growth considered below. Environmentally driven crack growth, which has been evaluated primarily, if not exclusively, by uniaxial stressing, clearly increases flaw sizes as the rate of crack growth increases, stressing rate decreases, and strength increases (e.g. with decreasing *G*). However, intrinsic increases in SCG as *G* decreases (Figure 2.7) would, at least partially, if not completely, counter this trend. As *G* increases down the finer *G* branch, the resulting decreased strength will decrease the contribution of slow crack growth, hence also adding to the slope of the finer *G* branch. As strengths approach those of the intersection on with the larger *G* branch, i.e. with the grain dimensions approaching those to contain the failure causing flaw in a single grain, crack growth rates will transition from those for polycrystalline samples to those of single crystals or of grain boundaries. Besides possibly further changing crack growth rates, this may result in a change in fracture mode as discussed in Chapter 2, Section III.A.

Kirchner and Ragosta [281] calculated that small (e.g. 10 μm) flaws within single Al_2O_3 grains would not lead to catastrophic failure at single crystal K_{IC} values unless *G* was ≥ 100 μm and loading rates high (e.g. 10^4 MPa/s) to limit environmentally induced slow crack growth. They concluded that cracks would otherwise be arrested as the grain boundary and failure would be determined by the polycrystalline K_{IC}. While these calculations show that large Al_2O_3 grains could be failure sources at single crystal K_{IC} values, their assumptions lead to more restricted calculated growth. They assumed that the transition from single to polycrystalline K_{IC} values occurs at the first grain boundary rather than over a range of e.g. 2–6 grains, as indicated for materials investigated [28] (Figure 2.16). They also assumed that TEA stresses were not a factor in such slow crack growth, i.e. flaws would not grow in the absence of external mechanical stressing, which has received little direct study. However, McMahon's [276] and Rice's [282] Al_2O_3 studies show that this either does not occur or readily saturates early in the life of a specimen. That the latter may be the case is shown by observations of Hunter et al. [283, 284] that microcracking of pure HfO_2 upon cooling in a sealed furnace only began occurring upon opening the furnace to air and continued for several tens of hours before saturation.

Gruver et al. [128] showed similar 96% Al_2O_3 origins from isolated large grains over a range of temperatures. Using 1/2 these G_m values as *c* for the eight large grain (of 20) fracture origins in liquid N_2 gave K_{IC} 3.9 ± 0.6 MPa·m$^{1/2}$, and seven of 20 in the 95% (and two of 20 hot pressed) Al_2O_3 having large grain origins at 22°C gave K_{IC} 3.2 ± 0.3 MPa·m$^{1/2}$. While these all generally agree with the polycrystalline K_{IC}, all gave at least one value ≥ two standard deviations below the average, suggesting that in those cases the grains were not the complete flaw, i.e., c extended into the surrounding average grain structure, so plotting σ value for those G_m values is questionable. Similarly, it cannot be ruled out that large grains giving the highest calculated K_{IC} values may have resulted in failure before the flaw reached the full grain size, i.e. the critical flaw size was $<G$.

Chantikul et al.'s [131] model based on crack bridging is an interesting extension of the basic fracture mechanics-c/G model. It also predicts two branches, with branch intersections at $c \sim 2-3G$. They showed the larger G slope decreasing as G increases, indicating a transition to single (or bi-) crystal strengths. Although it predicts a zero slope for the finer G branches, factors leading to a nonzero slope in this region could be included. Use of a grain boundary K_{IC} of 2.75 MPa·m$^{1/2}$ Al_2O_3 is questionable in view of K_{IC} for many common orientations of sapphire being ~ 2 MPa·m$^{1/2}$, and grain boundary values commonly being ~1/2 such values (Chapter 2, Section III.E). However, there are more fundamental issues concerning their model. It is based on TEA stresses, yet the same type of σ–$G^{-1/2}$ behavior is observed for cubic materials not having any TEA stresses. Although bridging has been observed in cubic as well as noncubic materials, there are a variety of concerns regarding the applicability of bridging to normal strength behavior, as discussed earlier and later.

VII. EVALUATION OF THE Σ–G$^{-1/2}$ MODEL PARAMETERS

A. Overall Review and Assessment

The primary failure mechanism in most ceramics at or near room temperature is brittle failure from preexisting flaws per Fig. 3.1. The first of two deviations from this is failure from a substantial density of microcracks that are preexisting, developing, or both relative to stressing to failure. These lead to rapid initial strength decreases (commonly closely paralleling those of the elastic moduli) with increasing G when G_s for substantial microcracking is reached or exceeded. Presumably strength shows normal G dependence for the preexisting (e.g. machining) flaw population at finer G ($< G_s$) with a transition to microcrack determined failure that depends on the G distribution and TEA, and possibly EA, specifics. The other deviation is microplastic initiation of flaws and failure in some ceramics. This becomes more marginal as stresses for microplasticity increase due to intrinsic effects at any temperature, decreasing temperatures, or inhibition of the microplasticity by impurities or additives. Higher stress for microplasticity makes this failure mechanism more dependent on better surface finish, larger G, and higher temperatures. The $G = \infty$ strength projection and intercept from microplastic failure is typically the stress to activate the microplasticity, which is usually ~ the strength of single crystals oriented for easiest activation of the microplasticity determining failure of polycrystalline bodies. Such extrapolation of polycrystalline to single crystal strengths, rather than polycrystalline strengths falling well below single crystal strengths, is a basic difference respectively between microplastic and preexisting flaw failure. Increasing competition of preexisting flaw versus microplastic failure as microplastic stresses increase commonly results in a switch from microplastic to preexisting failure as strengths increase with decreasing *G*. The grain size of this transition should incrase as the surface finish quality decreases and the stress for the microplasticity increases, and again appears to occur when *c*~ *G*/2. Thus all three failure mechanisms will commonly have a finer *G* branch due to preexisting flaw failure.

The primary factors controlling tensile strength in brittle failure are typically the initial flaw, any slow crack growth, and the toughness controlling its propagation to failure. Conventional fracture mechanics approaches to the *G* dependence of strength have focused on toughness, neglecting possible variations of flaw character, i.e. size and location relative to the grain structure, and thus does not provide guidance for understanding or predicting the slopes and intersection(s) of the finer and larger *G* branches. The slopes of these branches are important first and foremost since this is a primary tool for estimating the benefits of reducing *G* and secondarily since most proposing use of G_m assumed (based on limited data) a zero slope of such branches. However, data clearly shows variable, often substantial, positive slopes with some possibility of zero or even negative slopes, all of which are consistent with expectations of varying flaw and other changes with *G*. A key factor in the finer *G* branch slopes is that while, to a first approximation, flaw size is independent of *G*, there are second-order variations. These are indicated for machining flaws by the following representative theoretical equation predicting effects of material properties on machining flaw sizes [286]:

$$c \propto \left(\frac{E}{H}\right)^{1/3}\left(\frac{L}{K}\right)^{2/3} \tag{3.2}$$

where *E*=Young's modulus, *H*=hardness, *L*=load, and *K*= the appropriate toughness. For $c < G$ the toughness should be the appropriate single crystal or grain boundary values, while that for $c > G$ should be a polycrystalline value for small cracks, i.e. typically no *R*-curve or related effects.

Consider first effects of the above flaw size variations as a function of *G*. In a given material *E* typically has no *G* dependence, and *K* often has limited or no *G* dependence at finer *G* but clearly decreases for *c* approaching *G*/2. *H* measurably increases with decreasing *G* in the finer *G* region (Chap. 4, Sec. II), thus indicating some limited reductions in *c* with decreasing *G*. These effects increase strength as G decreases, giving a positive finer G slope. Different materials with different E and K values would shift flaw sizes, and thus also strengths, some in the finer G branches, with differing G dependences of K varying finer G branch slopes. Note also that the minimum in H (Chap. 4) occurs when the indent and grain sizes are similar, which may be a factor in the grain size where $c \sim G/2$.

Similar trends are expected for as-fired surfaces. Thus a decrease in flaw sizes in as-fired surfaces as G decreases is also expected, since decreasing grain sizes are typically obtained by reduced temperature–time exposure of the sample, e.g. of fibers. Such reduced exposure reduces the extent of grain boundary grooving associated with finer G, and the finer G itself increases the tortuosity of the sequence of grain boundary grooves acting as a flaw, decreasing severity of the resultant flaw, often probably more than for machined flaws, indicating higher positive finer G slopes.

Turning to experimental evaluations of flaw populations, much of this must be inferred from the G dependence of strength, particularly for as-fired surfaces, since there is no independent flaw data for them. This is also partly true for machined sample, since while there is considerable flaw data for them, detailed comparisons of different materials for varying G as a function of machining parameters are limited. However, extensive studies of machining flaws, mainly by Rice and colleagues [12, 13, 15–22], show that machining flaw sizes for failure for a broad range of ceramics with typical moderate to finer grit diamond grinding do not vary widely, e.g. c generally is in the 20–50 μm range. However, materials of higher than normal machined strengths such as WC-Co, TZP, and some Si_3N_4 bodies (Fig. 3.34) have finer machining flaw sizes controlling failure, e.g. more commonly in the 10–20 μm range (Fig. 3.34). Thus such bodies with high toughness at small crack sizes, typically in finer G bodies with effective toughening at finer G, have smaller flaw sizes, which are less likely to be affected significantly by R-curve and other related large crack effects but may experience more benefit from surface compressive stresses than is typical for most machining.

Besides statistical variations of factors noted above, there are other important sources of differing flaw dependences on G at finer G. The first of three to be noted is preferred grain orientation, local or global, especially if it changes with G. This clearly occurs, e.g. in BeO [169, 170], where extrusion results in some preferred orientation of substantially nonequiaxed powder particles, with the orientation then increasing with grain growth, apparently due to preferential growth of oriented grains. The σ–$G^{-1/2}$ for such bodies, though limited, indicates a negative finer G branch slope which is consistent with the data showing orientation increasing with G, which could counter the normal strength decrease as G increases.

A second source of changing finer G slopes is changing flaw shape as c increases to approach $G/2$. Strength anisotropy studies as a function of stress versus machining directions originally did not focus on effects of G on the extent of

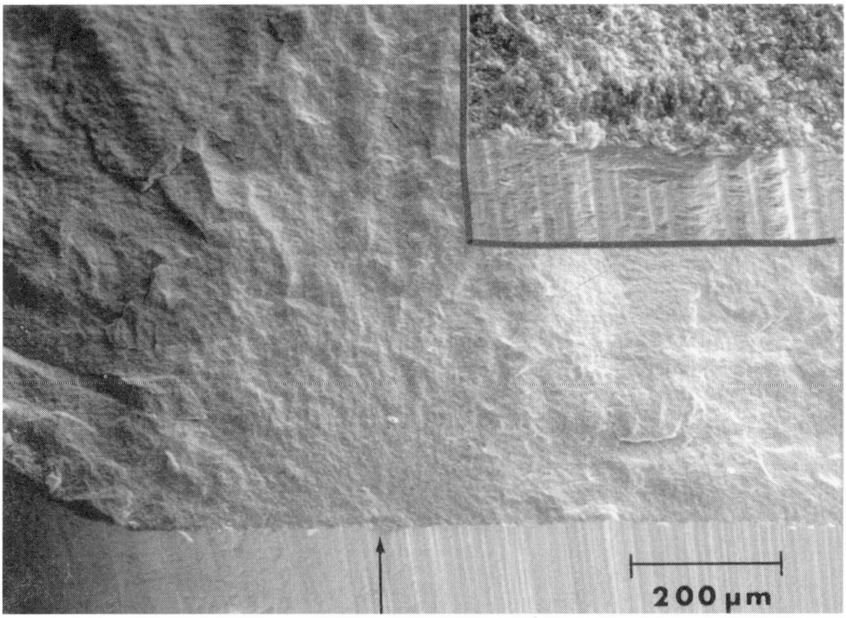

FIG. 3.34 SEM fractograph showing the fracture origin (arrow) in a dense hot pressed, higher strength Si_3N_4 from a typically smaller than normal machining flaw (shown at ~ 10-fold higher magnification in the insert).

flaw elongation parallel with abrasive motion. However, evidence now shows that flaw elongation is reduced as the largest flaw dimension, its length in this case, approaches the dimensions of the surface grains being machined [143, 144, 273]. As G increases toward and beyond the flaw dimensions, grains first begin to constrain flaw elongation, then the smaller flaw dimension, i.e. its depth. Further G increase then progressively relaxes these constraints so that anisotropic strength as a function of machining direction increases with further G increase. This is extensively shown by studies of strength anisotropy as a function of machining and stressing directions (Fig. 3.33) and is supported by some direct fractographic evidence. The progressive reduction in flaw elongation as G increases clearly reduces slopes of finer G branches for specimens tested with the stress axis normal to the machining direction. Machining studies indicate that finer G branch slopes for such specimens are often ~ 0 and may actually become negative as grain sizes approach the flaw depths [17].

A third possible source of changing finer G slopes is changing mechanisms involved in the failure process. An example of this is increasing contribution of microcracking as G increases. Thus, Tomaszewski's data for Al_2O_3 [134] shows evidence of microcracking increasing as G increased via decreased E (from strength not vibrational measurements). Such increasing microcracking as G increases should increase the finer G slope as indicated by σ–$G^{-1/2}$ trends for microcracking materials (Fig. 3.23) as well as similar associated decreases in E. Such microcrack contribution to the finer G slope could well be the source of the substantial finer G slopes commonly seen for Al_2O_3 (Figs. 3.15–3.17). Other possible combined effects are those due to effects of EA, TEA, or both. While these, especially TEA, can be important in microcracking, they may also have effects at grain sizes below those for microcracking, but at reduced failure stress levels, as is discussed for the larger G branches below.

Consider next the intersections of the finer and larger G branches. While there can be multiple finer G branches for a given material reflecting different finer G flaw populations, all will join a single larger G branch for that material when the flaw dimensions for each finer G branch ~ equals $G/2$. Flaw variations in the finer G branches will lead to corresponding variations of the G range of each branch intersection. However, there are also intrinsic variations to the G of such intersections due to statistical effects on forming flaws of about the same dimensions as those of the grains as discussed earlier. Some variable contributions due to factors such as EA and TEA may also occur.

A key factor in both the intersection of the finer and larger G branches and in larger G branch slopes is the extent of subcritical crack growth whether due to environmental effects or intrinsic growth. As previously noted, Kirchner and Ragosta [281] calculated that small (e.g. 10 μm) flaws within single Al_2O_3 grains undergoing environmentally driven SCG could lead to catastrophic failure at single crystal K_{IC} for $G \geq 100$ μm and loading rates high (e.g. 10^4 MPa/s) instead of arresting at the grain boundary. However, this must be an upper limit, since they assumed that the single- to polycrystal K_{IC} transition occurs at the first grain boundary rather than over multiple, e.g. 2–6, grains, indicated experimentally for most ceramics investigated [28] (Figure 2.16) and that TEA stresses were not a factor in such slow crack growth, which is contrary to some results [283, 284]. More fundamentally, there is direct experimental support for fracture occurring from flaws whose initial size is smaller than the grain in which they are located at or before their reaching the boundaries of this grain. This is shown by cases of fracture occurring from a single larger grain, with calculations frequently showing that the failure causing flaw size was smaller than the grain dimensions. More specific demonstration of this is given by results reported for $MgAl_2O_4$, and especially Y_2O_3, [2, 28] showing that failure frequently became catastrophic before the flaw reached the first grain boundary. Further, some origins from larger grains show single crystal mist-hackle features [2, 3, 12, 13, 21] within individual grains for failure from preexisting flaws (e.g. Figs. 3.7, 3.12) as well as slip nucleated failure (Fig. 3.4 B–D). Since mist boundary to flaw size ratios for single crystals are at least 3 to 1, and probably closer to those for glasses and polycrystals (~ 10 to 1), these observations set severe limits on the extent of subcritical crack growth. Thus fracture had to be critical before the flaw boundary reached the grain boundaries of the large grains in which failure initialed.

Consider in more detail the slopes of the larger G branches, where it should again be noted that the slopes of these branches on σ–$G^{-1/2}$ plots need to be multiplied by $(2)^{-1/2} \sim 0.71$ to obtain K since G is measured by a diameter and flaws by a radius. The K values from these slopes should be between those for easier grain boundary or single crystal fracture at one extreme and the polycrystalline values for a given material at the other extreme. Singh et al. [287] considered the issue of intragrain flaws first propagating at the single crystal K_{IC} then possibly arresting at the grain boundary without environmental effects. The first of three regions identified was the very large G region in which growth of the initial flaw cannot be arrested, and hence failure is determined by the initial flaw size and the appropriate single crystal K_{IC}, with no dependence on G. The transition from the larger G to the finer G branch occurs when $c \sim G/2$ if there is a step function change between single and polycrystalline K_{IC} values, and when $c = 3G$ for a more gradual K_{IC} transition. For the G range between these two extremes, i.e. the larger G region, they concluded that σ would be controlled by the polycrystalline K_{IC} and $c>G/2$ but with the specific c depending upon the type of K_{IC} transition. Evans [288] subsequently showed, based on a dimensional analysis, that the large G region could exhibit an intrinsic $G^{-1/2}$ dependence (independent of the original c, but with a slope intermediate between the single and polycrystalline K_{IC} values). The conditions cited for this were that the stress–crack length relationship have a maximum, and polycrystalline K_{IC} have no (or limited) G dependance. Subsequently Virkar et al. [289] combined and refined these two analyses, noting the need for better definition of the local K_{IC} values e.g. their dependence on c. The more complex case of crack propagation and arrest along grain boundaries has not been considered, with or without environmental effects.

The portion of the basic model for which there is little data is the transition from the larger G branch to single crystal strengths. Substantial data exists showing extension of strengths of both cubic and noncubic materials to and well below strengths of single crystals of the same material with the same machining finish (Figs. 3.10–3.13, 3.16–3.18, 3.20, 3.21, 3.24). Thus there must be some transition from these larger G strengths to those of the weaker single crystal orientations as sketched in Fig. 3.1. Besides the clear implications of this transition by strengths of larger G and single crystals, there is some limited data in this area. Thus there are a few strengths for (mainly fusion cast, optical grade) CaF_2 with G in the range of a few hundred to ~ 1000 μm that are similar or somewhat < those for comparably finished single crystal specimens [59]. Similarly Gentilman's [92] strengths of specimens from fusion-cast, transparent 2 $Al_2O_3 \cdot 1MgO$ with large (2–5 mm) grains tested with grain boundaries normal to the tensile surface are close to those of similarly finished single crystals. On the other hand, while there is no single crystal data, polycrystalline strengths for bars with the substantially oriented SbSI grains normal to the stress show strengths constant at low values (~ 15 MPa) for grains with lengths and diameters respectively of 300, 55 and 6000, 600 μm (Fig. 3.32). Finally, Mar and Scott [290], who fabricated sapphire bicrystals of controlled orientation, showed that when the *c*-axis misorientation across the bicrystal (tilt) boundary was > 36°, the bicrystals had spontaneous fractures from the TEA, hence essentially zero strength, but pure twist boundaries were mechanically sound.

The above sapphire bicrystal observations, as well as general considerations, clearly show that there can be substantial variations in strengths along multiple paths, transitioning from larger *G* to single crystal strengths dependent on material, specimen/grain size, and test method (Fig. 3.1). Thus specimens of Al_2O_3 with specimen-to-grain-size ratios that approximate tests of one or more bicrystals in series will have strengths ranging from zero to whatever value of strength the weakest boundary allows due to its TEA stresses. Tests of specimens with only one grain boundary across the specimen cross section, i.e. a single bicrystal, will vary from zero strength to higher values than for specimens with more boundaries along the length of the specimen, but each boundary extending across the specimen cross section. Similarly, larger specimens that have more than one grain boundary across a specimen cross section, i.e. have some boundaries intercepting parallel stress paths, would have lower maximum strengths, but less frequent zero strengths, since when some boundaries have zero strength, others having some strength could carry part of the load giving strengths > 0. Similar but generally significant effects due to EA are also likely to occur but often to be less severe than those of TEA.

Besides the above direct experimental evidence of TEA effects on sapphire bicrystal strengths, there is additional evidence of effects of TEA stresses affecting strengths not only when fracture of a single boundary is involved but also when fracture involves many grains. Thus Rice et al. [101, 102, 234, 291, 292] showed that as the flaw-to-grain-size decreased for noncubic materials due to increasing contributions of TEA or transformation stresses to failure, their contribution manifests itself as a reduction of the fracture toughness or energy calculated from the applied stress at failure and the flaw dimensions, geometry, and location. These stress contributions extrapolate to or above typical strengths of ceramics studied (Fig. 3.35) consistent with strengths going to zero in many sapphire bicrystals. However, effects of TEA stresses can impact strengths where multigrain flaws are involved, as revealed by measurements of the sizes of fracture mist, hackle, or macrocrack branching boundaries in materials with moderate to high TEA and related stresses, e.g. Al_2O_3 and B_4C to Pyroceram, graphite, and PZT [102]. This shows that effects of such stresses are manifesting themselves in the fracture processes occurring in these materials at ≥10 times the flaw

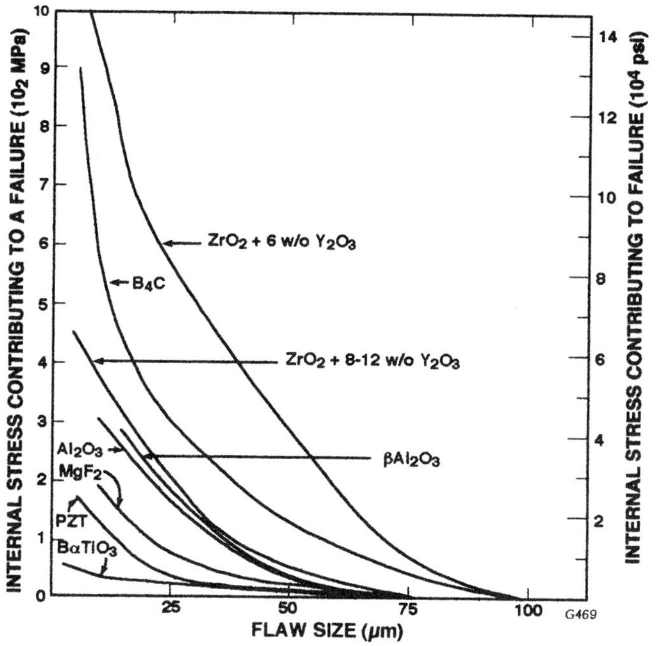

FIG. 3.35 Plot of TEA and related microstructural (internal) stress contribution to failure as a function of flaw size extrapolated to the range of the size of the grains. (From Rice et al. [291, 292], published with the permission of Plenum Press.)

sizes, e.g. starting at crack sizes of 1–4 mm. Again similar, but lesser, effects are expected from EA, for which observations on other mechanical properties such as spontaneous fracture and hardness have been indicated (Chap. 2, Sec. III.C and Chap. 4, Sec. II.D).

B. Quantitative Evaluation of the $\sigma - G^{-1/2}$ Model Parameters

Numerical evaluation of the basic fracture mechanics parameters underlying the $c - G$ relations of the model (Fig. 3.1) based upon the survey $\sigma - G^{-1/2}$ data is summarized in Table 3.1. Using the σ values for failure at the indicated large G–fine G branch intersections and pertinent toughness (K) values, the c values are calculated by

$$c = 0.63(K/\sigma)^{1/2} \tag{3.3}$$

assuming a half penny flaw. About 60% of the resultant c values are within a factor of 2–3 of the observed G value at the transition, in spite of uncertainties both of measurement of G values and of which G and K values to use (an average G was used in most cases). Further, several of the cases where c is $>3G$ at the intersection based on a half penny crack would be reduced for a more elongated crack (a slit crack reduces the numerical factor in Eq. (3.3) from 0.63 to 0.25, making the majority of these remaining intersections fall within $c \leq 3G$). This transition from larger to finer G branches at $c > \sim G/2$ is consistent with Singh et al.'s model and data indicating the single-to-poly-crystal K_{IC} transition occurs over a few to several grains rather than 1 [287]. The calculated large G slopes range from somewhat below single crystal to ~ polycrystalline K_{IC} values (as do such slopes for other materials, e.g. Cr_3C_2) HfC, TaC, WC, and ZrC). While three of these cases are due to investigators using G_m (Alford et al. [27] and Virkar and Gordon [195]) three are related to studies using an average G. One of the latter ($MgAl_2O_3$) shows the single crystal–polycrystalline K_{IC} transition over ~ 1 grain [28], hence consistent with the large G slope being ~ K_{IC}. While the range of slopes in part reflects uncertainties in the G values to use, it is consistent with crack growth modeling noted earlier, especially when the possibility of grain boundary K_{IC}'s substantially (e.g. twofold) lower than single crystal K_{IC}'s is considered.

Numerical evaluations corroborate two basic expectations of the model, namely that larger G branch slopes are < the polycrystalline toughness values, and that intersections of finer and larger G branches occur when flaw and grain dimensions are about the same and typically reasonably within the bounds of $c \sim G/2$ to $3G$ (Table 3.1). Further, there is no apparent difference in such comparisons whether the materials are cubic or noncubic, again arguing against TEA stresses, which occur only in noncubic materials, determining $\sigma - G$ behavior. Thus TEA stresses (and EA stress concentrations) do not determine the basic behavior of the model, but can impact behavior and model parameters, e.g. possibly modifying slopes, and especially impacting the transition from larger G to single crystal behavior. Also there is some indication of the calculated flaw sizes for failure of bodies with as-fired surfaces being somewhat larger than the grain dimensions at the larger–finer G branch intersections (e.g. data of Charles, Hing, and Alford et al. in Table 3.1). While this clearly needs more evaluation, it is consistent with expectations, since grain boundary grooves are probably not as severe flaws as cracks, which is consistent with their effective flaw size typically being a fraction of G, e.g. the calculations in Table 3.1 assume a sharp flaw which then predicts a flaw size larger than G. Finally, while the TZP and PSZ data sets, especially the latter, are not as adequate as desired, numerical evaluations of the limited data suggests that projected larger–finer G branch intersections may give similar results to the other ceramics.

It is important to recall that in calculations concerning large grains as flaws the small dimension of the grain on the fracture is the flaw size, not the larger one [which enters via the flaw geometry factor, Y of Eq. (2.2)], again arguing against plotting data versus G_m. For example, Prochazka and Charles [96] plotted σ (<140 to ~350 MPa) vs. the length (L) of the large tabular (exaggerated) α-grains at fracture origins in the β-SiC matrix, obtaining a slope of 3.9 MPa·m$^{1/2}$ with a G= ′ intercept at ~ 70, not 0, MPa). However, use of a G value based on the equivalent circular area (assuming the grain width ~ 10% of their length, per their micrographs) lowers the slope to ~ 2.3 MPa·m$^{1/2}$, which is more consistent with other data.

VIII. CORRELATION OF TOUGHNESS, TENSILE STRENGTH, AND YOUNG'S MODULUS

A key question to be addressed is the toughness–strength–reliability relations, the latter as commonly measured by the Weibull modulus. This arises since toughening mechanisms such as transformation toughening and crack branching, bridging, and related phenomena, and possibly microcracking, have been cited as means of increasing reliability, often with limited loss of, or increased, strength. Unfortunately, many investigators have assumed this is correct, and thus only measured toughness, neglecting strength or Weibull evaluations or both. However, there are some studies that provide further information as well as broader evaluations of self consistancy of toughness and strength that provide insight. These in fact show first that toughness can increase while strength decreases. Second and more broadly significant, increased tougheness leads to increased reliability in only limited cases where significant toughening occurs at sufficiently fine grain size.

Consider first evaluation of the self-consistency of the microstructural dependence of toughness with that of tensile strength, which provides by far the largest data base and a clear picture of broad discrepancies. Comparison of the porosity dependence of toughness and strength has clearly shown that high toughness from phenomena such as crack bridging associated with large cracks usually has no relation to strengths typically controlled by much smaller cracks [5, 6]. Similar comparison of the grain dependence of toughness and strength shows the same general discrepancy, as can readily be seen by comparing behavior in Chap. 2 and this chapter. Consider first noncubic materials, which provide more opportunity for toughening via crack branching and bridging due to TEA stresses, which have been cited as a source of such toughening and resultant σ–G behavior [131]. Thus many studies of noncubic materials show K_{IC}, and especially γ, independent of G at finer G, then rising and passing through a maximum as G increases [293] (Figs. 2.15 and 2.16). This is basically inconsistent with essentially all σ – G data (except possibly one set of Chandler and colleagues [169, 170], extruded BeO data having increasing orientation with increasing G), since none show the corresponding increasing σ with increasing G. The resolution of this dilemma is the recognition that much crack complexity (e.g. deflection, branching, bridging and related R-curve effects) that accounts for such K_{IC} and γ increases with G occurs on a substantially larger (e.g. mm) scale than normal strength controlling flaws, which are commonly a few tens of microns. Thus there is progressively less opportunity for the crack complexity observed with large cracks to be operative as $c \rightarrow G$. Further, since failure is a weak link process, it wll occur from the flaw having the lowest surrounding K_{IC}, i.e. the least opportunity for crack complexity. This is consistent with R-curve effects, and other extremes of K_{IC} (e.g. in PSZ) not controlling the initial σ, but with σ retained after thermal shock [293], i.e. with larger scale crack propagation.

Consider next cubic materials, where crack bridging has also been reported (Chap. 2, Sec. III.E). Most fracture toughness (K_{IC}) or energy (γ) data for them shows little or no G dependence over the ranges studied (e.g. > 1 μm to ~ 200 μm) and hence is not necessarily inconsistent with σ–G behavior [293, 294]. However, recent cubic (β) SiC data shows a K_{IC} maximum at G~0.5 μm [258] (Figure 2.14), which is consistent with earlier data for very fine (0.01–0.1 μm) G, CVD SiC (though complicated by residual stresses) indicating a lower K_{IC} than for normal G SiC (i.e. with $G \geq 1$ μm).

The first of two other evaluations of the relation of toughness and strength is to compare both toughness (K) and strength (σ) to Young's modulus (E). It is readily argued that barring some enhancement of toughness beyond simple elastic behavior, i.e. due to effects such as transformation toughening or crack branching/bridging, toughness will scale with E per Eq. (2.2). Various authors have shown that most ceramics exhibit such K–E correlation (Fig. 3.36A). Data points scattered below the average trend (i.e. $MgAl_2O_4$, MgO, B, B_4C, SiC, and to a lesser extent TiO_2) all reflect extensive transgranular fracture and known, or expected, easy cleavage on one or more sets of planes that are multiple in occurrence, e.g. {100} or {110} representing respectively three and four sets of planes [295]. This is in contrast to many, possibly all, of the materials above the average trend having fewer or less easy cleavage planes or both. Thus beta alumina has only basal cleavage, i.e. only one plane, and Al_2O_3 and TiB_2 are not known to have any cleavage and to have some, to substantial, intergranular fracture.

The counterpart of the above comparison is evaluation of strength trends as a function of E. Since $\sigma \sim K_{IC}(c)^{-1/2}$, and polycrystalline K_{IC} values tend to be proportional to E, comparison of strengths for similar c and G values for each material should also correlate with E and give insight to $\sigma - K_{IC}$ relations. Since the size and shape of machining flaws do not vary much for a given machining direction, much data satisfies this condition, especially at the same finer G(~ 4 μm in Fig. 3.36B) where effects of G on flaw geometry are not a factor and the polycrystalline K is pertinent. Such σ–E correlation is clearly better than the K–E (Fig. 3.36A). The three low, but not extreme, deviants from this trend, $CaZrO_3$, BeO, and TiB_2, probably reflect less development relative to most of the

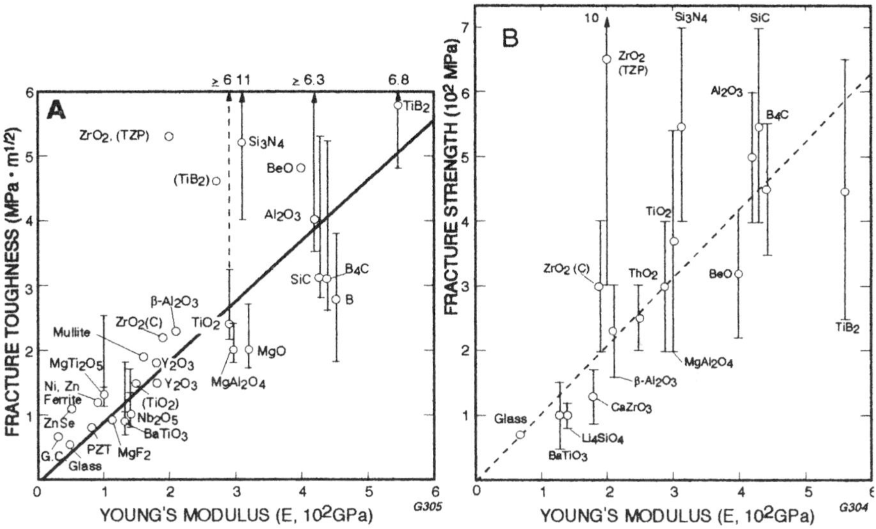

FIG. 3.36 Plots showing the correlations of (A) fracture toughness (*K*) and (B) tensile strength, both versus Young's modulus (*E*) for various ceramics respectively. (From Rice [293]), published with the permission of the *Journal of Materials Science.*)

other materials. The extreme high deviant is tetragonal ZrO_2 (TZP), which is consistent with transformation toughening increasing *K* above levels expected from those correlating with *E*. The next highest deviant from the $\sigma - E$ trend is Si_3N_4, which is consistent with the lower, but not the upper, portion of its $K - E$ trend. Similarly, note that the $\sigma - E$ trends do not correlate with the higher *K* values for TiO_2, BeO, Al_2O_3, and TiB_2. Similar issues of basic differences between σ and *K* results will be shown for particulate composites, and serious differences have been shown as a function of porosity [6, 293].

Note that the mechanical properties of highly oriented CVD graphite in the three directions relative to its orientation (Table 2.4), while falling above trends of Figs. 3.36A and 3.36B, show similar trends. However, the orientation with delamination normal to the crack deviates the most, again particularly for $K - E$ correlation, consistent with much of the higher large crack *K* not manifesting itself in strength.

Per the above theoretical and experimental correlation of strength and Young's modulus it is useful to consider $\sigma - G^{-1/2}$ data normalized by Young's modulus. Doing this using *E* values in Table 3.2 brings most materials' $\sigma - G^{-1/2}$ behavior closer (Fig. 3.37). This results in no distinction between cubic and non-

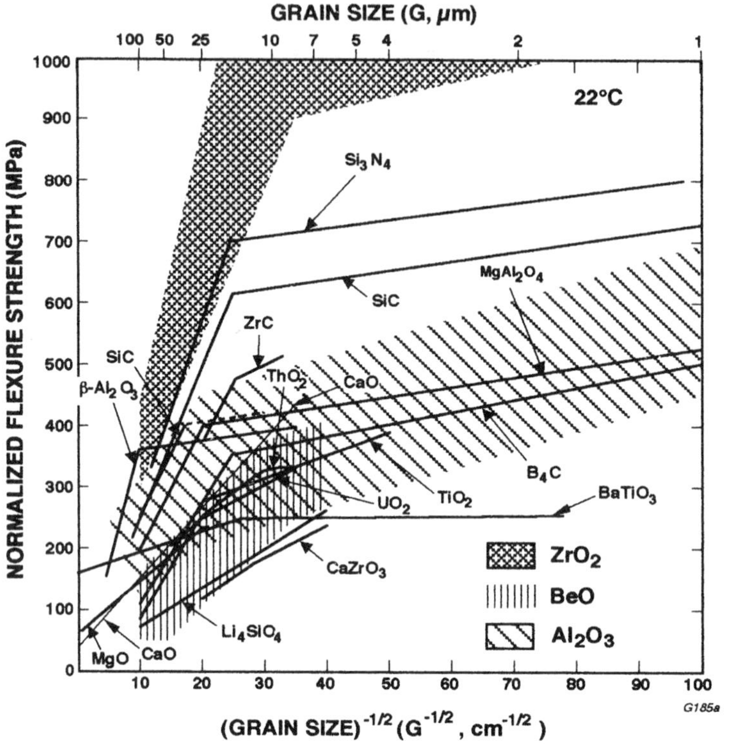

FIG. 3.37 Normalized σ–$G^{-1/2}$ data for various ceramics at 22°C to the same Young's modulus as for Al_2O_3 (~ 400 GPa, i.e. the σ plotted is that of the material × times $400/E_m$, where E_m = Young's modulus of the material, from Table 3.2). Note that more recent data has increased both the ZrO_2 and Si_3N_4 data, and that diamond data (per Fig. 3.14, but not included for clarity) falls in the main body of data, e.g. in the range for Al_2O_3 and BeO.

cubic materials, indicating that such structural differentiation does not correspond to any basic $\sigma - G^{-1/2}$ trends. Further, there is no clear difference for materials where microplasticity can determine strength, e.g. CaO, MgO, and $BaTiO_3$. The one pronounced differentiation in Fig. 3.37 is that of transformation toughened ZrO_2, especially TZP. Another is Si_3N_4 densified with oxide additives, especially since recent developments have raised this curve closer to or above the toughened ZrO_2 curve, which has also increased. These two outstanding material systems reflect higher toughnesses with finer *G*, so much of this toughness also impacts strength, both by limiting flaw sizes, as was noted earlier, and by inhibiting flaw failure.

TABLE 3.2 Young's Moduli (E in GPa) Used to Normalize Ceramic σ–G Relations

Material	E	Material	E	Material	E	Material	E
Al_2O_3	400	TiO_2	285	$BaTiO_3$	190	$MgAl_2O_4$	294
BeO	395	ThO_2	250	β-Al_2O_3	200	Diamond	1050
CaO	200	UO_2	230	$CaZrO_3$	180	SiC (β)	400
MgO	355	ZrO_2	230	Li_4SiO_4	138	ZrC	400

IX. TENSILE STRENGTH VERSUS TOUGHNESS AND RELIABILITY

An important extension of the above evaluations is directly to compare σ as a function of K for a given material, and hence also usually as a function of G, which sheds light on important exceptions, at least more extreme ones of TZP and in situ toughened Si_3N_4. Such K–σ deviations were first shown by Swain and Rose [296] for PSZ showing strength increasing with increasing toughness but then reaching various process (hence microstructure) dependent maxima, beyond which strengths decrease despite further increase in K (though this generally also begins to decrease with increased G, but often at larger G values). Various portions of such σ–K curves for various TZP materials are shown in Fig. 3.38, and for various dense Si_3N_4 bodies in Fig. 3.39. In TZP materials there is a general correlation of strength and toughness till substantial toughnesses are reached, with similar G dependence of strength (Fig. 3.21). The peaks in strengths and subsequent decreasing strength with further increase in toughness may correspond respectively to the intersection of the finer and larger G branches and the larger G branch itself. However, a study to verify this and to find whether there are differences from nontoughened bodies has not been done. Si_3N_4 σ–G behavior appears consistent with toughening and strengthening in finer G bodies but is not sufficiently detailed to confirm this. However, it clearly shows strength increasing as toughness increases, reaching a probable maximum and then decreasing. A key reason for strengths ceasing to increase with increasing toughness is that a sufficiently stressed region of sufficient scale and weakness will become the origin of failure in tensile stressing as a natural consequence of its meeting the failure criteria. On the other hand, propagation of a crack, especialy a larger one, arbitrarily introduced elsewhere in the body into such a region, can readily result in crack branching, bridging, or other toughening mechanisms. In toughened zirconia bodies, the decrease in strength may be precipitated by microcracking, usually of larger size and greater extent due to larger grain, precipitate size, or both, or by increased grain boundary phase (in PSZ). In Si_3N_4, strength maxima are usually the result of failure initiation from larger isolated or clustered grains, especially in self-reinforced bodies

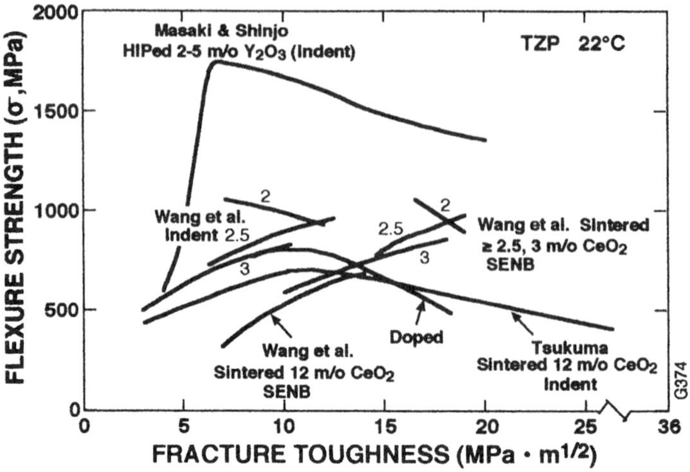

FIG. 3.38 Flexure strength versus toughness of various TZP bodies [179–181, 183, 302]. Note (1) stabilization (mostly Y_2O_3) shown in m/o, and (2) that such curves are similar to those shown for PSZ materials by Swain and Rose [296], but these for TZP are often stretched out over a larger *K* range. The three data points for TZP of Theunissen et al. [303] fall in the same range as the upper portion of the data of Ruiz and Readey [304]. Note similarity to Fig. 3.39.

[297–300]. These observations are consistent with observations of Becher et al. [301] that uncontrolled development of larger, elongated grains increases toughness but often reduces strength, while control of the development of the larger, elongated grain structure, e.g. by seeding, results in better toughness and strength.

Quantitative correlation of such σ–*K* maxima with grain sizes in Si_3N_4 is complicated by the fact that high toughnesses typically occur in bodies with mixtures of larger, substantially elongated grains (often with some excess oxide additive phase) in a finer, and often less elongated, grain matrix. However, while specific quantitative characterizations of such grain structures are at best difficult and uncertain, they have been qualitatively characterized by investigators, as is shown in Table 3.3, which shows strengths reaching maxima at intermediate *G* values, often sooner than the maxima for *K*. This data further shows strengths typically reaching a maximum versus toughness, as well as a similar maximum of Weibull moduli (*m*). Limited data also indicates a similar but more limited trend for Al_2O_3, quite possibly associated with the limited grain elongation [308]. However, again note that this occurrs at relatively modest *G* values. Similarly some strength benefits of increased toughnesses in some finer *G* SiC and TiB_2

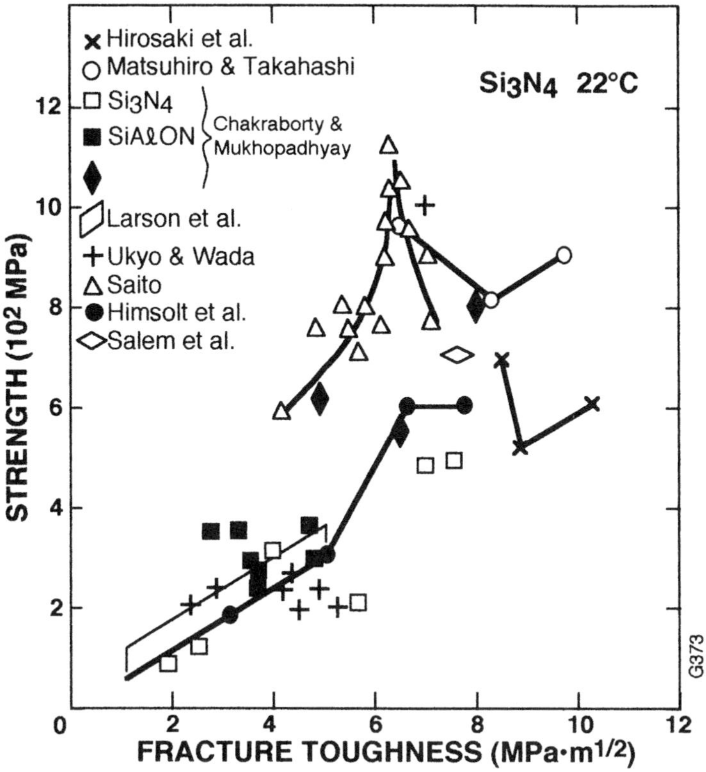

Fig. 3.39 Flexure strength versus toughness of various Si_3N_4 bodies [100, 239, 242, 293, 297, 298, 305–307]. Note that such curves are similar to those shown for PSZ materials (Fig. 3.38)

bodies may be occurring, but much more research is needed. One component of this should be evaluation of the fracture mode, since, as shown in Chapter 2, Section III.E, transgranular fracture typically substantially reduces or eliminates crack branching and bridging effects. Again these observations must be made over a range of properties, grain sizes, and crack sizes to assess adequately such effects as were discused above.

The complexity of sorting out the processing-microstructural issues are illustrated by the work of Zhang et al. [309] on α-sialons with varying amounts of $3Dy_2O_3{\cdot}5Al_2O_5$ (Dy garnet, DG) addition using either hot pressing or HIPing. Hot pressing resulted in strength decreasing modestly from 600 to 550 MPa as the DG content increased from 0 to 5 w/o, while IF toughness showed a similar decrease from 5.4 to 5.1 MPa·m$^{1/2}$. Such decreases were accompanied by some

Table 3.3 Fracture Toughness, Flexure Strength, and Weibull Moduli Trends with Grain Structure for Si_3N_4 and Al_2O_3[a]

		Fine *G*			Medium *G*			Larger *G*		
Material	Investigator	*K*	σ	*m*	*K*	σ	*m*	*K*	σ	*m*
Si_3N_4	Matsuhiro and Takahashi [297]	6.5	0.96		8.3	0.81		9.7	0.9	
Si_3N_4[b]	Himsolt et al. [298]	3.2			5.0			8.0		
			0.5			0.7			0.6	
Si_3N_4	Tani et al. [241]	4.5	0.6		6.0	0.55		6.0	0.55	
Si_3N_4	Kim et al. [300]				7.0[c]	0.88[c]		8.0[d]	0.68[d]	
Si_3N_4	Hoffmann [299]		1.1	14–20		0.9	46		0.81	18
Si_3N_4	Li et al. [244]			20			33			22
Al_2O_3[e]	Price et al. [308]				3.7	0.25	9			
					4.7	0.26	24	5.1	0.26	13

[a] *G* = grain size, *K* = fracture toughness in $MPa{\cdot}m^{-1/2}$, σ = flexure strength in GPa, and *m* = Weibull modulus.
[b] *K* and σ data shown on separate lines to reflect that the two properties were measured on bars from somewhat different processing, as opposed to all measurements being from bars from the same processing, as is the case where all data for a given *G* range are on the same line.
[c] Average grain diameter and length: 1 ± 0.3 and 5.3 μm.
[d] Average grain diameter and length: 2 ± 1 and 12 μm.
[e] The top line gives results for *G* = 4 μm and an aspect ratio of 1.5, while the lower line is for bodies respectively with *G* =4 μm and 5 μm, both with aspect ratios of 2.6 due to slightly different compositions of the grain boundary glass in these 96% alumina bodies.

modest coarsening of the fine equiaxed grain structure, which was also corroborated by some decrease in H_V (100 N) from 18.6 to 17.6 GPa. On the other hand, the HIPed bodies showed strength increasing from 420 to 550 MPa and toughness similarly increasing from 5 to 5.3 $MPa{\cdot}m^{1/2}$ as the DG content increased from 0 to 5 w/o. Increased addition, toughness, and strength in these samples were accompanied by greater coarsening and elongation of the grains, which was again reflected in lower hardnesses decreasing from 18.3 to 17.3 GPa as the addition increased. Thus opposite strength and toughness trends occurred for hot pressing versus HIPing as a function of additive content, showing important differences for two densification processes that often give the same results. However, the changes were consistent with each other, since strengths from both processes tracked with their toughness and reasonably well with the microstructural changes, i.e. higher strength from finer grain structure, with some countering of decreases from coarsening with development of an elongated but still reasonably fine grain structure. The corroboration of strength and toughnesses with microstructural changes by H trends is an example of the benefits of more property measurements.

Turning directly to the issue of reliability, Table 3.3 clearly shows that the Weibull modulus can reach substantial levels in Si_3N_4 with development of intermediate grain structures consisting of finer, often more equiaxed, grains of a few microns diameter combined with similar, or somewhat larger, diameter, but more elongated, grains. Such *m* maxima do not correlate with strength maxima, and probably not with toughness maxima, but do appear to correlate with either preceding, or the onset of, larger grains or clusters of them acting as fracture origins. This raises the issue of whether the increase in *m* is due to crack branching and bridging or to a more uniform population of grains to initiate failures, or some combination of these two effects. Detailed study of fracture surfaces may help resolve this issue, as was previously discussed [293], but detailed study involving correlation with other factors is also needeed.

One of the few other materials for which there is similar, though very limited, data is Al_2O_3 (Table 3.3), which suggests similar effects but on a far more modest scale, especially of strength. The limited strength levels raise some question whether such effects are part of the normal $\sigma - G$ behavior, since, as was noted earlier, $K - G$ behavior is contrary to the $\sigma - G$ behavior. No measurements of *m* as a function of grain size over any significant *G* range are known, but *m* can be calculated from the average strength (*A*) and the standard deviation (*S*) per Ref. 310 as

$$m = 1.21(A/S) - 0.47 \tag{3.4}$$

However, $\sigma - G$ data commonly entails limited (e.g. 5–10 or fewer) measurements per grain size, severely limiting the accuracy of such calculations. Further, even with more data points, limited shifts in grain structure can shift Weibull moduli. Thus Ting et al. [133], who directly measured *m* for their sintered alumina bodies, showed *m* increasing from 7 to 8 with narrower *G* distributions from ~0.5 to ~4.5 μm to $m \sim 11 - 13$ with *G* distributions of ~1 to ~8 μm. While this is consistent with data of Price et al., variations in these results again show the importance of having a reasonable data base for calculations of *m* as a function of *G* whether calculated directly or via Eq. (3.4).

The alumina data set of Tomaszewski [134] (Fig. 3.14) meets the above criteria, consisting of 15–25 measurements per data point for at least eight different *G* values, with ~ four on the finer *G* branch and four on the larger *G* branch. These give an average *m* of 9 ± 4 for the larger *G* branch and 26 ± 14 for the finer *G* branch, indicating higher *m* values at finer, not at larger, *G* where greater toughening due to branching and bridging should occur. Thus while limited possible effects of increased toughness increasing reliability have been indicated by Price et al. [308] and possibly by Ting et al. [133], this clearly is not a large effect and is not associated with important $\sigma - G$ changes, especially the change from finer *G* branch to larger *G* branch behavior, e.g. as suggested by Chantikul et al. [131].

Another material in which some data exists on toughness and strength changes with microstructure changes, especially with elongated grains, is SiC. Thus data of Cho and colleagues for in situ toughened silicon carbide [311, 312] shows strengths decreasing with any increase in toughness above the normal level (~ 3 MPa·m$^{1/2}$) (Fig. 3.40) in bodies sintered with 12 and 8 wt% respectively of Al_2O_3 and Y_2O_3 with or without 1% of α-SiC seeds. They showed that 5, 10, and 20% amounts of the combined additives, which resulted in the grain diameter respectively decreasing from 1 to 0.6 μm and aspect ratio increasing from ~ 6 to ~ 11, gave little change of strengths (~ 600 MPa) or toughness 5.6–5.8 MPa·m$^{1/2}$), i.e. consistent with the middle portion of the trend in Fig. 3.40. While apparently no direct measurements of Weibull moduli have been made in such studies of SiC, estimates of *m* from Eq. (3.4) indicate that if there are any changes in *m* as toughness increases, it is for *m* to decrease not increase with increased *K*.

The above review shows that increased toughness can lead to increased strength and reliability, i.e. increased Weibull modulus in some cases. These are mainly ZrO_2 toughened bodies and in situ toughened Si_3N_4 bodies, but even in these bodies improved strengths and reliability become self-limiting with continued increased toughness, i.e. there is an optimum balance between these proper-

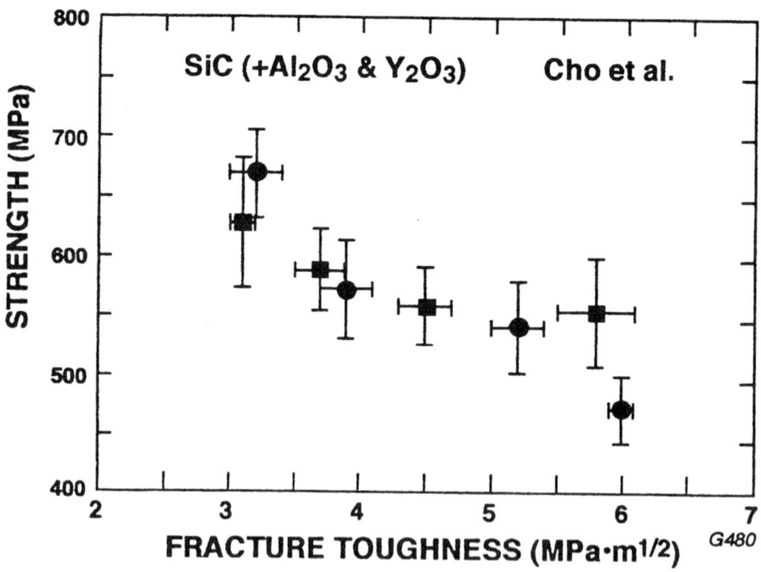

FIG. 3.40 Strengths versus toughness from tests at 22°C of in situ toughened SiC sintered with respectively 12 and 8 w/o of Al_2O_3 and Y_2O_3 with or without 1% of α- SiC seeds by Cho et al. [311, 312]. (Published with the permission of the *Journal of Materials Science.*)

ties that depends on the microstructure and the crack-microstructure scale. Such benefits are likely to be dependent on both the size and shape of the component and the nature of the stress-crack situations to be addressed. Further, such effects show limited or no translation of these benefits to the few other materials investigated, i.e. Al_2O_3 and SiC, indicating effects that probably depend on combinations of grain boundary phase-grain interactions and of the properties of the grains, e.g. of their fracture characteristics.

Finally, note two important opportunities to understand better the relation of crack propagation, especially toughness, tests to normal strength behavior. The first is the study of composites, addressed in Chapters 8–12, which again shows some of the same basic differences between toughness and strength, again questioning the applicability of much large crack data to normal strength. The second is various aspects of fractography, which is often more successful with monolithic versus composite ceramics and is probably basic to determining the apparently limited contributions of bridging and related *R*-curve effects to normal strength behavior. One basic aspect is corroboration of microstructural aspects of crack propagation in the fracture origin area, such as fracture mode, e.g. intergranular fracture favorable to bridging, and detailed comparison of matching fracture surfaces, e.g. for evidence of crack bridging. Another potentially important aspect is comparison of fracture mirror and related dimensions as a function of flaw, microstructure, and property changes. As noted earlier, such dimensions show clear reductions as flaw sizes decrease in bodies with substantial TEA stresses reflecting their contributions to failure [102]. Since such dimensions also reflect the toughness controlling fracture, they should thus correspondingly show increases with bridging and related *R*-curve effects, i.e. if these increase the toughness controlling failure this should be reflected in the mirror and related dimensions. Though such specific studies have not been conducted, qualitative evaluation of existing data does not show obvious evidence of such increases, i.e. no increase in mirror size other than the normal decrease with decreasing flaw size. Further, specific comparison of strength–mirror size data for hot pressed Al_2O_3 ($G\sim$ 1–2 μm) and commercial sintered aluminas ($G\sim$ 4–10 μm) shows them being essentially idntical [313], contrary to expected increasing bridging effects as G increases.

X. SUMMARY, DISCUSSION, CONCLUSIONS, AND RECOMMENDATIONS

Alkali halides, which even in polycrystalline form often show some macroscopic yield, have a single σ–$G^{-1/2}$ branch with a moderately positive σ–$G^{-1/2}$ slope (depending on material composition and temperature) extrapolating to single crystal yield stresses at $G=\infty$. Similarly, some somewhat harder, more refractory materials with known or expected microplasticity in single crystals such as CaO,

MgO, and $BaTiO_3$ typically show a single branch, at least for large to moderate G, with $\sigma > 0$ extrapolation at $G=\infty$ to ~ the single crystal failure stress, again indicating microplastically controlled strengths. Limited yield-extrapolated σ differences are attributed to purity differences between single crystals and polycrystalline grains. Other manifestations of microplastic behavior are σ increases with alloying (e.g. as for CaO, MgO, and $BaTiO_3$) and preferred orientation (e.g. of MgO, despite preferred orientation of {100} for easier fracture), as well as possible limited increases in strengths and internal failures as a result of surface work hardening from machining. Strengths, slopes, and intercepts decrease with increasing test temperature, moderately at first (reflecting decreases in both yield stresses and elastic moduli) and then more rapidly (e.g. above 1000°C or more commonly due to increasing grain boundary sliding) (Chap. 6).

Materials of intermediate hardness such as MgO and $BaTiO_3$ clearly show a change to greatly reduced, possibly zero G dependence of σ at finer G, with this transition being a function of surface finish. More severe machining, e.g. use of coarser grits or machining transverse, as opposed to parallel, to the tensile axis results in lower strengths and a transition to substantially reduced G dependence of σ at finer G. While there is not yet any direct fractographic evidence, such finer G failure is attributed to generating flaws sufficiently large to cause failure at lower stresses than for microplastic induced failure. Microplastic controlled failure is typically associated with failure initiating from a single grain, implying a transition to flaw failure when $c \sim G$. Such a transition to flaw controlled failure at finer G from microplastic control of σ at larger G is exactly the opposite of that proposed by Carniglia [8, 9]. There is also some evidence that very coarse machining, e.g. sawing, of MgO leads to strengths below the normal large G σ–$G^{-1/2}$ trends, possibly due to failure from larger flaws or failure via the slip-assisted crack growth. Fairly extreme finishing is needed to observe such machining effects, since unless fairly deep cracks are introduced in the surface, their role in failure is probably limited by surface work hardening. Note that even with slip-induced failure, mist and hackle are generated within larger grains, e.g. of CaO (Figs. 3.4B–D) again showing cracks having reached catastrophic propagation before reaching a grain boundary, as in failure from preexisting flaws.

The great bulk of refractory ceramic materials, which are of moderate, and most commonly substantial, hardness, and which fail from preexisting flaws, show a two-branch σ–$G^{-1/2}$ behavior over the normal G range (e.g. ~ 1–100 μm). The larger G branch shows a substantial positive slope, which is typically between single crystal (or grain boundary) and polycrystalline K_{IC} values, reflecting varying extents of subcritical growth of flaws initially < G. Such $c < G$ subcritical crack growth may be due to environmental effects or simply to propagation starting at single crystal (or grain boundary) K_{IC} values, but being arrested by surrounding grains, with resultant failure occurring at higher, e.g. polycrystalline, K_{IC}. However, even if c does not = $G/2$ at failure, there can still be measurable, frequently intrinsic $G^{-1/2}$, σ dependence. Similarly, observations of single crystal mist and hackle patterns in large grains show limits on flaw growth prior to failure (e.g. Figs. 3.7 and 3.12), and evidence of fracture related to single crystal instead of polycrystalline toughnesses. Thus substantial further work documenting local K values and understanding their effects on crack propagation and strength is needed. While this larger G branch may frequently extrapolate to $\sigma = 0$ at $G = \infty$, this is both uncertain and probably irrelevant since there must be a transition to single crystal (or bicrystal; i.e. grain boundary) strengths. Larger G strengths can frequently extend to a fraction, e.g. 1/5 to 1/2, of typical single crystal strengths for weaker orientations with comparable surface finishes. While some lower polycrystalline strengths can be due to defects, much of this appears to be intrinsic, e.g. due to lower grain boundary strengths/toughnesses as well as probable ontributions of TEA and possibly EA stresses to failure. Such factors may also mean that the larger G branch slope is not constant.

The transition from the larger G to the finer G region of $\sigma - G^{-1/2}$ behavior can intrinsically cover a range of c/G ratios (e.g. from 2 to 1/3) not just 2 (due to c being a radius and G a diameter). This approximate equivalence of c and G at the transition is born out by most data. It also implies less subcritical crack growth, and less effects of machining direction on σ. Data does show much more limited effects of the range of normal machining variables, e.g. machining direction or grit size on σ, when $G \sim$ the size of the flaws for the specific machining, with increasing effects at larger G (i.e. greater for single crystals) and finer G (where they tend to have the greatest effect).

The finer G branches result from $c > G$, so σ is controlled by polycrystalline K_{IC} (typically with limited or no contribution of R-curve). There may be multiple finer G branches, or more scatter in them, with the number or width being dependent on different or varying flaw populations, e.g. as occur from significantly different machining conditions, or varying types of flaws. Each finer G branch intersects the larger G branch when $c \sim G$ (with variations noted earlier). Finer G branches always have substantially lower slopes than the larger G branch. However, the assumption made by some (mainly those using G_m), that the slope of these finer G branches is intrinsically 0, is not supported by the majority of the data, nor with a variety of mechanisms that are known or expected to introduce some limited G dependence in this region. Several mechanisms that can give positive slopes are (1) reductions in machining flaw sizes or severity of grain boundary grooves (on as-fired or annealed surfaces) as G decreases, (2) possible grain-related cracks around pores, (3) decreasing contributions of EA, TEA, and related stresses as G decreases (relative to c) (4) increased grain boundary concentration of impurities or additives as G increases, and (5) reduced SCG as G decreases. Preferred grain orientation (e.g. from green body extrusion) can also affect this slope. This may give a negative slope due to increasing orientation with increasing G, as indicated for some BeO. Finer G

slopes of ~ 0 could occur by effects yielding negative slopes being balanced by those giving positive slopes. It is also possible that K_{IC} and σ begin to decrease intrinsically at very fine *G*, e.g. nm, beyond probable reductions in the crack deflection, branching, bridging, and related microstructural scale mixed mode effects. Such decreases may occur due to more disordered grain boundary structure becoming a measurable fraction of the specimen volume, i.e. such a fraction of the specimen so it has more disordered structure, e.g. like a glass, causing strength to decrease intrinsically. However, observed strength reductions at *G* ~ 1 μm or less, e.g. in oxides and MgF_2 (Fig. 3.24), are primarily or exclusively due to anion impurities.

The question of what *G* value to use in plotting σ is important, since it impacts the slope of both larger and finer *G* branches, as well as their intersection and hence quantitative evaluation of mechanisms involved, and the benefits to be achieved from *G* refinement. From a procedural standpoint, the *G* value should be consistent with the σ value used. The most effective way of doing this is via fractography, in order to determine not only the specific *G* value pertinent to the fracture origin but also the location of the fracture origin and hence any possible σ corrections. If fractography is not successful or is simply not conducted, use of G_m is not appropriate unless there is reasonable evidence for it being significant, since the maximum σ (which is all that is normally directly available from a flexure test) is more appropriate to G_a than G_m. Also, applicability of G_m will generally be substantially less for the finer than for the larger *G* branch.

For measuring appropriate *G* values, use of the common linear intercept method is not recommended. Instead, it is suggested that an average *G* diameter be measured for each grain along some, e.g. random, lines (preferably on the fracture surface) and averaged. Such a surface average grain diameter should be more appropriate for fracture, and both individual and average values can be compared to different *G* (e.g. G_m) values and weighted, to account for *G* distribution. With heterogeneous grain structures, the spatial and size distributions need to be considered. Such *G* measurement can, at least partially, account for grain shape effects, but they may also require more detailed evaluation of grain shapes. Grain orientation is often important, especially on a global scale, and may increase significantly with increasing *G;* it needs to be directly addressed and often is related to grain shape effects.

Another very important set of parameters are those of testing. Thus larger *G*, e.g. G_m, values may be more pertinent to larger specimens and stressed volumes (e.g. uniform tension tests) and less so for smaller specimens and stressed volumes (e.g. flexure, especially three point). Test atmosphere and temperature can provide important clues to subcritical crack growth, and probable *G* effects on this (as can machining effects as a function of *G*). Higher temperature tests, e.g. at 500–1000°C, can be important for differentiating microplastic and brittle failures (Chap. 6). Thus for both a σ decrease similar to that of *E* with temperature would be expected, e.g. is indicated for microplastic materials, e.g. CaO and MgO, but for brittle failure there can be a temperature dependence to EA, and there clearly is for TEA. This can lead to, at least initially, either less decrease of σ or a limited σ increase with temperature, especially in the *G* regime where these effects are most important, as is indicated for several orders.

A variety of other parameters need to be considered in accessing σ–*G* behavior. These include residual porosity (whose effects may not be fully independent of *G*), machining, whose effects can clearly change with *G*, and other body constituents. Constituents in solid solution generally have limited effect on failure from preexisting flaws or microcracks (e.g. via EA or TEA) but can have substantial effect on failure due to microplasticity. Constituents as second phases (temporarily or permanently) affect σ mainly via *G* (often significantly reducing it) but may have various other positive or negative effects on σ.

In a broader perspective, analysis of $\sigma - G^{-1/2}$ behavior and application of fractography have proved valuable for defining failure mechanisms, especially when combined with other studies such as of machining and different materials. Thus the first of two broad basic results are the fundamental differences in primarily the grain size and secondarily the shape and orientation dependence of much large crack toughness and normal strengths, which are often opposite, i.e. increasing toughness with decreasing strength. This results from two factors, namely from larger versus small crack effects as discussed in Chap. 2 and from changing *c/G* ratios (Figure 2.16) and some dependence of flaw sizes, especially from machining, on *G* via local values of elastic moduli, hardness, and toughness. This importance of local properties impacting the introduction of flaws and resultant body strengths is contrary to the typical focus on explaining ceramic strength behavior based on large crack toughness values with limited, or often no, attention to flaws causing failure and their microstructural dependence. This reduced relevance of large crack toughness values to normal strength behavior is controlling propagation of small cracks versus the importance of material and microstructure on properties controlling machining flaw generation is the second broad result [294]. However, such tests are relevant to strengths controlled by larger cracks, e.g. from serious thermal shock (Chap. 6) or impact damage. These basic results are corroborated and reinforced by similar evaluation of ceramic composites (Chaps. 8, 9, and 12).

REFERENCES

1. F. P. Kundsen. Dependence of Mechanical Strength of Brittle Polycrystalline Specimens on Porosity and Grain Size. J. Am. Cer. Soc. 42(8):376–388, 1959.
2. R. W. Rice. Review, Ceramic Tensile Strength: Grain Size Relations: Grain Sizes, Slopes, and Branch Intersections. J. Mat. Sci. 32:1673–1692, 1997.
3. R. W. Rice. Microstructure Dependence of Mechanical Behavior of Ceramics. Treatise Mat. Sci. Tech., Properties and Microstructure 11 (R. C. McCrone ed). Academic Press, New York, 1997, pp. 199–381.

4. R. W. Rice. Evaluation and Extension of Mechanical Property-Porosity Models Based on Minimum Solid Area. J. Mat. Sci. 31:102–118, 1996.
5. R. W. Rice. Comparison of Physical Property-Porosity Behavior with Minimum Solid Area Models. J. Mat. Sci. 31:1509–1528, 1996.
6. R. W. Rice. Porosity of Ceramics. Marcel Dekker, New York, 1998.
7. E. Ryshkewitch and D. W. Richerson. Oxide Ceramics, Physical Chemistry and Technology. 2d ed., General Ceramics, Inc., Haskell, NJ, 1985.
8. S. C. Carniglia. Petch Relation in Single-Phase Oxide Ceramics. J. Am. Cer. Soc 48(11):580–583, 1965.
9. S. C. Carniglia. Reexamination of Experimental Strength-vs.-Grain-Size Data for Ceramics. J. Am. Cer. Soc. 55(5):243–249, 1972.
10. H. P. Kirchner and R. M. Gruver. Strengthening Oxides by Reduction of Crystal Anisotropy. Cer. Fin. Co., Report No. 6 for Contract N00014-66-C0190, 1972. (See also H. P. Kirchner. Strengthening of Ceramics: Treatments, Tests, and Design Applications. Marcel Dekker, New York, 1979.)
11. R. W. Rice. Strength/Grain-Size Effects in Ceramics. Proc. Brit. Cer. Soc. 20:205–213, 1972.
12. R. W. Rice. Fractographic Identification of Strength-Controlling Flaws and Microstructure. Frac. Mech. Cer. 1 (R. C. Bradt, D. P. H. Hasselman, and F. F. Lange, eds.). Plenum Press, New York, 1974, pp. 323–343.
13. R. W. Rice. Machining Flaws and the Strength-Grain Size Behavior of Ceramics. The Science of Ceramic Machining and Surface Finishing II (B. J. Hockey and R. W. Rice, eds.). NBS Special Pub. 562, 1979, pp. 429–454.
14. S. M. Wiederhorn, B. J. Hockey, and D. E. Roberts. Effect of Temperature on the Fracture of Sapphire. Phil. Mag. 28(4): 783–795, 1973.
15. R. W. Rice. The Effect of Grinding Direction on the Strength of Ceramics. The Science of Ceramic Machining and Surface Finishing (S. J. Schneider and R. W. Rice, eds.). NBS Special Pub. 348, US Govt. Printing Office, Washington DC 1972, pp. 365–376.
16. R. W. Rice. Processing Induced Sources of Mechanical Failure in Ceramics. Processing of Crystalline Ceramics (H. Palmour III, R. F. Davis, and T. M. Hare, eds.). Plenum Press, New York, 1978, pp. 303–319.
17. R. W. Rice. Machining of Ceramics. Proc. of 2d Army Mat. Tech (J. J. Burke, A. E. Gorum, and R. N. Katz, eds.). Metals and Ceramic Info. Center, Columbus, OH, 1974, pp. 287–343.
18. R. W. Rice and J. J. Mecholsky, Jr. The Nature of Strength Controlling Machining Flaws in Ceramics. The Science of Ceramic Machining and Surface Finishing II (B. J. Hockey and R. W. Rice, eds.). US Govt. Printing Office, Washington DC, 1979, pp. 351–378.
19. R. W. Rice, J. J. Mecholsky, Jr., and P. F. Becher. The Effect of Grinding Direction on Flaw Character and Strength of Single Crystal and Polycrystalline Ceramics. J. Mat. Sci. 16:853–862, 1981.
20. R. W. Rice, J. J. Mecholsky, Jr., S. W. Freiman, and S. M. Morey. Failure Causing Defects in Ceramics: What NDE Should Find. NRL Memo. Report 4075, 10/30/1979.
21. R. W. Rice. Ceramic Fracture Features, Observations, Mechanisms, and Uses.

Fractography of Ceramic and Metal Failures (J. J. Mecholsky, Jr. and S. R. Powell, Jr., eds.). ASTM, STP 827, 1984, pp. 5–103.

22. R. W. Rice. Perspective on Fractography. Adv. Cer. 22, Fractography of Glasses and Ceramics (V. D. Frechette and J. R. Varner, eds.). Am. Cer. Soc., Westerville, OH, 1988, pp. 3–56.
23. R. L. Coble. Thermal Grooving in Polycrystalline Ceramics and Initiation of Brittle Fracture. J. Am. Cer. Soc. 54(1):59–60, 1971.
24. R. E. Tressler, R. A. Langensiepen, and R. C. Bradt. Surface-Finish Effects on Strength-vs.-Grain-Size Relations in Polycrystalline Al_2O_3. J. Am. Cer. Soc. 57(5):226–227, 1974.
25. R. C. Bradt, J. L. Dulberg, and R. E. Tressler. Surface Finish Effects and the Strength-Grain Size Relationship in MgO. Acta Met. 24:529–534, 1976.
26. D. C. Cranmer, R. E. Tressler, and R. C. Bradt. Surface Finish Effects and the Strength-Grain Size Relation in SiC. J. Am. Cer. Soc. 60(5–6):230–237, 1977.
27. N. McN. Alford, K. Kendall, W. J. Clegg, and J. D. Birchall. Strength/Microstructure Relation in Al_2O_3 and TiO_2. Ad. Cer. Mat's. 3(2):113–117, 1988.
28. R. W. Rice, S. W. Freiman, and J. J. Mecholsky, Jr. The Dependence of Strength-Controlling Fracture Energy on the Flaw-Size to Grain-Size Ratio. J. Am. Cer. Soc. 63(3–4):129–36, 1980.
29. B. R. Emrich. Technology of New Devitrified Ceramics—A Literature Review. Technical Doc. Rpt. No. ML-TDR-64-203 to AFML, 9/1964.
30. W. H. Rhodes and R. M. Cannon. Microstructure Studies of Polycrystalline Refractory Compounds. Summary Report for Contract N00019-C-0376, US Naval Air Syst., Washington, DC, 1974.
31. R. J. Stokes and C. H. Li. Dislocations and the Strength of Polycrystalline Ceramics. Materials Science Research 1 (H. Stadelmaier and W. Austin, eds.). Plenum Press, New York, 1963, pp. 133–157.
32. P. F. Becher and R. W. Rice. Strengthening Effects in Press Forged KCl. J. Appl. Phys. 44(6):2915–2916, 1973.
33. M. W. Benecke, C. R. Porter, and D. W. Roy. R and D on the Application of Polycrystalline Zinc Selenide and Cadmium Telluride to High Energy IR Laser Windows. Report No. AFML-TR-72-177, 1972.
34. J. C. Wurst. Thermal, Electrical, and Physical Property Measurements of Laser Window Materials. Proc. Conf. On High Power Infrared Laser Window Materials (C. A. Pitha ed.). AQFCRL-TR-73-0372(1), Special Report No. 162, Air Force Cambridge Research Labs 2:565–592, 6/1973.
35. F. Buch, and C. N. Ahlquist. The Yield Strength of Polycrystalline CdTe as a Function of Grain Size. Mat. Sci. Eng. 13:191–196, 1974.
36. M. Weinstein. Preparation of Fine-Grained PbTe by Ultrasonic Agitation of a Solidifying Melt. Trans. Met. Soc. AIME 230:321, 1964.
37. H.-E. Kim, and A. J. Moorhead. Effect of Doping on the Strength and Infrared Transmittance of Hot-Pressed Cesium Iodide. J. Am. Cer. Soc. 74(1):161–165, 1991.
38. D. W. Roy and P. E. Natale. Processing and Properties Study of IR Windows. Final Report No. AFAL-TR-66-349, for Contract AF 33(615)-2733, 9/1966.

39. S. W. Freiman, J. J. Mecholsky, Jr., and R. W. Rice. Fracture of ZnSe and As_2S_3 Laser Window Materials. Proc. 4th Annual Conf. on Infrared Laser Window Materials (C. R. Andrews and C. L. Stecker eds.). Air Force Matls. Lab WPAFB, Ohio, 1975, pp. 697–715.
40. R. C. Pohanka, R. W. Rice, and B. E. Walker, Jr. Effect of Internal Stress on the Strength of $BaTiO_3$. J. Am. Cer. Soc. 59(1–2):71–74, 1976.
41. B. E. Walker, Jr., R. W. Rice, R. C. Pohanka, and J. R. Spann. Densification and Strength of $BaTiO_3$ with LiF and MgO Additives. J. Am. Cer. Soc. 55(3):272–277, 1976.
42. R. C. Pohanka, S. W. Freiman, and B. A. Bender. Effect of the Phase Transformation on the Fracture Behavior of $BaTiO_3$. J. Am. Cer. Soc. 61(1–2): 72–75, 1978.
43. R. C. Pohanka, S. W. Freiman, and R. W. Rice. Fracture Processes in Ferroic Materials. Ferroelectrics 28:337–342, 1980.
44. V. C. S. Prasad and E. C. Subbarao. Deformation Studies on $BaTiO_3$ Single Crystals. Appl. Phys. Let. 22(8):424–425, 1973.
45. Y. V. Zabara, A. Y. Kudzin, and O. I. Fomichev. Dislocations in $BaTiO_3$ Single Crystals. Sov. Phys. Solid State 15(9):1852–1853, 1974.
46. R. W. Rice. CaO: II, Properties. J. Am Cer. Soc. 52(8):428–436, 1969.
47. R. W. Rice. Deformation, Recrystallization, Strength, and Fracture of Press-Forged Ceramic Crystals. J. Am. Cer. Soc. 55(2):90–97, 1972.
48. R. W. Rice, J. G. Hunt, G. I. Friedman, and J. L. Sliney. Identifying Optimum Parameters of Hot Extrusions. Boeing Co. Final Report for NASA, Contract NAS 7-276, 1968.
49. R. W. Rice. Machining, Surface Work Hardening, and Strength of MgO. J. Am. Cer. Soc. 56(10):536–541, 1973.
50. R. W. Rice. Strength and Fracture of Dense MgO. Ceramic Microstructures 76 (R. M. Fulrath and J. A. Pask, eds.). John Wiley, New York, 1968, pp. 579–587.
51. R. W. Rice. Strength and Fracture of Hot-Pressed MgO. Proc. Brit. Cer. Soc. 20:329–363, 1972.
52. R. W. Rice. Effects of Hot Extrusion, Other Constituents, and Temperature on the Strength and Fracture of Polycrystalline MgO. J. Am. Cer. Soc. 76(12):3009–3018, 1995.
53. R. M. Spriggs and T. Vasilos. Effect of Grain Size on Transverse Bend Strength of Alumina and Magnesia. J. Am. Cer. Soc. 46(5):224–228, 1961.
54. T. Vasilos, J. B. Mitchell, and R. M. Spriggs. Mechanical Properties of Pure, Dense Magnesium Oxide as a Function of Temperature and Grain Size. J. Am. Cer. Soc. 47(12):606–610, 1964.
55. A. G. Evans and R. W. Davidge. The Strength and Fracture of Fully Dense Polycrystalline Magnesium Oxide. Phil. Mag. 162(20):373–388, 1969.
56. W. B. Harrison. Influence of Surface Condition on the Strength of Polycrystalline MgO. J. Am. Cer. Soc. 47(11):574–578, 1964.
57. A. Nishida, T. Shimamura, and Y. Kohtoku. Effect of Grain Size on Mechanical Properties of High-Purity Polycrystalline Magnesia. Nippon Seramikkusu Kyoki Gakuyustu Konbunshi 98(427):412–415, 1978.
58. R. Burns and G. T. Murray. Plasticity and Dislocation Etch Pits in CaF_2. J. Am. Cer. Soc. 45(5):251–252, 1962.

59. R. Newberg and J. Pappis. Fabrication of Fluoride Laser Windows by Fusion Casting. Proc. 5th Annual Conf. on Infrared Laser Window Materials (C. R. Andrews and C. L. Stecker, eds.). Air Force Materials Lab WPAFB, Ohio, 1976, pp. 1065–1078.
60. S. K. Dickinson. Infrared Laser Window Materials Property data for ZnSe, KCl, NaCl, CaF_2, SrF_2, BaF_2. Air Force Cambridge Res. Lab. Report AFCRL-TR-75-0318, 6/6/1975.
61. S. W. Freiman, P. Becher, R. Rice, and K. Subramanian. Fracture Behavior in Alkaline Earth Fluorides. Proc. 5th Annual Conf. on Infrared Laser Window Materials (C. R. Andrews and C. L. Stecker, eds.). Air Force Matls. Lab WPAFB, OH, 1976, pp. 519–533.
62. R. M. Spriggs, L. A. Brissette, and T. Vasilos. Pressure Sintered Nickel Oxide. Am. Cer. Soc. Bul. 43(8):572–577, 1964.
63. W. B. Harrison. Fabrication and Fracture of Polycrystalline NiO. Honeywell Res. Cen. Third Interim Report for Contract DA-11-022-ORD-3441, 3/1965.
64. R. M. Spriggs, T. Vasilos, and L. A. Brissette. Grain Size Effects in Polycrystalline Ceramics. Materials Science Research, The Role of Grain Boundaries and Surfaces in Ceramics 3 (W. W. Kriegel and H. Palmour III, eds.). Plenum Press, New York, 1966, pp. 313–353.
65. F. P. Knudsen. Dependence of Mechanical Strength of Brittle Polycrystalline Specimens on Porosity and Grain Size. J. Am. Cer. Soc. 42(8):376–388, 1959.
66. C. E. Curtis and J. R. Johnson. Properties of Thorium Oxide Ceramics. J. Am. Cer. Soc. 40(2):63–68, 1957.
67. R. W. Rice. Unpublished data.
68. W. H. Rhodes, G. C. Wei, E. A. Trickett, M. R. Pascucci, S. Natansohn, and C. Brecher. Lanthana-Doped Yttria as an Optical Ceramic. GTE Lab. Annual Report TR 0095 05-90-879 for contract N60530-86-C-0022, 6/15/1990.
69. C. R. Kennedy and G. Bandyopadhyay. Thermal-Stress Fracture and Fractography in UO_2. J. Am. Cer. Soc. 59(3–4):176–177, 1976.
70. M. D. Burdick and H. S. Parker. Effect of Particle Size on Bulk Density and Strength Properties of Uranium Dioxide Specimens. J. Am. Cer. Soc. 39(5):181-, 1956.
71. F. P. Knudsen, H. S. Parker, and M. D. Burdick. Flexural Strength of Specimens Prepared from Several Uranium Dioxide Powders; Its Dependence on Porosity and Grain Size and the Influence of Additions of Titania. J. Am. Cer. Soc. 43(12):641–647, 1960.
72. A. G. Evans, and R. W. Davidge. The Strength and Fracture of Stoichiometric Polycrystalline UO_2. J. Nuc. Mat. 33:249–260, 1969.
73. R. F. Canon, J. T. A. Roberts, and R. J. Beals. Deformation of UO_2 at High Temperatures. J. Am. Cer. Soc. 54(2):105–112, 1971.
74. R. W. Rice, and W. J. McDonough. Ambient Strength and Fracture of ZrO_2. Mechanical Behavior of Materials. Soc. Mat. Sci., Japan 4:394–403, 1972.
75. R. W. Rice. Strength–Grain Size Behavior of ZrO_2 at Room Temperature. J. Mat. Sci. Lett. 13:1408–1412, 1994.
76. J. W. Adams, R. Ruh, and K. S. Mazdiyasni. Young's Modulus, Flexural Strength, and Fracture of Yttria-Stabilized ZrO_2 versus Temperature. J. Am. Cer. Soc. 80(4):903–908, 1997.

77. K. F. H. Ahlborn, Y. Kagawa, and A. Okura. Observation of the Influence of Microcracks on the Crack Propagation Inside of Transparent ZrO_2. Ceramics Today—Tomorrow's Ceramics 66C (P. Vincenzini, ed.). Elsevier, 1991, pp. 1857–1864.
78. J. R. Hague. The Effect of Stabilizer on Properties of ZrO_2. Am. Cer. Soc. Bul. 45(9):826, 1966.
79. A. G. King and Fuchs. Effect of Composition on the Strength of Magnesia Stabilized Zirconia. Am. Cer. Soc. Bul. 47(4):427, 1968.
80. R. P. Ingel, D. Lewis, B. A. Bender, and R. W. Rice. Temperature Dependance of Strength and Fracture Toughness of ZrO_2 Single Crystals. J. Am. Cer. Soc. 65(9):C-150–152, 1982.
81. R. C. Garvie, R. R. Hughan and R. T. Pascoe. Strengthening of Lime-Stabilized Zirconia by Post Sintering Heat Treatments. Processing of Crystalline Ceramics, Mat. Sci. Res. 11 (H. Palmour III, R. F. Davis, and T. M. Hare eds). Plenum Press, New York 1978, pp. 263–272.
82. J. F. Jue, and A. V. Virkar. Fabrication, Microstructural Characterization, and Mechanical Properties of Polycrystalline *t*-Zirconia. J. Am. Cer. Soc. 73(12):3650–3657, 1990.
83. R. W. Rice, K. R. McKinney, and R. P. Ingel. Grain Boundaries, Fracture, and Heat Treatment of Commercial Partially Stabilized Zirconia. J. Am. Cer. Soc. 64(12):C-175–177, 1981.
84. J. T. Bailey and R. Russell, Jr. Preparation and Properties of Dense Spinel Ceramics in the $MgAl_2O_4$-Al_2O_3 System. Trans. of Brit. Cer. Soc. 68(4):159–164, 1969.
85. J. T. Bailey and R. Russel, Jr. Magnesia-Rich $MgAl_2O_4$ Spinel Ceramics. Am. Cer. Soc. Bul. 50(5):493–496, 1971.
86. R. W. Rice and W. J. McDonough. Ambient Strength and Fracture Behavior of $MgAl_2O_4$. Mechanical Behavior of Materials, Soc. Mats. Sci. Japan 4:422–431, 1972.
87. D. M. Chay, H. Palmour III, and W. W. Kreigel. Microstructure and Room-Temperature Mechanical Properties of Hot-Pressed Magnesium Aluminate as Described by Quadratic Multivariable Analysis. J. Am. Cer. Soc. 51(1): 10–16, 1968.
88. W. H. Rhodes, P. L. Berneburg, and J. E. Niesse. Development of Transparent Spinel. AVCO Corp. Final Report for Army Matls. and Mechanics Research Center contract DAAG-46-69-C-0113, 1970.
89. W. G. Jacobs. Synthesis of $MgAl_2O_4$ Spinel, M. S. thesis, Rutgers State University, 1976.
90. W. T. Bakker and J. G. Lindsay. Reactive Magnesia Spinel, Preparation and Properties. Am. Cer. Soc. Bul. 46(11):1094–1097, 1967.
91. T. Kanai, Z.-E. Nakagawa, Y. Ohya, M. Hasegawa, and K. Hamano. Effect of Composition on Sintering and Bending Strength of Spinel Ceramics. Report of the Res. Lab. of Eng. Mat., Tokyo Inst. of Tech 13:75–83, 1987.
92. R. L. Gentilman. Fusion-Casting of Transparent Spinel. Am. Cer. Soc. Bul. 60(9):906–909, 1981.
93. T. I. Hou and W. M. Kriven. Mechanical Properties and Microstructure of Ca_2SiO_4- $CaZrO_3$ Composites. J. Am. Cer. Soc. 77(1):65–72, 1994.
94. J. S. Moya, P. Pena, and S. De Aza. Transforming Toughening in Composites Containing Dicalcium Silicate. J. Am. Cer. Soc. 68(9):C-259-62, 1985.

95. J. A. Coppola and R. C. Bradt. Measurement of Fracture Surface Energy of SiC. J. Am. Cer. Soc. 55(9):455–460, 1972.
96. S. Prochazka and R. J. Charles. Strength of Boron-Doped Hot-Pressed Silicon Carbide. Am. Cer. Soc. Bull. 52(12):885–891, 1973.
97. T. D. Gulden. Mechanical Properties of Polycrystalline β-SiC. J. Am. Cer. Soc. 52(11):585–590, 1969.
98. S. G. Seshardi, M. Srinivasan, and K. Y. Chia. Microstructure and Mechanical Properties of Pressureless Sintered Alpha-SiC. Cer. Trans. 2, Silicon Carbide '87 J. D. Cawley and C. E. Semler, eds.). Am. Cer. Soc., Westerville, OH, 1989, pp. 215–226.
99. Y. Takeda, and K. Nakamura. Effects of Additives on Microstructure and Strength of Dense Silicon Carbide. Proc. *23d.* Japan Congress on Materials Science. Soc. Mat. Sci. Japan, 1980, pp. 215–219.
100. D. C. Larsen, J. W. Adams, L. R. Johnson, A. P. S. Teotia, and L. G. Hill. Ceramic Materials for Advanced Heat Engines, Technical and Economic Evaluation. Noyes, Park Ridge, NJ, 1985.
101. R. W. Rice, S. W. Freiman, R. C. Pohanka, J. J. Mecholsky, Jr., and C. Cm. Wu. Microstructural Dependence of Fracture Mechanics Parameters in Ceramics. Frac. Mech. Cer. 4 (R. C. Bradt, D. P. H. Hasselman, and F. F. Lange, eds.). Plenum Press, New York, 1978, pp. 849–876.
102. R. W. Rice. Fractographic Determination of K_{IC} and Effects of Microstructural Stresses in Ceramics. Fractography of Glasses and Ceramics. Ceramic Trans. 17 (J. R. Varner and V. D. Frechette, eds.). Am. Cer. Soc., Westerville, OH, 1991, pp. 509–545.
103. G. V. Srinivasan and V. Venkateswaran. Tensile Strength Evaluation of Polycrystalline SiC Fibers. Cer. Eng. Sci. Proc. 14(7–8):563–572, 1993.
104. M. J. Koczak, K. Prewo, A. Mortensen, S. Fishman, M. Barsoum, and R. Gottschall. Inorganic Composite Materials in Japan: Status and Trends. ONRFE Sci. Monograph M7, 11/1989.
105. T. Shimo, I. Tsukada, T. Seguchi, and K. Okamura. Effect of Firing Temperature on Thermal Stability of Low-Oxygen SiC Fiber. J. Am. Cer. Soc. 81(8):2109–2115, 1998.
106. A. Elkind and M. W. Barsoum. Grain Growth and Strength Degradation of SiC Monofilaments at High Temperatures. J. Mat. Sci. 31:6119–6123, 1996.
107. X.-J. Ning, P. Pirouz, and S. C. Farmer. Microchemical Analysis of the SCS-6 Silicon Carbide Fiber. J. Am. Cer. Soc. 76(8):2033–2041, 1993.
108. R. T. Bhatt and D. R. Hull. Microstructural and Strength Stability of CVD SiC Fibers in Argon Environments. Cer. Eng. Sci. Proc. 12(10):1832–1844, 1991.
109. R. T. Bhatt and D. R. Hull. Strength-Degrading Mechanisms for Chemically-Vapor-Deposited SCS-6 Silicon Carbide Fibers in Argon Environments. J. Am. Cer. Soc. 81(4):957–964, 1998.
110. S. R. Nutt and F. E. Wawner. Silicon Carbide Filaments: Microstructure. J. Mat. Sci. 20:1953–1960, 1985.
111. R. W. Rice, and K. R. McKinney. Residual Stresses and Scaling CNTD SiC to Larger Sizes. J. Mat Sci. Lett. 1:159–162, 1982
112. S. Dutta, R. W. Rice, H. C. Graham, and M. C. Mendiratta. Characterization and Properties of Controlled Nucleation Thermochemical Deposition (CNTD)- Silicon Carbide. J. Mat. Sci. 15:2183–2191, 1980.

113. D. B. Miracle, and H. A. Lipsitt. Mechanical Properties of Fine-Grained Substoichiometric Titanium Carbide. J. Am. Cer. Soc. 66(8):592–597, 1983.
114. V. P. Bulychev, R. A. Andrievskii, and L. B. Nezhevenko. Theory and Technology of Sintering, Thermal, and Chemicothermal Treatment Process, The Sintering of Zirconium Carbide. Transl. from Poroshkovaya Metallurgiya 4(172):38–42, 4/1977.
115. J. A. Savage, C. J. H. Worst, C. S. J. Pickels, R. S. Sussmann, C. G. Sweeney, M. R. McClymont, J. R. Brandon, C. N. Dodge, and A. C. Beale. Properties of Free-Standing CVD Diamond Optical Components. Window and Dome Technologies and Materials 3060. SPIE Proc. (R. W. Tustison, ed.), 4/1997, pp. 144–159.
116. A. Wronski and A. Fourdeux. Slip-Induced Fracture of Polycrystalline Cr, Mo, and W: The Group Via BCC Transition Metals. Intl. J. Fract. Mech. 1(2):73–80, 1965.
117. J. R. Stevens. Effect of Surface Condition on the Ductility of Tungsten. High Temperature Materials II (G. M. Ault, W. F. Barclay, and H. P. Munger, eds.). Interscience, New York, 1962, pp. 125–137.
118. R. N. Orava. The Effect of Grain Size on the Yielding and Flow of Molybdenum. Refractory Metals and Alloys IV, Research and Development 1 (R. I. Jaffee, G. M. Ault, J. Maltz, and M. Semchyshen, eds.). Gordon and Breach, New York, 1966, pp. 117–140.
119. R. H. Forster and A. Gilbert. The Effect of Grain Structure on the Fracture of Recrystallized Tungsten Wire. J. Less Com. Met. 20:315–325, 1970.
120. A. P. Valintine, and D. Hull. Effect of Temperature on the Brittle Fracture of Polycrystalline Tungsten. J. Less Com. Met. 17:353–361, 1969.
121. D. B. Binns and P. Popper. Mechanical Properties of Some Commercial Alumina Ceramics. Proc. Brit. Cer. Soc. 6:71–79, 1966.
122. B. R. Steele, F. Rigby, and M. C. Hesketh. Investigations on the Modulus of Rupture of Sintered Alumina Bodies. Proc. Brit. Cer. Soc. 6:83–93, 1966.
123. H. Neuber and A. Wimmer. Experimental Investigations of the Behavior of Brittle Materials at Various Ranges of Temperature. AFML-TR-68-23, 1968.
124. R. F. Gruszka, R. E. Mistler, R. B. Runk. Effect of Various Surface Treatments on the Bend Strength of High Alumina Substrates. Am. Cer. Soc. Bul. 49(6):575–579, 1970.
125. A. G. Evans and G. Tappin. Effects of Microstructure on the Stress to Propagate Inherent Flaws. Proc. Brit. Cer. Soc. 20:275–297, 1972.
126. M. J. Hanney, and R. Morrell. Factors Influencing the Strength of a 95% Alumina Ceramic. Proc. Brit. Cer. Soc. 32:277–290, 1982.
127. M. McNamee and R. Morrell. Textural Effects in the Microstructure of a 95% Alumina Ceramic and Their Relationship to Strength. Sci. of Cer. 12:629–634, 1984.
128. R. M. Gruver, W. A. Sotter, and H. P. Kirchner. Fractography of Ceramics. Department of the Navy Report, Contract N00019-73-C-0356, 1974.
129. W. B. Crandall, D. H. Chung, and T. J. Gray. The Mechanical Properties of Ultra-Fine Grain Hot Pressed Alumina. Mechanical Properties of Engineering Ceramics (W. W. Kriegel and H. Palmour III, eds.). Wiley Interscience, New York, 1961, pp. 349–376.
130. E. M. Passmore, R. M. Spriggs, and T. Vasilos. J. Am. Cer. Soc. 48(1):1–7, 1965.

131. P. Chantikul, S. J. Bennison, and B. R. Lawn. Role of Grain Size in the Strength and R-Curve Properties of Alumina. J. Am. Cer. Soc. 73(8):2419–2427, 1990.
132. L. Sztankovics. The Effect of Grain Size and Porosity on the Bending Strength and Wear Resistance of Alumina Ceramics. Epitoanyag 42(3):88–95, 1990.
133. J.-M. Ting, R. Y. Lin, and Y.-H. Ko. Effect of Powder Characteristics on Microstructure and Strength of Sintered Alumina. Am. Cer. Soc. Bul. 70(7):1167–1172 (1991).
134. H. Tomaszewski. Influence of Microstructure on the Thermomechanical Properties of Alumina Ceramics. Cer. Intnl. 18:51–55, 1992.
135. T. Koyama, A. Nishiyama, and K. Niihara. Effect of Grain Morphology and Grain Size on the Mechanical Properties of Al_2O_3 Ceramics. J. Mat. Sci. 29:3949–3954, 1994.
136. D. Kovar and M. J. Readey. Grain Size Distribution and Strength Variability of High-Purity Alumina. J. Am. Cer. Soc. 79(2):305–312, 1996.
137. A. Krell and P. Blank. Grain Size Dependence of Hardness in Dense Submicrometer Alumina. J. Am. Cer. Soc. 78(4):1118–1120, 1993.
138. R. W. Rice. Specimen Size–Tensile Strength Relations for a Hot-Pressed Alumina and Lead Zirconate/Titanate. Am. Cer. Soc. Bul. 66(5):794–798, 1987.
139. G. E. Gazza, J. R. Barfield. and D. L. Preas. Reactive Hot Pressing of Alumina with Additives. Am. Cer. Soc. Bul. 48(6):605–610, 1969.
140. C. O. McHugh, T. J. Whalen, and J. Humenik, Jr. Dispersion-Strengthened Aluminum Oxide. J. Am. Cer. Soc. 49(9):486–491, 1966.
141. D. T. Rankin, J. J. Stiglich, D. R. Petrak, and R. Ruh. Hot-Pressing and Mechanical Properties of Al_2O_3 with an Mo-Dispersed Phase. J. Am. Cer. Soc. 54(6):277–281, 1971.
142. R. W. Rice. Effects of Ceramic Microstructural Character on Machining Direction–Strength Anisotropy. Machining of Advanced Materials (S. Johanmir, ed.). NIST Special Pub. 847, US Govt. Printing Office, Washington DC, 1993, pp. 185–204.
143. R. W. Rice. Correlation of Machining–Grain-Size Effects on Tensile Strength with Tensile Strength–Grain-Size Behavior. J. Am. Cer. Soc. 76(4):1068–1070, 1993.
144. R. W. Rice. Porosity Effects on Machining Direction–Strength Anisotropy and Fracture Mechanisms. J. Am. Cer. Soc. 77(8):2232–2236, 1994.
145. J. B. Wachtman, Jr., and L. H. Maxwell. Strength of Synthetic Single Crystal Sapphire and Ruby as a Function of Temperature and Orientation. J. Am. Cer. Soc. 42(91:432–433, 1959.
146. A. H. Heuer and J. P. Roberts. The Influence of Annealing on the Strength of Corundum Crystals. Proc. Brit. Cer. Soc. 6:17–27, 1966.
147. G. Orange, D. Turpin-Launay, P. Goeuriot, G. Fantozzi, and F. Thevenot. Mechanical Behavior of a Al_2O_3-AION Composite Ceramic Material (Aluminalon). Sci. Cer. 12:661–666, 1984.
148. A. Harabi and T. J. Davies. Mechanical Properties of Sintered Alumina-Chromia Refractories. Trans. Brit. Cer. Soc. 94(2):79–84, 1995.
149. R. J. Charles. Static Fatigue:Delayed Fracture. Studies of the Brittle Behavior of Ceramic Materials, Technical Report No. ASD-TR-628. Aeronautical Systems Division, Wright Patterson AFG, OB, 1962, pp. 370–404.

150. J. E. Bailey and H. A. Barker. Ceramic Fibers for the Reinforcement of Gas Turbine Blades. Ceramics in Severe Environments, Materials Science Researchs (W. W. Kriegel and H. Palmour III, eds.). Plenum Press, New York, 1971, pp. 341–359.
151. H. D. Blakelock, N. A. Hill, S. A. Lee, and C. Goatcher. The Production and Properties of Polycrystalline Alumina Rods and Fibers. Proc. Brit. Cer. Soc. 15:69–83, 1970.
152. P. Hing. Spatial Distribution of Tungsten on the Physical Properties of Al_2O_3-W Cermets. Sci. of Cer. 12:87–94, 1984.
153. S. Hori, R. Kurita, M. Yoshimura, and S. Somiya. Suppressed Grain Growth in Final-Stage Sintering of Al_2O_3 with Dispersed ZrO_2 Particles. J. Mat. Sci. Lett. 4:1067–1070, 1985.
154. F. F. Lange and M. M. Hirlinger. Hindrance of Grain Growth in Al_2O_3 by ZrO_2 Inclusions. J. Am. Cer. Soc. 67(3):164–168, 1984.
155. N. D. Nazarenko, V. F. Nechitailo, and N. I. Vlasko. The Manufacture and Properties of Oxide Fibers. Soviet Pwd. Metall. 4:265–267, 1969.
156. F. H. Simpson. Continuous Oxide Filament Synthesis (Devitrification). Boeing Co. Final report for contract AFML-TR-71-135, 1971.
157. B. H. Hamling. Metal Oxides. British Patent No. 1,144,033, 1969.
158. J. D. Birchall. The Preparation and Properties of Polycrystalline Aluminum Oxide Fibers. Trans. J. Br. Cer. Soc. 82:143–145, 1983.
159. A. K. Dhingra. Advances in Inorganic Fiber Developments. Contemporary Topics in Polymer Science, 5 (E. J. Vandenberg, ed.). Plenum Press, New York, 1984, pp. 227–260.
160. J. C. Romine. New High-Temperature Ceramic Fiber. Cer. Eng. Sci. Proc. 8(7–8):755–765, 1987.
161. M. H. Stacey. Developments in Continuous Alumina-Based Fibers. Trans. J. Br. Cer. 87:168–172, 1977.
162. R. N. Fetterolf. Development of High Strength, High Modulus Fibers. Babcock & Wilcox Report No. 7953 for contract AFM:-TR-70-197, 1970.
163. H. G. Sowman and D. D. Johnson. Ceramic Oxide Fibers. Cer. Eng. Sci. Proc. 6(9):1221–1230, 1985.
164. Method and Apparatus for Making Fibers, Patent Specifications. British Patent No. 1,141,207, 1969.
165. K. Jakus and V. Tulluri. Mechanical Behavior of a Sumitomo Alumina Fiber at Room and High Temperature. Cer. Eng. Sci. Proc. 10(9–10):1338–1349, 1989.
166. J. T. A. Pollock and G. F. Hurley. Dependence of Room Temperature Fracture Strength on Strain-Rate in Sapphire. J. Mat. Sci. 8:1595–1602, 1973.
167. P. Shahinian. High-Temperature Strength of Sapphire Filament. J. Am. Cer. Soc. 54(1):67–68, 1971.
168. S. A. Newcomb and R. E. Tressler. Slow Crack Growth of Sapphire at 800 to 1500°C. J. Am. Cer. Soc. 76(10):605–612, 1993.
169. B. A. Chandler, E. C. Duderstadt, and J. F. White. Fabrication and Properties of Extruded and Sintered BeO. J. Nuc. Mat. 8(3):329–347, 1963.
170. R. E. Fryxell and B. A. Chandler. Creep, Strength, Expansion, and Elastic Moduli of Sintered BeO as a Function of Grain Size, Porosity, and Grain Orientation. J. Am. Cer. Soc. 47(6):283–291, 1964.

171. G. G. Bentle and R. N. Kniefel. Brittle and Plastic Behavior of Hot-Pressed BeO. J. Am. Cer. Soc. 48(11):570–577, 1965.
172. K. Veevers. Recrystallization of Machined Surfaces of Beryllium Oxide. J. Australian Cer. Soc. 5(1):16–20, 1969.
173. J. Greenspan. Development of a High-Strength Beryllia Material. Army Materials and Mechanics Research Center Report No. AMMRC-TR-72-16, 1972.
174. J. S. O'Neill and D. T. Livey. Fabrication and Property Data for Two Samples of Beryllium Oxide. United Kingdom Atomic Energy Authority Research Group Report AERE-R4912, 1965.
175. N. A. Hill, J. S. O'Neill, and D. T. Livey. The Properties of BeO containing up to 15 w/o SiC and up to 2 w/o MgO. Proc. Brit. Cer. Soc. 7:221–232, 1967.
176. H. P. Kirchner and R. M. Gruver. Strength-Anisotropy-Grain Size Relations in Ceramic Oxides. J. Am. Cer. Soc. 53(5):232–236, 1970.
177. K. Tsukuma and M. Shimada. Strength, Fracture Toughness and Vickers Hardness of CeO_2-Stabilized Tetragonal ZrO_2 Polycrystals (Ce-TZP). J. Mat. Sci. 20:1178–1184, 1985.
178. R. C. Garvie. Partially Stabilized Zirconia Ceramics. US Patent 4,279,655, 1981.
179. J. Wang, M. Rainforth, and R. Stevens. The Grain Size Dependence of the Mechanical Properties in TZP Ceramics. Br. Cer. Trans. J. 88(1):1–6, 1988.
180. J. Wang, X. H. Zheng, and R. Stevens. Fabrication and Microstructure-Mechanical Property Relationships in Ce-TZPs. J. Mat. Sci. 27:5348–5356, 1992.
181. T. Masaki. Mechanical Properties of Toughened ZrO_2-Y_2O_3 Ceramics. J. Am. Cer. Soc. 69(8):638–640, 1986.
182. N. L. Hecht, S. M. Goodrich, D. E. McCullum, and P. P. Yaney. Characterization Studies of Transformation-Toughened Ceramics. Am. Cer. Soc. Bul. 71(6): 955–959, 1992.
183. K. Tsukuma. Mechanical Properties and Thermal Stability of CeO_2 Containing Tetragonal Zirconia Polycrystals. Am Cer. Soc. Bul. 65(10):1386–1389, 1986.
184. C. Puchner, W. Kladnig, and G. Gritzner. Mechanical Properties of CeO_2-doped ZrO_2. J. Mat. Sci. Lett. 9:94–96, 1990.
185. M. Kihara, T. Ogata, K. Nakamura, and K. Kobayashi. Effects of Al_2O_3 Addition on Mechanical Properties and Microstructure of Y-TZP. J. Cer. Soc. Jpn., Intl. Ed. 96:635–642, 1998.
186. D. B. Marshall, F. F. Lange, and P. D. Morgan. High-Strength Zirconia Fibers. J. Am Cer. Soc. 70(8):C-187–188, 1987.
187. J. Lankford. Plastic Deformation of Partially Stabilized Zirconia. J. Am. Cer. Soc. 66 (11):C-212–213. 1983.
188. J. Lankford. Deformation and Fracture of Yttria-Stabilized Zirconia Single Crystals. J. Mat. Sci. 21:1981–1989, 1986.
189. D. B. Marshall, M. C. Shaw, R. H. Dauskardt, R. O. Ritchie, M. J. Readey, and A. H. Heuer. Crack-Tip Transformation Zones in Toughened Zirconia. J. Am. Cer. Soc. 73(9):2659–2666, 1990.
190. P. F. Becher, M. V. Swain, and M. K. Ferber. Relation and Transformation Temperature to the Fracture Toughness of Transformation-Toughened Ceramics. J. Mat. Sci. 22:76–84, 1987.

191. A. G. Evans. Toughening Mechanisms in Zirconia Alloys. In: Fracture in Ceramic Materials. Noyes, 1984, pp. 16–55.
192. A. H. Heuer. Transformation Toughening in ZrO_2-Containing Ceramics. J. Am. Cer. Soc. 70(10):689–698, 1987.
193. A. G. Evans. Perspective on the Development of High-Toughness Ceramics. J. Am. Cer. Soc. 73(2): 187–206, 1990.
194. P. F. Becher, K. B. Alexander, A. Bleier, S. B. Waters, and W. H. Warwick. Influence of ZrO_2 Grain Size and Content on the Transformation Response in the Al_2O_3-ZrO_2 (12 mol% CeO_2) System. J. Am. Cer. Soc. 76(3):657–663, 1993.
195. A. V. Virkar and R. S. Gordon. Fracture Properties of Polycrystalline Lithia-Stabilized β″-Alumina. J. Am. Cer. Soc. 60(1–2):58–61, 1977.
196. W. J. McDonough, D. R. Flinn, K. H. Stern, and R. W. Rice. Hot Pressing and Physical Properties of Na Beta Alumina. J. Mat. Sci. 13:2403–2412, 1978.
197. G. J. May, S. R. Tan, and I. W. Jones. Hot Isostatic Pressing of Beta-Alumina. J. Mat. Sci. 15:2311–2316, 1980.
198. R. Stevens, Strength and Fracture Mechanisms in Beta-Alumina. J. Mat. Sci. 9:934–940, 1974.
199. J. N. Lingscheit, G. J. Tennenhouse, and T. J. Whalen. Compositions and Properties of Conductive Ceramics for the Na-S Battery. Am. Cer. Soc. Bull. 58(5):536–538, 1979.
200. T. L. Francis, F. E. Phelps, and G. MacZura. Sintered Sodium Beta Alumina Ceramics. Am. Cer. Soc. Bull. 50(7):615–619, 1971.
201. C.-Y. Chu, K. Bar, J. P. Singh, K. C. Goretta, M. C. Billone, R. B. Poeppel, and J. L. Routbort. Mechanical Properties of Polycrystalline Lithium Orthosilicate. J. Am. Cer. Soc. 72(9):1643–1648, 1989.
202. J. A. Kuszyk and R. C. Bradt. Influence of Grain size on Effects of Thermal Expansion Anisotropy in $MgTi_2O_5$. J. Am. Cer. Soc. 56(8):420–423, 1973.
203. J. J. Cleveland and R. C. Bradt. Grain Size/Microcrack Relations for Pseudobrookite Oxides. J. Am. Cer. Soc. 61(11–12):478–481, 1978.
204. W. D. Scott. Purification, Growth of Single Crystals, and Selected Properties of MgF_2. J. Am. Cer. Soc. 45(12):586–587, 1962.
205. W.-Y. Lin, M.-H. Hon, and S.-J. Yang. Effects of Grain Growth on Hot-Pressed Optical Magnesium Fluoride Ceramics. J. Am. Cer. Soc. 71(3):C-136–137, 1988.
206. H. H. Rice and M. J. Garey. Sintering of Magnesium Fluoride. Am. Cer. Soc. Bul. 46(12):1149–1153, 1967.
207. D. A. Buckner, H. C. Hafner, and N. J. Kriedl. Hot-Pressing Magnesium Fluoride. J. Am. Cer. Soc. 45(9): 435–438, 1962.
208. V. Mandorf and J. Hartwig. High Temperature Properties of Titanium Diboride. High Temperature Materials Il (G. Ault, W. Barclay, and H. Munger, eds.). Interscience, New York, 1963, pp. 455–467.
209. R. A. Alliegro. Titanium Diboride and Zirconium Diboride Electrodes. Encyclopedia of Electrochemistry (C. Hampell, ed.). Reinhold, New York, 1964, p. 1125.
210. R. D. Holliday, R. Mogstad, and J. L. Henry. Gas Evolution During Consolidation of Carbothermic TiB_2 Powder to High Density. Electrochem. Tech. 1(5–6):183, 1963.
211. H. R. Baumgartner and R. A. Steiger. Sintering and Properties of Titanium Di-

boride Made from Powder Synthesized in a Plasma-Arc Heater. J. Am. Cer. Soc. 67(3):207–212, 1984.
212. P. F. Becher, C. B. Finch, and M. K. Ferber. Effect of Residual Nickel Content on the Grain Size Dependent Mechanical Properties of TiB_2. J. Mat. Sci. Lett 5:195–197, 1986.
213. R. Telle and G. Petzow. Strengthening and Toughening of Boride and Carbide Hard Material Composites. Mats. Sci. Eng. A 105/106:97–104, 1988.
214. T. Watanabe and S. Kouno. Mechanical Properties of TiB_2-CoB-Metall Boride Alloys. Am. Cer. Soc. Bul. 61(9):970–973, 1982.
215. M. K. Ferber, P. F. Becher, and C. B. Finch. Effect of Microstructure on the Properties of TiB_2. J. Am. Cer. Soc. 74(1):C-2–4, 1983.
216. H. Itoh, S. Naka, T. Matsudaira, and H. Hamamoto. Preparation of TiB_2 Sintered Compacts by Hot Pressing. J. Mat. Sci. 25:533–536, 1990.
217. E. S. Kang, C. W. Jang, C. H. Lee, and C. H. Kim. Effect of Iron and Boron Carbide on the Densification and Mechanical Properties of Titanium Diboride Ceramics. J. Am. Cer. Soc. 72(10):1868–1872, 1989.
218. J. Matsushita, H. Nagashima, and H. Saito. Preparation and Mechanical Properties of TiB_2 Composites Containing Ni and C. J. Cer. Soc. Jpn., Intl. Ed. 99:1047-, 1991.
219. A. D. Osipov, I. T. Ostapenko, V. V. Tarosov, R. V. Tarasov, V. P. Podtykan, and N. F. Kartsev. Effect of Porosity and Grain Size on the Mechanical Properties of Hot-Pressed Boron Carbide. Porosh. Met. 1:63–67, 1982.
220. G. deWith. High Temperature Fracture of Boron Carbide: Experiments and Simple Theoretical Models. J. Mat. Sci. 19:457–466, 1984.
221. B. Champagne and R. Angers. Mechanical Properties of Hot-Pressed B-B_4C Materials. J. Am. Cer. Soc. 62:149–153, 1979.
222. M. Beauvy. Propriétés Mécaniques du Carbure de Bore <<Fritte>>. Rev. Int. Hautes Tempér Refract. 19: 301–310, 1982.
223. C. C. Seaton and S. K. Dutta. Effect of Grain Size on Crack Propagation in Thermally Shocked B_4C. J. Am. Cer. Soc. 57(5):228–229, 1974.
224. T. Vasilos and S. K. Dutta. Low Temperature Hot Pressing of Boron Carbide and Its Properties. Am. Cer. Soc. Bull. 53(5):453–454, 1974.
225. N. D. Rybal chenko, A. G. Mironova, V. P. Podtykan, I. T. Ostapenko, A. D. Osipov, and R. V. Tarasov. Effect of Conditions of Hot Pressing on the Structure and Mechanical Properties of Boron Carbide. Poroshkovaga Metallurgiya 8(248):39–43, 1983.
226. M. Bougoin, F. Thevenot, J. Dubois, and G. Fantozzi. Synthèse et Caractérisation de Céramiques Denses en Carbure de Bore. J. Less Com. Met. 114:257–271, 1985.
227. K. A. Schwetz, and W. Grellner. The Influence of Carbon on the Microstructure and Mechanical Properties of Sintered Boron Carbide. J. Less-Com. Met. 82:37–47, 1981.
228. K. A. Schwetz, W. Grellner, and A. Lipp. Mechanical Properties of HIP Treated Sintered Boron Carbide. Inst. Phys. Conf. Ser. No. 75, 1986, pp. 413–426.
229. J. Dubois, G. Fantozzi, M. Bougoin, and F. Thevenot. Microstructure et Propriétés Mécaniques de Materiaux Céramiques de Type B_xC/SiC, Frittes sans Charge. J. Phys. 2:C1-75, 1986.

230. G. A. Gogotsi, Y. A. L. Groushevsky, O. B. Dashevskaya, Y. U. G. Gogotsi, and V. A. Laytenko. Complex Investigation of Hot-pressed Boron Carbide. J. Less Com. Met. 117:225–230, 1986.
231. D. Kalish, E. V. Clougherty, and J. Ryan. Fabrication of Dense Fine Grained Ceramic Materials. ManLabs, Inc. Final Report for US Army Materials Research Lab. Contract DA-19-066-AMC-283(x), 11/1966.
232. K. Niihara. Mechanical Properties of Chemically Vapor Deposited Nonoxide Ceramics. Am. Cer. Soc. Bul. 1160, 1983.
233. K. Niihara. A. Nakahira, and T. Hirai. The Effect of Stoichiometry on Mechanical Properties of Boron Carbide. J. Am. Cer. Soc. 67(1) C-13–14, 1984.
234. R. W. Rice. Effects of Thermal Expansion Mismatch Stresses on the Room-Temperature Fracture of Boron Carbide. J. Am. Cer. Soc. 73(10):3116–3118, 1990.
235. B. Roebuck and E. A. Almond. Deformation Fracture Processes and the Physical Metallurgy of WC-Co Hardmetals. Intl. Mats. Reviews 33(2):90–110, 1988.
236. B. Roebuck. The Tensile Strength of Hardmetals. J. Mat. Sci., 14:2837–2844, 1979.
237. W. Rafaniello. Development of Aluminum Nitride: A New Low-Cost Armor. Dow Chem. Co. Final Report for US Army Research Office Contract DAAL03-88-C-0012, 12/1992.
238. K. Hiruta, H. Hirotsuru, R. Terasaki, and Y. Nakajima. Influence of Powder Characteristics on Flexural Strength of Aluminum Nitride Ceramics. Intl. Conf. on Aluminum Nitride Ceramics, Le Maridien Pacific, Tokyo, Japan, 3/8–11/98.
239. A. K. Mukhopdhay, D. Chakarborty, and J. Mukerji. Fractographic Study of Sintered Si_3N_4 and RBSN. J. Mat. Sci. Lett. 6:1198–1200, 1987.
240. R. K. Govila. Fracture Phenomenology of a Sintered Silicon Nitride Containing Oxide Additives. J. Mat. Sci. 23:1141–1150, 1988.
241. E. Tani, S. Umebayashi, K. Kishi, K. Kobatashi, and M. Nishijima. Effect of Size of Grains with Fiber-Like Structure of Si_3N_4 on Fracture Toughness. J. Mat. Sci. Lett. 4:1454–1456, 1985.
242. J. A. Salem, S. R. Choi, M. R. Freedman, and M. G. Jenkins. Mechanical Behavior and Failure Phenomena of an Insitu Toughened Silicon Nitride. J. Mat. Sci. 27:4421–4428, 1992.
243. A. J. Pyzik and D. F. Carroll. Technology of Self-Reinforced Silicon Nitride. Ann. Rev. Mater. Sci., Annual Reviews Inc. 24:189–214, 1994.
244. C.-W. Li, S.-C. Lui, and J. Goldacker. Relation Between Strength, Microstructure, and Grain-Bridging Characteristics in Insitu Reinforced Silicon Nitride. J. Am. Cer. Soc. 78(2):449–459, 1995.
245. J. E. Brocklehurst. Fracture in Polycrystalline Graphite. Chem. Phys. Carbon 13 (P. L. Walker and P. A. Thrower, eds.). Marcel Dekker, New York, 1977, pp. 145–279.
246. R. H. Knibbs. Fracture in Polycrystalline Graphite. J. Nuc. Mat. 24:174–187, 1967.
247. M. Wakamatsu, S. Ishingo, N. Takeuchi, and T. Hattori. Effect of Firing Atmosphere on Sintered and Mechanical Properties of Vanadium-Doped Alumina. J. Am. Cer. Soc. 74(6):1308–1311, 1991.
248. R. D. Bagley, I. B. Cutler, and D. L. Johnson. Effect of TiO_2 on Initial Sintering of Al_2O_3. J. Am. Cer. Soc. 53(3):136–141, 1970.

249. F. A Kröger. Enrichment of Titanium at Grain Boundaries in Al_2O_3. J. Am. Cer. Soc. 58(7–8): 355–356, 1975.
250. V. Jayaram, B. J. Dalgleish, and A. G. Evans. Some Observations of Microstructural Changes in Alumina Induced by Ti Inhomogeneities. J. Mat. Res. 3(4):764-, 1988.
251. E. Ryshkewitch and D. W. Richardson. Oxide Ceramics, Physical Chemistry and Technology. General Ceramics, Inc., Haskell, NJ, 1985, p. 217.
252. T. M. Clarke, D. L. Johnson, and M. E. Fine. Effect of Oxygen Partial Pressure on Precipitation in Titanium-Doped Aluminum Oxide. J. Am. Cer. Soc. 53(7):419–420, 1970.
253. M. Blanc, A. Mocellin, and J. L. Strudel. Observation of Potassium β''' -Alumina in Sintered Alumina. J. Am. Cer. Soc. 60(9–10):403–409, 1977.
254. R. W. Rice. The Effect of Gaseous Impurities on the Hot Pressing and Behavior of MgO, CaO and Al_2O_3. Proc. Brit. Cer. Soc. 12:99–123, 1969.
255. R. W. Rice. Ceramic Processing: An Overview. AICHE J. 36(4):481–510, 1990.
256. T. A. Wheat and T. G. Carruthers. The Hot Pressing of Magnesium Hydroxide and Magnesium Carbonate. Science of Ceramics (G. H. Steward, ed.). Brit. Cer. Soc., 1968, pp. 33–51.
257. P. W. Montgomery, H. Stromberg, and G. Jura. Solid Surfaces and the Gas-Solid Interface. Am. Chem. Soc.: Advances in Chemistry Series 33, 1961, p. 18.
258. H. Kodama and T. Miyoshi. Study of Fracture Behavior of Very Fine-Grained Silicon Carbide Ceramics. J. Am. Cer. Soc. 73(10):3081–3082, 1990.
259. D. Stoyan and H.-D. Schnabel. Description of Relations Between Spatial Variability of Microstructure and Mechanical Strength of Alumina Ceramics. Cer. Int'l 16:11–18, 1990.
260. T. Sakai. Effect of Oxygen Composition on Flexural Strength of Hot-Pressed AIN. J. Am. Cer. Soc. 61(9–10): 460–461, 1978.
261. K. Komeya and H. Inoue. The Influence of Fibrous Aluminum Nitride on the Strength of Sintered AIN-Y_2O_3). Trans. J. Brit. Cer. Soc. 70(3):107–114, 1971.
262. M. Yasuoka, K. Hirao, M. E. Brito, and S. Kanzaki. High-Strength and High-Fracture-Toughness Ceramics in the $Al_2O_3/LaAl_{11}O_{18}$ Systems. J. Am. Cer. Soc. 78(7):1853–1856, 1995.
263. H.-D. Kim, I.-S. Lee, S.-W. Kang, and J.-W. Ko. The Formation of $NaMg_2Al_{15}O_{25}$ in an α-Al_2O_3 Matrix and Its Effect on the Mechanical Properties of Alumina. J. Mat. Sci. 29:4119–4124, 1994.
264. J. A. Salem and J. L. Shannon, Jr. Crack Growth Resistance of Textured Alumina. J. Am. Cer. Soc. 72(1): 20–27, 1989.
265. F. V. DiMarcello, P. L. Key, and J. C. Williams. Preferred Orientation in Al_2O_3 Substrates. J. Am. Cer. Soc. 55(10):509–514, 1972.
266. D. K. Smith, Jr., and S. Weissmann. Residual Stress and Grain Deformation in Extruded Polycrystalline BeO Ceramics. J. Am. Cer. Soc. 5(6):330–336, 1968.
267. H. Tagai, T. Zisner, T. Mori, and E. Yasuda. Preferred Orientation in Hot-Pressed Magnesia. J. Am. Cer. Soc. 50(10):550–551, 1967.
268. J. E. Dykins. Tensile Properties of Sea Ice Grown in a Confined System. Physics of Snow and Ice, International Conf. on Low Temperature Science, Vol. 1, Part 1 (H. Oura, ed.). Hokkaido University 1967, pp. 5–36.

269. K. Yon, M. Hahn, R. W. Rice, and J. R. Spann. Grain Size Dependence of SbSI with Oriented Grains. Adv. Cer. Mats. 1(1):64–67, 1986.
270. T. Ohji, K. Hirao, and S. Kanzaki. Fracture Resistance Behavior of Highly Anisotropic Silicon Nitride. J. Am. Cer. Soc. 78(11):3125–3128, 1995.
271. C. Cm. Wu, K. R. McKinney, and R. W. Rice. Strength and Toughness of Jade and Related Natural Fibrous Materials. J. Mat. Sci. 25:2170–2174, 1990.
272. J. S. Reed and A.-M. Lejus. Effect of Grinding and Polishing on Near-Surface Phase Transformations in Zirconia. Mat. Res. Bull. 12:949–954, 1977.
273. R. W. Rice. Effects of Ceramic Microstructural Character on Machining Direction–Strength Anisotropy. Machining of Advanced Materials, (S. Johanmir, ed.). NIST Special Pub. 847, US Govt. Printing Office, Washington DC, 1993, pp. 185–204.
274. J. T. Chakraverti and R. W. Rice. Strengthening of Zirconia-Toughened Materials by Grit Blasting. J. Mat. Sci. 16: 404–405, 1997.
275. R. W. Rice and J. T. Chakravarti. Non-Machining Surface Strengthening of Transformation Toughened Materials. US Patent 5,228,245, 7/20/1993.
276. C. C. McMahon. Relative Humidity and Modulus of Rupture. Am. Cer. Soc. Bull. 58(9):873, 1979.
277. U. R. A. Lino and H. W. Hubner. Effect of Surface Condition on the Strength of Aluminum Oxide. Sci. Cer. 12:607–612, 1984.
278. A. Zimmermann, M. Hoffman, B. D. Flinn, R. K. Bordia, T.-J. Chuang, E. R. Fuller, Jr., and J. Rödel. Fracture of Alumina with Controlled Pores. J. Am. Cer. Soc. 81(9):2449–2457, 1998.
279. A. Zimmermann and J. Rödel. Generalized Orowin-Petch Plot for Brittle Fracture. J. Am. Cer. Soc. 81(10):2527–2532, 1998.
280. R. W. Rice. Pores as Fracture Origins in Ceramics. J. Mat. Sci. 19:895–914, 1954.
281. H. P. Kirchner and J. M. Ragosta. Crack Growth from Small Flaws in Larger Grains in Alumina. J. Am. Cer. Soc. 63(9–10):490–495, 1980.
282. R. W. Rice. Effects of Environment and Temperatures in Ceramic Tensile Strength–Grain Size Relations. J. Mat. Sci. 32:3071–3087, 1997.
283. S. L. Dole, O. Hunter, Jr., and D. J. Bray. Microcracking of Monoclinic HfO_2. J. Am. Cer. Soc. 61(11–12):486–490, 1978.
284. O. Hunter, Jr., R. W. Scheidecker, and S. Tojo. Characterization of Metastable Tetragonal Hafnia. Ceramurgica, Intl. 5(4):137-, 1979.
285. D. Lewis and R. W. Rice. Comparison of Static, Cyclic, and Thermal-Shock Fatigue in Ceramic Composites. Cer. Eng. Sci. Proc. 3(9–10):714–721, 1982.
286. D. B. Marshall. Failure from Surface Flaws. Fracture in Ceramic Materials—Toughening Mechanisms, Machining Damage, Shock (A. G. Evans, ed.). Noyes, New York, 1984, pp. 190–220.
287. J. P. Singh, A. V. Kirkar, D. K. Shetty, and R. S. Gordon. Strength–Grain Size Relations in Polycrystalline Ceramics. J. Am. Cer. Soc. 62(3–4):179–182, 1979.
288. A. G. Evans. A Dimensional Analysis of the Grain-Size Dependence of Strength. J. Am. Cer. Soc. 63(1–2):115–116, 1980.
289. A. V. Virkar, D. K. Shetty, and A. G. Evans. Grain-Size Dependence of Strength. J. Am. Cer. Soc. 64(1):56–57, 1981.

290. H. Y. B. Mar and W. D. Scott. Fracture Induced in Al_2O_3 Bicrystals by Anisotropic Thermal Expansion. J. Am. Cer. Soc. 53(10):555–558, 1970.
291. R. W. Rice, R. C. Pohanka, and W. J. McDonough. Effect of Stresses from Thermal Expansion Anisotropy, Phase Transformations, and Second Phases on the Strength of Ceramics. J. Am. Cer. Soc. 63(11–12):703–710, 1980.
292. R. W. Rice, S. W. Freiman, R. C. Pohanka, J. J. Mecholsky, Jr., and C. Cm. Wu. Microstructural Dependence of Fracture Mechanics Parameters in Ceramics. Fracture Mechanics of Ceramics—Crack Growth and Microstructure 4 (R. C. Bradt, D. P. H. Hasselman, and F. F. Lange, eds.). Plenum Press, New York, 1978, pp. 849–876.
293. R. W. Rice. Microstructural Dependence of Fracture Energy and Toughness of Ceramics and Ceramic Composites Versus That of Their Tensile Strengths at 22°C. J. Mat. Sci. 31:4503–4519, 1996.
294. R. W. Rice. Machining Flaw Size-Tensile Strength Dependence on Microstructure of Monolithic and Composite Ceramics. To be published.
295. R. W. Rice. Ceramic Fracture Mode-Intergranular vs. Transgranular Fracture. Ceramic Transactions 64: Fractography of Glasses and Ceramics 3 (J. R. Varner, V. D. Frechette, and G. D. Quinn, eds.). Am. Cer. Soc., Westerville, OH, 1996, pp. 1–53.
296. M. S. Swain and L. R. F. Rose. Strength Limitations of Transformation-Toughened Zirconia Alloys. J. Am. Cer. Soc. 69(7):511, 1986.
297. K. Matsuhiro and T. Takahashi. The Effect of Grain Size on the Toughness of Sintered Si_3N_4. Cer. Eng. Sci. Proc. 10(7–8):807–816, 1989.
298. G. Himsolt, H. Knoch, H. Huebner, and F. W. Kleinlein. Mechanical Properties of Hot-Pressed Silicon Nitride with Different Grain Structures. J. Am. Cer. Soc. 62(1–2):29–32, 1979.
299. M. J. Hoffmann. Analysis of Microstructural Development and Mechanical Properties of Si_3N_4. Tailoring of Mechanical Properties of Si_3N_4 Ceramics (M. J. Hoffmann and G. Petzow, eds.). Kluwer Academic Publishers, The Netherlands, 1994, pp. 59–72.
300. Y.-W. Kim, M. Mitomo, and N. Hirosaki. *R*-Curve Behavior and Microstructure of Sinterted Silicon Nitride. J. Mat. Sci. 30:5178—5184, 1995.
301. P. F. Becher, E. Y. Sun, K. P. Plunckett, K. B. Waters, C. G. Westmoreland, E.-S. Kang, K. Hiro, and M. E. Brito. Microstructural Design of Silicon Nitride with Improved Fracture Toughness, Part I: Effects of Grain Shape and Size. J. Am. Cer. Soc. 81(11):2821–2830, 1998.
302. T. Masaki and K. Shinjo. Mechanical Behavior of ZrO_2-Y_2O_3 Ceramics Formed by Hot Isostatic Pressing. Advs. Cer. 24: Sci. Tech. Zirconia III. Am. Cer. Soc., Westerville, OH, 1988, pp. 709–720.
303. G. S. A. M. Theunissen, J. S. Bouma, A. J. A. Winnubst, and A. J. Burggraaf. Mechanical Properties of Ultra—Fine Grained Zirconia Ceramics. J. Mat. Sci. 27:4429–4438, 1992.
304. L. Ruiz and M. J. Readey. Effect of Heat Treatment on Grain Size, Phase Assemblage, and Mechanical Properties of 3 mol% Y-TZP. J. Am. Cer. Soc. 79(9):2331–2340, 1996.

305. N. Hirosaki, Y. Akimune, and M. Mitomo. Effect of Grain Growth of β-Silicon Nitride on Strength, Weibull Modules, and Fracture Toughness. J. Am. Cer. Soc. 76(7):1892–1894, 1993.
306. Y. Ukyo and S. Wada. High Strength Si_3N_4 Ceramics. J. Cer. Soc. Jpn., Intl. Ed. 97:858–859, 1989.
307. S. Saito. Fine Ceramics. Elsevier, New York, 1988, pp. 182–183.
308. D. B. Price, R. E. Chinn, K. R. McNerney, T. K. Borg, C. Y. Kim, M. W. Krutyholowa, N. W. Chen, and M. J. Haun. Fracture Toughness and strength of 96% Alumina. Bul. Am. Cer. Soc. 76(5):47–51, 1997.
309. C. Zhang, W. Y. Sun, and D. S. Yan. Fabrication of Dy-—Sialon Ceramics. J. Mat. Sci. Lett. 17:583–586, 1998.
310. J. T. Neil. Calculating Weibull Modulus from Average and Standard Deviation. GTE Labs. Inc. Report TM-0135-07-89-066, 7/1989. See also J. Gong and Y. Li. Relationship Between the Estimated Weibull Modulus and the Coefficient of Variation of the Measured Strength of Ceramics. J. Am. Cer. Soc. 82(2):449–452, 1999 for a similar expression.
311. D.-H. Cho, Y.-W. Kim, and W. J. Kim. Strength and Fracture Toughness of In-Situ-Toughened Silicon Carbide. J. Mat. Sci. 32:4777–4782, 1997.
312. Y.-W. Kim, W. J. Kim, and D.-H. Cho. Effect of Additive Amount on Microstructure and Mechanical Properties of Self-Reinforced Silicon Carbide. J. Mat. Sci. Lett. 16:1384–1386, 1997.
313. H. P. Kirchner and R. M. Gruver. Fracture Mirrors in Alumina Ceramics. Phil. Mag. 27(6):1433–1446, 1973.

4

Grain Dependence of Indentation Hardness at ~ 22°C

I. INTRODUCTION

This chapter addresses the grain (mainly size) dependence of indentation hardness *(H)*. Indenter geometries are either spherical or pyramidal, the latter primarily Vickers or Knoop (designated respectively by subscripts V and K), which are dominant in ceramic testing and the focus of this chapter. Scratch hardness is also used some, but primarily as a test to simulate machining or wear; hence it is covered in conjunction with these topics in Chapter 5. Though grain size *(G)* is commonly the dominant parameter, grain shape and orientation, and especially their combination, can be important, and while studied little, are noted and discussed to the extent that data allows.

The grain dependence of hardness arises primarily from its impact on plastic deformation, the primary mechanism of forming permanent indentations. While in the past there was substantial controversy about the mechanisms of indentation, especially in very hard ceramics, it is now accepted that such crystalline materials do undergo local deformation to form indents primarily by dislocation slip mechanisms, as in metals. Such deformation in ceramics, which is more restricted, approximately in inverse proportion to the material hardness, has been extensively verified by two sets of observations and their self-consistency. The first is direct observations of dislocations and slip by optical birefringence, etching, microradiography, and especially by direct transmission electron microscopy. The second is by hardness behavior itself, especially the relation of hardness anisotropy of single crystals as a function of crystal structure orientation relative to the indenter geometry as a function of the degrees of activation of differing slip systems [1–5] (Sec. II.E). (Note, in glasses, not addressed here, other deformation mechanisms, e.g. densification may occur—see Chap. 10.) The grain size, shape, and orientation dependence of hardness thus results from the same grain structure constraints on yield stress as given in Eq. (3.1), with two modifications. First, σ_f, which is usually slightly > the yield stress for the easiest activated slip system in testing the strengths of single crystals, and was used as a convenient approximation for the actual yield stress, is replaced by the yield stress, σ_y. Second, σ_y is the general yield stress for the required deformation and is not necessarily that for only the easiest activated slip system [6]. Thus a Hall–Petch type dependence on grain size, *G* (with some impacts of grain shape and orientation) is a major, or the total, factor in the *G* dependence of *H*. Although much hardness data is given with no grain (or other material or test) characterization, there is substantial data providing support for such an H–$G^{-1/2}$ dependence, especially at mainly finer *G* [7], as will be extensively shown.

However, some data did not necessarily show *H* decreasing with increasing *G*, mainly at intermediate *G*, and some results showed the opposite *G* dependence, i.e. *H* increasing with increasing *G* at larger *G* [7–10]. Thus Armstrong et al. [8] reported a reverse BeO H_V–G dependence, i.e., decreasing from single crystal values (on {0001} and $\{10\bar{1}0\}$ planes) with decreasing *G* (Fig. 4.3). Similarly, Sargent and Page [9], though not giving specific *H* values, reported MgO single crystal H_V higher (on {100} planes) than for dense, hot pressed polycrystalline MgO (G = 130 or 10 μm). While not encompassing single crystal tests, Tani et al.'s [10] H_V (2 N, ~ 200 gm load) data for Al_2O_3 shows a marked decrease in *H* with decreasing *G* (from ~ 60 to ~ 6 μm, see Fig. 4.2) in contrast to the opposite trend for their Y_2O_3 data, thus showing that the different *H*–*G* trends are not due entirely to differing techniques of different investigators. Though these variations were limited, e.g. due to the frequently limited *G* range covered, especially at larger *G*, such variations are an underlying factor in H–$G^{-1/2}$ data not necessarily extrapolating to single crystal values. Thus some, e.g. Niihara and Hirai [11], considered a G^{-1} dependence of *H*, since this resulted in their fine *G* polycrystalline values more closely extrapolating to single crystal data (Fig. 4.13).

A recent review [7] showed that the above variations are manifestations of a very common, but variable, indent associated deviation below a simple Petch relation at intermediate *G* whose recognition was typically seriously restricted by insufficient *H*–*G* data. Thus much more extensive *H*–*G* data presented previously [7] and here clearly shows frequent deviation through an indent-*G* dependent *H* minimum at intermediate *G*, which is attributed to observed indent associated cracking. While the cracking from indent vertices used for toughness and strength evaluations (mainly in finer *G* bodies) may be a factor in this, the observed indent associated cracking of interest here is more confined around the indent, consisting mostly of a number of spall-lateral cracks (Sec. II.D). Such cracking increases as the indent and grain dimensions approach one another and decreases as the indent size gets larger or smaller than the grain size. The result is a material-, body-, and test-dependent lowering of hardness values below those expected from the Hall–Petch relation as cracking increases, which thus typically occurs at varying intermediate *G* values that shift with indent size and hence material, grain structure, and indent types and loads. Thus hardness first commonly decreases from single crystal values as *G* decreases, reaches a minimum, and then progressively increases with further decreasing *G* at finer *G*, ~ as the Hall–Petch relation at finer *G*. While there is no quantitative theory for these deviations, the substantial experimental data showing the trends for indent type, load, and material dependence is reviewed. Note that as the indent size approaches the grain size, increased scatter of hardness values is expected, since there is increased dependence of each value measured on the parameters of individual or adjoining grains indented.

The subsequent review of the grain dependence of hardness provides a more comprehensive data base and perspective than a more limited, earlier survey of the *G* dependence of *H* and related properties [12] by extensively drawing upon and extending a more recent and more extensive evaluation [7]. H_K and H_V data are reviewed first for single oxides, then for mixed oxides, then for borides, carbides, and nitrides, in alphabetical order, giving Knoop results first if sufficient data is available. Then limited data for spherical indenters and other materials is briefly addressed. Key trends to observe are that the more limited data for most softer, less refractory materials indicates a simple Hall–Petch relation, but almost all harder, more refractory materials (and a few softer, less refractory materials) have a superimposed indent–grain size dependent *H* minimum, mostly at intermediate *G* values. The limited data for bodies with nanoscale grains is shown to be generally consistent with data for bodies of the same composition but more normal *G* (i.e. ≥ 1 μm), but some nanograin bodies are shown to have *H* decreasing as *G* decreases, i.e. opposite from the normal behavior at finer *G*.

This review of H–$G^{-1/2}$ data is followed first by a review of indent character trends showing that the minimum in *H* values is associated with increasing local cracking, which is typically a maximum when the indent dimensions are ~ those of the grains under and around the indentation. Then observations of extrinsic effects, mainly grain boundary phases or impurities and indent type and load on such cracking, are discussed and summarized followed by discussion of possible contributions of factors such as TEA and EA. [Note that data also shows the commonly observed load dependence of both H_V and H_K, and $H_V < H_K$ at 100 gm, but the reverse tends to occur at 500 gm (Table 4.1).] This is followed by discussion of the limited data on effects of grain

TABLE 4.1 Approximate Average Hardnesses (GPa)[a]

	100 gm		500 gm	
Material	H_K	H_V	H_K	H_V
Al_2O_3	26	23	18	19
MgO	10.5	9.5	8	7
ZrO_2	15+	14	12-	13
$MgAl_2O_4$	18	17	13.5	15
SiC	37	34.5	23	27.5

[a]*Source:* After Rice et al. [7], published with the permission of the *Journal of the American Ceramic Society.*

shape and orientation, and especially implications on the latter from substantial single crystal data. Note that additional data on the room temperature hardness of both single and polycrystals is found in Chapter 7, Section II, on the temperature dependence of H.

Before proceeding to the data review, it is important to recall some of the factors that can impact hardness and thus potentially complicate accurate evaluation of its grain size dependence, considering first exclusively body aspects. Besides basic body chemistry, the presence of impurities or additives, as well as stoichiometry and varying crystal structure, and especially porosity, are commonly also important. While stoichiometry may often manifest much of its effects via its frequent impacts on grain parameters, especially size (as shown later), typically resultant lattice defects can directly impact dislocation motion to varying extents, hence directly impacting hardness in view of its basic dependence on local plastic deformation. Additionally, both stoichiometry and impurities or additives can change grain boundary strengths as well as stresses, e.g. those from TEA and EA, and thus the extent and impact of local cracking. Examples of enhanced local cracking due to grain boundary phases from (fluoride) additives are illustrated later, and possible impurity effects in some nanograin bodies are discussed. Changes in or mixes of different crystal structures of the same composition may have limited to substantial effects, e.g. by up to ~ 25% in Si_3N_4 [13,14] (see Fig. 4.13). Porosity can have varying and substantial effects on hardness, as is extensively addressed elsewhere [7,15,16] with much of its effect being due to the amount and character of porosity, e.g. via the exponential, e^{bP}, dependence of the ratio of H at some volume fraction porosity P to that at $P = 0$ over a reasonable fraction of the porosity range, e.g. to $P = 0.2$–0.5. The parameter b is commonly in the range of 3–7 for hardness, depending on the type of porosity.

Factors dependent both on the body character and on test factors are grain and hardness measurements. As noted in Chapter 1, Section IV.B, measurement of grain size is complicated by several factors, including typically neglected and more difficult to quantify aspects of grain shape and orientation, but also by both actual measurement, including, even for random equiaxed grain structures, what the grain size should be, e.g. a two-dimensional one (i.e. the size of the grains at their intersection with a surface) or their "true" three-dimensional size. While the latter is probably more pertinent to H, there is no clear guidance, but these issues are commonly overridden by the very loose measurement of most G values, including specifics of the measurements and any factors to convert the common linear intercept values to some "true" G commonly not being given. These commonly result in uncertainties in G values of ± 50%, and sometimes substantially more. Actual hardness measurements for a given body also depend on indent load [7,17–19] and on surface character, with the two being partly interrelated [19]. Thus H values typically increase below indent loads of ~ 1 to a few kilograms (i.e. ~ 10 or more N), with the increases being at first modest, but accelerating as the load decreases. Surface character is important for its impact on clarity of the indent dimensions, e.g. due to the degree of surface smoothness. However, mechanical surface finishing, which is almost exclusively used, also introduces surface cracks that may impact indent formation, and more fundamentally introduces extensive deformation in the surface [19,20]. The latter commonly is sufficient to work harden the surface, to depths dependent on the material, the grain structure and the nature of the final machining (commonly polishing), and commonly some on previous machining. Such surface work hardening is commonly a factor, often an important one, in the load dependence of H since higher loads cause more indent penetration into less work hardened material.

II. DATA REVIEW

A. Oxides

Brookes and Burnand's [3] H_K (500 gm load) data for the $(1\bar{1}00)$ plane of sapphire averaging 16 ± 2 GPa is somewhat lower than Rice et al.'s data [7] (Fig. 4.1) and considerably lower, as expected, than Becher's [19] H_K (100 gm) values of 22–30 GPa for the (0001) basal plane [and 22–35 GPa on the (1120) plane] from the five-fold higher load. Becher's data, where most of the variations are due to different surface finishes, is in good agreement with the current data. The substantial single- and polycrystal data of Rice et al. [7] suggests a limited minimum in H_K (100 gm) at intermediate G (~ 50 μm) i.e., H_K first decreases some, then increases with decreasing G. Their data shows a clearer and larger H_K minimum at G ~ 50μm for the 500 gm load. Al_2O_3 H_K–$G^{-1/2}$ (400 gm) data of Skrovanek and

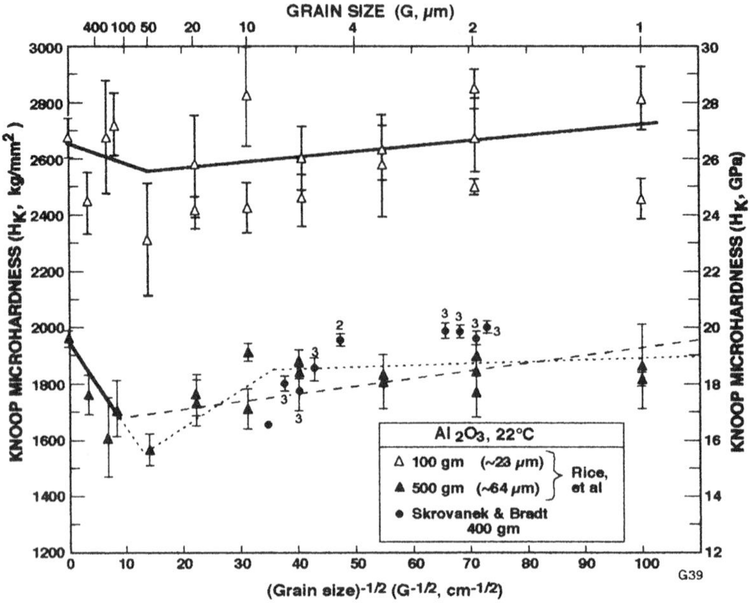

FIGURE 4.1 H_K vs. $G^{-1/2}$ for single crystal (sapphire) and polycrystalline (mainly hot pressed) Al_2O_3 of Rice et al. [7] (100 and 500 gm loads, giving typical indent diagonals respectively of ~ 8 and ~ 23 μm) and Skrovanek and Bradt [21] (400 gm load, sub- or superscripts the number of tests) at ~ 22°C. Vertical bars = standard deviation. Note (1) a probable H_K minimum in the 100 gm data, at G ~ 50 μm, and a definite minimum in the 500 gm load data at G ~ 50–100 μm (depending upon the exact trend of the finer G data, e.g., per two alternatives shown by dashed lines), (2) good agreement between data sets for 400 and 500 gm loads which, when combined, favors a greater G dependence in the smaller G regime. (From Ref. 7, published with the permission of the *Journal of the American Ceramic Society.*)

Bradt [21] agrees well with that of Rice et al. (500 gm load) but shows a greater H decrease with increasing G for G > ~ 5–8 μm (Fig. 4.1).

Kollenberg's [22] H_V (200 and 400 gm) values, averaging ~ 19 GPa for three different sapphire planes, are in reasonable agreement with the sapphire H_V of Rice et al. [7], which in turn is in reasonable agreement with Becher's data [19] considering possible surface finish and orientation differences. Rice et al.'s substantial data versus G does not show a clear minimum for 100 gm load but could be consistent with one at G ~ 20 μm. Their data at 500 gm load indicates a minimum at G ~ 60 μm. Limited H_V data of Lankford [23] for dense sintered Al_2O_3 (Lucalox, loads: 200–800 gm) agrees fairly well with Rice et al.'s data for the lower (200 gm) load, i.e. falling mainly between the 100 and 500 gm data, but is high for the higher loads. Tani et al.'s [10] H_V (2 N, ~ 200 gm) data, showing a substantial H decrease from their largest G (~ 60 μm) to their smallest G (~ 6 μm) and hence a probable H minimum, varies from being consistent in H values with those of Rice et al. to some substantial difference, but not total disagreement. In view of other data for alumina and other materials, their data implies a probable H_V minimum, but with a shift to finer G relative to other data, which may well be due to other factors such as G measurement, e.g., if they used as-measured linear intercept values, these would be less (e.g. ~ 50%) than the average (surface) diameters used in the present study.

Alpert, et al.'s [24] H_V (10 N, ~ 1 kg) data for sapphire and dense sintered larger grain Al_2O_3 (Vistal) and finer grain but slightly less dense 99.9% Al_2O_3 agrees with that of Rice et al. [7], but their values for 96% and 90% alumina are progressively lower (Fig. 4.2). Their three data points for dense Al_2O_3, by themselves, would indicate no G dependence of H_V, but they are consistent with the H_V minimum of Rice et al. Similarly, Clinton and Morrell's [25] earlier as-measured H_V (1 kg) sapphire values are also generally consistent with trends of the other data, but their values for various commercial sintered (95–99%) aluminas (with P to ~ 0.05) are lower. However, approximate correction to dense, pure alumina values by extrapolation as a function of alumina content and the exponential porosity dependence with a b value of 6 [7,15,16] brings such data into much more reasonable absolute agreement with data for dense pure aluminas, e.g. at 500 gm. More recent data of Krell` and Blank extending from ~ 4 μm down to ~ 0.4 μm [26] generally agrees in trend and absolute values with the other data (recognizing the higher, ≥ 10 N, loads), clearly showing a substantial increase in H as G decreases.

Turning to other single oxides, Rice et al.'s [7] H_V data for the (0001) surface of BeO crystals was essentially identical to that of Armstrong, et al.'s [8], but the latter's data on $(10\bar{1}0)$ is somewhat higher than the former data (Fig. 4.3). Tests on $\{11\bar{2}0\}$ and $\{10\bar{1}0\}$ were essentially the same as for the {0001} plane [7]. While the H_V values for both studies are also similar for polycrystalline values over the G range (~ 15–100 μm) in common and both studies show H_V decreasing over the larger G, range, Rice et al. showed H_V increasing with decreasing G at finer G in contrast to Armstrong et al.'s continuous H decrease. Whether this difference (at the finest two G's) is due to differences in G or H measurements or the specific specimens cannot be determined, but H differences due to differences in porosity could be a factor.

Armstrong and Raghuram's [2] H_K (100 gm) of ~ 8 ± 3 GPa for {100} surfaces of MgO crystals averages below, but clearly overlaps with, the projection of the data of Rice et al. [7] at $G = \infty$ (Fig. 4.4A). Rice et al.'s polycrystalline data [7] clearly indicates similar, limited H_K minimums for 100 and 500 gm loads at G ~ 15 and 30 μm respectively and an indicated limited H_V minimum at

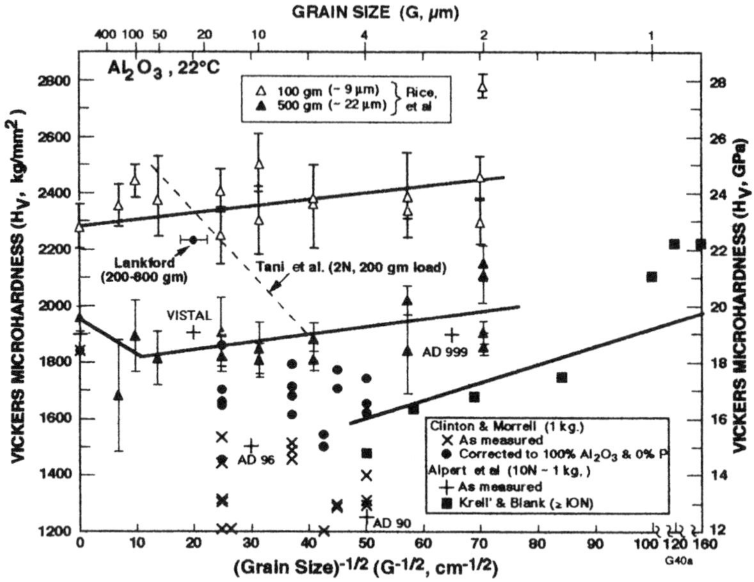

Figure 4.2 H_V vs. $G^{-1/2}$ for single crystal (sapphire) and polycrystalline (mainly hot pressed) $A1_2O_3$ of Rice et al. [7] (100 and 500 gm loads, vertical bars = standard deviation), giving typical indent diagonals respectively of ~ 9 and ~ 22 μm) and other data [24–26] at ~ 22°C. Note (1) no apparent H_V minimum at 100 gm load, (2) a probable H_V minimum at G = 50–100 μm at 500 gm load, (3) reasonable agreement of data for experimental and commercial bodies [24,25] with general trends, and often absolute values (especially with correction for other constituents and porosity), and (4) Krell and Blank [26] higher load (≥ 10 N) data extending from ~ 4 μm down to ~ 0.4 μm generally agreeing with other data and showing a substantial increase in H as G decreases. (From Ref. 7, published with the permission of the *Journal of the American Ceramic Society*.)

G ~ 30 μm for the 100 gm load and show a definite and more substantial minimum at G ~ 20 μm at the 500 gm load (Fig. 4.4B).

TiO_2 data is restricted mainly to single crystals and submicron grain sizes (Fig. 4.5). Becher's single crystal (rutile) data on {100} surfaces is one of the most comprehensive, reflecting H_V indent loads of 300, 200, and 100 gm (~ 3–1 N) and chemically polished, sanded, or diamond ground surfaces [19]. Hardness values increased (with coefficients of variation of < 5%) from averages of 8.7 to 12.9 GPa in the order of loads and surface finishes listed with the greatest increases occurring on going from 200 to 100 gm loads and from chemically polished, to sanded, to ground surfaces. His data is consistent with the other comprehensive study of Li and Bradt [27] as a function of load (50, 100, 200,

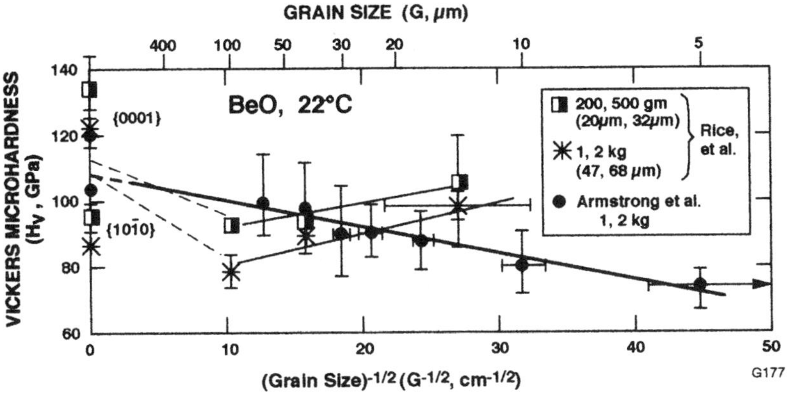

Figure 4.3 H_V vs. $G^{-1/2}$ for single and polycrystal (sintered, P ~ 0.05) BeO at 0.2, 0.5, 1, and 2 kg loads of Armstrong et al. [8] and (1 and 2 kg) loads of Rice et al. [7] at ~ 22°C. Respectively vertical and horizontal bars are standard deviation, and the lower ends of vertical bars represent the average value for the lower load, and the upper end the higher load value (coefficient of variations varied for < 1% to ~ 9%, averaging 3% for 4–6 measurements over all loads). Note the general agreement of both single and polycrystalline values between the two studies, except at finer *G*. (From Ref. 7, published with the permission of the *Journal of the American Ceramic Society*.)

and 300 gm) and orientation on {100}, {110}, and {111} surfaces of rutile crystals with mechanically polished surfaces giving Knoop hardnesses of 10.5 to 16 GPa, mostly ≤ 14 GPa. The estimated H_V value of Mayo et al. [28] of ~ 11 GPa from nanoindent tests on {100} mechanically polished rutile surface is also consistent with the above data. Data of Averback et al. [29] (see also Andrievski [30]) for H_V (load unspecified) of two TiO_2 (mainly rutile) bodies with $P = 0.06$ and 0.11 with G ~ 0.15 to 0.5 μm both show H decreasing with increasing G, with the rate of decrease clearly accelerating above G ~ 0.35 μm. While the actual hardness values are mostly below those for single crystals, correcting them for porosity would bring them at the minimum to about the mean of single crystal values, and more likely mostly somewhat above the upper range for single crystal values, i.e. very similar to other data (e.g. Figs. 4.1–4.4). (Thus the minimum b value of e^{-bP} for correcting for porosity would move H values for G ~ 0.04 μm to 10–12 GPa. However, Averback et al.'s data indicates a more likely b value of ~ 7.5, which appears more consistent with Mayo et al.'s [28] data. This b value would move the above H values up to ~ 15 GPa.) The two H_V data points (2 N load) of Guermazi et al. [31] at G ~ 0.015 and 5 μm (the former with P ~ 0.15 and the latter from further sintering of the former and unspecified P, but

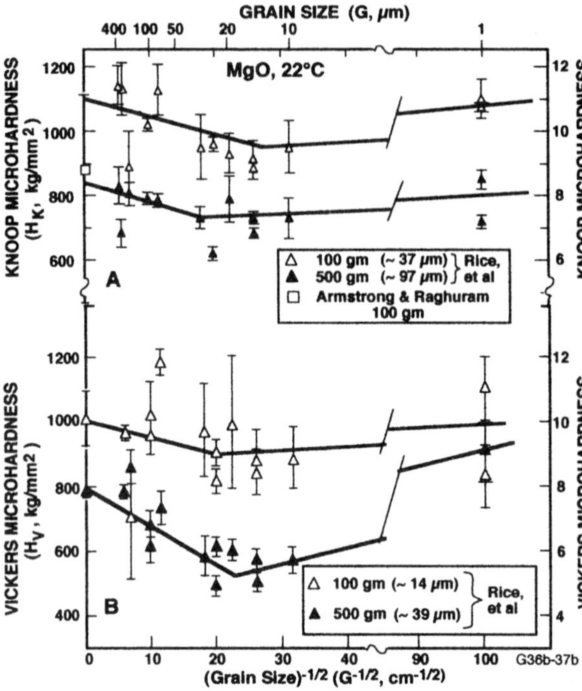

Figure 4.4 Hardness vs. $G^{-1/2}$ for hot pressed and single crystal MgO of mainly Rice et al. [7] at ~ 22°C. (A) H_K data (100 and 500 gm load) giving typical indent diagonals respectively of ~ 37 and ~ 97 μm, showing probable H_K minima at G ~ 15 and ~ 30 μm respectively for 100 and 500 gm loads. (B) H_V (100 and 500 gm loads, giving typical indent diagonals respectively of ~ 14 and ~ 39 μm) showing a probable H_V minimum at G ~ 30 μm for 100 gm load, and (2) definite H_V minimum at G ~ 20 μm at 500 gm load. Vertical bars = standard deviation. (From Ref. 7, published with the permission of the *Journal of the American Ceramic Society*.)

probably ~ 0) appear to be fairly consistent with the other data, e.g. correction of the finer G body with P ~ 0.15 with b = 4–7.5 gives H values of ~ 10–17 GPa. Thus the TiO_2 data, while having uncertainties of load and porosity character and correction, appears consistent with data for other materials over more typical G ranges with P ~ 0. Whether the greater rate of decrease at G > ~ 0.35 μm is the decrease toward a minimum as found for most other materials is somewhat uncertain, but given the low load–small indent size of the nanoindentations, this may be a reasonable possibility.

Tani et al.'s [10] H_V (2 N ~ 200 gm) data for dense hot pressed Y_2O_3 (Fig. 4.6) shows a substantial (> 20%) decrease in H from the finest (~ 0.2 μm) to the

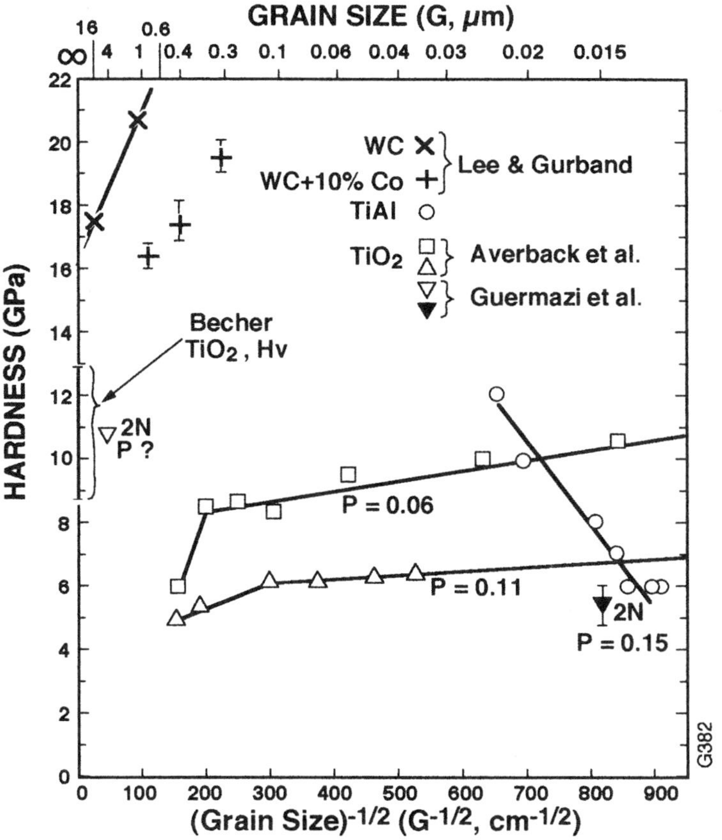

Figure 4.5 H_V $G^{-1/2}$ data for single and polycrystalline (rutile) TiO_2 [19,29,31], WC [30,80,81], and TiAl [99] at ~ 22°C, mostly for submicron grains.

largest (214 μm) G (in contrast to the significant H increase with increasing G they observed for their dense Al_2O_3, Fig. 4.2) but is considerably below most of Rice et al.'s [7] data, and slightly below Fantozzi et al.'s [33] limited (≤ 10 N ~ 1 kg, G = 1 and 10 μm) data. Limited larger G (~20–500 μm, 100 gm-30 kg) data of Rhodes [34] and Cook [35] (which showed little or no load dependence) lies somewhat (e.g. 10%) above Tani et al.'s data, while the limited (100–1000 gm, G ~ 2–500 μm) data of the present study lies 15–25% above Tani et al.'s data. However, some of the materials of these latter three studies contained ~ 10% ThO_2 or a La_2O_3 rich second phase (G = 500 μm material) which could increase H values. Tani et al.'s 200 gm H_K data questions any deviations from a simple

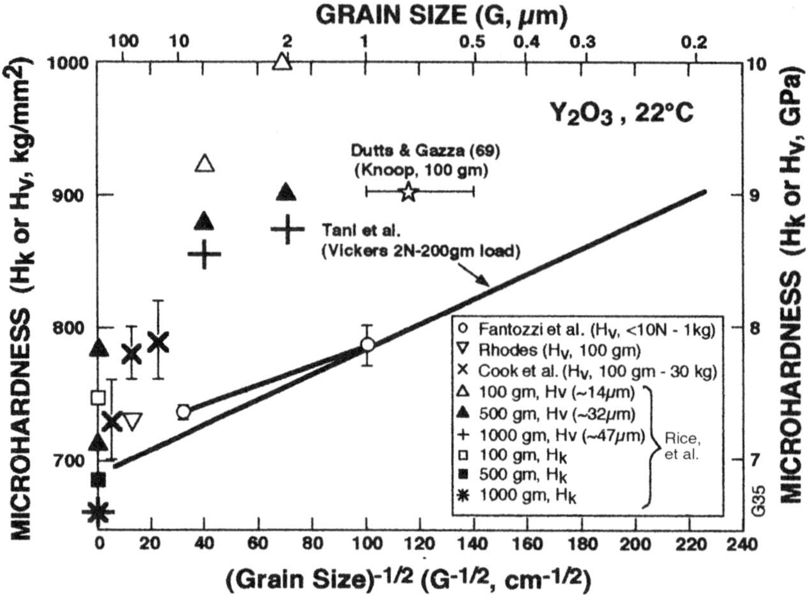

Figure 4.6 Hardness–$G^{-1/2}$ data for single and polycrystalline Y_2O_3. H_K of Rice et al. [7] (100 and 500 gm loads), and one data point (100 gm load) of Dutta and Gazza [32] at ~ 22°C. H_V data of Rice et al. (100, 500, and 1000 gm loads, giving typical diagonals respectively of 14, 32, and 47 μm), Tani et al. [10] (~ 200 gm), Fantozzi et al. [33] (~ 1 kg), Rhodes [34] (100 gm load, dense sintered, optical material), and Cook et al. [35] (100 gm–30 kg loads). Vertical bars = standard deviation. (From Ref. 7, published with the permission of the *Journal of the American Ceramic Society*.)

Hall–Petch relation. Heavier load data is too limited to be definitive but also questions deviations from a simple Hall–Petch relationship. This, combined with the significantly less cracking [and the earlier noted apparent avoidance of grain boundary fracture [7] (Chap. 2, Sec. II.D)], suggests there can be exceptions to the general deviations to a minimum below a Hall–Petch relation, especially for harder materials, based on material character, not just load and indent geometry.

Pajarec et al.'s [36] H_K (100 gm ~ 1 N) data for cubic (9.4 m/o Y_2O_3) ZrO_2 crystals averaging ~ 14.5 GPa for various orientations on {100} planes agrees with data of Rice et al. [7] for similar (but generally lower Y_2O_3 content) crystals (Fig. 4.7A). Cochran et al.'s [37] H_K (100 gm) of 11.8–13.0 GPa for ZrO_2 (+ 9.8 m/o Y_2O_3) crystals of unknown orientation is somewhat lower. The single crystal H_K data shows no clear dependence on G, which may be due to limited differences of composition and porosity levels but also likely reflects greater plastic ac-

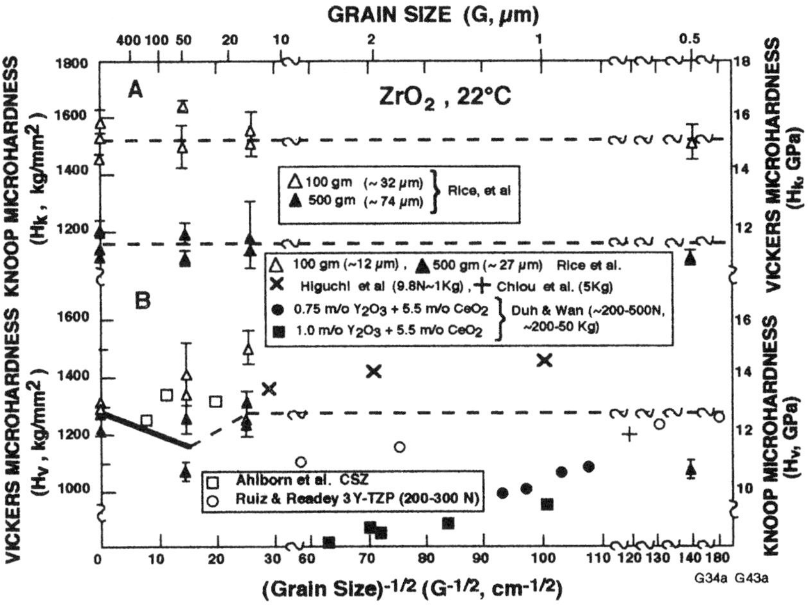

Figure 4.7 Hardness vs. $G^{-1/2}$ for ZrO_2. at ~ 22°C. (A) H_K (100 and 500 gm loads, giving typical indent diagonals respectively of ~ 32 and 74 μm) of Rice et al. [7] indicate no apparent *G* dependence, (B) H_V data of Rice et al. [7] (100 and 500 gm loads, giving typical indent diagonals of respectively ~ 12 and ~ 27 μm) and others [40–46]. Some data, especially that of Duh and Wan [44], shows a clear *G* dependence. Vertical bars = standard deviation. (From Ref. 7, published with the permission of the *Journal of the American Ceramic Society.*)

commodation often feasible in (partially stabilized) ZrO_2. Pajarec et al.'s [36] {100} H_V of 16.0 ± 0.1 GPa (200 gm ~ 2 N), independent of indenter orientation on the plane, is somewhat higher, but it is still in reasonable agreement with the present data (Fig. 4.7B). Watanabe and Komaki's [38] H_V (30 kg) data for ZrO_2 (2–4 m/o Y_2O_3) crystals of unknown orientation of ~ 12.0 ± 0.5 GPa is in excellent agreement with the present data and with earlier data of Wu and Rice [39]. Higuchi et al.'s [40] H_V (1 kg) data for hot pressed ZrO_2 (+ 8 m/o Y_2O_3 + 0.1–0.2 w/o C) is somewhat higher than, but in reasonable agreement with, Rice et al.'s [7] and Chiou et al.'s [41] (5 kg) data for sintered bodies (Fig. 4.7B). Data of Ramadas et al. [42] for sintered ZrO_2 (0.5 to 7.5 mol%) Y_2O_3 bodies (at 10 kg load), corrected to zero porosity [16,43], also generally agreed with their data. Higuchi et al.'s [40] data suggests greater *H* dependence at fine *G*, which is probably more realistic, since their hot pressed samples had constant composition. Rice et al.'s samples generally increased in Y_2O_3 content with increasing *G* and were sintered, having some (e.g., a few %) porosity (except for the single crystals). Their data indicates a limited H_V minimum at $G \sim 50$ μm, but no clear G dependence at finer G. However, this may be due to limited composition and porosity effects, since other data suggest H_V increasing with decreasing G at finer G. Duh and Wan's [44] data (for Y_2O_3 + CeO stabilizers and high loads) clearly shows the normal H–$G^{-1/2}$ dependence and some, but limited, load dependence. Ruiz and Readeys' data for 3 mol% Y-TZP (0.13–0.33% porosity) also showed a modest decrease from 12.7 GPa at the finest G (0.3 μm) to 11.3 GPa at the largest G (5 μm) [45]. Ahlborn et al.'s data for dense (transparent) fully stabilized ZrO_2 (with 10 mol% TiO_2 + 7 mol% Y_2O_3) suggested a slight decrease in H_V (unspecified load) from 13.1–13.4 GPa for $G = 25$ and 90 μm to 12.3 GPa at $G = 130$ μm [46].

Akimune and Bradt's [47] H_K (100 gm) data for stoichiometric $MgAl_2O_4$ crystals, {100} planes, averaging 16.5 ± 1.5 GPa for various orientations and just over 15 GPa for the {111} plane, is quite consistent with data of Rice et al. [7], while manufactured H_K (100 gm) data for commercially produced stoichiometric crystals, averaging 13.5 ± 1.5 GPa, is somewhat lower but not grossly inconsistent with other data (Fig. 4.8). Rhodes et al.'s [48] H_K (100 gm) for polycrys-

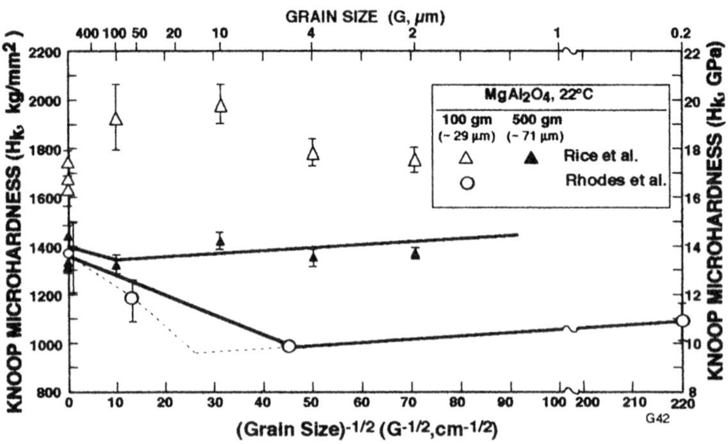

Figure 4.8 H_K vs. $G^{-1/2}$ for single and polycrystalline $MgAl_2O_4$ at ~ 22°C. Data of Rice et al. [7] (100 and 500 gm loads, giving typical indent diagonals respectively of ~ 29 and ~ 71 μm) and of Rhodes et al. [48] (100 gm load). Rice et al.'s 100 gm data, taken by itself, would suggest a possible H_K (probably anomalous) maximum at intermediate G, especially when effects of increasing (to several %) porosity with decreasing G are considered. Other data all indicates an H_K minimum. Data of Rhodes et al. suggests a definite H_K minimum between $G \sim 5$ and 20 μm. Vertical bars = standard deviation. (From Ref. 7, published with the permission of the *Journal of the American Ceramic Society*.)

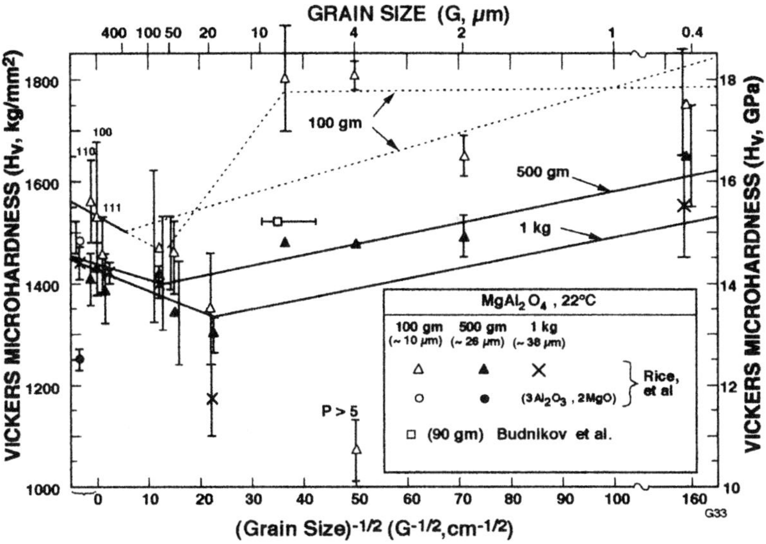

Figure 4.9 H_V vs. $G^{-1/2}$ data for $MgA1_2O_4$ at ~ 22°C of Rice et al. [7] (100, 500, and 1000 gm loads giving typical indent diagonals respectively of ~ 10, 26, and 38 μm). Note data for three principal planes for stoichiometric (Czochralski) and random planes of alumina rich (Vernulil) single crystals. Vertical bars = standard deviation. Horizontal bar = G range for Budnikov et al.'s [49] 90 gm data. (From Ref. 7, published with the permission of the *Journal of the American Ceramic Society.*)

talline $MgAl_2O_4$ specimens (press forged or hot pressed, typically with about 1% Li-based additions), is considerably below the present data (Fig. 4.8) but clearly suggests that hardness first decreases with increasing G and later increases with increasing G as one approaches single crystal values, i.e., a H_K minimum at G 5–15 μm at 100 gm load. While Rice at al.'s 100 gm data is uncertain, their 500 gm data showed a minimum at G ~ 100 μm, e.g., in view of some porosity in especially the finest G bodies. The more extensive single crystal H_V, including lower H_V of alumina rich versus stoichiometric crystals, and polycrystalline data of Rice et al. (Fig. 4.9), shows distinct H_V minima at G ~ 50–400, 50, and 20 μm, respectively for loads of 100, 500, and 1000 gm. Budnikov et al.'s [49] $MgAl_2O_4$ H_V (~ 90 gm) data lies below the other data but is reasonably consistent with it. Note (1) that the greater uncertainty in the degree and extent of G dependence at finer G is probably due to the limited porosity in finer G specimens, and (2) that both sets of Rice et al.'s data involve some specimens of varying stoichiometry (e.g. some of the bodies in Fig. 3.10), which probably adds to the variations.

Okazaki and Nagata's H_V (unspecified load) data for PZT, though covering only a limited G range (with $P \sim 0.05$) [12,50], indicates little or no minimum (Fig. 4.14).

B. Borides, Carbides, and Nitrides

Limited TiB_2 H_V data of Rice et al. [7] and the one H_V literature data point [51] agree but are insufficient to indicate whether deviations from a simple Hall–Petch H–G trend occur, but the limited data indicate H_K first decreases substantially and then increases substantially as G decreases (Fig. 4.10). The 100 gm data of other investigators substantially reinforces this possible trend, e.g., a H_K minimum at $G \sim 10$ µm. Although scattered, the 100 gm H_K data of Vahldiek and Mersol [53], Flinn et al. [54], McCawley et al. [55], and Nakano et al. [56] generally agree with each other, and with Rice et al.'s data, as does the unspecified H (2 N load) of Miyamoto and Koizumi's [57] reaction processed TiB_2 ($G \sim 5$ µm, $P \sim 0.05$). Similarly, 200 N H_V data of Watanabe and Kouno [58] for TiB_2 processed with various types and amounts of additives with 0 or a few percent porosity (data

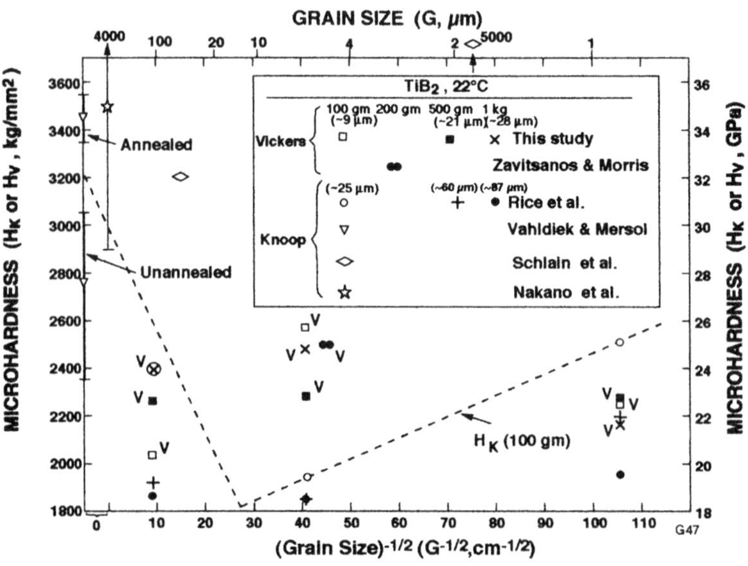

Figure 4.10 Hardness vs. $G^{-1/2}$ for single- and polycrystal TiB_2 at ~ 22°C. Rice et al.'s H_K data [7] (100, 500, and 1000 gm loads giving typical indent diagonals respectively of 25, 60, and 87 µm). Also shown are limited H_K data of Koester and Moak [52], Vahldiek and Mersol [53], Schlain et al. [54,55], and Nakano et al. [56] (100 gm load); also H_V data of Rice et al. (100 and 500 gm loads), and Zavitsanos and Morris [51] (200 gm load). Vertical bars = standard deviations. (From Ref. 7, published with the permission of the *Journal of the American Ceramic Society*.)

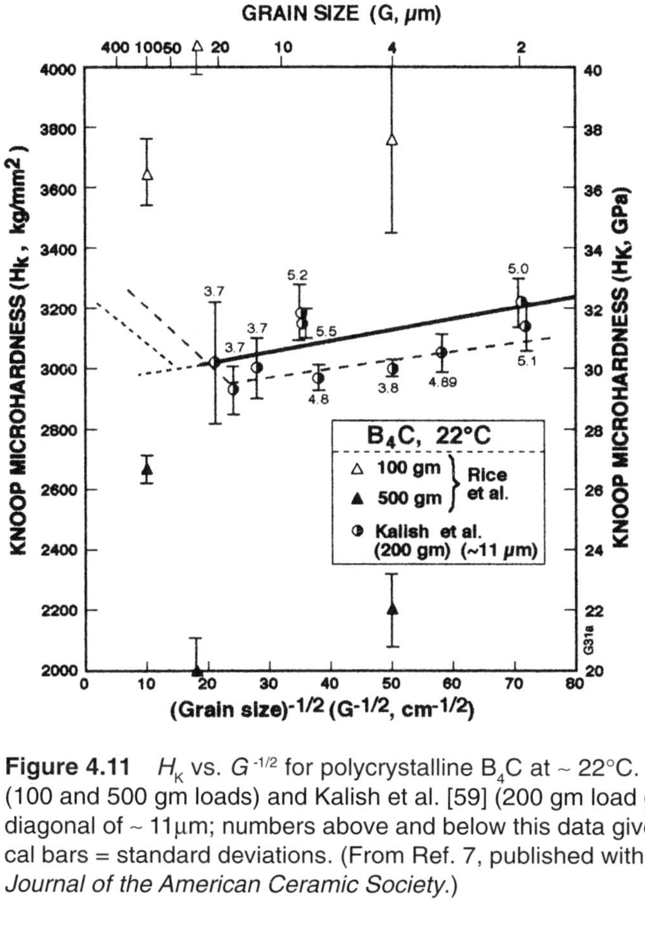

Figure 4.11 H_K vs. $G^{-1/2}$ for polycrystalline B_4C at ~ 22°C. Data of Rice et al. [7] (100 and 500 gm loads) and Kalish et al. [59] (200 gm load giving a typical indent diagonal of ~ 11μm; numbers above and below this data give the B/C ratio). Vertical bars = standard deviations. (From Ref. 7, published with the permission of the *Journal of the American Ceramic Society*.)

corrected to 0 porosity) for G ~ 2–7 μm giving values of ~ 21–23 GPa, with a limited decrease as G increases, is in good agreement with data of Fig. 4.10.

Kalish et al.'s [59] H_K (200 gm) data for hot pressed B_4C with B/C carbon ratios ranging from 3.7 to 5.5 is consistent with that of Rice et al. [7], i.e., lying in the appropriate range between the 100 gm and 500 gm load (Fig. 4.11). Their data show a general decrease in H_K with increasing G over the range observed (~ 2 to 23 μm), indicating that G was a more important factor in changing H than was the B/C ratio itself (though G was probably highly influenced by the B/C ratio). This data does not show a clear H minimum, but data for B/C = 3.7–3.8 may indicate one at G ~ 20 μm. The limited data of Rice et al. [7] would not indicate an H_K minimum at 100 gm load (in fact, if taken literally, it would imply a maximum) but would indicate an H_K minimum at G ~ 25 μm (500 gm load). Data of

Munir and Veerkamp [60] on their hot pressed B_4C showed H_V (unspecified load) decreasing from ~ 48.5 GPa at G ~ 2 μm to a minimum of ~ 42 GPa at G ~ 17 μm in reasonable agreement with H_K data of Fig. 4.11.

Niihara's [61] substantial tests of 6H SiC crystals shows 100 gm $H_{\underline{K}}$ ~ 24–27.8 GPa as a function of orientation on (0001), 26.9–31.5 GPa on {10$\bar{1}$0}, and 21.9–29.8 GPa on {11$\bar{2}$0} planes, in good agreement with Sawyer et al.'s [62] data for the same α crystal planes of respectively 29.5, 21.3–27.6, and 23.9–27.6 GPa at 100 gm and more limited 300 gm data of Adewoye and Page [63] of 27.8–31.3 GPa and of Fujita et al.'s [64] 500 gm values of 20–28 GPa (see also McColm [65] for some values). Page and colleagues' 1 kg polycrystalline SiC data [62,63,66,67] by itself suggests H_K first decreases and then increases as G decreases, i.e., an H_K minimum at G ~ 10 μm (Fig. 4.12). The limited (500 gm) data is consistent with Page's 1 kg data. While the limited 100

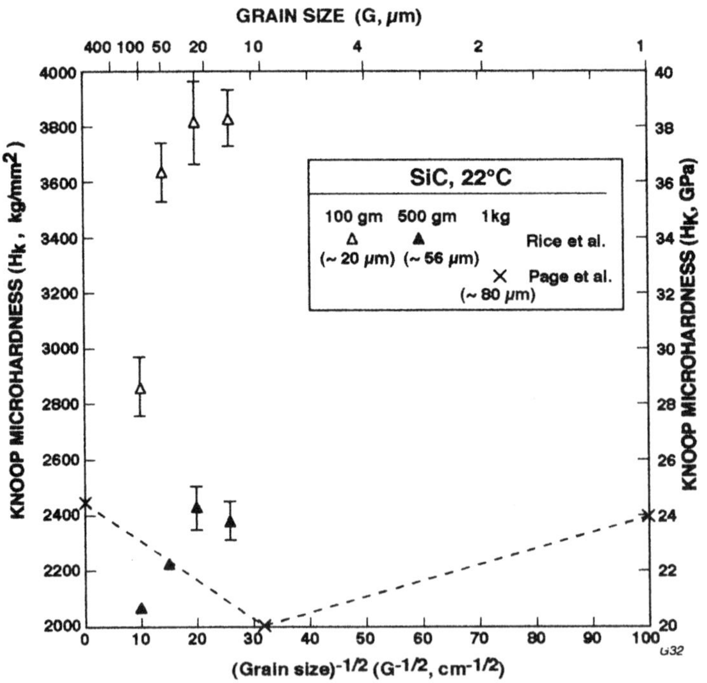

FIGURE 4.12 H_K–$G^{-1/2}$ data for single and polycrystalline SiC at ~ 22°C. Data of Rice et al. [7] (100 and 500 gm loads, giving typical diagonals respectively of ~ 7 and 20 μm) and of Page et al. [62,63,66,67] (1 kg). Vertical bars = standard deviations. (From Ref. 7, published with the permission of the *Journal of the American Ceramic Society*.)

gm SiC H_V data is insufficient to show any trend, higher load data of Rice et al. [7] and the literature [7,23,62] suggest an H_V minimum e.g. at G ~ 30 μm (500 gm). The phase content of some of the polycrystalline bodies varies or is not specified for some bodies, so some of the variation probably reflects varying α and β contents, which does not appear to have near as marked H impact as for Si_3N_4.

H_V SiC data shows similar trends, e.g. 100 gm values on the same crystal planes of ~ 27–32 GPa of [68], and 500 gm values of 30.2–32.5 and 37.7–46.7 GPa at 100 gm [69]. The limited 100 gm polycrystalline data of Rice et al. [7] was again scattered (but in a different fashion from the H_K data on the same specimens indicating specimen variations), but their 500 gm load data indicates a probable H_V minimum, as did data of Page et al., but at uncertain G values. More recent data on CVD SiC of Kim et al. [70] and Kim and Choi [71], though complicated by incomplete grain characterization, e.g. of varying degrees and types of grain orientation and colony structure, supports H increasing as G decreases at finer G. Again, more limited effects of α and β contents is indicated. More recent data at nanoscale G is generally consistent with H for normal G values (see Sec. III.A).

While TiC data is generally limited in the G range, especially for any one investigator, collectively there is considerable data (mostly for $G \geq 10$ μm), which is generally reasonably consistent for a given load range and indenter between different investigators. Collectively, this data is consistent with an H_K, and especially an H_V minimum at G ~ 20 μm). The single crystal H_K data of Rice et al. [7] (for material believed to have a Ti/C ratio of ~ 1) and the single crystal data of Rowcliffe and Hollox [72] and Hannick et al. [73] agree fairly well. Also, the highest Ti stoichiometry (0.93) data point of Miracle and Lipsitt [74] (1 kg load) agrees reasonably well with the similar G data of Rice et al. [7]. Similarly, there is reasonable agreement of H_V crystal data of Rice et al. with that of Kumashiro et al. [75] and that of Shimada et al. [76] (1 kg load) on {100}, which gave 26–28 GPa, as well as the large G (~ 400 μm) data of Cadoff et al. [77]. Data of Yamada et al. [78] also indicate H increasing with the C/Ti ratio and agrees with H_V values of Rice et al. H_V data of Yamada et al. and of Leonhardt et al. [79] both indicate H increasing at finer G, as overall does the other data.

Limited data for pure WC of Lee and Gurland [80] clearly indicates a marked increase in H_V (15 kg load) as G decreases from ~ 4 to ~ 1 μm (Fig. 4.5). Limited data of McCandlish et al. [81] (see Andrievski [30]) with somewhat smaller (nanoscale) grains in WC + 10 vol% Co also shows H markedly increasing with decreasing G, consistent with most other materials. Most, and probably all, of the lower H values for the WC-Co body versus those for pure WC of the same G are probably due to the effects of the Co, i.e. this data appears consistent with the model of Lee and Gurland [80] for effects of the Co boundary phase on the hardness of such bodies.

While there has been extensive development of AlN, hardness data is limited, especially over a reasonable G range. However, Rafaniello's more extensive data clearly shows H_V (9.8 N) decreasing with increasing G, e.g. decreasing ~ 10% between G ~ 2+ and 12 μm from ~ 10+ to ~ 9- GPa [82] (Fig. 5.4).

McColm [65] has compiled much of the limited BN data, giving H_V for hexagonal material ~ 2 GPa versus ~ 46 GPa for cubic material for loads of ~ 0.25–1.5 N with unspecified G. He also reports values of ~ 35 and 42 GPa for bodies with G ~ 3 μm with respectively 10 and 85% cubic content, thus providing extreme examples of the impact of phase content on H. He further compiled H_K (~ 4.9 N load) data on bodies of G = 1–3, 1–5, 3–9, and 10 μm giving respectively ~ 39, 34, 33, and 30 GPa, thus implying a typical H decrease with increasing G at finer G, but there are uncertainties due to differing binder compositions and amounts as well as probable variations in residual porosity. Single crystal values of 43 GPa at the same 4.9 N load again suggest an H minimum. [This H_K value is consistent with another (1 kg load) of 45 GPa [3].]

Polycrystalline (mainly hot pressed) Si_3N_4 data of Rice et al. [7] is generally consistent with other data [13,14,23,65,83–85] for similar bodies, i.e. those with substantial β-Si_3N_4 content (Fig. 4.13). This data is also consistent with that of Mukhopadhyay et al. [86,87] (H ~ 12–15 GPa for G ~ 1–4μ, data not shown in Fig. 4.13 to avoid confusion). Niihara and Hirai's H_V (100 gm) data for CVD (pyrolytic) polycrystalline SiC [11], along with (α) single crystals [62,63,66,88,89], is clearly above almost all hot pressed Si_3N_4 data. Grain boundary phases from additives and impurities in hot pressed samples probably contribute to their lower values. However, significantly lower β- versus α-Si_3N_4 polycrystalline or single crystal values are again shown, as previously reported [13], so variable β contents in hot pressed samples are an important reason for their lower and variable values. Distinctly lower H for β- vs. α-Si_3N_4 was shown by Greskovich and Yeh [13] and more recently by Ueno [90]. Both the hot pressed and especially the CVD (pyrolytic) data clearly suggests H first decreasing with increasing G (as also reported by Mukhopahyay et al. [86,87]) and then subsequently increasing, with single crystal values being higher than at least most of the range of polycrystalline values, i.e. a distinct H_V minimum at G ~ 20–100 μm at 100 gm load for CVD material and also possibly for hot pressed material, since it typically contains considerable β-phase. This trend for an H minimum provides an alternate explanation for the failure of Niihara and Hirai's data to extrapolate to single crystal values, which is consistent with a broad range of behavior of other materials, seriously questioning their proposed G^{-1} dependence of H. Note also that the data collectively indicates a probable H minimum at higher, e.g., 500–1000 gm, loads.

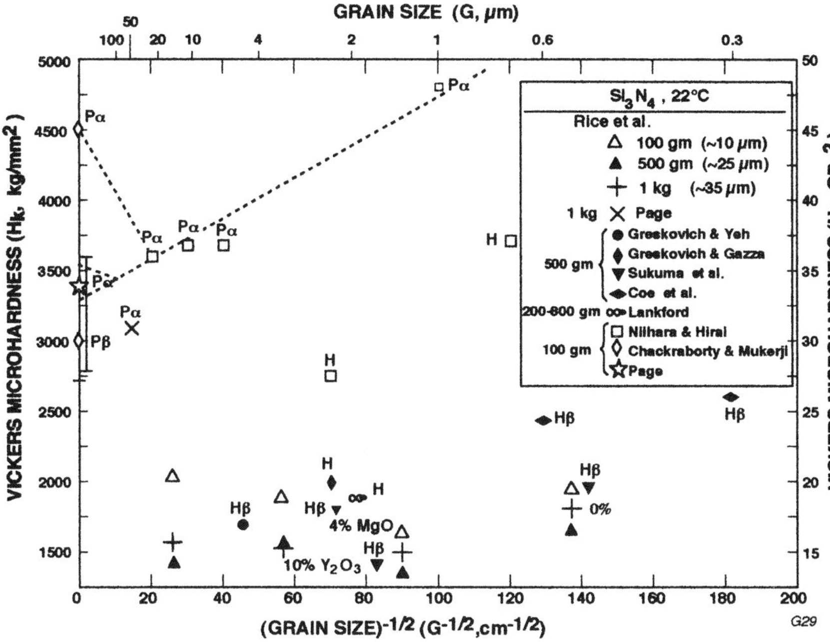

FIGURE 4.13 H_V vs. $G^{-1/2}$ for single and polycrystalline Si_3N_4 at ~ 22°C. Data of Rice et al. [7] (100 and 500 gm loads) and other studies at 100, 500, and 1000 gm loads as shown. Note that *P* indicates pyrolytic, i.e., CVD, Si_3N_4, *H* hot pressed, and α and β the predominance of these respective phases. Where specific data on additive content for hot pressed material is available, it is shown next to the data point (one data point for 0% additives is the result of high pressure hot pressing [84]). Vertical bars = standard deviations. (From Ref. 7, published with the permission of the *Journal of the American Ceramic Society*.)

C. Effects of Indenter Geometry and Load, and Other Constituents and Materials

Pyramidal indenters such as Knoop and Vickers have planar sides that make uniformly shaped indentations that vary only in relative size with indent load. However, such indenters give different hardness values due to differences in interaction with differing surface layers and microstructures as well as their not being defined on the same basis, i.e. Knoop and Vickers hardnesses are defined respectively by the maximum cross-sectional area and the actual surface contact area of the indenter with the material [91]. Table 4.1 summarizes the approximate average hardness values for common ceramic materials with substantial data for Knoop and Vickers hardness for 100 and 500 gm loads. This shows that while $H_K > H_V$ at a 100 gm load, often substantially so, the reverse is generally true at 500 gm loads. Since indent dimensions are inversely related to H values, and the H minimums commonly observed are related to indent-to-grain-size ratios due to cracking, as discussed in the next section, this indicates some relative shifts in H minimum with indent geometry and load and related cracking and the grain sizes at which they occur.

While Knoop and Vickers indentations are dominant for ceramics, some, mostly older, ceramic data has been taken with spherical indenters due for example to wider use of such indenters for metals. Only a fraction of this limited ceramic data is available as a function of G, but spherical indenter tests are common in metals where a Petch-type relation is generally followed, e.g. Bunshah and Armstrong [92]. Floyd's Rockwell 45N scale data for a 96% alumina composition with G varying from 6 to 25 μm clearly also follows a Petch-type relation [93], as does the Rockwell A hardness of WC-12 wt% Co (G ~ 1–6 μm) [94]. Given the typical large size of such ball indenters, and the high loads commonly used, a possible H minimum could occur with such indents but would be expected at much larger grain sizes [i.e. use of different size indenters could be useful in better defining the occurrence and mechanism(s) of such H minima along with load dependences].

Consider next the effects of other constituents on H, a large subject that can be only outlined here, highlighting possible impacts on G dependence of H. (See also Chap. 7, Sec. II for data on temperature dependence of hardness and Chaps. 10 and 11 for hardness of ceramic composites.) Impurities or additives that are in solid solution can affect H values. While solid solutions can reduce H, e.g. as indicated in alumina rich $MgAl_2O_4$ crystals (Fig. 4.9), it often increases H, e.g. as illustrated by alumina data. Thus Belon et al. [95] showed that ~ 2 mol% additions of isomorphous single (Cr_2O_3, Ti_2O_3, V_2O_3, or Ga_2O_3) or mixed oxides ($MgTiO_3$) individually increased H_K values (unspecified load) of $A1_2O_3$ by 15–25%, while simultaneous additions of Cr_2O_3 with Ti_2O_3 or Ga_2O_3 increased H values 35–40%. All additions passed through H maxima at 1.5 (Cr_2O_3) to 3 ($MgTiO_3$) m/o additions (in bodies made via vacuum melting, with additive levels determined after fusion). These $A1_2O_3$ results are similar to those of Albrecht [96] showing an H maximum at ~ 3 mol% Cr_2O_3. These Cr_2O_3 addition results are in contrast to those of Bradt [97] showing a linear H_V (500 gm) increase of nearly 1% per m/o Cr_2O_3 to at least 12 m/o in samples hot pressed at various temperatures (but with the reduced surface layer removed). Shinozaki et al. [98] showed similar, but varying, results with H_V (200 gm) increasing by 10–15%, peaking at ~ 20 m/o (with higher H for firing at 1600°C and somewhat lower H peaking at ~ 15 m/o for firing at 1700°C, both under reducing conditions). These different results probably reflect different rates and extents of Cr_2O_3 solution due to differing degrees of mixing, reduction, and firing, which have received little or no attention, but which may impact H directly or via G. However, Shinozaki et al.'s results imply a lower *H* maximum at lower Cr_2O_3 levels with larger *G* expected from higher firing. Bradt clearly showed lower *H* due to larger *G* (e.g. ~ 10 versus ~ 2 μm), but details of composition-processing-microstructure-hardness need further clarification.

The above variations in part indicate two sets of complexities of studying composition effects on *H*. The first set is due to additions of other materials impacting both sintering and grain growth, so truly valid comparisons can only be made by comparing bodies of sufficiently similar microstructures directly or by correcting results for such effects (e.g. using results of this chapter). These complexities are compounded by the extent of homogeneous mixing achieved, which is a complex function of the uniformity and scale of initial mixing and the temperature–time (and possibly atmosphere) of densification or additional heat treatment. The firing atmosphere can be important, since the stoichiometry of the additives, matrix, or both may be altered, impacting the extent and character of the solid solution. Such effects are closely related to those of stoichiometry of even a single constituent body, e.g. effects in bodies more susceptible to stoichiometric variations such as Cr, Ti (e.g. note effects on crack propagation, Fig. 2.5), and Zr oxides and B, Ti, and Zr carbides. Thus again note that major effects of stoichiometry are often primarily via effects on *G*, but there is some residual effect of the stoichiometry alone, e.g. B_4C (Fig. 4.11) and TiC [7,74]. The second set of complexities, stoichiometry variations of the body, additive(s), or both, may often vary with depth from the surface, which may vary with grain size due to differing temperature giving different *G* values, and to effects of grain boundary diffusion, which may be dependent on *G* values. For the above reasons, especially the latter one, single crystal data for the stoichiometry or composition considered are a valuable guidepost to compare with polycrystalline results, e.g. as extensively shown by results reviewed in this chapter. Thus data for carbide crystals of differing stoichiometry clearly show differing *H* values.

Effects of additives or impurities on *H* values as second phases may differ, often substantially, from those found with solid solution, with two extremes often having differing results. The first extreme, often with the greatest differences and variations, is formation of a second phase along grain boundaries. As briefly noted earlier (Fig. 4.5, see also Chap. 10), a softer bonding boundary phase such as Co for WC reduces hardness in proportion to both the hardnesses of the two phases, their volume fractions, and their microstructural character, i.e. grain size of the matrix and mean free path between binder phase boundary "sheets." On the other hand, residuals from LiF additions in MgO (Fig. 4.21) and $MgAl_2O_4$ greatly exacerbate cracking around indents due to their enhancing intergranular fracture, i.e. local indent effects similar and related to the macrofracture and crack propagation effects noted for such additives (Fig. 2.12). Similarly, the significant decrease in *H* with decreasing *G* at fine *G* in the limited TiAl data of Chang et al. [99] (see Andrievski [30]) indicating opposite trends from almost all other materials (Fig. 4.5) may reflect grain boundary impurities left from species absorbed on the nanoparticle similar to those indicated as sources of low strengths in nanoceramics (Chap. 3, Sec. IV.A).

The other extreme of second-phase distribution, which again reflects a composite structure (Chaps. 10 and 11), is homogeneous precipitation. Thus while effects will vary with the amount and the chemical and physical nature of the precipitate versus those of the matrix, precipitation commonly increases hardness values over those for solid solutions of the same composition. For example, studies of TiO_2 precipitation in Al_2O_3 of Bratton [100], show that the precipitation of the modest levels of TiO_2 solid solution achievable (~ 0.1 wt%) increases H_K values by 10–20%. Similarly, it is well known that precipitation in alumina rich $MgAl_2O_4$ crystals can increase H values, e.g. by ~ 15%, but with varying dependence on heat treatments, e.g. Lewis et al. [101]. However, again note that in differing polycrystalline compositions varying spinel stoichiometry has much of its effects on H values via its impact on G.

ZrO_2 bodies with varying additives for partial or full stabilization of the cubic or tetragonal phases can reflect various combinations of the above second-phase distributions. However, while such effects must be factors in the variations (and probably the indications of limited G dependence), the data, mainly for different types and levels of Y_2O_3 stabilization, does not appear to be a large factor in the H–G dependence, nor of ZrO_2-Y_2O_3 crystals [7,39]. The consistency of the ZrO_2 trends with other data, despite variable Y_2O_3 compositions, reinforces the limited effect of Y_2O_3 content.

Consider now other, primarily softer, less refractory materials, starting with limited halide data from a previous compilation [12] (Fig. 4.14), where, while not always specified, loads were commonly ~ 100 gm. Both the KCl and the BaF_2 H_V data, while each consisting of one data set for single crystals and a few or one for the respective polycrystals (G ~ 10 μm), indicate little or no opportunity for an intervening H minimum. Such proscription of an H minimum is uncertain for the one H_K data set for MgF_2 crystals and two data sets for fine-grain polycrystals.

H_V and H_K data on various rocks of nominally single-phase composition of Brace [102,103] all indicate that any intervening H minimum must be at $G >$ several hundred microns (Fig. 4.14). However, the apparent high loads (to 150 kg) would indicate that any minimum would be at such high G values as would the lower hardness of most of these materials, especially the limestone. Note that the higher H values for dolomite in part reflect solid solution effects, since it is a solution of Mg and Ca carbonates. (Note also that Brace's data is based on using the maximum G values (with measurement specifics not given), which increases the slopes on Petch plots to the extent of the average to maximum G ratio, which most likely varies some for the different rocks.)

Finally, consider data for cubic ZnS and ZnSe, which have been devel-

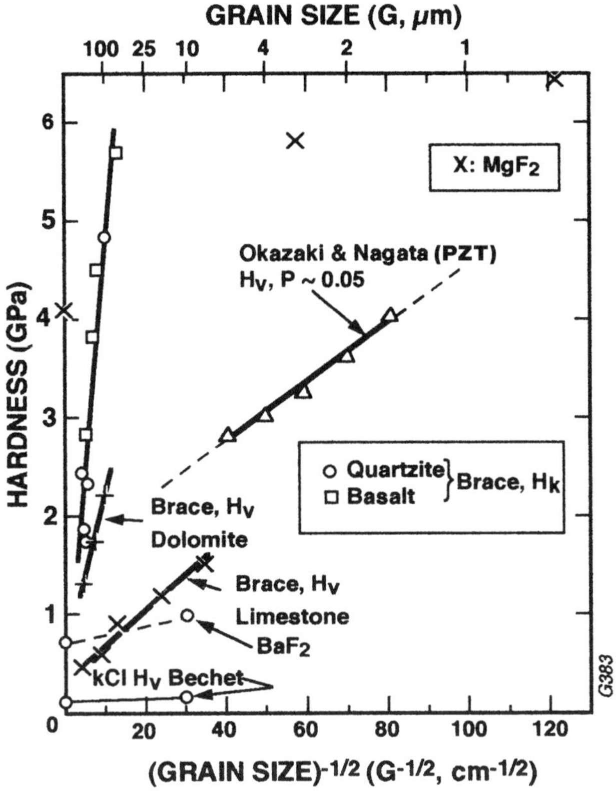

Figure 4.14 H vs. $G^{-1/2}$ for KCl, MgF_2, BaF_2 [12] and PZT [12, 50], and for some nominally single-phase rocks after Brace [102] at ~ 22°C. Note that the latter used the maximum G values, which increase the slopes by unknown and probably different amounts.

oped for IR window applications. Most extensive is H_V (10 N, ~ 1 kg and 100 N ~ 10 kg) data on CVD ZnS [104–106], covering a substantial G range, which clearly shows H first substantially decreasing with increasing G and then increasing again, so that projected single crystal values would be higher than a fair range of the polycrystalline values (Fig. 4.15). The projected H_V value is in good agreement with H_V values of ~ 1.8 ± 0.3 GPa from studies of Lendvay and Fock [107] on single crystals with similar multikilogram loads. This data is also consistent with limited, previously compiled [12] polycrystalline data (apparently with low, e.g. 100 gm, loads) for G ~ 3 and 0.8 μm, giving respective H_K values of ~ 2 and ~ 3 GPa. More limited H_K (50 gm load) data for isostructural and similarly prepared CVD ZnSe of Swanson and Pappis [108] for G = 30–100

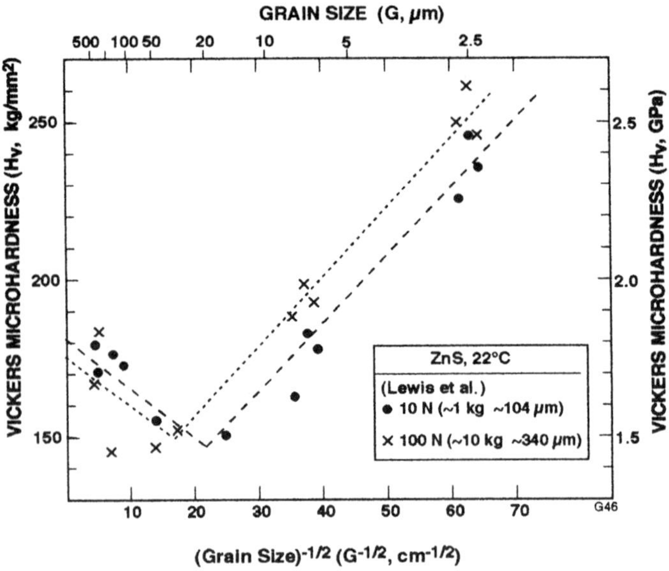

Figure 4.15 H vs. $G^{-1/2}$ for ZnS. Vickers hardness data for CVD-ZnS [106–108] at 10 N (~ 1 kg) and 100 N (~ 10 kg) loads at ~ 22°C. Note the distinct H_V minima at $G \sim 20$ µm and ~ 150 µm respectively. (From Ref. 7, published with the permission of the *Journal of the American Ceramic Society.*)

µm fit a Petch relation: $H_K \sim 0.8 + 1.7G^{1/2}$ GPa (for G in µm). It is not clear whether this data reflects an approach to an H minimum (i.e. at $G \sim$ or > 30 µm) or the absence of one given the lower hardness and load of ZnSe and load used.

D. Occurrence and Character of Indent-Related Cracking Associated with *H* Minima

As shown earlier, H minima were generally observed with most harder, more refractory ceramics, and also some softer, less refractory materials. It was also shown that the extent of deviation of the H minima below the normal Petch dependence and the G values of the H minima, while both material dependent, both generally increased as the indent load increased. It was also noted that these H minima were associated with a maximum of the amount and complexity of cracking. This section documents this cracking and its association with H minima and shows that such minima occur when the indent and grain sizes are about the same; but this cracking can be seriously exacerbated by grain boundary phases enhancing intergranular fracture. Thus in contrast to a continuous increase in H with decreasing G, when such H minima and associated cracking occur, H values first decrease from single crystal values (i.e. at $G = \infty$) as G decreases until the indent size is on ~ the same scale as the G and then increases as G decreases further, following a modified Petch relation (i.e. extrapolating to < single crystal H values at $G = \infty$).

First consider the G dependence of such minima, which is shown by both direct correlation with G values, and subsequent direct observations of the cracking associated with the H minima. Correlation with G data is provided by Table 4.2, which shows the ratio of the grain size at the H minima to the indent diagonal length (d) for materials having clear minima at one or more test loads. This shows that, while there is variation (discussed later), the H minima occur when G/d is ~ 2–5 for Vickers and ~ 1 for Knoop indents. (This trend is also indicated in many of the preceding H–$G^{-1/2}$ plots, where an average value of d is given for some of the loads and materials.)

Next consider the character and occurrence of such cracking of interest here, which is not the commonly reported ~ linear cracks extending out from one or more of the indent main diagonals often used for measuring toughness. The cracking of interest here is instead more localized around the indent, often of a spalling character, and in polycrystalline bodies it occurs extensively via intergranular fracture (Figs. 4.16–4.20). This localized cracking may exist mainly or exclusively by itself (e.g. Figs. 4.16B–D, 4.17D, 4.18BC, and 4.20), or in combination with the more commonly reported ~ linear cracking (e.g. Figs 4.17D and 4.19B). Again, much of the cracking of interest is of a spalling character and is often intergranular, especially when the indent size is similar to G. As G becomes larger than the indent size, cracking again diminishes towards that of single crystals, where cracking is generally less than in large grain bodies, and is

TABLE 4.2 Grain Size/Indent Diagonal Ratios at Hardness Minima[a]

	H_V			H_K	
Material	100 gm	500 gm	1 kg	100 gm	500 gm
Al_2O_3		50–100/23		50/24	50–100/68
BeO		100/26	100/58		
MgO	20/14	25/43		15/39	30/100
ZrO_2	50/12	50/29			
$MgAl_2O_4$	50–200/11	50/26	20/37		100/74
TiB_2				15/28	
ZnS			25/110		
Average	5 ± 4	2 ± 1	1 ± 0.9	1 ± 0.9	1 ± 0.5

[a]Values in μm. *Source:* After Rice et al. [7], published with the permission of the *Journal of the American Ceramic Society*.

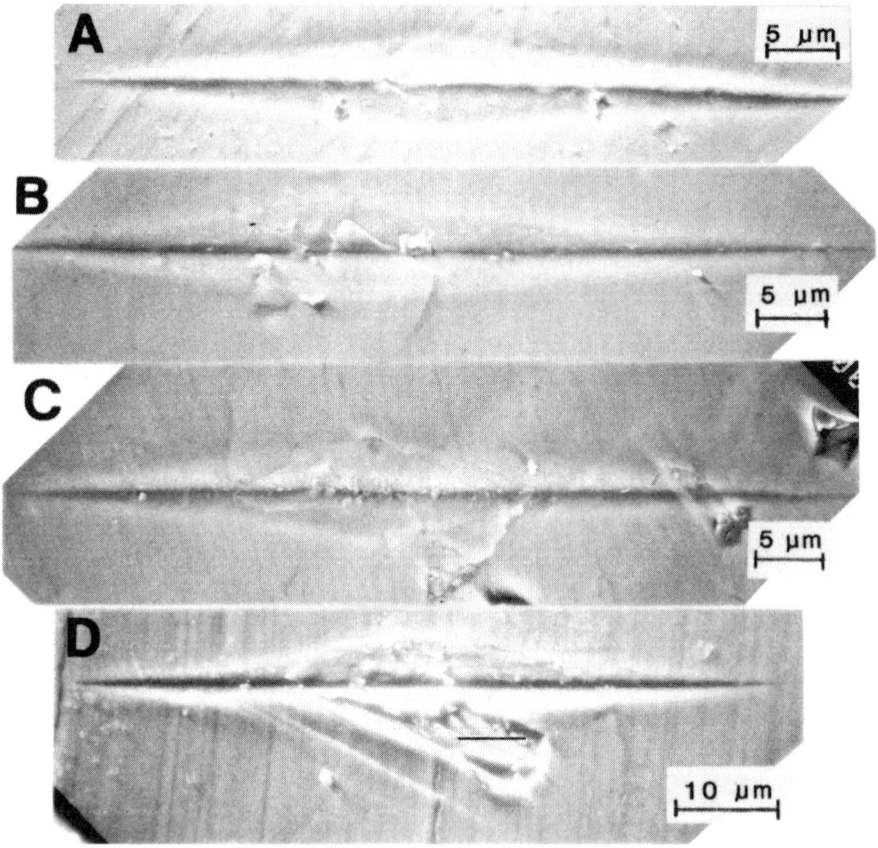

Figure 4.16 Knoop indents in polycrystalline $A1_2O_3$ at 500 gm with increasing *G*. Nominal *G* was (A) 6 μm, (B) 11 μm, (C) 3, 11 μm (bimodal), and (D) 40 μm. Note the general tendency for greater cracking, especially locally around the indent as *G* increases. (After Rice et al., [7], published with the permission of the *Journal of the American Ceramic Society*.)

more ordered (i.e., some along preferred fracture on cleavage planes), hence providing less opportunities for spalling (e.g., Fig. 4.19). Similar indent-induced cracking in dense sintered Al_2O_3 (*G* 20–30 μm) was shown by Lankford [109] at similar loads (e.g. 0.6 kg) at both room and modest elevated temperatures, and Anstis et al [110] reported such cracking in a dense Al_2O_3 with *G* ~ 20 μm (50 N load), but not finer *G* (~ 3 μm) or sapphire.

There are three intrinsic factors and one extrinsic factor that vary the above cracking. The first two and most distinct are test factors of load and indent type.

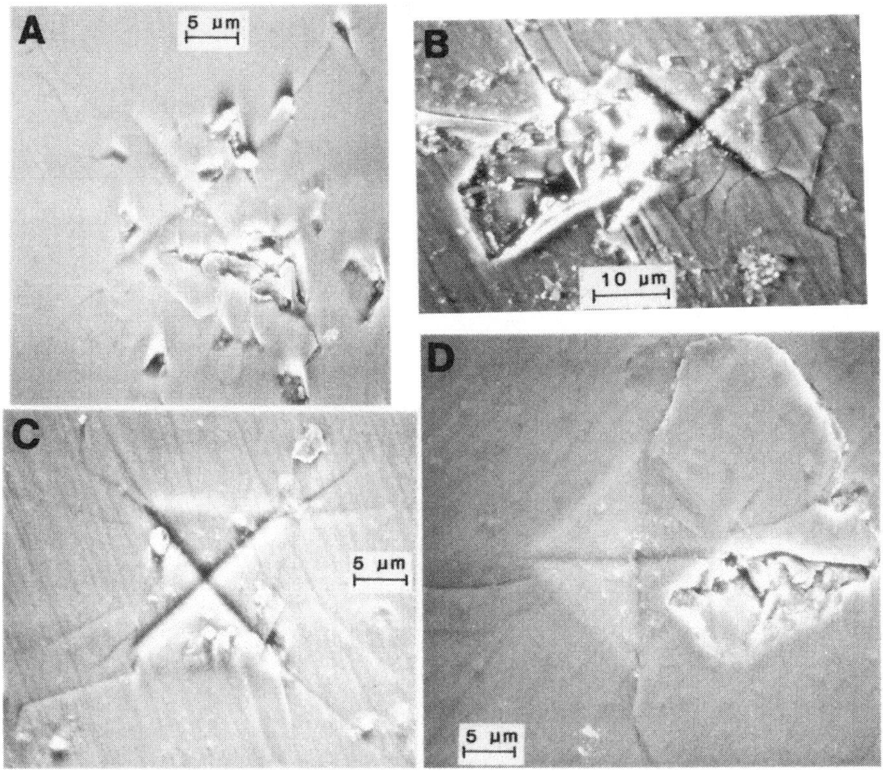

Figure 4.17 Vickers indents (500 gm) in hot pressed polycrystalline $A1_2O_3$ of varying *G*. (A) *G* ~ 6 μm; (B) *G* ~ 40 μm, (C) *G* ~ 40 μm, and (D) *G* ~ 50 μm. Note the general tendency for both more cracking and greater complexity of cracking, especially near the indent, as *G* increases. (From Ref. 7, published with the permission of the *Journal of the American Ceramic Society*.)

Thus such cracking and associated *H* minima, which probably first begin at some threshold load and then increase as indent load increases, so that while there can be considerable variation at a given load, the frequency and scope of this localized cracking clearly increased with increasing load (e.g., Figs. 4.18–4.20). The second factor is indent geometry, i.e. less cracking with Knoop vs. Vickers indents (e.g., Fig. 4.16 vs. 4.17) attributed to the long narrow character of the Knoop vs. the Vickers indenter (e.g. the length affecting the *G* value where cracking is a maximum, but the narrowness limiting the actual extent of cracking). The third factor, a body one, is the grain size (and also grain shape and orientation as noted in the next section). As shown by the H–$G^{-1/2}$ plots and Table

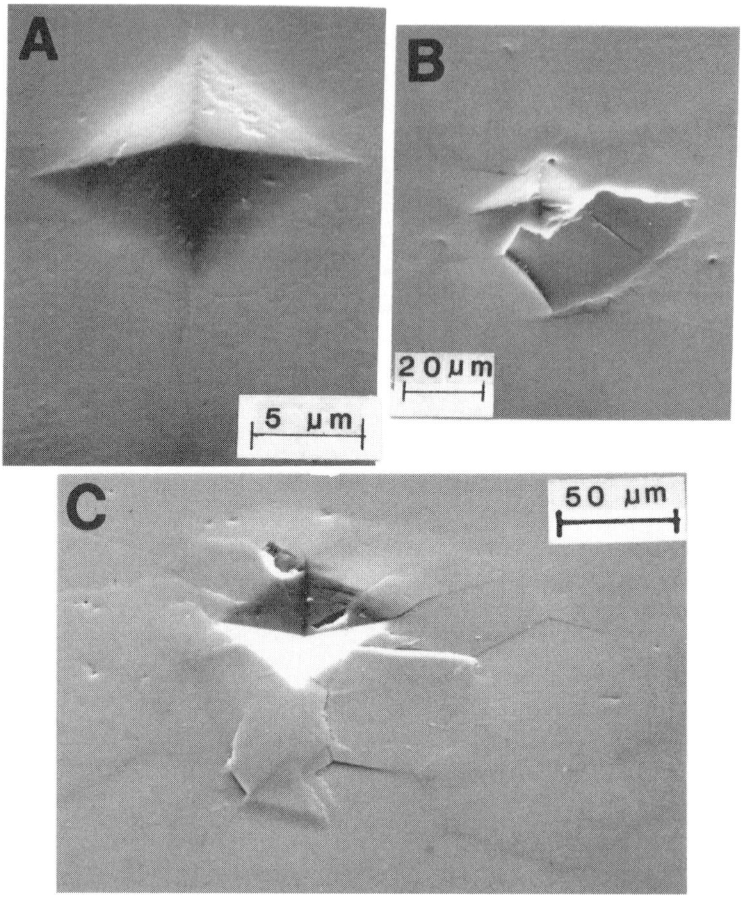

Figure 4.18 Vickers indents as a function of load in hot pressed and annealed MgO with $G \sim 100$ μm. Loads were (A) 100 gm, (B) 500 gm, and (C) 2 kg. Note the increase in extent and complexity of cracking, especially around the indent and its association with the grain structure with increased load. (From Ref. 7, published with the permission of the *Journal of the American Ceramic Society*.)

4.2, the H minima and associated cracking reach maxima at material and load-dependent G values. Clearly, in view of the G dependence, variations in G will vary the maxima in cracking, e.g. observations on a dense alumina with a bimodal G distribution (of ~ 3 and 11 μm) showed cracking more like the larger than the smaller G value (Fig. 4.16C), as might be expected by fracture being a weak link process and the larger G being closer to the G for most H minima in

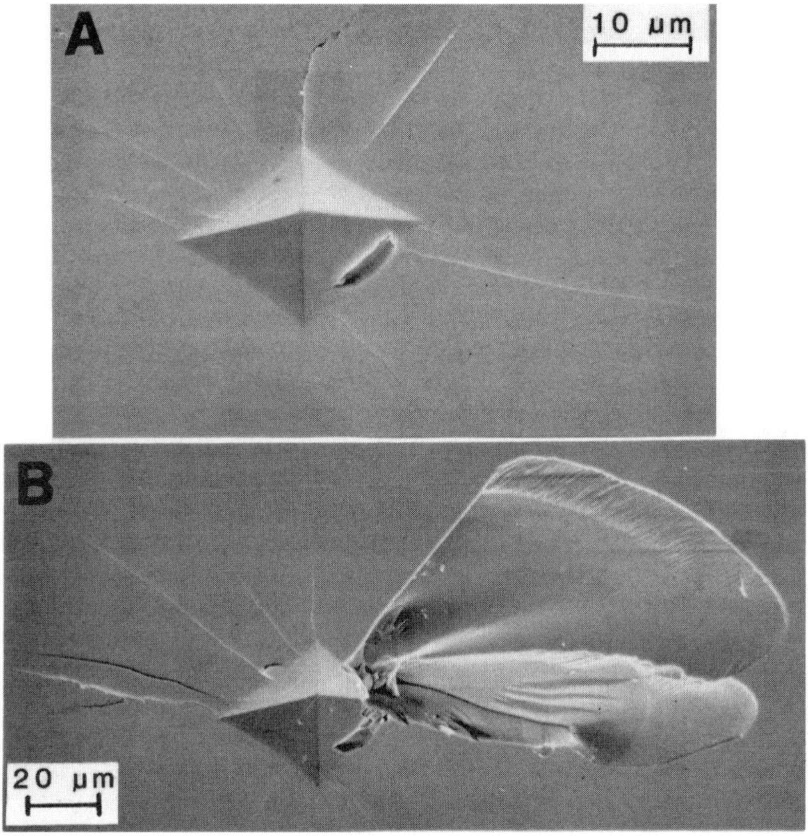

Figure 4.19 Vickers indents on {110} surfaces of stoichiometric $MgA1_2O_4$ crystals. (A) 500 gm load, and (B) 2000 gm load. Note increased extent and complexity of cracking at the higher loads. (From Ref. 7, published with the permission of the *Journal of the American Ceramic Society.*)

most materials and tests. On the other hand, though opportunities to observe this have not apparently occurred, a mixture of grains larger than the G for maxima cracking would be expected to give cracking more consistent with the smaller G (but also impacted by its volume fraction).

The fourth and sporadic, but important extrinsic, body factor is the accumulation of additive or body constituents at grain boundaries when they significantly enhance intergranular fracture, thus enhancing grain spalling, in proportion to their amount, distribution, and effects on fracture. Thus contrast Fig. 4.18 (MgO hot pressed with no additives) and Fig. 4.20 (MgO hot pressed

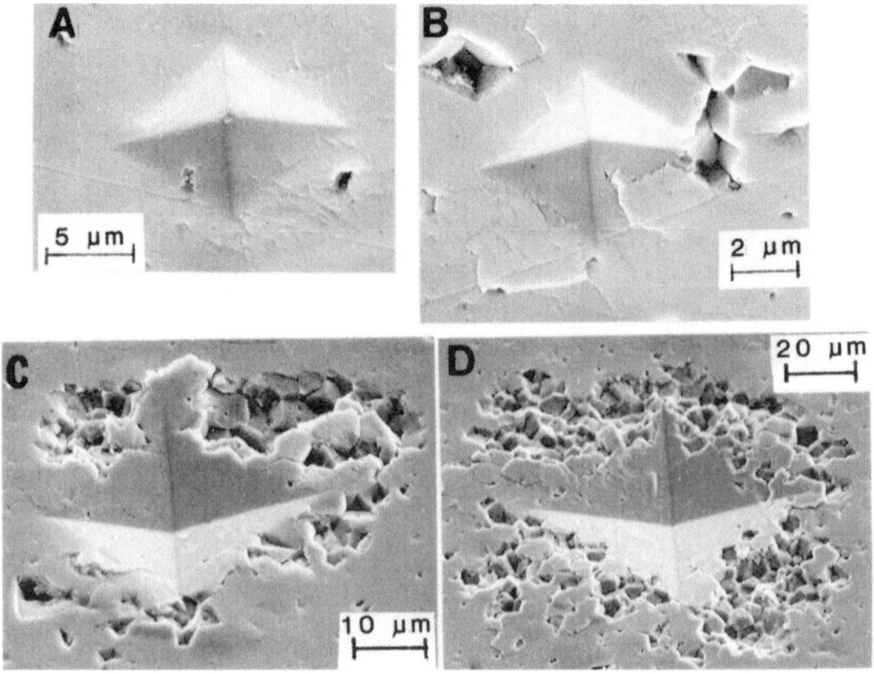

Figure 4.20 Vickers indents at various loads in MgO hot pressed (with LiF), but unannealed, having $G \sim 5$ µm. Loads were (A) 50 gm, (B) 100 gm, (C) 500 gm, and (D) 1 kg. Note increasing grain boundary cracking as load increased. (From Ref. 7, published with the permission of the *Journal of the American Ceramic Society.*)

with LiF, but unannealed). $MgAl_2O_4$ hot pressed with similar additives also showed distinctly more local cracking and spalling vs. sintered material with no additive, despite much larger *G* in both bodies [7]. Such effects in both MgO and $MgAl_2O_4$ are consistent with reduced fracture energies/toughnesses for such bodies and increased intergranular fracture on a macro scale (Fig. 2.12).

A set of tests performed by Sperisen et al. [111], though limited, are very suggestive in their results and implications for future testing. They used a Vickers indenter mounted on the cross-head of a conventional testing machine so the rate of loading and unloading, as well as the loading level and time, could be independently controlled along with two supplemental testing aspects. The first was the use of acoustic emission detection during indentation, and second the use of a specimen holder allowing the surface to be indented to be covered with a liquid; in their case by oil or water for an inert versus active fluid. Using two dense, pure hot pressed alumina bodies ($G \sim 1$ and ~ 6 µm) they observed effects of these parameters on both cracking from the indent corners (as for toughness measurements) and local spalling cracks, almost exclusively via intergranular fracture, in and around indents. While the cracking process was complex and somewhat variable, there were important trends, one of the most basic being that acoustic emission showed that most cracking occurred in three stages during loading, but some limited cracking occurred during two unloading stages. The nature and extent of these stages varied with G and the fluid environment of the indent. Indent cracks, e.g. for toughness measurement, which were mostly intergranular, appeared to initiate as Palmquist cracks that subsequently joined, with subsequent crack-to-indent-size ratios being the same for water and oil environments, reaching a constant level by loads of ~ 15 N for G ~ 1 μm, but were still not fully independent of load for G ~ 6 μm ~ 15 N load. The intergranular spall cracking was found to have no apparent threshold, i.e. observed at the lowest load tested (~ 10^{-2} N), and to be substantially greater for the larger G and with water versus oil. Though limited and raising issues of the relative effects of the fluids as chemical species versus lubricants, such tests are suggestive of important complexities that have been almost totally neglected and deserve much more use and evaluation.

E. Effects of Grain Shape and Orientation, and Other Factors

There is very little data directly on the effects of preferred grain orientation on hardness, and as is unfortunately so common, very limited or no quantitative characterization of the degree and character of preferred orientation in specimens. This is a serious lack, since local or global orientation (or both) can frequently be substantial and often, but not necessarily, occur with grain elongation. While either or both local and global orientation occur in vapor deposited (e.g. CVD) coatings or bodies [e.g. 68], some orientation can be much more common than is often noted in a variety of bodies, e.g. various sintered ones. However, there is substantial data on the effects of indent orientation on various single crystal surfaces that provides clear and detailed information on the limits of polycrystalline orientation effects on hardness. Such data is also a major demonstration of plastic deformation as the determining factor in hardness indentation of even the hardest materials. Thus, as noted in the introduction, the dependence of hardness on differing crystal surfaces and on indenter geometry and its orientation relative to crystal directions on a given crystallographic surface has been extensively and consistently shown to be due to varying activation of differing crystal slip systems [1–4]. There are substantially differing effects of different indenters, but there is a simple approximate interrelation between Knoop and Vickers [2] (the latter having less pronounced anisotropy). There are also the usual effects of load on hardness levels, and there may be some changes in the nature of the orientation dependence with differing loads due to some shifts in the balance of activation of differing slip systems. (There are also significant effects of temperature, e.g. Fig. 7.1, and differing effects of tensile versus compressive stresses [3] that can be pertinent to details of the relation of hardness and compressive strength, Chap. 5, Sect. II.B.) The orientation dependence of hardness on major, i.e. low index, crystal surfaces, is typically approximately a sinusoidal wave, often with some modifications, e.g. Fig. 4.21 (see also Fig. 7.1).

As reference for the extremes in the orientation dependence of hardness of polycrystalline ceramics, an outline of the single crystal orientation dependence is presented in Table 4.3 showing representative values of the maximum and minimum hardness and their ratio. Materials with extremely anisotropic structures (commonly materials with platy structures such as graphite, hexagonal BN, and mica) have extremes of hardness anisotropy and hence significant effects of preferred orientation on polycrystalline hardnesses (though these materials also commonly present challenges in obtaining good hardness values, especially with the loading direction parallel with the plane of the platy structure). Though not as extreme, other ceramics often have substantial hardness anisotropy, which often varies as much as or more for ceramics of various cubic versus those of noncubic structures (Table 4.3). Substantial other data is also available, e.g. for TiO_2 [27], $MgAl_2O_4$ [47], SiC [63,66–69], other carbides [72,73,112], Si_3N_4 [69,113], SiO_2 [69], MgO and LiF [114], and other oxides [36,115,116]. (Note that Ref. 113 is one of the few articles discussing composition–structure effects that may vary with details of grain/crystal growth that can be important in some materials such as SiC and Si_3N_4.)

These ceramic trends are similar to, but possibly less extreme than, those

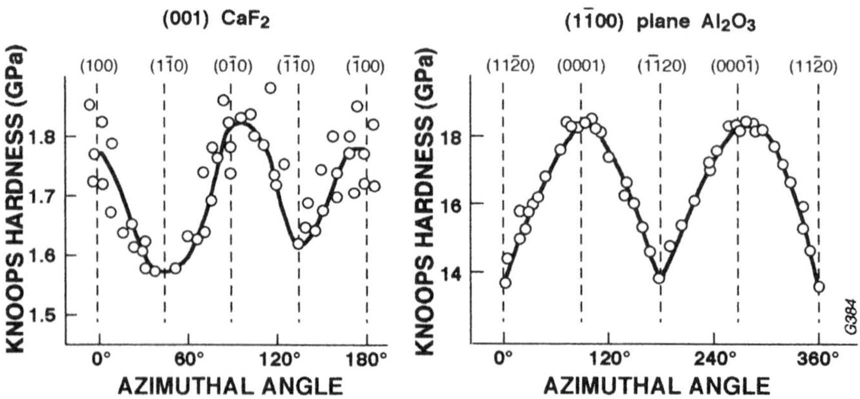

Figure 4.21 Knoop hardness anisotropy on a (001) surface of CaF_2 (A) and a $(1\bar{1}00)$ surface of Al_2O_3 (B) (presumably at a 100 gm load) after Brookes and Burnand [3]. (Published with the permission of the American Society of Metals.)

TABLE 4.3 Summary of Hardness Anisotropy of Some Ceramic Crystals[a]

Material	$^{H}H_{K}$ (GPa)	$^{L}H_{K}$ (GPa)	$^{H}H/^{L}H$
(A) NaCl Cubic Structure			
NaCl	0.20	0.18	1.1
LiF	0.97–.20	0.87–0.96	1.1–1.25
MnS	1.42–1.96	1.19–1.62	1.2
MnO	2.87	2.52	1.14
MgO	8.0	4.0	2.0
TaC	16.5	15.0	1.1
ZrC	22.5	19.8	1.14
HfC	25.0	18.5	1.35
$VC_{0.84}$	26.0	20.7	1.26
$TiC_{0.80}$	23.0	20.0	1.15
TiC	27.5	20.2	1.36
(B) Other Cubic Structures			
CaF_2	1.78	1.57	1.13
Diamond	96.0	69.0	1.39
(C) Hexagonal Structures			
Al_2O_3	18.0	14.0	1.29
Mo_2C	15.8	15.4	1.02
SiC	29.5	20.3	1.45

[a] ^{H}H and ^{L}H are respectively the high and low, i.e. maximum and minimum, values for different Knoop indenter orientations on differing crystal surfaces; values in GPa. A range of values reflects two or more differing sets of data.
Source: Ref. 3.

for metals (Table 4.4). Thus single crystal anisotropy in hardness is a basic guide to the orientation dependence of hardness in polycrystalline bodies, but the grain size dependence (which may vary some with at least higher levels of preferred orientation) must be accounted for. An additional complication in the grain size dependence of ceramic hardness is the grain-related cracking that commonly occurs, as was extensively shown earlier. This is likely to be affected more (mainly reduced) by increasing preferred orientation, but no data or models are available for guidance. Grain shape can also be a factor in the orientation dependence of H, e.g. Chakraborty and Mukerji [69] showed that H_V (100 gm) on prismatic planes doubled from 20 to 40 GPa as the aspect ratio of the grains (crystals) increased from 1 to ~ 5 (beyond which H was ~ constant). However,

TABLE 4.4 Summary of Hardness Anisotropy of Some Metal Crystals[a]

Material	$^{H}H_{K}$ (GPa)	$^{L}H_{K}$ (GPa)	$^{H}H/^{L}H$
(A) Face Centered Cubic (FCC) Structure			
Al	0.47	0.14–0.17	1.35–1.47
Cu	0.47	0.34	1.38
Ni	1.15	0.72	1.60
(B) Body Centered Cubic (BCC) Structure			
Nb	0.81	0.59	1.37
V	1.03	0.79	1.30
Cr	1.39	1.08	1.47
W	4.09–4.45	3.37–3.6	1.2–1.5
(C) Hexagonal Close Packed (HCP) Structure			
Zn	0.44	0.15	2.93
Mg	0.32–0.36	0.13–0.15	2.40
Ti	1.17–1.32	0.36–0.56	1.4–3.0[b]
Co	2.50	1.80	1.39
Zr	2.71	1.15	2.42

[a]^{H}H and ^{L}H are respectively the high and low, i.e maximum and minimum, values for different Knoop indenter orientations on differing crystal surfaces; values in GPa. A range of values reflects two or more differing sets of data.
[b]Most values 2.4–3.0.
Source: Ref. 3.

H_V was independent of aspect ratio at ~ 20.5 GPa. Another complication, as in hardness itself, is load dependence of H anisotropy. Though not extensively documented, Pajarec et al.'s data for the H_K anisotropy of cubic ZrO_2 crystals shows the maximum anisotropy decreasing from 1.31 with a 0.5 N load to 1.12 at a 2 N load, indicating an increased role of slip requiring higher stress activation [36].

Note first additional sources of H anisotropy data and second studies of Westbrook and Jorgensen [117]. The latter not only are an additional source of data on H anisotropy of ceramic crystals but also showed that adsorbed moisture on crystal surfaces can measurably alter their hardness values and thus also the measured anisotropy (see also Chap. 7, Sec. II). Finally note that, besides effects of electric fields and poling on cracks and related effects in ferroelectric and piezoelectric materials (Chap. 2, Sec. III.I), hardness can also be affected. Thus Park et al. [118] showed that while H_K (0.25–1 N) was isotropic for both single-

and dense polycrystals ($G \sim 60$ μm), H_K with the long axis of the Knoop indentation normal to the poling direction resulted in lower H and that parallel to the poling direction higher H (i.e. 4.42 and 4.69 GPa, which approximately straddle the unpoled, isotropic value of 4.53 GPa). A similar anisotropy was found in single crystals, i.e. respectively 4.06 and 4.35 GPa. Note that single crystal value is somewhat higher than the large G polycrystalline body, which indicates that H of $BaTiO_3$ may not have a minimum, e.g. suggesting effects of twinning.

III. GENERAL DISCUSSION

A. Variations, Uncertainties, Consistency, and *G* Range of Data

The substantial data reviewed earlier clearly shows that a Petch-type, i.e. a $G^{-1/2}$, dependence of ceramic hardness occurs over the entire, or more commonly the finer, portion of the grain size range. The deviation from a pure Petch relation (i.e where all H values increase from those of the average crystal values in a linear fashion as a function of $G^{-1/2}$) is associated with the extensive trend for a hardness minimum at some intermediate grain size due to cracking. However, before discussing such cracking it is useful to note variations commonly impacting data, which arise to some extent from test method or more extensively from the character of the specimen, its preparation, and its characterization.

The primary test factors, indenter type (i.e. shape) and load, which can significantly impact H values (e.g. especially for < 500 gm loads, as is well documented), are respectively usually and commonly given, making substantial data useful, but some of limited or no value. Loading (and unloading) rate, time, and environment, which can have some, often varying effects, are widely neglected, as is resultant indent character, e.g., symmetry and cracking. The latter can be a particular problem, since irregular indents are typically discarded rather than being seen as an indication of irregularities in the sample, especially locally around and under the indents. Such neglect can bias results toward a more favorable rather than a true hardness evaluation. Use of recording indenters [111] should be valuable to determine the nature and extent of effects of loading (and unloading) rates, times, environment, and local variations, especially for better documentation and understanding of indent associated cracking, as was indicated earlier. A specimen preparation factor impacting test factors is the load dependence of H not only on the final surface finish but also on earlier surface finish steps if the extent of cracking and especially surface work hardening of earlier finishing steps are not fully removed by subsequent finishing steps [19].

The first of the more extensive specimen factors impacting hardness is composition. Beyond the fundamental factor of the basic chemical composition of the body, e.g. NaCl versus TiC, are variations in composition, due both to stoichiometry and to other (added or accidental) constituents. Though commonly neglected, it was shown earlier that variations in stoichiometry often have their major effects via impacts on *G*, e.g. B_4C (Fig. 4.11). However, it appears there is also an intrinsic effect of stoichiometry beyond that via effects on *G*. This is more strongly indicated in TiC, where there is again some tendency for *G* to decrease with increasing C/Ti ratio [74]. A related compositional factor is phase content, which can be quite important (e.g. in Si_3N_4, Fig. 4.13), but substantially less in other materials, e.g. SiC.

The presence of added constituents of the same or different species of the body can also be important, with the former being an important factor in stoichiometry. Greater effects are indicated when the added constituents result in a second phase versus being in solid solution. Thus modest effects are indicated by solid solutions with Al_2O_3 [95–98], and possible greater effects from precipitates [100], e.g. due to excess Al_2O_3 in $MgAl_2O_4$ precipitating out in single crystals [101]; but much more study is needed. Similarly, effects of grain boundary phases can be significant, e.g. as in Co bonded WC (Fig. 4.5). Note that precipitation may also preferentially occur at free surfaces, as indicated in $MgAl_2O_4$ [101] and can also preferentially occur at or near grain boundaries. The above phenomena may interact with the *G* dependence of *H* in different intrinsic and extrinsic fashions. Thus both stoichiometric variations and second phase versus solid solution are typically in part (nonuniquely) related to grain size due to interactive effects of starting particle size and temperature on *G*. Totally unexplored is these compositional effects on hardness-related cracking and its *G* dependence.

One of the major limitations of *H–G* data, as in all microstructural property dependence, is microstructural characterization beyond that of composition noted above. Basic to this is the measurement of *G* itself, which presents three problems: (1) considerable *H* data is limited in its use because no specific *G* values are given, (2) besides uncertainties of factors such as grain shape, difficulties arise due to different (often unspecified) measurement methods, and (3) conversions to a "true" grain size, with both the definition of "true" and the conversion factors commonly not being specified. A fundamental issue is that the slip aspect of the *G* dependence of *H* would indicate use of a true three-dimensional or volume measurement of *G*, while the associated cracking may be more consistent with a two-dimensional surface grain size, requiring more basic understanding. Whether *G* represents an average as-measured grain diameter or a linear intercept value, or such values multiplied by a factor to give a "true" three-dimensional *G*, *G* values commonly have differences of 50–100% and possibly as much as 200%. Further, some idea of the *G* variation, i.e., range or standard deviation, is not given. (Ideally, *G* would be measured where the indent is made.)

Similarly, limits in other microstructural characterization, e.g. of grain shape and orientation, and of porosity and other constituents, are a probable source of data variations. Porosity is clearly of importance (e.g. Figs. 4.2 and

4.5) [12,15,16] but again presents a number of problems of incomplete characterization. While the amount of porosity is often given, its homogeneity, size, and relation to the grain structure are often not given. These latter issues can be important, since indents cover small areas, so heterogeneities in porosity can give "bad" (typically discarded) indents, biasing results. Such problems are more likely to occur with limited levels of porosity and medium to larger pores.

Most data is in the conventional *G* range of 1 micron and larger. There can be contamination problems in obtaining dense bodies with nanoscale grains (e.g. Chap. 3, Sec. V.A), which often get more severe as *G* decreases, which may be the source of *H* decreasing with decreasing nanoscale *G* for TiAl (Fig. 4.5). However, other bodies with such fine *G* are reasonably consistent with larger *G* body data. This is clearly the case for WC+ 10% Co [80,81] with *G* down to 300 nm (Fig. 4.5) and appears consistent with TiO_2 data [29,31] when corrected for porosity. More recent data of Vaßen and Stöver [119] for SiC (mainly β) with *G* down to ~ 80 nm with limited oxide boundary impurities and > 95% theoretical density showed H_V (10 N) increasing from 23–24 GPa at *G* ~ 0.3–5 μm to 24–27 GPa at *G* ~ 100 nm (= 0.1 μm) and also supports continuation of *H*–*G* trends to finer *G* with suitable purity and density. Their earlier data [120] covered a broader range and clearly followed a $G^{-1/2}$ dependence.

B. Basic Mechanisms of Cracking

Two closely related key issues are the *H* minima and related indent cracking as a function of *G*. Materials of moderate to high hardness for which the most comprehensive range of *H*–*G* data are available, e.g. Al_2O_3, (Figs. 4.1 and 4.2), MgO (Fig. 4.4) , $MgAl_2O_4$ (Figs. 4.8 and 4.9), and ZnS (Fig. 4.15), almost always have an *H* minimum at intermediate *G*, e.g., in the 10–50 μm range, especially as indent load increases. Despite more limited data, this trend for *H* is also indicated for hard nonoxides, e.g. TiB_2 (Fig. 4.10), SiC (Fig. 4.12), and Si_3N_4 (Fig. 4.13) extending the *H* and bonding range of the minima. Thus as *G* decreases from single crystals ($G = \infty$), *H* first decreases (instead of a steady increase for a simple Petch relation) and then increases with further decreasing *G*. Single crystal or large *G* values are generally higher in comparison with intermediate and often some finer *G* values though the single crystal values are therefore typically obtained on only one or two low index crystallographic planes, which usually do not represent the highest crystal *H* values. Such minima tend to be more pronounced with higher loads, and probably Vickers vs. Knoop indents. The occurrence and nature of such minima thus explain what were previously seen as anomalous hardness dependence, e.g. of Armstrong, et al.'s [8] BeO data (Fig. 4.3).

The correlation of cracking with the *H* minima suggests it causes the minima, e.g. when *G* ~ 1–3 times the indent diagonal (Table 4.2), superimposed on the normal Hall–Petch H–$G^{-1/2}$ relationship based on microplasticity. Though there is considerable variation in the scope and character of such cracking, grain boundary fracture is dominant whenever boundaries are near the indents indicating varying (statistical) grain misorientation, residual boundary porosity, and phases being factors. The importance of grain boundaries as well as of impurities is clearly shown by effects of grain boundary phases known to enhance intergranular fracture and lower fracture energy/toughness, i.e. residues from use of LiF in densifying some MgO and $MgAl_2O_4$ greatly enhance grain boundary fracturing around indents (Fig. 4.20). The broad ocurrence of such cracking over various material types and characters, e.g. from some softer oxides to much harder oxides and nonoxides, further reinforces the roll of cracking, again generally along grain boundaries. This correlation of cracking, especially along grain boundaries, is reinforced by such cracking being a maximum when the indent size is ~ G, i.e. when indent stresses are high on local grain boundaries of larger grains. Limited other observations support such cracking, e.g. when the indent size was ~ G in Al_2O_3 [121]. Very limited data indicates cracking occurs mainly on loading versus unloading, consistent with the former presumably having greater H effect; but this is another area for further research.

The importance of grain boundaries in such cracking suggests factors affecting grain boundary fracture as basic material properties impacting the process. In harder materials this suggests factors such as the degrees of elastic anisotropy (EA) and thermal expansion anisotropy (TEA), and deformation–boundary effects. Thus the apparent absence of such cracking in Y_2O_3 may reflect material property effects (e.g. as suggested by apparent avoidance of intergranular fracture in it, Chap. 2, Sec. II.D [122]), as does the apparent absence of such cracking in many softer materials. However, data indicates a probable absence of such cracking in less refractory, much softer materials such as some halides and chalcogenids, as might be expected from the easier and generally greater plastic flow in softer materials. While these cannot be fully sorted out, the following evaluation and summary in Table 4.5 provide some guidance.

Both TEA and EA are probable factors, since they can lead to considerably varying stress concentrations at grain boundaries. TEA clearly begins at material/body-dependent G values, and EA is believed to have similar dependences [123], explaining the apparent absence of indent-related cracking at fine G. TEA, which occurs only in noncubic materials and clearly commonly leads to grain boundary fracture (Chap. 2, Sec. II.C), is probably a factor in conjunction with local indent-induced stresses. Thus the more extreme H minima indicated for TiB_2 versus Al_2O_3 and BeO, which have lower TEA, suggest TEA as a factor. However, αSiC (noncubic) has low TEA and β (cubic) SiC no TEA [121] but a substantial H minima. On the other hand, SiC and other cubic materials, e.g. MgO, $MgAl_2O_4$, and ZnS, all have substantial EA (as does TiB_2) [124,125] and substantial H minima indicating a nonexclusive role of TEA due to EA impact.

TABLE 4.5 Hardness Minima, Associated Grain Size, and Related Material Properties

Material	G (μm)	H xl[a]	Hmin[a] (load, kg)	H xl/ H min	EA[b] (%)	TEA ($\Delta\alpha$ max)[c]
KCl	—	0.01 (V)	—	—	12	0
BaF_2	—	0.7 (V)	—	—	0.1	0
ZnSe	—	0.8 (V)	—	—	11.7	0
ZnS	20–30	1.8 (V)	1.4 (1-10)	0.78	8.4	0
MgF_2	?	4 (K)	?			0.4
MgO	12	11 (K)	9.8 (0.1)	0.89	} 2.3	0
MgO	30	8.5 (K)	7 (0.5)	0.82		
MgO	30	10 (V)	9.2 (0.1)	0.92		
MgO	25	8 (V)	5.5 (0.5)	0.69		
Y_2O_3	—	7 (V)			?	0
BeO	100	10.8 (V)	8 (1,2)	0.74	0.4	1.1
ZrO_2	50	13 (V)	11.5 (0.5)	0.88	6-10	0
$MgAl_2O_4$	100	14 (K)	13.5 (0.5)	0.96	} 7	0
$MgAl_2O_4$	5–14	13.8 (K)	9.6 (1)	0.70		
$MgAl_2O_4$	60–400	14.5 (V)	14 (0.5)	0.96		
$MgAl_2O_4$	20–60	14.5 (V)	13.2 (1)	0.91		
Al_2O_3	50	26.6 (K)	24.5 (0.1)	0.92	} 1.7	0
Al_2O_3	60–100	19.7 (K)	15.8 (0.5)	0.80		
Al_2O_3	—	22.8 (V)	— (0.1)	—		
Al_2O_3	100	19.5 (V)	17 (0.5)	0.87		
TiB_2	14	32 (K)	18 (0.1)	0.56	0.5	3
α-Si_3N_4	18	40 (V)	36 (0.1)	0.90	?	0.1 (β = 0.5)
SiC	10	24.5 (K)	20 (1)	0.82	7.3	0 (α - 0.4)

[a]Hardnesses of xl = single crystal and min. = polycrystalline minimum, both in GPa, from Figs. 4.1–4.15.
[b]Values from Refs. 124 and 125.
[c]Values from Refs 123, 126, and 127.

However, a single direct correlation with EA is questioned by the apparent absence of an H minimum in ZnSe, which has one of the highest EA levels, e.g. ~ 70% > than for ZnS. However, the lower hardness of ZnSe relative to ZnS may enhance plasticity in ZnSe relative to ZnS to the extent that it suppresses grain boundary cracking in ZnSe [whose strength may be controlled by microplasticity (Fig. 3.2)] but not in ZnS. One further factor that is likely to affect such local indent cracking is the anisotropy of slip itself, since fewer slip systems, i.e. greater slip anisotropy (e.g. in MgO [6]), usually result in more stress concentration at blocked slip bands at grain boundaries. This would also be another reason for reduced local cracking as G decreases below G values for the H minima. Thus four factors are seen as probable material parameters in such local indent cracking. The first three are increased slip anisotropy and resultant boundary stress concentrations and the enhancement of these due to TEA in noncubic materials, and especially and more generally EA in all materials, but often more pronounced in cubic materials. Countering these three stress concentration factors is increased plastic flow to relax stress concentrations, with this generally correlating with lower hardness.

Again, both expectations and limited data indicate that much of the cracking occurs during the indent formation, rather than during or after unloading. However, this is an important area for further study, since understanding when cracking is occurring is important ultimately to fully interrelate *H* with other physical properties, and it may have important implications for wear, particle erosion, and machining phenomena in ceramics in view of their close relation to indentation effects. The *H* minimum due to local cracking clearly has important implications regarding use of indent flaws for fracture toughness testing, i.e., potentially precluding use of this over a range of intermediate grain sizes, and it may be related to changing flaw sizes from machining as grain size changes (Chaps. 3, 8, and 12).

IV. SUMMARY AND CONCLUSIONS

The room temperature H_V–H_K–$G^{-1/2}$ trends of a variety of generally dense oxide and nonoxide ceramics covering considerable *H* and *G* ranges (including single crystals where feasible) show two related material trends. First is the expected basic Petch-type $G^{-1/2}$ dependence that is commonly found mainly in softer materials, e.g. alkali halides. The other is a deviation from the Petch relation via a superimposed *H* minimum at intermediate grain sizes. Thus *H* instead initially decreases from single crystal or large *G* values with decreasing *G* vs. the generally accepted trend for *H* to increase continuously with decreasing *G* (e.g. a $G^{-1/2}$ dependence), which is approached at finer grain sizes in this case. This second case is most commonly found in most, but not necessarily all, harder ceramics. The *H* minimum at intermediate *G*, which is dependent some on indent geometry and increases in extent and probably in the *G* at which it occurs with indent load, explains anomalous, non-Petch *H*–*G* trends previously observed. The overall *H*–$G^{-1/2}$ dependence in all materials as well as single crystal hardness anisotropy are both consistent with plastic deformation by slip as the fundamental mechanism of forming hardness indents.

The *H* minimum at intermediate *G* is associated with a maximum of indent related cracking, often of a spalling character along grain boundaries, so the latter is the probable cause of the former. Such cracking tends toward a maximum, i.e. an *H* minimum, when the indent and grain sizes are similar. The extent of this *H* minimum tends to be greater for Vickers vs. Knoop indents and as the load increases. It also shows considerable variability, e.g. a tendency to be exacerbated by residual grain boundary additives, impurities, and porosity, and a probable dependence on local statistical variations of grain orientations and factors such as elastic anisotropy (EA) and thermal expansion anisotropy (TEA). More fundamentally, the *H* minima and related local cracking are probably due to combinations of TEA, EA, and slip anisotropy driving the process (mainly via grain boundary stress concentrations), and the extent of deformation (generally inversely related to *H*) and related boundary stress relaxation limiting its occurrence. Other parameters such as stoichiometry may or may not be important; they cannot be sorted out without comparing microstructures, since compositional changes commonly change *G*, which can be the major mechanism of their affecting *H*.

Thus a variety of often incompletely determined factors impact *H–G* relations and need better definition and understanding. Effects of indent type are better, but not fully, understood. Load dependence and its relation to surface finish and subsurface effects (cracking and especially work hardening) need further attention. The mechanism of indent related cracking, i.e. both the issues of where it occurs in the indenting cycle and its intrinsic and extrinsic parameter dependence need much more attention. More thorough characterization of materials is clearly needed to assure full utility of *H* data and better understanding. Better *G* measurements are a key need, along with more attention to grain shape and orientation (though single crystal data provides important information on limits of the latter).

REFERENCES

1. B. C. Wonsiewicz and G. Y. Chin. A Theory of Knoop Hardness Anisotropy. The Science of Hardness Testing and Its Research Applications (J. H. Westbrook and H. Conrad, eds.). American Society for Metals, Metals Park, OH, pp. 167–173, 1973.
2. R. W. Armstrong and A. C. Raghuram. Anisotropy of Microhardness in Crystals. American Society for Metals, Metals Park, OH, pp. 174–186.
3. C. A. Brookes and R. P. Burnand. Hardness Anisotropy in Crystalline Solids. American Society for Metals, Metals Park, OH, pp. 199–211.
4. D. I. Golland and G. Mayer. The Characterization of Deformation Bands in Iron Single Crystals by Microhardness Analysis. American Society for Metals, Metals Park, OH, pp. 212–222.
5. A. G. Atkins. High-Temperature Hardness and Creep. American Society for Metals, Metals Park, OH, pp. 223–240.
6. J. J. Gilman. Hardness—A Strength Probe. American Society for Metals, Metals Park, OH, pp. 51–74.
7. R. W. Rice, C. Cm. Wu, and F. Borchelt. Hardness-Grain-Size Relations in Ceramics. J. Am. Cer. Soc. 77(10): 2539–2553, 1994.

8. R. W. Armstrong, E. L. Raymond, and R. R. Vandervoort. Anomalous Increase in Hardness with Increase in Grain Size of Beryllia. J. Am. Cer. Soc. 53(9):529–530, 1970.
9. P. M. Sargent and T. F. Page. The Influence of Microstructure on the Microhardness of Ceramic Materials. The Mechanical Engineering Properties and Applications of Ceramics (D. J. Godfrey, ed.). British Ceramic Society, Stoke on Trent, UK, 1978, pp. 209–224.
10. T. Tani, Y. Miyamoto, M. Koizumi, and M. Shimada. Grain Size Dependencies of Vickers Microhardness and Fracture Toughness in Al_2O_3 and Y_2O_3 Ceramics. Ceram. Intl. 12:33–37, 1986.
11. K. Niihara and T. Hirai. Chemical Vapour-Deposited Silicon Nitride. J. Mat. Sci. 12:1243–1252, 1977.
12. R. W. Rice. Microstructure Dependence of Mechanical Behavior of Ceramics. Treatise on Materials Science and Technology 2 (R. C. McCrone, ed.). Academic Press, New York, 1977, pp. 199–238.
13. C. Greskovich and H. C. Yeh. Hardness of Dense α-Si_3N_4. J. Mat. Sci. Lett. 2:657–659, 1983.
14. C. Greskovich and G. E. Gazza. Hardness of Dense α- and β-Si_3N_4 Ceramics. J. Mat. Sci. Lett. 4:195–196, 1985.
15. R. W. Rice. Porosity of Ceramics. Marcel Dekker, New York, 1998.
16. R. W. Rice. Comparison of Physical Property–Porosity Behavior with Minimum Solid Area Models. J. Mat. Sci. 31:1509–1528, 1996.
17. H. Li and R. C. Bradt. The Microhardness Indentation Load/Size Effect in Rutile and Cassiterite Single Crystal. J. Mat. Sci. 28:917–926, 1993.
18. J. B. Quinn and G. D. Quinn. Indentation Brittleness of Ceramics: A Fresh Approach. J. Mat. Sci. 32:4331–4346, 1997.
19. P. F. Becher. Surface Hardening of Sapphire and Rutile Associated with Machining. J. Am. Ceram. Soc. 57(2):107–108, 1974.
20. R. W. Rice. Machining, Surface Work Hardening, and Strength of MgO. J. Am. Cer. Soc. 56 (10):536–541, 1973.
21. S. D. Skrovanek and R. C. Bradt. Microhardness of a Fine-Grain-Size Al_2O_3. J. Am. Ceram. Soc. 62 (3–4):213–214, 1979.
22. W. Kollenberg. Plastic Deformation of Al_2O_3 Single Crystals by Indentation at Temperatures up to 750°C. J. Mat. Sci. 23:3321–3325, 1988.
23. J. Lankford. Compressive Strength and Damage Mechanisms in Ceramic Materials. Southwest Research Institute Interim Technical Report, Office of Naval Research, Contract N00014-75-C-068, 2/15 1980.
24. C. P. Alpert, H. M. Chan, S. J. Bennison, and B. R. Lawn. Temperature Dependence of Hardness of Alumina-Based Ceramics. J. Am. Cer. Soc. 71(8):C-371–373, 1988.
25. D. J. Clinton and R. Morrell. The Hardness of Alumina Ceramics. Proc. Brit. Cer. Soc. 34:113–127, 1984.
26. A. Krell` and P. Blank. Grain Size Dependence of Hardness in Dense Submicrometer Alumina. J. Am. Cer. Soc. 78(4):1118–1120, 1993.

27. H. Li and R. C. Bradt. Knoop Microhardness Anisotropy of Single-Crystal Rutile. J. Am. Cer. Soc. 73(5):1360–1364, 1990.
28. M. J. Mayo, R. W. Siegel, A. Narayanasamy, and W. D. Nix. Mechanical Properties of Nanophase TiO_2 as Determined by Nanoindentation. J. Mater. Res. 5(5):1073–1082, 1990.
29. R. S. Averback, H. J. Hofler, H. Hahn, and J. C. Logas. Sintering and Grain Growth in Nanocrystalline Ceramics. Nanostruct. Mater. 1:172– , 1992.
30. R. A. Andrievski. Review: Nanocrystalline High Melting Point Compound-Based Materials. J. Mat. Sci. 29:614–631, 1994.
31. M. Guermazi, H. J. Höfler, H. Hahn, and R. S. Averbach. Temperature Dependence of the Hardness Dependence of Nanocrystalline Titanium Dioxide. J. Am. Cer. Soc. 74(10):2672–2674, 1991.
32. S. K. Dutta and G. E. Gazza. Transparent Y_2O_3 by Hot-Pressing. Mat. Res. Bull. 4:791–796, 1969.
33. G. Fantozzi, G. Orange, K. Liang, M. Gautier, J. P. Duraud, P. Marie, and C. E. Le Gressus. Effect of Nonstoichiometry on Fracture Toughness and Hardness of Yttrium Oxide Ceramics. J. Am. Ceram. 72(8):1562–1563, 1989.
34. W. Rhodes. GTE Corp. Private communication, 1990.
35. R. J. Cook. Lateral Cracks and Microstructural Effects in the Indentation Fracture of Y_2O_3. J. Am. Cer. Soc. 73(7):1873–1878, 1990.
36. A. Pajarec, F. Guiberteau, A. Dominguez-Rodriguez, and A. H. Heuer. Microhardness and Fracture Toughness Anisotropy in Cubic Zirconium Oxide Single Crystals. J. Am. Ceram. 71(7):C-332–333, 1988.
37. J. K. Cochran, K. O. Legg, and G. R. Baldau. Microhardness of N-Implanted Yttria Stabilized ZrO_2. Emergent Process Methods for High-Technology Ceramics. Plenum Press, New York, 1984, pp. 549–557.
38. M. Watanabe and K. Komaki. NGK Spark Plug Co. Private communication, 1981.
39. C. Cm. Wu and R. W. Rice. Wear and Related Evaluation of Partially Stabilized ZrO_2. Cer. Eng. and Sci. Proc. 6(7–8):1012–1022, 1985.
40. S. Higuchi, Y. Takeda, K. Maeda, and T. Miyoshi. Effect of Reducing Grain Size on Mechanical Properties of Stabilized ZrO_2 Ceramics. J. Ceram. Soc. Jpn., Intl. Ed. 96:979–984, 1988.
41. B. Chiou, J. J. Hwang, and J. Duh. The Influence of MgO Addition on the Mechanical Properties of Yttria Stabilized Zirconia. High Tech. Cer. (P. Vincenzini, ed.). Elsevier, Amsterdam, The Netherlands, 1987, pp. 1225–1232.
42. R. Ramadas, S. C. Mohan, and S. R. Reddy. Studies on the Metastable Phase Retention and Hardness in Zirconia Ceramics. Mat. Sci. Eng. 60:67–72, 1983.
43. R. W. Rice. Comment on "Studies on the Metastable Phase Retention and Hardness in Zirconia Ceramics." Mat. Sci. Eng. 73:215–217, 1985.
44. J. G. Duh and J. U. Wan. Developments in Highly Toughened CeO_2-Y_2O_3-ZrO_2 Ceramic System. J. Mat. Sci. 27:6197–6203, 1992.
45. L. Ruiz and M. J. Readey. Effect of Heat Treatment on Grain Size, Phase Assemblage, and Mechanical Properties of 3 mol% Y-TZP. J. Am. Cer. Soc. 79(9):2331–2340, 1996.

46. K. F. H. Ahlborn, Y. Kagawa, and A. Okura. Observation of the Influence of Microcracks on the Crack Propagation Inside of Transparent ZrO_2. Ceramics Today—Tomorrow's Ceramics 66C (P. Vincenzini, ed.). Elsevier Science Publishers B.V., 1991, pp. 1857–1864.
47. Y. Akimune and R. C. Bradt. Knoop Microhardness Anisotropy of Single-Crystal Stoichiometric $MgAl_2O_4$ Spinel. J. Am. Ceram. Soc. 70(4):C-84–86, 1987.
48. W. H. Rhodes, P. L. Berneburg, and J. E. Niesse. Development of Transparent Spinel ($MgAl_2O_4$). Avco Corp. Report for Army Contract AMMRC-CR-70-19, 10/1970.
49. P. P. Budnikov, F. Kerbe, and F. J. Charitonov. The Hot Pressing of Aluminum-Magnesium Spinel ($MgAl_2O_4$). Sci. Cer. 4:69–78, 1968.
50. K. Okazaki and K. Nagata. Effects of Density and Grain Size on the Elastic and Piezoelectric Properties of $Pb(Zr\text{-}Ti)O_3$ Ceramics. Mechanical Behavior of Materials, Proc. Intl. Conf. on Mechanical Behavior of Materials 4, Kyoto, 8/15–20/1971. Soc. Matls. Sci., Kyoto, Japan, 1972, pp. 404–412.
51. P. D. Zavitsanos and J. R. Morris, Jr. Synthesis of Titanium Diboride by a Self-Propagating Reaction. Cer. Eng. Sci. Proc. 4(7–8):624–633, 1983.
52. R. D. Koester and D. P. Moak. Hot Hardness of Selected Borides, Oxides and Carbides to 1900°C. J. Am. Cer. Soc. 6:290–296, 1967.
53. F. W. Vahldiek and S. A. Mersol. Slip and Microhardness of IVa to VIa Refractory Materials. J. Less Com. Met. 55:265–278, 1977.
54. D. R. Flinn, J. A. Kirk, M. J. Lynch, and B. G. Van Steatum. Wear Properties of Electrodeposited Titanium Diboride Coatings. US Dept. of the Interior, Report of Investigations, 1981.
55. F. X. McCawley, D. Schlain, and G. R. Smith. Electrodeposition of Titanium Diboride. Am. Cer. Soc. Bull. 5(4):349, 1972.
56. K. Nakano, H. Matsubara, and T. Imura. High Temperature Hardness of Titanium Diboride Single Crystal. Jp. J. Appl. Phys. 13(6):1005–1006, 1974.
57. Y. Miyamoto and M. Koizumi. High Pressure Self-Combustion Sintering for Ceramics. J. Am. Cer. Soc. 67(11):C-224–223, 1984.
58. T. Watanabe and S. Kouno. Mechanical Properties of TiB_2-CoB-Metall Boride Alloys. Am. Cer. Soc. Bul. 61(9):970–973, 1982.
59. D. Kalish, E. V. Clougherty, and J. Ryan. Fabrication of Dense Fine Grained Ceramic Materials. ManLabs, Inc. Final Report for US Army Materials Research Laboratory, 11/1966.
60. Z. A. Munir and G. R. Veerkamp. An Investigation of the Parameters Influencing the Microstructure of Hot-Pressed Boron Carbide. Univ. of Calif. Dept. Mechan. Eng., Davis, CA, Progress Report, 11/1975.
61. K. Niihara. Slip Systems and Plastic Deformation of Silicon Carbide Single Crystals at High Temperatures. J. Less Com. Met. 65:155–166, 1979.
62. G. R. Sawyer, P. M. Sargent, and T. F. Page. Microhardness Anisotropy of Silicon Carbide. J. Mat. Sci. 15:1001–1013, 1980.
63. O. O. Adewoye and T. F. Page. Anisotropic Behavior of Etched Hardness Indentations. J. Mat. Sci. 11:981–984, 1976.
64. S. Fujita, K. Maeda, and S. Hyodo. Anisotropy of High-Temperature Hardness in 6H Silicon Carbide. J. Mat. Sci. Lett. 5:450-452, 1986.

65. I. J. McColm. Ceramic Hardness. Plenum Press, New York, 1990.
66. M. G. S. Naylor and T. F. Page. Microhardness, Friction and Wear of SiC and Si_3N_4 Materials as a Function of Load, Temperature and Environment. First Annual Tech. Rept. US Army Grant DA-ERO-78-G-010, 10/1979.
67. T. F. Page, G. R. Sawyer, O. O. Adewoye, and J. J. Wert. Hardness and Wear Behavior of SiC and Si_3N_4. Proc. Brit. Cer. Soc. 26:193–208, 1978.
68. T. Hiari and K. Niihara. Hot Hardness of SiC Single Crystal. J. Mat. Sci. 14:2253–, 1979.
69. D. Chakraborty and J. Mukerji. Effect of Crystal Orientation, Structure and Dimension of Vickers Microhardness Anisotropy of β, α-Si_3N_4, α-SiO_2 and α-SiC Single Crystals. Mat. Res. Bull. 17:843–849, 1982.
70. Y.-W. Kim, S.-W. Park, and J.-G. Lee. Composition and Hardness of Chemically Vapour-Deposited Silicon Carbide with Various Microstructures. J. Mat. Sci. Lett. 14:1201–1203, 1995.
71. D.-J. Kim and D.-J. Choi. Microstructure and Surface Roughness of Silicon Carbide by Chemical Vapour Deposition. J. Mat. Sci. Lett. 16: 286–289, 1997.
72. D. J. Rowcliffe and G. E. Hollox. Hardness Anisotropy, Deformation Mechanisms and Brittle-to-Ductile Transition in Carbides. J. Mat. Sci. 6:1270–1276, 1971.
73. R. H. J. Hannick, D. L. Kohlstedt, and M. J. Murray. Slip System Determination in Cubic Carbides by Hardness Anisotropy. Proc. Roy. Soc. Lond. A. 326:409–420, 1972.
74. D. B. Miricale and H. A. Lipsitt. Mechanical Properties of Fine-Grained Substoichiometric Titanium Carbide. J. Am. Cer. Soc. 66(8):592–597, 1983.
75. Y. Kumashiro, A. Itoh, T. Kinoshita, and M. Sobajima. The Micro-Vickers Hardness of TiC Single Crystals up to 1500°C. J. Mat. Sci. 12:595–601, 1977.
76. S. Shimada, J. Watanabe, and K. Kodaira. Flux Growth and Characterization of TiC Crystals. J. Mat. Sci. 24:2513–2515, 1989.
77. I. Cadoff, J. P. Nielsen, and E. Miller. Properties of Arc-Melted vs. Powder Metallurgy Titanium Carbide. Warmfeste und Karrosconfestandige Sinterwerkstoffe, Springer-Verlag, Vienna, Austria, 1956, p. 712.
78. O. Yamada, Y. Miyamoto, and M. Koizumi. High-Pressure Self-Combustion Sintering of Titanium Carbide. J. Am. Cer. Soc. 70(9):C-206–208, 1987.
79. A. Leonhardt, D. Schlafer, M. Seidler, D. Selbmann, and M. Schonherr. Microhardness and Texture of TiC Layers on Cemented Carbides. J. Less Com. Met. 87:63–69, 1982.
80. H. C. Lee and J. Gurland. Hardness and Deformation of Cemented Tungsten Carbide. Mat. Sci. Eng. 33:125–133, 1978.
81. L. E. McCandlish, B. H. Kear, and B. K. Kim. Processing and Properties of Nanostructured WC-Co. Nanostrucrured Matls. 1:119–122, 1992.
82. W. Rafaniello. Development of Aluminum Nitride: A New Low-Cost Armor. Dow Chem. Co. Final Report for US Army Research Office Contract DAAL03-88-C-0012, 12/1992.
83. R. F. Coe, R. J. Lumby, and M. F. Pawson. Some Properties and Applications of Hot-Pressed Silicon Nitride. Properties and Applications of Silicon Nitride,

Special Ceramics No. 5 (P. Popper, ed.). British Ceramic Society, Stoke on Trent, UK, 1977, pp. 361–376.

84. K. Tsukuma, M. Shimada, and M. Koijumi. Thermal Conductivity and Microhardness of Si_3N_4 with and Without Additives. Am. Cer. Bull. 60(9):910–912, 1981.

85. Y. Miyamoto, M. Koizumi, and O. Yamada. High-Pressure Self-Combustion Sintering for Ceramics. J. Am. Cer. Soc. 67(11):C-224–233, 1984.

86. A. K. Mukhopadhyay, S. K. Datta, and D. Chakraborty. Hardness of Silicon Nitride and Sialon. Cer. Intl. 17:121–127, 1991.

87. A. K. Mukhopdhay, S. K. Dutta, and D. Chakrborty. On the Microhardness of Silicon Nitride and Sialon Ceramics. J. Eur. Cer. Soc. 6:303–311, 1990.

88. P. J. Burchill. Hardness Anisotropy of α-Si_3N_4 Single Crystal. J. Mat. Sci. 13:2276–2278, 1978.

89. D. Chakraborty and J. Mukerji. Characterization of Silicon Nitride Single Crystals and Polycrystalline Reaction Sintered Silicon Nitride by Microhardness Measurements. J. Mat. Sci. 15:3051–3056, 1980.

90. K. Ueno. Microstructure Dependence of Fracture Toughness and Hardness of Silicon Nitride. Yogyo-Kyokai-Shi 97(1):85–87, 1989.

91. M. C. Shaw. The Fundamental Basis of the Hardness Test. The Science of Hardness Testing and Its Research Applications (J. H. Westbrook and H. Conrad, eds.). Am. Soc. for Metals, Metals Park, OH, 1973, pp. 1–11.

92. R. F. Bunshah and R. W. Armstrong. Continuous Ball Indentation Test for Examining Hardness Dependence on Indentor Size, Indentation Size, and Material Grain Size. Am. Soc. for Metals, Metals Park, OH, 1973, pp. 318–328.

93. J. R. Floyd. Effects of Firing on the Properties of Dense High-Alumina Bodies. Trans. Brit. Cer. Soc. 64:251–265, 1965.

94. H. E. Exner and J. Gurland. A Review of Parameters Influencing Some Mechanical Properties of Tungsten Carbide Cobalt Alloys. Powder, Metallurgy 13(25):13031, 1970.

95. L. Belon, H. Forestier, and Y. Bigot. The Hardness of Some Solid Solutions of Alumina. Special Ceramics 4 (P. Popper, ed.). British Ceramic Research Association, Stoke on Trent, 1968, pp. 203–209.

96. F. Albrecht. Anisotropy of the Hardness of Synthetic Corundum. Z. Krist. 106:183–190, 1954.

97. R. C. Bradt. Cr_2O_3 Solid Solution Hardening of Al_2O_3. J. Am. Cer. Soc. 50(1):54–55, 1967.

98. K. Shinozaki, Y. Ishikura, K. Uematsu, N. Mozutani, and M. Kato. Vickers Micro-Hardness of Solid Solution in the System Cr_2O_3- Al_2O_3. J. Mat. Sci. Lett. 15:1314–1316, 1980.

99. H. Chang, H. J. Höfler, C. J. Altstetter, and R. S. Averback. Synthesis, Processing and Properties of Nanophase TiAl. Scripta Met. Mat. 25:1161–1166, 1991.

100. R. J. Bratton. Precipitation and Hardening Behavior of Czochralski Star Sapphire. J. Appl. Phys. 42(1):211–216, 1971.

101. D. Lewis, B. A. Bender, R. W. Rice, J. Homeny, and T. Garino. Precipitation and Toughness in Alumina-Rich Spinel Crystals. Fracture Mechanics of Ceramics 8.

Microstructure, Methods, Design, and Fatigue (R. C. Bradt, A. G. Evans, D. P. H. Hasselman, and F. F. Lange, eds.). Plenum Press, New York, 1986, pp. 61–67.
102. W. F. Brace. Dependence of Fracture Strength of Rocks on Grain Size. Penn. State U. Mineral Experimental Station Bull. 76:99–103, 1963.
103. W. F. Brace. Brittle Fracture of Rocks. Intl. Conf. on States of Stress in the Earth's Crust—Preprints of Papers (W. R. Judd, ed.). Rand Corp. Report, 1963.
104. K. L. Lewis, J. A. Savage, K. J. Marsh, and A. P. C. Jones. Recent Developments in the Fabrication of Rare-Earth Chalcogenide Materials for Infrared Optical Applications. New Opt. Mat., SPIE Conf. Proc., 400, Soc. of Photo-Optical Instrumentation Enginers, Bellingham, WA, 1983.
105. K. L. Lewis, A. M. Pitt, J. A. Savage, J. E. Field, and D. Townsend. The Mechanical Properties of CVD-Grown Zinc Sulphide and Their Dependence on the Conditions of Growth. 9th Intl. Conf. on CVD, ECS Proc. 84(6):530–545, 1984.
106. D. Townsend and J. E. Field. Fracture Toughness and Hardness of Zinc Sulphide as a Function of Grain Size. J. Mat. Sci. 25:1347–1352, 1990.
107. E. Lendvay and M. V. Fock. Microhardness Anisotropy in Cubic and Hexagonal ZnS Single Crystals. J. Mat. Sci. 4:747–752, 1969.
108. A. W. Swanson and J. Pappis. Application of Polycrystalline ZnSe Prepared by Chemical Vapor Deposition to High Power IR Laser Windows. (Raytheon Co.) AFML Tech. Report TR-75-170, 1975.
109. J. Lankford. Comparative Study of the Temperature Dependence of Hardness and Compressive Strength in Ceramics. J. Mat. Sci. 18:1666–1674, 1983.
110. G. R. Anstis, P. Chantikul, B. R. Lawn, and D. R. Marshall. A Critical Evaluation of Indentation Techniques for Measuring Fracture Toughness: I. Direct Crack Measurements. J. Am. Cer. Soc. 64(9):533–538, 1981.
111. T. Sperisen, C. Carry, and A. Mocellin. Microfracture Behavior of Fine Grained Alumina Studied by Indentation and Acoustic Emission in Various Environments. Fracture Mechanics of Ceramics, Microstructure, Methods, Design, and Fatigue (R. C. Bradt, A. G. Evans, D. P. H. Hasselman, and F. F. Lange, eds.) 8. Plenum Press, New York, 1986, p. 69–83.
112. Y. Kumashiro and E. Sauma. The Vickers Micro-Hardness of Non-Stoichiometric Niobium Carbide and Vanadium Carbide Single Crystals up to 1500°C. J. Mat. Sci. 15:1321–1324, 1980.
113. J. Dusza, T. Eschner, and K. Rundgren. Hardness Anisotropy in Bimodal Grained Gas Pressure Sintered Si_3N_4. J. Mat. Sci. Lett. 16:1664–1667, 1997.
114. D. G. Rickerby. Observations of the Hardness Anisotropy in MgO and LiF. J. Am. Cer. Soc. 62(3–4):222, 1979.
115. T. W. Button, I. J. McColm, and S. J. Wilson. Hardness Anisotropy and its Dependence on Composition in Sodium Tungsten Bronzes and Rhenium Trioxide Single Crystals. J. Mat. Sci. 14:159–164, 1979.
116. A. M. Lejus, D. Ballutaud, C. R. Kha, and J. Solide. Microdureté de Monocristaux de V_2O_3. Mat. Res. Bull. 15:95–102, 1980.
117. J. H. Westbrook and P. J. Jorgensen. Effects of Water Desorption on Indentation Microhardness Anisotropy in Minerals. Am. Min. 53:1899–1909, 11–12/1968.

118. E. T. Park, J. L. Routbort, Z. Li, and P. Nash. Anisotropic Microhardness in Single-Crystal and Polycrystalline $BaTiO_3$. J. Mat. Sci. 33:669–673, 1998.
119. R. Vaßen and D. Stöver. Manufacture and Properties of Nanophase SiC. J. Am. Cer. Soc. Submitted for publication.
120. R. Vaßen and D. Stöver. Properties of Silicon-Based Ceramics Produced by Hot Isostatic Pressing Ultrafine Powders. Phil. Mag. B 76(4):585–591, 1997.
121. G. R. Anstis, P. Chantikul, B. R. Lawn, and D. B. Marshall. A Critical Evaluation of Indentation Techniques for Measuring Fracture Toughness: I. Direct Crack Measurements. J. Am. Cer. Soc. 64(9):533–538, 1981.
122. C. Cm. Wu, R. W. Rice, and P. F. Becher. The Character of Cracks in Fracture Toughness Measurements of Ceramics. Fracture Mechanics Methods for Ceramics, Rocks and Concrete (S. W. Freiman and E. R. Fuller, Jr., eds.). ASTM STP 74, ASTM, Philadelphia, PA, 1982, pp. 127–140.
123. Z. Li and R. C. Bradt. Thermal Expansion and Thermal Expansion Anisotropy of SiC Polytypes. J. Am. Cer. Soc. 70(7):445–448, 1987.
124. R. W. Rice. Possible Effects of Elastic Anisotropy on Mechanical Properties of Ceramics. J. Mat. Sci. Lett. 13:1261–1266, 1994.
125. D. H. Chung and W. R. Bussem. The Elastic Anisotropy of Crystals. Anisotropy in Single-Crystal Refractory Compounds (F. W. Vahldiek and S. A. Merson, eds.). Plenum Press, New York, pp. 217–245, 1968.
126. R. W. Rice and R. C. Pohanka. Grain Size Dependence of Spontaneous Cracking in Ceramics. J. Am. Cer. Soc. 62(11–12):559–563, 1979.
127. E. C. Skaar and W. J. Croft. Thermal Expansion of TiB_2. J. Am. Cer. Soc. 56(1):45, 1973.

5

Grain Dependence of Compressive Strength, Wear, and Related Behavior at ~ 22°C

I. INTRODUCTION

This chapter addresses first the grain (mainly size) dependence of compressive strength, primarily uniaxial, σ_C, but data on compressive strengths with superimposed hydrostatic compression is also presented. This is followed first by a review of the very limited data on grain effects on ballistic, i.e. armor, performance; then the more substantial, but still limited, data is presented first on erosion by impacting particles and resultant body strength and then on sliding wear. Since there are many manifestations of wear due to the variety of loading and environmental conditions, the focus will be on basic grain effects rather than on details of variations in wear tests and use that can occur. Then grain size effects on abrasive machining of ceramics are also discussed.

The first of two common themes among the properties considered is that all involve substantial compressive stress with varying local tensile stresses (which are a factor in differences between the different properties). The second relation is that all these properties have considerable correlation with hardness, which has some relation to elastic moduli, e.g. Young's and bulk moduli [1,2]. The utility of these property correlations is that H is obtained by a simple test rapidly performed on small samples at moderate cost. While this correlation is both partly compromised by the often complex and variable dependence of hardness as a function of factors such as surface finish, microstructure, and indenter load and configuration, it can also be an aid in sorting out similar dependences of wear, erosion, and machining.

Consider now the basic mechanism(s) of the grain dependence of these properties, i.e. in crystalline materials, since they are the ones having their volume nearly or fully occupied by grains. A key factor is again local plastic deformation as associated with hardness indentations (Chap. 4). While there is some uncertainty in the scope and specifics, e.g. changes and details of its occurrence in compressive strength, erosion, wear, and machining, local plastic flow is clearly a factor in all of these. While this is known empirically, it also stems from similar mechanisms of these properties and hardness, since erosion, most machining, and much wear entail particles or asperities penetrating surfaces. Such penetration causes local plastic flow and fracture similar to that involved in hardness indentation (Chap. 4), which thus entail local H values and E and K values respectively, as well as load dependence for both phenomena. While ballistic performance entails some penetration and compressive strength does not, both have some correlation with hardness [1–5] (hence also generally with elastic moduli [1,2]). The correlation of these properties, especially compressive strength, with hardness stems from the fact that hardness is a measure of the local yield stress (Y), which for many materials obeys the simple relation $Y = H/C$, where C is referred to as the constraint factor and is typically (for ceramics) 2.5 to 3 [2,6]. (Where there are significant differences in the yield stress for deformation, i.e. slip or twinning systems, that do not allow general deformation, e.g. do not represent five independent slip systems, then the values of C may be different [2], especially where stresses are limited, e.g. in tension. Also, in materials such as silicate glasses and polymers, where different deformation mechanisms may occur, their "yield" stresses also tend to scale with E [6–8]). Thus the H–σ_c correlation is attributed to the yield stress ~ $H/3$ being the upper limit to the compressive strength of dense crystalline ceramics [3–5]. This correlation and that of other properties to H is a factor in their G dependence, but there are variations for each property.

While there is a basic correlation of σ_c with H, there are potential differences in their G dependence, as there are substantial differences and some similarities of σ_c with tensile strength, σ_T, as shown by the general character of the mechanisms involved. Though not documented or understood in detail, it is now generally recognized that in well conducted compressive tests, generation of substantial local tensile stresses occurs due to, and on the scale of, the microstructure, especially pores and probably grains, and second phase particles or regions. These local tensile stresses generate and grow cracks but at higher stresses than in tensile testing since the individual cracks are typically semistable in the macro compressive stress field. The high stresses reached allow microplastic processes to generate cracks, and possibly contribute some to their stability, growth, or both, thus providing the basis for correlation with H.

Further, while ultimate compressive failure is typically somewhat explosive due to the high stored elastic energy of the high stresses involved, the local, stable tensile cracking leads to cumulative failure, as opposed to much more immediate catastrophic failure from rapid growth of a single crack in tensile loading. Thus compressive failure is seen as the progressive generation, growth, and coalescence of small, initially micro, cracks, generated and grown by local tensile stresses due to stress heterogeneities, microplastic processes, or both, culminating in brittle failure, e.g. as indicated by acoustic emission [5,9–13]. These processes are the probable source of the *G* dependence of σ_c and the similarities and differences of its *G* dependence versus those of *H* and tensile strength, since these entail respectively more concentrated local deformation, cracking, or both and a lesser role of larger scale propagation of single cracks. Also, the load dependence of *H* and its local cracking appear to be sources of differences in the *G* dependences of *H* and σ_c, e.g. higher load *H* values are probably more pertinent, but *H* minima from local surface cracking probably have limited or no pertinence. Note also that there are various parasitic effects, e.g. end crushing or cracking due to end strain incompatibilities between loading and specimen surfaces, that can lead to premature compressive failure in testing and may vary with *G*.

The field of ballistic impact on ceramic materials is complicated first by complexities of the process and second by the fact that many of the test results are classified because of the important application to military armor [14,15]. Complexities of the process arise most fundamentally from the speed and resultant very high strain rates of the process that is typically controlled by shock wave phenomena that alter or preclude more normal mechanical behavior. Further complexities arise since there are various types of projectile threats that involve different degrees or types of various mechanisms, with each impacted by how well the ceramic is packaged in terms of both backing as well as, mainly for the most demanding applications, side and frontal containment of the ceramic. However, there is one basic aspect of the process that is clear, namely that there are two basic stages in what happens to the ceramic. The first stage, which is the key one in defeating the projectile, is the generation and initial propagation of a compressive shock wave from the projectile impact as both the projectile and the ceramic armor under it are shattered into fine fragments. The second stage is the propagation of the compressive stress wave through the ceramic to its boundaries where it reflects as a tensile stress wave that causes much further damage in the ceramic, which is a dominant factor in the extent to which there may be some second hit capability of the remaining ceramic in stopping a second projectile. Projectile penetration is defeated by erosion and fragmentation of the receding penetrator nose by the multitude of ceramic microfragments through which the projectile moves.

Erosive wear also generally correlates with hardness and its *G* dependence as well as via indentation cracking effects from the particle impacts. Though not directly modeled, such G dependence can in principle be at least approximately obtained from empirical or analytical models via their dependence on underlying physical properties such as K and H, e.g. an empirical model for the erosive wear rate of ceramics (W_E) from particle impact gives [16]

$$W_E \propto v_0^d R^e \rho^f K^g H^h \tag{5.1}$$

where v_0 = particle initial velocity, R = particle radius, ρ = particle density, and K and H are local values of fracture toughness and hardness of the body being eroded. For two models the exponents have respective values of $d = 3.2$ or 2.4, $e = 3.7$ (for both), $f = 1.3$ or 1.2, $g = -1.3$ (for both), and $h = -1.25$ or $+ 0.11$. Thus the G dependence of W_E is reflected in the G dependence of fracture toughness and hardness to the extent that they are suitably accurate in reflecting the material behavior over the G regime of interest. A review of these models revealed that while they generally approximately fit data, there were discrepancies, mainly greater than predicted dependence on H and K, which were attributed to microstructural aspects of erosion not being adequately accounted for in the models [16].

Similarly, modeling of the introduction of cracks from sharp indentations has been used as a basis of modeling particle impact damage, wear processes, and material removal processes in abrasive machining. Resultant models for the resultant crack size, c, give [17]

$$c \propto \left[\left(\frac{F}{K}\right)\left(\frac{E}{H}\right)^{1/2}\right]^{2/3} \tag{5.2}$$

where F = the load on the indenter or abrasive particle and E, K, and H are local, not necessarily global, values of Young's modulus, toughness, and hardness. The G dependence of the resultant crack size arises via effect of H and K since E normally does not depend on G.

Wear is even more complex because of the broader diversity of phenomena involved, which typically include varying mixes of indent fracture from asperities or abrasive particles, plastic flow (locally around indentations and in thin surface layers), and chemical reaction. Effects of these phenomena are further varied by frictional, e.g. stick–slip and heating, effects. A partial separation of wear behavior is between abrasive and sliding wear, the former involving more indentation fracture and the latter more frictional and chemical effects.

While the penetration of individual erosive or abrasive particles or of individual wear surface asperities would indicate the use of load-dependent H values reflecting similar penetration, there is substantial uncertainty in this due to both multiple and varying asperities in wear processes, and generation of new, rough surface in wear, erosion, and machining versus a single uniform indentation on a fixed surface for hardness measurements. Even greater uncertainty exists for the pertinent K values, since in wear, erosion, and machining, the crack scale will typically be much smaller than for normal fracture toughness testing, the larger scale of the latter being a major factor in significant variations in K–G relations (Chap. 2). Local temperature and environmental effects add further uncertainty.

While the diversity of wear presents challenges, it generally reflects important G dependences stemming from not only correlations with H but also indent-fracture and plastic deformation, as well as possible K–G dependence. Though such G dependence can depend on grain boundary character and grain size (and its distribution), shape, and orientation, it may aid in sorting out the varying mechanisms and is clearly important in developing, selecting, and applying wear resistant ceramics.

The subsequent sections review the G dependence of the above properties in the order: compressive strength, ballistic performance (briefly in view of very limited data) and erosion, wear, and machining. (No data on grain shape and orientation are known.) This review provides a more comprehensive data base and perspective on the G dependence of these properties (complementing and extending more limited earlier surveys of their G dependence [3,4,18]. This review also clearly shows the critical need for more study, especially of a more comprehensive nature.

II. GRAIN DEPENDENCE OF COMPRESSIVE STRENGTH

A. Grain Size Dependence

Four previous surveys of the G dependence of compressive strength of ceramics [3–5,18] provide the background for much of the following review, which addresses first compressive strength without superimposed confining stresses and then with such stresses. Much, especially earlier, data is for bodies with some to substantial porosity, especially at finer grain sizes, thus requiring correction of compressive strengths to $P = 0$, e.g. as extensively discussed elsewhere [1] to expand the limited data base. However, despite the limited data base, there is a fairly clear G dependence consistent with expectations based on earlier reviews.

Limited data of Alliegro [18,19] for hot pressed ZrB_2 ($P\sim$ 0.01–0.09) clearly show, even with no or conservative P correction, substantial σ_c levels with a simple linear decrease as a function of $G^{-1/2}$ extrapolating to substantial values at $G^{-1/2} = 0$, i.e. for single crystals (Fig. 5.1). Data for similarly hot pressed TiB_2 of Alliegro [18,19], (corrected for P = 0–0.11, mostly $\leq$ 0.05), though of more limited G range, indicate similar G dependence, but at somewhat higher σ_c levels (> 2 GPa) and is consistent with the one data point of Mandorf and Hartwig (extrapolated to $P = 0$) [20]. Other data for TiB_2 + 20%

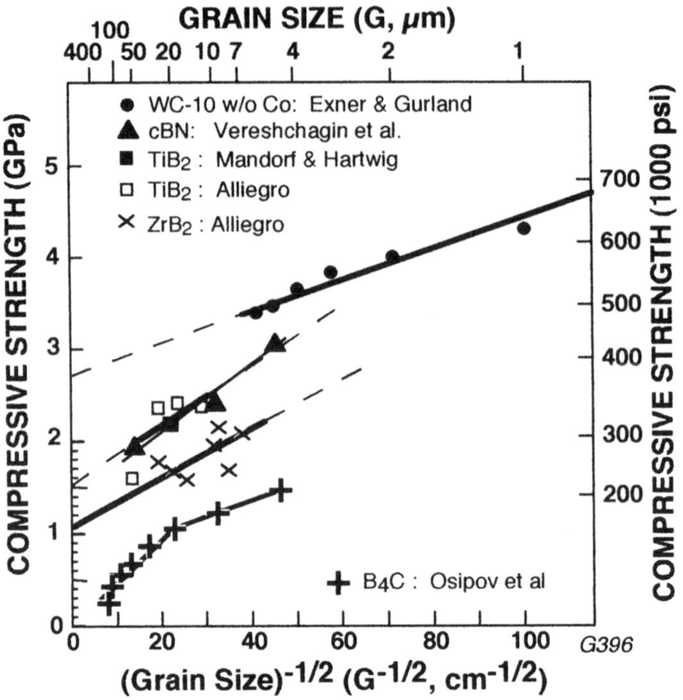

Figure 5.1 Compressive strength versus $G^{-1/2}$ for ZrB_2 [19], TiB_2 [19,20], and WC + 10 w/o CoO [23], and cubic BN (cBN) [24] at ~ 22°C. Note the high strength levels relative to those for tensile strength, and that despite these values being low relative to theoretical expectations, they clearly show substantial G dependence; also that some data points reflect more than one test, e.g ~ 5 for cubic BN. (From Ref. 3, published with the permission of Plenum Press.)

SiC was reported to be > 3 GPa for G = 5–10 μm [21], but more recent tests of pure hot pressed TiB_2, G~ 10 μm, giving σ_c = 5.7 ± 0.2 GPa, are seen as closely approaching the theoretical compressive strength [14,15,22]. Theoretical compressive strength is commonly taken as ~ $E/10$ or $H/3$, the former not being load dependent, but not reflecting any G dependence, which is at least partly reflected in $H/3$ (where H should probably be at higher loads where there is little load dependence, but not reflecting surface cracking effects).

Data for WC + 11 w/o Co with little or no porosity reported by Exner and Gurland [23] shows very similar G dependence, but at somewhat higher σ_c levels. Data of Tracy and colleagues [14,15,22] for hot pressed (G~ 1 μm and sintered (~ 4 G μm) SiC also indicates some G dependence, since the former had σ_c

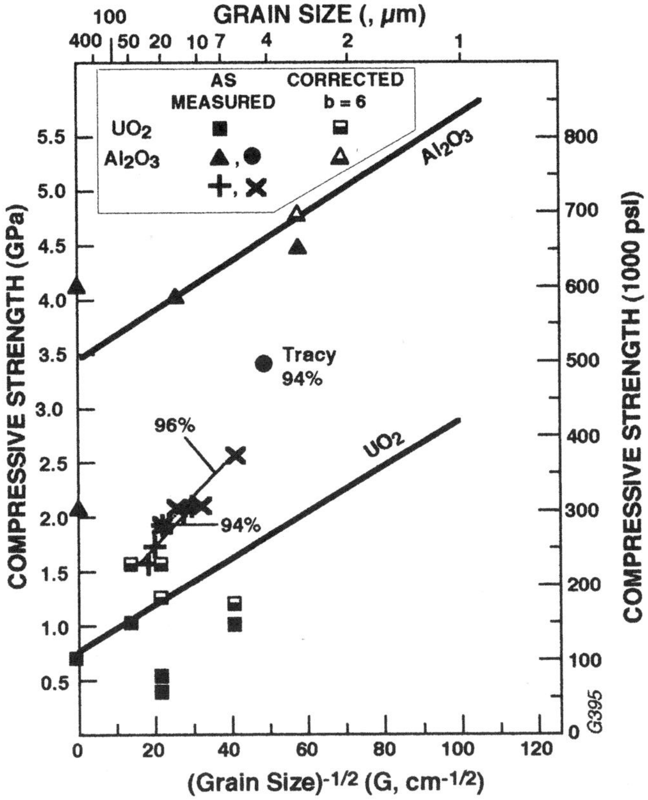

FIGURE 5.2 Compressive strength versus $G^{-1/2}$ for 94% alumina [4,26], higher purity aluminas [3], and sapphire (stressed perpendicular and parallel with the *c*-axis) [27], and from a more recent test a commercial 94% alumina [22,28], along with data for UO_2 polycrystals [30] and single crystals [31] at ~ 22°C.

~ 6.3 and the latter ~ 4.6 GPa, though respective percent porosities of 1–2 and 3–4 are a factor in this difference. Data of Vereshchagin et al. [24] on cubic BN is limited but clearly indicates a Petch relation. B_4C data of Osipov et al. [25] shows similar behavior but is bilinear, i.e. very similar to their tensile strength behavior (Fig. 3.1) but with higher slopes and compressive strengths ~ 3 times tensile strengths. The bilinear character and low multiple of compressive versus tensile strength are consistent, i.e. serious parasitic tensile stresses are the likely source of both, as are the lower strength levels and scatter in much of the other compressive strength data.

There is more extensive data for Al_2O_3. Earlier data of Floyd [4,26] for commercial 94 and 96% alumina bodies, both as-produced and with added heat treatment to increase *G* (which also decreased density some) clearly showed a *G* dependence of their compressive strengths that closely paralleled that of their *H*. Strengths were somewhat higher for 96 versus 94% alumina as expected, and Floyd's tests on the same bodies in flexure clearly showed less *G* dependence of tensile versus compressive strengths (especially for tensile strengths of the finer *G* branch, Chap. 3), which is also shown by comparing the *G* dependence of both strengths for other ceramics). More limited data on mostly purer, e.g. 99.9% alumina bodies, gives higher strengths, which are approached when the 94 and 96% alumina bodies' strengths are corrected for limited porosity and silicate-based glass content. Tests of Tracy et al. [22,28] on 94% alumina are seen as closely approaching the theoretical compressive strength (again taken as $H/3$ or $E/10$ as noted earlier) for such a body of 3.4 ± 0.2 GPa. These higher strengths are also consistent with their probable extrapolation between the compressive failure stresses of ~ 4.2 and 2.1 GPa for sapphire tested respectively with the stress axis parallel and normal to the *c*-axis [27]. Castaing et al. [29] showed apparent yield of sapphire at compressive stresses of ~ 6 GPa normal to the *c*-axis with 1.5 GPa confining pressure and a strain rate of 2×10^{-5}/s but noted that this was due to delayed fracture from creep rupture at 25°C, not dislocation glide till test temperatures ≥ 200°C.

The limited compressive strength data for ThO_2 of Knudsen [32] and of Curtis and Johnson [33] corrected for porosity shows a substantial linear trend as a function of $G^{-1/2}$ and a substantial extrapolation of ~ 1.4 GPa at $G^{-1/2} = 0$ [3]. Similarly, UO_2 data of Burdick and Parker [30] and Igata et al. [31] extrapolates to ~ 0.7 GPa at $G^{-1/2} = 0$, consistent with Igata et al.'s compressive strength of ~ 0.7 GPa for UO_2 crystals. The implication of plastic flow in compressive failure by the extrapolation of polycrystalline compressive strengths to those of single crystals is reinforced by observation of slip in UO_2 by Yust and Hargue [34]. Paterson and Weaver [35] measured compressive strengths of three MgO bodies of *G* ~ (1) 10–15 μm (*P* ~ 0), hot pressed with 3% LiF and then annealed at 1300°C, (2) 30 μm, isopressed and sintered at 1700°C, *P* = 0.03, and (3) 300–500 μm, core drilled from skull melted ingots. Recognizing that their finest *G* body had lower strength (attributed to residue from the LIF addition) than their two remaining data points for the intermediate and large grains showed some measurable decrease in σ_C as *G* increased, and on extrapolation to $G = \infty$. The latter is consistent with it being ~ the yield stress for single crystals stressed in the <111> direction, i.e. ~ 0.4 GPa versus that for <100> stressing ~ 0.15 GPa of Copley and Pask [36].

Paterson and Weaver [35] also measured compressive strengths of the same three above MgO bodies at 22°C, but with superimposed hydrostatic pressure (Fig. 5.3). They used two different types of rubber jackets to prevent fluid

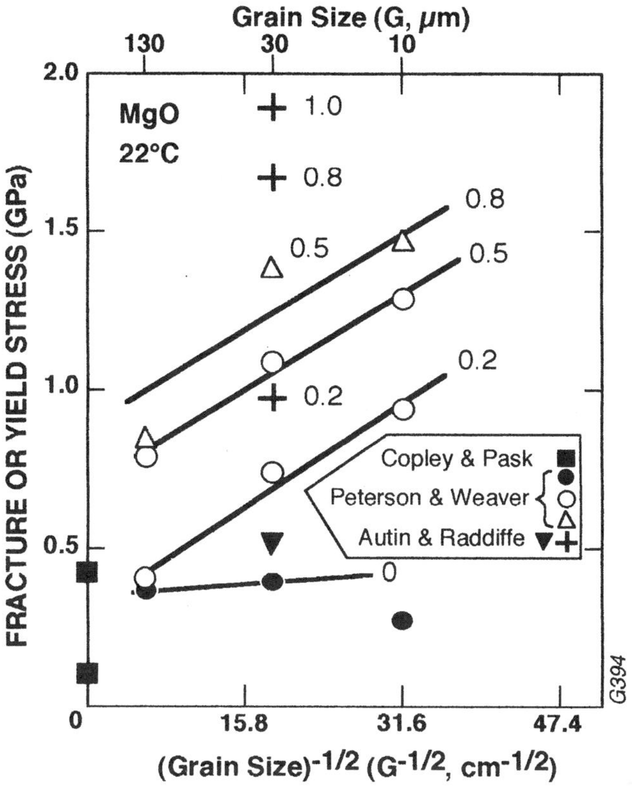

FIGURE 5.3 Compressive strength of MgO versus $G^{-1/2}$ at 22°C with superimposed hydrostatic compressive stresses of 0 (i.e. none), 0.2, 0.5, 0.8, and 1 GPa from Paterson and Weaver [35] and Auten and Radcliffe [37] along with single crystal yield stresses (hence ~ compressive strengths) of Copley and Pask [36] for <100> and <111> stress axes.

intrusion into cracks, with some differing effects of the two jacket materials. All tests with the application of such pressure showed higher G dependence of σ_C, with higher values for extrapolation to $G = \infty$ for the two highest confining pressures. (Note that the finer G specimens, those made with LiF, are much more consistent with values for the other bodies made without LiF in contrast to tests with no confining pressure.) Auten and Radcliffe [37] also measured compressive strengths of one polycrystalline MgO material ($G \sim 30$ μm) prepared very similarly to that of Paterson and Weaver's intermediate G body, but conducted their tests without a rubber jacket on the samples. As shown in Fig. 5.3,

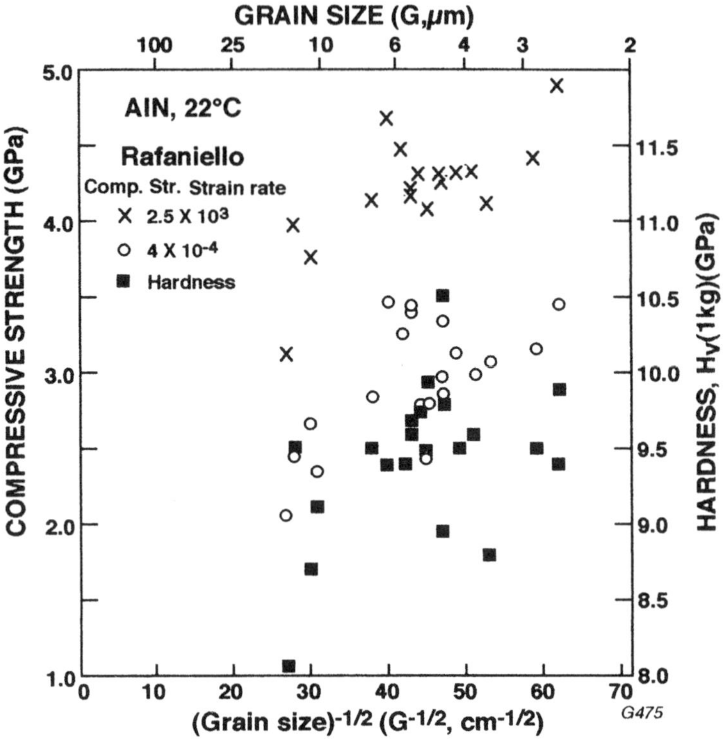

FIGURE 5.4 Compressive strength and hardness, H_V(9.8 N) at ~ 22°C for dense (< 3% porosity) AlN sintered with different amounts and types of oxide sintering aids [43]. Note that compressive tests used dumbbell specimens after Tracy and were conducted at three strain rates (the highest and lowest rates are shown) with all showing strength increasing as G decreased and strain rate increased. Also note the similar trend for hardness and its being ~ 3 times compressive strength.

their results showed very similar trends with confining pressure but were somewhat higher in fracture or yield stress.

Extrapolation of the polycrystalline data to $G = \infty$ falls between the values of Copley and Pask [36] for MgO single crystals stressed in <100> and <111> directions but closer to the substantially higher value for yield/fracture in the <111> direction as expected. Both Weaver and Patterson [38] and Auten et al. [39] measured compressive strengths of MgO single crystals at 22°C along <100> loading axes as a function of confining pressure using respectively rubber jackets and no jackets. Both obtained compressive strengths with substantial single crystal deformation. Patterson and Weaver's values for jacketed samples were ~ 120 MPa without confining pressure and increasing, mostly modestly, with increasing pressure. Auten et al.'s values for unjacketed samples were ~ 80 MPa without confining pressure and decreased modestly with increasing pressure. However, the averages of these two sets of tests were ~ independent of confining pressure and agreed fairly well with values of Copley and Pask. (Auten et al. noted that a probable reason for their lower results was surface cracking and resultant loss of surface chips without jackets versus such cracking and especially loss of surface chips being constrained with jackets on the specimens.)

Limited data of Brace [3,40,41] for nominally single phase limestone ($CaCO_3$) and dolomite ($CaCO_3$ + $MgCO_3$) follows a $G^{-1/2}$ dependence and parallels his data for high load hardness (both using the maximum rather than the average *G* (Fig. 4.14) [3]. Fredrich and Evans [42], besides reviewing the limited compressive data and models for rocks as a function of *G* (generally showing Petch-type behavior), also presented their own data for four marble limestones with *G* from ~ 6–1800 μm, which also clearly followed a Petch relation. They further showed that tests with confining pressures of 5 or 10 MPa resulted in essentially the same relation, except for a possible upward shift of the data by a few percent as the confining pressure increased.

Finally consider more recent AlN data of Rafaniello [43], which over a limited *G* range (~ 2+ to 12 μm) showed a substantial decrease with increasing *G* consistent with a $G^{-1/2}$ dependence (Fig. 5.4). Note that compressive strength increased with increasing strain rate, which was corroborated by tests at 0.6 sec^{-1} being in between the two sets of results in Fig. 5.4 but closer to those for the lowest strain rate.

B. Other Microstructural and Testing Effects, *H* Correlation, and Failure Mechanisms

No data specifically on effects of grain shape or orientation on compressive strength is known. However, the correlation of σ_C and *H*, and the substantial data on the dependence of *H* on crystal orientation, provide guidance, as does limited data on effects of grain shape on indentation fracture (Chap. 4, Sec. II.E). There is also limited single crystal data, e.g. that for sapphire compressively stressed parallel or normal to the *c*-axis (Fig. 5.2). There are also some theoretical estimates of some, e.g. diamond structure, crystals that provide some guidance, such as that of Nelson and Rouff [44] predicting theoretical strengths of diamond of 1.8 GPa for [100] loading and 4.1–4.8 GPa for [111] loading.

The first of two important loading parameters not addressed in the above review is strain rate. The limited data has been reviewed by Lankford [45,46] along with presenting his own data for sintered, nearly dense SiC (*G*~ 3 μm) and dense sintered Al_2O_3 (*G* ~ 25 μm). He showed that there was an ~ 30% increase in compressive failure strength of Al_2O_3 over the strain rate range of ~ 10^{-5} to > 10^2 sec^{-1}, while there was none for SiC. He observed that in both materials microcrack nucleation appeared to be athermal. While strain rate effects on slip and probably twin deformation are typically thermally activated, this may not always be true at high strain rates, so further study is needed. Lankford also showed a marked rate of increase in compressive strengths at strain rates > 10^2 sec^{-1} indicating a fundamental change in mechanism for both materials for which there is some theoretical basis in inertial effects on crack initiation and propagation. He noted that extrapolation of the high strain rate increase in compressive strengths indicates reaching theoretical strengths (~ $E/10$) at rates of ≥ 10^6 sec^{-1}.

The issue of loading rate also raises the issue of possible environmental effects, especially of slow crack growth, particularly due to moisture. Lankford [10] reported that the strain rate dependence of the compressive strength in his testing of (Lucalox) Al_2O_3 corresponded to n values for slow crack growth for tensile strength, i.e. values of ~ 50 (hence implying a G dependence per Fig. 2.7A). He subsequently noted similar correspondence of compressive and tensile slow crack growth in SiC and Si_3N_4, but not in Mg PSZ [12]. Similarly Nash [47] reported that a 97.5% and a 99.7% alumina (both noted as having average G of a few microns) showed reductions of σ_C from respectively ~ 1.7 and 2.2 GPa to ~ 1.2 and 1.5–1.7 GPa in air and Ringer's solution (and other water-based solutions for the 99.7 body) in testing at a strain rate of 0.55×10^{-3}/sec. However, Ewart and Suresh [48] ruled out environmental effects in their studies of compressive fatigue crack propagation in Al_2O_3.

While slow crack growth must be considered in view of the local nucleation, growth, and coalescence of microcracks under local tensile stresses, there are two sets of issues that raise serious questions about such growth in true compressive failure. The first is the uncertainties in n values in crack propagation under macro tensile stresses, in particular the indicated significant G dependence (Fig. 2.7A). This for example raises the question of whether there is a similar G dependence in compression and whether the tensile n values used for comparison have sufficiently similar G values for adequate comparison. The second and more fundamental question is that of the extent and time of environmental access to the crack tips. While crack propagation of a single surface connected crack allows ready access of the test environment to the crack tip, in well conducted tests, compressive failure is seen as a process of cumulative nucleation, growth, and coalescence of microcracks throughout the bulk of the sample. Thus over much of this process many, probably most, microcracks are not accessible by the environment over much of the stage of the compressive failure process. This raises the question of how effective the environment can be in the probably limited portion of the failure process during which it has considerable access into the sample and the crack tips. On the other hand, in less well conducted compressive tests, local tensile stresses, e.g. often from end constraints, commonly result in "compressive" failure due to propagation of one or a few surface connected cracks. Clearly both the surface connected and commonly resultant macro nature of these cracks and their dependence on substantial parasitic tensile stresses provide much more opportunity for environmental effects on their growth and the resultant reduced "compressive" strength. This probable role of such surface connected cracks also raises the issue of whether there is a specimen size effect on compressive failure, since many microcracks appear to form at the surface, which could allow them to grow preferentially in from the surface. Thus what might start as a true compressive failure could become dominated by cracks propagating in from the surface, but with this being increasingly favored as specimen size decreases. This basic issue of the extent of intrinsic versus extrinsic environmental effects on compressive failure is an important scientific issue. However, it is also potentially a very important one in the engineering design of ceramic components for substantial compressive loading, since such loading is likely to be plagued by more parasitic tensile stresses than in most compressive testing. Another complication in considering environmental effects on compressive strength is the possibility that the fluids involved may act as lubricants in addition to, or instead of, environmental effects, since crack friction is seen as playing an important role in compressive strengths.

The second stressing aspect to consider is repeated loading, i.e. compressive fatigue. While this has been studied substantially less than tensile fatigue (with very limited attention to grain effects in both), there is some limited guidance mainly in a review by Suresh [49] and especially tests by Ewart and Suresh [48]. They studied cyclical fatigue crack growth from machined notches in two high-purity Al_2O_3 bodies (AD 995 and 999 with respective G of ~ 18 and 3 μm) and one orientation of sapphire. The latter resulted in no stable crack propagation, while the two polycrystalline bodies showed similar overall crack growth characteristics, but the finer G material required ~ 50% higher compressive stress for stable crack propagation, i.e. consistent with higher compressive strengths at finer G. (Note that they observed an enhanced extent of crack propagation near the machined surfaces pertinent to the issue of surface and related finishing effects on crack propagation discussed in Chap. 2, Sec. III.B.) The absence of stable crack propagation in sapphire versus stable propagation in the polycrystalline bodies via almost exclusively intergranular fracture shows that such crack propagation in polycrystalline bodies is clearly associated with grain boundaries, and it further questions conventional slow crack growth (which occurs with tensile loading in sapphire, Chap. 2, Sec. III.B) in compressive loading. They concluded that large compressive stresses induce inelastic deformation in the vicinity of notch roots that can result in residual tensile stresses at the notch tip due to permanent strains left upon unloading. They saw their results as not necessarily being consistent with microplasticity as a mechanism of compressive failure and saw probable contributions of TEA stress and crack face friction effects (but did not consider possible effects of EA).

Before considering mechanisms and correlations, consider variations in compressive strength values. Though almost never, or never, used, Weibull moduli are as pertinent to compressive as to tensile strengths (as they are not only to other mechanical and nonmechanical properties but also to microstructural parameters such as grain size, shape, and orientation). While Weibull moduli (m) values have not been directly measured, compressive strength data of Tracy and colleagues [15,22], besides generally being the most free of serious parasitic stresses, also commonly made several, typically 10, measurements for each material, thus providing reasonable data to calculate m values via Eq. (3.4). As shown in Table 5.1, Weibull moduli calculated for σ_C (m_C) roughly follow trends for tensile moduli (m_T) they directly measured for flexural strengths, σ_T, but with $m_C > m_T$ by ~ 40–400%. While 10 tests per material is a limited data base on which to calculate m values, the consistency of m_C being > m_T, and the fact that higher m_C than m_T values were found across six materials indicates a trend that is highly probable. The higher moduli in compressive failure are consistent with expectations of Weibull moduli being less sensitive to microstructural heterogeneities in compressive loading than in tensile loading, but the differences may be limited by the greater G dependence of σ_C versus that of σ_T.

Consider now the correlation of hardness and compressive strength. Rice, in his review of ceramic compressive strength [3], showed that it correlated with hardness, i.e. the upper limit was typically $H/3$ (Fig. 5.5), though other constraint factors than three may occur, as also discussed by Lankford [5]. Deviations below this limit were attributed to, roughly in order of decreasing scope, impact, or both, (1) frequent and often substantial parasitic stresses in testing, (2) material defects and heterogeneity, and (3) other, lower stress mechanisms of failure, e.g. twinning. Lankford accepted this H correlation and further documented and extended it. A summary of results shown in Fig. 5.6 clearly shows that (1) more recent results from generally better conducted tests, often with denser specimens,

TABLE 5.1 Comparison of Weibull Moduli for Flexure (Tensile, m_T) and Compressive (m_c) Failure of Ceramics[a]

Material	m_T	m_c	Material	m_T	m_c
Al_2O_3 (94%)	—		SiC-S	10.8	14
TiB_2-HP	29	41	SiC-HP	9.6	31
B_4C-HP	5.2	18	Al_2O_3/SiC_w	5.1	25

[a] m_T values measured in studies published by Tracy and colleagues [15,22], while m_c values were calculated from their compressive strength data per Eq. (3.4). HP = hot pressed, S = sintered, and SiCw = a SiC whisker composite.

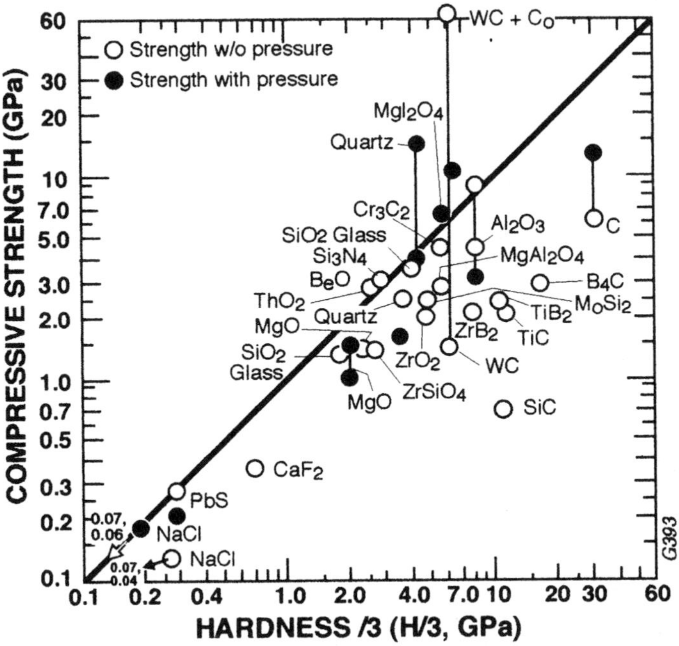

FIGURE 5.5 Compressive strength versus *H*/3 for various ceramics from a survey of Rice (previously unpublished) from a variety of sources, including Patterson and Weaver [35] (circular points).

give a generally closer approach of σ_C to $H/3$, and that (2) data for bodies tested with superimposed hydrostatic loads commonly equal or exceed $H/3$ as the hydrostatic pressure increases. Data of Bairamashvili et al. [51] on hot pressed B_4C, B_6O, and SiB_4 with $P \sim 0.1$–0.04 and $G \sim 1$–3 μm may also support the H–σ_C correlation, since it appears that their σ_C values may be shown as 10 times their true value. If this is the case, their respective values of σ_C for the above respective order for the materials versus those of the corresponding $H/3$ values (in parentheses for H with 1–2 N load) are 7.9 (11.7), 6.3 (10.7), and 5.0 (9.0). This correlation of σ_C with $H/3$ and the typical association of $H/3$ with the general yield stress of a material is a strong indicator of yielding as a basic mechanism controlling compressive strength in well conducted tests of samples of reasonable to good quality.

The σ_C–H correlation is quite suggestive, but it does not fully prove a microplastic mechanism of failure (discussed further below.) Thus the correlation with H may reflect more basic correlations with elastic moduli in general [1,2], and shear modulus specifically, e.g the theoretical compressive strength limit

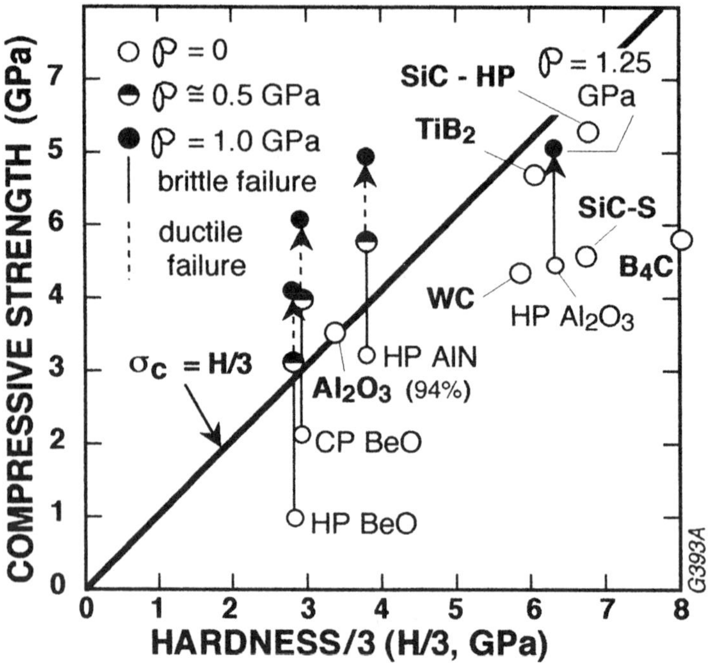

FIGURE 5.6 Compressive strength versus $H/3$ for various ceramics from better conducted tests with and without superimposed hydrostatic stress levels (shown in GPa) at 22°C. (Data from Refs. 5,14,15,22,28,50.)

being ~ 80% of the shear moduli for [111] loading of crystals of Ge, Si, and diamond [44]. Further, the substantially increased data, especially for the G dependence of H, also raises some questions, and the load and indenter shape dependence of H raises the issue of what H values are appropriate. While higher load H values appear more appropriate, as noted earlier, there are uncertainties regarding which indenter shape is more pertinent, e.g. possibly Vickers over Knoop, but with the merits of spherical indenters relative to these unknown. Further, the use of any indenter geometry at a fixed load means that the resultant indents are changing in the number of grains involved as G (or the G range) of a given sample changes, raising the issue of to what extent such changing grain-to-microplastic-zone size ratios in measuring H correspond to such ratios of possible microplasticity in compressive failure. There are also issues raised by the σ_C–G data that do not appear to correlate well in detail with H–G data, e.g. though limited, the σ_C–G data shows no evidence of σ_C minima at any intermediate G, as H often does. Most likely the H minimum found in more comprehensive H testing of a number of ceramics is not directly pertinent to compressive failure, since this entails free surface cracking versus bulk, i.e. much more constrained, cracking in compressive failure (though the possible mechanisms of generating the surface cracking–spalling in H testing may be pertinent, as is discussed below). However, neglecting the H minimum exacerbates another possible problem in detailed correlation of H–G and σ_C–G data, namely that σ_C–G data often shows more change over the same G range than does H. Thus while there is a very useful general correlation between H and σ_C, the specifics of the mechanisms and their impact on the correlation through factors such as the G dependences of H and σ_C are uncertain.

Consider now the evidence for microcrack formation and coalescence as the basic mechanism of compressive failure. Though again the data is generally limited, there is considerable support for such cracking, mainly from four sources, but very limited information on most details, which are generally important. The first source is extensive modeling showing that compressive loading results in cracks parallel with the applied uniaxial stress initiating from pores (see Adams and Sines [52] and Sammis and Ashby [53]). The second source is direct observation of such cracks from both large artificial pores in glass [53] and fine natural pores in relatively dense polycrystalline ceramics such as Al_2O_3 [9] and SiC [11]. Further and very important is direct observation of individual cracks or reduced transparency of single crystals of MgO and Al_2O_3 [29,38,39]. The third source is acoustic emission data, e.g. as for the previous Al_2O_3 [9] and SiC [11]. While some emission may be from other sources, much of it, e.g. of Nash [47], is highly likely to reflect cracking. The fourth source is interrupted compressive loading followed by subsequent tensile loading normal to the original compressive loading axis by Sines and Taira [54]. They did such testing on a commercial Al_2O_3, reaction bonded SiC, and HIPed Si_3N_4 and showed that such compressive loading prior to tensile testing in the normal direction resulted in the onset of tensile strength degradation with prior compressive loading above ~ 40% of the compressive strength and further degradation approaching zero tensile strength at ~ double the level of compressive loading relative to the ultimate compressive strength in Al_2O_3 and SiC, which had porosity and coarser microstructures. However, no degradation from such prior compressive loading was found in the nominally pore free, finer G Si_3N_4 tested. Thus there is considerable data showing microcracking from compressive loading, but much remains to be determined about the specifics of the onset, growth, and coalescence of such cracking. It is also important to recognize that the extent and cause(s) of such crack processes may change with material, microstructure, loading, and temperature. Another indication of crack generation during compressive stressing is the results of Stucke and Wronski [55] showing progressive reductions in tensile (flexure) strength as superimposed hydrostatic pressure on three MgO, two Al_2O_3, and one UO_2 bodies was increased.

Turning to the mechanism of microcrack generation, growth, and coalescence, it is highly probable that there are multiple mechanisms with varying contributions as a function of which stage of compressive failure is being addressed and what the material, microstructure, and loading conditions are. While microcracking from pores is clearly an important source, there must be other sources of microcracks for two reasons. First, similar compressive strength behavior is seen for ceramics with little or no porosity and those with more substantial porosity, e.g. between sintered and hot pressed Al_2O_3, but corrected to $P = 0$. The second is the extrapolation of polycrystalline compressive strengths with little or no porosity, or corrected to zero porosity, to the range of single crystal compressive strengths, e.g. for UO_2 and Al_2O_3 (Fig. 5.2). For other mechanisms of generating microcracks than from pores it is logical to consider the same candidate mechanisms as for indent-related cracking and resultant H minima due to cracking from deformation, especially slip anisotropy, and TEA and EA stresses. Clearly there are important differences between the two microcracking situations, i.e. indent cracking is mostly or exclusively surface connected and occurs at or near the edge of the zone of ~ hydrostatic compression, while compressive microcracking occurs in conjunction with local tensile stresses in an overall compressive stress field. The latter is the logical cause of compressive microcracking presumably occurring at much higher stresses. TEA and EA are clearly possible sources of microcracks, since they can result in significantly increased local stress that could exceed the local stress for nucleation of microcracks. Some or all of the G dependence of compressive strengths would thus be due to the probable G dependence of such microcracking from TEA and EA [56,57]. Note that the stress concentrations due to EA can be substantial, e.g. commonly to 1.5–3 [58]. However, it varies substantially not only with the degree of EA but also interactively and significantly with grain shape, i.e. aspect ratio, and with the orientation of the grain elongation relative to both the stress axis and the relation of the grains or regions of them with higher versus lower local elastic moduli from the EA per modeling by Hasselman [58].

While EA and TEA are highly probable factors in compressive microcracking, it is likely that microplastic deformation due to slip, twinning, or both is also a factor for several reasons. These include the clear occurrence of slip at similar stress levels under indenters, the implications of yielding at ~ $H/3$, and the correlation of this with σ_C. Another indicator of plasticity is the G dependence of σ_C following a Petch relation and thus correlating with single crystal behavior where there is no TEA or EA. While some might argue that the Petch relation is misleading, as it was for the finer G branch for tensile strengths (Fig. 3.1), it should be noted that there is no clear evidence for a larger G branch for σ_C as for tensile strength, and stresses for σ_C are much more consistent with those for plastic deformation than those of the finer G branch for tensile strength. Two other important, but widely neglected, factors are the transitions from failure with no to substantial supporting hydrostatic pressure and the changes in *H* and especially σ_C with modest differences in temperature (Chap. 6).

Additional support for plasticity as an important factor in compressive failure is provided by specific observations. While Nash's transgranular fracture mode of what is probably rhombohedrahl cleavage in Al_2O_3 may be evidence of twinning, his TEM observations of twins, along with some of the substantial acoustic emission he observed probably being from twinning, are reasonable indicators of twinning in Al_2O_3 [47]. An important observation is the demonstration of deformation accompanying compressive testing of Al_2O_3, MgO, and UO_2 crystals, as well as considerable evidence of this in polycrystalline MgO. Another is Lankford's observation of crack initiation by blockage of slip or twin bands or both in compressive, as well as tensile, stressing of dense sintered Al_2O_3 [9,10], and both microscopic and macroscopic evidence of plastic deformation in compressive testing of PSZ bodies, including a clear yield point and substantial subsequent plastic strain [12,13]. While it is legitimately argued that each of these is to some extent a special case, they clearly show that plasticity can occur in compressive failure. More importantly, they show that there are different options for plasticity in ceramics at these stress levels, which is a reminder that thinking of a single mechanism is probably an inappropriate and simplistic approach. Further, the extent of compressive plasticity in the MgO and especially PSZ cases clearly implies that other cases of progressively less plasticity are likely to occur and be a factor.

It should also be noted that some of the arguments raised against microplasticity by some are at best uncertain. Thus the argument that common intergranular fracture in compressive failure or crack propagation is contrary to microplasticity is on shaky grounds, since slip and twinning, while sometimes leading to transgranular fracture, result in intergranular fracture e.g. in Al_2O_3 [9,10] and MgO (Chap. 3, Sec. III). Wiederhorn et al.'s [59] observation via transmission electron microscopy (TEM) that there are no dislocations associated with tips of cracks introduced into Al_2O_3 by indents or thermal shock is of uncertain applicability. These observations are most likely of the tips of arrested cracks rather than of their initiation, thus showing only that crack growth and arrest was not associated with slip at or near the arrested crack tips. Further, the argument that such TEM is necessary to confirm the role of plasticity in compressive failure is potentially misleading, since it neglects the issue of the number and scale of microcrack nuclei that may be needed, especially for earlier stages of ultimate compressive failure, and the feasibility of finding these by TEM. Thus, for example, crack nucleation due to blockage of one slip or twin band at the boundary with an adjacent grain occurring at one in a thousand grains may well be a substantial number of crack nuclei in a body, but they clearly provide a very low probability of suitable detection by TEM. It is much more logical to recognize that there is probably a substantial range of plastic contributions, many on a very modest scale, so a variety of self-consistent tests and evaluations are needed to address this issue. Such more comprehensive evaluations should preferably entail testing of a variety of materials, with a variety of microstructures, loading conditions, test temperatures, and associated evaluations, e.g. acoustic emission, microscopies, and radiography.

III. BALLISTIC STOPPING CAPABILITIES OF CERAMIC ARMOR, GRAIN AND OTHER DEPENDENCE

Ceramics are often used as armor against bullets and other, often heavier, projectiles, since they typically have the best stopping power, especially as a function of weight. As noted earlier, much of the physics of failure in such impact regimes is different from that of conventional failure due to the much high loading rates of such impact. This commonly occurs at rates that exceed the speed of sound in both the impacting and the impacted bodies, so stresses propagate in both as shock waves, since the normal elastic distribution of stress is exceeded, which means that any portion of the target does not experience stress until the shock wave reaches it. Again, the initial shock wave is compressive and then becomes tensile on reflection from the ceramic surfaces, with the latter causing much of the fracturing, e.g. the typical "spider web" crack pattern with normal bullets and granulation from special armor piercing (i.e. long rod, kinetic energy) projectiles after the projectile has been destroyed.

At the low end of projectiles in terms of size, hardness, and velocities there are limited tests indicating some improved stopping power as *G* decreases and purity increases in alumina bodies. Thus tests by Ferguson and Rice [60] on alumina bodies of varying *G* and purity, with limited or no porosity (including sapphire, showed some *G* dependence, especially in less pure bodies (Fig. 5.7). While these tests were with smaller, softer projectiles (nonarmor piercing, that were used to simulate energetic metal fragments), they have some value as a starting point.

Rafaniello's [43] .30 caliber AP data for dense sintered AlN, though scattered, clearly also shows that ballistic stopping ability, i.e. ballistic limit velocity, decreases as *G* increases (Fig. 5.8). However, his data for .50 caliber AP stopping ability (using depth of penetration into the Al backing material as the measure of this) is more widely scattered and is at best only somewhat suggestive of decreasing stopping ability as *G* decreases. The apparent progressive decrease in the clarity, extent, or both of stopping ability increasing as *G* decreases for the above .22, .30, and .50 caliber tests is consistent with even less indication of *G* effects against heavier projectile threats. Thus as one proceeds progressively to more serious projectiles ranging from .30 and .50 caliber to various types of armor piercing projectiles that differ greatly in density, hardness (e.g. of W or WC), size, and velocity (e.g. $\leq$ 1000 versus to 2000 m/s) [14,15], correlation

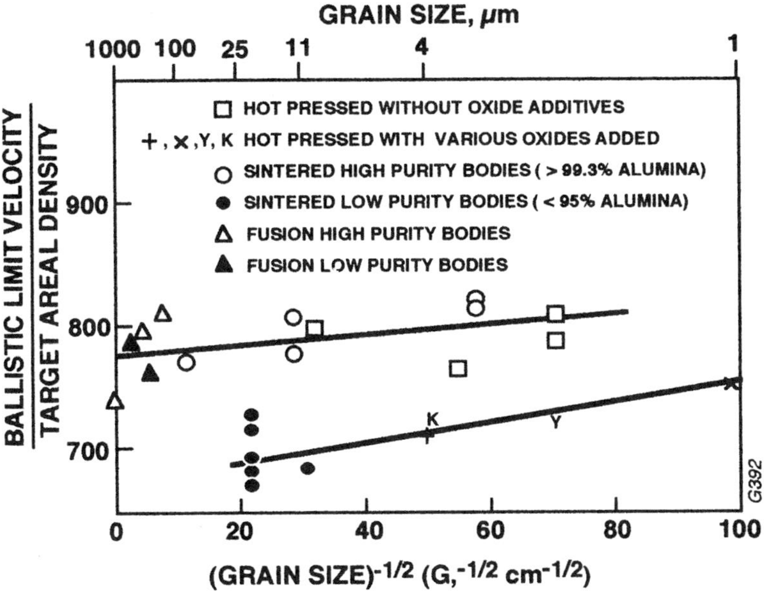

FIGURE 5.7 Ballistic limit velocity (i.e. the velocity at which half the projectiles penetrate the target and half fail to do so) normalized by the areal density of the target (i.e. the weight per unit area) versus the inverse square root of grain size *(G)* for tests of various alumina bodies at ~ 22°C with high power .22-caliber ductile bullets used to simulate fragment stopping capabilities of armor. Note the two lines, the upper one for pure alumina bodies (including sapphire) and the lower one for less pure aluminas (i.e. with less stopping ability). (From Ref. 60, published with the permission of Plenum Press.)

with *G* reduces significantly. This changes material rankings; B_4C is commonly the most weight efficient ceramic armor against .30 and .50 caliber bullets, but it is less effective against more severe armor piercing projectiles.

While the specifics of the material properties and hence microstructures that determine the range of behavior, especially stopping different armor piercing projectiles, are not fully understood, there are some guidelines [14,15]. The first and most fundamental one is that ballistic projectile stopping power is probably dependent on more than one property and changes with the nature of the projectile, so microstructural needs probably change. Second, properties controlling normal tensile fracture, i.e. fracture toughness and tensile strength, have no positive correlation, and some have negative correlations, with armor performance. For example, grain boundary phases that often increase fracture toughness by enhancing intergranular fracture and crack bridging and branching (at

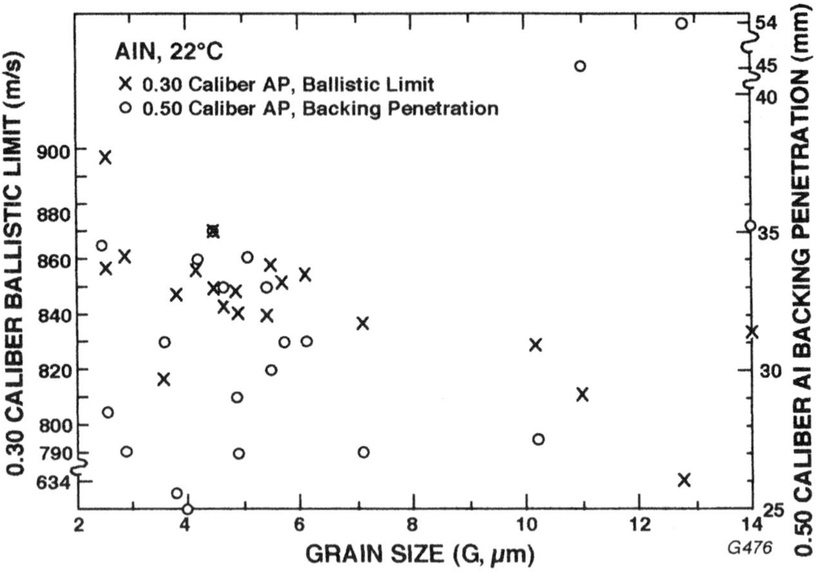

FIGURE 5.8 Ballistic stopping ability of dense (~ 3 to 0.3% porosity) sintered AlN (the same bodies as in Figs. 3.28 and 5.4) versus average *G* for .30 and .50 caliber AP (armor piercing) projectiles, using respectively ballistic stopping velocity and depth of penetration into the Al backing for the ceramic target as the measures of stopping ability.

least as measured with large scale cracks, Chap. 2, Sec III) typically give poorer armor performance. Thus many ceramic composites are less desired, and larger *G* bodies that may give more concentration of boundary phases are typically less desired materials.

There are some rough indicators of armor performance, since successful ceramics typically have high Young's moduli and indentation hardnesses, moderate to low densities (at low to zero porosity), low Poisson's ratios, and moderate to fine *G*. However, the specific dependences on these properties and the above noted microstructures are not established, due not only to the complexity and possible multiplicity of processes but also to the probability that in many cases there is a threshold requirement, beyond which the property or related microstructural dependence decreases. Thus armor must have a hardness greater than that of the projectile to stop the latter effectively, but beyond that there is only a rough correlation with hardness. It is also known that the presence of preexisting flaws in ceramic armor reduces its performance, but the mechanism of flaw generation during impact, e.g. due to EA and TEA and associated grain size, shape, and orientation, is unknown. The negative effects of preexisting flaws suggest correlation with Weibull moduli, e.g. some possible rough correlation with such tensile moduli has been noted [14,15]. However, while there is no clear correlation with even well measured compressive strengths, it seems more likely that such moduli for compressive failure, which roughly correlate with tensile moduli (Table 5.1), may be better candidates to reflect flaw density. Thus there are likely to be useful correlations of properties and their microstructural dependences that can benefit ceramic armor selection and fabrication, but they must require balances between various factors, as was noted above, with the balances probably shifting with the nature of the ballistic threat. Further testing of microstructural factors, as addressed in Figs. 5.7 and 5.8, with progressively higher-velocity and harder threats, with the recognition that various and changing property correlations are involved, seems needed. While hardness and compressive strength correlations suggest a possible dependence on grain size, i.e. better ballistic performance, hence "stopping power," at finer G, there are many uncertainties and very few tests (especially in the unclassified literature).

Finally it should be noted that while there has been little study of plastic flow in ballistic impact on ceramics, there is clear evidence of substantial plastic flow in the debris near the impact zone. Given the complexity of the ballistic impact, including its very short duration and multiple interactions, there is uncertainty as to the extent to which the post-failure tensile shock wave contributes to this. However, the observed yielding under shock wave conditions, i.e. the Hugoniot Elastic Limit [10], indicates that such yielding is an important factor in ballistic stopping power.

IV. GRAIN DEPENDENCE OF EROSION WEAR AND RESIDUAL CERAMIC STRENGTH

Consider now the grain dependence of both erosion due to impact of particles on the surfaces of ceramics and the resultant ceramic strength. There is substantial interest in the effects of solid particles because erosive wear by them is an important issue in many applications components, e.g. potentially in engines where particles are ingested or generated, or in equipment to use streams of abrasive particles, e.g. grit blast nozzles. However, high-velocity impacts of liquid, especially rain, drops are also important for some aircraft and especially missile components, particularly microwave and IR windows and domes. Effects of the impacts of solid particles or liquid drops is a complex subject, but considerable modeling and testing has taken place to give at least partial insight into the mechanisms and parameter dependences. While, as is unfortunately very common, issues of microstructural, e.g. grain size, effects on erosion and induced damage of impacted bodies have received limited attention, there is some information on grain effects.

Consider first more general information on particle erosion, especially the general correlation of increasing *H* with increasing erosion resistance, or conversely increasing rates of erosion with decreasing *H*. As part of their particle erosion studies of ceramics (which are among the most extensive), Wada and Watanabe [61] showed that the erosion rate was inversely proportional to the ratio of the hardness of the impacted surface to that of the impacting particle, i.e. a linear dependence on a log–log plot. Similarly, others have shown correlations of erosion with *H*, e.g. of the onset of serious rain erosion with the hardness of the impacted ceramic [16,62,63]. Both the correlation of erosion with *H* and the generally recognized increase of *H* as *G* decreases suggest reduced erosion as *G* decreases, as did qualitative observations of lower erosion at finer *G*, e.g. of Wada and Watanabe [61] in Si_3N_4.

Consider now the limited experimental data on *G* dependence of particle erosion of ceramic surfaces. A study of SiC erosion using Al_2O_3 particles (37–270 μm dia.) at velocities of ~ 108 m/s by Routbort and Scattergood [64] provided some direct indication of a grain size effect. They showed that the erosion rate of reaction bonded (RB) SiC (i.e. with large *G*, *P*~ 0, and excess Si) progressively increased substantially as the size of the impacting particle increased, as expected from theory. However, similar tests of hot pressed SiC (*G*~ 2–4 μm), while being ~ consistent with their RB SiC at small particle sizes, progressively deviated to less erosion as the size of the eroding alumina particles increased, with erosion rates for the largest alumina particles being < for the finer particles, i.e. contrary to all models. They noted however that the damage zone sizes for single particle impacts was ~ 0.2–0.3 times the particle size for normal incidence, so the finer impacting particles created damage on the scale of individual grains, while larger impacting particles created damage zones covering more grains as the particle size increased. These observations were supported by their fractography showing that whole grains were often removed. The implication is that erosion is a function of grain size, with possible maxima of erosion when the damage zone and *G* are ~ equal, i.e. similar to the conditions for *H* minima as a function of *G*.

More quantitative data on the *G* dependence, though not a focus of the substantial study of Wiederhorn and Hockey [16], can be extracted from it. They measured the erosion rate (mass loss of a variety of ceramics using 150 μm SiC abrasive particles carried at controllable velocities by injection of a particle stream into an air stream normal to the specimen flat surface. While their data shows general correlations of erosion rates with inverse functions of hardness and fracture toughness per Eq. (5.1), they made measurements on two dense, pure polycrystalline bodies, a sintered one (*G*~ 30 μm) and a hot pressed one (*G*~ 3-4 μm), as well as on a $\{10\bar{1}1\}$ surface of sapphire. Their data was used to obtain the erosion resistance (the reciprocal of the erosion rate), which has been here normalized by dividing all resistance values by that for sapphire under the specified conditions (allowing direct comparison of data at their three test temperatures of 25, 500, and 1000°C). Plotting this normalized erosion resistance versus $G^{-1/2}$ for their various particle velocities (37–125 m/s) and test temperatures shows that all data follows the same trend as a function of G (Fig. 5.9).

Breder and Giannakopoulos [65] conducted a narrower study of particle erosion (and subsequent target strength, discussed below), focusing on five commercial alumina bodies, all ≥ 99.5% purity except for a 90% alumina (with the second finest G of ~ 4 μm). They utilized four grit sizes of SiC particles having average diameters of 360, 460, 550, and 1300 μm and employed a slinger arm to generate particle velocities of 37–106 m/s at normal incidence. An equation similar to that considered by Wiederhorn and Hockey [16] was modified to allow for a variable toughness due to varying extents of R-curve behavior by introducing various weightings of a parameter m in each exponent in Eq. (5.1) such that for $m = 0$ the toughness was constant, i.e. independent of crack size as in the original models. They showed, using such a modified equation based on inclusion of their parameter m in the exponents, that there was no correlation of erosion with R-curve behavior based on analysis of the slopes of log–log plots of eroded volume (V) versus the kinetic energy of the eroding particles (U). However, they

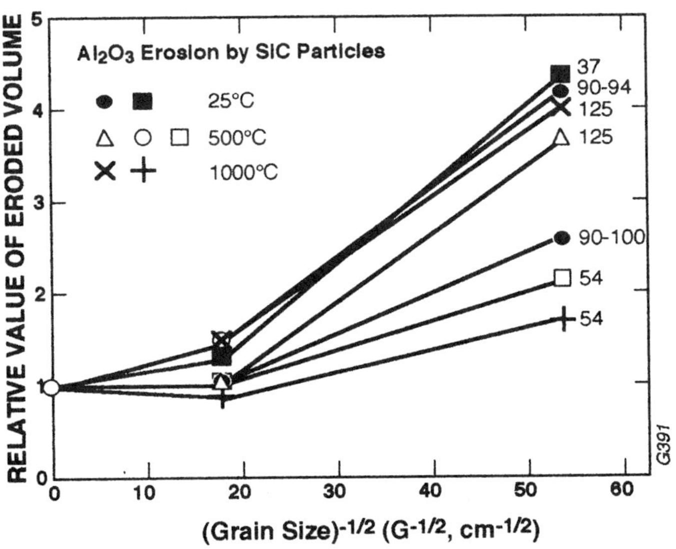

FIGURE 5.9 Relative erosion of sapphire and two dense polycrystalline bodies versus the inverse square root of the grain size *(G)* for the various temperatures and particle velocities shown (at the right of each line in m/s). (Data from Ref. 16.)

did not attempt any correlations of erosion with the grain structure, which progressed from the above two finer *G* bodies to ones of 17, 22, and 38 μm (the latter reflecting a bimodal grain structure averaging ~ 8 and 60 μm). In further analysis of their data for indications of a *G* dependence, this author found none for the slopes of their *V–U* curves or of their toughness values but did find a definite correlation of their calculated *m* values and *G*. Thus their *m* values progressively decreased from 0.33 at the largest average *G* to values of 0.28 and 0.22 for *G* ~ 22 and 17 μm respectively, and *m* = 0.16 and 0.08 respectively at *G* ~ 2 and ~ 4 μm. The lowest *m* value at *G* ~ 4 μm might suggest a minimum, but this low value likely reflects the impact of the ~ 10% glass content of the 90% alumina body. The exact meaning of the above correlation is uncertain, but it clearly serves as a reminder that neglect of grain structure in analysis of data can be a serious limitation on understanding. Routbort and Matzke [66] similarly noted apparent failure of available models to reflect accurately the microstructural effects in their study of erosion of various SiC bodies.

The above study of Wiederhorn and Hockey, which includes data for sapphire specimens, illustrates the value of single crystal tests in general and especially for studying effects of grain structure. Data of mainly Rickerby and Macmillan [67,68] provides useful detailed studies of MgO crystal erosion. Unfortunately there is limited direct comparison of these studies with their study of a polycrystalline MgO body [69], though some differences are indicated, but their crystal data provides a basis for future comparisons. Similarly, data of Schoun and Subramanian [70] on single crystals of LiF, NaCl, KCl, and CaF_2 and of Routbort et al. [71] on crystals of Si provide an opportunity for comparison to polycrystalline erosion and hence added insight into effects of grain structure.

Consider now the effects of erosion of ceramic surfaces on the resultant strengths of the eroded specimens tested with the eroded surface as the tensile surface in uniaxial or biaxial flexure (the latter for disk or plate samples). Ritter [72,73] has evaluated the residual strength after erosion of the same five commercial alumina bodies studied by Breder and Giannakopoulos [65], apparently using the same apparatus, SiC abrasive particle sizes, and velocity range. Biaxial flexure strengths of eroded and uneroded specimens, the latter with Vickers indents, mainly with 100 N loads, showed that large grain size bodies gave rise to crack resistance and resultant increased resistance to indent and erosion induced strength loss. This was based on using a similar analysis to that of Breder and Giannakopoulos to allow for variable toughness, but in an equation predicting residual strength rather than erosion rates based on the same underlying indentation–crack concepts giving higher *m* values, since *m* correlated very similarly for both indented and eroded strengths as for the erosion rate, as noted earlier. However, the actual strengths of both indented and eroded disks were highest at intermediate *G* (~ 17 μm), not at the maximum *G*, i.e. clearly differing from both the normal σ_T–G trend (Fig. 3.1) and the predicted strength trend. Further confusion arises since Breder and Giannakopoulos concluded that *R*-curve effects were not operative in the erosion of these alumina bodies under the same test conditions. Whether enhanced material removal occurs when the impact crater is ~ *G* is unknown. Similar comparison was made using three commercial Si_3N_4 bodies [74], which, though more limited in the range of finer *G* values and complicated by bimodal grain structures, showed a similar trend for *m* to increase with *G* at finer *G*.

Besides the obvious very limited data on, and especially attention to, *G* effects on erosion and resultant strengths, there are two sets of issues that need to be noted for further attention. The first set is microstructural, i.e. not only *G*, but its variations along with grain shape and orientation. While the effects of *G* have been barely addressed, the other grain parameters have not been addressed at all. As for other properties, single crystal data provides insight to the limits of grain orientation effects, but such crystal data is very limited for erosion and apparently nonexistent for residual strengths from erosion.

Turning to other erosion studies, there can be added complexity. For individual particle impacts there may be a material-dependent velocity threshold for the onset of strength reduction [75]. For multiple particle impacts, erosion may not decrease strengths, and in some cases erosion can actually increase strengths. Thus Wada et al. [76] and subsequently Chakraverti and Rice [77,78] showed that abrasive grit blasting can measurably increase resultant strengths of TZP materials. This may be seen as an isolated case due to transformation toughening, which in part it is (e.g. PSZ bodies also show greater erosion resistance relative to their hardness, at least partly correlating with their higher relative toughness). However, Gulden [80] reported that while RBSN eroded with various size and velocity quartz particles had reduced strengths, such erosion of a commercial sintered 95% alumina and especially a commercial hot pressed Si_3N_4 showed a probable increase in strength. Similarly Laugier [81] eroded surfaces of a sialon, Al_2O_3 + 4 wt% ZrO_2, and an Al_2O_3 + TiN +TiC cutting tool bodies with Al_2O_3 grits and then measured fracture toughness via Vickers indents in the as-eroded surfaces and after polishing various depths of material from the eroded surfaces. He reports that all three materials showed substantially increased toughness in the eroded surfaces, e.g. respectively nearly fivefold, about fourfold, and nearly fourfold. However, the increased toughness rapidly diminished as material was removed from the eroded surface, implying that the surface toughening was due to a fairly shallow layer (e.g. ~ 20–40 μm) of compressive stress (e.g. from impact generated slip).

In addition to the complications of the absence of strength degradation, or more seriously strength improvements as a result of surface erosion, there are more general issues. These include the transition from a few to many impacts, i.e. from isolated impacts and damage areas to overlapping impacts, damage areas, or both, which has been only partially addressed. There is also the issue of environmental effects on both erosion and subsequent strengths. While strain rates in impact loading may limit effects of environment, there may still be environmental effects in play due to residual stresses, and in subsequent strength testing. The possible G dependence of such environmental effects (Fig. 2.7) is an added complication that needs to be addressed in a broader evaluation of G effects. Finally, there is a serious need better to evaluate the nature of erosive impact damage and its impact on both erosion and resultant strengths in terms of the nature of the cracks, their scale and character relative to the grain size, and the relevance of macro crack propagation tests to such cracking. Thus impact cracks are as complex or more complex than many indent cracks, consisting of various combinations, natures, and extents of cone, lateral, and median cracks. While such more complex crack patterns are likely to reduce somewhat the stress intensity on the resultant crack that propagates, e.g. in a post erosion strength test, there is at best a very uncertain and probably limited relation between the effect of such stress intensity reduction and toughness increases due to crack bridging and branching in large scale crack propagation. The latter have commonly been indirectly asserted via assumed contributions of R-curve effects, but the issues of both crack scale and their nature (as just noted) need to be recognized and objectively evaluated.

V. GRAIN DEPENDENCE OF WEAR AND MACHINING

A. Grain Dependence of Scratch Hardness and Abrasive Wear

Two of the common methods of studying wear are its direct study by various tests of two bodies against one another and by simulating the effects of a single abrasive particle or surface asperity impressed into and translated over the test surface under varying conditions. The grain effects on the former test are discussed in the following section and the latter, which is based on scratch hardness testing, is the subject of this section.

Wu et al. [82] conducted scratch hardness tests as a function of G for several ceramics using a conical diamond point with an included angle of 90° and a tip radius of ~ 15 μm with loads of 0.5, 1, and 1.5 kg on the tip as the specimen is rotated under it at ~ 0.1 cm/s to make a circle ~ 1 cm diameter. Such tests are referred to as diamond pin-on-disk (DPOD) tests. The wear (W) was taken as the average of the wear track cross-sectional area in planes normal to the wear surface; the reciprocal ($1/W$), i.e. the wear resistance, was plotted versus $G^{-1/2}$, since this is analogous to such plots of indentation hardness. Data for dense hot pressed polycrystalline Al_2O_3 of various grain sizes down to ~ 3 μm was all below the data for sapphire (Fig. 5.10), which may indicate effects of grain boundaries and related TEA and EA stresses. Subsequent results for a commercial 85%

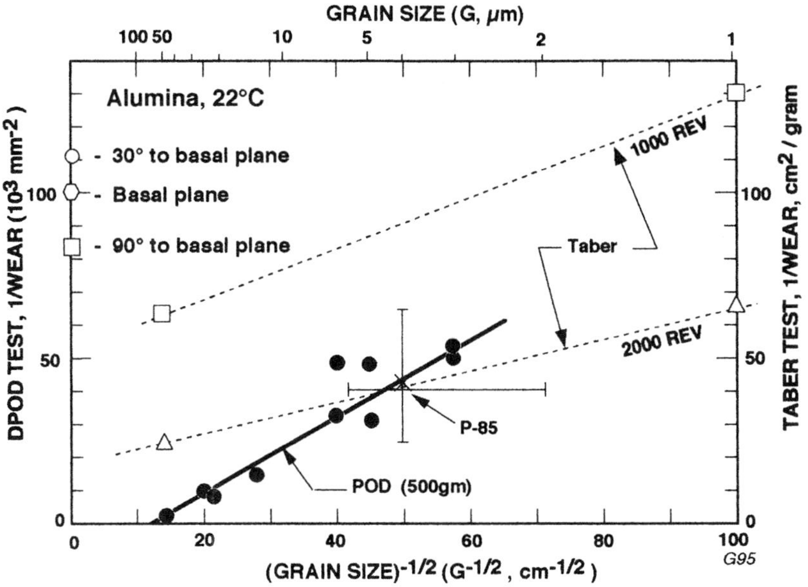

FIGURE 5.10 Diamond pin-on-disk scratch tests of hot pressed and an 85% commercial alumina (P-85) Al_2O_3, as well as three orientations of sapphire with a 0.5 kg load and for Taber wear tests of two similar aluminas at 1000 and 2000 revolutions. Note that both tests clearly show wear resistance decreasing as G increases and that the DPOD results are not only well below the sapphire data but also extrapolate to a negative intercept at $G^{-1/2} = 0$. (From Refs. 82, 83, published with the permission of Ceramic Engineering and Science Proceedings.)

alumina [83] agreed reasonably well with the hot pressed data. The polycrystalline data clearly showed substantial decrease in wear resistance as G increases, not only extrapolating well below those for three orientations of sapphire but also to a negative intercept at $G^{-1/2} = 0$. This suggests that it has a distinct minimum similar to, but probably more extreme than for, indentation hardness (e.g. Figs. 4.1,4.2).

Wu et al. [84] also made DPOD tests on several commercial SiC bodies of varying additive amount and type, and frequently with substantial G ranges, again plotting $1/W$ versus $G^{-1/2}$. Their H_V (0.5 kg load) showed similar trends with their DPOD data at 0.5 and 1 kg (Fig. 5.11). Both the DPOD tests extrapolate to positive intercepts at $G^{-1/2} = 0$, but the 1 kg load tests cover a reduced range of $1/W$. Both $1/W$ and H_V plots indicate that in bodies with substantial ranges of grain sizes, their wear behavior is more consistent with data for bodies

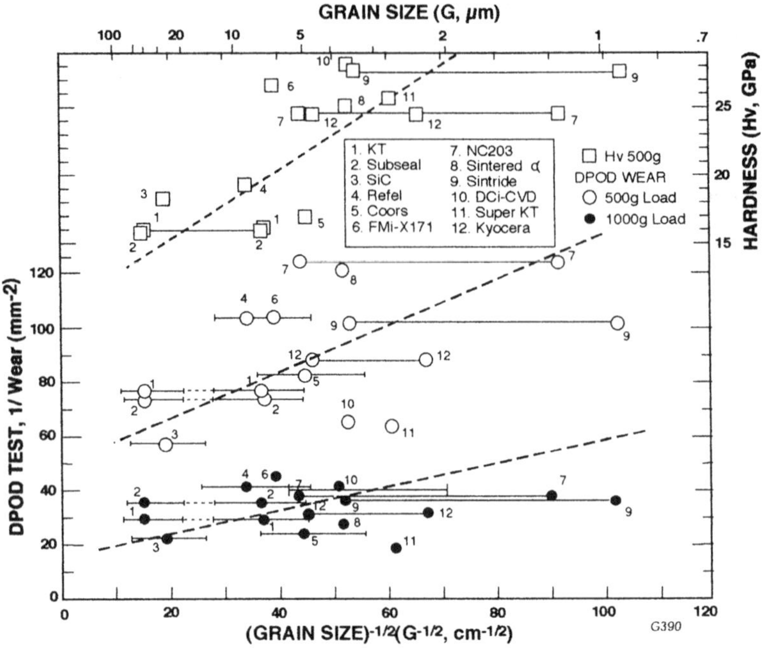

FIGURE 5.11 Plot of H_V (0.5 kg load) and the reciprocal of the DPOD wear (0.5 and 1 kg loads) versus $G^{-1/2}$ for the same set of commercial SiC bodies. Note that both sets of 0.5 kg data extrapolate to positive intercepts at $G^{-1/2} = 0$, that the 1 kg data is consistent with this despite its compression, and that data for bodies with wide G ranges agree with other data toward or at their larger G values. (After Wu et al. [84], published with the permission of Ceramic Engineering and Science Proceedings.)

with narrower G ranges when plotted versus the larger G values, i.e. the larger G values dominate such wear behavior.

Wu et al. [82] also made DPOD tests on (1) various hot pressed MgO bodies along with data on a {100} MgO crystal surface, (2) various experimental and commercial PSZ and TZP bodies (all with Y_2O_3 partial stabilization, except two with MgO) and on ~ {110} surfaces of crystals with 20 and 12 wt% Y_2O_3, and (3 mostly dense $MgAl_2O$ bodies and {100}, {110}, and {111} surfaces of stoichiometric crystals (Fig. 5.12). Again plots of $1/W$ versus $G^{-1/2}$ all showed $1/W$ decreasing as G increased, with many polycrystalline values substantially below single crystal values and extrapolating near, or below, the lower end of the single crystal values. Again, there was compression of the range of wear resistance at higher loads, so polycrystalline values tended to extrapolate closer to

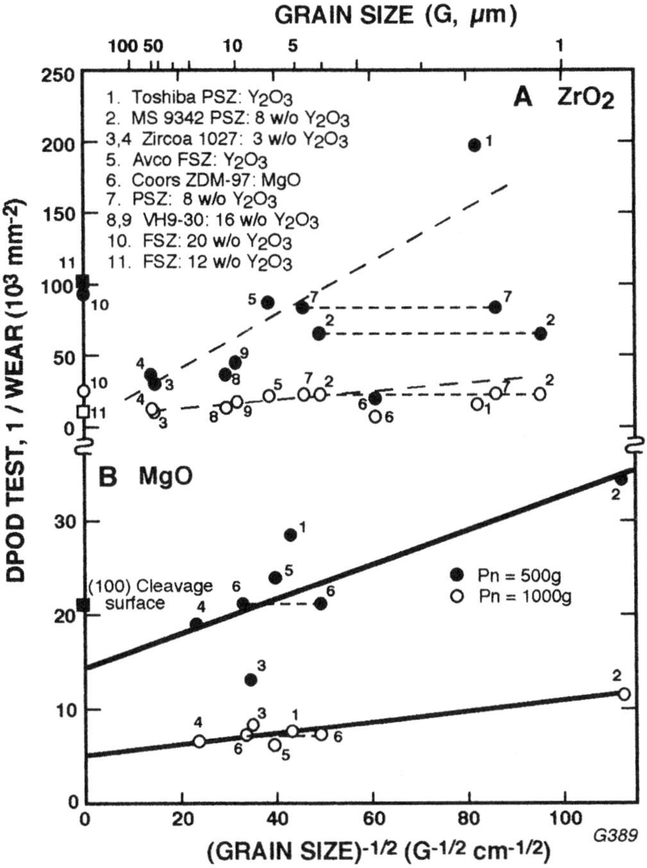

FIGURE 5.12 Plot of $1/W$ versus $G^{-1/2}$ for (A) ZrO_2 and (B) MgO for loads of 0.5 and 1 kg. Note again that (1) where bodies have a large range of grain sizes, results are more consistent with the larger G rather than the finer G, (2) polycrystalline trends extrapolate to below single crystal values, indicating a minimum in the wear resistance analogous to H minima frequently observed, and (3) compression of the data range at higher loads. (From Ref. 82, published with the permission of Ceramic Engineering and Science Proceedings.)

single crystal values at 1 versus 0.5 kg loads. All three of these data sets could be consistent with a minimum of $1/W$ at some intermediate G value as for indentation hardness, since there is either no data at larger G, e.g. > 20–50 μm for ZrO_2 and MgO, or almost no data between 0.5 and 50 μm for $MgAl_2O_4$ (which may be why it showed less G dependence than the other two materials). Also note that

TABLE 5.2 Diamond POD Wear of Various Nonoxide Ceramics of Varying Grain Sizes at Three Loads

Ceramic (~ *H*)[a]	Fabrication[b]	Grain size (*G*, μm)	Diamond POD wear track cross-sectional area (m)^2 loads 0.5 kg	1 kg	1.5 kg
AlN [12]	S	6 ± 1	0.06 ± 0.02	0.18 ± 0.06	0.28 ± 0.15
	HP	8 ± 3	0.08 ± 0.1	0.18 ± 0.1	0.26 ± 0.15
	HP	16 ± 3	0.08 ± 0.07	0.30 ± 0.20	0.59 ± 0.25
Si_3N_4 [18][c]	HP[d]	0.5	—	2.3 ± 0.6	21 ± 2
	HP	1.3	—	20 ± 3	48 ± 3
	HP	15	—	7 ± 2	45 ± 7
TiC [26]	HP	16	8.7 ± 0.9	22 ± 5	39 ± 5
TiB_2 [23]	HP	0.9	6.7 ± 1.2	24 ± 5	51 ± 12
	S	6	11 ± 2	25 ± 5	51 ± 12
	S	120	5.8 ± 1.2	52 ± 27	138 ± 75

[a]Approximate H_V (500 gm load) values.
[b]S = sintered, HP = hot pressed, all to low, e.g. ~ 0, porosity.
[c]NC 132.
Source: Ref. 82.

separate DPOD and H_V tests of ZrO_2 crystals as a function of Y_2O_3 content indicate that both are essentially independent of Y_2O_3 content until it gets below 5 w/o, i.e. where significant monoclinic content begins to appear, in which case *H* progressively decreases and wear substantially increases [82].

More limited data of Wu et al. [83] from their DPOD tests on various nonoxide materials of low to ~ zero porosity, though scattered some, also shows a general trend for increased wear (i.e. decreased wear resistance) as *G* increased (Table 5.2). Scatter is attributed to factors such as residual porosity, heterogeneities of additive distribution (and possibly to some extent the type and amount of additive), and variations in grain structure. One important example of this was the observation of significant relic structure from spray drying in a sialon body. Microcracking, which is another source of higher wear rates, is suspected in the large *G* TiB_2.

An important question regarding such scratch tests is how similar or different they are relative to both indentation hardness (Chap. 4) and other wear tests (considered in the next section). Scratch tests are very similar to indent tests in that a tip, commonly a sharp one, is pressed into the surface to make an impression, but in the scratch hardness test the tip must also plow through the material in order to make its track. Direct comparison of both tests is feasible by plotting scratch hardness versus H_V at the same loads on the same or similar materials and surface finishes for various ceramics for the same *G* (e.g. for G

~ 15 μm) where there is more data [82–84]. This comparison shows there is a general correspondence between the two tests, as might be expected, but with two sets of deviations. The first and most obvious was two materials having much higher scratch hardness versus H_V than the other materials, i.e. Si_3N_4 and especially AlN. Some difference might be expected, since while indent tests have varying extents of interactions with grain structure and hence with grain boundaries (Chap. 4), scratch tests inherently have more such interaction. Thus where impressions in either test are smaller than the grain size, the indent test will reflect more limited effect of the grain structure, while the scratch test is literally forced into successive grains and resultant broader interaction with the grain structure. The combination of the wear track character (considered next), and the fact that fracture is often a more serious source of wear, suggest that the deviations of Si_3N_4 and especially AlN are due to greater plastic flow. The second difference between indent and scratch tests is that while most materials compared in the two tests show reasonable correlation, this correlation is not constant as a function of *G*, since variations decrease as load increases. Thus comparison at another *G* result in different scatter, slopes, or both, as is also indicated by the two tests generally having different slopes as a function of *G*, which is again consistent with more interaction of scratch tests with the grain structure.

SEM examination of the wear tracks showed a broad trend for the track surface to fall in the range from a relatively smooth surface, with some undulations and striations parallel with the track at fine *G*, to more, or exclusively, fractured surface, usually with substantial to exclusive intergranular fracture at larger *G* (Figs. 5.13,5.14). The smoother surfaces with striations are typically interpreted as evidence of plastic flow. A frequent intermediate step between striations and fracture is the occurrence of partial, then complete, approximately periodic transverse interruptions of the smooth striated track surfaces as *G*, or other parameters presumably enhancing stick–slip conditions, increases, which again indicates plastic flow. The occurrence of plastic flow is supported by TEM studies of wear and abraded surfaces, e.g. Hockey et al. [86]. Though not directly studied, the intergranular fracture mode appears to commence at a finer *G* and be more complete at larger *G* and in bodies with weaker grain boundary phases, as is probably true for both MgO (Fig. 5.13) and $MgAl_2O_4$ (Fig. 5.14). Further support for the indications of plastic flow in such track formation and its role in limiting wear rates, and fracture enhancing them, is the greater extent, smoothness, and persistence to larger *G* in bodies showing lower wear, e.g. Si_3N_4 and AlN.

Mukhopadhyay and Mai [87] did similar scratch tests on three sintered, dense, pure alumina bodies [G = 0.7, 5, and 25 μm, having H_V (500 gm load) of 19, 16, and 11 GPa] using a conical diamond with a 100 μm tip radius with loads of 8–40 N. Tangential forces increased with load but were mixed as a function of

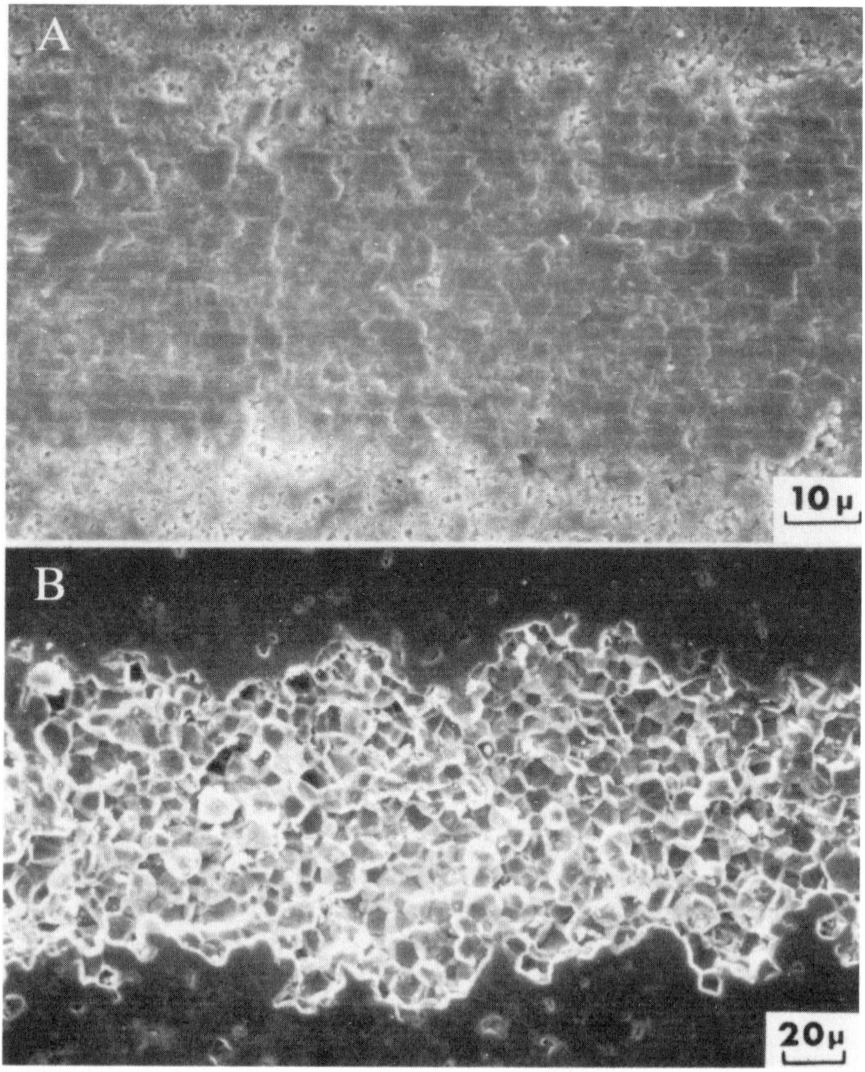

FIGURE 5.13 SEM photos of DPOD scratch wear tracks under a 1 kg load on hot pressed MgO with *G* = (A) 0.7 μm and (B) 8 μm. Note the smoother surface at finer *G* with striations parallel with the scratch path indicating plastic flow (but some possible cracking, probably along grain boundaries, under the translucent surface layer) and the essentially complete intergranular fracture at larger *G*. More cracking at such finer *G*s is attributed to residual boundary phases. (From Ref. 82, published with the permission of Ceramic Engineering and Science Proceedings.)

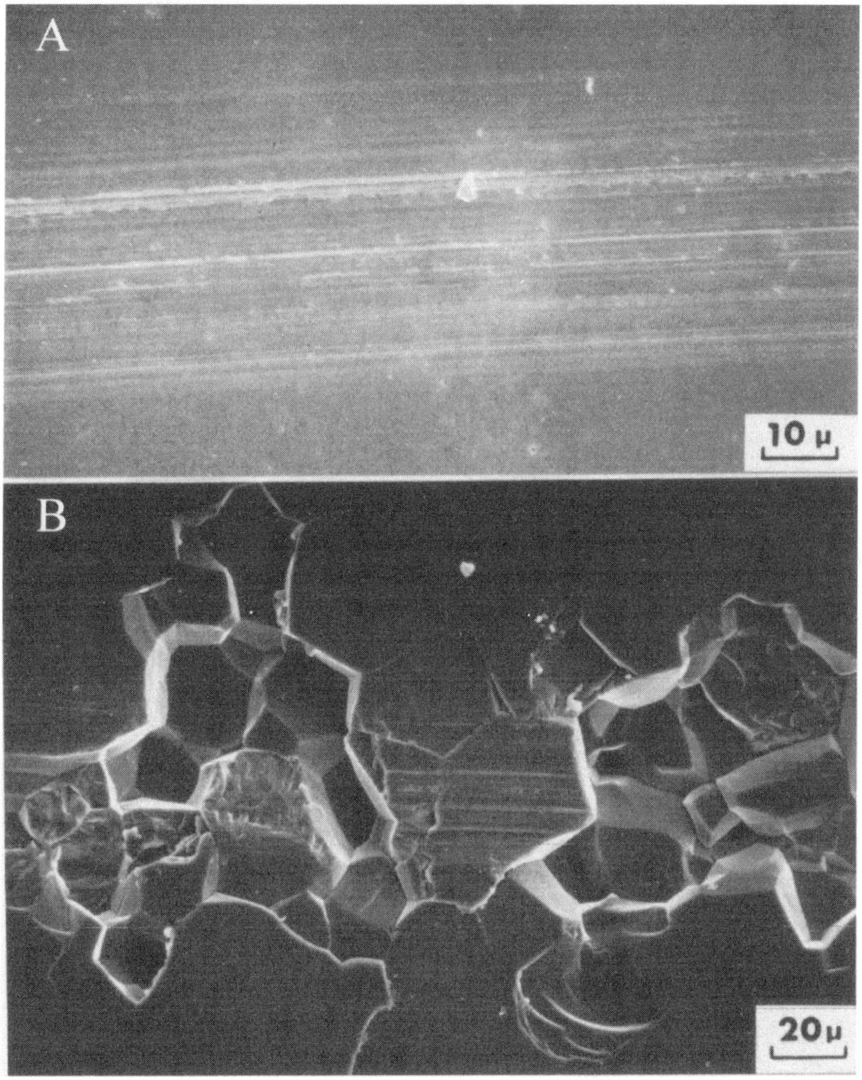

FIGURE 5.14 SEM photos of DPOD scratch wear tracks under a 1 kg load for $MgAl_2O_4$ with G = (A) 0.4 μm and (B) 50 μm. Note the smooth track surface except for some striations parallel with the track at finer G and the marked increase in microfracture, mainly intergranular at the larger G, and that this often occurs over a wider region than the track width. (From Ref. 82, published with the permission of Ceramic Engineering and Science Proceedings.)

G, except for a distinctly higher rate of increase with load for the finest G body at higher load for two passes versus one pass of the diamond tip. Friction coefficients tended to decrease with increasing load, more so for two passes versus one pass, except again for the finest G body, where friction markedly increased with load. Groove widths increased with both load and G, and wear rate initially increased rapidly and then more slowly or not at all as load increased with significant G dependence. The finest G body had the greatest changes in, but the lowest, wear rate, and the two larger G bodies, lower rates of increase (and only at lower loads) with overall higher wear rates, i.e. the intermediate G body had nearly, and the largest G body over, an order of magnitude higher wear rate. Track examination showed that the finest G wear appeared to be mainly plastic deformation to ~ 10 N loads, which then transitioned to mixed plastic deformation and intergranular fracture above 10 N, while the two larger G bodies, especially the largest G body, was mainly controlled by inter- and transgranular fracture. They noted that their results did not support the concept of material removal by joining of lateral and radial cracks based on indentation fracture and attributed the substantial G dependence to a combination of the G dependence of H and TEA stress effects.

Xu and colleagues [88–90] conducted similar scratch tests, i.e. using a conical diamond tip with an included angle of 120° and a tip radius of ~ 10 μm under loads of 10–40 N (i.e. 1–4 kg) on dense sintered Al_2O_3 bodies of G = 3, 9, 15, 21, and 35 μm. They reported scratch hardness values with 10 and 40 N loads decreasing from ~ 25 to ~ 21.5 GPa as G increased from 3 to 9 μm, i.e. similar relative changes as found by Wu et al. [82]. Xu et al.'s [89] scratch hardness data for G 15–35 μm was presented as then being constant at ~ 21 GPa, but their data would also suggests a possible minimum at ~ 20 GPa at G = 21 μm. Though scattered, some their data indicated values ~ 5% lower for 40 versus 10 N loads.

The scratch wear track results of Xu and colleagues [88–90] generally corroborate and extend those of Wu et al. [82], e.g. increasing intergranular microfracture as G increases, and provide very useful additional information, e.g. showing slip or twin bands, or both, involved in both the wear track forming process and the microfracture process. They showed that for single scratches the volume of removed material increased as the square of G (Fig. 5.15), and that the net volume of material removed in sequentially forming two parallel scratches was distinctly dependent on G. This G dependence resulted from, first, the extent of lateral damage due to cracking increasing as G increased due to its increasingly extending beyond the width of the indenter diameter at the specimen surface and, secondly, the interactions of this lateral cracking as a function of the separation of two parallel scratches. Thus they clearly showed that there were critical scratch separation distances for the onset and maximum of interaction to enhance material removal by microfacture that increased with both load and G. They showed that the scratch separation for the onset of inter-

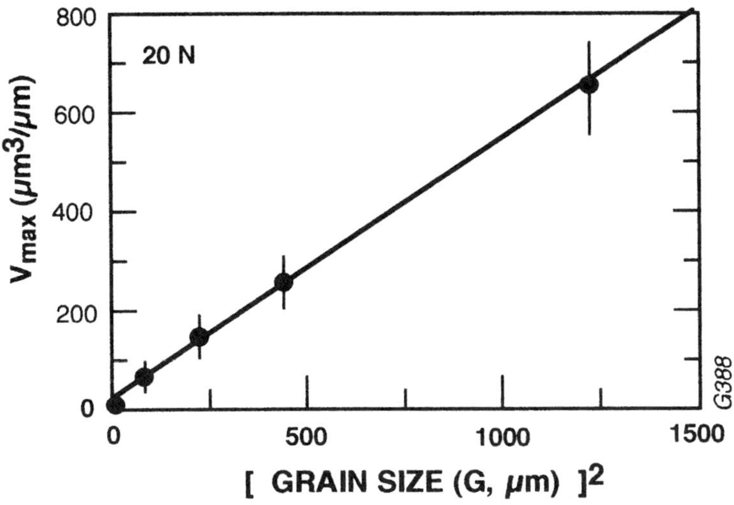

FIGURE 5.15 Plot of the volume material removed versus the grain size *(G)* squared for dense sintered alumina bodies with a 20 N load on the scratch indenter tip. (Data from Ref. 90, published with the permission of the *Journal of the American Ceramic Society*.)

action was ~ 3+ times *G* and that for maximum interaction was ~ 2- times at a 20 N load, with the scratch separations for these interactions increasing linearly with grain size (Fig. 5.16). (The *G* multiples are not totally constant for the onset and maximum interaction, since the plots of scratch separations for these interaction stages versus *G* do not go through the origin.) The material volume removed increased not only as the square of *G* (Fig. 5.15) but also as the square of the load. Thus progressively less material is removed as the scratches start to and then fully overlap. However, they also showed that the volume removed also increased ~ linearly as the number of repeat scratches in the same track increases, but with the rate of increase again being a significant function of both load and *G*.

Rice [91] summarized many of the known and expected effects of microstructure, especially *G*, on ceramic wear, noting for example that grain size effects on wear can arise not only from the size of the surface indentation relative to *G* and the subsequent motion of the indenter across grain boundaries but also from the depth of penetration into the grain. The probable role of EA and TEA on wear were noted, and that of the transition of fracture energy and toughness from polycrystalline to single crystal or grain boundary values (Fig. 2.15) has been previously noted [92], as has the greater influence of TEA (and probably EA) as the crack size approaches *G* [93]. Xu et al. [89] also noted probable

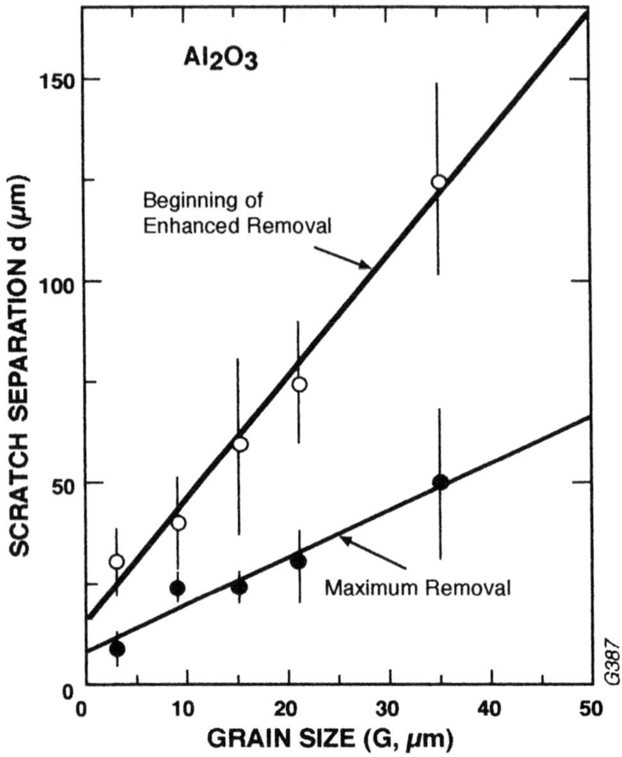

Figure 5.16 Plot of the separation (d) of two parallel scratches on the surfaces of a dense sintered alumina for enhanced removal as a function of the grain sizes of the alumina. Vertical bars are standard deviations. Note the ~ constant multiple of separation for initiation and maximum of enhanced removal due to scratch interactions, and that the curves do not go through the origin. (Data from Ref. 90, published with the permission of the *Journal of Materials Science.*)

contributions of TEA to their alumina wear results and cited them as a reason that use of substantially lower fracture toughness values may be more appropriate than normal polycrystalline values.

Diamond POD tests (2 kg load) of fibrous grain structures and their orientation effects were conducted by Wu [94] on the same jade samples discussed earlier (Table 2.3). The Guatemalan jadeite with the lowest toughness of ~ 3 MPa·m$^{1/2}$, coarser G (dia. ~ 40 μm with an aspect ratio of ~ 2), and ~ random grains had the smallest width (~ 64 μm) and cross section of the track from the diamond, but subsurface crack damage often extended to twice the track width.

The Siberian hornblende with somewhat higher toughness (~ 3.5 MPa·m$^{1/2}$) and finer G (~ 5 μm with an aspect ratio of ~ 10) with ~ random grain orientation had ~ twice the track cross section and ~ 20% greater track width, but with a similar extent of subsurface crack damage. The LB hornblende with finer G (~ 1 μm dia with an aspect ratio of ~ 50), substantial orientation, and associated toughness anisotropy ranging from ~ 2–5 times the other jades had anisotropic track character. This character was similar for any direction of fibers parallel with the plane of the track, since the diamond tip samples all orientations in the plane of its circular path. This gave track cross sections ~ 2 times those for the jadeite, but track and damage widths ~ the same as for the other hornblende. However, when the fibrous direction was normal to the surface and plane of the track, the second lowest track cross section and track width were obtained, indicating a definite anisotropy of response. An uncertainty in these tests was the hardness values for these materials and a measure of the depth of associated damage below the track (as opposed to the extent of damage near the surface outside of the track). However, these results show a definite anisotropy due to substantial preferred orientation, as might be expected, but pronounced effects of whether damage was much worse for motion normal to the fibers, as might be expected for some fibrous materials, was apparently not observed.

As with other properties, single crystal wear results are an important reference point for grain structure, especially orientation, effects. Though used more in other properties reviewed in previous sections and chapters, there is some data on friction and wear for single crystals beyond that previously cited. Thus, for example, Bowden and Brookes [95] showed that sliding friction on {100} surfaces of LiF, MgO, and diamond was the greatest in <100> and the least in <110> directions, with the extent of the anisotropy increasing as the sharpness of the slider tip increased. This anisotropy was related to the depth of penetration due to increased activation of slip and resultant cracking, especially in LiF and MgO, which was consistent with the frictional anisotropy reflecting the hardness anisotropy (Fig. 4.21 and Tables 4.3 and 4.4), especially as the sharpness of the slider increased. Buckley [96–98] reported friction coefficients for some planes on some single crystals, e.g. on sapphire varying from 0.5 to 1 depending on the surface and direction of motion. Similarly he showed the rate of wear as a function of crystal plane and direction for rutile similar to that of hardness anisotropy, but of much greater amplitude, e.g. peak-to-valley ratios of up to 7 for wear rate.

Turning to polycrystalline results, consider first abrasive wear, e.g. measured by weight loss, of a sample rotated under an abrasive wheel loaded against the sample by a fixed deadweight load so that the abrasive wheel is free to rotate about its axis, which is normal to that of the specimen rotation. This test (a commercial manifestation of which is the Taber Wear Tester) is thus similar to a pin-on-disk test with multiple contacts of the abrasive wheel replacing the single contact of the pin against the rotating disk sample. Wu et al. [82] conducted some Taber tests on two bodies of $G \sim 1$ and 50 μm from the same or similar alumina used for their DPOD tests giving a clear increase in wear as G increases and hence decreasing wear resistance as G increases, but at a lower rate than in the DPOD tests (Fig. 5.10). However, again note larger G polycrystalline results below those for single crystals. Taber testing of the same SiC specimens from the DPOD tests (Fig. 5.11) resulted in a separation of the data into two groups, one of RBSC specimens that had larger G, Si second phase, higher wear rates for a given G, and a lower slope versus $G^{-1/2}$, and the other sintered or hot pressed SiC with finer G, lower wear rates at a given G, and a higher slope versus $G^{-1/2}$ (Fig. 5.17).

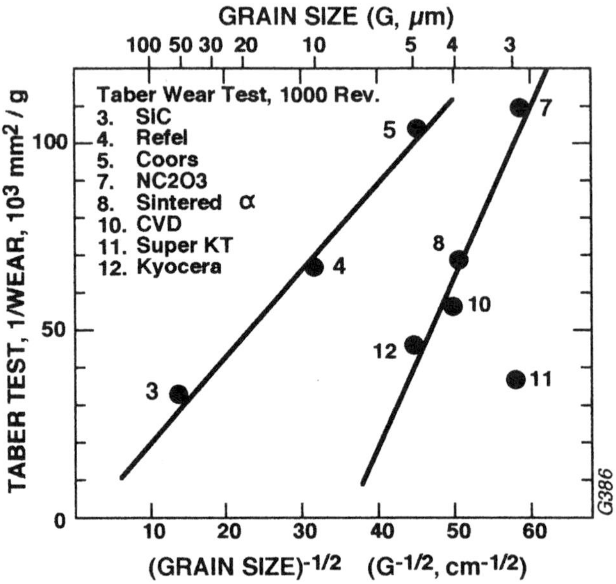

FIGURE 5.17 Wear rate (i.e. inverse of wear weight loss per unit wear area) versus the inverse square root of grain size for Taber wear tests of various RBSC and sintered and hot pressed SiC bodies, the same ones as were shown in Fig. 5.11. Note that this wear test separates the bodies into two groups, one of RBSC with more wear and a lower slope versus $G^{-1/2}$ and the other the sintered and hot pressed bodies with a higher slope versus $G^{-1/2}$, but displaced to the right, i.e. to lower wear. (From Ref. 84, published with the permission of Ceramic Engineering and Science Proceedings.)

B. Grain Dependence of Sliding and Related Wear

Turning to predominately sliding wear, Cho et al. [99] conducted some pin-on-disk tests of three of the same or similar Al_2O_3 bodies used by Xu and colleagues where the "pin" was a 12 mm diameter Si_3N_4 ball (rather than a rod with a spherical end) rotated at 100 rpm under a load of 450 N with paraffin oil lubricant. The widths of the resultant wear track used as a measure of the wear as a function of test time showed an incubation period of low wear for all samples and then a change to a much more rapid wear rate that began at a specific time for each body, with the time varying inversely with *G*, i.e. the time for onset of more rapid wear increasing substantially as *G* decreased. They observed corresponding wear track development as being only smooth striations before significant wear onset and microfracture, mostly intergranular after the onset of increased wear. The occurrence of plastic deformation was confirmed by TEM showing "severe accumulation of dislocation pileups and twins with strong crystallographic features, characteristic of abrasion damage in ceramics." They presented a model based on TEA stresses and resultant cracking for their observed *G* dependence of the onset of significant wear, predicting a dependence close to, but not exactly of, a Hall–Petch form. Liu and Fine [100] also modeled their data based on microcracking from TEA stresses, with similar agreement between their model and the data. However, neither group addressed the issue of whether such wear behavior is unique to noncubic materials, as might be implied by a direct dependence on TEA.

Gahr et al. [101] studied unlubricated sliding and oscillating wear using a hollow cylinder whose outer surface was rotated against a thin block, or oscillated over a plate, of the same ceramic, using respectively a 10 N load and a speed of ~ 0.8 m/s and a 100 N load and a speed of 0.02 m/s at 20 Hz. Both a dense sintered, pure alumina ($G \sim 0.8$–12 μm) and three mol% Y TZP ($G \sim 0.3$–2 μm were used. Both *H* and K_{IC} were measured (10 N load) showing *H* decreasing from 19.5 to 16 GPa for Al_2O_3 and ~ 13.8 to 12 GPa for the TZP as *G* increased, and K_{IC} increasing with *G* and then being ~ constant at higher *G*, i.e. respectively from ~ 3.7 to ~ 4.1 MPa·m$^{1/2}$ by $G \sim 4$ μm and 5 for Al_2O_3 to nearly 10 MPa·m$^{1/2}$ by $G \sim 1.2$ μm for ZrO_2. They also showed that surface roughness was low and constant over the limited *G* range for the TZP, and for the alumina started from the levels achieved for TZP (~ 0.1+ μm) at fine *G* but increased to ~ 0.7 μm at *G* ~ 12 μm. They showed that the unidirectional sliding coefficient of friction for the alumina was nearly independent of *G* at ~ 1, while that for the TZP increased from ~ 0.5 at the finest *G* to ~ 1 by $G = 0.6$–0.8 μm and constant beyond. For unidirectional sliding the wear resistance of both the alumina and the TZP varied as $G^{-1/2}$, with that for TZP being over an order of magnitude lower, but for reciprocating sliding the *G* dependences were respectively G^{-2} and $G^{-1/3}$, so the larger *G* wear rates for the alumina were about the same as for the TZP. Their examination of the wear tracks showed the same general trend, namely more plastic flow at finer G and more intergranular fracture as G increased, especially in alumina over its larger G range. Their results clearly showed there is no correlation of their indent K_{IC} values and wear results.

He et al. [102] conducted similar pin-on-disk tests of fine grain Y-TZP using a 4 mm diameter SiC ball under a load of 8 N oscillated over the surface at 8×10^{-2}m/s in dry N_2. They found that wear decreased linearly with $G^{-1/2}$ to G = 0.7 μm and then transitioned to a higher rate of wear, varying as G^{-1} for $G > 0.9$ μm. The higher wear regime was characterized by delamination, accompanying grain pullout, and transformation to monoclinic ZrO_2.

While the above results all show substantial decrease in wear as G decreases, there are tests showing opposite trends. Thus Xiong et al. [103] used a pin-on-disk test of Al_2O_3 disks and pins with G = 4, 8, or 12 μm with a 10 N load on the pin as the disk was rotated or oscillated at 1000 rpm and ~ 1 m/s in air. A key difference was that they used flat rather than rounded pin tips. Their wear rates (based on weight loss) for both the disk and the pin of the same G clearly showed wear of both linearly decreasing as G increased, e.g. by ~ 3/4 from G = 4 to G = 14 μm, i.e. opposite of the G dependence in other tests. They reported that the wear surfaces of the finer G test samples showed severe damage manifested as extensive intergranular fracture and substantial debris, while the largest G samples showed some intergranular fracture but no wear debris. They suggested that the opposite G dependence from that normally found versus that of their tests resulted from the use of a flat pin generating more debris at finer G that could not be removed because of the flat versus normal rounded pin tip.

There is other evidence for an opposite dependence of wear on G, i.e. for wear to increase rather than decrease as G decreases. A more detailed study of this is the wear of Ni-Zn ferrite head materials due to the passage of CrO_2 or Fe_2O_3 recording tapes over them. Thus Kehr et al. [104] showed that the amount of ferrite wear decreased by 10–20% as G increased from 3+ to ~ 39 μm for both tapes, but with about threefold higher absolute wear levels for the CrO_2 tape. The wear trends for as-machined or annealed (1 hr in air at 900°C) ferrite surfaces were the same, though there was a trend for the machined surfaces to have somewhat lower wear, at least against the Fe_2O_3, as expected from probable work hardening of such surfaces. This trend for less wear at larger G occurred with a small reduction in porosity from ~ 0.3 to ~ 0.1%, but this was not seen as sufficient to explain the difference (though there may be an effect of incorporating pores within grains with grain growth on wear behavior, as is noted below). That the amount of porosity decreased by itself is not sufficient to explain the increase in wear with increased G, but it is consistent with normal decreases in both hardness and flexure strength as G increases (both linearly versus $G^{-1/2}$, with somewhat higher strengths for annealed versus machined samples). Note, however, that similar ferrites also showed more transgranular fracture with SCG and more intergranular fracture in fast fracture than most materials (Chap. 2, Sec. III.B).

There is further precedence for wear apparently decreasing as G increases in various industrial wear tests and applications, at least of alumina bodies [105]. For example, many 99% commercial alumina products are sintered with MgO additions to control grain size (e.g. to ~ 8 μm) to give better mechanical performance, including wear resistance for various applications. However, for some wear applications, better performance is found by leaving out the MgO, which about doubles the grain size from the same or similar firing with MgO additions. The resulting larger grain bodies of very similar density as those sintered with MgO gives better wear resistance in some tests (e.g. rotating specimens in a water–sand slurry) and applications but worse performance than its finer grain counterpart in other applications and tests, e.g. of grit blast resistance. It has been suggested that the better performance of the larger G bodies may reflect probable increased intra- versus intergranular pores, but these results are a clear indication of the complexities of microstructural interactions in wear, i.e. similar to those, but possibly even more complex than, for other properties, and the need for more comprehensive testing and characterization. It should also be noted that such industrial tests also generally show many of the wear and erosion trends indicated earlier, e.g. the common superiority of Si_3N_4 (and some toughened bodies, Chap. 10, Sec. III.C).

Another important mode of wear behavior is rolling contact fatigue (RCF), since this is the method of selecting ceramics for ball and roller bearing applications. In such tests a ceramic rod with a high-quality, bearing grade surface finish is rotated at high speed while held between three balls in a plane normal to the rod axis with high contact pressures between the balls and the rod. In such tests and applications what is sought are materials and finishes that result in noncatastrophic failure with long incubation times for formation of small local spalls which can in turn lead to more serious damage and vibration. There has apparently not been any study of G dependence in Si_3N_4, which dominates the ceramic bearing field, in part since G commonly does not vary widely in Si_3N_4, and, though no clear correlation apparently exists, generally finer G bodies are common in this application. However, Rice and Wu [85] conducted limited RCF tests on four ZrO_2 materials, a PSZ single crystal, and three polycrystalline TZP bodies. Despite some limited porosity in the latter, they all showed longer RCF lives than the single crystal, indicating better performance with finer G, especially at the finest G (~ 0.25 μm), where the life was over an order of magnitude higher. Failure was by spalling, which in the polycrystalline bodies was typically from larger, mostly isolated, pores or pore clusters, indicating further potential advantage of G with improved body quality, especially less and smaller pores.

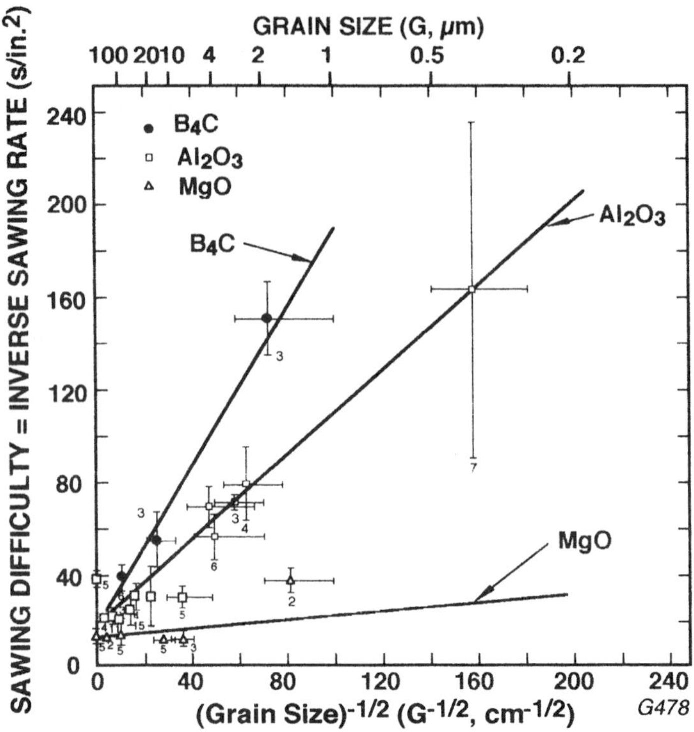

FIGURE 5.18 Difficulty of sawing with a standard commercial diamond blade with a fixed horizontal force moving the specimen into the blade versus the inverse square root of the grain size for the three ceramics shown. The correlation coefficients were reported to be 0.99, 0.98, and 0.6 for B_4C, Al_2O_3, and MgO respectively. (From Ref. 106, published with the permission of the *Journal of the American Ceramic Society*.)

VI. GRAIN DEPENDENCE OF MACHINING RATES AND EFFECTS

While the preceding sections reviewed material removal and damage by erosion and wear, there is also interest in and need for documenting and understanding corresponding grain effects on machining of ceramics. However, while there is common interest in limiting the extent of surface damage in both cases, it is desired to have higher machining rates rather than lower ones as typically sought with wear and erosion. Clearly both fixed and free abrasive machining involve similar basic mechanisms found in erosion and wear, so the topics are related.

Rice and Speronello [106] made a fairly comprehensive study of the rate of diamond sawing and of diamond grinding of various ceramics and grain sizes under a fixed horizontal force driving the work piece into the saw blade or grinding wheel. Thus they measured the times to saw through a given area or remove a given volume of material using commercial diamond blades and grinding wheels respectively. Their sawing tests for 12 ceramics (six with reasonable to substantial G range) and grinding tests of seven of these ceramics (three with reasonable to substantial G range) clearly showed that the difficulty of machining (i.e. the reciprocal of the sawing or grinding rate) was a linear function of $G^{-1/2}$ (Fig. 5.18). They further showed that more extensive sawing data correlated with representative Vickers hardnesses of the ceramics (i.e. an average value for each material neglecting G dependence) whether the slope of the inverse sawing rate versus $G^{-1/2}$ plot was used or the sawing difficulty at a fixed G was used (Fig. 5.19).

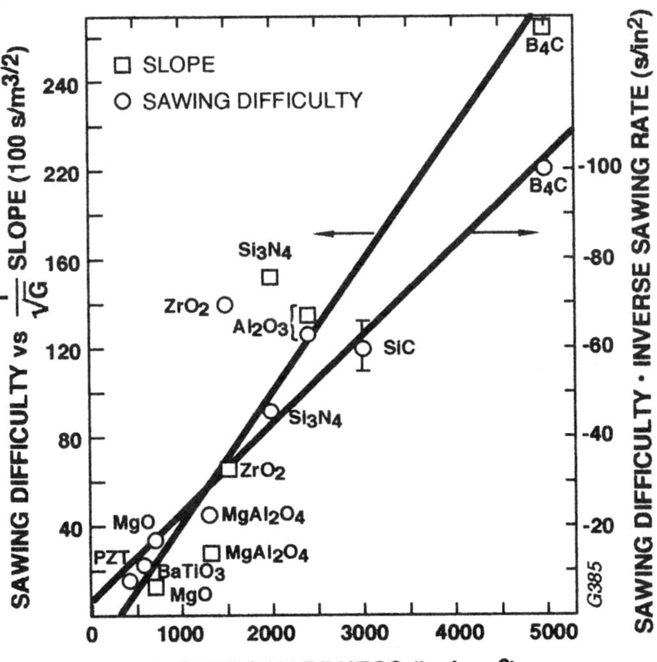

FIGURE 5.19 Plot of the slope of the sawing difficulty versus $G^{-1/2}$ slope (left scale) and the actual sawing difficulty at a fixed G (50 μm here), versus representative Vickers hardnesses showing a correlation of sawing and hardness (e.g. correlation coefficients of 0.91 to 0.95). (From Ref. 106, published with the permission of the *Journal of the American Ceramic Society*.)

Earlier work by Goyette et al. [107] showed that the vertical forces in single point diamond machining of five dense sintered Al_2O_3 bodies decreased significantly, e.g. ~ twofold, from the smallest to the largest G (2–40 μm) for a fixed rpm over the > threefold rpm range used. Similarly, Marshall et al. [108] showed that normal grinding forces increased with depth of cut as expected and that for any fixed depth of cut was substantially higher for finer G (~3 μm) than coarser G (20 and 40 μm) high purity alumina bodies. However, a 96% alumina with intermediate G (~ 11 μm) had higher forces, sapphire still higher, and a 90% alumina (G ~ 4 μm) still higher forces. Thus they noted that purity was also a factor: some less pure materials had higher forces. Normal fracture toughness values from conventional tests, i.e. with cracks large with respect to most microstructures, showed an inverse correlation with grinding forces (hence grinding resistance), but there was a better, though scattered, correlation with toughness values projected to finer crack sizes, some of these approaching single crystal values.

Xu et al. [89] showed similar effects of G on grinding results in the alumina bodies used in their earlier summarized studies. Normal and tangential grinding forces both increased with depth of cut in respectively a nearly linear and a linear fashion, with substantially higher rates of increase for G = 3 μm, but no significant differences for G = 9–35 μm. Average surface roughness progressively increased, but with diminishing increases as G increased; the number, and hence also the area, of microcracks (intergranular) was zero for G = 3 μm, but beginning at G = 9 μm they increased at a substantial, ~ linear, rate as a function of $G^{-1/2}$.

VII. DISCUSSION AND SUMMARY

The properties considered in this chapter are commonly more varied and complex than those of Chapters 2 and 3, and less well documented, for their dependence on G and other material, body, and test parameters. All have several aspects in common, i.e. compressive stresses, as for hardness, and all have at least a general correlation with hardness and involve some degree of plastic deformation, which varies, often in incompletely known fashions, for the different properties. Grain parameters play some, commonly an important, role in all of them, but only grain size has been addressed to any, but still limited, extent (sometimes not explicitly), so other grain effects offer important opportunities not only for study and understanding of their roles in each of the properties but also for better and broader understanding of the properties themselves. For example, a number of materials and bodies studied most likely have measurable but varying degrees of grain elongation, e.g. Si_3N_4, α-SiC, Al_2O_3, and TiB_2, and various bodies, e.g. hot pressed ones, commonly have some preferred orientation, so these unaddressed factors are probably factors in data scatter, and especially uncertainties in property correlations. Thus better attention to grain and other microstructural effects will significantly increase understanding of each of the properties.

An important factor limiting correlation of properties of this chapter with those of earlier chapters, and hence with better known microstructural correlations, is the common correlation with multiple properties. Properties in previous chapters commonly have strong correlations with two properties, namely Young's modulus and fracture energy (which are also related), though there are variations and uncertainties, especially in the latter (Chap. 2, Sec. III; Chap. 3, Sec. VII). These uncertainties and those of the specific roles of TEA and EA are compounded for the properties in this chapter due to the uncertainties, and often more complex dependence of their properties.

Three factors illustrate the added uncertainties in the dependence of properties of this chapter on other more basic properties and hence on their grain structure dependence. First, the correlation of H with properties in this chapter is more uncertain because the load, indenter geometry, surface condition, strain rate, and environment, hence the grain structure dependences probably vary in uncertain fashions, even for those properties where loading is localized to specific areas as in erosion and wear similar to indentation. However, even for these cases, there are differences and complexities, e.g. scratch hardness and wear inherently involve more interaction with the grain structure than in at least some ranges of load, geometry, and grain structures than for indentation hardness. Similarly, both erosion and especially ballistic performance involve effects of the hardness of both the impacting object and the target. The second factor is the high strain rates of some of these processes such as many erosion situations (e.g. for some missile domes and windows) and especially armor applications, particularly against kinetic energy penetrators.

The third factor is uncertainty in the local toughness to use and impacts of TEA and EA stresses on the local fracture process. The most pervasive issue is that of toughness associated with cracks on a scale <, equal to, or only somewhat > the local G, i.e. similar to situations discussed for tensile failure (Chap. 2, Sec. III; Chap. 3, Sec. VII). Note first that addressing this issue in the terms of short versus long crack behavior is misleading, since this implies that even a short crack, which can still have a length large in comparison to the grain size, would have behavior substantially closer to that of a long crack in comparison to a crack with both its length and depth similar to or < G. Further, the treatment in terms of long and short cracks has focused on the baseline toughness being that for easier single crystal fracture, instead of also including that for grain boundary fracture, which may be similar, or substantially lower, depending on boundary character and on how many boundary facets are involved. For erosion and wear, and possibly compression, effects of environment may also be factors, though for compression this may be more of a factor for failure involving more surface connected cracks due to nonuniform stressing. Also such possible effects in compression may be more due to lubrication effects between cracks than to true slow crack growth, since compressive failure typically appears to involve varying friction between crack faces.

Next consider the related issues of self-consistency of the results and their completeness, since the two are closely related, i.e. fewer results mean less opportunity to evaluate self-consistency. An important aspect of self-consistency is consideration of data scatter, and its consistency with both the property and the mechanisms involved as well as the microstructural variations and their impact and correlation with property variations. Thus, for example, though widely neglected, the coefficients of variation, hence also the Weibull moduli, of well conducted compression tests are higher, commonly substantially so, in comparison to such values on the same materials in tension (Table 5.2). This is consistent with compressive failure being a process of cumulative failure from the nucleation and growth of many small cracks in the body versus tensile failure reflecting failure due to the most severe flaw developed or existing in the region of high stress. Similarly, with more data it may be feasible to correlate variations in erosion and wear with factors such as the distribution of asperities and surface grain sizes, shapes, and orientations.

Despite the limitations and uncertainties outlined above, consider the grain size trends demonstrated beyond the general decrease with increasing G. Compressive strengths, which extrapolate to the range of differing single crystal values (indicating probable effects of grain shape and elongation), generally decrease more with increasing G and show less variation (for well conducted tests), than tensile strengths; but many issues remain. These include specifics and variations of plastic deformation; crack nucleation and growth; strain rate, environment, and size effects. Though not addressed very extensively, grain size probably plays a role in ballistic stopping abilities of ceramics, e.g. as indicated by limited data with smaller, slower, softer bullets; but much more study is needed to address the complexities due to the high strain rates, differing projectiles, and apparent accentuation of other microstructural variables such as grain boundary phases and bonding. Differentiation between microstructural and property effects as a threshold requirement versus a continuous factor seems important. Erosion shows higher material removal as G increases and indicates some possible correlation with the G dependence of H, e.g. possibly via reduction of the resistance to erosive material removal at intermediate G. There are also uncertainties in the G dependence of strengths remaining after some erosion, with an important question being what fracture toughness values to use as a function of cracks approaching, or becoming <, G. Again the terminology of long versus short cracks does not adequately reflect this issue of crack size relative to G, and toughness values for fracture along a few or one grain boundary facet need to be considered; these K values may be substantially < single crystal values.

Scratch hardness, which can simulate aspects of a single wear asperity, clearly has similarities with indentation hardness, e.g. Fig. 5.11 with values over a significant *G* range below single crystal values indicating possible effects of TEA and EA and of grain shape and orientation. This data also indicates the bias of results toward better correlation with the sizes of larger grains where there is a substantial *G* distribution. This is similar to the role of larger grains in tensile failure, probably partly reflecting fracture as part of the scratch track generation mechanisms. However, there must also be differences between scratch and hardness, and especially tensile strength, tests. Thus, for example, the motion of the scratch indenter increases interaction with the grain structure over that for hardness indentation only, as indicated by materials such as Si_3N_4 and AlN deviating significantly from the correlation of the two tests and these deviations apparently being associated with greater plastic flow (which presumably also has significant *G* dependence). An important factor in repeated scratching is the interaction of adjacent scratches by enhanced fracture between the scratches, initiating and reaching maxima respectively at ~ 3 and ~ 2 times *G*, showing another important impact of *G* (and suggesting similar studies of indent-related cracking) (Fig. 5.16). The preeminent difference of tensile failure and scratch fracture is that the former is usually dominated by a single weak link, and thus is impacted by the statistical aspects of flaw location, e.g. from machining, and a single large grain or cluster of them. On the other hand, scratches cross many grains and can interact with each other, both of which can be impacted by the character of many individual larger grains or grain clusters. Thus the role of grain size distribution on wear is a potentially important but generally unaddressed topic.

Wear, despite many variations in conditions, usually correlates with hardness and thus commonly increases with *G*, i.e. wear rate decreases as *G* decreases, as do machining rates (Figs. 5.18 and 5.19). However, opposite wear trends with *G* were noted, e.g. with effects of residual porosity changing from primarily intergranular to some intragranular locations as grains grow, being suggested in one of the cases of opposite *G* dependence. In the other case, the greater difficulty of forming or removing wear debris was suggested as a possible cause of the opposite *G* dependence from that commonly found. These clearly highlight the need for further study, especially with more comprehensive testing and characterization, and evaluation of possible contributions of a broader range of material properties such as TEA and EA, as is true for all properties in this chapter. Impacts of TEA and EA in wear are indicated by the frequent increase in intergranular fracture as *G* increases (i.e. opposite of most fracture trends, Fig. 2.5 in both noncubic materials (with TEA and EA) and cubic materials (with EA only), e.g. Figs., 5.13, 5.14.

Finally, four key factors, especially about erosion and wear, deserve added emphasis. First, they clearly depend on *G* via its impacts on both plastic deformation and fracture, with various probable environmental, reaction, and interactive effects. Second, there is a probable role of EA and TEA in these behaviors, which probably underlie material-dependent G levels for onset or changes in mechanisms and results, e.g. to intergranular fracture. Third, a major issue out of several that need much more study, but deserves particular note, is the transition from individual impacting particle or asperity indentations or scratches to many interactive ones, and the probable enhancement of effects of such impacts or tracks when their dimensions are in a range similar to that of the grains. Fourth, both the importance and the diversity of wear make it an important area for further study.

REFERENCES

1. R. W. Rice. Porosity of Ceramics. Marcel Dekker, New York, 1998.
2. J. J. Gilman. Hardness-A Strength Probe. The Science of Hardness Testing and Its Research Applications (J. H. Westbrook and H. Conrad, eds.). Am. Soc. for Metals, Metals Park, OH, 1973, pp. 51–74.
3. R. W. Rice. The Compressive Strength of Ceramics. Ceramics in Severe Environments, Materials Science Research 5 (W. W. Kriegel and H. Palmour III, eds.). Plenum Press, New York, 1971, pp. 195–227.
4. R. W. Rice. Microstructure Dependence of Mechanical Behavior of Ceramics. Treatise on Materials Science and Technology 2 (R. C. McCrone, ed.). Academic Press, New York, 1977, pp. 199–238.
5. J. Lankford. The Compressive Strength of Strong Ceramics: Microplasticity Versus Microfracture. J. Hard Mat. 2(1–2):55–77, 1991.
6. M. C. Shaw. The Fundamental Basis of the Hardness Test. The Science of Hardness Testing and Its Research Applications (J. H. Westbrook and H. Conrad, eds.). Am. Soc. for Metals, Metals Park, OH, 1973, pp. 1–11.
7. R. W. Rice. Discussion of The Fundamental Basis of the Hardness Test. The Science of Hardness Testing and Its Research Applications (J. H. Westbrook and H. Conrad, eds.). Am. Soc. for Metals, Metals Park, OH, 1973, pp. 12–13.
8. D. M. Marsh. Plastic Flow in Glass. Proc. Roy. Soc. A 279:420–435, 1964.
9. J. Lankford. Compressive Strength and Microplasticity in Polycrystalline Alumina. J. Mat. Sci.12:791–796, 1977.
10. J. Lankford. Temperature-Strain Rate Dependence of Compressive Strength and Damage Mechanisms in Aluminum Oxide. J. Mat. Sci. 16:1567–1578, 1981.
11. J. Lankford. Uniaxial Compressive Damage in α-SiC at Low Homologous Temperatures. J. Am. Cer. Soc. 62(5–6):310–312, 1979.
12. J. Lankford. Plastic Deformation of Partially Stabilized Zirconia. J. Am. Cer. Soc. 66(11):C-212–213, 1983.
13. J. Lankford. Deformation Mechanisms in Yttria-Stabilized Zirconia. J. Mat. Sci. 23:4144–4156, 1988.

14. D. Viechnicki. W. Blumenthal, M. Slavin, C. Tracy, and H. Skeele. Armor Ceramics—1987. Proc. 3rd TACom Armor Coord. Conf. for Light Combat Vehicles. Secret 17–19 Feb., Vol. 2, pp. 27–54, 1987.
15. C. A. Tracy, M. Slavin, and D. Viechnicki. Ceramic Fracture During Ballistic Impact. Advances in Ceramics 22: Fractography of Glasses and Ceramics (V. D. Frechette and J. R. Varner, eds.). American Cer. Soc., Westerville, OH, 1988, 99 295–306.
16. S. M. Wiederhorn and B. J. Hockey. Effect of Material Parameters on the Erosion Resistance of Brittle Materials. J. Mat. Sci. 18:166–180, 1983.
17. D. B. Marshall. Failure from Surface Flaws. Fracture in Ceramic Materials—Toughening Mechanisms, Machining Damage, Shock (A. G. Evans, ed.). Noyes, Park Ridge, NJ, 1984, pp. 190–220.
18. R. W. Rice. Strength/Grain Size Effects in Ceramics. Proc. Brit. Cer. Soc. 20:205–257, 6/1972.
19. R. A. Alliegro. Titanium Diboride and Zirconium Diboride Electrodes. Encyclopedia of Electrochemistry (C. A. Hampel, ed.). Reinhold, New York, 1964, pp. 1125–1130.
20. V. Mandorf and J. Hartwig. High Temperature Properties of Titanium Diboride. High Temperature Materials 11 (G. M. Ault, W. F. Barclay, and H. P. Munger, eds.). Interscience, New York, 1963, pp. 455–467.
21. J. V. E. Hansen. The Norton Co. Private communication, 1970.
22. W. A. Dunlay, C. A. Tracy, and P. J. Perrone. A Proposed Uniaxial Compression Test for High Strength Ceramics. US Army Materials Technology Lab. Report MTL-TR-89-89, 9/1989.
23. H. E. Exner and J. Gurland. A Review of Parameters Influencing Some Mechanical Properties of Tungsten Carbide-Cobalt Alloys. Powder, Metallurgy 13(25):13–31, 1970.
24. L. F. Vereshchagin, I. S. Gladkaya, and V. N. Slesarev. Compressive Strength and Thermal Stability of Polycrystalline Cubic BN. Izvestiya Akademiya Nauk SSSR, Neorganicheskie Materialy 13(6):1022–1024, 1977.
25. A. D. Osipov, I. T. Ostapenko, V. V. Tarosov, R. V. Tarasov, V. P. Podtykan, and N. F. Kartsev. Effect of Porosity and Grain Size on the Mechanical Properties of Hot-Pressed Boron Carbide. Porosh. Met. 1:63–67, 1982.
26. J. R. Floyd. Effects of Firing on the Properties of Dense High-Alumina Bodies. Trans. Brit. Cer. Soc. 64:251–265, 1965.
27. P. W. Bridgman. Studies in Large Plastic Flow and Fracture. McGraw-Hill, New York, 1952, p. 120.
28. C. A. Tracy. A Compressive Test for High Strength Ceramics. J. Testing Eval. 15(1):14–19, 1987.
29. J. Castaing, J. Cadoz, and S. H. Kirby. Prismatic Slip of Al_2O_3 Single Crystals Below 1000°C in Compression Under Hydrostatic Pressure. J. Am. Cer. Soc. 64(9):504–511, 1981.
30. M. D. Burdick and H. S. Parker. Effect of Particle on Bulk Density and Strength Properties of Uranium Dioxide Specimens. J. Am. Cer. Soc. 39(5):181–187, 1956.
31. N. Igata, R. R. Hasiguti, and K. Domoto. Micro-Plasticity of Uranium Dioxide at

Room Temperature. Proc. 1st Intl. Conf. on Fracture 2. Japanese Soc. Strength and Fracture Materials, 1966, pp. 883–898.
32. F. P. Knudsen. Dependence of Mechanical Strength of Brittle Polycrystalline Specimens on Porosity and Grain Size. J. Am. Cer. Soc. 42(8):376–388, 1959.
33. C. E. Curtis and J. R. Johnson. Properties of Thorium Oxide Ceramics. J. Am. Cer. Soc. 40(2):63–68, 1957.
34. C. S. Yust and C. J. Hargue. Deformation of Hyperstoichiometric UO_2 Single Crystals. J. Am. Cer. Soc. 54(12):628–635, 1971.
35. M. S. Paterson and C. W. Weaver. Deformation of Polycrystalline MgO Under Pressure. J. Am. Cer. Soc. 53(8):463–471, 1970.
36. S. M. Copley and J. A. Pask. Deformation of Polycrystalline MgO at Elevated Temperatures. J. Am. Cer. Soc. 48(12):636–642, 1965.
37. T. A. Auten and S. V. Radcliffe. Deformation of Polycrystalline MgO at High Hydrostatic Pressure. J. Am. Cer. Soc. 59(5-6):249–253, 1976.
38. C. W. Weaver and M. S. Paterson. Deformation of Cubic-Oriented MgO Crystals Under Pressure. J. Am. Cer. Soc. 52(6):293–301, 1969.
39. T. A. Auten, S. V. Radcliffe, and R. B. Gordon. Flow Stress of MgO Single Crystals Compressed Along [100] at High Hydrostatic Pressure. J. Am. Cer. Soc. 59(1–2):40–42, 1976.
40. W. F. Brace. Dependence of Fracture Strength of Rocks on Grain Size. Penn. State Univ. Mineral Experimental Station Bull. 76:99–103, 1963.
41. W. F. Brace. Brittle Fracture of Rocks. Intl. Conf. on States of Stress in the Earth's Crust—Preprints of Papers (W. R. Judd, ed.). Rand Corp. Report, 1963.
42. J. T. Fredrich and B. Evans. Effect of Grain Size on Brittle and Semibrittle Strength:Implications for Micromechanical Modeling of Failure in Compression. J. Geophysical Res. 95:10907–10920, 7/1990.
43. W. Rafaniello. Development of Aluminum Nitride:A New Low-Cost Armor. Dow Chem. Co. Final Report for US Army Research Office Contract DAAL03-88-C-0012, 12/1992.
44. D. A. Nelson, Jr., and A. L. Rouff. The Compressive Strength of Perfect Diamond. J. Appl. Phys. 50(4):2763–2764, 1979.
45. J. Lankford. Mechanisms Responsible for Strain-Rate-Dependent Compressive Strength in Ceramic Materials. J. Am. Cer. Soc. 64(2):C–33–34, 1981.
46. J. Lankford. High Strain-Rate-Dependent Compression and Plastic Flow of Ceramics. J. Mat. Sci. 15:745–750, 1996.
47. A. Nash. Compressive Failure Modes of Alumina in Air and Physiological Media. J. Mat. Sci. 18:3571–3577, 1983.
48. L. Ewart and S. Suresh. Crack Propagation in Ceramics Under Cyclic Loads. J. Mat. Sci. 22:1173–1192, 1987.
49. S. Suresh. Fatigue Crack Growth in Brittle Materials. J. Hard Materials 2(1–2):29–54, 1991.
50. H. C. Heard and C. F. Cline. Mechanical Behavior of Polycrystalline BeO, Al_2O_3, and AlN at High Pressure. J. Mat Sci. 15:1889–1897, 1980.
51. I. A. Bairamashvili, G. I. Kalandadze, A. M. Eristavi, J. Sh. Jobava, V. V. Chotu-

lidi, and Yu. I. Saloev. An Investigation of the Physicomechanical Properties of B_6O and SiB_6. J. Less Com. Met. 67:455–461, 1979.
52. M. Adams and G. Sines. Crack Extension from Flaws in a Brittle Material Subjected to Compression. Tectonophy. 49:97–118, 1978.
53. C. G. Sammis and M. F. Ashby. The Failure of Brittle Porous Solids Under Compressive Stress States. Acta Metall. 34(3):511–526, 1986.
54. G. Sines and T. Taira. Tensile Strength Degradation of Ceramics from Prior Compression. J. Am Cer. Soc. 72(3):502–505, 1989.
55. M. S. Stucke and A. S. Wronski. The Dependence of the Strength of Several Oxide Ceramics on Hydrostatic Pressure. Proc. Brit. Cer. Soc. 25:109–126, 1975.
56. R. W. Rice and R. C. Pohanka. Grain Size Dependence of Spontaneous Cracking in Ceramics. J. Am. Cer. Soc. 62(11–12):559–563, 1979.
57. R. W. Rice. Possible Effects of Elastic Anisotropy on Mechanical Properties of Ceramics. J. Mat. Sci. Lett. 13:1261–1266, 1994.
58. D. P. H. Hasselman. Single Crystal Elastic Anisotropy and the Mechanical Behavior of Polycrystalline Brittle Refractory Materials. Anisotropy in Single-Crystal Refractory Compounds (F. W. Vahldiek and S. A. Merson, eds.). Plenum Press, New York, 1968, pp. 247–265.
59. S. M. Wiederhorn, B. J. Hockey, and D. E. Roberts. Effect of Temperature on the Fracture of Sapphire. Phil. Mag. 28(4):783–796, 1973.
60. W. J. Ferguson and R. W. Rice. Effect of Microstructure on the Ballistic Performance of Alumina. Materials Research 5, Ceramics in Extreme Environments (W. W. Kriegel and H. Palmour III, eds.). Plenum Press, New York, 1971, pp. 261–270.
61. S. Wada and N. Watanabe. Solid Particle Erosion of Brittle Materials (Part 3)—The Interaction with Material Properties of Target and that of Impingement on Erosive Wear Mechanism. Yogyo-Kyokai-Shi 95(10):573–578, 1987.
62. S. Wada and N. Watanabe. Solid Particle Erosion of Brittle Materials (Part 1)—The Relation Between Erosive Wear and Properties of Gas Pressure Sintered Si_3N_4. Yogyo-Kyokai-Shi 94(11):1157–1163, 1986.
63. A. Swanson, P. Miles, and J. Paoois. Chemically Vapor Deposited Semi-Conductors for Lasers and Infrared Window Applications. Raytheon Co. Final Technical Report AFML-TR-77-34, 7/1977.
64. J. L. Routbort and R. O. Scattergood. Anomalous Solid-Particle Erosion Rate of Hot-Pressed Silicon Carbide. J. Am. Cer. Soc. 63(9–10):593–595, 1980.
65. K. Breder and A. E. Giannakopoulos. Erosive Wear in Al_2O_3 Exhibiting Mode-I R-Curve Behavior. Cer. Eng. Sci. Proc. 11(7–8):1046–1060, 1990.
66. J. L. Routbort and Hj. Matzke. On the Correlation Between Solid-Particle Erosion and Fracture Parameters in SiC. J. Mat. Sci. 18:1491–1496, 1983.
67. D. G. Rickerby, B. N. Pramilabai, and N. H. Macmillan. The Effect of Crystallographic Orientation on Damage in MgO Due to Spherical Particle Impact. J. Mat. Sci. 14:1807–1816, 1979.
68. D. G. Rickerby and N. H. Macmillan. The Effect of Approach Direction on Damage in MgO Due to Spherical Particle Impact. J. Mat. Sci. 15:2435–2447, 1980.

69. D. G. Rickerby and N. H. Macmillan. Erosion of MgO by Solid Particle Impingement at Normal Incidence, J. Mat. Sci. 16:1579–1591, 1981.
70. S. R. Schuon and K. N. Subramanian. Microstructural Aspects of Impact Erosion in Single Crystals of LiF, NaCl, and CaF_2 by Blunt Projectiles. J. Mat. Sci. 18:732–740, 1953.
71. J. L. Routbort, R. O. Scattergood, and E. W. Kay. Erosion of Silicon Single Crystals. J. Am. Cer. Soc. 63(11–12):635–640, 1980.
72. J. E. Ritter. Effect of Microstructure on the Impact Damage of Ceramics. Ceramics Today—Tomorrow's Ceramics (P. Vincenzini, ed.). Elsevier, 1991, pp. 2733–2742.
73. J. E. Ritter. Effect of Microstructure on the Impact Damage of Ceramics. Cer. Intl. 17:165–170, 1991.
74. J. E. Ritter, S. R. Choi, K. Jackus, P. J. Whalen, and R. J. Rateick, Jr. Effect of Microstructure on the Erosion and Impact Damage of Sintered Silicon Nitride. J. Mat. Sci. 26:5543–5546, 1991.
75. Y. Akimune, T. Akiba, and T. Ogasawara. Damage Behavior of Silicon Nitride for Automotive Gas Turbine Use when Impacted by Several Types of Spherical Particles. J. Mat. Sci. 30:1000–1004, 1995.
76. S. Wada, N. Watanabe, and H. Hasegawa. Solid Particle Erosion of Brittle Materials (Part 10)—Strengthening by Erosion in Tetragonal ZrO_2 Ceramics. J. Cer. Soc. Jpn., Intl. Ed. 97:128–132, 1989.
77. J. T. Chakraverti and R. W. Rice. Strengthening of Zirconia-Toughened Materials by Grit Blasting. J. Mat. Sci. Let. 16: 404–405, 1997.
78. R. W. Rice and J. T. Chakraverti. Non-Machining Surface Strengthening of Transformation Toughened Materials. US Patent No. 5,228,245, 7/20/1993.
79. S. Srinivasan and R. O. Scattergood. Erosion of Mg-PSZ by Solid Particle Impact. Adv. Cer. Mat. 3(4):345–352, 1988.
80. M. E. Gulden. Study of Erosion Mechanisms of Engineering Ceramics, Static Fatigue of Ceramics. Solar Turbines Intl. Final Report for 1/4/1973–1/1/1980 for ONR Contract N00014-73-C-0401, 7/1980.
81. M. T. Laugier. Surface Toughening of Ceramics. J. Mat. Sci. Lett. 5:252, 1986.
82. C. Cm. Wu., R. W. Rice, D. Johnson, and B. A. Platt. Grain Size Dependence of Wear in Ceramics. Cer. Eng. Sci. Proc. 6(7–8):995–1011, 1985.
83. C. Cm. Wu., R. W. Rice, C. P. Cameron, L. E. Dolhert, J. H. Enloe, and J. Block. Diamond Pin-on-Disk Wear of Al_2O_3 Matrix Composites and Nonoxides. Cer. Eng. Sci. Proc. 12(7–8):1485–1499, 1991.
84. C. Cm. Wu., R. W. Rice, B. A. Platt, and S. Carrle. Wear and Microstructure of SiC Ceramics. Cer. Eng. Sci. Proc. 6(7–8):1023–1039, 1985.
85. R. W. Rice and C. Cm. Wu. Wear and Related Evaluations of Partially Stabilized ZrO_2. Cer. Eng. Sci. Proc. 6(7–8):1012–1022, 1985.
86. B. J. Hockey. Observations by Transmission Electron Microscopy on the Subsurface Damage Produced in Aluminum Oxide by Mechanical Polishing and Grinding. Proc. Brit. Cer. Soc. 20:95–115, 6/1972.
87. A. K. Mukhopadhyay and Y.-W. Mai. Grain Size Effect on Abrasive Wear Mechanisms in Alumina Ceramics. Wear 162–164:258–268, 1993.

88. H. H. K. Xu, and S. Jahanmir. Microfracture and Material Removal in Scratching of Alumina. J. Mat. Sci. 30(4):2235–2247, 1995.
89. H. H. K. Xu, L. Wei, and S. Jahanmir. Influence of Grain Size on the Grinding Response of Alumina. J. Am. Cer. Soc. 79(5):1307–1313, 1996.
90. H. H. K. Xu, S. Jahanmir, and Y. Wang. Effect of Grain Size on Scratch Interactions and Material Removal in Alumina. J. Am. Cer. Soc. 78(4):881–891, 1995.
91. R. W. Rice. Micromechanics of Microstructural Aspects of Ceramic Wear. Cer. Eng. Sci. Proc. 6(7-8):940–958, 1985.
92. R. W. Rice, S. W. Freiman, and J. J. Mecholsky, Jr. The Dependence of Strength-Controlling Fracture Energy on the Flaw-Size to Grain-Size Ratio. J. Am. Cer. Soc. 63(3-4):129–136, 1980.
93. R. W. Rice, R. C. Pohanka, and W. J. McDonough. Effect of Stresses from Thermal Expansion Anisotropy, Phase Transformations, and Second Phases on the Strength of Ceramics. J. Am. Cer. Soc. 63(11-12):703–710, 1980.
94. C. Cm. Wu. Wear of Ceramic Natural Fiber Composites. Engineered Materials for Advanced Friction and Wear Applications, Proc. Intl. Conf., AST Intl., Metals Park, OH, 1988, pp. 79–84.
95. F. P. Bowden and C. A. Brookes. Frictional Anisotropy in Nonmetallic Crystals. Proc. Roy. Soc. A, 294(1442):244–258, 1966.
96. D. H. Buckley. Friction and Wear of Ceramics. Am. Cer. Soc. Bull. 51(12):884–905, 1972.
97. D. H. Buckley. Friction and Wear of Glasses and Ceramics. Surfaces and Interfaces of Glasses and Ceramics, Materials Science Research 7 (V. D. Frechette, W. C. LaCourse, and V. L. Burdick, eds.). Plenum Press, New York, 1974, pp. 101–126.
98. D. H. Buckley and K. Miyoshi. Friction and Wear of Ceramics. Wear 100:333–353, 1984.
99. S.-J. Cho, B. J. Hockey, B. R. Lawn, and S. J. Bennison. Grain Size and R-Curve Effects in the Abrasive Wear of Alumina. J. Am. Cer. Soc. 72(7):1249–1252, 1989.
100. H. Liu and M. E. Fine. Modeling of Grain-Size-Dependent Microfracture-Controlled Sliding Wear in Polycrystalline Alumina. J. Am. Cer. Soc. 76(9):2393–2396, 1993.
101. K.-H. Z. Gahr, W. Bundschuh, and B. Zimmerlin. Effect of Grain Size on Friction and Sliding Wear of Oxide Ceramics. Wear 162–164:269–279, 1993.
102. Y. He, L. Winnubst, A. J. Burggraaf, H. Verweij, P. G. Van der Varst, and B. de With. Grain-Size Dependence of Sliding Wear in Tetragonal Zirconia Polycrystals. J. Am. Cer. Soc., 79(12):3090–3096, 1996.
103. F. Xiong, R. R. Manory, L. Ward, M. Terheci, and S. Lathabai. Effect of Grain Size and Test Configuration on the Wear Behavior of Alumina. J. Am. Cer. Soc. 80(5):1310–1312, 1997.
104. W. D. Kehr, C. B. Meldrum, and R. F. M. Thornley. The Influence of Grain Size on the Wear of Nickel-Zinc Ferrite by Flexible Tape. Wear 99:109–117, 1975.
105. J. T. Chakraverti and K. Anderson. Diamonite Products, Shreve, OH, private communication.

106. R. W. Rice and B. K. Speronello. Effect of Microstructure on Rate of Machining of Ceramics. J. Am. Cer. Soc. 59(7–8):330–333, 1976.
107. L. F. Goyette, T. J. Kim, and P. J. Gielisse. Effect of Grain Size on Grinding High Density Aluminum Oxide. Am. Cer. Soc. Bull. 56(11):1018, 1977.
108. D. B. Marshall, B. R. Lawn, and R. F. Cook. Microstructural Effects on Grinding of Alumina and Glass-Ceramics. J. Am. Cer. Soc. 70(6):C-139–140, 1987.

6

Grain Effects on Thermal Shock Resistance and Elevated Temperature Crack Propagation, Toughness, and Tensile Strength

I. INTRODUCTION

Previous chapters have addressed the grain, mainly size, dependence of mechanical properties nominally at 22°C. This chapter complements those by addressing the changes in the grain dependence of tensile properties as a function of temperature. Hardness, compressive strength, and wear and related behavior at elevated temperatures are addressed in the next chapter. Some attention is given to effects of temperatures < 22°C, but the primary focus is on effects of elevated temperatures, mainly where brittle fracture is still dominant. Thus while high-temperature creep, stress rupture, and related slow crack growth processes are addressed to a limited extent, they are large complex topics, commonly with even less specific attention to specific grain structure effects. Therefore the focus is on the grain dependence of mechanical properties in the range where fracture is still brittle or transitioning to nonbrittle processes such as creep rupture, where grain boundary sliding is commonly the dominant process. This restricted focus is still complex since (1) while there is more, but still limited, information on grain dependence, most data is more limited than at ~ 22°C, and (2) the transition to nonbrittle processes is often subtle, not well defined, and can vary substantially with test parameters, e.g. stress/strain rates, the nature and amount of even limited grain boundary phases, grain structure, temperature, and the interactions of these.

This chapter also briefly reviews effects of test environment and (mainly moderate) temperature on the flexure strength (σ) of polycrystalline ceramics as a function of grain size (G), in part complementing and extending discussion of slow crack growth (SCG) in Chap. 2. Additional insight is sought by comparing these polycrystalline strength trends with the behavior of single crystals (where available) and of Young's modulus. This chapter draws substantially on a more recent review [1] and complements this and other reviews of σ–$G^{-1/2}$ behavior at 22°C, updating and extending them [1–3]. The purpose of this chapter is to consider these factors as a guide to understand better the grain dependence of mechanisms of brittle fracture as a function of T. Thus while some behavior in the 1000–1500°C range is noted, the focus is on the –200 to 1000°C range; creep and high-temperature stress-rupture are only briefly considered. Similarly, while environmental effects are further considered, the focus is on their impact on σ–G relations. While useful information and implications regarding σ–G relations are obtained, uncertainties and inadequacies are shown that provide guidance for improved studies.

Properties are treated in the order of crack propagation and fracture toughness followed by tensile strength and then thermal stress and shock failure, for which there is limited data. This chapter shows that strengths can increase substantially more at $T < 22$°C than expected from increases in E due to suppression of SCG and that there is limited data at $T > 22$°C and < 800–1000°C, apparently due to the common assumption that no significant property changes occur in this temperature range. This is often found to be incorrect by limited more comprehensive tests. Comparison of the relative temperature dependence of Young's modulus, toughness, and tensile strength is used to indicate mechanisms and changes. Some important strength changes in this intermediate range are shown, e.g. especially in Al_2O_3, where corresponding changes in hardness and compressive strength are shown in the next chapter. In general, more consistency is found between toughness and tensile strength at $T > 22$°C.

II. MODELS AND CONCEPTS

A. Young's Modulus, Toughness, and Tensile (Flexure) Strength

Modeling of properties as a function of temperature is typically based on accessing the temperature dependence of the key parameters in the mechanism that is dominant in the temperature regime considered. Where there is a transition between mechanisms, some interpolation between the expectations of the different mechanisms may be sought for guidance. Where other mechanisms become operative, they must be addressed, with creep being a fundamental one, as is discussed below.

First consider the temperature dependence of elastic moduli, which are basic parameters, hence correlators, with properties discussed in previous chapters. While elastic moduli normally do not depend on grain parameters, comparison of their temperature dependence with that of other mechanical properties can indicate changes in mechanisms. This is particularly so for fracture toughness and related crack propagation and tensile strength, since they depend directly on Young's modulus *(E)* [Eq. (2.2)]. This comparison with the temperature dependence of elastic moduli is aided by their typically better documented, better understood, and simpler decrease with increasing temperature due to their reflecting basic atomic bonding and the resultant decreases of this with temperature. Decreases of 1 to 2% per 100°C temperature rise are common. Thus Anderson [4] showed that Wachtman's empirical equation [5] for Young's modulus *(E)* at any temperature *(T)*,

$$E = E0 - ATe^{-(T_0/T)} \tag{6.1}$$

where E_0 = Young's modulus at absolute zero temperature (T_0) and A a is constant, was theoretically correct, i.e. could be derived from the Mie–Gruneisen equation of state if Young's modulus was replaced by the bulk modulus. The equation was also theoretically correct for Young's modulus provided that the temperature dependence of Poisson's ratio is small, which is true for some, but not all, ceramics. Others have corroborated and extended these results, e.g. Ref. 6. Such intrinsic changes in elastic moduli with temperature are an important factor in the temperature changes of other properties that scale with elastic moduli, so deviations of other mechanical properties from such decreases with temperature are an important indication of changes in mechanisms. Greater deviations of polycrystalline versus single crystal behavior from the temperature dependence of E are also an important sign that microstructural effects are changing with temperature, as is shown in Sec. IV. Thus while grain boundary sliding (e.g. whose earlier onset may be indicated by internal friction) [7–11] may also affect the apparent elastic moduli if the stressing period is sufficiently long relative to the rate of creep such that some measurable nonelastic strain occurs during the stress cycle; such sliding typically has greater effect on other mechanical properties.

Other property changes with temperature may cause changes in the temperature dependence of other mechanical properties directly or via effects on elastic properties. A prime example of the latter is effects of temperature changes on microcracking from thermal expansion differences, e.g. anisotropy (TEA) between grains, which decreases as temperature increases and vice versa and also commonly depends on G, as was discussed in Chap. 2, Sec. III.C. As is discussed elsewhere [8], the significant decreases that substantial microcracking can make in elastic and other mechanical (and nonmechanical, e.g. thermal) properties can be reversed due to microcracks generally decreasing in size, number, or both, with many closing and partly or fully healing, as temperature substantially increases. Thus the normal decrease of elastic (and other mechanical) properties with increasing temperature may be temporarily reduced or reversed till further temperature increases result in the normal decrease in properties, with the reverse trend on cooling. Similarly, though less established and understood, elastic anisotropy (EA) may affect or cause such cracking, aiding or countering effects of TEA in noncubic materials, or by itself in cubic materials [8]. However, in contrast to the more consistent temperature dependence of TEA, that of EA may decrease, be nearly constant, increase, or show combinations of these as temperature increases, making evaluations material specific (Fig. 7.14).

Besides the above effects of temperature on properties via changes underlying those of interest, there are also the mobility and activity of water as the most pervasive sources of slow crack growth (SCG) via chemical effects that greatly decrease as temperature increases. Thus in materials such as Al_2O_3, which experience substantial SCG, their strengths at lower temperatures (e.g. commonly –196°C in liquid N_2) are increased far more than expected from the 1–2% increase in *E*. Similarly, temperature increases, especially above 100°C, reduce the amount of moisture at crack tips and hence resultant SCG from it. The other important change in crack propagation and resultant mechanical properties arises from grain boundary sliding and its effects on both the crack size and character at failure and the energetics of propagation including impacts of possible effects on crack bridging and branching. However, all of these again raise questions and complications of the impact of crack size effects in crack propagation tests versus in the properties to be predicted such as tensile strength.

Beyond the above effects of temperature on underlying parameters such as Young's modulus and crack size, e.g. the latter due to localized grain boundary sliding at and near crack tips, there are the broader mechanisms of nonelastic, i.e. creep or other deformation processes. These become of increasing significance, commonly on both a broader scale spatially in the sample or component and in terms of materials as temperatures increase and strain rates decrease. While there can be important differences between tensile and compressive stressing, for reference and guidance the following representative equation for the strain rate $(d\in/dt)$ in creep is

$$\frac{d\varepsilon}{dt} = \frac{AM_s}{kT}\left(\frac{b}{G}\right)^m\left(\frac{\sigma}{M_s}\right)^n \tag{6.2}$$

where A = a constant, M_S = the shear modulus, k= Boltzmann's constant, T = absolute temperature, b = the Burgers vector, G = grain size, σ = stress, and m and n are constants for a given body and condition, which are respectively typically

0–3 and 1–5. Thus even in the case of the least dependence on G, strain rate increases as G^{-1}, i.e. greater than the G dependence for brittle fracture of $G^{-1/2}$. Such dependence of creep rate is essentially in the opposite direction from that of brittle fracture, since higher creep strain rates correspond to failure in faster times, at lower stresses, or both as G decreases versus increased strength in brittle fracture as G decreases.

B. Thermal Stress, Shock, and Fatigue Failure

The classical way of addressing thermal stress and shock failure was to solve the boundary value problems, or more commonly use existing compilations of solutions, e.g. Ref. 12. Such solutions almost invariably assumed isotropy of the body and that properties were temperature independent, since to do otherwise substantially complicates the problem. Application of these solutions, which are typically quite dependent on component geometry and boundary conditions, is by comparing the resulting thermal stresses to the known, or expected, failure stresses of the material, e.g. via use of Weibull statistics of failure [13]. Such solutions and experimental results yielded a number of thermal stress failure criteria depending on the material parameters and the thermal environment [14–18]. These basically fall into two broad categories of thermal stress failure resistance, i.e. where (1) no significant crack generation or growth occurs, so there is no loss of strength, or (2) thermal stresses are so extreme that crack generation, or more commonly growth, cannot be avoided but can sometimes be limited so that some reasonable level of strength remains.

The maximum tensile stress σ_t from a rapid temperature change is

$$\sigma_t = E\alpha\,(\Delta T)(1-\nu)^{-1} \tag{6.3}$$

Where E = Young's modulus of the body, α = its thermal expansion, ΔT = the temperature difference (typically in quenching), and ν = Poisson's ratio. Various modifications of this are made, e.g. to account for time dependence of establishing this stress. Thus a factor reflecting the ratio of the time-dependent maximum stress to the theoretical maximum stress, i.e. of ≤ 1, is often multiplied into the right side of Eq. (6.3) to account for this [19]. Similarly, the right side of Eq. (6.3) may be multiplied by the thermal conductivity to reflect the higher temperature gradients from lower conduction in the body. Such equations are readily solved for the critical ΔT, i.e. ΔT_C, and hence the critical quench temperature, T_C, into a given environment, commonly a water bath, that causes measurable strength degradation. Thus though there are uncertainties or refinements such as specifics of the heat transfer from the specimen to the fluid [20] and effects of the bath temperature on this [21], such standardized quench tests are widely used to compare and rank materials.

Modeling of thermal stress and shock effects was advanced by analysis of bodies with arrays of cracks based on previous mechanical analysis of such bodies. Models based on such crack arrays were developed, applied, and extended by Hasselman [14–18], who showed their consistency with fracture mechanics formulation [18], providing direct analysis of the effects of microcracks [22,23]. Hasselman also noted that thermal stresses could also cause failure of function due to excessive buckling or fracture from bending stresses [24], with criteria for resistance to these being respectively (1) low thermal expansion and a low aspect ratio of the body, and (2) a high body aspect ratio and value of $\sigma^2(\alpha E^2)^{-1}$. Some analysis of thermal fatigue has also been made [25].

The changes in mechanisms and underlying mechanical properties with increased temperatures noted earlier can be factors in the thermal stress or shock failure in ceramics of high resistance to such failure. However, their impact is typically limited, since thermal stress failure, while determined by the extreme high temperature involved, is usually controlled by fracture in the cooler regions. This results from the fact that the main sources of flaws for failure in ceramics are typically in or near the body surface. Thus in rapid cooling of heated bodies, or of localized heating of part of a piece of ceramic, the coolest area(s) include much or all of the free surface, where many of the flaws for failure are located. Besides the above simpler model and more complex ones, there are now a variety of computer programs for thermal stress and shock, with some allowing consideration of temperature dependence of properties, which can be important in some cases.

III. FRACTURE TOUGHNESS AND RELATED CRACK PROPAGATION AS A FUNCTION OF G AND TEMPERATURE

A. Fracture Toughness as a Function of G and Temperature

This section addresses fracture toughness (and hence also energy) and related *R*–curve effects; high temperature SCG due to grain boundary sliding is addressed in the following section, as are effects of environments such as H_2O on SCG at $T < 22°C$ and possible effects of G on them, since they are typically evaluated by flexure measurements. Unfortunately there is very little direct information on the G dependence of crack propagation behavior for these two sections due to combinations of the limited G range of materials available (e.g. SiC to some extent, and especially Si_3N_4), the lack of tests on bodies of various G, and the lack of specifying the G values for bodies studied. However, limited data allowing comparison of single- and polycrystal results exists, as does somewhat more data comparing fracture toughness and elastic moduli as a function of test temperature. Such comparison indicates some effects of G, and effects of grain boundary phases, which are a factor in grain effects, as is also shown by some direct studies.

Consider first the fracture toughness and *R*–curve effects in Al_2O_3, for which there is more data than for other materials. Thus the data and survey for sapphire of Iwasa and Bradt [26] show considerable variation as a function of orientation, as does other data [27,28], but with a reduction in toughness anisotropy as the test temperature increases (Fig. 6.1). Their evaluation showed much greater decrease in the fracture toughness for basal plane fracture than expected elastically, indicating an increasing role of nonelastic processes as *T* increases, so such basal fracture goes from being by far the least preferred to the most preferred as temperature increases. This is consistent with indicated toughness minima, e.g. due first to increased ease of slip or twinning aiding crack growth and subsequent crack blunting as such deformation becomes more extensive, and especially direct TEM observations showing increasing slip deformation as *T* increases (Wiederhorn et al [29]).

There is a fair amount of fracture toughness data for polycrystalline

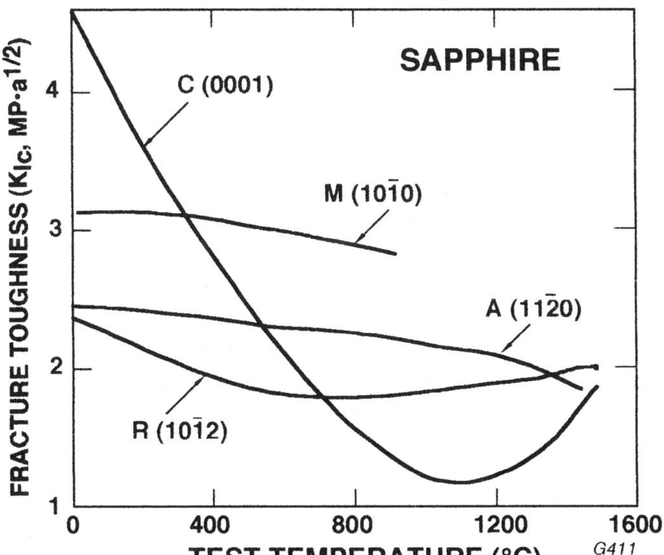

FIGURE 6.1 Fracture toughness of sapphire (mainly from indent tests) on various indicated fracture planes as a function of test temperature. Note that (1) data could indicate a less rapid decrease in (0001) fracture than shown to ~ 200°C, and (2) the indicated expectation for much greater decrease of basal fracture, i.e. on (0001) planes, is based on the temperature dependence of *E* (close to results for the A and M planes). (From Ref. 26, published with permission of the American Ceramic Society.)

Al_2O_3 to compare with the single crystal results. Most of this shows toughness nearly constant as a function of T or more commonly decreasing over most or all of the temperature range, with the rate of decrease increasing at higher temperatures. Kobayashi et al.'s [30] data for a commercial alumina via a wedge loaded DCB test is an extreme example of the former, showing little change from 22 to 1000, then to 1600°C. De With's [31] data for dense alumina sintered with MgO additions showed, via NB tests, ~ linear toughness (and strength) decreases to the limits of their tests, ~ 1200°C, that were ~ those expected for E for both G ~ 23 and ~ 36 μm (the latter obtained with 40 ppm CaO addition). He also showed that while toughness levels were lower than data from other investigators for similar aluminas, they had similar slopes, as was expected for the temperature dependence of E. The values for the finer G body were ~ 15% > for the larger G body, i.e.~ 3.8 and 3.3 MPa·m$^{1/2}$ respectively at 22°C, indicating a G dependence different from that of toughness independent of or increasing with G at 22°C (Chap. 2, Sec. III.F). Also, his more extensive data for the finer G body would be consistent with a toughness minimum at ~ 700°C and a maximum at ~ 1100°C.

Dalgleish et al. [32] tested three commercial (95, 97.5, and 97.7%) aluminas in the G range of respectively ~ 4, 24, and 2 μm and P ~ 0.06, 0.09, and 0.01 to 1000°C showing toughness minima at 300–600°C and then maxima at ~ 800°C, respectively ~ 10% below and 10–30% above values at 22°C. No significant change in acoustic emissions from each alumina occurred to 650°C, beyond which emission, attributed to flow of the glassy grain boundary phase, increased substantially, e.g. as indicated by intergranular fracture increasing with T. The most probable microstructural correlation was for lower toughness at larger G, e.g. by 40%, i.e. consistent with de With's data above. Others also showed toughness decreasing with increasing T, e.g. Tai and Watanabe [33] by ~ 30% to 700°C, and Moffatt et al. [34] showed a similar trend with accelerating decreases of ~ 70–80% at 1400°C. Jakus et al. [35] evaluated an 85% commercial alumina (G ~ 5 μm) from 1150 to 1275°C showing that bridging of the crack by glassy ligaments occurred.

Xu et al. [36] reported grain bridging of a pure, dense alumina (G ~ 10.5 μm) based on R–curve effects from indentation cracks (mostly ~ 100–300 μm) with modest toughness increases, e.g. ~ 30%, with increasing crack size, with similar rates of increase for all test temperatures (25, 400, 700, 1000, and 1300°C). The decrease in toughness of ~ 35% over the temperature range, which is somewhat greater than expected from decreases in E (i.e. ~ 27%), occurred mostly by 400°C. They analyzed their data based on a bridging model assuming that toughness is the sum of an intrinsic toughness, K_0, and bridging contributions, giving K_0 values decreasing from 2.2 to 1.4 MPa·m$^{1/2}$ from 25 to 1300°C. However, such values appear low in comparison to averaging single crystal values in Fig. 6.1, which would be for transgranular fracture with a crack size of ~

G rather than much larger cracks used. While fracture mode was not addressed, intergranular fracture is expected, especially at higher temperatures, which could lower K_0 values, for cracks approaching or on the scale of G, but again such low values would appear to be low for cracks encompassing 5–10 grains.

Grimes et al. [37] studied toughness and R–curve behavior of a commercial, 99.5 pure (AD-995) alumina made with a SiO_2 sintering aid giving $G = 19 \pm 4$ μm (range 2–40 μm). They showed that while there was some variation for the two NB tests used (chevron and straight notch), as expected, there was little or no decrease in toughness values from 4 to 4.5 MPa·m$^{1/2}$ from 22°C to 650°C and then an accelerating decrease to ~ 1 MPa·m$^{1/2}$ at 1400°C. They observed limited change in the fracture mode of mainly intergranular fracture for $G < 15$ μm and mostly transgranular fracture for $G > 20$ μm to 950°C, except for some reduction in the transgranular fracture, which accelerates at higher temperatures. By 1200°C, some crack branching was observed, with this increasing as temperature increased. Renotching of the specimens confirmed that the R–curve effects, which were similar in relative proportion to the toughness at each test temperature, were due to effects in the wake zone, which were attributed to bridging (and at higher temperatures, branching).

As was noted above, individually data of de With [31] and Dalgleish et al. [32] each indicated toughness decreasing as G increases. These data sets are also generally consistent with each other as well as other limited data for ≥ 95% alumina bodies where G values are available (Fig. 6. 2). Thus despite differing tests, limited differences in purity, and residual porosity, these data sets collectively reinforce a decrease in toughness as G increases, but with this G dependence decreasing as T increases.

Another material for which there is both single- and polycrystal data as a function of T is $MgAl_2O_4$. Stewart and Bradt [38,39] showed that the IF toughness of stoichiometric $MgAl_2O_4$ crystals stressed on <100>, <110>, <111> axes all showed limited decreases similar to, but less than, the expected decrease in E, to ~ 900–1200°C, but then increased at an increasing rate with further T increases to the limit of testing (1500°C). White and Kelkar [40] and others [41–44] showed that the toughness of a large G (~ 100 μm), transparent $MgAl_2O_4$ hot pressed with LiF additions (that reduce toughness at room temperature, Fig. 2.13) initially decreased from ~ 2.4 MPa·m$^{1/2}$ at 22°C slower than E but then accelerated its decrease as the melting temperature (800+°C) of LiF was reached and then very modestly increased to the limit of testing at 1500°C (Fig. 6.3). To ~ 800°C the fracture mode was ~ 50% intergranular and then increased, consistent with indicated and expected effects of LiF and possible increased grain bridging as intergranular fracture increased.

Baudin et al. [43], using $MgAl_2O_4$ sintered from 99.6% pure spinel powder to ~ 2.5% porosity with $G = 1.5 \pm 0.8$ μm, showed E decreasing from 205 GPa at 22°C to ~ 170 GPa at 1300°C, i.e. ~ a 10% decrease. Toughness decreased from

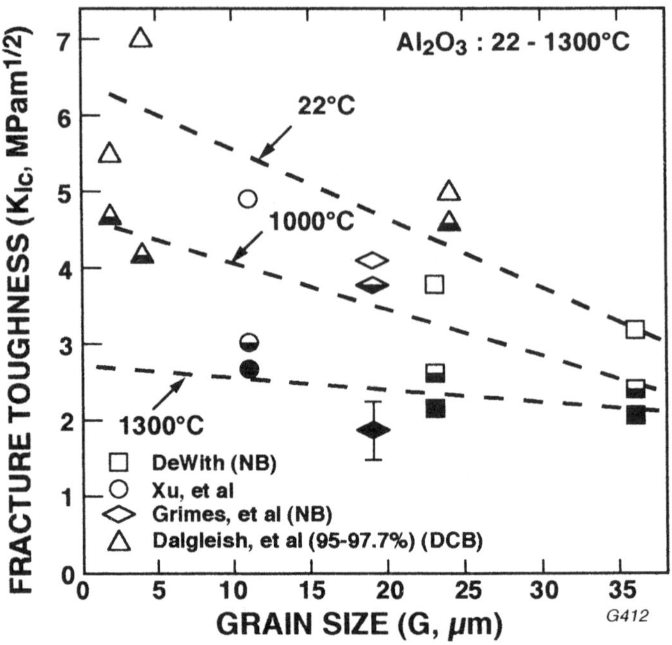

FIGURE 6.2 Fracture toughness versus grain size *(G)* for ≥ 95% alumina bodies at 22, 1000, and 1300°C, where *G* values were available from Xu et al. [36], Grimes et al. [37], de With [31], and Dalgleish et al. [32]. (Note that the latter data was corrected for $P = 0.01$, 0.06, and 0.09 respectively for $G = 2$, 4, and 24 μm using e^{-4P}, but trends would still be similar without such corrections.) Despite differences in measurement techniques (indicated for toughness, *G* determinations not always specified) and some differences in fabrication, purity, and residual porosity, the data consistently indicates a decrease in toughness as *G* increases and an overall decrease in both toughness and its *G* dependence as *T* increases.

3 MPa·m$^{1/2}$ at 22°C to a minimum of 2.3 MPa·m$^{1/2}$ at 800°C and then increased to a maximum of 2.8 MPa·m$^{1/2}$ at 1200°C. Strength behavior was opposite; starting at 175 MPa at 22°C, it increased to a maximum of 200 MPa at 800°C and then decreased to 160 MPa at the limit of testing of 1300°C. They also showed that toughness at 1200°C decreased ~ 20% (along with increasing transgranular fracture) with increasing strain rate, similar to the ~ 30% decrease for single crystals [38], but the former may reflect more reduction in bridging than in slip effects, which are the expected source of single crystal decreases. Baudin and Pena [44] subsequently showed that the 1200°C toughness of a stoichiometric spinel and two alumina rich ones, all having similar microstruc-

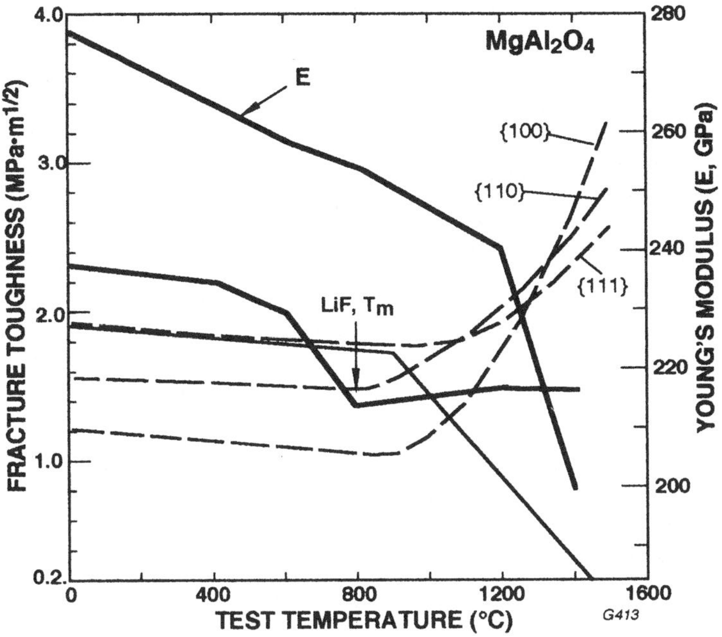

FIGURE 6.3 Comparison of the temperature dependences of fracture toughnesses of polycrystalline $MgAl_2O_4$ hot pressed with LiF additions, $G \sim 100$ μm, $P \sim 0$ [40], $MgAl_2O_4$ sintered without additives, $G = 1.5 \pm 0.8$ μm, $P \sim 0.02$ [39] and stoichiometric crystals of three orientations (dashed lines), along with Young's modulus (*E*) data for a dense polycrystalline body. Note the (1) limited toughness differences for the two polycrystalline bodies and the highest crystal values measured in different tests to several hundred degrees, (2) more rapid decrease of the body made with LiF as the LiF melting point is approached, then little or no decrease at higher temperatures in contrast to the lack of such a drop for the body made without LiF, but then a greater decrease for the latter at higher temperatures, and (3) similar initial rates of decrease for the three crystal orientations, reaching a minimum and then increasing substantially (but presumably decreasing again at higher temperatures).

tures, exhibited respectively intergranular fracture with bridging and mostly transgranular fracture.

Inoue and Matzke [45] measured toughness of sintered ThO_2 ($G \sim 20$ μm with much of the ~ 8% porosity being intragranular and a micron or more in dia.) using Hertzian cracks from spherical indenters of Al_2O_3 or steel from 22 to 388°C. These indenters gave toughness values of respectively 1.22 and 1.07 at 22°C decreasing to ~ 0.77 MPa·m$^{1/2}$ for both at 388°C, i.e. severalfold times the

~ 5% decrease in *E*. On the other hand, Ohnishi et al. [46] reported that sintered (~ 99% dense) mullite toughness decreased similar to *E*, i.e. by ~ 5% from 22 to 500°C, beyond which toughness slowly, and then more rapidly, increased with increasing *T* till slowing its increase by the limit of testing of 1400°C (Fig. 6.4). Strengths showed similar trends, but less, and more complex, increases at higher temperature (in contrast to no change till 1200°C and then only a slight decrease for hot pressed mullite). Baudin [47] reported lower *E* values but with a similar relative decrease with increasing *T*, and a similar range of toughnesses and strengths values, but decreasing ~ 10% to minima at ~ 800°C, then increasing to maxima ~ 20% > values than at 22°C at ~ 1400°C. Mah and Mazdiyasni [48] reported fracture toughness calculated from their strengths and fractography of ~ 1.8 MPa·m$^{1/2}$ at 22°C, decreasing ~ 20% to a minimum of ~ 1.5 MPa·m$^{1/2}$ at ~ 1100°C and then increasing substantially as temperature further increased, apparently due to slow crack growth that was observed beginning at ~ 1300°C and was attributed to limited amounts of a glassy grain boundary phase observed in TEM.

Consider now the behavior of nonoxides, starting with SiC. Henshall et al. [49] measured toughness of 6H α–SiC crystals using NB tests with the notch parallel with {11$\bar{2}$0}for propagation in the <1$\bar{1}$00> direction, obtaining 3.3 MPa·m$^{1/2}$ from 22°C to 1000 K and then 5.8 MPa·m$^{1/2}$ at 1773 K. However, Naylor and Page [50], using an indentation technique, reported toughness decreasing from 4.6 MPa·m$^{1/2}$ at 22°C to 2.0 MPa·m$^{1/2}$ at 800°C for fracture on (0001)

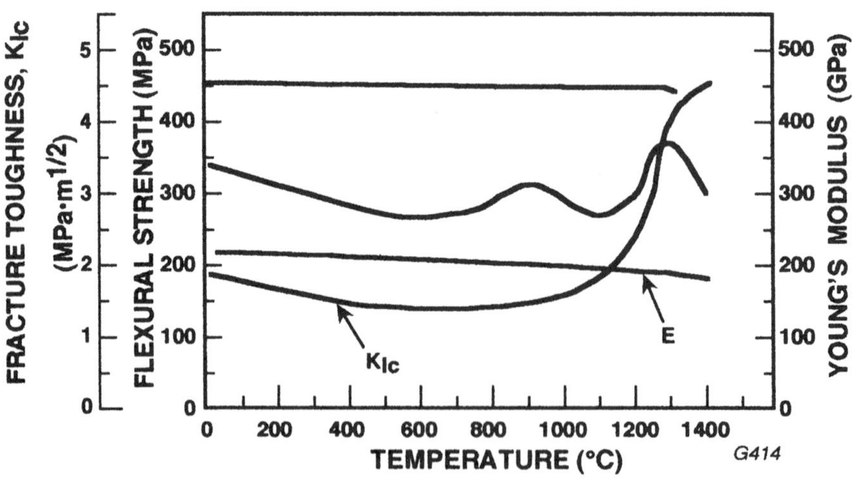

FIGURE 6.4 Young's modulus, toughness, and two strength curves for mullite versus test temperature. (From Ref. 46.) See also Figure 6.19.

planes. While such room temperature values are high in comparison to values from fractography by ~ 60% and 100% respectively [51] (Chap. 2, Sec. III.D), the temperature dependences are of prime interest here. Guillou et al. [52] also measured (Berkovich) indentation toughness of 6H SiC crystals for indents on (0001) planes with cracks in <10$\bar{1}$0> directions showing ~ 40% decrease by 600°C and then an ~ bottoming out at ~ 1.6 MPa·m$^{1/2}$, as with Naylor and Page (Fig. 6.5). Thus there is agreement and there are similarities and differences in the SiC data.

Evans and Lange [53] showed that the DT toughness of commercial SiC hot pressed with Al_2O_3 addition was ~ constant at ~ 4 MPa·m$^{1/2}$ to ~ 1100°C and then decreased substantially, i.e. to ~ 2.8 MPa·m$^{1/2}$ at 1400°C. Henshall et al. [54] subsequently evaluated the NB toughness (and delayed failure) of the same SiC (G ~ 1.5 μm), giving toughness of ~ 6.1 MPa·m$^{1/2}$ to ~ 1000°C and then decreasing to ~ 4 MPa·m$^{1/2}$ by 1773 K. While their absolute values are about 50% higher, the relative change in toughness with temperature were very similar. These changes correspond to a change from mixed inter- and transgranular fracture mode at 300 K to all intergranular fracture at > 1373 K, and the onset of de-

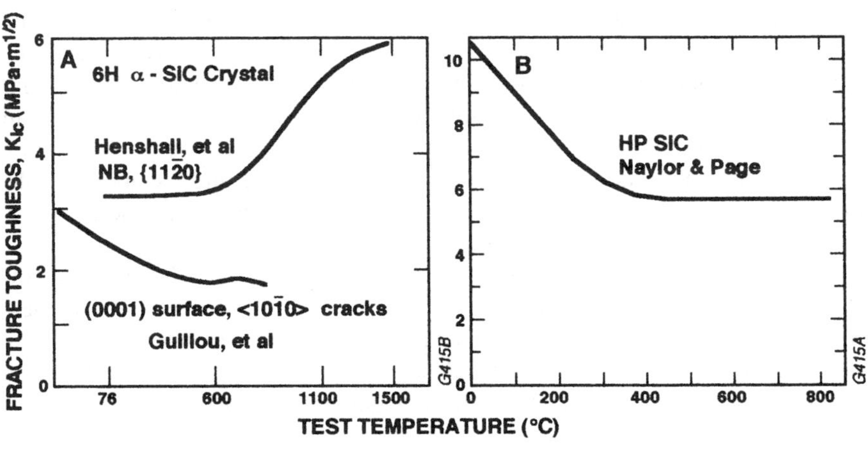

FIGURE 6.5 Fracture toughness of SiC as a function of test temperature. (A) For crystals: NB data of Henshall et al. [49] and indent data of Guillou et al. [52] for different (shown) crystal orientations. (B) Indent data of Naylor and Page [50] for hot pressed SiC. Note the (1) broad range of values at 22°C, which are somewhat to substantially higher in comparison to single crystal values from fractography and from comparison to polycrystalline values [51], (2) differing temperature trends in (A), limited in crystal values at 22°C, but they are nearly as high as most polycrystalline values, but with opposite temperature trends, and (3) the unusually high polycrystalline values in (B).

layed failure at ≥ 1273 K. Naylor and Page's [50] indent tests of hot pressed SiC from the same source (and Al_2O_3 additions), while much higher in values, show a similar temperature trend.

In contrast to the above SiC hot pressed with Al additions, SiC densified with B, B+C, or B_4C additions (i.e. commonly commercially sintered α=SiC) shows much less or no change of toughness (and often strength) to temperatures of ~ 1500°C. Thus, Evans and Lange [53] showed that DT toughness of commercial sintered α-SiC increased slightly [~ 10% to the limit of testing (1500°C) in contrast to SiC hot pressed with Al_2O_3 additions, as was noted above]. Similarly, Ghosh et al. [55] showed toughness measured by three techniques on the same commercial sintered α=SiC all being constant between 3 and 4 MPa·m$^{1/2}$ over the 20–1400°C range. Srinivasan and Seshadri [56] also showed (NB) toughness constant at ~ 4.8 MPa·m$^{1/2}$ over this range when tested unoxidized in an inert atmosphere, but with varying increases in toughness above 600 –1000°C depending on testing in air or with prior oxidation of the samples (the latter giving greater increases). Popp and Pabst [57] corroborated that the toughness of commercial sintered α–SiC was nearly constant to the limits of their testing of 1200°C (actually showing a few percent decrease from the value of ~ 3.5 MPa·m$^{1/2}$ at 22°C) with negligible difference as a function of strain rates in an air atmosphere. This limited change was noted as correlating with the predominant transgranular fracture mode reported to at least ~ 1400°C. However, they showed that while a reaction-processed SiC with ~ 13% residual Si had a similarly constant toughness at ~ 4 MPa·m$^{1/2}$ over the same temperature range in air tested at high strain rates (1260 mm/min), it markedly increased by ~ 200% from ~ 900–1200°C at low strain rates (0.024 mm/min). These and the above results are generally consistent with those for high-temperature slow crack growth, which is discussed in the next section.

Kriegesmann et al. [58] corroborated that sintered and hot pressed SiC having mainly transgranular fracture at elevated temperatures retains toughness (and strengths) there, i.e. showed no decrease from values at ~ 22°C to the limit of testing of 1400°C. On the other hand, such bodies having mainly intergranular fracture at elevated temperatures, while showing toughnesses (and strengths) essentially the same as at 22°C at 800°C, showed decreases of 20–40% at 1400°C. The bodies showing decreases included hot pressed material having the highest strengths at 22 °C (by 12–16% versus the other two hot pressed materials and by ~ 70% versus the sintered SiC) and included bodies made from either α= or β= SiC powder. Bodies derived from either α or β powder showed that the SiC crystal phase was not the determinant of the retention or loss of toughness or strength. While the specific densification aids used were not disclosed, they did note that the use of B or B-based aids corresponded with transgranular fracture and toughness (and strength) retention at elevated temperatures versus aluminum-based aids correlated with increased intergranular fracture and toughness (and strength) decreases at elevated temperatures. These authors briefly discussed and showed micrographs of the microstructures of the bodies studied, showing that all four bodies had G values averaging ~ 4 μm, but with different distributions of sizes, shapes, or both. Thus of the two hot pressed bodies, both from α powder and with similar nominally equiaxed toughness (and strengths), the one with larger isolated, equiaxed grains (e.g. to ~ 5+ μm) had somewhat lower toughness (~13%) but higher strength (~ 4%) at 22°C and no loss of toughness (or strength) at elevated temperature. However, the body hot pressed from β powder that had tabular grains (~ 2 μm dia. and ~ 5–10 μm long) had the greatest strength at 22°C but ~ 20% loss at 1400°C (toughness not measured).

B_4C hot pressed without additives (P ~ 0.08, G ~ 5 μm) has been reported to have toughness decreases of ~ 30% from 22 to 1200°C by Hollenberg and Walther [59]. While their toughness values appear low by ≥ 50%, the decrease with temperature is similar to, but possibly somewhat greater than, the decrease in E.

Next consider results for dense Si_3N_4. Naylor and Page [50], using an indentation technique, reported toughness of CVD material (i.e. made without additives, but quite possibly having some columnar or oriented grain structure or both) decreasing by nearly 50%, e.g. from ~ 5.6 MPa·m$^{1/2}$ at 22°C to ~ 2.9 MPa·m$^{1/2}$ at 800°C, i.e. greater than expected for E. Data for dense sintered or hot pressed Si_3N_4 and sialons also generally showed lower toughness at 1000–1100 than at 22°C, usually consistent with decreases in E, i.e. 10–20% [60,61]. However, tests continued to higher temperatures typically show a subsequent increase to toughness maxima at 1200–1400°C [63], with the temperature and extent of such maxima dependent on the type of toughness test, the strain rate of the test, and the nature of the material. Ohji et al. [63], using Si_3N_4 hot pressed with 3 and 5 wt% respectively of Al_2O_3 and Y_2O_3, showed the common tendency for toughness to decrease little or not at all to > 1000°C and then to show a modest maximum at ~ 1200°C (while tensile strength had a small maximum at ~ 1000°C); it had a substantial loading rate dependence, e.g. a maximum at a displacement rate of 10^{-3} mm/min at 1260°C.

Finally, consider the temperature dependence of toughness of other single crystals than Al_2O_3, $MgAl_2O_4$, and SiC presented earlier. NB data of Ingel et al. [64] for Y-PSZ and Y-CSZ showed the former decreasing to a minimum at ~ 1000°C while the CSZ crystals probably reached a minimum at ~ 600°C, both subsequently increasing substantially with further temperature increases beyond the minima (Fig. 6.6). The latter increase was attributed to plastic deformation, e.g. as shown by macroscopic deformation, especially in the CSZ, and higher toughness values at higher strain rates at high temperatures. This is in contrast to a similar but continuing decrease of the toughness of a commercial Mg-PSZ (and associated intergranular fracture). In all three cases the decreases are more than expected from E decreases, but possibly more for the two PSZ materials, which

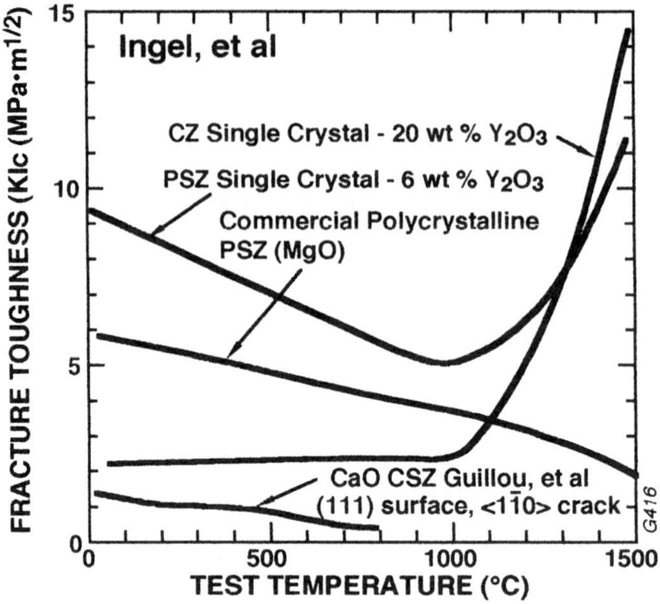

FIGURE 6.6 Fracture toughness (NB) of ZrO_2 single crystals partially and fully stabilized with Y_2O_3 (stress axis probably ~ <110>), as well as for a commercial Mg-PSZ material, versus test temperature. (From Ingel et al. [64], published with permission of the *Journal of the American Ceramic Society.*) Also note indent data of Guillou et al. [52].

must reflect added decreases due to decreasing effects of transformation toughening. Note also (1) Ingel et al. reported DCB tests at 22°C giving toughnesses of 3–6 and ~ 1.5 MPa·m$^{1/2}$ for respectively PSZ and CSZ crystals, and (2) Guillou et al. [52] showed indentation toughness of Ca-CSZ crystals (14 m/o CaO, indented on {111} planes with <1$\bar{1}$0> cracks decreasing from 1.32 to 0.48 MPa·m$^{1/2}$ for $T = 22$ to 800°C, i.e. consistent in this trend with the Y-PSZ crystal data).

Ball and Payne's [65], NB tests of quartz crystals of differing orientations showed toughness decreasing from ~ 0.8 MPa·m$^{1/2}$ to a minimum of ~ 0.6 MPa·m$^{1/2}$ at ~ 200°C and then increasing with further temperature increase (indicating limited effects of differing orientations, Chap. 2, Sec. III.B). Guillou et al. [52] also measured Vickers indentation toughness of MgO crystals for {100}<100> fracture showing a decrease from 1.7 MPa·m$^{1/2}$ at 22°C to a minimum at ~ 250°C and then a maximum of ~ 1.3 MPa·m$^{1/2}$ at ~ 250°C and then a maximum of ~ 1.9 MPa·m$^{1/2}$ at ~ 350°C with an ~ 20 N load, with less pronounced changes, higher minimum, and lower (e.g. 10%) maximum values at an ~ 3 N load. Mah and Parthasarathy [66] showed that SENB toughness of YAG single crystals increased from 2.2 MPa·m$^{1/2}$ at 22°C to 4.5 and 5.5 MPa·m$^{1/2}$ at 1600°C respectively in air and vacuum, with most of the increases above 1200°C and no significant orientation dependence.

The above review shows that crystal toughnesses often increase substantially at high temperatures due to plastic flow and less disparity between different fracture toughness tests and some of these with strength results, especially in their *G* dependence, as is commonly found in testing at or near room temperature. However, as is discussed later, this often has little or no relevance to high-temperature toughness of polycrystalline bodies, where grain effects, especially grain boundary sliding and failure, dominate, e.g. as indicated by the continued decrease in polycrystalline PSZ toughness (Fig. 6.6). Toughness and strength testing of bodies, especially single crystals showing substantial crack tip and even bulk plasticity (e.g. Fig. 6.6), indicate more consistent agreement regarding the onset and effects of such plasticity. However, caution and the need for more comparative testing is indicated by results of Hirsch and Roberts [67], who used toughness tests to determine the ductile–brittle transition in single crystals of Si. They found that this transition varied by up to 250°C between DCB and indentation fracture tests. The presence of dislocations at or near the crack tip due to the indent was an important factor in lowering the ductile–brittle transition temperature, e.g. polishing off most of the indent raised the transition temperature 50°C and abrading the area lowered the temperature 40°C.

B. Crack Propagation as a Function of G and Temperature

Slow crack growth (SCG) measurements at elevated temperatures typically use the same types of tests as are used for SCG due to environmental effects at room temperature and the same equation, i.e. Eq. (2.3), with the exponent *n* thus again being used as a value to characterize the process, as was discussed by Evans [68]. However, the first and most general of two aspects of such data is serious limitations of specific effects as a function of *G* or other grain parameters, and quantitative effects of grain boundary phases. Second is the transition from one mechanism to another as *T* increases, which, while aided by the common use of the same equation for various mechanisms of slow crack growth, has been widely neglected. Thus data is almost exclusively for room temperatures or at high temperatures where SCG due to grain boundary sliding dominates. However, Evans and Lange [53] showed that DT toughness of commercial SiC hot pressed with Al_2O_3 addition (having room temperature SCG due to moisture giving $n \sim 80$ with activation energies characteristic of stress corrosion processes) was disappearing, i.e. $n > 200$, at 600°C. On the other hand, SCG was again clearly evident by 1400°C ($n \sim 21$), but of a different sort, i.e. characterized by much higher activation energies characteristic of plastic flow via grain boundary sliding.

The above high-temperature SCG is a highly thermally activated process, as is indicated in Fig. 6.7, which shows the range of commonly encountered *n* values. A greater degree of oxidation in more oxidizing atmospheres for oxidizable materials also increases the rate of crack growth. The substantial changes with temperature, and sometimes atmosphere, commonly mask effects of *G* and related parameters, especially when they do not encompass substantial changes. However, the extent of slow crack growth in a given specimen can often be clearly revealed by the predominant intergranular fracture mode on the subsequent fracture surface where the remaining fracture mode is partly, often mostly, transgranular. Some of this demarcation by fracture mode change may be visible on elevated temperature fracture, if not obscured by oxidation of nonoxide fractures. However, it is commonly particularly pronounced on fractures that were later completed at lower, especially room, temperature after the high-temperature slow crack growth [63]. Fig. 6.8 gives examples of this for SiC. Note also

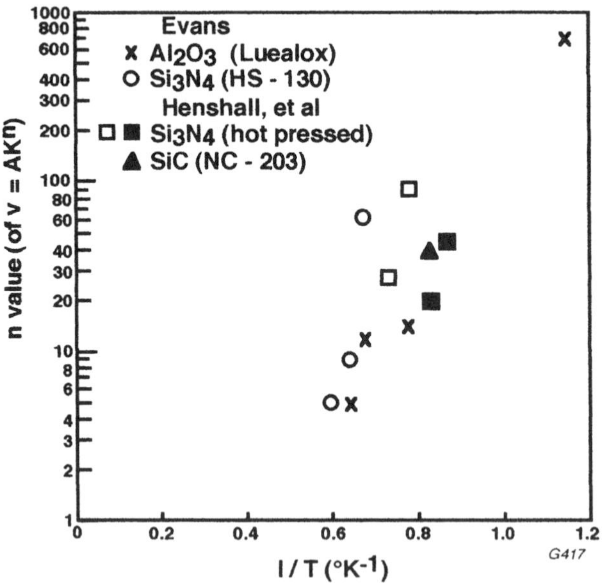

FIGURE 6.7 Plot of high-temperature slow crack growth *n* values [per Eq. (2.3), $v = AK^n$] versus the inverse test temperature (in *K*). Note the effect of test atmosphere, mainly the degree of its oxidation potential (solid symbols for simulated turbine atmosphere, others for non-oxidizing atmosphere). (Original data from Evans [68] and Henshall et al. [54]).

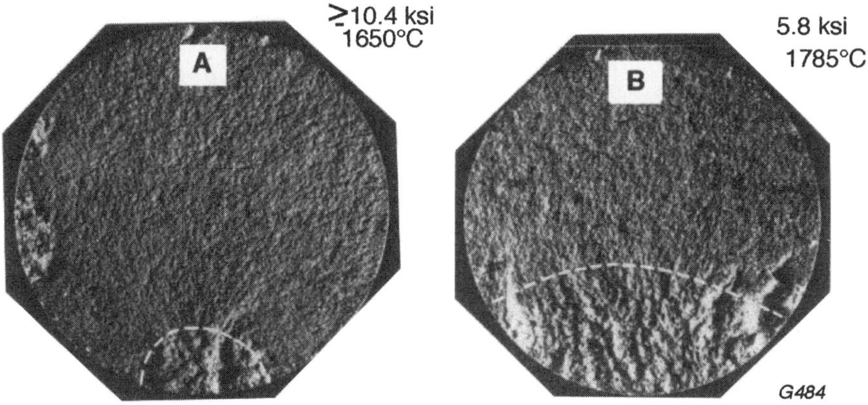

FIGURE 6.8 Examples of high-temperature slow crack growth in true tensile tested hot pressed SiC. SEMs showing regions of high temperature SCG (rougher areas outlined by dashed white lines) exposed by subsequently completing fracture at room temperature, with indicated temperatures and stresses. (From Ref. 69, published with permission of Brook Hill Publishing Company.)

that examining specimen surfaces after high-temperature stressing can also reveal other sites and approximate extents of high-temperature slow crack growth.

Turning to specific cases, static and cyclic crack growth studies of a commercial (AD-998) alumina ($G \sim 5$ µm) at 1200°C by Lin et al. [70] again showed mainly intergranular fracture, usually along a single crack due to SCG, but under some conditions growth of more than one crack was indicated. They also indicated some contribution of glassy phase bridging despite the very low level of such material in this 99.8% pure body. Horibe and Sumita [71] studied high temperature static and dynamic crack propagation fracture stress of two SiC bodies densified with B + C additions for two powders differing mainly in their contents of SiO_2 (0.34 vs. 0.53 w/o) and Al (0.03 vs. 0.39 w/o). While some results are complex, their overall conclusion was that the higher Al content correlated with greater creep and lower strength at 1500°C, i.e. consistent with effects of Al_2O_3 additions in the previous section.

Baumgartner reported SCG of dense sintered TiB_2 in molten Al at ~ 970°C (of interest for possible use of TiB_2 as electrodes for Al refining) based on both DT [72] and dynamic fatigue [73] tests (see Fig. 20 for σ–$G^{-1/2}$ data). DT tests showed no SCG in purer, finer G (< 10 µm, $P \sim 0.01$) but definite SCG in larger G (~ 20 µm, $P \sim 0.02$) with more impurities (especially ~ 0.3 w/o oxygen), where the intergranular penetration of Al was accompanied by cell impurities of

Fe, Si, and P. Dynamic fatigue tests conducted on higher purity TiB_2 [74] as a function of a wider range of *G* showed limited intergranular penetration of molten Al into the TiB_2, but no intergranular crack growth, i.e. transgranular fracture from surface connected, e.g. processing, flaws. However, dynamic fatigue tests were consistent with changing behavior as a function of *G*, showing generally increasing negative *n* values [i.e. of Eq. (2.3)] of ~ 30–80–100 as *G* increased from ~ 1–3 μm and then becoming increasingly positive values of 44 and 53 as *G* increased to ~ 10 and then 17 μm. Baumgartner argued that the failure process was liquid metal embrittlement, i.e. a lowering of strength due to lowered toughness from the presence of liquid Al at the crack tips, instead of stress corrosion, which entails lowering of strength due to SCG. He also suggested that the trend in *n* values could be explained by possible competition of plastic blunting of cracks decreasing as *G* increased, i.e. mainly in finer *G*, stronger bodies, versus liquid metal embrittlement via reduced toughness at the crack tip. Thus the latter would be mainly operative in the absence of crack tip blunting in larger *G* bodies, where microcracking from TEA stresses was also present and probably contributing, especially in the largest G body.

Of the few studies of crack propagation and fracture energy and toughness in graphite, one examined crack propagation as a function of temperature and environment (e.g. atmospheres of H2O, CO, or He). Thus Freiman and Mecholsky [75] showed that both an isotropic (POCO-AXF-5Q) and an anisotropic (ATJ-S) graphite exhibited stable crack propagation to respectively ~ 800 and 1600°C in H_2O, CO, or He atmospheres. Above these temperatures crack propagation was catastrophic. They attributed the stable crack propagation to stress corrosion due to H_2O in the pores (hence explaining no influence of the external test atmosphere, and consistent with strength results). They showed fracture energies in both materials increasing by of the order of 50% and fracture toughnesses by about 20% from 22 to 1400–1600°C. Similarly Sato et al. [76] showed fracture toughnesses of their three ~ isotropic (molded) and one anisotropic (extruded) graphites increasing by respectively ~ 50 (for the anisotropic, extruded material) and > 80% to maxima at ~ 2100°C or at, or beyond, their maximum test temperature of 2600°C. These changes are substantially more than they found for Young's modulus of these same graphites (and about the same or intermediate for tensile strength), but similar in overall trend. Extruded material had the lowest toughness and the least increase, consistent with its measurement being with the oriented grains versus those of the molded graphites being across the grain orientation.

Turning to discussing the grain structure dependences of toughness and crack propagation as a function of temperature, there is limited data on the basic, intrinsic factors of this, namely grain size *(G)*, as well as shape and orientation. Further, such possible grain effects are compromised by the frequent dominant effect of grain boundary effects, since small amounts of grain boundary phases can dominate higher temperature crack propagation and toughness, usually enhancing the former and limiting the latter. Such phase effects can readily mask effects of substantial changes in *G* since, for a fixed volume of grain boundary phase, its amount along an average grain boundary scales with the *G*, i.e. if *G* doubles so does the amount of the boundary phase along an average grain boundary. However, the first and clearest experimentally of four factors is that grain structure still has an important role in these processes and the substantial dependence of tensile strengths on basic grain parameters shown in the next section. This reflects both the more extensive strength data and the frequent differences, mainly due to crack scale effects, especially relative to the microstructure, between crack propagation and related measurements versus those in determining strengths. Second is the clear effects of different crystal planes, which, while often diminishing with temperature, are clearly significant, and sometimes complex (e.g. Fig. 6.1), which clearly imply basic effects of grain shape and especially orientation if not masked or overridden by grain boundary phase effects. Third are differences between the temperature dependences of *E* and *K*, which mainly aid confirmation of grain boundary effects. Fourth, and also basic, is expectations from other models and behavior, namely fracture at lower temperatures, e.g. 22°C, as is extensively discussed in Chap. 4, and high temperature behavior, e.g. as reflected in the *G* dependence in Eq. (6.2).

IV. EFFECTS OF GRAIN SIZE AND TEMPERATURE ON TENSILE (AND FLEXURE) STRENGTHS

A. Effects of Environment on the *G* and *T* Dependence of Strength, Including at $T < 22°C$

As discussed in Chap. 2, Sect. III.B, slow crack growth can occur at room temperature due to the effects of an active fluid, especially gaseous, species that alters or breaks crack tip bonds, with H_2O being one of the most severe and prevalent of such species. Thus strength testing at lower temperatures reduces the activity and mobility of active species, especially if they are solidified, i.e. strength testing in liquid nitrogen, hence at its boiling point of –196°C, essentially halts most SCG, including that due to H_2O. This allows a ready test for SCG and assessment of the extent of SCG by comparing strengths at –196°C and a higher temperature, usually ~ 22 °C, since the ~ 1–4 % increase in strength due to increases in *E* at –196 versus 22°C is negligible to most strength increases due to essentially eliminating SCG at –196°C which are typically 10 to > 50%. Further, such tests often allow changes in flaw sizes due to SCG to be observed by subsequent fractography or implied by changes in strengths and toughnesses. Such tests also indicate that the occurrence of SCG is either (1) intrinsic, if the failure initiating flaw (and usually also surrounding) fracture mode is mainly or exclusively transgranular, or (2) extrinsic, due to grain boundary character, usually phases, if the failure initiating flaw (and also possibly surrounding) fracture mode is mainly or exclusively intergranular as shown by Rice and Wu [77]. Thus they showed that such tests, while corroborating SCG in silicate glasses and Al_2O_3, showed increasing SCG in CeO_2, Y_2O_3, ZrO_2, and TiO_2 respectively, while no SCG occurred in refractory borides, carbides, and nitrides such as TiB_2, ZrB_2, SiC, TiC, ZrC, AlN, and Si_3N_4, which in pure form (e.g. from CVD) exhibited more or exclusive transgranular fracture, unless there was sufficient oxide containing grain boundary phase and resultant substantial to exclusive intergranular fracture. There can also be special cases where there are environmental interactions with slip or twinning that may affect strengths, but these are mostly uncertain and limited.

Turning to specific σ–$G^{-1/2}$ data, Figure 6.9 summarizes much alumina data at –196 versus 22°C [2,3,31,78–85], which while scattered shows a trend for higher strengths at –196°C, especially when data from the same investigators and bodies are compared, e.g. that of Charles [86] and Gruver et al. [84]. Overall this shows (1) the same two-branch σ–$G^{-1/2}$ behavior, (2) both with finer grain size σ–$G^{-1/2}$ slopes > 0. (3) single crystal strengths > many polycrystalline samples with similar surface finishing [2,78–82], and (4) greater single- and polycrystal strengths at –196°C versus 22°C. While some data, e.g. for press forged Al_2O_3 [3,83], does not clearly show increases at –196°C vs. 22°C due to scatter and the limited extent of the data, specific comparisons more clearly show single- and polycrystal strength increases. Thus Heuer and Roberts [80,87] showed sapphire strength increasing ~ 35–50% in liquid N_2 (–196°C) vs. 22°C in air for various surface finishes. Other investigators [89–91] showed similar increases, but Charles [86] showed a 75% increase. For dense hot pressed Al_2O_3 tested at –196°C, Rice [2] showed a 30% strength increase for most grain sizes but a 45% increase for G = 1–2 μm. Similarly, Charles showed ~ 20% strength increase for lamp envelope Al_2O_3 (G ~ 6–150 μm), Neuber and Wimmer [81] a ~ 30% increase for 99.5% Al_2O_3 (porosity, P, ~ 0.05, G ~ 35 μm), Davidge and Tappin [82] a ~ 25% strength increase for 95% Al_2O_3, P ~ 0.07, G ~ 8 μm, and Gruver et al. [84] a ~ 30% increase for 96% Al_2O_3, P ~ 0.05, G ~ 7 μm (Fig. 6.9) in liquid N_2 vs. air at 22°C. Overall the polycrystalline strength increase is probably less than for sapphire (except possibly at G ~ 1–2 μm), reinforcing sapphire strengths being even > many polycrystalline values at –196°C versus 22°C. Tests in the absence of H_2O at 22°C (e.g. in vacuum) showed that much, but not all, of the increase in strength at –196°C is due to the elimination of slow crack growth (SCG). Thus Charles showed sapphire strength increased only ~ 17% at –196 vs. 22°C but decreased ~ 50% in wet air vs. vacuum at 22°C, while lamp envelope Al_2O_3 (G ~ 40 μm) showed only about an 8% increase and an ~ 44% decrease respectively; i.e. indicating less increase in liquid N_2 but similar decrease in wet air to that of sapphire. He also showed an ~ 20% increase in 22°C (air) strength for

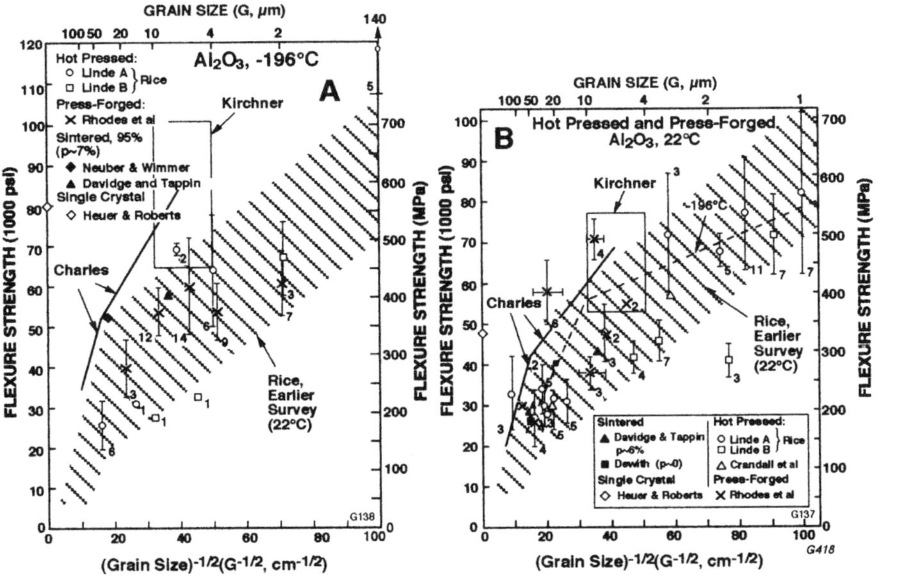

FIGURE 6.9 Comparison of σ–$G^{-1/2}$ data, mainly for hot pressed and pressed forged, Al_2O_3, at (A) – 196°C and (B) 22°C. For reference, the range of data from an earlier survey [1] of data at 22°C is shown in (A) and the mean trend line for actual data at –196°C from (A) is shown in (B) as a dashed line. Note the (1) generally lower strengths of the author's specimens made from Linde B versus Linde A powders, (2) greater scatter and possible lower strength level of the pressed forged vs. hot pressed Al_2O_3 , the latter mainly at finer *G* (~ 10 μm), (3) single crystal strengths being higher than much of the polycrystalline data at –196 and 22°C, and (4) direct comparison of Charles [86] and Gruver et al.'s [84] data (the latter labeled Kirchner) at both temperatures. (From Rice [1], published with permission of the *Journal of Materials Science*.)

a substantial *G* range (~ 6–150 μm) at a strain rate of 2.7 × 10^{-4} vs. 1.4 ×10^{-2}/min. The lower strengths of bodies made from Linde B powder may reflect more anion impurities [1], while some lower strengths and greater scatter of press forged specimen strengths probably reflects more variation in grain size shape and orientation, and possibly of residual stresses.

Assessing whether the absence of SCG at –196°C shifts the *G* dependence of strength, e.g. due to possible effects of *G* on SCG (Fig. 2.8), is difficult due to the variations noted above, as well as complexities of the actual SCG. Thus the occurrence of SCG in single crystals, at rates generally similar to those of polycrystalline Al_2O_3 [92] (Chap. 2, Sec. III.B) indicates that transgranular

SCG can occur in polycrystalline materials and thus may not impact the single–polycrystal strength balance. While SCG occurs in sapphire, the first of two complications is that sapphire SCG has been measured on only a few fracture planes, and the extent of less, or no, SCG on other planes is not known. Second and more basic is that SCG in polycrystalline Al_2O_3 is mainly or exclusively by intragranular fracture (which may reflect the preceding complication), especially at finer grain sizes (Fig. 2.6) in contrast to more transgranular fracture commonly observed in fast fracture, and may decrease with decreasing *G* (Fig. 2.8). Changing SCG as a function of *G* could change the intersection and slopes of the two branches of the σ–$G^{-1/2}$ behavior.

McMahon [93] showed strength of sintered, high Al_2O_3 bars at 22°C being a function only of surface finish and relative humidity (to 70%) during the test, and not of prior humidity exposure. He showed that the relative level of strength and its decrease due to H_2O varied as follows for different specimen surface conditions: (1) as-fired surfaces gave the lowest strength and the least (~ 5%) strength decrease, (2) surfaces ground perpendicular to the specimen axis gave intermediate strengths and the greatest (~ 15%) decrease, and (3) surfaces ground parallel with the specimen length gave the highest strength and an intermediate strength decrease with increasing relative humidity. Thus the relative strength decrease with increasing humidity was a function of surface finish as well as moisture content. Rice [80] showed that dense, hot pressed Al_2O_3 averaged ~ 20% strength decrease on testing in distilled H_2O vs. air at 22°C, but that retesting in air of bar sections previously tested in H_2O (after drying) returned them to their original air strength. These two studies show that strength degradation due to SCG occurs during actual loading and is a function of the environment only during stressing. This implies that SCG either does not occur due to microstructural (e.g. thermal expansion anisotropy, TEA) stresses or that it saturates (at least for typical multigrain size flaws) after initial exposure.

SCG also occurs in polycrystalline BeO [94] where the specifics of the SCG fracture mode are poorly documented (the overall fracture mode for tests in air at 22°C is predominantly transgranular). While SCG does not occur in some single crystals such as MgO and apparently ZrO_2 [77], it can occur intergranularly in MgO, as shown by Rhodes et al. [95]. More SCG in finer versus larger grain MgO (~ 25 μm vs. ~ 45 μm) is uncertain because of impurity differences but may imply a grain size effect in view of there typically being thicker grain boundary phase as grain size increases. Whether there are intrinsic differences in SCG rates between materials exhibiting only intergranular versus at least some transgranular SCG is unknown. In contrast to the above oxides, fast fracture in a Mn-Zn ferrite [96] was mainly by intergranular fracture, while SCG occurred mostly by transgranular failure, especially with *G* ~ 45 μm and somewhat less with *G* ~ 35 μm with more grain boundary (e.g. Ca) phase, again suggesting possibly greater effects at finer grain size (Chap. 2, Sec. III.B and Chap. 2 end note).

Li ferrites show SCG, which has also been reported to be sensitive to losses of Li on firing [97], which may imply gradients of stoichiometry between grain boundaries and the rest of the grain, which could be a factor in changing fracture modes and in possible grain size effects. Significant decreases in Young's modulus and internal friction increases of HfO_2 [98] occurred upon opening the vacuum furnace (after sintering or heat treating for grain growth), saturating after only ~ 2 days; thus indicating microcracking from TEA stresses alone and SCG saturates in limited time in the absence of an applied stress.

Though noted in Chap. 2, Sec. III.B, it is again worth noting that corroding species which do not lead to SCG may also reduce strengths. Thus CaO and MgO crystals do not exhibit SCG due to H_2O, which attacks them independent of stress to produce hydroxides, whose expanded volume, when this occurs in constrained locations, especially pores or cracks, results in fracture due to repeated stages of stress buildup and release by cracking over days to weeks for CaO and much longer for MgO. Similar effects occur in polycrystalline bodies, where pores accessible from the surface are also important sources of this hydration damage. Possibly more extreme is the substantial to complete degradation of some TZP bodies over certain ranges of Y_2O_3 contents and G (Fig. 2.9).

Regarding nonoxides, SCG has been shown in some halide single crystals, e.g., AgCl and CaF_2, the latter also showing probable effects of slip limiting the extent of SCG, e.g. via easier arrest of cracks [92]. SCG in polycrystalline MgF_2 and ZnSe being 100% intergranular (whereas fast fracture is essentially 100% transgranular) indicates grain boundary control of SCG in these materials. McKinney et al. [99] reported essentially no SCG with large-scale cracks, e.g. DCB or DT tests, in various Si_3N_4 materials and no small-scale SCG (i.e., no delayed failure in pure Si_3N_4, made by either CVD or reaction sintering), but clear delayed failure in Si_3N_4 made with oxide additives (with the extent of SCG generally increasing with the amount of oxide additive) via all intergranular fracture. They attributed this large vs. small crack behavior to oxide distribution along grain boundaries, i.e. maintaining that of the many flaws available on the surface for SCG, at least one could always be found that had sufficient contiguity of grain boundary oxides for sufficient SCG to lower strength. On the other hand, large cracks, as used in a DCB test, covered too broad a range of grain boundaries, many of which may not have sufficient contiguity of oxide content to allow continuous SCG. Recently SCG has been reported (via essentially 100% intergranular fracture) in AlN [100,101] on a similar or lower level than in Al_2O_3, with the extent of SCG apparently correlating with the residual oxide grain boundary content [92]. While there appears to be intrinsic SCG in carbon materials [75,92], SCG does not appear to occur in carbides, e.g. B_4C, SiC, TiC, and ZrC (or borides, e.g., TiB_2 and ZrB_2) unless sufficient grain boundary phase (e.g. oxide) is present to provide the material and path for SCG [77,99]. Thus SiC made with oxide additives shows SCG, but not SiC made with B-C additions or by CVD (i.e. without additives), i.e., paralleling the Si_3N_4 results. This is corroborated by such materials showing no SCG exhibiting predominant to exclusive.

As test temperatures increase, e.g. to and beyond a few hundred degrees C, the amount of active species such as H_2O is commonly reduced, reducing the amount of SCG. However, in some cases active species may be contained in pores, causing SCG at high temperatures independent of the external test environment, as was indicated earlier in graphites. On the other hand, at higher temperatures other species may cause corrosive damage or SCG, e.g. due to their liquefaction (hence mobility), increased reactivity, or both. As noted in the previous section, oxidation of nonoxide materials can be an important manifestation of this, especially if either there is a grain boundary phase that will combine with the oxidation product, e.g. to form a softer, more reactive glass, or the nonoxide produces oxide liquid or glassy phases, e.g. B_2O_3 or SiO_2.

B. Effects of Grain Size on Strength of Al_2O_3 (and Ice) at Elevated Temperatures

Increasing temperatures above 22°C in air generally decreased sapphire strength [26,29,85–91,102–104] often drastically, e.g. losing 1/3 to 3/4 of its strength at 22°C upon reaching a minimum at 400–600°C depending on orientation (Fig. 6.1), surface finish, and test environment. Hurley [103] observed a rapid strength decrease from 22 to ~ 400°C for both, <11$\bar{2}$0>and *c*-axis (<0001>) filaments and then a plateau to ~ 700 and 900°C respectively before rapidly decreasing again. (However, compression testing of sapphire rods of the same orientations showed respectively a slow decrease, similar to that for Young's modulus, and then a very rapid decrease starting at ~ 800°C.) The level and especially the temperature of the strength minimum can be affected by other parameters. Charles [86] showed a strength minimum at ~ 900°C for sapphire tested in air as-annealed (1200°C) vs. 400–600°C for mechanically finished surfaces. These tests in various atmospheres showed sapphire strength decreasing by ~15% to a minimum at ~ 600°C in vacuum with less decrease in dry or wet H_2 (but strength ~ 20% lower in dry H_2 than in vacuum and ~ 20% lower for wet versus dry H_2) before all merging together at ~ 900°C. Iwasa and Bradt's [26] (indentation-fracture) fracture toughness tests of sapphire oriented for basal or rhombohedral fracture showed similar trends; i.e. decreasing ~ 25 and 75% to minima at ~ 800 and 1000°C respectively (Fig. 6.1). (Their K_{IC} tests of sapphire oriented for fracture on A or M planes follow the decreases of Young's moduli with increasing temperature.) Less strength decrease, i.e. a higher minimum strength (but at a somewhat lower temperature) was indicated in one [89] but not another [91] test of Cr doped sapphire. However, Sayir et al. [104], who observed strength minima at 300°C and maxima at 900°C in undoped sapphire, reported that 500 ppm MgO or TiO_2 (separately or combined) doping eliminated the minima and maxima.

Carniglia's surveys [105,106] of the σ–$G^{-1/2}$ behavior of Al_2O_3 showed strengths of finer grain size, dense bodies at 400°C ~ the same as at 22°C and then decreasing at a moderate rate up to 1000–1200°C and more rapidly beyond 1200°C. Differentiation of strength as a function of temperature in the larger grain size region was even more moderate. (Correcting for Carniglia's failure to plot all data at 22°C and erroneously plotting some data at higher strength reduces the limited differentiation his plot showed between fine grain bodies at 22 and 400°C.) Charles' [86] testing of lamp envelope Al_2O_3 (G ~ 40 μm) showed strength ~ constant from ~ 200–600°C and then dropping gradually (e.g. ~ 5%/100°C) in vacuum, while tests in dry and wet H_2 (the latter again at lower strength levels as for sapphire) showed a strength minimum at ~ 400°C and a maximum at ~ 1100°C. Neuber and Wimmer's [81] air testing of a ≥ 99.5% Al_2O_3 (P ~ 0.05, G ~ 35 μm) showed distinct strength minima (at ~ 400°C) and maxima (at ~ 800°C, Fig. 6.12) for each of four sets of rods having diameters of 2–8 mm, with the strength levels slightly lower for each increase in diameter. Kirchner et al. [107,108] also showed a definite strength minimum at ~ 400°C for their dense hot pressed Al_2O_3, tested as-polished, or strengthened by surface compression from quenching in silicone oil. The quenched material also showed a strength maximum at ~ 800°C; however, there was substantial scatter in both the maxima and minima for their bodies. While Jackman and Roberts [91] clearly showed such maxima and minima for single crystals, their tests of a 99.3+% Al_2O_3 (P ~ 0.05, G ~ 50 μm) showed only an uncertain indication of a strength minimum at ~ 500°C. Mizuta et al.'s [109] HIPed, transparent Al_2O_3 (uniform G ~ 1–2 μm) showed no maxima or minima; instead strength was ~ constant at ~ 780 MPa to > 1000°C and then dropped to ~ 700 MPa at 1100°C. Thus such minima, maxima, both, or a plateau at intermediate temperatures are shown in almost all [80,87,88,90,110] (Fig. 6.12) but not all [80,103] Al_2O_3 studies.

Al_2O_3 σ–$G^{-1/2}$ data [84,85,88,89,111–113] at 1200–1315°C (Fig. 6.10) shows similar two-branch behavior but with lower strength (e.g. ~ 50%, possibly more at fine grain size) than at 22°C, with reasonable agreement between different studies. Again, higher single crystal than many polycrystalline strengths are seen, as is a σ–$G^{-1/2}$ slope > 0 at finer grain size. While strength–temperature data for bodies of various grain sizes shows the overall expected strength decrease with increasing grain size, there is commonly a limited maximum, or at least ~ a strength plateau over a significant intermediate temperature range (Fig. 6.11).

Impurities or additives may or may not have significant effects in this temperature range. Thus there was no effect of AlON additions (other than via grain size) on strength (or K_{IC}) to at least 800°C [114] nor of CaO [31]. Crandall et al. [85] showed similar trends for Al_2O_3 hot pressed with or without 3% SiO_2 (Fig.

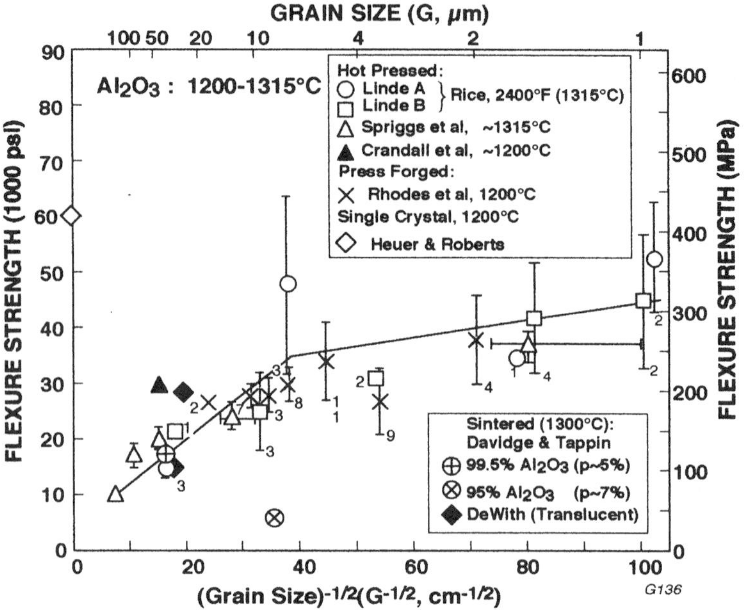

FIGURE 6.10 σ–$G^{-1/2}$ data, mainly for hot pressed and pressed forged Al_2O_3 , at 1200–1315°C. Note the general consistency of data from different sources and its indication of a two-branch σ–$G^{-1/2}$ relationship with the finer G branch having positive slope, and the generally lower strengths relative to those for single crystals. (Published with permission of the *Journal of Materials Science*.)

6.11). However, typical commercial (sintered) Al_2O_3 having an SiO_2-based (usually) glass phase commonly shows an intermediate (strain-rate-,composition- and possibly P-dependent) strength maximum at 700–1100°C, and then greater strength decreases [83,110] at higher temperature.

Al_2O_3 based polycrystalline fibers show similar strength–temperature trends. Tests of pure α–Al_2O_3 (Dupont FP) and Al_2O_3-SiO_2 fibers show the same strengths at 22° and 800°C, only moderate (~ 10%) decrease by 1000°C, and then a more rapid decrease [115-117] (Fig. 6.12). Al_2O_3-20% ZrO_2 fibers show ~ 10% higher strength at 800°C before dropping back to the same strength of 22°C at ~ 1000°C (and more rapid decrease at higher temperatures. Neither set of fibers was tested at 22°C < T > 800°C).

The above strength changes with increasing temperature *(T)* are put in broader perspective by comparing single- and polycrystal Al_2O_3 (including fiber) strength normalized by their values at 22°C, along with similar Young's modulus *(E)* and K_{IC} normalization (Fig. 6.12). This shows the well-known steady E–T

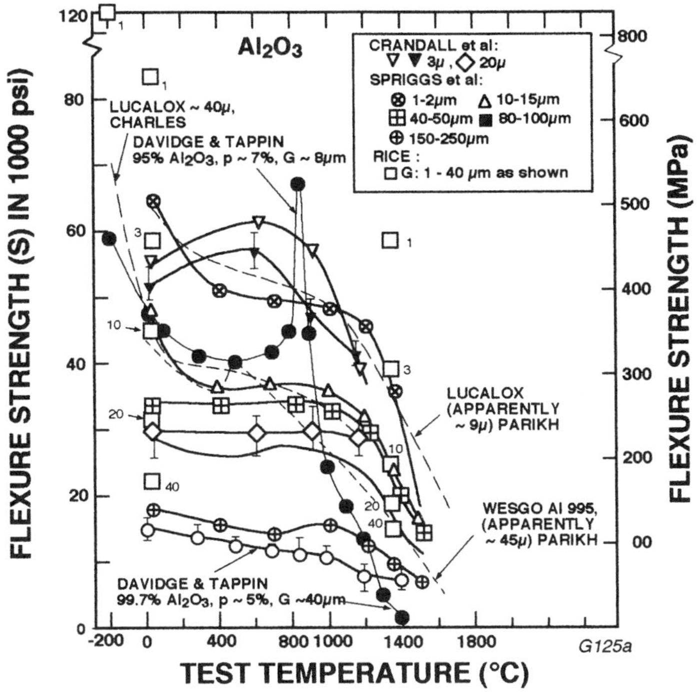

FIGURE 6.11 Flexure strength versus test temperature for different Al_2O_3 bodies reflecting primarily different grain sizes and secondarily some composition and processing differences. Note that the solid symbol of Crandall et al. [84] is for $Al_2O_3 + SiO_2$, and that the open symbol is for pure Al_2O_3, as is all other data except that of McLaren and Davidge [110]. (From Rice [1], published with permission of the *Journal of Materials Science*.)

decrease of 10–20% for both single- and polycrystals by 1200°C [108–120]. This is in marked contrast to a typically much faster initial decrease of both relative crystal K_{IC} and strength (typically oriented for basal or rhombohedrahl fracture), toward minima at ~ 400–800°C and then rising to pronounced strength maxima that can be ≥ to that at 22°C and then falling (rapidly). While absolute strength values vary as expected (e.g. with surface finish), these trends occur for crystals of various orientations [80,89,91] and machining [80,89], as well as as-grown (0°) crystal filaments [102]. Again, while sapphire strength values are higher when H_2O is not present, or with reduced activity, the trends are also relatively independent of the environment, since the basic trends are similar, whether the testing is done in vacuum or in air. Most polycrystalline tests at $T > 22°C$, <

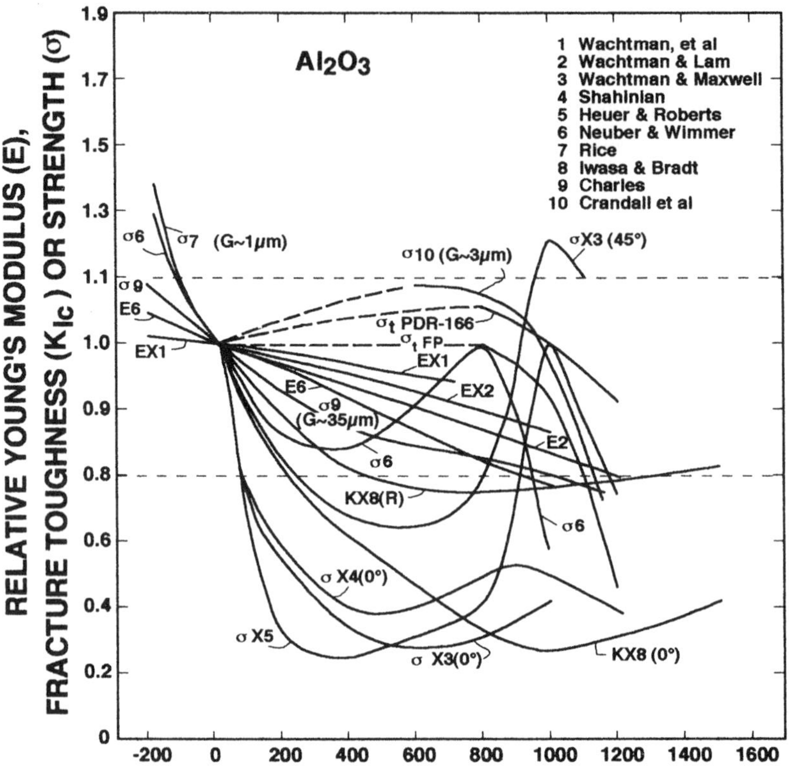

FIGURE 6.12 Relative Young's modulus *(E)*, fracture toughness (K_{IC}), and strength (σ) of single and polycrystalline Al_2O_3 versus test temperature (normalized by taking the property value at 22°C = 1). X following the property designation (*E*, K_{IC}, and σ) designates single crystal values (followed by the crystal orientation in ( ) if known). Numbers following the property and crystal designations give the source of the data (from listing, upper right). For polycrystalline values, grain sizes are shown in ( ) where pertinent and available. Curves designated σ_t are for true tensile testing of fibers (FP, a coarser grain pure alumina and PDR, a finer grain alumina-zirconia fiber). While most tests were in air, some were in vacuum or liquid N_2. Note the change in scale between relative values of 0.8 and 1.1 in order better to differentiate the data there, and that *E–T* trends, especially for single crystals, are a key basis of comparison. (From Rice [1], published with the permission of the *Journal of Materials Science*.)

800°C indicate a strength minimum at 400–600°C, and these and higher temperature tests showed little or no relative strength decrease from 22°C levels until ~ 800°C and may often show a limited maximum at 600–800°C (also observed for some fibers, tested in true tension, designated by σ_t in Figure 6.12, or flexure).

Typical ice, i.e. solid H_2O, has several properties, including strength, similar to Al_2O_3. This includes their single crystal basal slip and associated large strain rate sensitive yield drops, and scaling with the fraction of absolute melting point (i.e. homologous temperature), e.g. compare Higashi's [121] and Kronberg's [122] data and effects of solid solution species (e.g. Jones and Glen [123]). Thus Schulson et al.'s [124] strength for ice at –10°C (~ 96% of the absolute melting temperature) provides not only information for ice but also some guidance for Al_2O_3. Their results showed a Petch-type relation for samples with and without purposely introduced sharp cracks (Fig. 6.13), implying similar behavior at similar relative strain rates for Al_2O_3 without other significant effects,

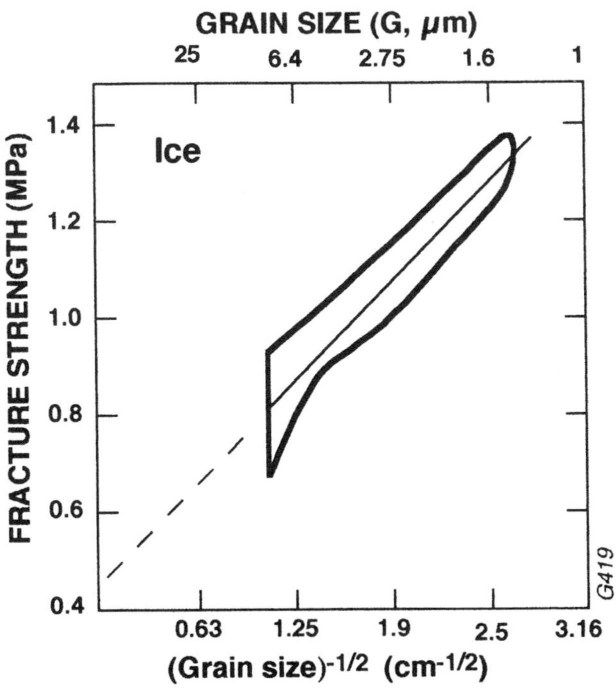

FIGURE 6.13 Tensile strength of ice at – 10°C with a strain rate of 10^{-3}/s. Note the Petch behavior despite the testing being at > 96% of the absolute melting temperature, implying analogous behavior in Al_2O_3. (From Ref. 124.)

e.g. from grain boundary phases. Thus their results indicate that Al_2O_3 can have brittle, Petch-type behavior at very high temperatures. Similarly, Dykins' [125] results show the ratio of tensile strengths of ice stressed parallel or perpendicular with growth dendrites to be ~ 3; with the latter strength decreasing from ~ 1.4 to 0.9 MPa as the test temperature increased from –27 to –3°C which implies that similar grain orientation effects can substantially impact the high temperature strength of polycrystalline Al_2O_3.

C. Effects of Grain Size on Strength of BeO at Elevated Temperatures

Bentle and Kneifel [126] showed polycrystalline BeO strength averaging ~ 10% greater at –196° than at 22°C in vacuum, suggesting a possible 15–30% increase for grain sizes > ~ 40 μm and a possible ~ 5–10% decrease at grain sizes of 20–40 μm. They also showed that testing in air or water vs. vacuum at 22°C reduced strengths of BeO (G ~ 20 μm) ~ 8–10% and 15–20% respectively. Similarly, Rotsey et al. [94] showed a strain-rate dependence of BeO (G ~ ≤ 3 μm, P ~ 0.04) indicating ~ 30% strength decrease in water vs. air at 22°C for circumferential ground (pressed) rods. Slightly higher strength but similar relative changes were found in air but no change in silicone oil. Annealed samples had higher strength and somewhat greater decreases (~ 40%) for testing in air but decreased nearly 10% for tests in silicone oil.

BeO shows a typical Young's modulus decrease with increasing temperatures, e.g. ~ 15% by 1200°C [118,119] (Fig. 6.14). Tests of (as-grown) single crystals in vacuum at 500°C and 1000–1800°C [127] showed slip only at ≥ 1000°C, with strengths following the decrease of Young's modulus with temperature fairly closely. Carniglia's earlier σ–$G^{-1/2}$ surveys [105,106] of BeO showed moderate or no strength decreases at fine grain sizes until > 800°C, with strength possibly increasing at intermediate temperatures. Though there was less differentiation of strengths versus temperature at larger grain sizes, there was even greater indication of strength, first increasing with increasing temperature and then decreasing. Bentle and Kniefel's [126] data for 500–1300°C almost always showed a strength maximum at 500–1200°C which was > strengths at 22°C (Fig. 6.14). Samples made with 1% MgO (G ~ 60–150 μm) showed substantially lower strength maxima. Such maxima were not seen for slightly less (~99.3%) dense samples (e.g. G ~ 45 and 60 μm) or with greater impurity contents (mainly 5000–7000 ppm F) than those shown in Fig. 6.14 (99.7–99.8% dense); instead strength was ~ constant (e.g. 400–800°C). The tests of Chandler et al. [129] at 300–1200°C in air showed moderate (~ 15–25%) relative strength maxima at 500–1000°C for 99.9% pure (UOX and HPA) (P ~ 0.02–0.03, G ~ 20 μm), while a less dense (99.7%) pure (AOX) BeO showed σ–T closely following E–T trends. However, the same AOX BeO with G ~ 50 μm (P ~ 0.04) showed a strength max-

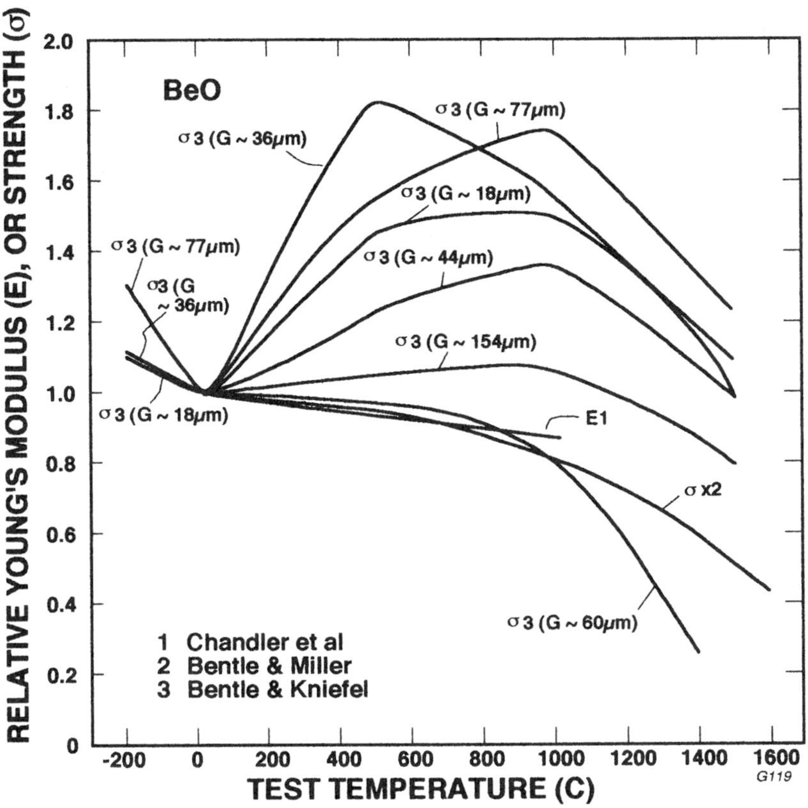

FIGURE 6.14 Relative Young's modulus *(E)* and flexure strength (σ) for (as-grown) single crystal and polycrystalline BeO (normalized by taking the value at 22°C = 1). Curve designations are analogous to those of Fig. 6.12. Note that Bentle and Miller's [127] and Bentle and Kniefel's [126] tests were in vacuum (*G* ~ 60 and 154 μm is with 1% MgO). Data of Fryxell and Chandler [128] is for both unoriented (from AOX powder) and oriented grains (from UOX powder), and that *E–T* trends, especially for single crystals, are a key basis of comparison. (From Ref. 1, published with the permission of the *Journal of Materials Science*.)

imum at 1000°C, 35% > at 22°C, and UOX BeO with 0.5% MgO rising to a slightly lower maximum. Fryxell and Chandler [128], using the same materials and process, showed all specimens having a relative strength maximum at 500–800°C with the level of the relative maximum increasing with increasing grain size from 7–10% (*G* ~ 20 μm) through 20% (>*G* ~ 50) to 40–43% (*G* ~ 90 μm). There was typically a tendency for lower relative strength maxima with

AOX BeO (no additive) than with UOX BeO (+0.5% MgO); the latter also showed preferred orientation increasing with increasing grain size. The absolute strength values were highest (~ 200 MPa) for G ~ 20 μm and intermediate for G ~ 50 μm bodies for both AOX and UOX, the latter showing ~ 50 and ~ 65% grain orientation for the two grain sizes respectively. The ~ 90 μm G bodies had strengths of ~ 100–130 MPa for AOX and ~130–175 MPa for UOX with ~ 80% grain orientation. Relative strength maxima at intermediate temperatures were also reported by Stehsel et al. [130] for three commercial cold pressed and one commercial slip cast and fired BeO and two (both commercial) of four hot pressed BeO samples tested. While the latter tests and those of Chandler and colleagues were in air, those of Bentle and Kniefel were in vacuum, indicating that these trends (e.g. the maxima) are not due solely, if at all, to environmental (e.g. H_2O) effects. On the other hand, Carniglia et al. [131] showed strengths (in vacuum) of dense hot pressed BeO being ~12% higher at –200°C vs. in air at 22°C (σ ~ 270 MPa), and 45, 51, and 27% higher respectively at ~ 550, 1000, and 1500°C.

D. Effects of *G* on Strength of CaO and MgO at Elevated Temperatures

As noted in Chap. 2, Sec. III.B, K_{IC} of MgO crystals increased in H_2O (but not DMF) [132]. Both Janowski and Rossi [133] and Rice [134] showed that MgO crystal yield stresses decreased ~ 20% and strength ~ 15% (but with greater ductility) in water versus in air at 22°C. Both showed that crystal pieces tested in water returned yield and fracture stresses back to their original air tested levels when retested (dried) in air. Thus SCG has not been observed in MgO single crystals, but yield and fracture stress reductions have been, indicating enhanced dislocation mobility (as does the increased toughness and ductility). On the other hand, similar CaO crystal tests showed yield and fracture stresses increasing respectively by ~ 5–25% and 5–35% in water versus air at 22°C [134]. Testing MgO crystals in liquid N_2 raised yield stresses ~ 80–130%, consistent with Copley and Pask's [135] (compression) and Thompson and Roberts's [136] tests and fracture stresses 10–15% versus in air at 22°C. Corresponding CaO crystal increases were ~ 105% and 90% respectively. However, long-term exposure of CaO crystals to liquid or vapor H_2O results in propagation of cleavage cracks attributed to the wedging action of resultant $Ca(OH)_2$ in preexisting cracks [134].

Polycrystalline MgO tests by Janowski and Rossi [133] and Rice [134] showed strengths lower (e.g. ~ 15%) in water than in air at 22°C; i.e. very similar to crystal tests. Both also showed recovery of the strength loss on drying and retesting in air. (Rice's tests covered G ~ 2–100 μm, showing no grain size trend.) However, strength in air was only ~ 10% lower than in liquid N_2, i.e. only ~ 10% of the difference found for single crystals. On the other hand, Rhodes et al. [95] reported delayed failure in polycrystalline MgO [G ~ 25–45 μm, P

0–0.007, and ≤ 0.02–0.6% impurities]. While the two finer grain bodies ($G = 26$ and 30 μm) showed delayed failure at ~ 50% of the inert strength (versus ~ 80 and 70% for $G = 46$ and 43 μm respectively), they were also the lowest purity (99.4 and 99.6 versus 99.98+ and 99.92% respectively). Thus they concluded that purity was the dominant variable in SCG, which is consistent with most of the impurities being at the grain boundaries with intergranular fracture (in contrast to mostly transgranular fracture in similar grain size bodies tested in air [137]). They also observed a possible fatigue limit (~ 80%) in the highest purity body, which they postulated to be due to the absence of a continuous grain boundary impurity film.

Recrystallized CaO crystal bars at ~ 1100 and 1300°C showed little or no strength decrease from 22°C. However, macroscopic yield frequently preceded brittle, almost exclusively transgranular fracture [134]. Limited polycrystalline MgO studies at moderate temperatures typically showed either an initial limited strength rise to a maximum at 400–700°C (especially as grain size increased) or a lower rate of decrease before more rapid strength decrease with increasing temperature. These polycrystalline strength trends are also supported by data of Evans et al. [138] ($G \sim 25$ and 150 μm, particularly for chemically polished samples). MgO single crystals recrystallized by pressed forging or hot extrusion [139,140] also showed little or no strength reduction with increasing temperature, some macroscopic yielding by ~ 1300°C and extensively at ~ 1500°C (but maintaining transparency and subsequent brittle, cleavage, fracture). While hot extruded MgO specimens from hot pressed and annealed billets showed similar strength for the same grain size as from recrystallized crystals at 22°C, the former showed a greater strength decrease at 1540°C, and the recrystallized crystals averaged ~ twice the strength as hot extruded, hot pressed MgO. Similarly, while the latter showed somewhat greater occurrence of grain boundary fracture at 22°C, it showed much greater frequency and amounts of this at higher temperature than the recrystallized crystals [139,140]. The above trends (which are consistent with those of Day and Stokes [141], $G \sim 100 \pm 50$ μm, $T \geq 1700$°C) are put in better perspective by plotting properties normalized by their 22°C values (Fig. 6.15). This shows (1) a moderate Young's modulus decrease of 10–15% by 1200°C, (2) substantially faster yield stress decrease (for <100> stressing), (3) strength ~ constant or a strength maximum between 400 and 800°C, and (4) a trend for less strength decrease and higher relative maxima at higher temperature as grain size increases.

E. Effects of Grain Size on Strength of Other Cubic Single Oxides ThO_2, UO_2, Y_2O_3, and ZrO_2 at Elevated Temperatures

ThO_2 shows positive σ–$G^{-1/2}$ slopes for finer grains at 22 and 1000°C but somewhat higher strength at 1000 vs. 22°C across the grain size range studied

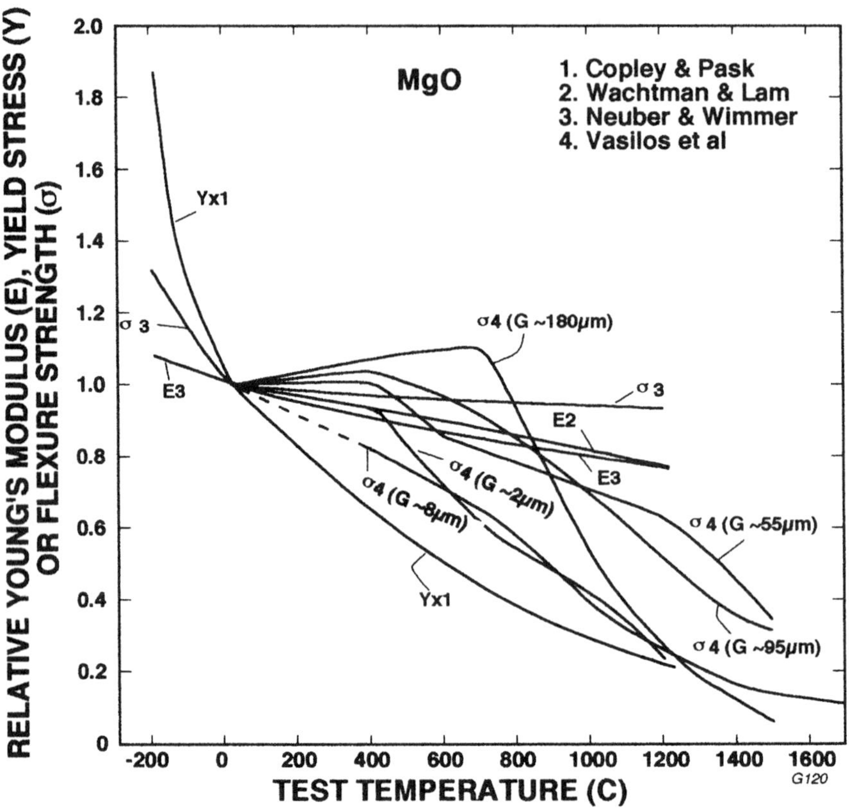

FIGURE 6.15 Relative Young's modulus *(E)*, yield stress (<100> single crystal, Y), and flexure strength (σ) of MgO versus test temperature (normalized by taking the property at 22°C = 1). Note that curve designations are analogous to those of Fig. 6.12, and that *E–T* trends, especially for single crystals, are a key basis of comparison. (After Rice [1], published with permission of Materials Science *the Journal of*).

[142,143] (Fig. 6.16). Collectively, UO_2 flexure data [144–148] is consistent with the basic σ–$G^{-1/2}$ model at 22 and 1000°C and indicates probable increased strength at 1000°C (Fig. 6.17). Diametral compression data [149] at 22°C also agrees with these trends. Individual data sets more clearly show strength increasing with temperature. Thus Burdick and Parker [144] showed UO_2 strength increased to a maximum at 700–1100°C with net increases of 20–35% for *G* ~ 20 μm (*P* ~ 0.15–0.22) and 50–70% at > *G* ~ 40 μm (*P* 0.08–0.12). Knudsen et al. [145] showed ~ 20% strength decrease for > *G* ~ 45 μm (*P* ~ 0.1) and a 5 to 75%

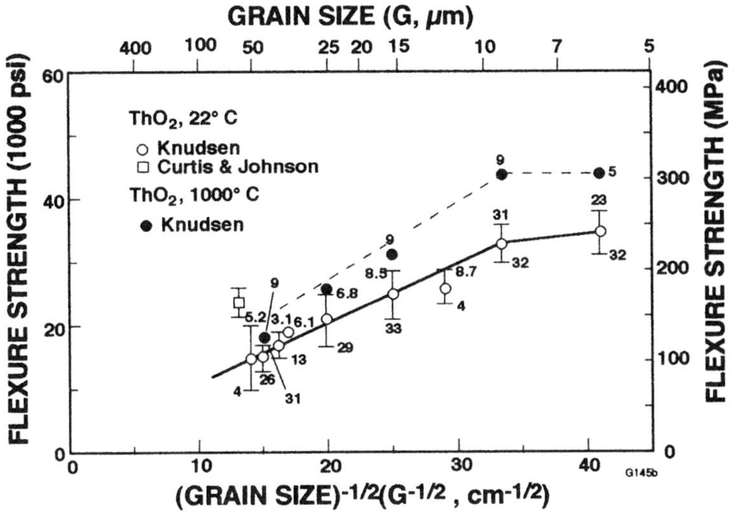

FIGURE 6.16 ThO_2 σ versus $G^{-1/2}$ at 22 and 1000°C. Data of Knudsen [142] corrected for variable porosity (superscripts in %) using $b = 4.2$ and 6.6 at 22 and 1000°C respectively per his analysis. Numbers with high temperature data points and below the error bars for tests at 22°C are the numbers of values averaged. Note the consistency of resultant corrected data despite quite variable P levels for different G bodies, and one data point of Curtis and Johnson [143]. (From Ref. 1, published with the permission of the *Journal of Materials Science.*)

increase for G = 20–25 μm (P ~ 0.08–0.24) between 22 and 1000°C. Evans and Davidge [146] showed no strength increase with initial temperature increase for their G = 8 μm UO_2 till ~ 500°C, and then a significant rise, peaking at ~ 800°C (a ~ 35% total increase) before decreasing again. Their ~ 25 μm G body showed a longer, slower strength rise, peaking at ~ 1100°C with a similar net increase before decreasing. Beals et al. [147] showed a similar σ–T increase and maximum strength (G ~ 25 μm, P ~ 0.03). Canon et al. [148] showed that strength increased slowly to a maximum (about 20% higher than at 22°C) at ~ 1400°C and then dropped sharply, for bodies with G = ~ 8, ~ 15, and ~ 31 μm.

While ZrO_2 single crystals (11.1 m/o, ~ 18.5 w/o Y_2O_3) show a typical Young's modulus decrease (e.g. ~ 1–2%/100°C) with increasing temperature to the limit of testing (700°C), polycrystalline behavior is more complex [150]. Polycrystalline ZrO_2 (CaO or MgO) stabilized showed somewhat greater Young's modulus decreases to ~ 400°C and then transitions to ~ an extrapolation of the above single crystal data [81,150]. Wachtman and Corwin [151] showed an internal friction peak in ZrO_2, generally in the 300–400°C range but decreasing some

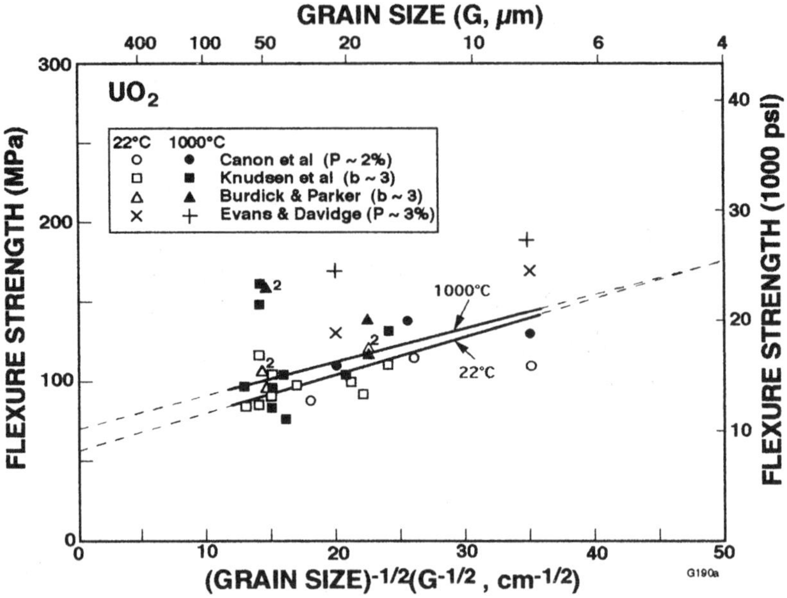

FIGURE 6.17 UO_2 σ vs. $G^{-1/2}$ at 22 and 1000°C. Data of Burdick and Parker [144], Knudsen et al. [145], Evans and Davidge [146], and Canon et al. [148] are for flexure at both temperatures. (Data of Kennedy and Bandyopandhyay [149], only at 22°C and from diametral compression, is not plotted, but showed a similar, lower trend.) Note data plotted as-measured with $P \sim 0.02$ for Canon et al. and ~ 0.03 for Evans and Davidge, while Knudsen et al's. data was corrected for P (= 0.05–0.24, mostly 0.05–0.1) and Burdick and Parker ($P = \sim 0.08$–0.12) using $b = 3$ (a 2 next to some data points indicates two identical points). Kennedy and Bandyopandhyay's data are plotted as-measured for $P = 0.03$–0.09 (shown next to data points). Note there is no trend for strength to decrease from 22 to 1000°C, and in fact strength appears to be greater at 1000 than at 22°C.

in magnitude and in the temperature of the maximum as the CaO content increased from 2 to 20%. Shimada et al. [152] showed an initial somewhat greater Young's modulus decrease to ~ 400°C and then (within the 650°C limit of testing) a similar transition as above for dense ($P \sim 0$) sintered ZrO_2 (+ 3 m/o, ~ 5.5 w/o Y_2O_3), as have Adams et al. [153] for sintered ($P \sim 0.02$–0.07, $G = 15$–50 μm) and hot pressed ($P \sim 0.005$–0.02, $G \sim 1$–3 μm) ZrO_2 (+ 6.5 m/o, ~ 11 w/o, Y_2O_3). Adams et al. also showed that their ZrO_2-Y_2O_3 and a commercial ZrO_2-Y_2O_3 body (with ~ 1 w/o SiO_2) had a much greater Young's modulus decrease from ~ 100 to ~ 400°C than three commercial ZrO_2-MgO bodies tested. Rapid initial E–T decreases have also been more recently reported for (1) ZrO_2 + 33m/o Tb_4O_7 between 200 and 500°C (but not with 33 m/o Pr_6O_{11}) [154], (2) 3 m/o

Y_2O_3 between ~ 22 and 300°C [9], and (3) 2 m/o Y_2O_3, + 8 m/o Y_2O_3, and 12 m/o CeO_2 (with respectively similar, greater, and smaller decreases, the latter when the CeO_2 was partly reduced, but no effect when it was fully oxidized) [10]. In the latter two cases, as well as that of Shimada et al., the anomalous Young's modulus decreases were associated with maxima or high levels of internal friction. Nishiyama et al. [11] also recently reported an internal friction peak at ~ 150°C in ZrO_2-2.8 m/o Y_2O_3. These ZrO_2 changes are corroborated by similar effects of Dole and colleagues [98] in HfO_2, i.e. an internal friction peak in unstabilized (monoclinic) HfO_2 at ~ 400°C, and drops in both Young's and shear moduli and an internal friction peak in HfO_2-20 m/o Er_2O_3.

While fully stabilized ZrO_2 crystals (22 w/o Y_2O_3) show essentially no strength changes till T ~ 1500°C [64], polycrystalline strengths show similar but greater deviations from E–T trends (Fig. 6.18). Thus the lack of significant single crystal strength changes between –196 and 22°C indicates limited, or no, single crystal SCG. Partially stabilized (6 w/o Y_2O_3) crystals (which start from about four-fold higher strength than fully stabilized crystals) show an initial strength decrease much greater than that of Young's modulus until ~ 500°C; then it levels off (at ~ twice the strength of fully stabilized crystals) until ~ 1500°C. Adams et al.'s tests of sintered and hot pressed ZrO_2 (+ ~ 11 w/o Y_2O_3) [153], though scattered, showed an average initial trend similar to the PSZ (6% Y_2O_3) crystals but continued to have much greater strength decrease than Young's modulus decrease, to a modest minimum at ~ 700°C. Drachinskii et al. [155] showed a slightly greater strength decrease to the limit of their tests (500°C, Fig. 6.18) in ZrO_2 + 4 m/o (~ 7 w/o) Y_2O_3 sintered and then annealed substantially. However, specimens with limited annealing after sintering dropped to a strength minimum of ~ 80% of their 22°C values at 100–200°C and then rose to a strength maximum at 300–400°C that could be similar or > (e.g. ~ 35%) strengths at 22°C. Higher temperature tests of some of these lesser annealed samples showed first a strength minimum at ~ 700°C (e.g. Adams et al.), but strength values ranged from 50% relative to 22°C down to ~ 30% relative to their strength maxima at ~ 300°C (i.e. in either case less relative decreases than for Adams et al.); then there was a strength maximum at ~ 1000°C. Such greater strength deviations and complexities are apparently not limited to ZrO_2-Y_2O_3 bodies, as shown by Neuber and Wimmer [81], who report greater strength decrease than of Young's modulus with increasing temperature, and probable inflections at ~ 300 and 800°C (Fig. 6.18) in ZrO_2 (with ~ 5 wt% CaO or MgO).

Fracture mode changes accompany the above ZrO_2 strength decreases with increased temperature (Fig. 18), e.g. Adams et al. [153] saw mostly transgranular fracture at 22°C, mixed trans- and intergranular fracture at 1000°C, and 100% intergranular fracture by 1500°C in their ZrO_2 (~ 11 w/o Y_2O_3) bodies, similar to PSZ (2.4 w/o MgO) [64]. Drachinskii et al. [155] observed transgranular fracture varying from 40 to 90% for specimens of various annealing in tests at 100°C, with the least transgranular fracture being for the lowest strength (180 MPa), but

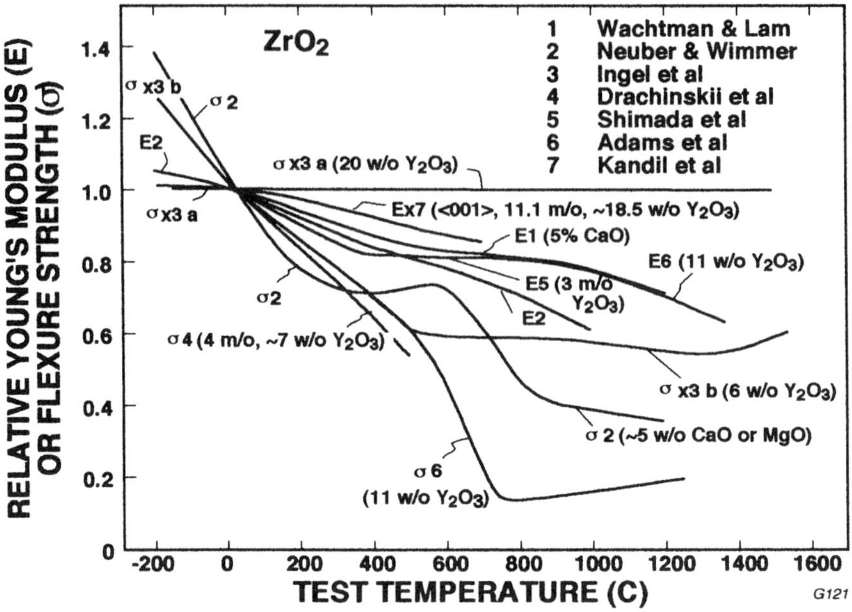

FIGURE 6.18 Relative Young's modulus *(E)* and flexure strength (σ) of single- and polycrystalline ZrO_2 versus test temperature (normalized by taking the property at 22°C = 1). Note the stabilizer in Neuber and Wimmer's ZrO_2 is not specified but is believed to be either CaO or MgO (~ 5 wt%, *P* ~ 0.13, ~ 25 μm) [81]. Note that curve designations are analogous to those of Fig. 6.12 (along with designation of some compositions) and that *E–T* trends, especially for single crystals, are a key basis of comparison. (From Ref. 1, published with the permission of the *Journal of Materials Science.*)

an intermediate % [153] for the highest strength (410 MPa), vs. 90% at 370 MPa. Rice [156] observed fracture initiation from grain boundaries surrounded entirely by 100% transgranular fracture not only in MgO, CaO, and $MgAl_2O_4$ but also in ZrO_2 (12.4 w/o MgO) and ZrO_2 (+11 w/o Y_2O_3, from the same processing as specimens used by Adams et al.). The substantial intergranular fracture at higher temperatures correlates with substantial grain boundary sliding creep, and even superplasticity found at ≥ 1000°C in fine grain TZP [156].

F. Effects of Grain Size on Strength of Mixed Oxides at Elevated Temperatures

As noted in the previous section, $MgAl_2O_4$ toughness decreased (~ 20%) to a minimum at ~ 900°C for {100} fracture (and less for other orientations) [38].

Toughness of hot pressed $MgAl_2O_4$ decreased slowly with increasing temperature (e.g. ~ 10% by ~ 900°C) for G = 5, 12, and 25 μm (but possibly less for G = 40 μm) and then much more rapidly [39], or was constant to ~ 800°C and then increased for G ~ 35 μm [41]. Overall strength behaved similarly, being the same at 22 and 200°C; it dropped by ~ 25% to a minimum at ~ 600°C or a plateau at ~ 400–800°C and then decreased slowly at higher temperature, i.e. similar to the temperature dependence of Young's modulus [39].

Penty [157] prepared and tested hot pressed mullite bodies ($P \leq 0.01$, G ~ 0.75–1.5, mostly > 1 μm) showing a strength minimum at ~ 700°C and a maximum at 1200°C, with some variation with stoichiometry (Fig. 6.19). Mah and Mazdiyasni [48] reported flexure strengths increasing from ~ 130 MPa at 22°C to ~ 145 MPa at 1500°C in dense, transparent, hot pressed mullite (61.9 mol% Al_2O_3, G ~ 3–9 μm). However, measurements at only 22, 1000, and 1200–1500°C missed possible intervening strength changes. Their lower strengths appear to be consistent with their larger grain size, though specimen size and surface finish are probably also factors. Fracture toughness calculated from fractography gave ~ 1.8 MPa·m$^{1/2}$ at 22°C, decreasing ~ 20% to a minimum

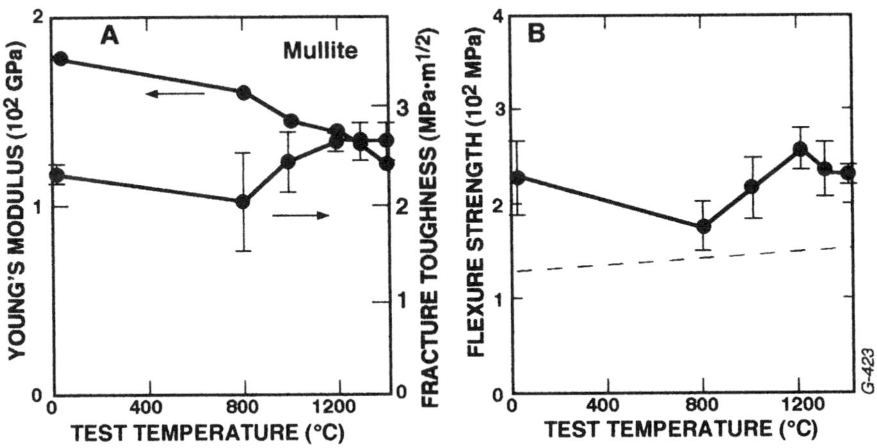

FIGURE 6.19 Mechanical properties of polycrystalline mullite versus test temperature for mullite. (A) Young's modulus and toughness data of Baudin [47]; (B) strength data of Baudin (solid line) [47] and Mah and Mazdiyasni (dashed line) [48], with the former being nearly identical to that of Penty [157] (not shown for clarity). Note measurements only at 22, 1000, and 1300–1500°C by Mah and Mazdiyasni missed seeing possible changes at intervening temperatures and that their larger G is probably an important factor in their overall lower strengths, despite having residual P ~ 0 versus 0.006 for Penty. Vertical bars = standard deviations. See also Figure 6.4.

of ~ 1.5 MPa·m$^{1/2}$ at ~ 1100°C and then increasing substantially as the temperature further increased, apparently due to slow crack growth that was observed beginning at ~ 1300°C and was attributed to limited amounts of a glassy grain boundary phase observed in TEM. Ohnishi et al. [46] showed a similar toughness trend with temperature, i.e. initially decreasing similar to E but then increasing substantially at higher temperature. Their strengths for sintered mullite, though higher, clearly showed more complex changes than a simple linear trend with increasing temperature. Baudin [47] reported lower E values but with a similar relative decrease with increasing T and a similar range of toughnesses and strengths values but decreasing ~ 10% to minima at ~ 800°C and then increasing to maxima ~ 20% > values at 22°C at ~ 1400°C.

G. Effects of Grain Size on Strength of Borides, Carbides, and Nitrides at Elevated Temperatures

While there is little or no σ–G data for most of these materials, there is some, as well as some other pertinent data on a few of them, e.g. TiB_2. Thus σ–$G^{-1/2}$ data for dense sintered TiB_2 tested at 970°C in argon or molten Al by Baumgartner [73] both showed two-branch behavior typical of brittle failure from flaws, except the larger G branch and hence the branch intersection may occur at somewhat finer G due to effects of probable microcracks in the largest G body (Fig. 6.20). The presence and effects of microcracks was shown by direct observations of Baumgartner and Steiger [74] in the largest G (~ 24 μm, P= 0.004) body and effects on properties in such bodies and those with finer G. Thus at 22°C Young's modulus of intermediate G (~ 11 μm) bodies was ~ 6%< that of the finest G bodies (~ 1 μm), and that of the largest G bodies was typically 20% < the intermediate G bodies despite porosities of ~ 2.4, 0.6, and 0.3% respectively in the finer through the largest G bodies. In the extreme, E of the largest G body at 22°C was as low as 270 versus 545 GPa for a finer G (~ 4 μm), and the thermal conductivity of the latter was ~ 25% > the former. Finer and intermediate G bodies showed E decreasing by ~ 24% from 22 to 1000°C, while the larger G bodies decreased by ~ 14%. These trends are supported by their strengths increasing in inert atmosphere testing as temperature increased by ~ 30 to 100% as G increased, with most (e.g. 80%) of the increase occurring by ~ 1000°C. Further, the load–deflection curves for the finer G bodies were linear to at least 1250°C, while those for heavily microcracked larger G bodies were nonlinear prior to fracture, which was attributed to additional stress-induced microcracking. These observations are supported by results of Mandorf and Hartwig [158], who showed that while their Young's moduli decreased less, e.g. by ~ 4% to ~ 1000°C (then accelerated in their decreases, especially with higher porosity), their flexure strengths increased more, e.g. by ~ 25% on reaching a maximum of ~ 305 GPa at ~ 1400°C (for G estimated at ~ 25 μm with a few

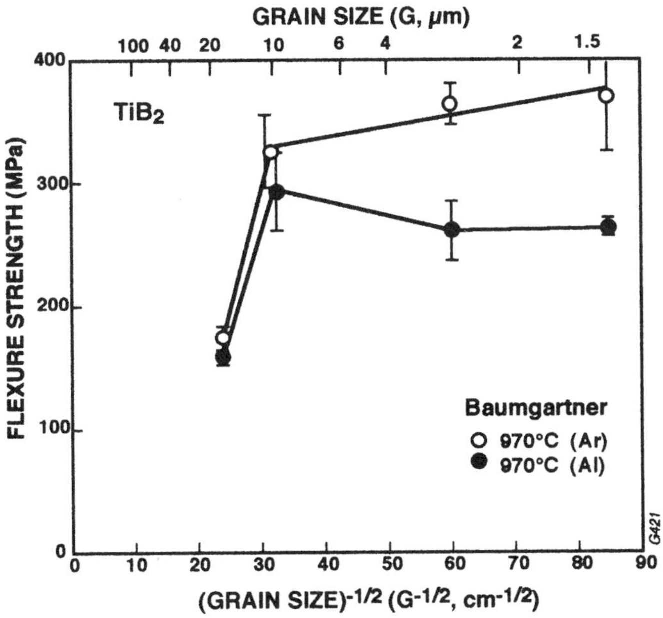

FIGURE 6.20 Flexure strength versus the inverse square root of grain size ($G^{-1/2}$) for dense sintered TiB_2 at ~ 970°C in argon (Ar) and molten aluminum (Al) from Baumgartner [72]. Note that (1) the finest *G* body had residual porosity of ~ 2.4% (~ 3–8 times the other bodies) and that the second finest *G* body was tested at 1000°C, so correction for these further increases their strengths, (2) the largest *G* body was microcracked giving lower strength, and (3) the data of Matsushita et al. [159] at finer *G* and that of Mandorf and Hartwig [158] at larger *G* are consistent with the above plot, though the latter indicates less larger *G* strength decrease, consistent with less apparent microcracking. Baumgartner attributed strength reduction in molten aluminum to liquid metal embrittlement, i.e. a reduction of crack tip toughness, instead of SCG, since fractography showed Al had limited penetration along grain boundaries but good penetration into surface connected processing flaws, there was no evidence of SCG, and fracture was transgranular.

percent porosity) indicating substantial microcrack closure. Data of Matsushita et al. [159] for dense sintered TiB_2 tested in Ar showed only a few % increase, mainly by 1000°C in testing to 1400°C (tests in air resulted in substantially greater increases, e.g. ~ 50% at 1000°C and then decreasing back to inert atmosphere levels at 1200 and 1400°C due probably to oxidation effects). The lack of significant increase in inert atmosphere is attributed, at least in part, to their finer *G*, estimated at ~ 5 μm. Limited tests of ZrB_2 at 1000°C showed no strength change from 22°C [160] or a maximum at ~ 300°C (~ 600°C for HfB_2, both in inert atmosphere) [161].

Limited B_4C data indicates limited changes from the σ–$G^{-1/2}$ behavior at 22°C. Thus de With [162] showed that the strength of the commercial hot pressed B_4C (G ~ 10 μm, P ~ 0) in dry nitrogen decreased from ~ 390 MPa at 22°C by only a few percent at ~ 600°C and then more rapidly to a minimum of ~ 300 MPa at ~ 1000°C, then increasing back to its 22°C level at ~ 1200°C. This was very similar in form, but larger in the extent of changes for toughness of the same material, with both strength and toughness tests giving ~ 100% transgranular fracture across the temperature range. Gogotsi et al.'s [163] behavior of a commercial dense hot pressed B_4C (G unspecified) showed very similar behavior in testing in Ar, except for starting from a strength of 300 MPa at 22°C (testing in air resulted in ~ 10% strength reduction by 600°C and then dropping to ~ 200 MPa at 1000 and 1200°C). Several investigations [164–166] showed strengths of B_4C decreasing very little until ~ 800°C (and limited decrease above 800°C) [162], which is consistent with K_{IC} trends.

Similar tests of SiC showed strength and K_{IC} maxima at ~ 1400°C [167], and considerable investigation of dense sintered and hot pressed SiC for engine and other applications commonly showed strength at 1000°C similar to that at 22°C or somewhat (e.g. ~ 20%) higher, typically for G ~ 2–10 μm [168]. Miracle and Lipsitt [169] showed limited (e.g. 10–20%) strength increases or decreases, or possibly no strength changes, in TiC from 22° to 600°C, and in some cases to 1000–1200°C depending on C/Ti ratios of 0.66, 0.75, 0.83, and 0.93 (G respectively 22, 21, 20, and 14 μm). Substantial strength decreases occurred at higher temperatures, with the earliest and greatest strength decrease for the C/Ti = 0.66, G ~ 22 μm body. Thus the strength change with increasing temperature generally did not follow the ~ 5% decrease of Young's modulus in this temperature range [170]. More extensive testing of dense sintered or hot pressed Si_3N_4, as well as less dense RSSN, showed that some bodies had lower strengths by 800–1000°C vs. 22°C, many had no decrease, and several increased (again by up to ~ 20%) [168]. This again shows strength not following the E–T trend (e.g. ≤ 5% decrease by 1000°C. Although such increases are most common for RSSN, they are not restricted to it (increases in strength can result from surface oxidation removing flaws in such nonoxides, especially RSSN).

V. GRAIN EFFECTS ON THERMAL SHOCK BEHAVIOR

Broader studies of thermal stress and shock resistance of ceramics support the general applicability of models discussed earlier, in particular confirming that fracture normally occurs on cooling rather than heating, since the former results in tensile stresses on the surface where fracture initiating flaws are typically much more prevalent. While such fracture thus occurs in material with less, possibly no, increase over ambient temperature, the temperature exposure may have effects via the thermal gradient impacts on stresses and properties, e.g. Young's modulus and thermal conductivity.

Data specifically on the grain, mainly size, dependence of thermal stress and shock resistance of ceramics, though quite limited, provides information to support the simple models, possible modification of them, and other insights to mechanical behavior, e.g. crack bridging effects. Gupta's [171,172] study of thermal shock failure of various alumina bodies of differing *G* with little or no porosity using the typical water quench test provides some of the clearest data. In this test, bars are heated in a furnace to a fixed temperature and then quenched into a water bath and subsequently tested for their resultant strength. This is repeated with other specimens with the post quench strength plotted versus the quench temperatures, with particular attention to the critical quench temperature, i.e. T_C, where strength decreases, often catastrophically. He showed that sapphire had the highest critical T_C, ~ 250°C (as well as the highest strength consistent with much data of Chap. 3, Sec. 1) but failed catastrophically, or nearly so, i.e. had zero or negligible strengths, after quenching at $\geq T_C$. Bodies with *G* of 10, 34, 40, and 85 μm, which had initial strengths decreasing from 340 to 160 MPa, all had a T_C of 200°C (Fig. 6.21). However, while the residual strengths after quenching well above T_C were ~ the same for all polycrystalline grain sizes, the strength retained for quenching at and somewhat above T_C increased with increasing *G*, and more importantly the strength decrease above T_C became gradual rather than abrupt (Fig. 6.21). Gupta [172] subsequently showed that plotting strength retained after quenching at T_C linearly increased with *G*, reaching 1 at *G* ~ 80 μm, consistent with his experimental results.

Tomaszeweski [173] conducted a more extensive study of *G* effects on thermal shock resistance of Al_2O_3 as an extension of his substantial study of effects of *G* on mechanical properties of Al_2O_3 (Fig. 2.11), covering the *G* range of ~ 4 to 600 μm. He showed a similar T_C of ~ 200°C with the extent of abrupt strength decrease again decreasing as *G* increased, so that it disappeared by *G* ~ 100 μm, beyond which a gradual strength decrease with increasing quench *T* occurred with the starting strength level and the decrease diminished with further *G* increase, with a very low strength level of ~ 10 MPa and no strength decreases occurring at *G* ~ 570 μm. These changes are related to linear reductions of the (static) Young's modulus with increasing *G*, i.e. from ~ 300+ to ~ 150 GPa at G 100+ μm and then to ~ 100 GPa at *G* ~ 460 μm due to microcracking.

Seaton and Dutta [174] showed some similar and different results for *G* effects on thermal shock of B_4C, which had very similar *E*, TEA, and toughness to Al_2O_3, but dominant transgranular rather than mixed or mainly intergranular fracture. They showed that, while starting strengths were somewhat higher for finer *G*, as was expected, the overall T_C was the same for *G* = 2 and 16 μm, as for Al_2O_3. However, the larger *G* showed strengths starting to decrease at < T_C,

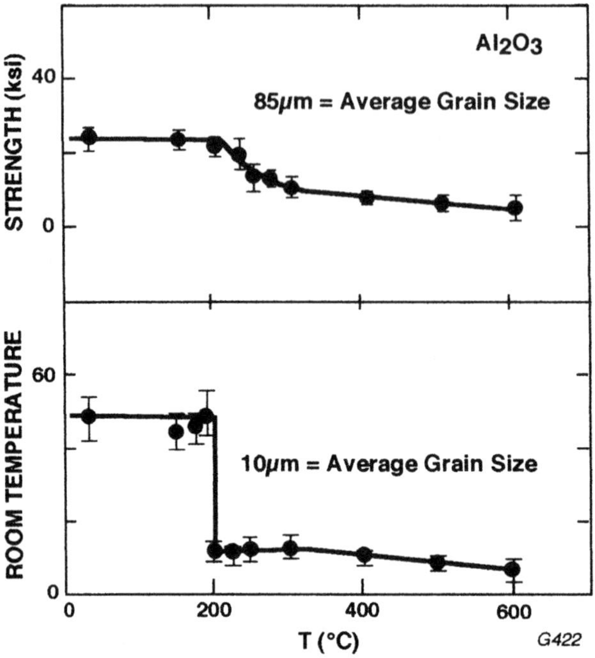

FIGURE 6.21 Flexure strengths at 22°C for quenched Al_2O_3 [172] with $G = 85$ μm and 10 μm. Note the similarity that while starting strengths increase with decreasing G, the overall ΔT values are independent of G, but residual strengths are somewhat higher, especially as a fraction of starting strengths at higher G for Al_2O_3, but not for B_4C, which reflects differences between these two materials, e.g. possibly of crack–grain bridging. (Published with the permission of the *Journal of the American Ceramic Society.*)

and then were lower at and somewhat above ΔT in contrast to the Al_2O_3 results (Fig. 6.21). The reasons for these differences are not known for certain, but they may reflect differing effects of crack–grain bridging effects. Thus as G increases in Al_2O_3, opportunity for such bridging increases and the resultant crack sizes involved in much of the thermal shock damage probably becomes large enough to involve a sufficient number of grains to limit strength loss. On the other hand, the predominant transgranular fracture in B_4C appears to limit possible crack bridging and thus the opportunity to mitigate strength losses in thermal shock.

Kennedy and Bandyopadhyay [149] conducted similar quench tests on four UO_2 bodies with $G \sim 2$–19 μm and porosity of ~ 9 to 3%, which generally decreased with increasing G. Again T_C was ~ independent of G (at ~ 100°C), but there was no clear improvement of the residual strength as G increased. In fact, strengths after quenching the two finer grain bodies did not reach those of the two larger G bodies at T_C until quench temperatures of 300–400°C.

Other tests show further complications and evaluation possibilities, e.g. in studies of Capolla and Bradt [175] of two recrystallized SiC refractory bodies with one having somewhat higher strength and WOF and somewhat lower thermal expansion with a substantial bimodal G distribution versus the other body with a grain structure of ~ uniform grains of about the size of, or somewhat >, the larger grains in the first body. The former body had a T_C of 350°C, slightly higher than the second body with its ~ 325°C, but it had a substantially lower rate of strength decrease and a higher retained strength than the second body. The property best correlating with these differences in thermal shock behavior was the WOF, but the underlying microstructural reasons for this are not clear, e.g. the body with somewhat finer large grains in a matrix of substantially finer grains, hence a substantially smaller average G, had substantially better performance. While this difference in grain structure is probably a factor, interaction between the grain structure and the substantial and similar levels of porosity in both bodies may be important, since there is clear precedent for porosity critically interacting with other microstructural aspects significantly to improve thermal shock [8]. Finally, these investigators [175] and others [167,168] have shown the value of using damping as a tool to monitor thermal shock damage, i.e. showing substantial increases in damping below and beyond T_C, then leveling off, with the overall levels being substantially higher with more thermal shock damage. Similarly, acoustic emission, though again used only a limited amount, can be a valuable indicator of thermal shock effects.

VI. DISCUSSION

A. Overall Strength–Grain Size Behavior as a Function of Temperature

As temperature increases, there must be a transition in the grain dependence of crack propogation and tensile strength, which is also significantly impacted by strain rate. This overall transition arises since the grain size dependence of these properties is fundamentally opposite at lower temperatures from what it is at sufficiently high temperatures, with the transition occurring at lower temperatures with lower strain rates and higher temperatures at higher strain rates. At lower temperatures, where brittle fracture initiation and propagation dominate, even where crack nucleation, growth, or both occur due to microplastic processes, strengths inherently increase with decreasing grain size, as was extensively shown in Chap. 3 and here. This chapter also clearly shows that such brittle fracture, most commonly manifested by the typical two-branch σ–$G^{-1/2}$ behavior found at ~ 22°C, commonly extends to ≥ 1000°C and sometimes to > 90% of the absolute melting point, as was shown for ice (Fig. 6.13). This is also supported by scaling of strengths with E commonly continuing to ≥ 1200°C (Figs. 6.22 and 6.23), similar to that found at 22°C (Fig. 3.37). This shows that while less σ–$G^{-1/2}$ data exists at elevated temperature, sufficient does exist (mainly at 1000–1300°C) to show basic similarities with behavior at 22°C, i.e. the common occurrence of two-branch σ–$G^{-1/2}$ curves, often higher strength levels for nonoxides, and limited differentiation of microplastic and flaw failure. It further indi-

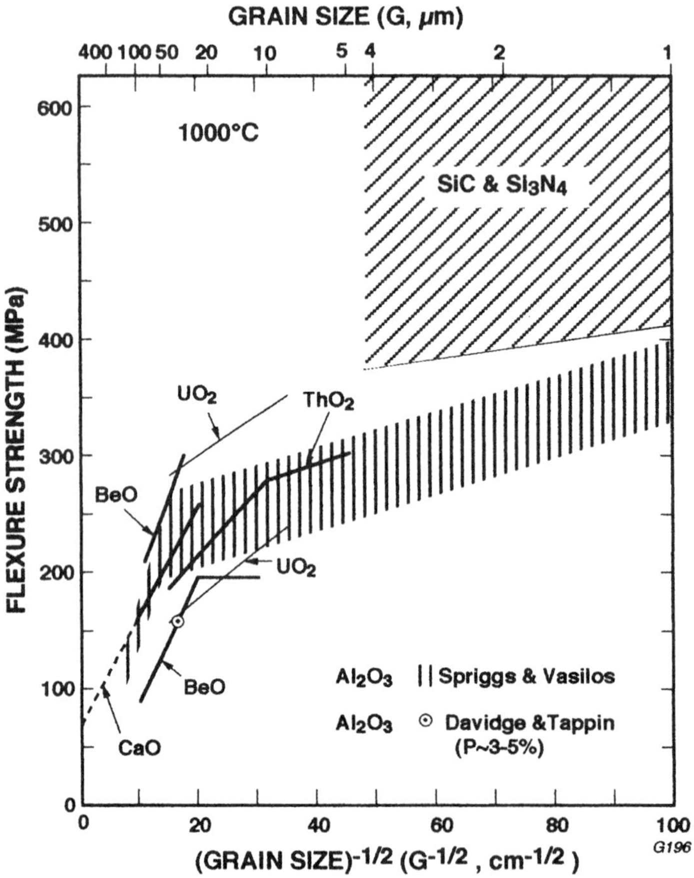

FIGURE 6.22 σ–$G^{-1/2}$ data for various ceramics at ~ 1000°C normalized to the same Young's modulus as Al_2O_3 (as in Chap. 3, Fig. 3.37 and Table 2). (From Ref. 1, published with the permission of the *Journal of Materials Science*.)

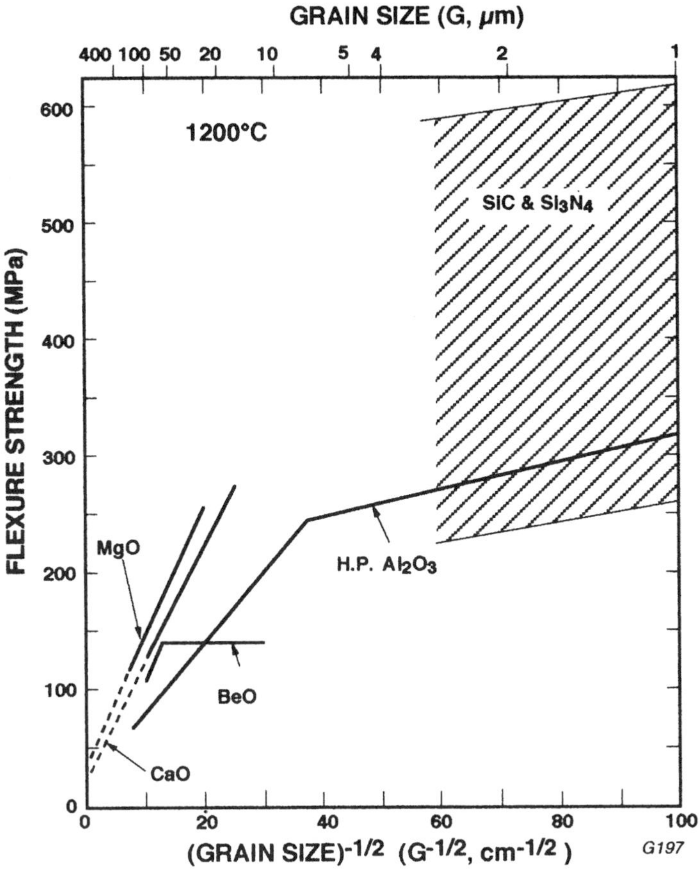

FIGURE 6.23 σ–$G^{-1/2}$ data for various ceramics at ~ 1200°C normalized to the same Young's modulus as Al_2O_3 (as in Figs. 3.37 and 6.22, and Table 3.2). (From Ref. 1, published with the permission of the *Journal of Materials Science*.)

cates that oxides tend to fall in a lower group and nonoxides such as SiC and Si_3N_4 in a higher group. This probably reflects some intrinsic as well as developmental effects. A possible difference is that oxides have more slip at elevated temperatures. Though this may temporarily increase strengths, e.g. as indicated for ThO_2 and UO_2, where strengths may become in part controlled by microplasticity (Figs. 6.16 and 6.17), much more study of dense quality bodies as a function of grain size and temperature is needed.

The extent to which brittle fracture extends to higher temperatures depends on the material, its microstructure, especially the amount and type of grain boundary phases, and test conditions, to some extent test atmosphere, but especially on strain rate. Again this higher temperature transition occurs due to opposite G dependences, i.e. finer grains result in easier grain boundary sliding and deformation and hence lower strengths as opposed to higher strengths at lower temperatures, i.e. indicated per Eq. (6.2), with strengths tending to be inversely proportional to the deformation strain rate. Greater deformation at finer G is shown by the occurrence of superplastic deformation in fine G ceramics similar to such behavior in metals [176,177].

The details of this transition, from lower temperature strengths being proportional to $G^{-1/2}$ and higher temperature strengths to various powers of $G \geq 1$, are complex and poorly defined. The general transition is fairly well indicated for several specific bodies, e.g. as reviewed by Quinn [178] and illustrated in Figure 6.24. However, what is almost totally missing is, especially quantitative, documentation of how such transitions depend on changes in grain size, shape,

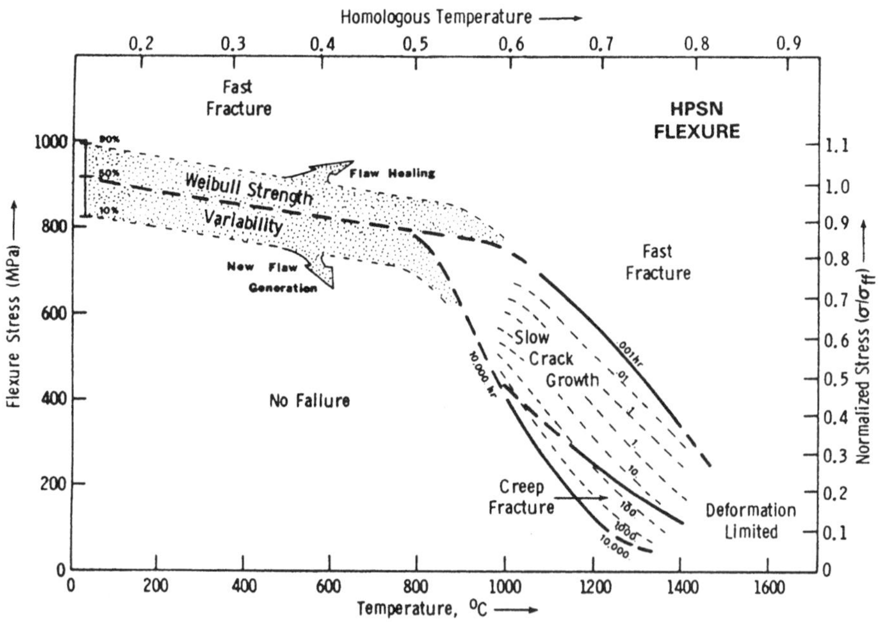

FIGURE 6.24 Plot of flexure stress versus test temperature for a specific commercial hot pressed Si_3N_4 for various strain rates, thus forming a useful failure mechanism map for such test variables. Unfortunately, detailed information on the effects of grain structure and differing amounts and types of grain boundary phases is not available. (From Ref. 178, published with the permission of the *Journal of Materials Science*.)

or orientation, as well as on the amount and character of grain boundary phases. The lack of such information arises from both the lack of adequate data as a function of T and the complexity of high-temperature behavior and the transition between this and lower temperature behavior. The complexity stems substantially from other changes and transitions in behavior that occur over intermediate temperature ranges. Unfortunately, many have focused on only the basic lower to higher temperature failure changes and thus do not begin higher temperature measurements till ≥ 1000°C, thus missing changes that often occur at lower temperatures. Frequent modest and sometimes extreme changes occur from the simple trend for toughnesses and especially strengths to decrease in proportion to the decreases in E till temperatures ≥ 1000°C, where grain boundary sliding or other deformation processes begin to occur depending on material and strain rates. Thus minima, maxima, or both of toughness, strength, or both commonly occur at $T \leq 1200$°C (e.g. Figs. 6.12, 6.14, 6.15, and 6.18) before continuous and accelerating decreases in strengths and polycrystalline toughnesses (disparities for single crystal toughnesses will be discussed later).

B. Property Changes Impacting Strength and Its Grain Size Dependence as a Function of Temperature

Young's modulus decrease with temperature again provides a baseline comparision of strength–temperature behavior, since it is a basic factor in strength changes with temperature. Such E–T trends, while not directly revealing grain structure dependence, can aid in this by considering other possible superimposed strength changes, especially those that are known or expected to depend on grain structure. Such comparison is the purpose of this section given the limited amount of experimental data reviewed earlier, starting with changes in slow crack growth.

The substantial occurrence of SCG due to environmental species, especially H_2O, clearly disappears as T decreases below 22°C for tests in liquid nitrogen at –196°C (Fig. 6.25). Thus if strengths and Young's moduli are plotted versus T starting well below room temperature, those materials experiencing SCG at modest T, e.g. due to H_2O, would show a deviation below strengths paralleling the decreases of E as T increases. However, such materials should also show a positive deviation of strengths versus T back toward the T dependence of E as T increases above 22°C due to reduced SCG from H_2O and other fluid species due to increased temperature reducing the amount of such species at the crack tip. Though not studied in detail, such a change was shown by Evans and Lange's [53] n values for SiC hot pressed with Al_2O_3 additions increasing with T, i.e. ~ 80 at 22°C and > 200 at 600°C. However, as temperature increases, other mechanisms of SCG can come into play, commonly due to grain boundary sliding, especially due to boundary phases. Again, Evans and Lange's [53] study of

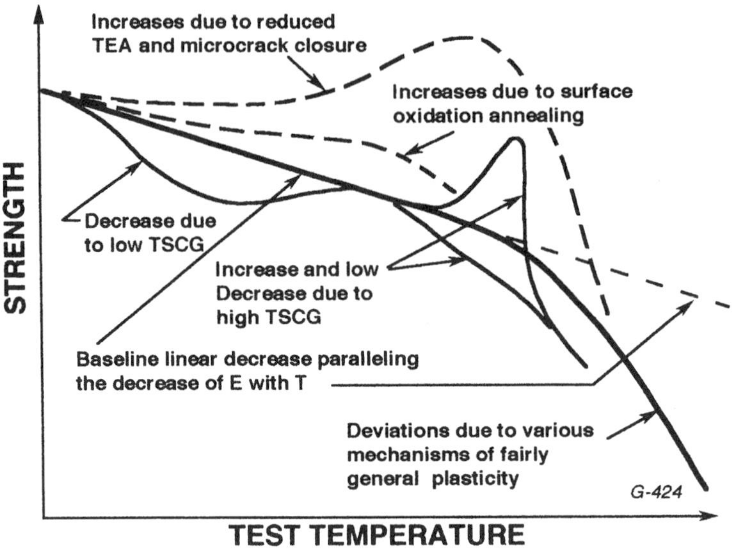

FIGURE 6.25 Schematic plot of the impact of various mechanisms for deviations from the overall temperature dependence of tensile strength from the inherent dependence of Young's modulus. Note that besides basic dependence on temperature (T), (1) the SCG deviations depend on test environment, strain rate, material, and often the amount and character of grain boundary phases, and also probably also on G (Fig. 2.8), (2) surface annealing and especially surface oxidation effects are respectively body and material and atmosphere specific, (3) TEA effects are material specific and clearly depend on grain structure, (4) higher temperature plastic effects are very strain rate dependent and may depend substantially on grain structure and boundary phases, and (5) effects of EA (not shown) are probably either positive or negative depending on their character and that of TEA and possibly plastic anisotropy and grain structure and boundary phases.

SiC hot pressed with Al_2O_3 additions showing SCG with $n \sim 21$ at $T = 1400°C$ illustrates this point.

Consider other changes as T increases that impact toughness and especially strength. Intrinsic changes include decreasing TEA as T increases, with such effects clearly being material and grain structure dependent (Chap. 2, Sec. II.C), and probably also by grain boundary phases. Thus increases in strengths of ZrB_2, HfB_2, and TiB_2 to maxima at intermediate T are probably due to reduction of TEA stresses as T increases. The increase in the relative strength changes as G increases in TiB_2 (while the overall strength decreased), as indicated by Baumgartner and Steiger [74]. Matsushita's [159] data also appears to be generally consistent with this, considering uncertainties in *G* and boundary phases. Such effects can be much more pronounced in very anisotropic materials, e.g. in BN and more extensively shown in graphites, where strengths can peak at 50% to threefold increases over those at 22°C at temperature of the order of 2500°C, e.g. Rice [8]. Strength increases in bodies with TEA can occur due to closing of microcracks from TEA as *T* increases, as appears to be at least part, if not all, of the cause of changes in larger *G* TiB_2. However, strengths can also increase due to increased *T* reducing the opportunuty for microcracks to develop due to the combination of TEA and applied stresses. This is indicated by the absence of pronounced effects of increasing temperature on conductivity of graphites, i.e. if the strength increases with increasing temperature in graphites were due mainly to closure of microcracks, similar effects would be expected on thermal conductivity. The absence of a large effect on thermal conductivity of graphites indicates that much of the strength effects occur due to reduced opportunity for stress-induced microcracking, which would be consistent with the conductivity results, sine there would be limited microcracking to affect conductivity as observed.

EA should also have some similar effects as TEA, including grain effects, but there are three sets of complications. The first is that EA stresses depend on both the extent and nature of the EA as well as on the stresses in the body, i.e. EA varies the stress in the body but does not generate stresses as TEA does. Thus EA effects would be expected to increase with increasing *G* similar to those for TEA, but the tensile stresses that can be sustained by larger *G* bodies are inherently limited by *G* as reviewed in this chapter and in Chapter 3. The second set of complications is that, as will be discussed in more detail later (e.g. Fig. 7.13), EA can increase or decrease with, or be ~ independent of, temperature, depending on the material and the temperature range considered. Thus while there is a consistent trend for a given material and temperature range, there is no general trend for all materials and temperatures as there is for TEA. The third complication is that EA effects are interactive with other anisotropies, i.e. of plastic deformation and resultant stress concentrations, of TEA, and of the grain shape and its orientation relative to the stress axes as a function of the nature of the EA, as modeled by Hasselman [179]. Thus while it is important to consider EA in evaluating mechanical behavior, its effects can be much more variable and complicated, e.g. since EA may vary in its crystallograpic dependence from the other anisotropies.

Other changes with increasing test temperature include reduction of residual (e.g. machining) stress, oxidation, and increasing plastic deformation. Residual stress changes are typically small and may often be more compressive from machining, so that their removal may give a limited decrease in strength. Oxidation of surfaces can consume and thus eliminate surface flaws, e.g. from machining, and thus increase strengths as indicated for TiB_2 [159]. However, such effects are complicated and limited by various factors such as the nature of the oxide coating (material and body dependent), the formation of bubbles or other defects, and residual stresses in the coating, effects of the latter depending on coating character and thickness.

Increasing plastic deformation due to twinning, slip, or both, as well as grain boundary sliding, plays a complex and often incompletely addressed role in the changes of strengths as temperature increases. Plasticity due to boundary sliding often occurrs at modest temperatures, is both material and body specific, and depends substatnially on strain rate. While this may cause a very temporary, strain-rate-, body-, etc. dependent increase in toughness, strength, or both, it more broadly leads to greater than normal progressive decreases with increasing temperature. While increasing crystal plasticity with increasing temperature may lead to resultant polycrystalline plasticity, this is often not the case or requires temperature and body character that are of limited interest. This results from the fact that increased plasticity in crystals often still leaves substantial anisotropy in the plasticity, i.e. limited ability to relieve arbitrary stress concentrations as will occur at grain boundaries of grains undergoing shape changes from deformation. In other words, plasticity of single crystals and hence individual grains often increases much faster than the occurrence of sufficient, i.e. typically five independent, slip or deformation systems for general ductility. This disparity in amount of deformation versus its general stress-relieving ability can result in increased strength decreases, commonly via intergranular failure. This is probably at least part of the cause of the transition to intergranular fracture of CaO and MgO (Chap. 6, Sec. IV.D), but increasing EA with increasing temperature is also probably a factor in this. Thus it is common for single crystals to show markedly increasing strain-rate-dependent toughnesses (e.g. Figs. 6.1, 6.3, and 6.6), with limited, no, or opposite effects of the toughness and strength with temperature. A more spectacular example of this negative effect of increased deformation on strength as temperature increases is the often neglected marked strength decreases in Al_2O_3 at only a few hundred degrees, that appear to be due to increased twinning, as was discussed in Sec. V.C.

Finer grain branch(es) show limited, and larger grain branches show substantial, grain size dependence of strength. For microplastic controlled strength, the larger grain branch ~ extrapolates to the single crystal strength, reflecting the easier modes of microplasticity activation [139–141]. For brittle fracture, the larger grain branch commonly extends, often substantially, below the lowest single crystal strength (as a function of orientation) for comparable surface finish. Where microplasticity occurs, it competes with flaw failure, with the balance between the two mechanisms often being shifted by specimen quality (i.e. processing defects), surface finish, temperature, and possibly test environment.

This survey shows that substantial strength changes can occur in the (often neglected) regime ≤ 1000°C. Thus significant changes of the relative single- and polycrystal strengths may occur, and there may also be variation of these changes with grain size. Parameters affecting such variations include not only environment (i.e. mainly H_2O here) and temperature but also surface finish (especially machining effects). Further, possible effects of material parameters (e.g., TEA and elastic anisotropy, EA) vary with temperature, microstructure, and possibly environment, as do effects of surface finish (environmental effects are also a function of temperature). However, effects of environment and surface finish can be at least partly separated out, though studies have not often done this.

C. Temperature Effects on the Grain Size Dependence Al_2O_3 and BeO

That the temperature dependence of σ–G relations can be complex and involves other effects can be better seen from the relative temperature dependence of Al_2O_3, BeO, MgO, and ZrO_2 (for which there is reasonable data, Figs. 6.12, 6.14, 6.15, and 6.18). Thus for $T < 600$–$800°C$, Al_2O_3 and BeO, both noncubic materials with similar, significant TEA, show opposite σ–T trends, i.e. BeO strength increases with temperature while Al_2O_3 strength decreases, especially for single crystals, and hence with no TEA.

The initial, substantial strength decreases in sapphire and polycrystalline Al_2O_3 had been speculated to be due to increasing crack tip microplasticity, i.e. slip or twinning, but was questioned by crack tip dislocations or twins not being found by Wiederhorn et al. [29]. However, a number of earlier observations suggested a possible explanation for the sapphire σ–T minimum based on twinning as follows. Heuer [180] reported twins introduced in sapphire by either surface scratching or fracture (e.g. rhombohedral twins at least as low as –196°C), possible cracks following twins, possible crack nucleation by twin–twin and twin–grain boundary intersections, and twins being thicker and larger above 600°C. Becher [181] showed both rhombohedral and basal twins introduced by surface abrasion and frequent association with resultant surface cracks. He subsequently indicated probable cracks along basal twin-matrix interfaces [182]. Scott and Orr [183] showed the resolved shear stress for rhombohedrahl twinning dropping from ~ 225 MPa at ~ 320°C to ~ 5 MPa by ~ 600°C and remaining constant thereafter to ≥ 1500°C. Though Scott and Orr's tests were in compression (requiring shortening of the specimen), thus not necessarily reflecting tensile behavior (requiring elongation), their changes closely mirrored the strength changes of sapphire, suggesting cause and effect i.e. similar twinning in tension. Alloying effects reported by Sayir [104] support this. K_{IC} results of Iwasa and Bradt [26] might appear to question this, but being obtained by the (Knoop) indentation-fracture tests, they are thus essentially a strength test, and indents are common sources of twins [181,184]. (Twin-matrix interfaces could have lower K_{IC}, and be preferred sites for SCG, e.g. be consistent with the marked strength drops in, at least machined, sapphire due to both increasing temperature and environment effects. Annealed surfaces may also have twin-flaw combinations, e.g. from previous machining or handling, but reduced in extent or severity, e.g. as possibly indicated by Charles' [86] data for annealed sapphire.) There is also evidence that twinning is associated with tensile failure in $BaTiO_3$ single- and polycrystals [185].

Recent research and development has confirmed that rhombohedral twinning is the source of the strength minimum in sapphire [186–192], despite the twinning being activitated in compression. Such confirmation resulted from both direct observation of the twinning and resultant failure as well as successful steps to suppress it and thus limit strength lossses. Both efforts were motivated by the severe weakening this mechanism causes in the use of sapphire as an ir-dome material for missiles, due to the resultant much easier thermal stress failure from aerodynamic heating. Thus Mecholsky's [186] fractographic studies showed sapphire flexure bars failed from twin crack nucleation due to compressive under the flexure loading points. (This is an atypical but not unique example of failure from compressive stresses in flexure. Failure from compressive stresses in flexure also often occurs in fiber composites.) Subsequently Harris [187], Savrun et al. [188], and Schmid and Harris [189] showed that the orientation–temperature dependence of sapphire flexure strength is due to rhombohedral twinning and resulatant crack nucleation, typically from twin–twin intersections. Such failure has been corroborated by reduction of failure of sapphire missile domes by minimizing compressive stresses on rhombohedral planes by orienting the domes relative to the asymmetrical aerodynamic heating. Corroboration is also supplied by both doping of sapphire [104,190,191] and especially a proprietary treatment [191,192] (speculated to be neutron irradiation), since both, especially the latter, reduce twinning. The doping (alloying) results are consistent with differing results cited for Cr_2O_3 doped sapphire, since such effects are a function of the dopant, its amount, test temperature, and sapphire orientation.

Twinning-induced fracture also appears consistent with Al_2O_3 σ–G effects via grain size limiting twin size less at moderate and large grain size, but more in the finer grain branch where too many grains are encompassed by the flaw size (c) for individual grain–twin interactions to be significant. Thus the substantial scatter of Kirchner and Gruver's hot pressed Al_2O_3 strength minima and maxima [107,108] with $C \sim 20$ μm and $G \sim 2$–5 μm may reflect effects of known grain heterogeneity. Also, Mizuta et al.'s [109] lack of a strength minimum is consistent with their apparently uniform, fine grain size. Al_2O_3 fibers, while not being tested as low as 400–500°C, would be consistent with no minimum due to the fine grain size (but a maximum at 800–1100°C). Neuber and Wimmer's [81] strength minima (and maxima) at intermediate grain size are consistent with such a twinning mechanism, as are Charles' [86]. (His larger grain, lamp envelope Al_2O_3 showing less of a strength minimum and at higher temperature suggest that environmental factors may also play a role in these σ–*T* minima and maxima.) Only a suggestion of a strength minimum in tests of Jackman and Roberts [91] (*G* ~ 50 μm) may be due to the probable larger pore size of the residual (~ 5%) porosity frequently being a key factor in failure. However, the role of TEA stresses cannot be neglected since, for example, large (e.g. isolated) grains are often preferred sources of failure in Al_2O_3 (and other ceramics) [78].

The subsequent significant strength upturn and resultant relative σ–*T* maximum of much of the Al_2O_3 data (e.g. at ~ 800–1,000°C) could reflect crack tip blunting due to plasticity in single crystals, since slip and twinning are clearly observed to occur to an increasing extent in this e.g. 600–1000°C, range, including at crack tips [29]. However, this is unlikely to be significant in polycrystalline Al_2O_3, especially as flaw size (*c*) becomes progressively > the grain size (*c* > *G*), since crack tip stress relief encompassing a number of grains is much less likely in view of the limited number of slip and twin systems. Instead of (or in addition to) such microplastic effects, reduction of TEA stresses [193] must be considered. The strength maximum occurs at, or close to, the temperature range at which such stresses are believed to disappear, e.g. based on spontaneous microcracking from such stresses. Evidence has been presented that such stresses increasingly directly contribute to failure at 22°C as the flaw size approaches the grain size [1,79,194] (Fig. 3.35), i.e. pertinent to much of the larger grain branch, with decreasing effects as grain size decreases along the fine grain branch. On the other hand, K_{IC} at 22°C (measured with large cracks) commonly shows a maximum at intermediate grain size, originally attributed to microcracking from TEA stresses [193] but now attributed more to *R*-curve effects (Figs. 2.16, 2.17). The latter effects are believed generally not to be pertinent, since flaws controlling strength are commonly not on a sufficient scale in the pertinent grain size range. However, the specifics of both of these mechanisms, their possible interactions, and their actual temperature dependence are, at best, limited.

Reduction in TEA stresses with increasing temperature is a possible mechanism for the BeO σ–*T* maximum, as originally suggested by Bentle and Kniefel [126] and Clarke [195], e.g. the temperature range of the maximum (500–1000°C) approaches that estimated for the disappearance of TEA stresses based on microcracking from such stresses [193]. Also, other factors, such as greater grain boundary stress relief due to higher stress in testing than for spontaneous cracking (i.e., with no external stressing), could reduce the temperature for maximum strength. Particularly supportive of such a stress relief mechanism is the absence of any apparent single crystal complications as for Al_2O_3. Again, the stress-relief mechanism should be dependent on *c* not being >> *G*, since the effect of such stresses goes to zero when averaged over many grains [2,78–80,194]. The indicated grain size dependence of the σ–*T* maxima (e.g. at *G* ~ 40–100 μm) supports this postulate. However, note that reduction of TEA stresses as an explanation of the σ–*T* maxima also means that SCG effects may be underestimated by tests in liquid N_2, since this increases TEA stresses, which would thus limit strength increases due to reduced SCG at $T > 22°C$.

Clearly grain boundary phases can play an important, but variable, role in the σ–*T* behavior, especially beyond ~ 600°C. Thus SiO_2-based grain boundary phases in Al_2O_3 can not only relieve TEA stresses but also lead to grain boundary sliding and attendant strain rate dependent maxima [82,110] (Fig. 6.11), as can grain boundary phases in other oxides and nonoxides (e.g. Si_3N_4). This is also shown by less pronounced maxima, or only an ~ strength plateau in BeO with additives or impurities [126]. Such differences probably reflect interrelated effects of the boundary phase and its degree of wetting, which can also be a function of processing, e.g. less SiO_2 wetting of Al_2O_3 under reducing condition [196], as indicated by differences between commercial (air) sintered ~ 95% Al_2O_3 (Fig. 6.11) and Al_2O_3 hot pressed with 3% SiO_2 [85].

D. Temperature Effects on the Grain Size Dependence on Other Ceramics, and Overall Mechanisms

The high EA of ZrO_2 bodies at modest *T* and its increase with increasing *T* (Fig. 7.13) suggest that it may be a factor in the transition from trans- to intergranular fracture at modest *T* (Chap. 2, Sec. III). Similarly, such a fracture mode transition at higher *T* in MgO has been suggested as reflecting its more modest but increasing EA with increasing *T* [197]. However, again other mechanisms may be involved, e.g. as indicated by decreases in *E* of ZrO_2 (Fig. 6.18), since this would presumably not occur due to EA unless it was causing microcracking (and then possibly only in tests with substantial applied stress, most likely static versus dynamic modulus measurements), but may be due to effects of lattice defect structures formed. Thus the often more extreme decrease of Young's modulus (and strength) at modest temperature in fully or partially stabilized ZrO_2, especially with Y_2O_3, also correlates with oxygen defects, e.g. forming anisotropic complexes as indicated by correlation of internal friction and other loss measurements via conductivity and dielectric tests [10,198]. This is corroborated by correlations of Young's modulus decreases (especially with Y_2O_3 or reduced CeO_2 additions) and variations with the stabilizer type and amount and reduction of CeO_2 [10]. While such effects of reduction have been neglected or associated with darkening attributed to other effects [199,200], this is likely to be important due to reducing conditions in hot pressing or HIPing samples, and especially high-temperature heat treatment of PSZ [201] (usually achieved via induction heating of carbon). Such defect effects have been indicated in ThO_2 [202] and are likely to occur in other materials, e.g. CeO_2 and $MgAl_2O_4$ [i.e. the latter *E–T* (Fig. 7.13) and σ–*T* jog at 500–700°C, Fig. 6.18]. While Young's modulus decreases would contribute to σ decreases, the latter are much larger, indicating an enhancement of the above oxygen defect mechanism or the addition of one or more other mechanisms.

Impurities, especially at grain boundaries, are a possible factor in the ZrO_2 σ–T decreases, especially in view of observed increased intergranular fracture initiation with temperature vs. mostly transgranular at lower temperature. However, it is not clear why ZrO_2 should be so much more sensitive to impurities, nor why they would be a factor at such low temperatures (e.g. 200–400°C). While, as noted earlier, SCG was not observed in Y_2O_3 fully stabilized ZrO_2 crystals, polycrystalline SCG via grain boundaries may be a possibility, but extensive transgranular fracture at and near 22°C argues against this. Destabilization of partially stabilized ZrO_2 by H_2O has also been observed, but only for a modest range of temperature, grain size, and Y_2O_3 content, not explaining similar effects for CaO, MgO, or Tb_4O_7 stabilization or full stabilization with Y_2O_3. Further, this effect appears to be a corrosion phenomenon [203–206], not SCG; i.e. degradation over the exposed area, not just at tips of sufficiently stressed cracks. Attributing moderate temperature decreases in ZrO_2 mechanical properties to attack of H_2O (or other species such as HCl [204,206]) also appears inconsistent with some similar strength trends for both $ZrO_2 + Y_2O_3$ single- and polycrystals (in view of probable association of this H_2O effect with grain boundaries, hence not pertinent to single crystals). This would also possibly imply some opposite effects of H_2O and boundary impurities, since the latter may often interfere with the reaction with H_2O. H_2O effects also appear to be inconsistent with many of the property changes continuing well beyond the temperature range of this destabilizing mechanism. Thus while H_2O effects may contribute to the E–T and especially σ–T changes, they cannot be the fundamental cause of them.

While EA may decrease or not change much with increasing temperature for some materials, it shows considerable increase with temperature for several materials recently reviewed [197], e.g. CaO, MgO, and ZrO_2. The latter shows EA increases significantly in the temperature range where Young's modulus and strength show marked decreases (Figs. 6.18, 7.13) and shows substantial composition dependence, implying even higher EA for partially stabilized materials (e.g. those of Drachinskii et al. [155].

The similarity of TEA and EA providing local (grain boundary) stress concentrations (the latter, only with an external stress applied to the body) might suggest EA as an analogous possibility for some (e.g. MgO) σ–T maxima at intermediate temperature, i.e. as for TEA as a possible cause of such maxima in Al_2O_3 and BeO. However, the common continued rise of EA with temperature noted above would appear to rule this out [197] (TEA stresses decrease with increasing temperature). On the other hand, increasing deformation with temperature combined with EA-T changes might be a possible mechanism. Such EA contribution would probably increase with grain size, analogous to the grain size dependence of spontaneous cracking from TEA (Chap. 2).

The marked EA of ZrO_2 may correlate with the occurrence of grain boundary fracture origins in larger grain bodies, fully and partially stabilized ZrO_2 [197]. The temperature rise of ZrO_2 EA may also contribute significantly to its higher temperature grain boundary sliding. Further, since EA increasing with temperature is very broad if not universal, its rise may be a factor in the *E–T* and σ–*T* jogs of $MgAl_2O_4$ noted earlier (at ~ 500–700°C), similar to, but less pronounced than, for ZrO_2. While the EA of MgO [197,207,208] is relatively low at 22°C, hence much less likely to be a factor at moderate temperature, its substantial EA levels at higher temperature [197], e.g. ~ 1200°C, may be related to increased fracture initiation from even relatively clean (i.e. recrystallized) grain boundaries at > ~ 1200°C [139,140]. Thus EA needs to be considered as another broad factor besides, or in addition to, grain boundary impurities in increasing intergranular failure with increasing temperature.

Other materials show little or no initial strength decrease until temperatures of ~ 1000°C or higher. Thus ThO_2 and UO_2 show higher strength at 1,000 vs. 22°C (Figs. 6.16 and 6.17). Whether such effects in ThO_2 are related to mechanical and electrical relaxation in the temperature range are unknown. Further, as noted earlier, nonoxides such as B_4C, SiC, and TiC, show limited, or possibly no, initial strength decrease, and in some cases possibly a slight increase with initial temperature increases, in contrast to the *E–T* decrease (typically a few to ~ 10% to 1000°C). Some of these differences could reflect reduction of TEA stresses, e.g. in B_4C, but in the case of B_4C, effects of substantial twinning, and in α-SiC of polytypes, are unknown. While oxidation and relief of surface compressive stresses from machining may also be factors, tests in neutral or reducing atmospheres show that these are, at best, partial factors.

The changes in strength with temperature, environment, and grain size of most ceramics are overall consistent with flaw induced failure. Thus slow crack growth is a well established adjunct to normal flaw failure, and microplastic nucleation of cracks, or assisting their growth, are accepted mechanisms interacting, and consistent, with conventional flaw failure. The same is true of changes in single crystal strengths and changes in grain boundary effects whether intrinsic, e.g. due to changes in TEA or EA stresses, or extrinsic, e.g. due to impurities. However, while the above concepts are known, fully effective quantification of the contributions to failure is generally not available.

Bridging, widely cited as an important factor in behavior of many ceramics, e.g. suggested [209] and questioned [210] in Al_2O_3 at lower temperatures, might be seen as enhanced at elevated temperatures due to increased intergranular fracture, but the issue of bridging effects at higher temperatures is at least as uncertain. This is due again to issues of observing bridging via arrested cracks along specimen surfaces, incompatible *G* dependences of large crack toughnesses and normal small crack strengths at lower temperatures applying at higher temperatures, as do effects of material and microstructural parameters effecting flaws, especially from machining, controlling strength as discussed in Chaps. 3 and 4 for room temperature behavior of monolithic ceramics and in Chaps. 8 and 9 for ceramic composites. Further, while increased intergranular fracture at higher test temperatures would be consistent with increased crack bridging, this corresponds to grain boundary weakening and associated decreased, not increased, strengths and toughnesses, which in fact may be responsible for some reduction of toughness–strength discrepancies. Though variability in brittle–ductile transitions and higher temperature strain-rate-dependent plastic deformation result in further toughness–strength differences at higher temperatures, there are other mechanisms that can dominate strengths of materials where bridging might occur.

Thus the strength minima and maxima observed with sapphire, as well as a number of (mainly larger grain) polycrystalline Al_2O_3 bodies, raise questions of how a single crystal mechanism, e.g. possibly twinning in sapphire as was noted earlier, impacts a polycrystalline body. Clearly, this can be the case if flaws causing failure are on the scale of one or a few grains, as was indicated earlier, but it seems unlikely that twinning could impact failure with flaw propagation over several to many grains, as is implied by crack scales needed for bridging, as is also implied by the absence of strength minima and maxima in finer grain Al_2O_3 bodies where cracks cover a number of grains. Again, the increased intergranular fracture with increased temperature over much of this range also raises questions about bridging in view of the strength decreases that occur [211].

The behavior of other materials also raises serious question regarding the role, if any, of bridging on their normal strength behavior. Thus BeO generally shows the opposite strength–temperature trend to ~ 1000°C but has similar Young's modulus, TEA, and slow crack growth to Al_2O_3, so at least one of these two materials would appear to be inconsistent with bridging. MgO shows similar though more moderate trends than BeO, but not greatly less, as would be expected if TEA stresses (absent in MgO) were a major factor in bridging, as is commonly proposed. ZrO_2 shows substantial strength decrease with initial temperature increases, which is accompanied by some increase in intergranular failure, which should aid bridging and hence limit strength decrease, i.e. the opposite of what appears to happen. Also, the decrease in Young's modulus, which appears to be due to lattice defects, raises further questions of how bridging could be a factor in associated strength changes.

E. Summary and Conclusions

Limited data on the grain dependence of thermal shock shows critical quench temperatures and retained strengths tending to increase some as G increases, but obviously at the expense of starting strength. Greater retained strengths may reflect benefits of possible crack bridging/R-curve effects, but this has not been investigated.

Single crystal toughnesses, e.g. of Al_2O_3 (Fig. 6.1), $MgAl_2O_4$ (Fig. 6.3), SiC (Fig. 6.5), ZrO_2 (Fig. 6.6), and MgO, while initially commonly decreasing with increasing temperature, typically exhibit a minimum, then a subsequent maximum, followed by a continued, probably accelerating, decrease. This reflects effects of overall increasing plastic deformation, primarily by slip, though in special cases, especially in sapphire, lower temperature twinning causes significant minima as a function of orientation, as implied by strength results (which implies a similar toughness maximum, probably due to increasing slip). However, while there is a corresponding minimum in the strength of sapphire, the temperature trends of toughnesses often do not correlate well with those of the corresponding polycrystals, due to grain boundary effects, which increase with temperature and many boundary phases.

Polycrystalline toughnesses, while sometimes showing minima and maxima (Fig. 6.4), which may not correlate well with those for single crystals, often show less variations from a continuous decrease as temperature increases (Fig. 6.3), especially at higher temperature, e.g $\geq$ 1000°C. *R*-curve effects are observed, e.g. due to glassy grain boundary ligaments, or more generally impurity enhanced or intrinsic increased intergranular failure as temperature increases. However, these are also often associated with lower toughness (Fig. 6.2), and especially lower strength, relative to purer bodies of the same material. While there continues to be some *G* dependence for some materials, this appears to diminish at higher, relative to lower, temperatures (Fig. 6.2).

Turning to σ–$G^{-1/2}$ behavior, this overall typically follows a two-branch behavior as at 22°C, i.e. limited grain size dependence at finer grain size due to $c < G$, and a substantial *G* dependence at larger grain size due to $c \leq G$. Such two-branch behavior occurrs at temperatures < 22°C (Fig. 6.9) and at higher temperatures, e.g. commonly to at least 1200–1300°C (Figs. 6.10, 6.22, and 6.23), though being material and strain rate dependent. An extreme of this in terms of the fraction of absolute melting temperature is indicated in ice (Fig. 6.13). All of this reinforces the dominance of flaw mechanisms of failure, as does the scaling with *E* (Figs. 6.22 and 6.23). Where microplastic failure occurs, mainly at medium and larger grain size, strength ~ extrapolates to the stress for the easiest activated mode of single crystal microplasticity. Higher relative σ of materials such as ThO_2 (Fig 6.16) and UO_2 (Fig. 6.17) at higher temperature may indicate increasing effects of microplasticity. Where flaw failure occurs, strengths at large grain size generally extend well below strengths for the weakest crystal orientation. No clear differentiation between cubic and noncubic materials failing from flaws was found, i.e. the mechanisms of failure are not primarily determined by structurally related effects. There is some indication of nonoxides such as SiC and Si_3N_4 (i.e. more covalently bonded) materials having higher relative strength, but the relative balance of intrinsic versus extrinsic reasons (e.g. more successful development) for this is not clear, showing that much remains to be documented and understood about σ–G–T behavior.

The need for further documentation and understanding is also shown by the fact that while flaw failure predominates, substantial complexity exists as reflected in significant deviations, especially from E–T behavior. Though more limited, there is sufficient data to show that a number of variations occur in the above trends, such as shifts in single- versus polycrystal strengths and probably between strengths for different grain sizes due to SCG and other effects, mainly at > 22°C and ≤ 1000°C, where testing is often particularly neglected. These variations are best seen for Al_2O_3 (Fig. 6.12), BeO (Fig. 6.14), MgO (Fig. 6.15), and ZrO_2 (Fig. 6.18) for which there is most data, including for E, whose general trends for different materials as well as for the specific material of interest is important.

Consider now a summary of the main variations, starting with sapphire, partly addressed earlier. Its strength drops rapidly from at least –196°C to a minimum at ~ 400–800°C and then rises to a maximum at 900–1100°C, before steadily decreasing at higher temperature. Polycrystalline Al_2O_3 often shows a similar, though usually less drastic, initial strength drop and may exhibit (1) a strength minimum, a subsequent maximum (similar to but less extreme than for single crystals), or both, or (2) an approximate strength plateau at intermediate temperature (e.g. 400–800°C). Both these trends appear to require sufficiently large grains and may be overridden by the presence of other sources of failure, e.g. pores. Both are also in contrast to the simple, steady, moderate decrease of Young's modulus (e.g. ~ 10–15% by 1200°C), which would also be the expected strength trend if only simple flaw failure were occurring. In contrast to this, neither BeO single- nor polycrystals show similar rapid initial strength drops at > 22°C that Al_2O_3 does, but crystals show simple σ–T and E–T trends, while polycrystals often show significant strength maxima at intermediate temperatures, with impurities (or additives) again limiting these. MgO, while having overall σ–T dependence consistent with slip induced fracture, shows intermediate temperature polycrystalline strength maxima (less pronounced than in BeO) or plateaus similar to BeO and Al_2O_3, despite the differences in underlying mechanisms. ZrO_2 shows polycrystalline E decreasing more rapidly with increasing temperature than single crystal Young's moduli, and even greater polycrystalline strength decreases. Other limited oxide and nonoxide data indicate some strength increases, or no decrease form 22 to ~ 1000°C (including in nonair atmospheres, ruling out surface oxidation effects), i.e. not following E–T decreases nor those expected due to relaxation of surface machining stresses.

Explanations for some, and known or probable factors for other, variations can be cited, the latter including environmental factors such as SCG, whose temperature dependence is poorly documented. At modest T, SCG effects (due often to H_2O) apparently occur only during external stressing, either not occurring, or (more probably) fairly rapidly saturating due to internal (e.g. TEA) stresses alone, and can occur transgranularly (especially in larger grains), but also intergranularly in polycrystalline bodies with or without single crystal SCG, e.g. due to grain boundary phases. SCG is affected by temperature and may be interactive with microplasticity, TEA and EA, and surface machining stresses. However, though the balance between increased reactivity versus reduced content as *T* increases, must ultimately cease SCG, intermediate trends are uncertain, as is the case of corrosion effects such as H_2O effects in some TZPs. Even less is known about high-temperature gas-driven SCG, e.g. as indicated in graphites, and liquid SCG, e.g. for TiB_2 in molten Al, where grain boundary (especially O_2) phases appear important. The second factor is changes of basic properties such as *E*, EA, and TEA as *T* increases. While TEA decreases with increasing *T* are fairly well known, its interactions with other factors such as grain orientation (e.g. in BeO, Chap. 2, Sec. III.H), boundary impurities, and EA are complications. EA has had much less attention, is probably dependent on grain shape and orientation and their changes with temperature, and varies widely with material, presenting difficulties of prediction. *E* normally decreases slowly, e.g. 1–2%/100°C, till 1000–1500°C, and hence individually and collectively *E* changes with *T* provide a useful reference point for comparing changes in other properties such as toughnes and strength. Some anomalous *E* changes do occur with increasing *T*, e.g. in $MgAl_2O_4$ and especially ZrO_2, where defect effects are probable factors via, or in addition to, EA changes (both have higher EA). Such defect (and related internal friction) effects probably extend to several other materials, e.g., CeO_2, ThO_2, and UO_2. Occasional phase transformations, e.g. at ~ 1200°C for PSZ and > 2000°C for BeO, can also be important factors. More generally, higher temperature environmental factors such as oxidation of nonoxides, or reaction or reduction of oxides, can become a critical factor due to changed surface flaw populations and possible microstructural changes (including pores in surface reaction phases).

The third and most specific and pervasive is plastic deformation, with slip or twinning at lower *T* being more limited and material specific. As noted earlier, sapphire's rapid strength drop with increasing *T* reflects failure from twin crack nucleation, and the subsequent strength maxima in Al_2O_3 (and BeO) probably reflect increased microplasticity to allow crack tip blunting. Other materials such as ThO_2 and UO_2 are probably more representative of typical effects of the onset of plastic deformation and CaO and MgO as examples of normal, gradual increases in deformation as *T* increases. The latter two clearly show effects of the deformation, including macroscopic deformation, but with brittle fracture (as for alkali halides), with these changes recognizable, but gradual, not nearly as spectacular as the onset of rhombohedral twinning failure in Al_2O_3 (which may be complicated by SCG from H_2O) . The other, much more pervasive, type of deformation is that due to first grain boundary sliding and then more general creep mechanisms as *T* increases, especially above 1200–1500°C. These can be substantially affected by grain size, shape, orientation, and boundary phases, e.g. SiO_2=based ones, but the latter effects can depend substantially on fabrication-wetting effects as in some Al_2O_3 bodies.

While existing data provides some insight, much more information is needed. Not only is there very little SCG information on single crystals (including materials for which crystals are readily available, e.g. TiO_2 and $MgAl_2O_4$), but the documentation in the most studied material, sapphire, is incomplete. Data for grain size effects in polycrystalline materials are even less well defined. There is reasonable evidence of TEA affecting strength, but specifics of this are still lacking, e.g., levels of these stresses, and how their effects depend on key parameters, e.g. flaw size. While significant EA increases with temperature in a number of, but not all, ceramics may cause increased grain boundary fracture initiation of many ceramics at higher temperatures, much less is known of its effects. Besides such direct polycrystalline studies, this also requires more single crystal elastic moduli–temperature data. Finally, an overall key need is for polycrystalline studies that explore enough variables, e.g. grain size, temperature, elastic moduli, and strength, that provide a reasonable opportunity of sorting out different factors. Narrow studies, focused on a single, often simplistic, approach or mechanism are of much less, if any, use.

REFERENCES

1. R. W. Rice. Review, Effects of Environment and Temperature on Ceramic Tensile Strength–Grain Size Relations. J. Mat. Sci. 3071–3087, 1997.
2. R. W. Rice. Microstructure Dependence of Mechanical Behavior of Ceramics. Treatise Mat. Sci. Tech., Properties and Microstructure 11 (R. C. McCrone, ed.). Academic Press, New York, 1977, pp. 199–381.
3. R. W. Rice. Strength/Grain-Size Effects in Ceramics. Proc. Brit. Cer. Soc. 20:205–213, 1972.
4. O. L. Anderson. Derivation of Wachtman's Equation for the Temperature Dependence of Elastic Moduli of Oxide Compounds. Phy. Rev. 144(2):553–557, 1966.
5. J. B. Wachtman, Jr., W. E. Teft, D. J. Lam, Jr., and C. S. Apstein. Exponential Temperature Dependence of Young's Modulus for Several Oxides. Phys. Rev. 122:1754–1759, 1961.
6. M. L. Nandanpawar and S. Rajagopalan. Wachtman's Equation and Temperature Dependence of Bulk Moduli in Solids. J. Appl. Phys. 49(7):3976–3979, 1978.
7. J. B. Wachtman, Jr. Mechanical and Electrical Relaxation in ThO_2 Containing CaO. Phys. Rev. 131(2):517–527, 1963.
8. R. W. Rice. Porosity of Ceramics. Marcel Dekker, New York, 1998.
9. M. Weller and H. Schubert. Internal Friction, Dielectric Loss, and Ionic Conductivity of Tetragonal ZrO_2-3% Y_2O_3 (Y-TZP). J. Am. Cer. Soc. 69(7):573–577, 1986.

10. Masakuni Ozawa, Tatsuya Hatanaka and Hideo Hasegawa. Internal Friction and Anelastic Relaxation of ZrO_2 Polycrystals Containing 2 mol% Y_2O_3, 8 mol% Y_2O_3 and 12 mol% CeO_2. J. Cer. Soc. Japan 99:628–632, 1991.
11. K. Nishiyama, M. Yamanaka, M. Omori, and S. Umekawa. High-Temperature Dependence of the Internal Friction and Modulus Change of Tetragonal ZrO_2, Si_3N_4 and SiC. J. Mat. Sci. Lett. 9:526–528, 1990.
12. B. A. Boley and J. H. Weiner. Theory of Thermal Stresses. John Wiley, New York, 1960.
13. S. Manson and R. W. Smith. Theory of Thermal Shock Resistance of Brittle Materials Based on Weibull's Statistical Theory of Strength. J. Am. Cer. Soc. 38(1):18–27, 1955.
14. D. P. H. Hasselman. Unified Theory of Thermal Shock Fracture Initiation and Crack Propagation in Brittle Ceramics. J. Am. Cer. Soc. 52(11):600–604, 1969.
15. D. P. H. Hasselman. Thermal Stress Resistance Parameters for Brittle Refractory Ceramics: A Compendium. Am. Cer. Soc. Bul. 49(12):1033–1037, 1970.
16. D. P. H. Hasselman. Figures-of-Merit for the Thermal Stress Resistance of High-Temperature Brittle Materials: A Review. Ceramurgia Intl. 4(4):147–150, 1978.
17. D. P. H. Hasselman. Thermal Stress Crack Stability and Propagation in Severe Thermal Environments. Ceramics in Severe Environments, Materials Science Research 5 (W. W. Kriegel and H. Palmour III, eds.). Plenum Press, New York, 1971, pp. 89–103.
18. D. P. H. Hasselman. Analog Between Maximum-Tensile-Stress and Fracture-Mechanical Thermal Stress Resistance Parameters for Brittle Refractory Ceramics. J. Am. Cer. Soc. 54(4):219, 1971.
19. T. Oztener, K. Satamurthy, C. E. Knight, J. P. Singh, and D. P. H. Hasselman. Effect of ΔT and Spatially Varying Heat Transfer Coefficient on Thermal Stress Resistance of Brittle Ceramics Measured by the Quenching Method. J. Am. Cer. Soc. 66(1):53–58, 1983.
20. H. Henecke, J. R. Thomas, Jr., and D. P. H. Hasselman. Role of Material Properties in the Thermal-Stress Fracture of Brittle Ceramics Subject to Conductive Heat Transfer. J. Am. Cer. Soc. 67(6):393–398, 1984.
21. P. F. Becher. Effect of Water Bath Temperature on the Thermal Shock of Al_2O_3. J. Am. Cer. Soc. 64(3):C-17–18, 1981.
22. D. P. H. Hasselman. Analysis of the Strain at Fracture of Brittle Solids with High Densities of Microcracks. J. Am. Cer. Soc. 52(8):458–459, 1969.
23. D. P. H. Hasselman and J. P. Singh. Analysis of Thermal Stress Resistance of Microcracked Brittle Ceramics. Am. Cer. Soc. Bul. 58(9):856–860, 1979.
24. D. P. H. Hasselman. Role of Physical Properties in Post-Thermal Buckling Resistance of Brittle Ceramics. J. Am. Cer. Soc. 61(3–4):178, 1977.
25. J. P. Singh, K. Niihara, and D. P. H. Hasselman. Analysis of Thermal Fatigue Behavior of Brittle Structural Materials. J. Mat. Sci. 16:2789–2797, 1981.
26. M. Iwasa and R. C. Bradt. Fracture Toughness of Single-Crystal Alumina. Advances in Ceramics 10, Structure and Properties of MgO and Al_2O_3 Ceramic (W. D. Kingery, ed.). Am. Cer. Soc., Columbus, OH, 1984, pp. 767–779.
27. P. F. Becher. Fracture-Strength Anisotropy of Sapphire. J. Am. Cer. Soc. 59(1–2):59–61, 1976.

28. B. N. Kim and T. Kishi. Estimation of Fracture Resistance of Al_2O_3 Polycrystals from Single-Crystal Values. Mat. Sci. Eng. A176:371–378, 1994.
29. S. M. Wiederhorn, B. J. Hockey, and D. E. Roberts. Effect of Temperature on the Fracture of Sapphire. Phil. Mag. 28(4):783–796, 1973.
30. A. S. Kobayashi, A. F. Emery, A. E. Gorum, and T. Basu. Fracture Toughness of Alumina at Room Temperature to 1600°C. ICM 3, Cambridge, UK, 8/1979, pp. 3–9.
31. G. de With. Fracture of Translucent Alumina: Temperature Dependence and Influence of CaO Dope. J. Mat. Sci. 19:2195–2202, 1984.
32. B. J. Dalgleish, A. Fakhr, P. L. Pratt, and R. D. Rawlings. The Temperature Dependence of the Fracture Toughness and Acoustic Emission of Polycrystalline Alumina. J. Mat. Sci. 14:2605–2615, 1979.
33. W.-P. Tai and T. Watanabe. Elevated-Temperature Toughness and Hardness of a Hot Pressed Al_2O_3-WC-Co Composite. J. Am. Cer. Soc. 81(1):257–259, 1998.
34. J. E. Moffatt, W. J. Plumbridge, and R. Herman. High Temperature Crack Annealing Effects on Fracture Toughness of Alumina and Alumina–Silicon Carbide Composite. J. Am. Cer. Soc. 95(10):23–29, 1996.
35. K. Jakus, J. E. Ritter, and R. H. Schwillinski. Viscous Glass Crack Bridging Forces in a Sintered Glassy Alumina at Elevated Temperatures. J. Am. Cer. Soc. 76(10):33–38, 1993.
36. H. H. Xu, C. P. Ostertag, and R. F. Krause, Jr. Effect of Temperature onToughness Curves in Alumina J. Am. Cer. Soc. 78(1):260–262, 1995.
37. R. E. Grimes, G.P. Kelkar, L. Guazzone, and K. W. White. Elevated-Temperature *R*-Curve Behavior of a Polycrystalline Alumina. J. Am. Cer. Soc. 73(5):1399–1404, 1990.
38. R. L. Stewart and R. C. Bradt. Fracture of Single Crystal $MgAl_2O_4$. J. Mat. Sci. 15:67–71, 1980.
39. R. L. Stewart and R. C. Bradt. Fracture of Polycrystalline $MgAl_2O_4$. J. Am. Cer. Soc. 63(11–12):619–623, 1980.
40. K. W. White and G. P. Kelkar. Fracture Mechanisms of a Coarse-Grained, Transparent $MgAl_2O_4$ at Elevated Temperatures. J. Am. Cer. Soc. 75(12):3440–3444, 1992.
41. A. Ghosh, K. W. White, M. G. Jenkins, A. S. Kobayashi, and R. C. Bradt. Fracture Resistance of a Transparent $MgAl_2O_4$. J. Am. Cer. Soc. 74(7):1624–1630, 1991.
42. D. W. Roy and J. L. Hastert. Polycrystalline $MgAl_2O_4$ Spinel for High Temperature Windows. Cer. Eng. Sci. Proc. 4(7–8):502–509, 1983.
43. C. Baudin, R. Martinez, and P. Pena. High-Temperature Mechanical Behavior of Stoichiometric Magnesium Spinel. J. Am.Cer. Soc. 78(7):1857–1862, 1995.
44. C. Baudin and P. Pena. Influence of Stoichiometry on Fracture Behavior of Magnesium Aluminate Spinels at 1200°C. J. Eur. Cer. Soc. 17:1501–1511, 1997.
45. T. Inoue and Hj. Matzke. Temperature Dependence of Hertzian Indentation Fracture Surface Energy of ThO_2. J. Am. Cer. Soc. 64(6):355–360, 1981.
46. H. Ohnishi, T. Kawanami, K. Miyazaka, and T. Hiraiwa. Mechanical Properties of Mullite. Ceramic Materials and Components for Engines, Proc. Second Intl.

Symp. (W. Bunk and H. Hausner, eds.). Verlag Deutsche Keramische Gesellschaft, 1986, pp. 6–39.

47. C. Baudin. Fracture Mechanisms in a Stoichiometric $3Al_2O_3{\cdot}2SiO_2$ Mullite. J. Mat. Sci. 32:2077–2086, 1997.
48. T.-Il. Mah and K. S. Mazdiyasni. Mechanical Properties of Mullite. J. Am. Cer. Soc. 66(10):699–703, 1983.
49. J. L. Henshall, D. J. Rowcliffe, and J. W. Edington. Fracture Toughness of Single-Crystal Silicon Carbide. J. Am. Cer. Soc. 60(7–8):373–375, 1977.
50. M. G. S. Naylor and T. F. Page. Microhardness, Friction and Wear of SiC and Si_3N_4 Materials as a Function of Load, Temperature and Environment. Annual Tech. Report for Contract DAERO-78-G-010, 10/1979.
51. R. W. Rice. Fractographic Determination of K_{IC} and Effects of Microstructural Stresses in Ceramics. Fractography of Glasses and Ceramics II, Ceramic Trans. 17 (J. R. Varner and V. D. Frechette, eds.). Am. Cer. Soc., Westerville, OH, 1991, pp. 509–545.
52. M. O. Guillou, J. L. Henshall, R. M. Hooper, and G. M. Carter. Indentation Hardness and Fracture in Single Crystal Magnesia, Zirconia, and Silicon Carbide. Special Ceramics 9, Proc. Brit. Cer. Soc. 49:191–202, 1992.
53. A. G. Evans and F. F. Lange. Crack Propagation and Fracture in Silicon Carbide. J. Mat. Sci. 10:1659–1664, 1975.
54. J. L. Henshall, D. J. Rowcliffe, and J. W. Edington. K_{IC} and Delayed Fracture Measurements on Hot-Pressed SiC. J. Am. Cer. Soc. 62(1–2):36–42, 1979.
55. A. Ghosh, M. G. Jenkins, K. W. White, A. S. Kobayashi, and R. C. Bradt. Elevated-Temperature Resistance of a Sintered α-Silicon Carbide. J. Am. Cer. Soc. 72(2):242–247, 1989.
56. M. Srinivasan and S. G. Seshadri. The Application of Single Edge Notched Beam and Indentation Techniques to Determine Fracture Toughness of Alpha Silicon Carbide. Fracture Mechanics Methods for Ceramics, Rock, and Concrete (S. W. Freiman and E. J. Fuller, Jr., eds.). ASTM STP 745, 1981, pp. 46–68.
57. G. Popp and R. F. Pabst. Effect of Loading Rate on Fracture Toughness of SiC at High Temperature. J. Am. Cer. Soc. 64(1):C-18–19, 1981.
58. J. Kriegesmann, A. Lipp, K. Reinmuth, and K. A. Schwetz. Strength and Fracture Toughness of Silicon Carbide. Ceramics for High-Performance Applications, III Reliability (E. M. Lenoe, R. N. Katz, and J. J. Burke, eds.). Plenum Press, New York, 1983, pp. 737–751.
59. G. W. Hollenberg and G. Walther. The Elastic Modulus and Fracture of Boron Carbide. J. Am. Cer. Soc. 63(11–12):610–613, 1980.
60. M. Lee and M. K. Brun. The High Temperature Fracture Toughness of SiAlON. Cer. Eng. Sci. Proc. 4(9–10):864–873, 1983.
61. D. Munz, G. Himsolt, and J. Eschweiler. Comparison of High-Temperature Fracture Toughness of Hot-Pressed Si_3N_4 with Straight-Through and Chevron Notches. J. Am. Cer. Soc. 63(5–6):341–342, 1980.
62. A. A. Wereszczak, M. K. Ferber, R. R. Sanders, M. G. Jenkins, and P. Khandelwal. Fracture Toughness (K_{IC} and γ–wof) of a HIPed Si_3N_4 at Elevated Temperatures. Cer. Eng. Sci. Proc. 14(7–8):101–112, 1993.

63. T. Ohij, S. Sakai, M. Ito, Y. Yamauchi, W. Kanematsu, and S. Ito. Fracture Energy and Tensile Strength of Silicon Nitride at High Temperatures. J. Cer. Soc. Jpn., Intl. Ed. 98:244–251, 1990.
64. R. P. Ingel, D. Lewis, B. A. Bender, and R. W. Rice. Temperature Dependence of Strength and Fracture Toughness of ZrO_2 Single Crystals. J. Am. Cer. Soc. 65(9):C-150–152, 1982.
65. A. Ball and B. W. Payne. The Tensile Fracture of Quartz Crystals. J. Mat. Sci. 11:731–740, 1976.
66. T.-II Mah and T. A. Parthasarathy. Effect of Temperature, Environment, and Orientation on the Fracture Toughness of Single-Crystal YAG. J. Am. Cer. Soc. 80(10):2730–2734, 1997.
67. P. B. Hirsch and S. G. Roberts. The Brittle-Ductile Transition in Silicon. Phil. Mag. A 64(1):55–60, 1991.
68. A. G. Evans. High Temperature Slow Crack Growth in Ceramic Materials. Ceramics for High Performance Applications (J. J. Burke, A. E. Gorum, and R. N. Katz, eds.). Brook Hill, Chestnut Hill, MA, 1974, pp. 373–396.
69. R. W. Rice, S. W. Freiman, J. J. Mecholsky, Jr., R. Ruh, and Y. Harada. Fractography of Si_3N_4 and SiC. Ceramics for High Performance Applications II (J. J. Burke, A. E. Gorum, and R. N. Katz, eds.). Brook Hill, Chestnut Hill, MA, 1978, pp. 373–396.
70. C.-K. J. Lin, D. F. Socie, Y. Xu, and A. Zangvil. Static and Cyclic Fatigue of Alumina at High Temperatures: II Failure Analysis. J. Am. Cer. Soc. 75(3):637–648, 1992.
71. S. Horibe and M. Sumita. Fatigue Behavior of Sinteres of SiC: Temperature Dependence and Effect of Doping with Aluminum. J. Mat. Sci. 23:3305–3313, 1988.
72. H. R. Baumgartner. Subcritical Velocities in Titanium Diboride Under Simulated Hall-Hetoult Cell Conditions. Am. Cer. Soc. Bull. 6(9):117–175, 1984.
73. H. R. Baumgartner. Mechanical Properties of Densely Sintered High-Purity Titanium Diboride in Molten Aluminum Environments. J. Am. Cer. Soc. 67(7):490–497, 1984.
74. H. R. Baumgartner and R. A. Steiger. Sintering and Properties of Titanium Diboride Made from Powder Synthesized in a Plasma-Arc Heater. J. Am. Cer. Soc. 67(3):207–212, 1984.
75. S. W. Freiman and J. J. Mecholsky, Jr. Effect of Temperature and Environment on Crack Propagation in Graphite. J. Mat. Sci. 13:1249–1260, 1978.
76. S. Sato, H. Awaji, and H. Akuzawa. Fracture Toughness of Reactor Graphite at High Temperatures. Carbon 16:95–102, 1978.
77. R. W. Rice and C. Cm. Wu. The Occurrence of Slow Crack Growth in Oxide and Non-Oxide Ceramics. To be published.
78. R. W. Rice. Ceramic Tensile Strength–Grain Size Relations: Grain Sizes, Slopes and Intersections. J. Mat. Sci. 32:1673–1692, 1997.
79. R. W. Rice. Fractographic Identification of Strength-Controlling Flaws and Microstructure. Frac. Mech. Cer. 1 (R. C. Bradt, D .P. H. Hasselman, and F. F. Lange, eds.). 1974, pp. 323–343.
80. R. W. Rice. Machining Flaws and the Strength Grain Size Behavior of Ceramics.

The Science of Ceramic Machining and Surface Finishing II (B. J. Hockey and R. W. Rice, eds.). National Bureau of Standards Special Publication 562 (US Government Printing Office), Washington, DC, 1979, pp. 429–454.
81. H. Neuber and A. Wimmer. Experimental Investigations of the Behavior of Brittle Materials at Various Ranges of Temperature. AFML–TR–68–23, 1968.
82. R. W. Davidge and G. Tappin. The Effects of Temperature and Environment on the Strength of Two Polycrystalline Aluminas. Proc. Brit. Cer. Soc. 15:47–60, 1970.
83. Reports of the AVCO Corp., Lowell, Mass., on Microstructure Studies of Polycrystalline Refractory Oxides for US Naval Air Systems Command contracts, 1966–1968, e.g., W. H. Rhodes, D. J. Sellers, R. M. Cannon, A. H. Heuer, W. R. Mitchell, and P. Burnett, Summary report AVSSD-0098-68-RR for Contract N000-19-67-C-0336, 1968.
84. R. M. Gruver, W. A. Sotter, and H. P. Kirchner. Fractography of Ceramics. Summary Report, Naval Air Systems Command, Report for Contract No. N00019-73-C-0356, Nov. 22, 1974.
85. W. B. Crandall, D. H. Chung, and T. J. Gray. The Mechanical Properties of Ultra-Fine Grain Hot Pressed Alumina. Mechanical Properties of Engineering Ceramics (W. W. Kriegel and H. Palmour III, eds.). Wiley Interscience, New York, 1961, pp. 349–376.
86. R. J. Charles. Static Fatigue: Delayed Fracture Studies of the Brittle Behavior of Ceramic Materials Technical Report No. ASD-TR-628, 370-404, Aeronautical Systems Division, Wright Patterson AFG, Ohio, 1962.
87. A. H. Heuer and J. P. Roberts. The Influence of Annealing on the Strength of Corundum Crystals. Proc. Brit. Cer. Soc. 6:17–27, 1966.
88. R. W. Rice. Machining of Ceramics. Ceramics for High Performance Applications (J. J. Burke, A. E. Gorum, and R. N. Katz, eds.). Brook Hill, Chestnut Hill, MA, 1974, pp. 287–343.
89. J. B. Wachtman, Jr., and L. H. Maxwell. Strength of Synthetic Single Crystal Sapphire and Ruby as a Function of Temperature and Orientation. J. Am. Cer. Soc. 42(9):432–433, 1959.
90. J. Congleton, N. J. Petch, and S. A. Shiels. The Brittle Fracture of Alumina Below 1000°C. Phil. Mag. 19(160):795–807, 1969.
91. E. A. Jackman and J. P. Roberts. On the Strength of Polycrystalline and Single Crystal Corundum. Trans. Brit. Cer. Soc. 54(7):389–398, 1955.
92. S. W. Freiman. Effect of Environment on Fracture of Ceramics. Ceramurgia International 2(3):111–118, 1976.
93. C. C. McMahon. Relative Humidity and Modulus of Rupture. Am. Cer. Soc. Bull. 58(9):873, 1979.
94. W. B. Rotsey, K. Veevers, and N. R. McDonald. The Effect of Strain Rate, Environment, and Surface Condition on the Modulus of Rupture of Beryllia. Proc. Brit. Cer. Soc. 7:205–219, 1967.
95. W. H. Rhodes, R. M. Cannon, Jr., and T. Vasilos. Stress-Corrosion Cracking in Polycrystalline MgO. Fract. Mech. Cer. 2 (R. C. Bradt, D. P. H. Hasselman, and F. F. Lange, eds.). Plenum Press, New York, 1973, pp. 709–733.

96. E. K. Beauchamp and S. L. Monroe. Effect of Crack-Interface Bridging on Subcritical Crack Growth in Ferrites. J. Am. Cer. Soc. 72(7):1179–1184, 1984.
97. A. Fort, D. Sharp, B. Ash, K. Papworth, and D. Reed. Influence of Liquid Environments on the Strength of Ferrite Memory Cores. J. Am. Cer. Soc. 55(6):329, 1972.
98. S. L. Dole, O. Hunter, Jr., F. W. Calderwood, and D. J. Bray. Microcracking of Monoclinic HfO_2. J. Am. Cer. Soc. 61(11–12):486–490, 1978.
99. K. R. McKinney, B. A. Bender, R. W. Rice, and C. Cm. Wu, Jr. Slow Crack Growth in Si_3N_4 at Room Temperature. J. Mat. Sci. 26:6467–6472, 1991.
100. M. E. O'Day and G. L. Leatherman. Static Fatigue of Aluminum Nitride Packaging Materials. Intl. J. Microcircuits Elect. Pack. 16(1):41–48, 1993.
101. C. Cm. Wu, K. R. McKinney, S. W. Freiman, G. S. White, R. W. Rice, L. E. Dolhert, and J. H. Enloe. Slow Crack Growth in AlN. To be published.
102. P. Shahinian. High-Temperature Strength of Sapphire Filament. J. Am. Cer. Soc. 54(1):67–68, 1971.
103. G. F. Hurley. Mechanical Behavior of Melt-Grown Sapphire at Elevated Temperature. Applied Polymer Symp. 21:121–130, 1973.
104. H. Sayir, K. P. D. Lagertof, M. R. DeGuire, and A. Sayir. Bend Strength of Undoped and Doped Laser Grown Sapphire Fibers. Presented in 16th Annual Conference and Exposition on Composites and Advanced Ceramics, Cocoa Beach, Florida, January 7–10, 1991.
105. S. C. Carniglia. Petch Relation in Single-Phase Oxide Ceramics. J. Am. Cer. Soc. 48(11):580–583, 1965.
106. S. C. Carniglia. Reexamination of Experimental Strength-vs.-Grain-Size Data for Ceramics. J. Am. Cer. Soc. 55(5):243–249, 1972.
107. H. P. Kirchner. The Elevated Temperature Flexural Strength and Impact Resistance of Alumina Ceramics Strengthened by Quenching. Mat. Sci. Eng. 13:63–69, 1974.
108. H. P. Kirchner. Strengthening of Ceramics, Treatments, Tests and Design Applications. Marcel Dekker, New York, 1979, p. 88.
109. H. Mizuta, K. Oda, Y. Shibasaki, M. Maeda, M. Machida, and K. Oshima. Preparation of High Strength and Translucent Alumina by Hot Isostatic Pressing. J. Am. Cer. Soc. 75(2):469–473 (1992).
110. J. R. McLaren and R. W. Davidge. The Combined Influence of Stress, Time and Temperature on the Strength of Polycrystalline Alumina. Proc. Br. Cer. Soc., Mechanical Properties of Cer. 2 (R. W. Davidge, ed.). 25:151–167, 5/1975.
111. N. M. Parikh. Factors Affecting Strength and Fracture of Nonfissionable Ceramic Oxides. Nuclear Applications of Nonfissionable Ceramics (A. Boltax and J. H. Handwerk, eds.). American Nuclear Soc., Hindsdale, IL, 1966, pp. 31–56.
112. R. M. Spriggs and T. Vasilos. Effect of Grain Size on Transverse Bend Strength of Alumina and Magnesia. J. Am. Cer. Soc. 46(5):224–228, 1963.
113. R. M. Spriggs, J. B. Mitchell, and T. Vasilos. Mechanical Properties of Pure, Dense Aluminum Oxide as a Function of Temperature and Grain Size. J. Am. Cer. Soc. 47(7):323–327, 1964.
114. G. Orange, D. Turpin-Launay, P. Goeuriot, G. Fantozzi, and F. Thevenot. Mechanical Behavior of a Al_2O_3–AlON composite Ceramic Material (Aluminalon). Sci. Cer. 12:661–666, 1984.

115. J. C. Romine. New High-Temperature Ceramic Fiber. Cer. Eng. Sci. Proc. 8(7–8):755–765, 1987.
116. M. H. Stacey. Developments in Continuous Alumina-Based Fibers. Br. Cer. Trans. 87:168–172, 1977.
117. J. E. Bailey and H. A. Hill. Ceramic Fibers for the Reinforcement of Gas Turbine Blades 5 (W. W. Kreigel and H. Palmour III, eds.). Plenum Press, New York, 1971, pp. 341–359.
118. J. B. Wachtman, Jr., W. E. Tefft, D. G. Lam, Jr., and C. S. Apstein. Exponential Temperature Dependence of Young's Modulus for Several Oxides. Phys. Rev. 122(6):1754–1759, 1961.
119. J. B. Wachtman, Jr., and D. G. Lam, Jr. Young's Modulus of Various Refractory Materials as a Function of Temperature. J. Am. Cer. Soc. 42(5):254–260, 1959.
120. R. Duff and P. Burnett. Microstructure Studies of Polycrystalline Refractory Oxides. Summary Report for Contract No. 3-65-0316-f, 1966.
121. A. Higashi. Mechanisms of Plastic Deformation in Ice Single Crystals. Physics of Snow and Ice, Intl. Conf. on Low Temperature Science, I. Conf. on Physics of Snow and Ice (H. Ôura, ed.). Bunyeido, Sapporo, Japan, 1967, pp. 277–289.
122. M. L. Kronberg. Dynamical Flow Properties of Single Crystals of Sapphire I. J. Am. Cer. Soc. 45(6):274–279, 1962.
123. S. J. Jones and J. W. Glen. The Effect of Dissolved Impurities on the Mechanical Properties of Ice Crystals. Phil. Mag. 19(157):13–24, 1969.
124. E. M. Schulson, S. G. Hoxie, and W. A. Nixon. The Tensile Strength of Cracked Ice. Phil. Mag. A 59(2):303–311, 1989.
125. J. E. Dykins. Tensile Strength of Sea Ice Grown in a Confined System. Physics of Snow and Ice, Intl. Conf. on Low Temperature Science, I. Conf. on Physics of Snow and Ice (H. Ôura, ed.). Bunyeido, Sapporo, Japan, 1967, pp. 523–537.
126. G. G. Bentle and R. M. Kniefel. Brittle and Plastic Behavior of Hot-Pressed BeO. J. Am. Cer. Soc. 48(11):570–577, 1965.
127. G. G. Bentle and K. T. Miller. Dislocations, Slip, and Fracture in BeO Single Crystals. J. Appl. Phys. 38(11):4248–4255, 1967.
128. R. E. Fryxell and B. A. Chandler. Creep, Strength, Expansion, and Elastic Moduli of Sintered BeO as a Function of Grain Size, Porosity, and Grain Orientation. J. Am. Cer. Soc. 47(6):283–291, 1964.
129. B. A. Chandler, E. C. Duderstadt, and J. F. White. Fabrication and Properties of Extruded and Sintered BeO. J. Nuc. Mat. 8(3):329–347, 1963.
130. M. L. Stehsel, R. M. Hale, and C. E. Waller. Modulus of Rupture Measurements on Beryllium Oxide at Elevated Temperatures. Mechan. Prop. of Eng. Cer. 16 (W. Kriegel and H. Palmour III, eds.). Interscience, New York, 1961, pp. 225–235.
131. S. C. Carniglia, R. E. Johnson, A. C. Hott, and G. G. Bentle. Hot Pressing for Nuclear Applications of BeO; Process, Product, and Properties. J. Nuc. Mat. 14:378–394, 1964.
132. D. A. Shockey and G. W. Groves. Effect of Water on Toughness of MgO Crystals. J. Am. Cer. Soc. 51(6):299–303, 1968.
133. K. R. Janowski and R. C. Rossi. Mechanical Degradation of MgO by Water Vapor. J. Am. Cer. Soc. 51(8):453–455, 1968.

134. R. W. Rice. CaO: II, Properties. J. Am. Cer. Soc. 52(8):428–436, 1969.
135. S. M. Copley and J. A. Pask. Plastic Deformation of MgO Single Crystals up to 1600°C. J. Am. Cer. Soc. 48(3):139–146, 1965.
136. D. S. Thompson and J. P. Roberts. Flow of Magnesium Oxide Single Crystals. J. Appl. Phys. 433–434, 1960.
137. R. W. Rice. Strength and Fracture of Hot-Pressed MgO. Proc. Brit. Cer. Soc. 20:329–364, 1972.
138. A. G. Evans, D. Gilling, and R. W. Davidge. The Temperature-Dependence of the Strength of Polycrystalline MgO. J. Mat. Sci. 5:187–197, 1970.
139. R. W. Rice. Deformation, Recrystallization, Strength, and Fracture of Press-Forged Ceramic Crystals. J. Am. Cer. Soc. 55(2):90–97, 1972.
140. R. W. Rice. Effects of Hot Extrusion, Other Constituents and Test Temperature on the Strength and Fracture of Polycrystalline MgO. J. Am. Cer. Soc. 76(12):3009–3018, 1993.
141. R. B. Day and R. J. Stokes. Mechanical Behavior of Polycrystalline Magnesium Oxide at High Temperatures. J. Am. Cer. Soc. 49(7):345–354, 1966.
142. F. P. Knudsen. Dependence of Mechanical Strength of Brittle Polycrystalline Specimens on Porosity and Grains Size. J. Am. Cer. Soc. 42(8):376–388, 1959.
143. C. E. Curtis and J. R. Johnson. Properties of Thorium Oxide Ceramics. J. Am. Cer. Soc. 40(2):63–68, 1957.
144. M. D. Burdick and H. S. Parker. Effect of Particle Size on Bulk Density and Strength Properties of Uranium Dioxide Specimens. J. Am. Cer. Soc. 39(5):181–187, 1956.
145. F. P. Knudsen, H. S. Parker, and M. D. Burdick. Flexural Strength of Specimens Prepared from Several Uranium Dioxide Powders; Its Dependence on Porosity and Grain Size and the Influence of Additions of Titania. J. Am. Cer. Soc. 43(12):641–647, 1960.
146. A. G. Evans, and R. W. Davidge. The Strength and Fracture of Stoichiometric Polycrystalline UO_2. J. Nuc. Mat. 33:249–260, 1969.
147. R. J. Beals, J. H. Handwerk, and G. M. Dragel. High Temperature Mechanical Properties of Uranium Compounds. In High Temperature Technology, Proc, Third Intl. Symp. on High Temp. Tech. in Pacific Grove, CA, Intl. Union of Pure & Applied Chem., Butterworths, London, 1968, pp. 265–278.
148. R. F. Canon, J. T. A. Roberts, and R. J. Beals. Deformation of UO_2 at High Temperatures. J. Am. Cer. Soc. 54(2):105–112, 1971.
149. C. R. Kennedy and G. Bandyopadhyay. Thermal-Stress Fracture and Fractography in UO_2. J. Am. Cer. Soc. 59(3–4):176–177, 1976.
150. H. M. Kandil, J. D. Greiner, and J. F. Smith. Single-Crystal Elastic Constants of Yttria-Stabilized Zirconia in the Range of 20° to 700°C. J. Am. Cer. Soc. 6(5):341–346, 1984.
151. J. B. Wachtman, Jr., and W. C. Corwin. Internal Friction in ZrO_2 Containing CaO. J. Res., National Bureau of Standards A. Physics and Chemistry 69A(5):457–460, 9–10/1965.
152. M. Shimada, D. Matsushita, S. Kuratani, T. Okamoto, M. Koizumi, K. Tsukuma and T. Tsukidate. Temperature Dependence of Young's Modulus and Internal

Friction in Alumina, Silicon Nitride, and Partially Stabilized Zirconia Ceramics. J. Am. Cer. Soc. 67(2):C-23–24, 1984.

153. J. W. Adams, D. C. Larsen, R. Ruh, and K. S. Mazdiyasni. Young's Modulus, Flexure Strength, and Fracture of Yttria-Stabalized Zirconia versus Temperature. J. Am. Cer. Soc. 80(4):903–908, 1997.
154. S. L. Dole. Elastic Properties of Hafnium and Zirconium Oxides Stabilized with Praseodymium or Terbium Oxide. J. Am. Cer. Soc. 66(3):C-47–49, 1983.
155. A. S. Drachinskii, V. A. Dubok, V. V. Lashneva, V. G. Vereshchak, and V. V. Kovylyaev. Features of the Temperature Dependence of Failure Stress for Zirconium Dioxide. Problemy Prochnosti 3:368–370, 1987.
156. R. W. Rice. Ceramic Fracture Features, Observations, Mechanisms, and Uses. Fractography of Ceramic and Metal Failure, ASTM STP 827 (J. J. Mecholosky, Jr., and S. R. Powell, eds.). 1984, pp. 5–103.
157. R. A. Penty. Pressure-Sintering Kinetics and Mechanical Properties of High-Purity, Fine-Grained Mullite. Ph.D. thesis. Lehigh University, 1972.
158. V. Mandorf and J. Hartwig. High Temperature Properties of Titanium Diboride. High Temperature Materials II (G. Ault, W. Barclay, and H. Munger, eds.). Interscience, New York, 1963, pp. 455–467.
159. J. Matsushita, H. Naggashima, and H. Saito. Preparation and Mechanical Properties of TiB_2 Composites Containing Ni and C. J. Cer. Soc. Jpn., Intl. Ed. 99:1047–1050, 1991.
160. R. A. Alliegro. Titanium Diboride and Zirconium Diboride Electrodes. The Encyclopedia of Electrochemistry (C. Hampell, ed.). Reinhold, New York, 1964, p. 1125.
161. D. Kalish, E. V. Clougherty, and K. Kreder. Strength, Fracture Mode, and Thermal Stress Resistance of HfB_2 and ZrB_2. J. Am. Cer. Soc. 52(1):30–36, 1969.
162. G. de With. High Temperature Fracture of Boron Carbide: Experiments and Simple Theoretical Models. J. Mat. Sci. 19:457–466, 1984.
163. G. A. Gogotsi, Yu. G. Gotosi, V. V. Kovylyaev, D. Yu. Ostrovoi, and V. Ya. Ivas`kevich. Behavior of Hot-Pressed Boron Carbide at High Temperature, II. Strength. Proshkovaya Met. 6(318):77–82, 1989.
164. M. Bougoin, F. Thevenot, J. Dubois, and G. Fantozzi. Synthèse et Characterization de Céramiques Denses en Carbure de Bore. J. Less Com. Met. 114:257–271, 1985.
165. G. A. Gogosti, Y. A. L. Groushevsky, O. B. Dashevskaya, Yu. G. Gogotsi, and V. A. Lavrenko. Complex Investigation of Hot-Pressed Boron Carbide. J. Less Com. Met. 117:225–230, 1986.
166. G. W. Hollenberg and G. Walther. The Elastic Modulus and Fracture of Boron Carbide. J. Am. Cer. Soc. 63(11–12):610–613, 1980.
167. Yukio Takeda and Kunihiro Maeda. Mechanical Properties of High Thermal Conductive SiC Ceramics. J. Cer. Soc. Jpn. 99:699–700, 1991.
168. D. C. Larsen, J. W. Adams, L. R. Johnson, A. P. S. Teotia, and L. G. Hill. Ceramic Materials for Advanced Heat Engines, Technical and Economic Evaluation. Noyes, Park Ridge, NJ, 1985.
169. D. B. Miracle and H. A. Lipsitt. Mechanical Properties of Fine-Grained Substoichiometric Titanium Carbide. J. Am. Cer. Soc. 66(8):592–597, 1983.

170. R. H. J. Hannink and M. J. Murray. Elastic Moduli Measurements of Some Cubic Transition Metal Carbides and Alloyed Carbides. J. Mat. Sci. 9:223–228, 1974.
171. T. K. Gupta. Strength Degradation and Crack Propagation in Thermally Shocked Al_2O_3. J. Am. Cer. Soc. 55(5):249–253, 1972.
172. T. K. Gupta. Critical Grain Size for Noncatastrophic Failure in Al_2O_3 Subjected to Thermal Shock. J. Am. Cer.Soc. 56(7):396–397, 1973.
173. H. Tomaszewski. Influence of Microstructure on the Thermomechanical Properties of Alumina Ceramics. Cer. Intl. 18:51–55, 1992.
174. C. C. Seatton and S. K. Dutta. Effect of Grain Size on Crack Propagation in Thermally Shocked B_4C. J. Am. Cer. Soc. 57(5):228–229, 1974.
175. J. A. Capolla and R. C. Bradt. Thermal Shock Damage in SiC. J. Am. Cer. Soc. 56(4):214–218, 1973.
176. T. G. Langdon. The Role of Grain Boundaries in High Temperature Deformation. Mat. Sci. Eng. A166:67–79, 1993.
177. Y. Maehara and T. G. Langdon. Review: Superplasticity in Ceramics. J. Mat. Sci. 25:2275–2286, 1990.
178. G. D. Quinn. Fracture Mechanism Maps for Advanced Structural Ceramics. J. Mat. Sci. 25:4361–4376, 1990.
179. D. P. H. Hasselman. Single Crystal Elastic Anisotropy and the Mechanical Behavior of Polycrystalline Brittle Refractory Materials. Anisotropy in Single-Crystal Refractory Compounds (F. W. Vahldiek and S. A. Merson, eds.). Plenum Press, New York, 1968, pp. 247–265.
180. A. H. Heuer. Deformation Twinning in Corundum. Phil. Mag. 13(122):379–393, 1966.
181. P. F. Becher. Abrasive Surface Deformation of Sapphire. J. Am. Cer. Soc. 59(3–4):143–145, 1976.
182. P. F. Becher. Fracture-Strength Anisotropy of Sapphire. J. Am. Cer. Soc. 59(1–2):59–61, 1976.
183. W. D. Scott and K. K. Orr. Rhombohedral Twinning in Alumina. J. Am. Cer. Soc. 66(1):27–32, 1983.
184. H. M. Chan and B. R. Lawn. Indentation Deformation and Fracture of Sapphire. J. Am. Cer. Soc. 7(1):29–33, 1988.
185. B. E. Walker, Jr., R. W. Rice, R. C. Pohanka, and J. R. Spann. Densification and Strength of $BaTiO_3$ with LiF and MgO Additives. Am. Cer. Soc. Bull. 55(3):274–276, 1976.
186. J. J. Mecholsky, Jr. Private communication, 1998.
187. D. C. Harris. Overview of Sapphire Mechanical Properties and Stratagies for Strengthening Sapphire. Proc. 7th DoD Electromagnetic Windows Symp., Johns Hopkins Applied Phys. Lab. (D. C. Harris, ed.). 5/1998, pp. 310–318.
188. E. Savrun, C. Toy, and W. D. Scott. Axial Compression Testing of Single Crystal Sapphire. 7th DoD Electromagnetic Windows Symp., Johns Hopkins Applied Phys. Lab. (D. C. Harris, ed.). 5/1998, pp. 358–364.
189. F. Schmid and D. C. Harris. Effects of Crystal Orientation and Temperature on the Strength of Sapphire. J. Am. Cer. Soc. 81(4):885–893, 1998.

190. E. Savrun, C. Toy, and W. D. Scott. Strengthening Sapphire by Microstructural Modifications. J. Am. Cer. Soc. 81(4):365–372, 1998.
191. D. C. Harris and L. F. Johnson. Navy Mechanical Test Results from the Sapphire Statistical Characterization and Risk Reduction Program. J. Am. Cer. Soc. 81(4):337–343. 1998.
192. T. M. Regan and D. C. Harris. High Temperature C-Axis Strengthened Sapphire. J. Am. Cer. Soc. 81(4):344–351, 1998.
193. R. W. Rice and R. C. Pohanka. Grain-Size Dependence of Spontaneous Cracking in Ceramics. J. Am. Cer. Soc. 62(11–12):559–563, 1979.
194. R. W. Rice, R. C. Pohanka, and W. J. McDonough. Effect of Stresses from Thermal Expansion Anisotropy, Phase Transformation and Second Phases on the Strength of Ceramics. J. Am. Cer. Soc. 63(11–12):703–710, 1980.
195. F. J. P. Clarke. Residual Strain and the Fracture Stress–Grain Size Relationship in Brittle Solids. ACTA Metallurgica 12:139–143, 1964.
196. M. K. Aghajanian, N. H. MacMillan, C. R. Kennedy, S. J. Luszcz, and R. Roy. Properties and Microstructures of Lanxide® Al_2O_3-Al Ceramic Composite Materials. J. Mat. Sci. 24:658–670, 1989.
197. R. W. Rice. Possible Effects of Elastic Anisotropy on Mechanical Properties of Ceramics. J. Mat. Sci. Lett. 13:1261–1266, 1994.
198. M. Weller and H. Schubert. Internal Friction, Dielectric Loss, and Ionic Conductivity of Tetragonal ZrO_2-3% Y_2O_3 (Y-TZP). J. Am. Cer. Soc. 69(7):573–577, 1986.
199. J. S. Moya, R. Moreno, and J. Requena. Black Color in Partially Stabilized Zirconia. J. Am. Cer. Soc. 71(11):C-479–480, 1988.
200. R. W. Rice. Comment on "Black Color in Partially Stabilized Zirconia." J. Am. Cer. Soc. 75(7):1745–1746, 1991.
201. R.W. Rice, K. R. McKinney, and R. P. Ingel. Grain Boundaries, Fracture, and Heat Treatment of Commercial Partially Stabilized Zirconia. J. Am. Cer. Soc. 64(12):C-175–177, 1981.
202. J. B. Wachtman, Jr. Mechanical and Electrical Relaxation in ThO_2 Containing CaO. Phys. Rev. 131(2):517–527, 1963.
203. J. J. Swab. Low Temperature Degradation of Y-TZP Materials. J. Mat. Sci. 26:6706–6714, 1991.
204. M. Hirano, T. Matsuyama, H. Inada, K. Suzuki, H. Yoshida, and M. Machida. Effect of Composition on Phase Stability Under Hydrothermal Conditions and Fracture Strength of Yttria- and Ceria-Doped Tetragonal Zirconia-Alumina Composites Fabricated by HIP. J. Cer. Soc. Jpn. 99(5):382–387, 1991.
205. I. Thompson and R. D. Rawlings. Effects of Liquid Environments on Zirconia-Toughened Alumina, Part I, Chemical Stability. J. Mat. Sci. 27:2823–2830, 1992.
206. I. Thompson and R. D. Rawlings. Effects of Liquid Environments on Zirconia-Toughened Alumina, Part II, Mechanical Properties. J. Mat. Sci. 27:2831–2839, 1992.
207. D. H. Chung and W. R. Buessem. The Elastic Anisotropy of Crystals. Proc. Intl. Symp. 2 (F. W. Vahldiek and S. A. Mersol, eds.). Plenum Press, New York, 1968, pp. 217–245.

208. G. Simmons and H. Wang. Single Crystal Elastic Constants and Calculated Aggregate Properties: A Handbook. MIT Press, Cambridge, Massachusetts, 1977.
209. P. Chantikul, S. J. Bennison, and B. R. Lawn. Role of Grain Size in the Strength and R-Curve Properties of Alumina. J. Am. Cer. Soc. 73(8):2419–2427, 1990.
210. R. W. Rice. Comment on "Role of Grain Size in the Strength and R-Curve Properties of Alumina." J. Am. Cer. Soc. 76(7):1898–1899, 1993.
211. R. W. Rice. Ceramic Fracture Mode-Intergranular vs. Transgranular Fracture. In Fractography of Glasses and Ceramics III (V. Frechette, J. R. Varner, and G. Quinn, eds.). Am. Cer. Soc., Westerville, OH, 1996, pp. 1–53.

7

Grain Dependence of Hardness, Compressive Strength, Wear, and Related Behavior at Elevated Temperatures

I. INTRODUCTION

This chapter complements the preceding one by completing the review of the grain dependence of mechanical properties of monolithic ceramics as a function of temperature. Hardness, compressive strength, wear, and related behavior are addressed, while the preceding chapter addressed crack propagation, toughness, and tensile strength. This chapter also compares the limited data on both tensile and compressive failure in the same material, i.e. the overall trend for the large difference between the two to disappear at higher temperatures, as in ductile metals, but for some differences to remain.

Unfortunately there is again a significant lack of data first simply as a function of only temperature, and especially also as a function of grain parameters (mainly size), though the extent of this limitation varies with the particular property. However, besides such direct data, there are three other sources of some information on the grain dependence as a function of temperature. First is the substantial data on such dependence of properties at ~ 22°C reviewed in Chapters 4 and 5, and the fact that properties generally undergo gradual changes with increasing T, so grain dependences are also expected to change gradually. Second is information on the grain dependence of one property versus temperature and the correlations of this property with other related ones, e.g. of E and H

with each other and with other mechanical properties as discussed in Chapters 2–6. Third is information on the temperature dependence of these properties for single crystals, since these reflect both the limit of *G* and of the orientation dependence of textured polycrystalline bodies.

Thus hardness, compressive strength (σ_C), wear, and related behavior are also affected by temperature-driven changes in underlying properties such as *E* and by the onset and increase in various deformation, e.g. creep, processes. However, there are other basic effects due to the inherent reduction of stresses for plastic deformation as temperature (*T*) increases due to two counter effects. On the one hand, increased plasticity will reduce local fracture and hence cracking associated with hardness indentations, compressive stressing, wear, and erosion. On the other hand, increased plasticity also means reduced hardness and more penetration of asperities into mating wear surfaces as *T* increases. Further, basic changes can occur in the mechanisms of failure, and hence in their *G* dependence. Thus, for example, increased plasticity in particulates causing erosion, or in the surfaces they impact, can change erosion, e.g. shifting of the angle for maximum erosion from 90° (i.e. normal) to the eroding surface for brittle processes to 30° for ductile erosion processes [1,2]. At higher temperatures, increased bonding (e.g. via welding) and increased chemical reaction can also become important factors in erosion, and especially wear. Similarly, repeated stressing, i.e. fatigue testing as temperatures increase, can be compounded by changes from mainly or exclusively brittle fracture to increasing effects of various nonelastic processes (again with an inverse effect of strain rate). Besides affecting properties, increased plastic deformation and reduction of brittle fracture as temperature increases also affects strength tests. Thus while alignment and interface stresses are important issues in compressive testing of brittle materials, they become less critical as the degree of plastic deformation increases as *T* increases. Similarly, though receiving almost no study, hardness-related cracking should decrease, i.e. as *T* increases.

While changes in properties considered in this and the previous chapter are generally gradual with increasing temperature, three factors should be noted. First, there can be important changes at modest temperatures, e.g. such as shown for the tensile strength of Al_2O_3 (Fig. 6.12), which are indicated as correlating with changes of hardness, and especially compressive strength, of Al_2O_3 (Fig. 7.6). Second, there can be substantial and rapid changes in properties as a function of temperature in the limited, but important, cases where properties have been measured through a phase transformation, e.g. as mainly seen in the more extensive *H–T* data. Third, though not commonly noted, as macroplasticity occurs in ceramics at higher temperatures, the differences in tensile versus compressive strengths decrease, especially for single crystals, as does the H/σ_C ratio.

II. GRAIN DEPENDENCE OF HARDNESS AS A FUNCTION OF TEMPERATURE

The unfortunate tendency of investigators to make measurements as a function of temperature on only a few bodies, often only one, with little or no indication of even G is particularly prevalent for hardness. Thus much of the G dependence of H as a function of T must be implied from the extensive demonstration of such dependence at 22°C (Chap. 4) and extrapolations of this to elevated temperatures. Such extrapolations are guided by the reasonable H–T data for one or a few bodies, usually of unspecified G, and single crystals as well as correlation with other related properties. The latter is a useful source of information given H–σ_C correlations, and more σ_C–T–G data in the next section. Thus the focus of this section is mainly on the T dependence of H for dense polycrystalline bodies, mostly of unknown or uncertain G, and of single crystals, as well as some comparison of the two, and in turn some comparison of these to E–T changes.

Consider first the issue of hardness anisotropy as a function of the crystallographic plane on which single crystal indents are made and the direction of the apexes of the indenter relative to the crystal directions on the indented plane. As shown in Chapter 4, there is frequently substantial hardness anisotropy at room temperature in many ceramics arising from activation of differing combinations of deformation modes, mainly slip systems, with anisotropy often nearly as, and in some cases more, in cubic versus noncubic crystal structures. Since anisotropy of hardness (or of any solid property) must disappear when the material melts, this suggests that the anisotropy may progressively decrease as temperatures increase. However, data for ice crystals [3,4], which have many similarities to α-Al_2O_3 on the basis of homologous temperatures, shows substantial relative H anisotropy, e.g. from –5 to –12°C a minimum ~ 35% < maximum H values. Such levels of anisotropy within 2% of the absolute melting point clearly show that H anisotropy does not gradually decrease to zero as T increases toward the melting point.

Other data clearly shows that anisotropy of hardness can vary substantially as a function of temperature for a given material and between different materials. Thus Atkins' review [5] showed that relative H anisotropy of TiC and VC, both cubic ceramics, respectively decreased substantially from 25 to 250°C (26% to 5%) and then increased some (~ 10%) by 610°C; it increased from ~ 5% at –196°C to ~ 7% and then to ~ 18% at 25 and 350°C (Fig. 7.1). Bsenko and Lundström [6] showed little or no change in the relative H_V anisotropy of arc melted, noncubic, B rich HfB_2, using a lower load (0.5 N) and the large G to make measurements within individual grains from 115 to 210°C, despite a nearly 20% decrease in average H_V values. Nakano et al. [7] reported H_K (0.1 kg load) for two directions on three planes of TiB_2 crystals at 22, 250, 500, 750, and 1000°C showing anisotropy disappearing at ~ 250°C and then increasing with further temperature increases (Fig. 7.2).

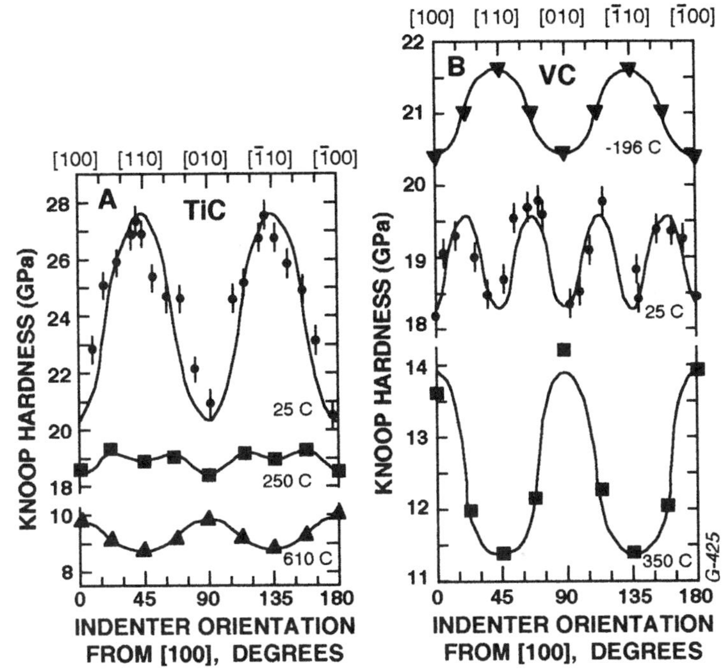

FIGURE 7.1 Knoop hardness anisotropy of cubic TiC and VC shown at three temperatures each. Note that while *H* anisotropy can decrease as *T* increases, e.g. in TiC at 250°C, it can also increase as *T* increases, e.g. in TiC at 610°C and VC at 350°C. See also Figure 7.2. (From Ref. 5, published with permission of the ASM.)

Similar crossover of H_K (0.1 kg) curves for differing planes and directions of α-SiC crystals, as shown above for TiB_2, was shown by Niihara [8], again at modest *T* (300–400°C). Overall *H* decreases of 60–80% were reported for different planes and orientations between 22 and 1600°C, with varying, moderate, but definitive degrees of deviation from linear decreases due to curvatures or inflections in the *H–T* curves indicating changes in plastic flow as a function of *T* related to slip changes, e.g. indicating substantial increases in plasticity above 800°C. Similar changes were indicated in H_K (0.5 kg load) tests of Fujita et al. [9] to 1400°C, again with crossover of *H* values for some planes at 300–500°C as well as at 1200°C. Hirai and Niihara [10] reported essentially linear decreases for their H_V (0.1 kg load) tests of SiC crystals to 1500°C with less anisotropy (as expected for H_V). However some clear changes, but no clear decreases, of *H* anisotropy were seen along with similar *H* decreases by 60% or

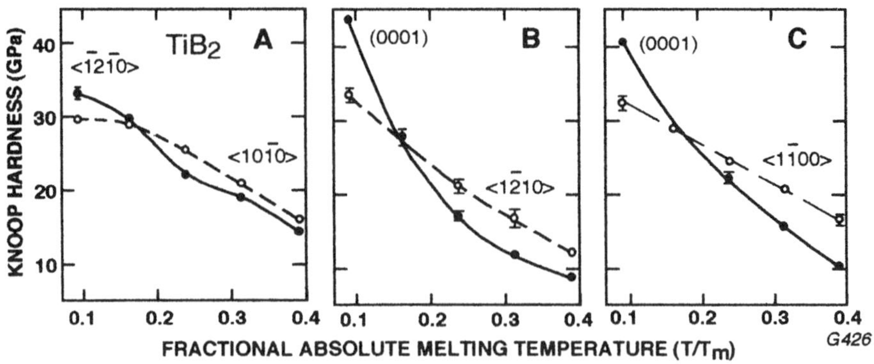

FIGURE 7.2 H_K (0.1 kg load) data as a function of the homologous temperature (i.e. fraction of the melting point taken as 2980°C) for two directions on three planes of TiB_2 crystals at 22, 250, 500, 750, and 1000°C showing anisotropy disappearing at ~ 250°C and then increasing with further temperature increases. (From Ref. 7, published with permission of the *Japanese Journal of Applied Physics.*)

more. Note that the lower temperature crossovers of *H* values for one crystal plane versus another commonly correspond to much less initial decrease in *H* on one plane. A frequent factor in this, besides differing changes in slip, can be desorption of moisture, as indicated by studies of Westbrook and Jorgensen [11] discussed below.

Consider now general *H–T* data trends for ceramic crystals, starting with cubic oxides. Guillou et al. [12] showed H_V of MgO crystals ({100} surfaces) and Ca-CZ (Berkovich hardness, H_B, on {111} surfaces) both decreasing to 30–40% of their 22°C values at 800°C (Fig. 7.3A), i.e. an order of magnitude or more than expected *E–T* decreases. Their MgO data is consistent with the mutual indentation (H_M) data for MgO crystals of Atkins and Tabor [13] and Westbrook's [14] data for indentations within single grains of polycrystalline MgO. Turning to noncubic ceramics, Kollenberg [15] showed H_V (2–3.9 N loads) for three sapphire orientations being similar to that of Alpert et al. [16] for 10 N load, showing similar decreases to ~ 26% of their 22°C values by respectively 750 and 1000°C (Fig. 7.1B). Kollenberg [17] also showed H_V (~ 1–2 N loads) for five orientations of hematite (Fe_2O_3) crystals (isomorphous with sapphire) substantially (~ 90%) decreasing with some variations in anisotropy and in overall *H* values with *T* in tests to 800°C (~ 0.6 T_m). These *H* decreases in single oxides such as Al_2O_3 and Fe_2O_3 are in marked contrast to a decrease to ~ 70% of their H_V (2 N load) value at 22°C by 1000°C in Kollenberg and Schneider's [18] measurements on (001) and (010) surfaces of mullite crystals. This decrease of <

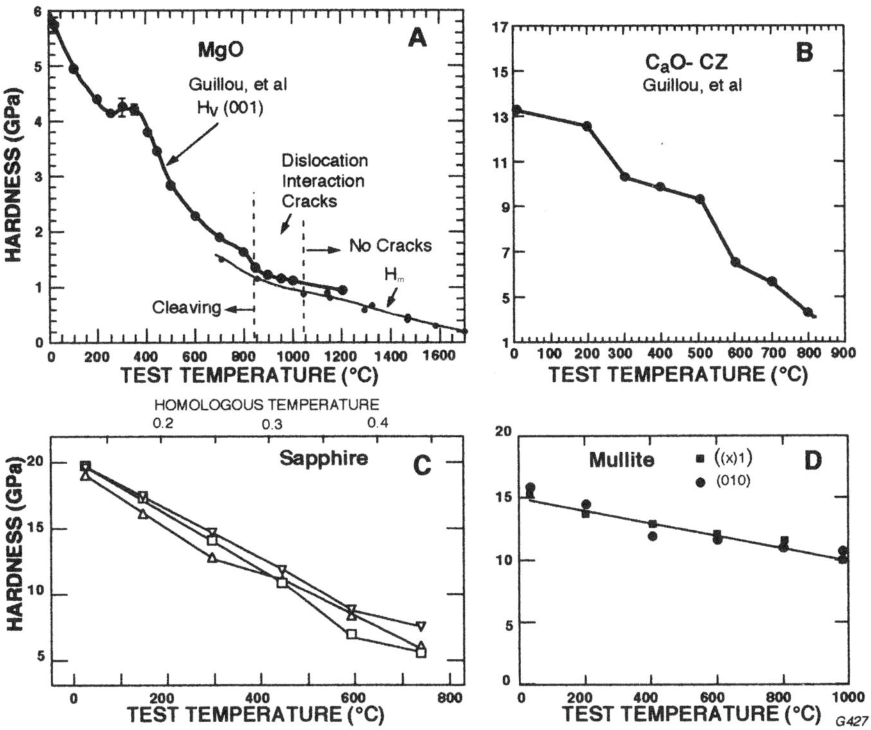

FIGURE 7.3 Hardness versus temperature for selected crystallographic surfaces on crystals of cubic ZrO_2-CaO and MgO [12], and noncubic Al_2O_3 [15,16] and mullite [18]. All hardnesses are Vickers with similar loads, except Berkovich (H_B) for the Ca-CZ and mutual indentation (H_M) for the higher T tests of MgO. Note reasonable agreement for the different MgO tests. (Published with permission of the British Ceramic Society, the *Journal of Materials Science,* and the *Journal of the American Ceramic Society.*)

½ that of Al_2O_3 over the same temperature range, despite mullite having a lower melting point, is attributed to greater difficulty of plastic deformation in such multiple constituent materials due to the requirement of cooperative motion of the added atomic species. Such results are consistent with those for other multiconstituent materials, e.g. as shown by Westbrook [14], as discussed below. Note that while several of these materials show essentially linear decreases in H over the limited temperature range, this is often not the case, as indicated by MgO and Ca-CZ, as well as a number of nonoxides discussed below.

Turning to nonoxide ceramics, deviations, often substantial, from simple linear decreases in H as T increases occur for both cubic and noncubic ceramics.

Thus, for example, Kumashiro et al. [19] demonstrated that their H_V measurements to 1500°C on {100} surfaces of cubic NbC, TaC, and ZrC crystals, as well as measurements of others on other cubic carbide crystals, showed various, often substantial, inflections in their *H–T* plots. While plotting data as a function of homologous temperature did not necessarily bring such inflections into coincidence, *H* values commonly decreased to ~ 25% of their values at 22°C by 1500°C. More extensive is earlier work of Kohlstedt [20] for his measurements on crystals of NbC, TiC, VC, and ZrC showing that all have overall ~ linear *H* decreases as a function of the fractional T_m (to values of 0.4–0.6), often with reasonable parallelness of some data. However, there were variations such as significant deviations from linearity, parallelness, or both above a fractional T_m of ~ 0.25. His measurements showed little or no deviations to less *H* increase as *T* approached ~ 20°C, which probably reflects measurement in vacuum and thus removal of much of the adsorbed water, as is discussed below. Guillou et al. [12] showed the hardness of α-SiC crystals [Berkovich, H_B, 2 N load on a (0001) surface] decreasing to ~ 40% of their 22°C values at 800°C (Fig. 7.4A) with a significant inflection between 300 and 400°C. Fujita et al. [9] showed similar as well as different rates of decrease, but similar total decrease for three orientations of H_K(500 gm) indents on 6H crystals (Fig. 7.4B). Sawyer et al. [21] showed decreases of ~ 65% from 22 to 800°C. Hirai and Niihara [10] studied H_V (100 gm) on three surfaces of 6H SiC crystals, showing modest separations (e.g. ~ 10%) that decreased ~ linearly by ~ 75% to 1500°C. Bsenko and

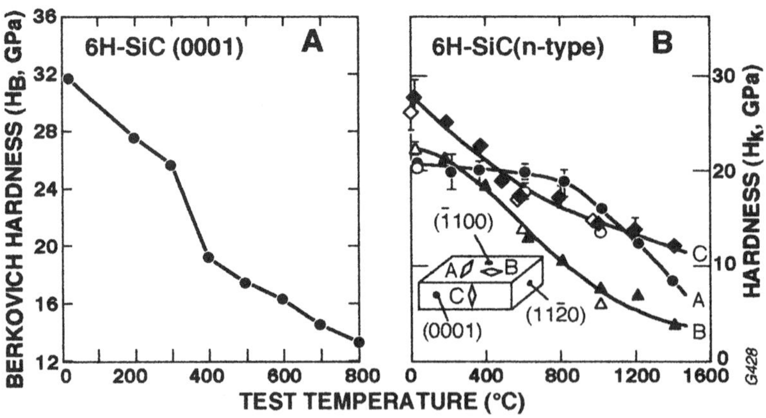

FIGURE 7.4 Hardness versus test temperature for 6H α-SiC crystals. (A) Berkovich hardness (2 N). (From Ref. 12, published with permission of the British Ceramic Society.) (B) Knoop hardness (500 gm). (From Ref. 9, published with permission of the *Journal of Materials Science*.)

Lundström [6] showed linear decreases in H_V of arc melted ZrB_2 and HfB_2, using a lower load (0.5 N) and the large G to make measurements within individual grains, thus giving an ~ single crystal value averaged over the random grain orientations.

Westbrook's [14] extensive testing and review of ceramic hardness versus T provides useful guidance in considering H–T–G relations. His H_V values were at low, 0.05–0.1 kg. loads so indents were typically small compared to the larger (unspecified) G values of his samples, so they are more representative of single crystal values averaged over all orientations. This correlation with single crystal data is shown by direct comparison. Westbrook's data commonly showed deviations at various temperatures in various materials similar to those noted above, but three additional sources of variations were identified. The first and most general was a trend for an increasing rate of H decrease with increasing T starting at ~ $\frac{1}{2}T_m$. The second and fairly general variation was a common lowering of H values at modest temperatures due to adsorbed moisture. Removal of this by heating in vacuum could increase the apparent H values, e.g. by 20–50% over the range of 50–300°C [11,14], as shown by retention of most or all of the higher H values on cooling and testing under vacuum. Due to differing adsorption characteristics by different crystal surfaces, desorption can change H anisotropy and may be a factor in other single crystal measurements, as was noted earlier. The third and more specialized variation was the significant changes that occur in H due to thermally driven crystal structure changes, e.g. increases of up to severalfold in H when the low-to-high quartz structure transition occurs in SiO_2 at 573°C [14,22]. This is accompanied by a change in the H anisotropy and a marked increase in the rate of H decrease with increasing T, with similar though less extreme effects of the same phase changes observed in isomorphous GeO_2 and $AlPO_4$ [22]. Additionally, Westbrook's measurements and survey showed on a homologous temperature basis that while binary compounds, e.g. NaCl structure oxides or carbides, may frequently have higher H values at lower T, their rate of H decrease with increasing T is typically substantially higher than that for most or all refractory ternary compounds, e.g. for mullite, spinel, and beryllium aluminate.

Consider now the limited data specifically on the effects of T on the G dependence of H. Though direct comparisons of H values for single- and polycrystals is limited, some does exist to further indicate G dependence as a function of temperature. Thus Alpert et al. [16] showed little or no difference between H_V for sapphire and two high-purity polycrystals (G ~ 3 and 20 μm), while showing expected progressively lower H_V for 96 and 90% (G ~ 11 and 4 μm) versus 99+% alumina bodies at lower T, but all merging together by T ~ 1000°C.

There is very limited polycrystalline data over substantial T ranges that can be directly compared with single crystal data, but a few possible comparisons are indicated. Thus Westbrook's [14] data for Al_2O_3 shows a decrease from ~ 20 to 6 GPa from 22 to 800°C, i.e. similar to Alpert et al., while his MgO data from 22 to

900°C shows H decreasing from ~ 8 to ~ 1.1 GPa, i.e. somewhat faster than single crystal values (Fig. 7.3). Additional examples of polycrystalline H values versus T for mostly nonoxides are shown in Figures 7.5 and 7.6. The former shows data for some CVD materials. The latter is of interest because of the H data being on the same samples tested for compressive strength, as is discussed in the next section. Note that, despite the limited T range, there is some tendency for more complex T dependences of polycrystalline H values. While some of this may reflect variations discussed above, e.g. effects of adsorbed water, some of this reflects additional complexities of polycrystalline bodies, e.g. of Al_2O_3, as is discussed in conjunction with compressive strength in the next section.

Lankford's [27] results for one dense body each of SiC, Si_3N_4, and Al_2O_3 are useful not only because they are better characterized (including G) but also

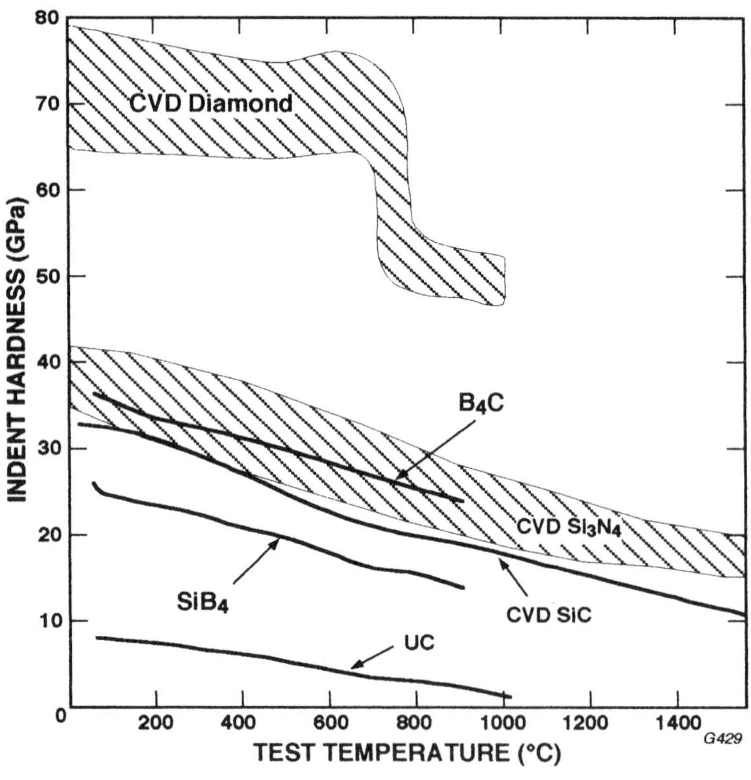

Figure 7.5 Temperature dependence of H of hot pressed B_4C and SiB_4 (9.8–19.6 N) (B_6O falls ~ halfway between these two) [23], CVD diamond (H_V, 7 N load) [24], arc-cast UC [25] H_V (5.5 kg load), and CVD SiC and Si_3N_4 [26].

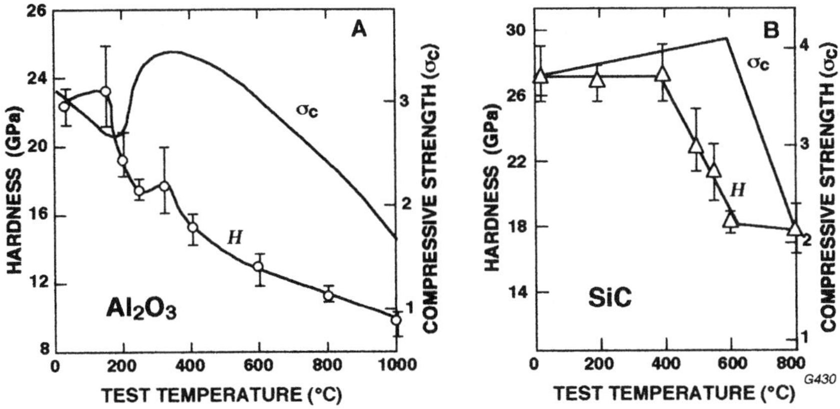

FIGURE 7.6 Hardness and compressive strength (σ_C) data of Lankford [27] versus test temperature for (A) Al_2O_3 and (B) SiC. (Published with permission of the *Journal of Materials Science.*)

because of corresponding compressive strength measurements on these, as is discussed in the next section. His hardness results show substantial inflections and changes as *T* increases for these three materials (Fig. 7.6), which are discussed in more detail in the next section in conjunction with the corresponding changes in compressive strength. However, it should be noted here that while almost no observations have been made on indent cracking as a function of temperature, i.e. much less than the limited observations at room temperature, Lankford [27] has noted this. Thus he reported that Vickers indentation (0.6 kg) generally reflected substantial increases in intergranular fracture around indents with substantial decreases in *H* as *T* increased. This was specifically illustrated for dense sintered Al_2O_3 showing more intergranular and less transgranular fracture at 1000 versus 22°C.

The role of grain boundaries, as indicated by Lankford's observations of increasing decreases in *H* with increasing *T* correlating with increased intergranular fracture, is supported by other observations. A basic one is generally less decrease in *H* at higher *T* for crystals versus polycrystalline bodies, e.g. as is indicated by the latter decreasing more rapidly near and above ~ $\frac{1}{2}T_m$, as noted by Westbrook [14]. This issue was also raised by Niihara's observation that elevated *T* decreases of *H* are substantially higher in most polycrystalline SiC and Si_3N_4 than for example in single crystals [26]. However, he presented data for CVD SiC and Si_3N_4 as well as Si_3N_4 hot pressed without additives showing much less *H* decrease with increasing *T* at higher *T*, thus focusing on the issue of the character of the grain boundaries. This is also indicated by effects of grain boundary phases on high temperature SCG and tensile strengths (Chaps. 5 and 6), and in compressive strength (next section), e.g. residues from the use of LiF for MgO and $MgAl_2O_4$, and oxide additions for SiC and Si_3N_4. However, much remains to be established, including effects of possible preferred orientation and grain elongation, e.g. in many CVD bodies.

III. GRAIN DEPENDENCE OF COMPRESSIVE STRENGTH AS A FUNCTION OF TEMPERATURE

Data for the dependence of compressive strength on grain size and temperature is almost opposite that of hardness. Thus while there was very little actual data as a function of *G* and *T*, but considerable data for single- and polycrystals of one (often unspecified) *G* as a function of *T* for *H*, there is very little single crystal data, but reasonable polycrystalline data as a function of *G* and *T*. Much of the data available is from earlier studies and has been discussed in three previous reviews, which serve as an important source for this section [28–30].

Reasonable earlier data exists for Al_2O_3 [28], which, though some requires correction to zero porosity for finer *G* bodies, shows progressive reduction of strength with increasing *G* (linearly with decreasing $G^{-1/2}$) and *T* (Fig. 7.7). This data also shows increasing indications of plastic deformation as *T* increases, e.g. above 1200°C, but dependent on strain rate as is shown and discussed further below. Another factor to note is that as *T* and resultant plastic deformation increase, the difference between compressive and tensile strengths decreases, especially in single crystals where there is no opportunity for differences in porosity generation from grain boundary sliding in tensile versus compressive stressing. Thus by or below 1600°C the $G^{-1/2} = 0$ intercept for dense Al_2O_3 is the same for tensile and compressive strengths at normal strain rates, but the σ–$G^{-1/2}$ slope is still substantially greater for compressive versus tensile strength (Fig. 7.8).

Consider now ice, a noncubic material that often has similar properties to α-Al_2O_3 on a homologous temperature basis as noted earlier (Chap. 6, Sec. IV.B; Chap. 7, Sec. II). Schulson clearly showed the typical $G^{-1/2}$ dependence of compressive strength to at least 96% T_m (Fig. 7.9) with no indication of bulk plastic deformation in stress–strain curves, but nevertheless with strain rate dependence [34]. While Schulson showed this data extrapolating to zero compressive strength at $G^{-1/2} = 0$, it could also be consistent with an intercept strength of ~ 0.4 MPa, as found for tensile strengths in similar testing (Fig. 6.13), again indicating convergence of compressive and tensile strengths at higher temperatures, especially for single crystals, though again the strengths and slope of the compressive testing at the same temperature and strain rate are higher than for tensile strength. Gold's [35] compressive creep studies of ice showed substantial cracking, with the amount of cracking as a function of strain showing maxima that increased in magnitude and occurred at lower strains as the test *T* increased.

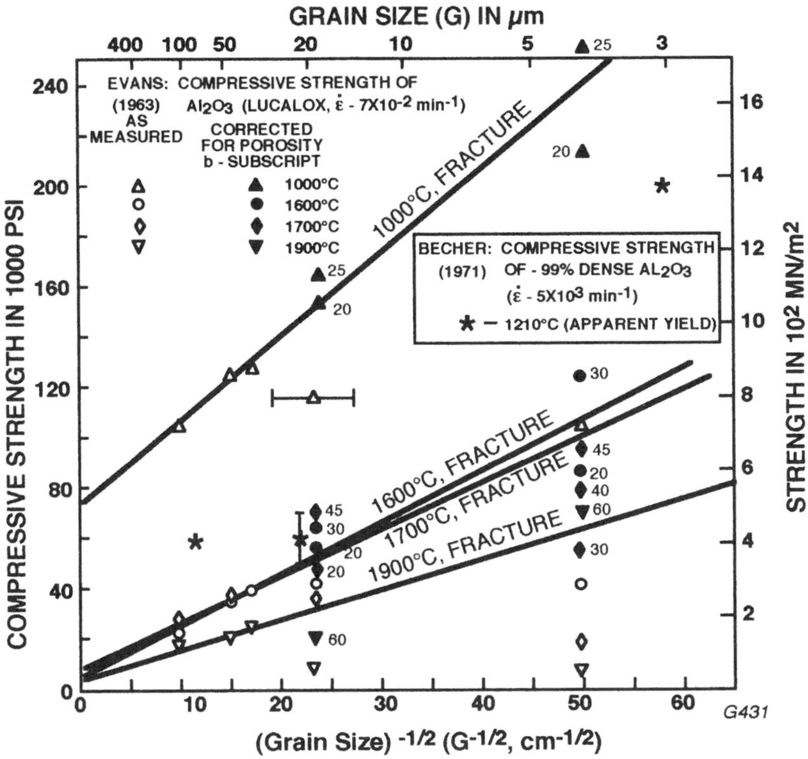

FIGURE 7.7 Compilation of compressive strength of Al_2O_3 versus $G^{-1/2}$ at various temperatures, from limited data of Becher [31] and more data from Evans [32], where data for finer *G* bodies has been corrected for the volume fraction porosity *(P)* to zero porosity per e^{-bP} for the *b* values shown. The *b* values were selected to give linear extensions of the σ_C–$G^{-1/2}$ for dense bodies at larger *G* and are consistent with other data and the expected increasing effect of porosity in limiting strength as *T* increases [33]. Note the consistency with other σ_C–$G^{-1/2}$ data for dense Al_2O_3 at 1600°C in Fig. 7.8. (After Rice [28], published with the permission of Academic Press.)

Cracks were on the scale of 1–2 grains, transgranular, ~ parallel to the compressive stress, and elongated in the direction of the substantial elongation of the columnar ice grains (oriented normal to the stress axis).

Turning to cubic oxides, there is a reasonable amount of data for dense MgO from Evans [32] and Copley and Pask [36] that shows similar behavior, especially when combined with each other and data for the yield stress of crystals stressed along a <111> axis, which is typically at or above the polycrystalline

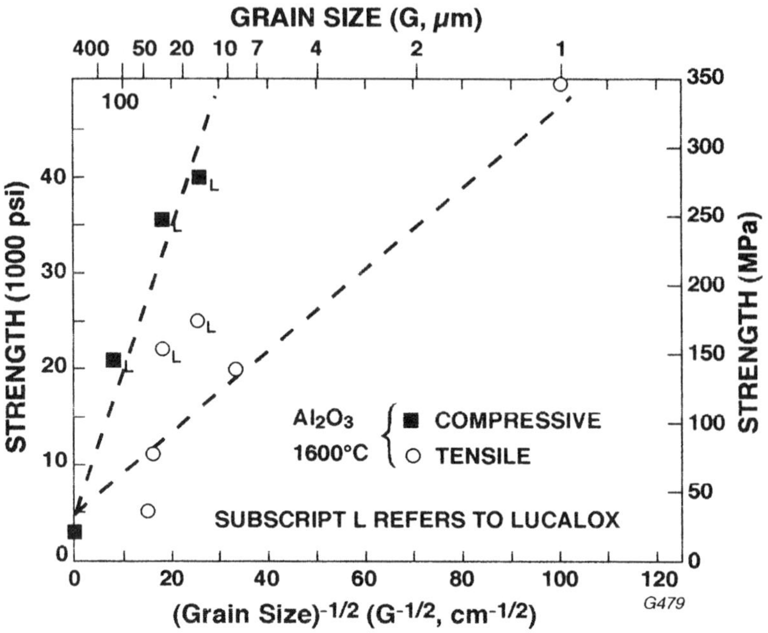

FIGURE 7.8 Compilation of some tensile and compressive strength versus $G^{-1/2}$ for dense Al_2O_3 at 1600°C. Note both strengths extrapolate to a common value at $G^{-1/2} = 0$, i.e. to ~ the compressive strength of one orientation of sapphire. (From Ref. 28, published with permission of Academic Press.)

failure stress [37] (Fig. 7.10). This again shows progressive reduction of strength with increasing G (linearly with decreasing with $G^{-1/2}$) and T, as well as the advantage of evaluating the self-consistency of different data sets. Langdon and Pask [37] also noted two other trends consistent with increased plasticity as T increased. The first was the change from mainly a single axial macrocrack for the macroscopic mode of compressive failure at lower temperatures to a more complex mode of conical fracture from the loading platens and lateral barreling as T increased. However, the general onset of bulk deformation at 800–1200°C depending on material and microstructure did not mean the immediate cessation of cracking, which first generally became more complex in terms of character and spatial character and then overall less in extent as T increased further. Second, transgranular fracture was dominant to ~ 1000°C, except for much more intergranular fracture in material made with LiF additions, again indicating effects of residual grain boundary phases. While attributed to somewhat greater CaO and SiO_2 impurity levels, the continued transgranular fracture may reflect more ef-

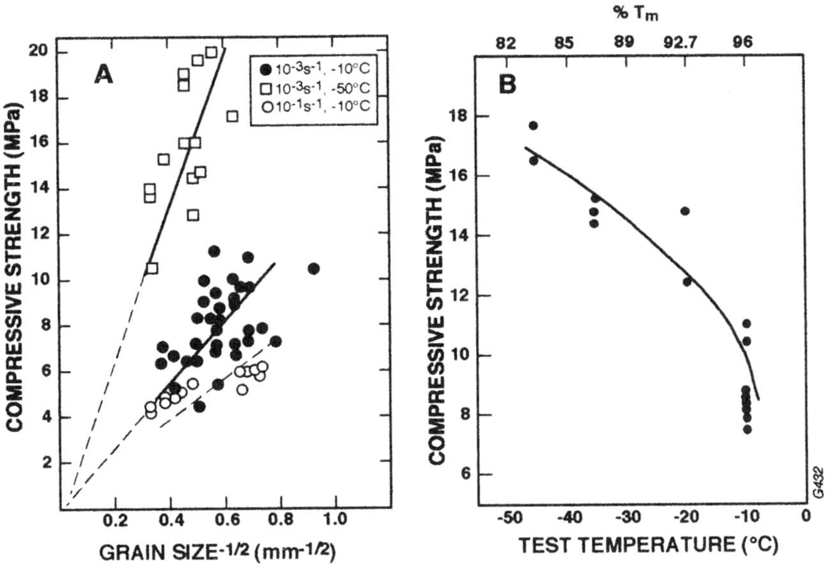

FIGURE 7.9 Ref. 34 ice compressive strength versus (A) $G^{-1/2}$ at two temperatures and two strain rates, and (B) test and homologous temperatures (T_m) (G 1.3 mm, strain rate 10^{-3}/s). (Published with permission of Philosophical Magazine.)

fect of the modest amount and size of residual pores along grain boundaries, and possibly the smaller ratio of such pore sizes to G, indicating the need for more detailed study. Thus while macroscopic ductility was generally substantial by 1200°C and further increased as T increased, there was still some cracking to at least 1400°C, again with some of it transgranular (in a body made without LiF additions).

Data of Ünal and Akinc [38] for one body of dense sintered Y_2O_3 (cubic, with $G \sim$ 5–40, average ~ 25 μm) showed similar results at a low strain rate (~ 6×10^{-6}/s). Thus bulk deformation, which occurred at 1200 but not at 1000°C, while terminating the axial macrofracture mode, did not immediately eliminate macrofracture; incompletely developed and connected axial cracks still formed, but they became progressively less extensive as temperature further increased. Both the compressive yield stress decreased more rapidly and true plastic deformation increased above 1200°C in comparison to brittle fracture at lower temperatures. While much of the fracture was intergranular, some transgranular fracture was observed to at least 1200°C.

Comparison of $H/3$ values and $G^{-1/2} = 0$ intercepts for compressive strengths of Al_2O_3 and MgO (Figs. 7.7 and 7.10) in Figure 7.11 shows that they

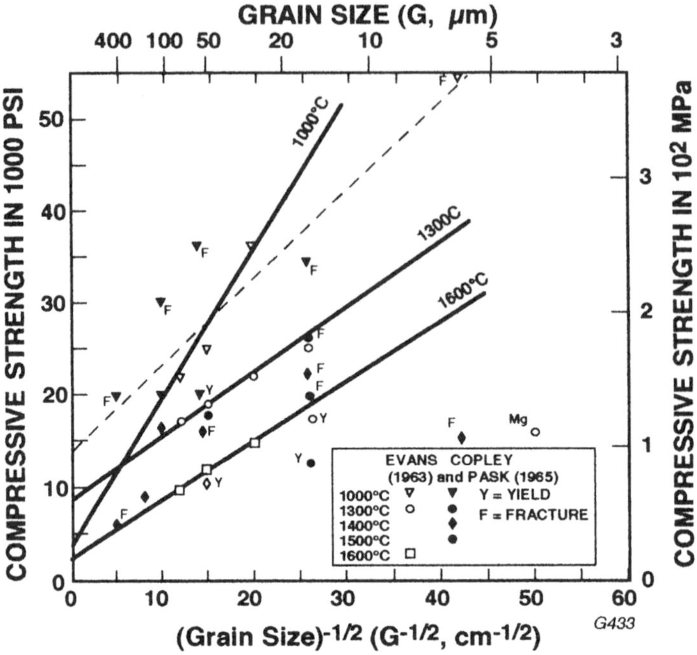

FIGURE 7.10 Compilation of compressive strength data for MgO versus $G^{-1/2}$ at various temperatures, data mainly from Evans [32] and Coply and Pask [36], after Rice [28]. However, data of Ref. 37 for the yield stress of MgO crystals stressed along <111> axes, which is ~ at or somewhat above polycrystalline yield stresses, is shown by arrows at the left axis (top to bottom) for 1000, 1300, and 1600°C. Slopes for the different test temperatures have been adjusted modestly to account for the single crystal data. Note the designation of whether a data point represents fracture (F) or yield (Y). (Published with the permission of Academic Press.)

generally parallel each other, as expected. However, $H/3$ for each material is higher, with the difference possibly increasing as T increases. This indicates the need for further evaluation of H–yield relations as discussed later. Also shown in Figure 7.11 is the tensile yield stress for MgO crystals as a function of T from Day and Stokes [39]. This again shows the convergence of tensile and compressive behavior, especially in single crystals, as plasticity becomes more extensive as T increases, i.e similar to data for Al_2O_3 (Fig. 7.7), as was also suggested above for ice.

Two other factors show that the general correlation of compressive strength and hardness continues to high temperatures, but that the relationship

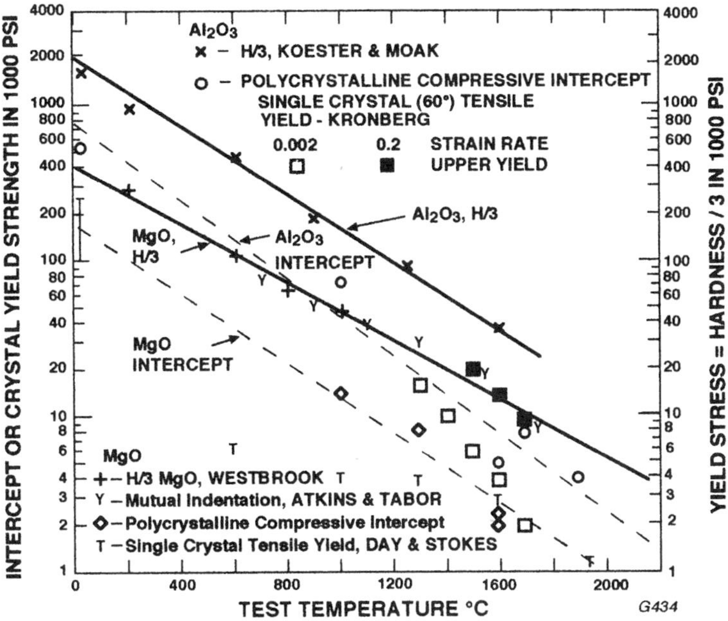

FIGURE 7.11 Stress at the $G^{-1/2} = 0$ intercept, actual single crystal yield, or *H*/3 for Al_2O_3 and MgO as compiled by Rice [28]. Note the convergence of the tensile and compressive yield stresses of MgO by ~ 1600°C, i.e. similar to Al_2O_3 data of Figure 7.7. (Published with permission of Academic Press.)

also must vary some from the simple *H*/3 correlation. The first is Lankford's [27] measurements of both properties on the same bodies, showing that the two properties do not follow parallel paths as a function of *T*, but compressive strengths decrease more rapidly with *T*. Thus the ratio of hardness to compressive yield stress varies and is often higher at higher *T* (Fig. 7.6). Second, the *G* dependence of compressive strength appears to be substantially greater (e.g. nearly twofold in Fig. 7.12 for TiB_2) than that of *H* at elevated *T*, as is also noted at room temperature. Kohlstedt [20] considered this issue in conjunction with his study of *H* of refractory NaCl structure carbide crystals discussed earlier, especially above the brittle–ductile transition where the discrepancy was greatest. He applied Marsh's [40] theory of indentation deformation, obtaining $H \sim 4.5Y$ instead of $\sim 3Y$ (see Sec. IV), which was in the right direction but still too low in comparison to the data. He cited higher strains and strain rates in hardness versus strength testing as a probable important factor in this difference, but it is clear that more study is needed to refine understanding of *H* and yield relations beyond the approximate, but useful, correlations observed.

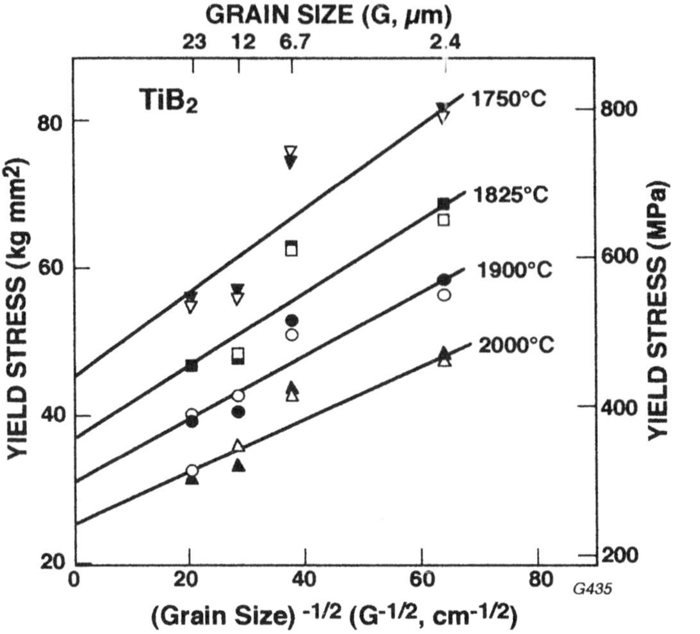

FIGURE 7.12 Compressive yield strength of dense sintered TiB_2 versus $G^{-1/2}$ at 1750–2000°C (~ 58–65% T_m) of Ref. 41. Their data and fitting the points for each temperature to a common activation energy of 0.8 eV/atom show a clear Hall–Petch relation. (Published with permission of the *Journal of Materials Science*.)

Ramberg and Williams' [41] study of the compressive strength of dense sintered TiB_2 in vacuum (5×10^{-4}/s strain rate) showed that macroscopic plastic deformation was seen only above 1700°C, where a clear Hall–Petch relation was shown (Fig. 7.12). They attributed the plastic deformation to slip, consistent with their evaluation that five or more independent slip systems can operate in TiB_2 at high T. Their tests of a single crystal at 2000°C gave a fourfold higher yield stress than indicated by projection of the polycrystalline data to $G^{-1/2} = 0$. This was attributed to the presence of laminar TiC precipitates hardening the single crystals.

The above studies, though useful and more substantial than most, reflect the common tendency to neglect measurements between 22 and ~ 1000°C. Though Lankford's studies of compressive strengths and hardness are for only one body (hence a given G) for each of three materials, Al_2O_3, SiC and Si_3N_4, his results (Fig. 7.6) are important for showing substantial changes at lower temperatures, direct correlations of H and σ_C for the same body versus temperature, and the impacts of test variables such as strain rate [42]. Consider first his Al_2O_3 data, which shows more extensive and complex changes, all of which are of some uncertainty in their origin but appear real and some related to other effects. Thus the initial increase in H to ~ 200°C may reflect water desorption effects as shown by Westbrook's results [11,14] of the previous section. The apparent H minimum at ~ 250°C probably correlates with the more definitive σ_C minimum at ~ 200°C, especially since the σ_C minimum shifts in temperature and magnitude with strain rate (Fig. 7.13), which increases its similarity to the larger, anomalous tensile strength minimum (Fig. 6.12). This correlation is further suggested by the fact that this minimum appears to be due to twinning, whose presence has been activated in (but not necessarily restricted to) compressive testing. Thus Lankford attributed the compressive strength peak at ~ 300°C to enhanced twinning [27]. This is not necessarily inconsistent with twinning being the cause of the σ_C minima at modestly lower T, since the temperatures of these minimum and maximum strengths are dependent on test conditions, especially strain rate (Fig. 7.13), and Lankford notes that twin thickness increases substantially as T increases over this range {42–45]. This increased twin thickness could substan-

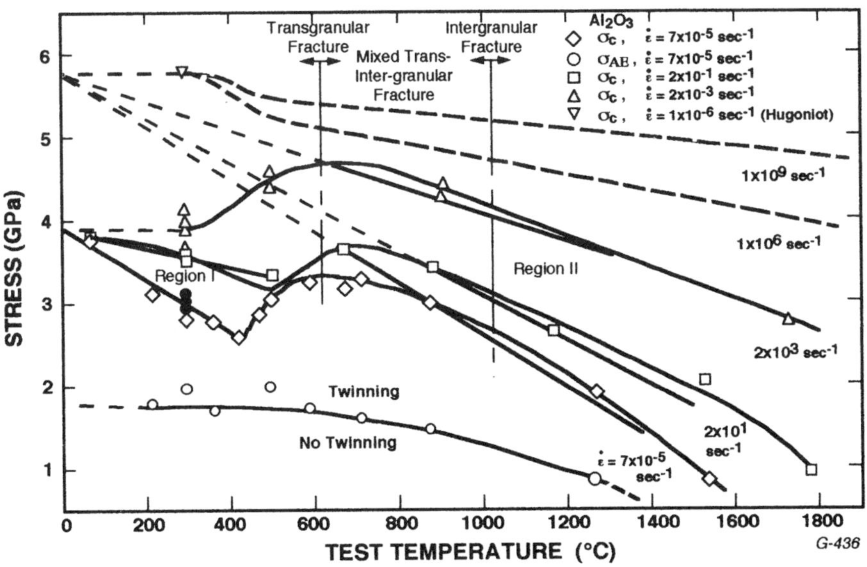

FIGURE 7.13 Compressive stress versus test temperature for a dense polycrystalline Al_2O_3 (G ~ 25 μm) for a wide range of strain rates. Note the designation of the general changes in fracture mode and in twinning and related acoustic emission. (From Ref. 42, published with permission of the *Journal of Materials Science.*)

tially change its effects as indicated. The differences between the *H* and σ_C minima may reflect differences in stress state, i.e. much of the stress around an indent is ~ hydrostatic, while that in compression is uniaxial, which is more effective for activating twinning.

Lankford [46,47] has also reported on the *T* dependence of compressive strengths of a Y-PSZ crystal and a commercial Y-TZP (fine grain) and Mg-PSZ (larger grain) to 800°C, with some observations on mechanisms, strain rate, orientation, and grain size effects. Thus crystals with 5 w/o Y_2O_3 stressed along <100> and <123> directions showed both differences in compressive ultimate strengths and deformation behavior at 22°C as well as in changes at higher *T*, e.g. 700°C. It was concluded that single crystals, which had ultimate strengths similar or > the polycrystals, reflected combinations of slip, transformation plasticity, and ferroelectric domain switching, while polycrystalline deformation at 22°C was by transformation plasticity and at ~ 800°C by forming unstable shear bands with flow via grain boundary sliding and cavitation.

Briefly turning to his SiC and Si_3N_4 data, the initially flat *H–T* trend may again reflect, at least partly, water desorption, while the increase in σ_C may reflect reduced brittleness, i.e. some limited plastic accommodation. Lankford observed that the rapid decreases in both hardness and compressive strength for SiC and Si_3N_4 corresponded to transitions from substantial transgranular to substantial intergranular fracture, as he also observed for Al_2O_3 [27].

Finally, note that work on superplastic flow in compressive deformation of ceramics of sufficiently fine *G* [48] indicates increased deformation of finer *G* bodies, at least qualitatively consistent with the *G* dependence, e.g. Eq. (6.2).

IV. GRAIN DEPENDENCE OF EROSION AND WEAR AS A FUNCTION OF TEMPERATURE

A basic change in erosion and its effects on residual strength due to impact of particles on ceramic surfaces as a function of temperature is increasing plasticity. This results in a basic change from brittle to ductile erosion, which varies as a function of particulate and target character, and especially particle velocity, hence strain rate. As noted earlier, this transition changes the maximum erosion from particle impacts normal to the surface for brittle erosion to an angle ~ 30° from the surface for ductile erosion [1,2].

Very few tests have been conducted as a function of temperature or at any elevated temperature, and there is thus little or no information on grain structure dependence. However, tests by Shockey et al. [49] on normal impact effects on dense Si_3N_4 (NC132, $G \sim 1$ μm) with either steel or WC spheres 2.4 mm dia. at 20 and 1400°C at velocities of ~ 20 to 200 m/sec indicate changes and complexities that can occur at higher temperatures. While the WC spheres produced elastic fracture (ring and cone cracks) at room temperature, they produced elastic-plastic fracture (plastic impressions and radial cracks) at 1400°C. The type and extent of impact damage from WC spheres at 1400°C appeared to be more deleterious than at 20°C under equivalent impact. On the other hand there was no change in the fracture pattern for steel sphere impact at the two temperatures, showing the importance of the impacting material and its character versus that of the target. Though the effects of a cold particle hitting a hot surface is unknown (but is often a real engineering issue), these limited tests indicate complexities even in the absence of varying target microstructures, indicating the need for substantial further study.

The situation with regard to wear is very similar to that for erosion. Thus dry sliding wear tests of dense sintered Al_2O_3 (G ~ 8 μm) plates against themselves of Xiao et al. [50] at 800–1200°C at velocities of 0.002 to 0.2 m/s under applied loads of 107–320 N giving nominal contact pressures of 0.4 to 1.2 MPa showed an important added complication that occurs at high temperatures. This was the formation of a very fine (e.g. nm scale) grain layer within the wear track with the thickness of the layer varying inversely, while the grain size in the layer increased, with the test temperature, e.g. thicknesses of several and ~ 1 μm and G ~ 0.1 and 1 μm respectively for 800 and 1200°C. They concluded that this layer formed by dynamic recrystallization. While the effects of the starting body grain structure on such recrystallization is unknown, it serves as a signal of added complexity that needs to be addressed in the complexities of wear at elevated temperatures.

V. DISCUSSION

A. Effects of Temperature and Elastic Anisotropy

The set of factors that may affect properties of this chapter are mainly those of the previous chapters, except that their temperature dependence must also be considered, as was partly done in the last chapter. Thus the temperature dependence of the various modes of plastic deformation, i.e. slip, twinning, grain boundary sliding, and diffusive creep, must be considered along with their strain rate dependences. A key example of this appears to be the probable contribution of twinning in Al_2O_3 to the significant flexure and tensile strength minima as discussed in Chap. 6 (e.g. Fig. 6.12). As noted earlier, the similar minima seen for H and σ_C in the present chapter (Fig. 7.6A) are probably due to the same cause, the differences arising from the differences in stress states and strain rates. The first of two important factors in the probable twinning effects in Al_2O_3 is the rapid reduction in the stress for twinning and then the thickening of the twins as T increases. The former appears to lead to the significant tensile strength and lesser compressive strength and especially hardness decreases as T increases, while the latter appears to be a factor in the subsequent property increases. The second factor is the indicated G dependence of these minima, i.e. the indicated diminution and then disappearance of them at finer G in flexure, which should be checked in hardness and compressive testing. Other more general changes in plastic deformation with increasing temperature are first the general increase in the extent of plastic deformation and the related reduction of stresses to activate it and second the changes in the anisotropy of such deformation, e.g. as revealed by hardness anisotropy and related wear anisotropy. However, such anisotropy clearly does not simply continuously decrease with increasing T; instead it often varies significantly (e.g. Figs. 7.1 and 7.2). While effects of this anisotropy are probably mitigated some by the increased plasticity as T increases, the issue of possible effects of variations in plastic anisotropy with T on other properties should not be neglected.

Two other anisotropies that need to be considered are TEA and EA along with their interactions with each other and with other factors such as anisotropies of plasticity and of single crystal (hence grain) fracture toughness. TEA stresses continuously decrease as T increases, going to zero at the stress relief or fabrication temperatures, and their effect becomes increasingly mitigated by increased plasticity. (Note that there can be some increasing TEA stress as T increases above the fabrication temperatures, where such temperatures are below stress relief temperatures, but this is limited to special low-temperature fabrication.) Further, effects of TEA stresses may be more limited in their effects on properties of this chapter, since they typically do not reflect failure from a single weakest link such as for tensile strength where TEA probably has more effect, as discussed in Chap. 6. However, TEA still probably plays an important role, e.g. in intergranular fracture in wear.

Turning to EA, its effects are probably more complex and pervasive, the latter arising from the fact that EA occurs in essentially all crystalline materials, frequently being as substantial or more in cubic versus noncubic materials, while TEA exists in only noncubic materials [51–53]. For cubic materials a common measure of EA is A*:

$$A* = \frac{3(A-1)^2}{3(A-1)^2 + 25A} \qquad \left(A = \frac{2C_{44}}{C_{11} - C_{12}} \right) \tag{7.1}$$

where the C_{ij} are the crystalline elastic constants and $A*$ is usually given as a percentage [51,52]. For noncubic crystals two expressions are required to define EA, but the one for $A*$ shear is most closely related to the above expression. Among cubic ceramics, ZrO_2, UO_2, $MgAl_2O_4$, β-SiC, ZnS, and ZnSe have high EA (e.g. 5–10%, which means that the ratios of maximum to minimum Young's moduli are ~ 1.5–2), i.e. $A*$ of ~ 8% means that the ratio of maximum to minimum Young's modulus is 2 and that $A*$ ~ 20% corresponds to a ratio of 3.

The complexity of EA effects arises from several sources. One is that the local microstructural stresses due to EA depend directly on the EA and the applied stress and hence generate the highest stresses at and near stress concentrations such as crack tips. The second reason is that EA can increase or decrease substantially, or not change much, as *T* increases, depending on the material (Fig. 7.14) (and possibly on the temperature range). This behavior is in contrast to that of TEA, which decreases with increasing *T*, as was noted above, and whose stresses are independent of, and hence additive with, the applied stresses. The effects of TEA on properties clearly depend on grain size (*G*), and some on shape and orientation relative to the applied stress, as EA effects also appear to, but the latter depend on grain shape and orientation of grain elongation relative to the stress axis as a function of the stress character per modeling by Hasselman [53]. Another general complication for all the various anisotropies is that they can be interactive, but their dependences of crystal orientation, while often the same or similar, can differ substantially.

EA was indicated as a contributor to intergranular fracture (e.g. Fig. 2.3), especially at very large *G*, SCG, possibly some *G* dependence of fracture en-

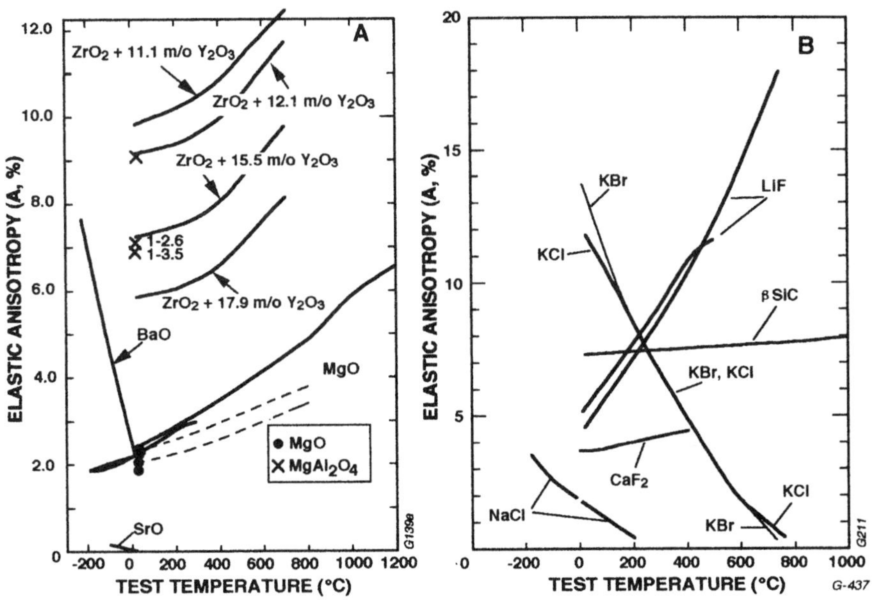

FIGURE 7.14 Sample plots of EA versus test temperature for (A) some oxides, with some compositional effects shown for some ZrO_2 and $MgAl_2O_4$ bodies (i.e. the latter designated in MgO/Al_2O_3 ratios) and (B) some halides and SiC. See also note on p. 253 of this book. (From Ref. 51, published with permission of the *Journal of Materials Science*.)

ergy/toughness–G dependence, hardness related cracking (Chap. 4, Sec. II.D), and related phenomena at room temperature. The differing trends of EA with T for different materials prevent general guidance on its effects by extrapolating those indicated at room temperature to higher temperatures, since the extrapolations would often be different, and commonly uncertain or unknown. Further, there are other factors increasing intergranular fracture as T increases besides possible increases in EA. However, consideration of specific materials where the EA–T dependence is known is at least suggestive, especially where it can be compared with trends for other similar materials but with different EA.

Thus the high EA of ZrO_2 bodies at modest T and its increases with increasing T (Fig. 7.14) suggest that it may be a factor in the transition from trans- to intergranular fracture at very modest T (e.g. Fig. 2.5, Chap. 6). Similarly, such a fracture mode transition at higher T in MgO has been suggested as reflecting its increasing EA with increasing T [51]. However, again other mechanisms may be involved, e.g. as indicated by decreases in E of ZrO_2, as shown in Figure 6.18, e.g. this would presumably not occur due to EA, unless it was causing microcracking (and then only in tests with substantial applied stress, most likely static versus dynamic modulus measurements), but it may be due to effects of lattice defect structures formed.

B. Mechanisms, Comparison with Tensile Strength and Self-Consistency

The temperature dependence of properties provides an important opportunity to corroborate mechanisms by comparing expected with observed temperature dependences of the same and related properties, i.e. evaluating self-consistency. Thus comparison of the temperature dependence of microplastically controlled compressive strength with actual data and with the temperature dependence of hardness is of value, as is comparison of the G and T dependences of compressive and tensile strengths. While data is limited, such comparisons are suggestive, as will be summarized below.

Comparison of measured or extrapolated (i.e. to $G = \infty$) single crystal compressive strengths, or preferably both, supports the concept that much compressive failure, even at room temperature, is controlled by microplastic processes of a more general nature than may be involved in some tensile failure. Thus values for MgO crystals of 50–120 MPa for tensile strength and 100–400 MPa for compressive strengths at ~ 22°C (Figs. 3.5,5.3) are consistent with higher stresses activating more deformations systems in compression. On the other hand, such values are respectively ~ 20 and ~ 60 MPa at 1300°C, which is at least approximately consistent with expected temperature reductions of stresses for slip. Also, where slip is the mechanism of both tensile and compressive failure, large differences in their strength–$G^{-1/2}$ slopes would not be expected (at least until grain boundary failure becomes important, as will be discussed below). Thus while there is considerable variation in the room temperature tensile data giving slopes of ~ 4–12 MPa·cm$^{1/2}$, that for compression at 22°C falls in the lower half of this range, i.e. reasonably consistent, and at 1300°C the slopes for both are ~ 4–5 MPa·cm$^{1/2}$. The even more limited data for UO_2 appears consistent with such trends, i.e. $G = \infty$/single crystal values of 700 and 60 MPa respectively in compression and tension at 22°C and slopes for both of ~ 2 MPa·cm$^{1/2}$, and no significant change for tensile values at 1000°C, thus supporting possible microplastic control of room temperature tensile strength of high-quality UO_2.

Comparison of $G = \infty$/single crystal values for compressive and tensile strengths of materials with established flaw initiated failure is less meaningful, since the two values reflect differing mechanisms. Similarly there are uncertainties in comparing their strength–$G^{-1/2}$ slopes. However, comparison of these values for compressive strengths between different materials and temperatures, though data is again very limited, is suggestive. Thus compressive data for TiB_2 giving extrapolated crystal values of only 500 MPa (Fig. 5.1) versus 3.5 GPa for sapphire (Fig. 5.1) and still ~ 450 MPa for TiB_2 at 1750°C (Fig. 7.12) all indicate that the earlier TiB_2 data at ~ 22°C is substantially low, e.g. due to parasitic testing stresses.

Finally, the comparison of Al_2O_3 and ice tensile and compressive strength data at higher temperatures is also suggestive, e.g. Figure 7.8 shows a common single crystal value at 1600°C. This is reasonable, since anisotropy of plastic flow generally decreases with increasing temperature, i.e. the yield stresses for more difficult to activate slip generally decrease faster than the average yield stress. Thus at higher temperatures the yield stress for the easiest and hardest slip systems will approach one another, as thus will the stresses for single crystal failure in tension versus compression, i.e. similar to ductile metals. However, the strength–$G^{-1/2}$ slope for compressive failure is higher than for tensile failure (e.g. a ratio of ~ 30). This may reflect greater ease and propensity for polycrystalline failure via grain boundary sliding mechanisms in tension than in compression, i.e. consistent with, and probably a precursor to, different types and character of creep failure in compression and tension.

VI. SUMMARY AND CONCLUSIONS

There is very little data on the G dependence of H as a function of T, leaving much of such dependence to be implied by H–G relations at 22°C and their extrapolation from H–T data for a single polycrystalline body (often of at best uncertain G) and for single crystals. There is reasonable data for the latter providing a data base to estimate limits of H anisotropy as a function of T for some important ceramics. Tests of both single- and polycrystals showed lower H due to adsorbed H_2O, and resultant reduced increases in H as T is increased to a few hundred degrees, which is a widely neglected factor in the *T* dependence of *H*. A more significant change in the *H–T* trends occurred in a medium *G*, dense Al_2O_3, namely an *H* minimum at ~ 200–300°C (Fig. 7.6A), which appears to correlate, with similar minima of increasing severity in respectively compressive and tensile (Figs. 6.12,7.6A) strengths. Beyond this, there are often changes in, e.g. inflections in, *H–T* relations observed in both single- and polycrystals as *T* increases, indicating changes in plastic flow, with differences between single- and polycrystals indicating differences in mechanisms that may also be dependent on *G*. Both *H* measurements on different crystal planes and of *H* anisotropy on individual planes as a function of *T* show that *H* anisotropy does not change in a simple similar pattern as a function of *T*. Data on ice shows substantial *H* anisotropy to ≥ 0.98 T_m. The very limited data on indent cracking shows this continuing in a medium *G*, dense Al_2O_3 to at least 1000°C, with increased intergranular fracture, as is common for other fracture as a function of *T*. These observations and other property correlations, especially with compressive strength, indicate mainly gradual if any changes in *H–G* relations as *T* changes, except where phase transformations occur, which can cause sharp, often large, *H* changes. However, while some *G* dependence of *H* is thus expected at higher *T* similar to that at lower *T*, the overall reduction of *H* with increasing *T* implies decreases differentiation of *H* as a function of *G* as *T* increases.

In contrast to *H* data, there is almost no single crystal data, but there is reasonable data on the *G* dependence of σ_C as a function of *T*, showing a normal Petch-type dependence to substantial *T*, e.g. > 0.96 T_m in ice and > 0.65 T_m in TiB_2. While compressive failure by macroscopic fracture generally ceases as macroscopic plastic deformation begins, there can often be continued, but diminishing, macrocracking as *T* increases. Correlation between *H* and σ_C continues as a function of *T*, including the occurrence of a strain-rate-dependent (Fig. 7.6A, 7.13) σ_C minimum at ~ 400°C that appears to correlate with one for *H* (Fig. 7.6A) and for tensile strength (Fig. 6.12). However, the H–σ_C correlation, i.e. the H/σ_C ratio, is more complex, e.g. involving a changing constraint factor, as shown by direct *H* and σ_C measurements on the same bodies as a function of *T* (Fig. 7.6). Extrapolations of tensile and compressive strengths to $G^{-1/2} = 0$ (i.e. implied or actual single crystal values) often become similar or the same as *T* further increases. This also indicates basic changes that probably underlie changing H/σ_C ratios, since much higher compressive versus tensile strengths at lower *T* are attributed to σ_C being driven by highly constrained local plastic deformation, while tensile strength is predominantly controlled by brittle failure from the most severe flaw. Thus while the H/σ_C ratio probably changes due to differing measurements, e.g. in terms of strain rates and environmental sensitivities, and changes in the extent and character of plastic flow, there is some useful, but variable, relation.

There is very little data on high-temperature erosion or wear, let alone on their G dependence. However, what limited data does exist clearly indicates that complex changes are probable in such properties as a function of T and possibly also G as T increases. Increased bonding, reaction, and deformation are all factors in these changes.

Besides limited or no data on the grain size dependence of these properties, there is much less, if any, data on effects of grain shape and orientation (though single crystal data does indicate the limits of the latter for H). Further, the properties of this and previous chapters are probably often affected by other properties and factors, e.g. TEA, EA, and plastic anisotropy, but limited data and probable complex interactions cloud such interactions, which appear to be an important area for further research. However, the effects of at least some of these other factors are also apparently complicated by their dependence on grain shape and orientation. Note that while TEA (which exists only in noncubic materials) decreases with increasing T, EA may increase substantially, decrease, or change little, depending on the material.

Finally, the first of two overall observations is that the transition from a Hall–Petch dependence with strengths increasing as G decreases to creep and related deformation with strengths decreasing with decreasing G as T increases is often pushed to high values of relative T_m for many materials at normal or higher strain rates. Studies of both creep and superplasticity support such changes, but many more details are needed. These opposite dependences on G pose serious challenges for processing, designing, and using ceramics at high temperatures. Second, the increased, often dominant, role of grain boundary phases, even in very limited amounts, also poses important challenges. These phases also pose serious challenges for processing, designing, and using ceramics at high temperatures.

NOTE ADDED IN PROOFS

Since completion of this chapter Palko et al. [54] have reported that the elastic anisotropy of yttria is ~ 1% at room temperature and does not change measurably up to limits of their testing of 1200°C. This is consistent with the suggestion that the absence of a hardness minimum for yttria (Fig. 4.6) and high transgranular fracture in crack propagation tests reflect low EA.

REFERENCES

1. A. W. Ruff and S. M. Wiederhorn. Erosion of Solid Particle Impact. Nat. Bureau of Standards Report, NSBIR 78-1575, 1/ 1979.
2. A. W. Ruff and L. K. Ives. Measurement of Solid Particle Velocity in Erosive Wear. Wear 35:195–199, 1975.
3. T. R. Butkovich. Hardness of Single Ice Crystals. Research Paper 9, Final Report for Project SIB 53-9, Corps of Engineers, US Army Snow, Ice, and Permafrost Research Establishment, Wilmette, IL, 1/1954.

4. E. L. Offenbacher and I. C. Roselman. Hardness Anisotropy of Single crystals of Ice Ih. Nature Physical Science 234:112, 12/1971.
5. A. G. Atkins. High-Temperature Hardness and Creep. The Science of Hardness Testing and Its Research Applications (J. H. Westbrook and H. Conrad, eds.). Am. Soc. for Metals, Metals Park, OH, 1973, pp. 223–240.
6. L. Bsenko and T. Lundström. The High-Temperature Hardness of ZrB_2 and HfB_2. J. Less Com. Met. 34:273–278, 1974.
7. K. Nakano, H. Matubara, and T. Imura. High Temperature Hardness of Titanium Diboride Single Crystal. Jpn. J. Appl. Phys. 13(6):1005–1006, 1974.
8. K. Niihara. Slip Systems and Plastic Deformation of Silicon Carbide Single Crystals at High Temperatures. J. Less Com. Met. 65:155–166, 1979.
9. S. Fujita, K. Maeda, and S. Hyodo. Anisotropy of High-Temperature Hardness in 6H Silicon Carbide. J. Mat. Sci. Lett. 5:450–452, 1986.
10. T. Hirai and K. Niihara. Hot Hardness of SiC Single Crystal. J. Mat. Sci. Lett. 14:2253–2255, 1979.
11. J. H. Westbrook and P. J. Jorgensen. Effects of Water Desorption on Indentation Microhardness Anisotropy in Minerals. Am. Min. 53:1899–1909, 11–12/1968.
12. M. O. Guillou, J. L. Henshall, R. M. Hooper, and G. M. Carter. Indentation Hardness and Fracture in Single Crystal Magnesia, Zirconia, and Silicon Carbide. Special Ceramics 9; Proc. Brit. Cer. Soc. 49:191–202, 1992.
13. A. G. Atkins and D. Tabor. Mutual Indentation Hardness of Single-Crystal Magnesium Oxide at High Temperatures. J. Am. Cer. Soc. 50(4):195–198, 1967.
14. J. H. Westbrook. The Temperature Dependence of Hardness of Some Common Oxides. Rev. Hautes Temp. Réfract. 3:47–57, 1966.
15. W. Kollenberg. Plastic Deformation of Al_2O_3 Single Crystals by Indentation at Temperatures up to 750°C. J. Mat. Sci. 23:3321–3325, 1988.
16 . C. P. Alpert, H. M. Chan, S. J. Bennison, and B. R. Lawn. Temperature Dependence of Hardness of Alumina-Based Ceramics. J. Am. Cer. Soc. 71(8):C-371–373, 1988.
17. W. Kollenberg. Microhardness Measurement on Haematite Crystals at Temperatures up to 900°C. J. Mat. Sci. 21:4310–4314, 1986.
18. W. Kollenberg and H. Schneider. Microhardness of Mullite at Temperatures to 1000°C. J. Am. Cer. Soc. 72(9):1739–1740, 1989.
19. Y. Kumashiro, Y. Nagai, and H. Katō. The Vickers Micro-Hardness of NbC, ZrC, and TaC Single Crystals up to 1500°C. J. Mat. Sci. Lett. 1:49–52, 1982.
20. D. L. Kohlstedt. The Temperature Dependence of Microhardness of the Transition-Metal Carbides. J. Mat. Sci. 8:777–786, 1973.
21. G. R. Sawyer, P. M. Sargent, and T. F. Page. Microhardness Anisotropy of Silicon Carbide. J. Mat. Sci. 15:1001–1013, 1980. See also M. G. S. Naylor and T. F. Page. Microhardness, Friction and Wear of SiC and Si_3N_4 Materials as a Function of Load, Temperature and Environment. First Annual Tech. Rept. US Army Grant DA-ERO-78-G-010, 10/1979.
22. J. H. Westbrook. Temperature Dependence of Strength and Brittleness of Some Quartz Structures. J. Am. Cer. Soc. 41(11):433–440, 1958.
23. I. A. Bairamashvili, G. I. Kalandadze, A. M. Eristavi, J. Sh. Jobava, V. V. Chotulidi,

and Yu. I. Saloev. An Investigation of the Physicomechanical Properties of B_6O and SiB_6. J. Less Com. Met. 67:455–461, 1979.
24. D. R. Mumm, K. T. Faber, M. D. Drory, and C. F. Gardinier. High-Temperature Hardness of Chemically Vapor-Deposited Diamond. J. Am. Cer. Soc. 76(1):238–240, 1993.
25. D. J. Brown and J. J. Stobo. Properties of Uranium Monocarbide. Trans. J. Brit. Cer. Soc. 62:177–182, 1963.
26. K. Niihara. Mechanical Properties of Chemically Vapor Deposited Nonoxide Ceramics. Am. Cer. Soc. Bull. 63(9):1160–1163, 1984.
27. J. Lankford. Comparative Study of the Temperature Dependence of Hardness and Compressive Strength in Ceramics. J. Mat. Sci. 18:1666–1674, 1983.
28. R. W. Rice. The Compressive Strength of Ceramics. Ceramics in Severe Environments, Materials Science Research 5 (W. W. Kriegel and H. Palmour III, eds.). Plenum Press, New York, 1971, pp. 195–227.
29. R. W. Rice. Microstructure Dependence of Mechanical Behavior of Ceramics. Treatise on Materials Science and Technology II (R. C. McCrone, ed.). Academic Press, New York, 1977, pp. 199–238.
30. J. Lankford. The Compressive Strength of Strong Ceramics: Microplasticity Versus Microfracture. J. Hard Mat. 2 (1–2):55–77, 1991.
31. P. F. Becher. Deformation Behavior of Alumina at Elevated Temperatures. Mat. Sci. Res. 5:315–329, 1971.
32. P. R. V. Evans. Studies of the Brittle Behavior of Ceramic Materials. N. A. Weil, ed. Armour Res. Foundation of Illinois Inst. Tech. Report ASD-TR-61-628, Part II for Air Force Contract AF33(616)-7465, 1963, pp. 164–202.
33. R. W. Rice. Porosity of Ceramics. Marcel Dekker, New York, 1998.
34. E. M. Schulson. The Brittle Compressive Fracture of Ice. Acta Metall. Mater. 38(10):1963–1976, 1990.
35. L. W. Gold. The Process of Failure of Columnar-Grained Ice. Phil. Mag. 26:311–328, 1972.
36. S. M. Copley and J. A. Pask. Deformation of Polycrystalline MgO at Elevated Temperatures. J. Am. Cer. Soc. 48(12):636–642, 1965.
37. T. G. Langdon and J. A. Pask. Effect of Microstructure on Deformation of Polycrystalline MgO. J. Am. Cer. Soc. 54(5):240–246, 1971.
38. Ö. Ünal and M. Akinc. Compressive Properties of Yttrium Oxide. J. Am. Cer. Soc. 79(3):805–808, 1996.
39. R. B. Day and R. J. Stokes. Mechanical Behavior of Polycrystalline Magnesium Oxide at High Temperatures. J. Am. Cer. Soc. 49(7):345–354, 1966.
40. D. M. Marsh. Plastic Flow in Glass. Proc. Roy. Soc. A 279:420–435, 1964.
41. J. R. Ramberg and W. S. Williams. High Temperature Deformation of Titanium Diboride. J. Mat. Sci. 22:1815–1826, 1987.
42. J. Lankford. Temperature-Strain Rate Dependence of Compressive Strength and Damage Mechanisms in Aluminum Oxide. J. Mat. Sci.16:1567–1578, 1981.
43. J. Lankford. Mechanisms Responsible for Strain-Rate-Dependent Compressive Strength in Ceramic Materials. J. Am. Cer. Soc. 64(2):C-33–34, 1981.
44. J. Lankford. High Strain-Rate-Dependent Compression and Plastic Flow of Ceramics. J. Mat. Sci. 15:745–750, 1996.

45. J. Lankford. Uniaxial Compressive Damage in α-SiC at Low Homologous Temperatures. J. Am. Cer. Soc. 62(5–6):310–312, 1979.
46. J. Lankford. Plastic Deformation of Partially Stabilized Zirconia. J. Am. Cer. Soc. 66(11):C-212–213, 1983.
47. J. Lankford. Deformation Mechanisms in Yttria-Stabilized Zirconia. J. Mat. Sci. 23:4144–4156, 1988.
48. Y. Maehara and T. G. Langdon. Review, Superplasticity in Ceramics. J. Mat. Sci. 25:2275–2286, 1990.
49. D. A. Shockey, D. C. Erlich, and K. A. Dao. Particle Impact Damage in Silicon Nitride at 1400°C. J. Mat. Sci. 16:477–482, 1981.
50. H. Xiao, T. Senda, and E. Yasuda. Dynamic Recrystallization During Sliding Wear of Alumina at Elevated Temperatures. J. Am. Cer. Soc. 79(120):3242–3249, 1996.
51. R. W. Rice. Possible Effects of Elastic Anisotropy on Mechanical Properties of Ceramics. J. Mat. Sci. Lett. 13:1261–1266, 1994.
52. D. H. Chung and W. R. Buessem. The Elastic Anisotropy of Crystals. Proc. International Symposium 2 (F. W. Vahldiek and S. A. Mersol, eds.). Plenum Press, New York, 1968, pp. 217–245.
53. D. P. H. Hasselman. Single Crystal Elastic Anisotropy and the Mechanical Behavior of Polycrystalline Brittle Refractory Materials. Anisotropy in Single-Crystal Refractory Compounds (F. W. Vahldiek and S. A. Merson, eds.). Plenum Press, New York, 1968, pp. 247–265.
54. J. W. Palko, S. Sinogeikin, A. Sayir, W. M. Kriven, and J. D. Bass. The Single Crystal Elasticity of Yttria to High Temperature, submitted to J. Appl. Phy.

8

Particle (and Grain) Effects on Elastic Properties, Crack Propagation, and Fracture Toughness of Ceramic Composites at ~ 22°C

I. INTRODUCTION

Chaps. 1 introduced grain and particle parameters that are important for mechanical properties, while Chapters 2–7 have addressed in detail the effects of primarily grain size and secondarily shape and orientation on nominally monolithic, i.e. single phase, ceramics. This chapter begins a similar review of primarily particle effects (i.e. of the dispersed phase) on mechanical properties by addressing elastic moduli, and crack propagation and fracture toughness of ceramic composites. Some observations are also made on the effects of the matrix grain size or other parameters, e.g. for noncrystalline or single crystal matrices on the properties covered in this chapter and subsequent ones.

A basic similarity of monolithic ceramics and ceramic composites is that properties affected by grain parameters in monolithic ceramics are typically also affected first by the particle parameters in composites, and also generally some by the grain parameters of the matrix. Another similarity is that many of the problems and uncertainties of crack propagation and fracture toughness and the relation of these to tensile strength of monolithic ceramics also occur for many ceramic composites. Important differences to note between the behavior of monolithic and composite ceramics are that while grain parameters typically have limited or no effect on elastic properties, thermal expansion, and electrical and thermal conductivity, these properties can be dependent on the particle parameters of the dispersed phase in ceramic composites. Again, as noted in Chapter 1, the term particle, while often used in the specific sense for composites of a matrix with dispersed (single crystal or polycrystalline) particles, is also used to include platelet, whisker, and (mainly short) fibers in composites, which are also addressed, though the latter only a very limited amount for comparative purposes.

It should be noted that it is only feasible to review selected aspects of papers in this extensive area of research. The goal is to provide a substantial summary of much of the pertinent data, and suitable background, on the mechanical behavior of ceramic composites with a focus on microstructural control of, or impact on, mechanical properties. This focus is intended to aid understanding of such composites and their design and processing from both scientific and engineering standpoints. Further note that it is important that material of this chapter on crack propagation be compared to that particularly of Chapter 9 on tensile strength, and secondarily to that of Chapters 10 and 11 on other mechanical properties and elevated temperature behavior. While this was also the case for monolithic ceramics in Chapters 2–5, it is even more important here due to less extensive microstructural evaluation of both crack propagation and strength behavior and especially of both on the same composite.

II. THEORETICAL AND CONCEPTUAL BACKGROUND

A. Elastic Properties

Since elastic properties of the two or more phases in a composite are seldom identical, and may in fact be significantly different, the elastic properties of such composites are very much a function of their composition and secondly of their microstructure. Elastic properties of dense monolithic ceramics, though having their complexities, are overall simpler since there is only one phase and there is no dependence on grain size [unless this correlates with microcracking, e.g. per Eq. (2.4)]. An important complexity of the elastic properties of monolithic ceramics shared by composites is that the elastic properties of both depend on the degree of preferred orientation of the phases involved, but composites are more complex in having two or more phases as opposed to one whose orientation must be considered. Further, elastic properties of composites can depend on the shapes of the matrix and second phase particles over and above effects of these shapes on preferred orientations of each phase. Elastic properties of composites may also depend on grain and particle sizes, e.g. as both such sizes and shapes may affect contiguity of one or more of the phases, which may affect elastic properties of the composite.

There has been substantial development of models for the elastic properties of composites because of both the importance and the complexity of the subject, but a detailed review of this development is not conducted here. While there is no single expression or family of expressions adequately predicting elastic properties of all ceramic composites of interest, only key points, reviews, and useful expressions will be summarized here for two reasons. First, first-order predictions of elastic properties available are generally adequate relative to the other uncertainties and needs to understand microstructural dependences of the mechanical properties of composites of primary interest here. Second, neither the models nor the composite characterization are sufficiently detailed and accurate to address fully many of the more detailed aspects of the elastic behavior of ceramic composites.

Consider first-order predictions of elastic moduli, especially Young's modulus, of ceramic composites based on bounding techniques. These yield upper and lower limits for the elastic properties based on the assumptions made, which, while broad, are often for simplified idealized systems. Thus the simplest and widest bounds are obtained from a model based on slabs of two isotropic materials, which for stressing parallel with the plane of the slabs gives a rule of mixtures upper bound for Young's modulus (E_{UC}):

$$E_{UC} = \phi E_P + (1-\phi)E_M \tag{8.1}$$

where ϕ = the volume fraction second (e.g. particulate) phase, E_P = the Young's modulus of the second (e.g. particulate) phase, and E_M= the Young's modulus of the other (e.g. matrix) phase. (This equation is commonly a good approximation for the modulus of fiber composites in their linear elastic region for stressing parallel with the fibers [1].) The lower bound (E_{LC}) from such a model is obtained when the parallel slabs are stressed normal to their planes, giving

$$E_{LC} = (E_P\, E_M)\, [\phi E_M + (1-\phi)E_P]^{-1} \tag{8.2}$$

These expressions, especially Eq. (8.1), are often suitable for a first-order estimate of many composites, especially for those with constituents whose moduli do not differ substantially, e.g. by a few fold or less (see Fig. 8.11).

Models for tighter bounds have been derived, with that of Hashin and Shtrickman [2] being well known. More recently Ravichandran [3] has presented a model based on an idealized composite structure of a uniform simple dispersion of identical isotropic cubic particles in a dense surrounding isotropic matrix (i.e. so only a single unit cell needs to be considered). He obtained for the upper (E_{UC}) and lower (E_{LC}) bounds respectively

$$E_{UC} = \{[cE_PE_M + E_M^2](1+c)^2 - E_M^2 + E_PE_M\}[(cE_P + E_M)(1+c)^2]^{-1} \tag{8.3}$$

$$E_{LC} = \{[E_PE_M + E_M^2\,(1+c)^2 - E_M^2](1+c)\}[(E_P - E_M)c + E_M\,(1+c)^3]^{-1} \tag{8.4}$$

where

$$c = \phi^{-1/3} - 1 \tag{8.5}$$

Similar closed form expressions are given for bulk and shear moduli, as well as Poisson's ratio. Good agreement was shown with data, e.g. for WC-Co bodies. While there are other bound expressions, it will be shown later that limited trials of the above-noted expressions give reasonable results (e.g. Fig. 8.11), as does the obvious use of the average of Eqs. (8.3) and (8.4), which is suitable for most present purposes.

Modeling to give even more rigorous expressions rather than bounds (even if relatively close) has been extensively conducted, but while substantial progress has been made, there are still important issues because of the simplifications and idealizations generally required to treat such complex problems rigorously. Thus models typically assume an isotropic matrix with isotropic inclusions of a single, simple shape, typically of uniform size, and no specification of its spatial distribution other than generally being uniform. Many models assume explicitly or implicitly a dilute dispersion of second phase, i.e. so interactions between adjacent particles can be neglected. Shapes are most commonly assumed to be spherical, but spheroids (and hence in the extreme rods) as well as platelets and needles have been considered, which can be important [4-6].

Many of the more rigorous models, which often do not give simple closed form expressions [e.g. as in Eqs. (8.1) through (8.4)], fall into one of the three following categories: (1) differential, (2) generalized self-consistent (GSC), and (3) Mori–Tanaka (M–T) approaches to the problem. Such models often give only bulk modulus rigorously, and sometimes also shear modulus, but generally Young's modulus and Poisson's ratio, and sometimes also shear modulus, can only be estimated, e.g. via bounding techniques. Christensen [7] reviewed the applicability of the above three model types for spherical particles, showing that while all three generally agreed with each other and data to $\phi = 0.2$–0.5, there was significant divergence at higher ϕ values, with the GSC method giving the best agreement with data. The other two models varied substantially in their degree of agreement and disagreement with each other, the GSC results, and data. Note that the advantages of the GSC method for such composites were achieved only after improvements to the forerunner self-consistent method (SCM) were made, since the SCM was not always accurate over the full range of ϕ. Much of the ground work for the SCM was laid by Hashin [8].

Further note that the above models and most or all other models are of uncertain applicability as the size of the filler particles is no longer constant, and especially as their shape changes from spherical to more irregular or elongated shapes, though again there are special models for important idealized shapes such as discs, rods (fibers), and needles. An important problem for some particulate, whisker, and especially platelet ceramic composites is the combined issues of shape, elastic anisotropy, and varying (usually poorly characterized) degrees of preferred orientation. Finally note that many models are primarily or exclusively applicable to two-phase composites, while many composites contain three or more phases (e.g. concrete and some composites noted later). While a number of models can be modified to make estimates for such multiphase composites, there are a number of models derived for multiphase bodies, including some older models that may be useful, e.g. those of Paul [9], Cohen and Ishai [10], Kerner [11], and Budiansky [12].

B. Crack Propagation and Fracture Toughness

The nature of crack propagation and its relation to fracture toughness (hence also fracture energy) in ceramics, especially ceramic composites, have been reviewed by Rice [13–18]. These reviews form a basis for much of this review, which focuses more on conceptual as opposed to detailed quantitative theories. The latter can be useful, but except for fiber composites, which is a large subject, only noted here, there are so many uncertainties in many of the models and their applicability that quantitative agreement is most likely more fortuitous than real. The reader is referred to other reviews addressing quantitative modeling of various proposed toughening mechanisms [19–24].

Consider first conceptual models mostly for composites containing nominally equiaxed, dispersed isotropic particles, starting with the most extensively verified but more restricted mechanism of transformation toughening. This has been used mainly for the martensitic tetragonal to monoclinic crystal phase transformation of ZrO_2, which is also applicable to HfO_2, but little work has been done on the latter, and efforts to find transformations in other materials suitable for similar use have not resulted in significant successes. The effectiveness of this process stems from first the fact that the transformation is diffusionless (i.e. requires no diffusion of atoms), so it can occur rapidly, i.e. in response to crack tip stress effects. It arises secondly due to the unusual character of the transformation in that the lower temperature, monoclinic, phase is less dense than the intermediate phase tetragonal ZrO_2 structure. As a result of the opposite phase density trend relative to the normal temperature trends of phases (i.e. higher temperature phases are normally of lower density than their lower temperature counterparts), ZrO_2 particles trapped in a matrix may not be able to transform due to the matrix constraining the expansion required to transform the trapped ZrO_2 particles. Microcracking can be a result if the ZrO_2 particles are too large, or the matrix too compliant per Eq. (2.4), and thus be a factor in the resultant mechanical behavior. However, the most fundamental effect of transformation toughening is when trapped tetragonal ZrO_2 particles or grains transform due to crack tip stresses relaxing the matrix constraint on their transformation. The resultant transformation results in compressive strains around transformed ZrO_2 grains or particles due to the shape and (~ 5%) volume increase of the resultant monoclinic phase over the original metastable tetragonal phase. These compressive strains from the transformation zone that results around the crack tip (Fig. 8.1A)

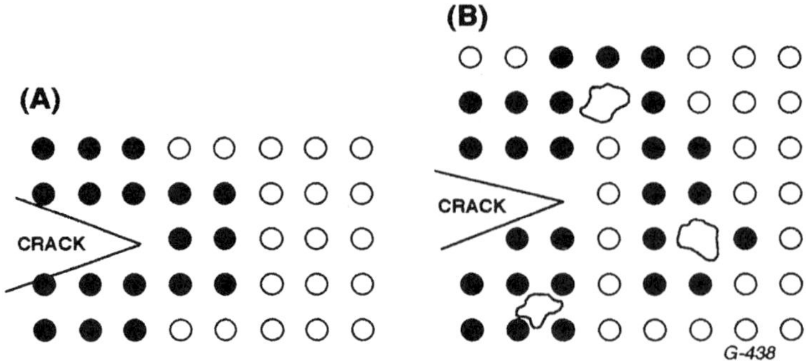

FIGURE 8.1 Transformation toughening from a dispersion of tetragonal ZrO_2 particles (open circles) in a matrix around a crack (normal to the page). Schematic of (A) the zone of tetragonal ZrO_2 particles transformed to the monoclinic phase (solid circles) not only at the crack tip but also along the crack faces (giving a crack wake zone similar to Fig. 8.2B), and (B) a speculated addition of microcracking sources to possibly extend the net transformation zone as discussed in the text. (From Refs. 13, 14. Published with permission of Ceramic Engineering and Science Proceedings.)

and along its faces (similar to Fig. 8.2B) thus partially counteract or shield the crack from the normal stress levels, making it more difficult for it to propagate [19]. It has been suggested that it might be feasible to extend the physical scope of the transformation zone around the crack by introducing sources of microcracking around the crack tip, which could in turn induce transformation of additional ZrO_2 particles or grains to extend the net zone size [13,14] (Fig. 8.1B). Some probable demonstration of this has been made, but more is needed.

Consider next microcracking, which as noted above can result from and accompany ZrO_2 transformation. This can also be an important factor in mechanical properties of monolithic ceramics with thermal expansion anisotropy [as a function of G and E per Eq. (2.4)], and possibly due to EA (Chap. 7, Sec. V.A), with resultant effects on thermal expansion and elastic properties, in addition to other mechanical properties. Ceramic composites with a dispersed phase or phases of differing thermal expansion from the matrix are also a source of microcracks, again per Eq. (2.4) and possible effects of elastic property differences. (Note that in either case, single crystal particles of noncubic structure may have more extreme effects than polycrystalline particles, since the crystalline anisotropy will often accentuate the particle–matrix property differences, especially if the matrix grain size is similar to or > the particle size.) Recall that the original concept of microcracking was for it to occur primarily in two lobes, one located above and one below the crack plane, both mostly somewhat ahead of the crack tip (Fig. 8.2A). It is now generally accepted that most microcracking occurs as a zone or sheath along the crack surfaces (Fig. 8.2B), i.e. similar to transformation toughening. Thus the effectiveness of this microcracking sheath around the crack surfaces, like the effects of transformation around a crack, are attributed to the strain expansion from microcracking and resultant compressive shielding of the crack tip from some of the tensile stresses driving its propagation. Based on the original modeling, it was predicted that the toughness increases would in turn increase as the distribution of microcrack size narrowed and approached the optimum size for microcracking, but how much impact these size effects have on the crack tip stress shielding has apparently not been addressed. Local concentration of microcracks, i.e. a designed heterogeneity of their spatial distribution by dispersing particles that can produce a higher density of microcracks than the matrix itself, has been proposed (e.g. Fig. 8.2C), which appears to have been demonstrated, as is discussed later.

Consider next crack deflection and branching, the former being one of the earlier nontransforming toughening mechanisms considered, which may be independent of each other and microcracking but can have various interrelations with one another. Modeling showed increased effect over that from increased

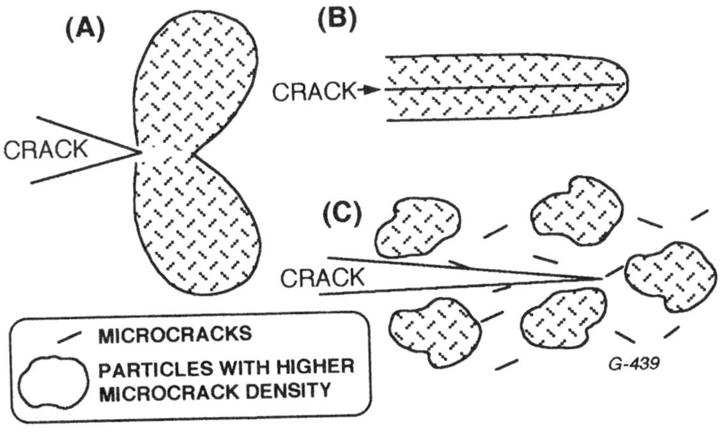

Figure 8.2 Schematic of microcrack toughening: (A) as originally proposed occurring in two lobes ahead of and above and below the crack, (B) primarily in the crack wake zones as more recently proposed and generally seen as much closer to actual occurrences, and (C) a proposed local concentration of microcracks to enhance their effectiveness by increasing their net concentration while limiting their opportunity for longer range linkage to enhance larger scale crack propagation. Views are normal to the crack plane. (C modified after Rice [13,14], published with permission of Ceramic Engineering and Science Proceedings.)

fracture area as from crack deflections from a plane path, and effects of the volume fraction of crack deflectors, their spacing and shape on toughening (Fig. 8.3). Certainly one, but not the only, way to obtain crack deflection is to have microcracking occur at and near the crack tip, e.g. similar to a proposed mechanism for intergranular fracture (Fig. 2.3). Another way is to introduce elongated particles, especially platelets with a highly preferred cleavage or preferred fracture surface, e.g. along its larger interfaces with the matrix.

A related, e.g. possibly more extreme, case of crack deflection is the line tension concept for toughening from crack pinning. This assumes that there are particles or other barriers to crack propagation that result in at least temporary pinning of the crack front at these points. Such pinning effects are commonly treated via line tension along the crack front, which, while raising some theoretical issues, has been used to yield quantitative relations for increases in fracture energy for idealized systems (Fig. 8.4A) [20,21]. Simple

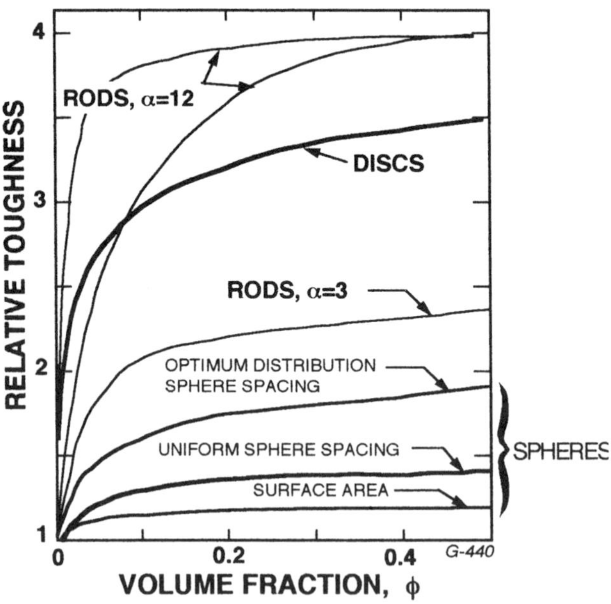

FIGURE 8.3 Summary plot of modeling results for increased fracture energy as a function of volume fraction of crack deflecting particles and their shape. Note progressively greater effects of rods (which increases with their aspect ratio,) versus spherical and platelet (discs), and increased effects of optimized distribution of spherical particles indicating possible benefits of this for rod and disc particles. (From Refs. 23,24,14. Published with permission of Ceramic Engineering and Science Proceedings.)

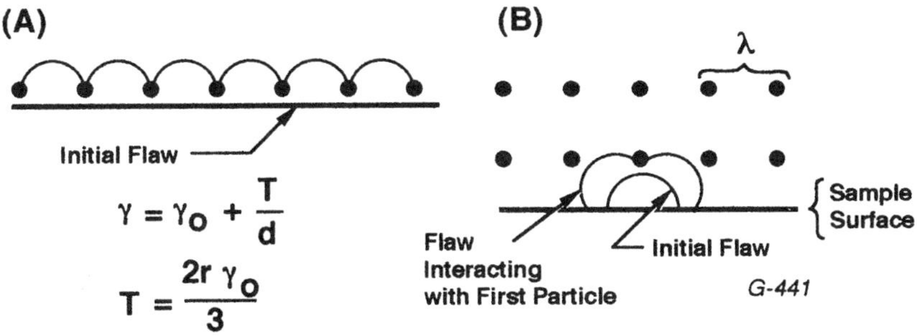

FIGURE 8.4 Schematic of (A) the basic line tension (T) model and increased fracture energy (γ), after Lange [20], and (B) diminished effects as the crack size approaches the pinning point separation. (After Rice [13], published with permission of Ceramic Engineering and Science Proceedings.)

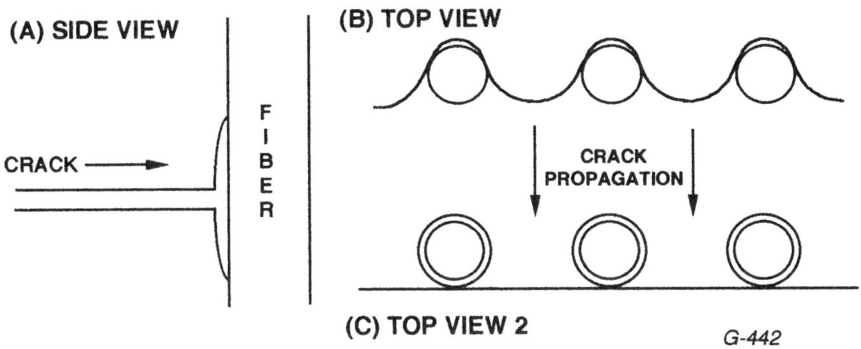

FIGURE 8.5 Schematic of the crack pinning in a fiber composite as a mechanism to aid fiber pullout. The side view (A) normal to the crack plane shows the crack held up at a fiber, while the top view 1 (B) (i.e. parallel with the crack plane) shows the crack pinned by three fibers in a row and top view 2 (C) shows the crack having advanced beyond the pinned fibers, leaving them with peripheral cracks that most likely enhance fiber pullout. (From Ref. 13. Published with permission of Ceramic Engineering and Science Proceedings.)

modification of this model indicates possible diminishing effects as crack sizes decrease to approach the pinning point spacing (Fig. 8.4B) [13]. Such modeling has also been extended to address anisotropy of shape of the pinning particles, but issues and limitations of resultant anisotropy appear not to have been fully evaluated [13,21]. Similar effects may occur with fiber composites (Fig. 8.5).

Crack branching, i.e. simply forming of one or more branch cracks along a crack front (Fig. 8.6), while clearly a possible source of increased toughness, has received little or no explicit modeling consideration. This appears to be due, at least in part, to the mistaken view that this is simply an extension of crack deflection. However, this view must be incorrect, since each branch may (and commonly does) have crack deflections. Thus while crack deflection may cause or be a factor in crack branching, and clearly complicates the quantatative evaluation of effects of branching, branching clearly results in energy dissipation and crack stress effects beyond those of a single crack alone with deflections. While crack branching may occur from natural crack bifurcations, it may also arise from crack deflections, whether or not they arise from microcracking, which may also lead directly to branching.

Crack wake bridging has become a widely cited mechanism of toughening in ceramic composites, as it has for monolithic ceramic bodies, based on ready observation of particles bridging the wake zone of cracks in composites (Fig. 8.7) or grains in monolithic ceramics. However, all the issues discussed in Chapter 2 regarding the implications of such observations with larger cracks propagated at limited velocities and then arrested for observations at the intersection of the large cracks with typically machined surfaces and their applicability to normally much smaller flaws controlling ceramic strengths apply here, as will be shown later. Again, the same uncertainties in the details of bridge formation apply, i.e. the extent to which some microcracking or crack branching initiation may occur at, just ahead of, or behind the crack tip to create bridging particle (grains) is also pertinent.

The last toughening mechanism of pullout is well recognized and established as the major mechanism in toughening of continuous fiber ceramic composites, which has been extensively analyzed [19], and whose validity is

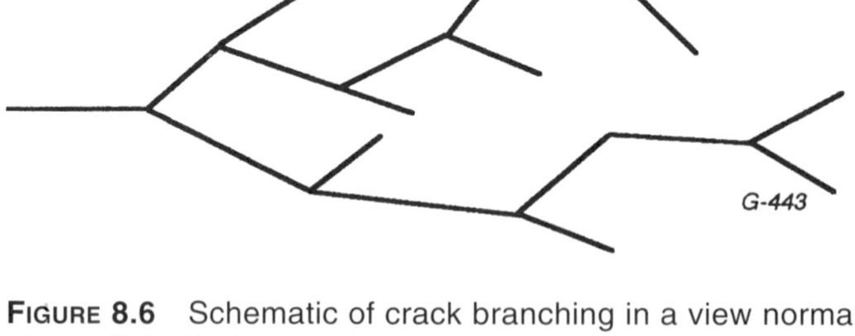

FIGURE 8.6 Schematic of crack branching in a view normal to the mean crack plane. Note that branching may occur naturally, e.g. due to differing orientations of preferred fracture planes at different positions along the crack front, due to other sources of crack deflections, or to microcracking, and can coexist with crack deflections and microcracking over and above that which may cause the crack branching.

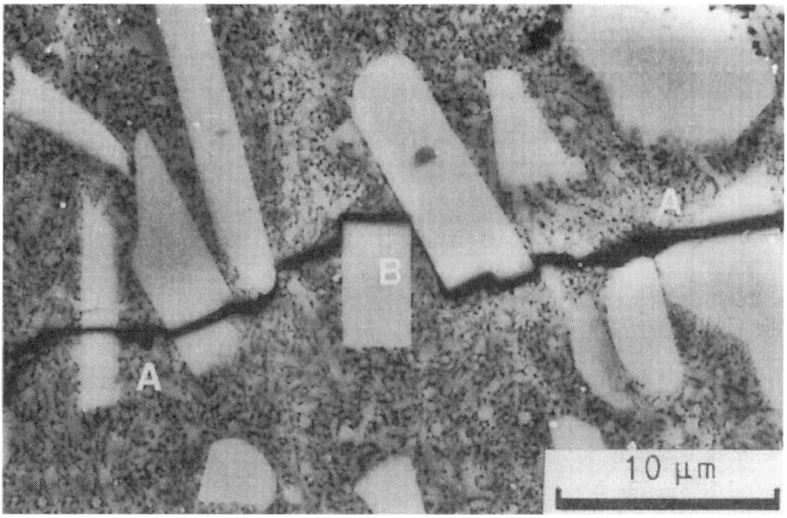

FIGURE 8.7 Fracture and bridging in the wake of a crack in a composite of SiC platelets in an SiC matrix. (A) refers to fractured platelets and (B) to bridging platelets. (From Ref 25. Published with permission of the Journal of Materials Science.)

clearly demonstrated by the effectiveness of fiber coatings to inhibit fiber–matrix bonding and hence enhance fiber pullout [26]. Besides such various direct observations of fiber pullout, pullout is also consistent with the larger scale of generally noncatastrophic propagating cracks (that may frequently be partly or fully arrested) in such fiber composites. However, its validity and applicability to progressively chopped (i.e. short) fiber, to whisker and platelet, and ultimately to normal particulate composites is progressively more uncertain. This arises in part due to the scales of possible pullout being so much less in such composites compared to continuous fiber composites, making the former difficult to distinguish clearly from the simple equivalent of intergranular fracture in such composites. Added uncertainties arise since in these other composites, pullout becomes similar to if not identical to crack bridging with all of its uncertainties.

Though often not emphasized or even explicitly identified, it is important to note the known or probable dependences of the various toughening mechanisms on microstructural parameters such as particle (grain) size, uniformity, orientation, and (where applicable) fracture mode. It is also important to address similarly known or possible effects of crack size. Both are summarized in Table 8.1, based in part on an earlier evaluation of Rice [15].

TABLE 8.1 Summary of Status and Microstructural and Crack Size Dependence of Toughening Mechanisms

Toughening mechanism	Verification	Particle (grain) size dependence	Crack size effect
Transformation	Substantial	Increased optimum size with increased stabilization and matrix *E*, and possibly decreasing volume fraction. Decreasing benefits as size distribution broadens	Some *R*-curve effects shown
Microcracking	Some	Optimum size, e.g. per Eq. (2.4), probable minimum size and increasing degradation at larger sizes. Decreasing benefits as size distribution broadens	Possible increased effectiveness as crack size increases
Crack deflection	Some	No clear direct size effect, but possibly some via spacing effects for a given volume fraction. Significant orientation effect. Size distribution effects uncertain	Probably increased effects, then saturation as crack size increases
Crack pinning	Limited	Increases with decreasing particle size, but probable limitations at small and large sizes. Probable significant orientation and size distribution effects	Reduced effects at finer, then saturated at larger, sizes
Crack branching	Limited	Probably depends on size and orientation, but in varying fashions as a function of possible contributing mechanism(s), e.g. crack deflection or microcracking	Probably greater effect and occurrence with increasing crack size
Crack bridging	Some	Generally increases as particle size increases. More effective for intergranular versus transgranular fracture	Increased effect, then saturation, as crack size increases
Pullout (especially of fibers)	High	Based on area dependence of frictional work in pullout, linearly increased effects with the inverse of fiber diameter. Significant orientation dependence	Not for finer fibers, but maybe for large fibers (filaments), and for whiskers and platelets

The first of two other important issues that are often not addressed is that of the extent to which cracks actually interact with the dispersed particles. This is generally not an issue in most fiber (and related directional solidified) composites, since the issue of fiber–crack orientations and opportunities for cracks avoiding the fibers is generally effectively zero. Such opportunity is also limited in composites of high-volume fractions of dispersed particles, whiskers, and platelets but is dependent on orientation effects in the order listed. However, as the volume fractions decrease, particularly for more equiaxed particulates, the degree of crack–particle interaction may decrease faster than expected. This may arise since while it is inviting to use straight lines on a photomicrograph to estimate the extent of crack–particle interactions, this may be very misleading, Thus as previously shown by Rice [13], limited curvature of cracks may allow them to avoid many particles, as is illustrated in Fig. 8.8. Further, the degree of crack–particle interaction can be dependent on crack velocity, a factor almost universally neglected, but limited study clearly shows a significant velocity dependence, as is discussed later.

The issue of possible crack–particle interaction has been addressed in some models as outlined in Fig. 8.9. This indicates that there is some but less interaction between a crack and a particle in hydrostatic tension, e.g. due to its having a thermal expansion greater than, and good bonding to, the matrix. On the other hand, the nature of the stress in the surrounding matrix for particles with lower thermal expansion than the matrix indicates stronger crack–particle interactions. However, such models neglect the changes in the local matrix stress states as the crack approaches the particle, as well as possible effects of crack velocity (which can vary the stresses ahead of the crack tip).

The second commonly neglected issue in most models (and many studies) is that of variations of the spatial distribution of the dispersed phase, especially serious heterogeneities of it. Thus an agglomerate or other accumulation of the matrix material often acts as a weaker source for easier local crack propagation, and particularly seriously, if large enough, as a source of failure. Equally, and often more seriously, is when there is a clustering or agglomeration of the dispersed particulate material, since this often can act as a defect, e.g. a flaw, and one which is often then surrounded by a region of reduced toughness due to its frequent lower concentration in the area surrounding the agglomeration (Fig. 8.10).

The above models have implications for crack propagation behavior beyond the measurement of fracture toughness, but they leave various uncertainties, some of which have been noted above. Another uncertainty is environmental effects, i.e. SCG. While the models provide some guidance of how SCG may progress, they by themselves provide no guidance as to the occurrence of SCG, which probably entails issues of the contiguity of dispersed particles or grain boundary phases subject to SCG, e.g. as noted for Si_3N_4 made with oxide additives (Chap. 2, Sec. III.B).

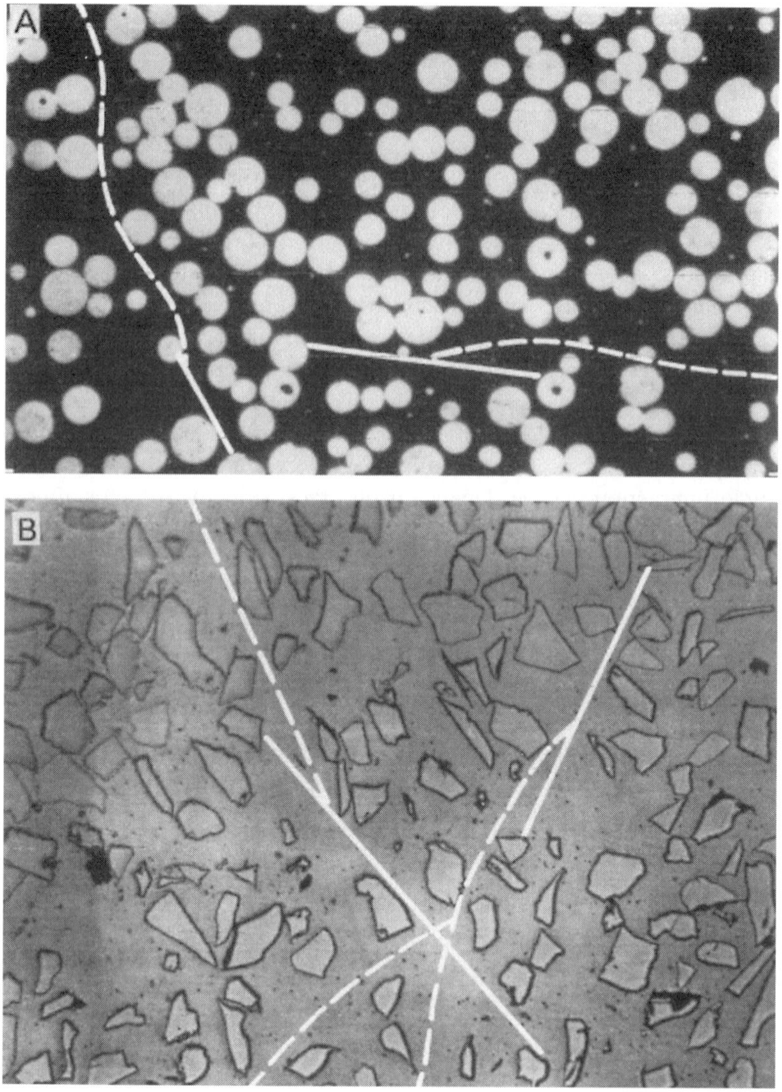

FIGURE 8.8 Photomicrographs of composites with (A) 40 v/o spherical particles ~ 25 μm dia. and (B) 30 v/o of irregular particles ~ 40 μm dia. While the solid white lines indicate limited mean free path lengths between particles, i.e. high crack–particle interaction, the dashed white lines show that significantly reduced crack–particle interaction may occur with limited change in crack trace shape. The extent to which this occurs is unknown, but it probably depends on factors such as interactive stresses between particles and cracks (e.g. Fig. 8.9) and crack velocities. (From Ref. 13. Published with permission of Ceramic Engineering and Science Proceedings.)

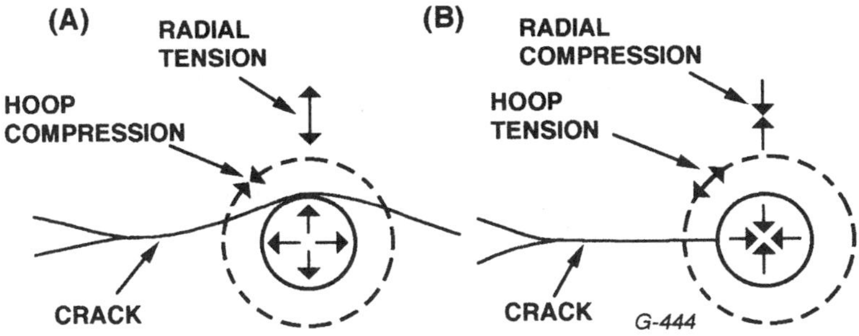

FIGURE 8.9 Schematic of idealized interactions of a matrix crack with (A) a particle in hydrostatic tension (e.g. due to higher expansion than, and strong bonding to, the matrix), and (B) in hydrostatic compression (e.g. due to lower thermal expansion than the matrix). (From Ref. 13. Published with permission of Ceramic Engineering and Science Proceedings.)

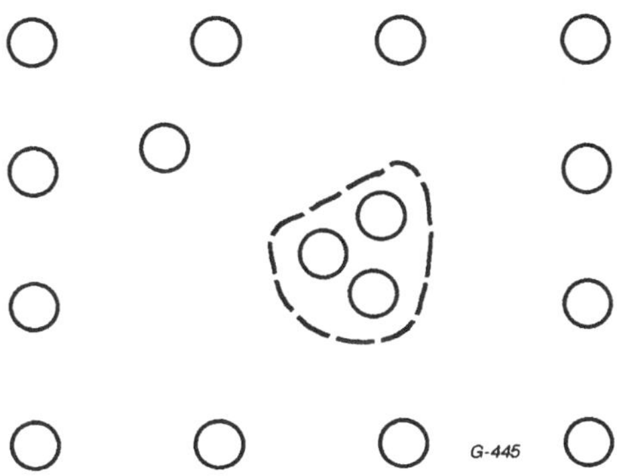

FIGURE 8.10 Schematic of a serious heterogeneity (enclosed by dashed line) of toughening particles (open circles), which may result in this acting not only as a flaw but also often as one with less local toughening. (From Ref. 13. Published with permission of Ceramic Engineering and Science Proceedings.)

Finally, it is important to note two points. First, a few combinations of mechanisms have been suggested (e.g. Figs. 8.1B and 8.2C) and others have been discussed [13,14], but most evaluations continue to assume that only one mechanism need be considered, usually without any justification. Second, in part as a corollary problem, while observations and analysis of the crack wake bridging concept are an important component of research in this area, it is also important to note briefly two negative aspects of this on research. One is the focus on wake bridging (enhanced by its ready observation) and an almost total lack of research on other mechanisms based on the premise that bridging is applicable and generally dominant, without clearly demonstrating its applicability to normal strength behavior. Another related negative effect is a common tendency simply to measure toughness and observe associated large-scale crack bridging, but to not measure strength, instead assuming that it follows the compostional, microstructural, and other dependences of toughness. As will be shown in Chap. 9, this assumption is often seriously incorrect, leaving the field with fragmented, incomplete, and uncertain data.

III. PARTICLE PARAMETER EFFECTS ON ELASTIC PROPERTIES

The primary parameters of the elastic properties of ceramic composites, as for any type of composites, are the composition, which determines the elastic properties of each of the constituents, and the volume fractions of the constituent phases. Secondary factors are the specific character of the second phase, such as its shape, orientation, and degree of interconnection (all of which are often related, e.g. interconnection of the second phase is related to volume fraction, size, and shape), and whether it occurs in the composite as a single crystal (hence being a second phase with elastic anisotropy that may be important). There is some, but not extensive, data on the elastic properties, most commonly Young's modulus of ceramic composites. Useful compilations of data for particulate [27] and whisker composites [28] are available. However, little comparison of results to models, especially comparison to two or more different models, has been made. The focus here is a brief summary indicating the status and reasonable approaches as an aid in addressing other mechanical properties.

Hasselman and Fulrath [29] evaluated Young's modulus of their composites of hot pressed sodium borosilicate glass with 10–50 v/o Al_2O_3 particles (~ 50 μm). They showed that use of the rule of mixtures upper bound gave average values that typically exceeded measured values by ~ 9–12%, while the average of the Hashin–Shtrikman bounds tended to fall below measured values by 0.5 to ~ 5% as the volume fraction alumina increased. Application of Eqs. (8.3) and (8.4) gave averages that ranged from ~ 9% high to < 2% low using their stated value of E for the alumina (which probably presents some uncertainty). Lange's [30] measured E values for 10, 25, and 40 v/o alumina particles (3.5, 11, 44 μm), which all gave the same values for a given v/o, in a similar glass gave average values of the upper and lower bounds [Eqs. (8.1) and (8.2)] that were ~ 8–14% higher than measured values (assuming E for the alumina to be (~ 400 GPa). Use of Eqs. (8.3) and (8.4) gave averages that ranged 5–8% higher than measured, i.e. about half the difference for the rule of mixtures bounds. Frey and Mackenzie [31] found that predictions used by Hasselman and Fulrath above worked even better on their composites of glass with Al_2O_3 or ZrO_2 ~ spherical particles (~ 125–150 μm), i.e. accurate to ± < 4%, while for alumina particles in a glass matrix of the same expansion, Binns [32] found that elastic properties agreed with the mean of Eqs. (8.1) and (8.2). Freiman and Hench reported < 3% error in using the Hashin–Shtrikman approach for crystallized glasses in the $LiO_2 \cdot 2SiO_2$ system [33]. Jessen et al. [34] reported that addition of spherical particles (44–75 μm dia.) of a Fe-Ni-Co alloy in a borosilicate glass matrix gave Young's moduli increasing per Eq. (8.1) at < 10 v/o addition and then transitioning to *E* closer to, and at or slightly below, that given by Eq. (8.2) by respectively 25 and 50–65 v/o metal. See also the note at the end of this chapter.

Donald and McMillan [35] made composites with varying contents (mostly up to 30 v/o) of chopped Ni wires (~ 3 mm long and 0.05, or mainly 0.125, mm dia.) by mixing them with powdered glasses and hot pressing. Composites with a glass matrix with an expansion 8.3 ppm/°C below that of Ni decreased from *E*~ 54 GPa at 0 Ni to ~ 44 GPa at 20–30 v/o Ni, while glasses with ~ the same expansion as Ni decreased from ~ 68 GPa at 0 Ni to a minimum of ~ 46 GPa at 20 v/o Ni and then rising to ~ 64 GPa at 40 v/o Ni. A glass matrix with an expansion ~ 1.7 ppm/°C > Ni showed a similar trend, i.e. a minimum of ~ 50 GPa at 20 v/o Ni and then increasing with more Ni, but less so than with matched expansion. In contrast to this, Zwissler et al. [36] added chopped (1.6 mm long) 304 stainless steel (SS) wires (6, 12, and 25 μm in dia.) to an FeO matrix by hot pressing and found *E* linearly increasing from 129 GPa for FeO in agreement with the rule of mixtures as the chopped wire content increased to the limits of their experiments of 10–15 v/o, depending on wire size. The lack of reductions in *E*, and hence apparently of microcracking, is attributed to the limited expansion difference (FeO ~ 2 ppm/°C > the SS) and the smaller sizes and lower v/o.

Turning to composites with crystalline matrices, Ono et al. [37] showed that the ratio of Young's modulus and density (E/ρ) for the composite system Al_2O_3-ZrO_2(+ 3 m/o Y_2O_3) was linear as a function of ZrO_2 w/o over the whole composition range. However their data with unstabilized ZrO_2, while starting with the same slope at low ZrO_2 additions, deviated significantly below the linear trend, especially from 10 to 20 w/o and 90 to 100 w/o ZrO_2, resulting in a value of ~ $^1/_2$ that of ZrO_2 with 3 m/o Y_2O_3 for the completely unstabilized ZrO_2, indicating serious microcracking. French et al. [38] similarly measured *E* across the complete range of Al_2O_3 and ZrO_2 contents, but with fully stabilized cubic ZrO_2 (with 8 m/o Y_2O_3). Their results showed a linear decrease of *E* between the two extremes of ~ 400 to ~ 240 GPa respectively for Al_2O_3 and ZrO_2, but without the deviations of Ono et al. with unstabilized ZrO_2.

Yuan et al's [39] E values decreasing by up to ~ 25% for mullite +0-25 v/o ZrO_2 (with varying Y_2O_3 levels) from the nearly identical values of mullite and ZrO_2 illustrates challenges of sorting out composite behavior. Corrections for the generally increasing 2.6–8.7% porosity as ZrO_2 levels increased via e^{-4P} indicates bodies with ~ 1 μm ZrO_2 particles had E values ~ 5–10% lower only with 20–25 v/o ZrO_2 additions, and those with 2 or 4 μm ZrO_2 particles had values ≥5% and 10–15% lower with respectively 5–10 v/o and 15–20 v/o

ZrO_2. ZrO_2 particle size dependences of E clearly indicate microcracking, as does effects of quenching, and data of Ishitsuka et al. [40] showing E for 50 v/o TZP in mullite being 150 GPa versus 235 and 215 GPa for the 2 constituents. Ruf and Evans [41] showed that additions of up to 40 v/o ZrO_2 to ZnO followed a rule of mixtures very closely, and extensions of this to 60 v/o were still close to a rule of mixtures. Limited deviations to values a few percent lower than the rule of mixtures at 40–60 v/o were attributed to limited microcracking.

Another important system is that of Al_2O_3-TiC, where while there is limited difference of constituent *E* values, *E* has been reported to increase ~ linearly from 393 to 415 GPa as the TiC content increased from 0 to 40 w/o (35 v/o) [42]. Small (1–5 v/o) additions of submicron particles of high modulus βSiC to lower modulus $BaTiO_3$ (Fig. 8.19) substantially increased the modulus, though there are some effects of a $BaTiO_3$ phase change [43]. Similarly, adding nanoscale βSiC particles to βSi_3N_4, while not increasing *E* at the 5 v/o level, linearly increased it over the 5–20 v/o range, approaching the investigators' (apparently rule of mixtures) expectations at the higher levels [44].

Other systems offering opportunities for evaluation are those where the second constituent may be present in solution or as a second phase, e.g. as for crystallized glasses and a few all crystalline constituent composites. AlN-SiC composites are an example of the latter, since the two end phases can form extensive solid solutions depending on particle sizes and times at temperatures of densification or post treatment; otherwise they yield two-phase bodies. Ruh et al. [45] showed that solid solution bodies had somewhat higher moduli over most of the composite range (and the Hashin–Shtriktman bounds following at or slightly above their solid solution trend) (Fig. 8.11). Estimating the moduli of their composites using $^1/_2$ the sum of values from Eqs. (8.3) and (8.4) and their values for pure AlN and SiC agrees fairly well with their solid solution data, adding to the question of the specific sources of differences between the two-phase and solid solution bodies. Comparison of this data with other data for composites having one constituent in common shows variation in the moduli for the same material, as is also common in the literature for ceramics in general. Thus Ruh et al. give *E* of dense SiC as 347 GPa, while Mah et al. [46] give it as ~ 427 GPa and the latter give *E* ~ 290 GPa for their Si_3N_4 matrix while Baril et al. [25] give this as ~ 313 GPa. While these differences are not large and are generally representative, they illustrate an important problem in predicting moduli of composites, since such values of constituents vary, and it is uncertain how much of such variation is due to measurement or material issues (and their interrelation, e.g. due to heterogeneities in the bodies).

Another type of composite of interest is one where the second phase is produced by an in situ reaction as was used by Chen and Chen [47] to produce La hexaluminate platelets in an Al_2O_3 matrix. They report *E* decreasing from ~ 420 GPa for pure Al_2O_3 to ~ 230 GPa for the pure aluminate, with greater decreases occurring from 80 to 100 v/o aluminate.

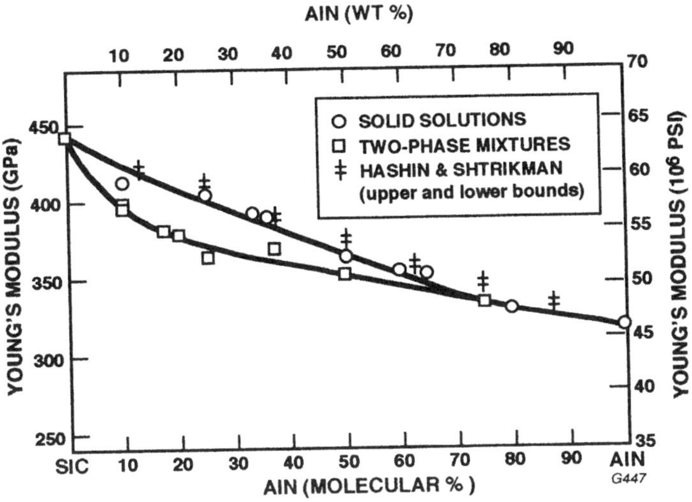

FIGURE 8.11 Young's modulus of SiC-AlN solid solutions or particulate composites versus AlN content. Note reduced Young's modulus of the particulate composite versus the solid solution. (From Ref. 44. Published with permission of the American Ceramic Society.)

Turning to two-phase crystalline composites of all nonoxide constituents, the most extensive data from a volume fraction standpoint (ϕ= 0 to 1) is that of Endo et al. [48] for SiC-TiC particulate composites, which they noted followed a linear relation as a function of weight fraction (Fig. 8.12). Taking half the sum of Eqs. (8.3) and (8.4) (using their values for the two end points of pure SiC and TiC) results in very good agreement with the mean of Endo et al.'s results. Similar comparison of values for Si_3N_4-TiC composites (ϕ= 0 to 0.5) of Mah et al. [46] shows more scatter. How much of this reflects inaccuracies in Eqs. (8.3) and (8.4), measurement variations, and actual material variations, e.g. due to heterogeneities of distribution of the second phase or variations in possible degree of preferred orientation of the matrix grains, the dispersed particles, or both, in the composite, cannot be ascertained.

Ferrari and Fillipponi [50] reviewed the application of effective medium theories, specifically earlier self-consistent scheme (SCS) and Mori–Tanaka (M–T), as well as Hashin–Shtrikman (H–S) bounds to ceramic composites of nonspherical Al_2O_3 or spherical particles in glass matrices (discussed earlier), as well as to pores in glass. This shows some of the problems even for simpler, more ideal composites. They showed that the data for nonspherical Al_2O_3 particles (0–50 v/o) was fitted by the H–S lower bound, which is also the M–T

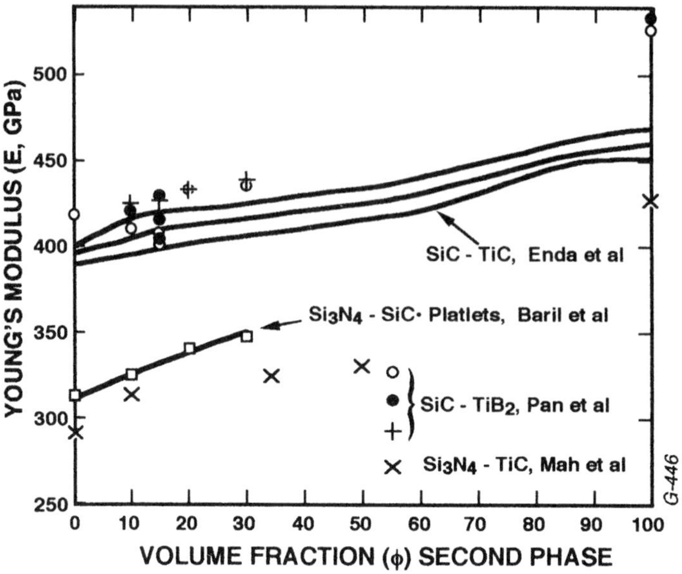

FIGURE 8.12 Young's modulus versus volume fraction second phase for particulate composites of SiC-TiC by Endo et al. [48] (where both the range and the mean are shown by upper, lower, and intermediate lines), of Si_3N_4-TiC by Mah et al. [41], and of SiC-TiB_2 by Pan et al. [49], as well as of Si_3N_4-SiC platelet composites by Baril et al. [25], Note that use of Eqs. (8.3) and (8.4) agrees almost identically with the mean trend for the data of Endo et al., while the data of Mah et al. shows measurable variations (scatter) between the data and such model predictions.

model for spherical particles, and the SCS model is also very close to the data. The dilute spherical particle models, while agreeing with data to ~ 20 v/o, fell progressively below data as the Al_2O_3 content increased, e.g. being 10–15% low at 50 v/o. Overall the dilute concentration model fitted better, though it was high at lower v/o Al_2O_3, e.g. by ~ 15% at 20 v/o, but it appears to be low at high v/o Al_2O_3. For glass with spherical *W* particles, the H–S lower bound and hence the M–T model for spherical particles were close, e.g. being ~ 5% low at 50 v/o W particles, while the SCS model was a good fit. Again the dilute spherical inclusion model was low, e.g. ~ 20% low at 50 v/o W. Their evaluations for voids in glass, being the extreme of low modulus inclusions, is indicative of problems for such composites, with a major one being the failure of the SCS and dilute concentration models, since they go to zero at 50 v/o pores. Data was close to the upper H–S bound, and still closer to the M–T model, but the latter fit was based on assuming an aspect ratio of 0.8 for the pores, which is really a curve fitting approach, since there is no experimental or theoretical basis for this, i.e. equilibrium pores in glass are spherical. Such fitting and the lack of explanation why in some cases the H–S lower and in others the H–S upper bound fitted better are illustrative of the limitations of current models as well as of characterization of composites. This is very similar to the problems of effects of porosity on elastic properties which are impacted not by only pore shape but also by how they are arrayed in the body, i.e degree of randomness versus the degree and type of ordered stacking, orientation of nonspherical pores and their contact or intersection (which is impacted by volume fraction, shape, orientation, and size [51]).

Data for platelet and whisker composites is often more uncertain, since this invariably entails varying, generally incompletely characterized degrees of preferred orientation and its variations in the bodies. This also presents the challenges of first knowing the crystal structure–platelet or whisker morphology and resultant elastic properties as a function of orientation in the platelets or whiskers. Thus, for example, the Young's modulus data for Si_3N_4-SiC platelet composite (ϕ= 0 to 0.3) of Baril et al. [25] (Fig. 8.12) faces these uncertainties for comparison to any model.

One set of platelet composites for which some limited, useful comparative data is available is those made with fine BN platelet particles dispersed in matrices of Al_2O_3 [52], mullite [52,53], SiC [54], and AlN [55]. Data for the relative Young's modulus (measured parallel to the hot pressing axis) versus v/o of BN shows a common trend for all three composites except for two deviations, one at higher BN loadings in SiC and one data point for mullite-BN produced by in situ reactions of B_2O_3 and AlN or Si_3N_4 [53] (Fig. 8.13). Collectively these suggest a consistent pattern based on the substantial preferred orientation of the BN platelets normal to the hot pressing axis, as clearly shown for Al_2O_3 - [52] and mullite - [53] based composites made by hot pressing of the matrix and BN powders, while the platelets are essentially randomly oriented when produced in situ via reaction hot pressing (Fig 11.5). Bodies made by the former process are anisotropic, e.g. having $E \sim 35\%$ higher in the plane of hot pressing versus normal to it [53], while the latter process gives essentially isotropic properties [54]. The high BN expansion of ~ 25 ppm/°C in the *c* direction (i.e. normal to the plane of the platelets, but only ~ 1 ppm/°C in the *a* direction) versus that of the alumina and mullite matrices (respectively ~ 9 and 5 ppm/°C) strongly suggests that separations between the BN platelet faces should occur, which is also indicated in TEM studies [56]. Thus data for *E* normal to the plane of oriented BN platelets should approach that for platelet pores as suggested by Lewis et al. [52], with a common trend for relative moduli irrespective of matrix, while data for isotropic, unoriented bodies should be higher due to less alignment of platelet–matrix separations in the direction of measurement, as shown in Figure 8.13. The deviation at higher BN loadings in Si_3N_4 may reflect less creep-orientation accommodation in it during hot pressing.

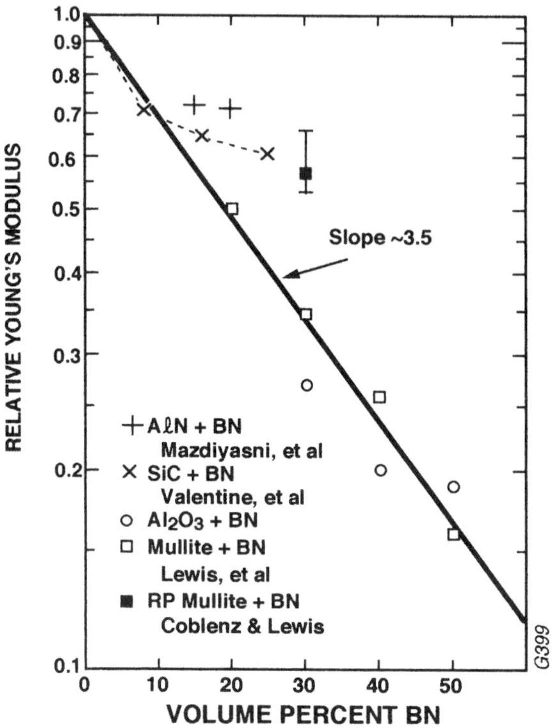

Figure 8.13 Relative Young's modulus at ~ 22°C as a function of v/o fine BN platelet particles in composites with Al_2O_3 [52], mullite [52,53], SiC [54], and AlN [55] matrices. Measurements parallel to the hot pressing axis, hence normal to the plane of preferred orientation of the BN platelets, for composites made by simply hot pressing the composite ingredients [52,54], but not in the essentially isotropic bodies made by forming the BN platelets in situ by reaction processing (RP) of B_2O_3 with AlN or Si_3N_4 [53]. Note (1) data for Al_2O_3 and mullite matrices does not give values for the matrix alone, so values of 410 and 220 GPa were used respectively, (2) the common trend for all three composites with preferred platelet orientation regardless of matrix, and the higher value for the mullite-BN reaction processed composite without significant BN platelet preferred orientation, and (3) that the common slope of ~ 3.5 is reasonably consistent with that for platelet pores oriented normal to the stress direction [52].

Turning to whisker composites, these offer all the complications of particulate and platelet composites, especially frequent substantial preferred orientation of the whiskers, though probably differing from the degree and character of orientation in platelet composites. No detailed studies of elastic properties of ceramic whisker composites have been made, but a few that are available indicate some of the variability and uncertainties. Consider first Ashizuka et al.'s [57] study of SiC whisker additions to cordierite-, anorthite-, and diopside-based glass systems, whose baseline *E* values were respectively ~ 90, 130, and 160 GPa. All *E* values increased ~ linearly by ~ 40 GPa to reach maxima at ϕ respectively of ~ 0.25, 0.3, and 0.4 and then dropped rapidly, with the extent of the decrease increasing in the reverse of the order listed. As expected, the percentage increase was greatest in the cordierite-based system having the lowest *E* value (Table 8.2). The decreases in *E* are attributed to microcracking, since anorthite has the closest thermal expansion to SiC, anorthite is higher, and cordierite is substantially lower, but the decrease is also dependent on ϕ, *E*, or both.

Consider next polycrystalline matrices for whisker composites, beginning with Wadsworth and Stevens' [58] addition of 0.3 volume fraction SiC whiskers hot pressed in a lower modulus matrix (cordierite) showing that *E* in the plane of pressing increased 24% from 140 to 192 GPa (Table 2). They showed that these results were nearly halfway between the bounds of Eqs. (8.1) and (8.2) using a value of $E = 440$ GPa for dense polycrystalline SiC as Yang and Stevens [59] did. Kumazawa et al. [60] showed that addition of 0–40 v/o of SiC whiskers to a mullite matrix increased *E* from ~ 220 GPa at 0 SiC to ~ 290 GPa, i.e by ~ 30%. Tamari et al. [61] reported that the Young's modulus of Al_2O_3 with mullite whiskers decreased linearly as ϕ increased from 0 to 0.3. Yang and Stevens' study of Al_2O_3-SiC whisker composites is one of the most studied systems if not the most studied. They showed that *E* in the plane of hot pressing (i.e. ~ parallel with the planar orientation of the whiskers) increased linearly as the volume fraction of SiC whiskers increased from 0 to 0.3. Correcting their results for limited porosity, they showed that *E* increased ~ 10% from ~ 397 to 408 GPa (Table 2) and that this trend was consistent with predictions from a rule of mixtures relation using the polycrystalline value of *E* for the SiC whiskers, a procedure of uncertain applicability to other bodies.

Fisher et al. [62] made a more detailed experimental and analytical study of elastic properties of composites of SiC whiskers in matrices of Al_2O_3 and Si_3N_4 with up to 30 v/o whiskers made via hot pressing. Although SEM examination showed considerable orientation of whiskers parallel to the plane of hot pressing, very little anisotropy in wave velocities was found. Thus while models for oriented whiskers were considered, behavior was nearly isotropic (which gave elastic moduli ~ the same as for stressing normal to aligned whiskers, e.g. supporting a model of random three-dimensional arrays of whiskers giving isotropic behavior). Shalek et al. [63] similarly measured *E* as a function of SiC

Table 8.2 Comparison of Mechanical Properties of Some SiC Whisker Composites with Different Matrices[a]

Matrix	E (GPa)	ΔE (%)	K ($MPam^{1/2}$)	ΔK (%)	σ (MPa)	$\Delta\sigma$ (%)	Investigator
Cordie-	—	—	1.25	330	85	365	Gadkaree [171]
rite	140	35	1.9	74	170	41	Wadsworth and Stevens [58]
	90	50	2.0	115	220	86	Ashizuka et al. [57]
Mullite			2.7	160 (85)	320	190 (122)	Wu et al. [172]
	210	—	2.0	40	440	31	Kumazawa [60]
			2.2	105	—	—	Becher et al. [163]
Si_3N_4	335	1–2	5.5 (7.1)	45 (46)	400 (650)	–25 (–30)	Shalek et al. [63]
		(ϕ=0.2)	6	–18	900	–38	Lundberg et al. [173]
			3.7 (4.7)	27 (36)	780	26	Buljan et al. [174]
Al_2O_3	395	3	4.5	27–56	560	18	Yang and Stevens [59]
			4.2	107	300	120	Becher et al. [163]
			4.0	75	460	60	Iio et al. [175]

[a] E = Young's modulus, K = fracture toughness, σ = flexure strength, Δ = the incremental change (in %), i.e if a value doubles its incremental change is 100%. Note some decreases in properties. All changes are for ϕ=0.3, except data of Lundberg et al. for ϕ=0.2.

whisker content in a Si_3N_4 matrix for different hot pressing conditions, showing differences of >10% with no whiskers. For the two hot pressing temperatures with the most data, one first decreased and one steadily increased, with both reaching maxima at ϕ=0.2, one at the *E* value for ϕ=0 and the other nearly 15% above the ϕ=0 value. Desmarres et al. [64] reported that addition of 30 v/o SiC whiskers to Si_3N_4 resulted in very little anisotropy, i.e. *E* was 348 and 340 GPa parallel respectively to the hot pressing axis and the plane of pressing, which represented an ~ 20% increase over the Si_3N_4 alone (280 GPa).

While composite moduli values can often be reasonably estimated from rule of mixtures, or more precise, bounds, there are various problems that can arise and limit the utility of more precise models. A basic problem is whether some, especially unknown, microcracking has occurred, e.g. as noted earlier for Al_2O_3-ZrO_2. Another important factor can be the degree of preferred orientation, especially of morphologically shaped single crystal dispersions such as whiskers or platelets, since these introduce significant (not always well documented) crystalline anisotropies and often incompletely characterized degrees and character of preferred orientation. Other issues in the measuring, and especially the estimation, of elastic moduli are the degrees of reaction and bonding between the matrix and the dispersed phase, accurate data for the properties of the phases involved, and the amount and character of any porosity in the bodies.

IV. PARTICLE (AND GRAIN) PARAMETER EFFECTS ON CRACK PROPAGATION

Of the four aspects of crack propagation of fracture mode, i.e. inter- or transgranular fracture (IGF or TGF), slow crack growth (SCG) due to environmental effects, more macro behavior such as crack branching, and fracture energy/toughness, only the latter has received substantial attention, which is discussed in the next section. This section addresses the limited information on the first three noted aspects of crack propagation.

The limited data on fracture mode in particulate and related composites is summarized here from the review by Rice [65] based on his own and literature observations. He noted that there was considerable TGF in many particulate and whisker composites fractured at nominally 22°C, with much of this in the matrix (commonly Al_2O_3, *G* ~1–10 μm), but also frequently of the dispersed phase. Thus laboratory and commercial Al_2O_3-TiC (*G* ~ 2–5 μm, strength >700 MPa) showed >50% TGF, which is consistent with Krell and Blank's [66] observation of considerable TGF in their Al_2O_3-TiC. While the ready cleavage of some dispersed phases such as TiC may be a factor in the often substantial TGF of the Al_2O_3 matrix in these composites, there seems to be a broader effect, since increased TGF is seen in other composites with less distinct cleavage of the dispersed phase. Thus Rice reports that Al_2O_3-ZrO_2 bodies (*G* ~ 2–5 μm, strength ~ 700 MPa) commonly showed 10–30% or more TGF, i.e. similar to, possibly more than in, bodies of either phase alone with the same *G* along with similar results of hot pressed ThO_2 (3 wt% Y_2O_3) with 25 m/o ZrO_2 (again with some reduction in grain size). Further, there clearly also was substantial TGF in the Al_2O_3 matrix with SiC whiskers, which frequently fractured both transversely and along the Al_2O_3-SiC interface (Fig. 8.14). While the TGF preferentially occurred in the larger grains on the fracture surface, as in monolithic ceramics, the degree of TGF in the composite matrix is typically higher than in the matrix alone (of the same *G*). This was clearly shown by the work of Baek and Kim [67] where the matrix alone ($G \sim 6$ μm) had predominantly IGF, but with 20% SiC whiskers the matrix (with ~ the same *G*) had mainly TGF. Dauskardt et al. [68] also reported essentially 100% TGF in the Al_2O_3 matrix ($G \sim 2$ μm) of Al_2O_3-SiC whisker composites fractured in fatigue tests.

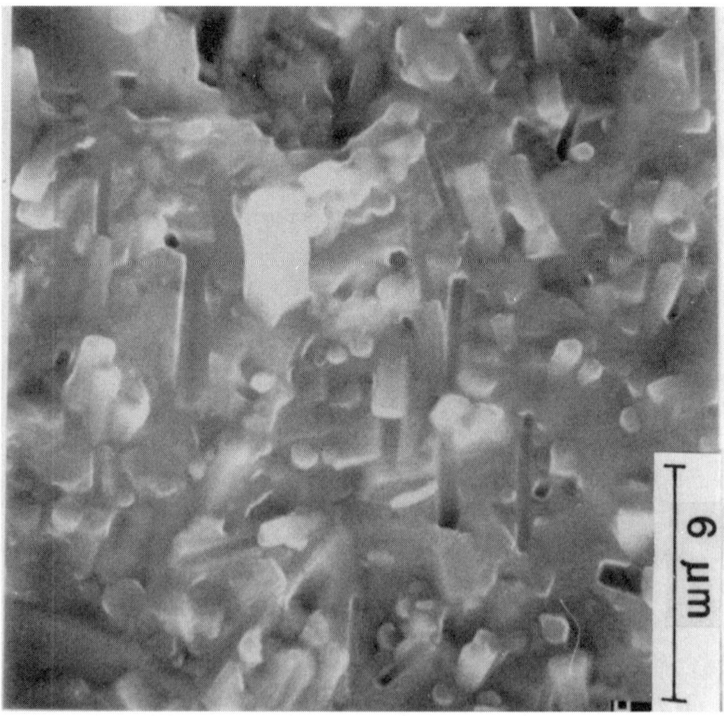

FIGURE 8.14 Flexure fracture of commercial (Cercom) Al_2O_3-SiC whisker composite at 22°C. Note substantial TGF in the Al_2O_3 and considerable transverse fracture of the whiskers, as well as considerable interfacial fracture (i.e. IGF) around part of many SiC whiskers.

WC-Co composites, which have been extensively studied, also show substantial TGF. Thus earlier flexure strength studies of Gurland and Bardzil [69] showed 50–75% TGF of WC grains for G ~ 1–4 μm, and while Murray [70] showed 10% for similar *G* material in K_{IC} tests, Hara et al. [71] reported a transition from IGF to TGF as *G* increased. Pickens [72], Pickens and Gurland [73], and Roebuck and Almond [74] concluded from their own studies and surveys of WC-Co data that TGF increases with *G* and the amount of Co. Pickens was more quantitative, indicating more IGF for $G < 2$ μm and more TGF for $G \geq 2$ μm; he noted more TGF of larger grains in a given body (which thus can vary TGF with *G* distribution) and cited possible effects of crack velocity and of *G* measurements. Comparison with the one data point for pure WC [75] suggests that WC-Co has more TGF at comparable *G*. An important factor in the fracture mode of such WC-Co composites is the degree of contiguity of the phases, with increased contiguity of the WC phase enhancing TGF of the WC [76].

The overall fracture mode provides some important indications about the fracture process, e.g. transgranular fracture of the dispersed phase implies less effect of crack bridging, since this is typically enhanced by intergranular fracture. However, other information is typically needed in conjunction with fracture mode, such as particle sizes, spacings, agglomeration, etc., important both generally, e.g. for effects on toughening mechanisms, and locally for effects on failure initiation. Such information is typically lacking, as is that on another factor that can be important, namely the effectiveness of the second phase in interacting with the crack, e.g. since, as illustrated in Fig. 8.8, limited crack variations can reduce crack–second phase particle interactions significantly. Though almost totally neglected, such interactions can vary as a function of a number of material and fracture parameters, e.g. crack velocity.

There has been limited study of SCG in ceramic composites, despite the important role that such crack growth can play in monolithic ceramics and the possible contribution that composite stresses between phases may contribute to such growth, at least on a local basis. From results of nominally single phase ceramics, composites with all phases susceptible to SCG are likely to have SCG, but affected by the microstructure and related stresses, and composites of phases having no SCG would have none (provided no grain boundary effects override such exclusion). The extent of SCG in composites having at least one phase susceptible to SCG probably depends on both the extent and the variation of contiguity of the SCG susceptible phase (e.g. as indicated in SiC and Si_3N_4 with oxide containing grain boundary phases, Chap. 2, Sec. III.B). However, there are a few studies of ceramic composites with oxide matrices susceptible to SCG that show that the composites can have significantly reduced SCG.

One set of materials on which there has been some SCG study are those composites that have a silicate glass matrix. Carlström et al. [77] reported that dynamic fatigue tests of four alumina-containing electrical porcelains gave *n* values of 26–35 and 31–48 at two different loading rate ranges. The ranges of these values thus encompass the *n* values that Soma et al. [78] measured on an electrical porcelain containing crystallites of mullite, quartz, and alumina. Smyth and Magida [79] using dynamic fatigue found $n \sim 30$ for a commercial crystallized glass (MACOR ®). Cook et al. [80] reported *n* values of 24–27 for three different bodies crystallized from LiO_2-SiO_2 glasses.

Jessen and Lewis [81] showed that composites of 1–10 v/o dispersed spherical particles of a Fe-Ni-Co alloy (~ 44–75 μm dia.) hot pressed in a borosilicate sealing glass of ~ matching thermal expansion exhibited SCG, i.e. *n* values of ~ 30 similar to the glass matrix itself (see Sec. 3 for elastic properties [34]). However, the stress intensities required for equivalent crack velocities in the composite as in the glass alone were increased by ~ 50%, so there was some reduction in the net SCG for a given stress intensity in the composite versus the glass alone. However, tests of composites made with 2.5 to 10 v/o preoxidized particles (giving an oxide layer 1–2 μm thick; the above original composites were made without any oxidation of the alloy particles) greatly increased the *n* values, e.g. to 400–1000 (while also increasing stress intensities required for equivalent crack velocities in the composite as in the pure glass matrix). Thus SCG was essentially stopped by controlling composite character, specifically by substantial particle–matrix bonding.

Jessen and Lewis also conducted other studies of crack propagation and fracture in their Fe-Ni-Co alloy particles–glass matrix with two additional important crack propagation results. First, they showed that lower velocity cracks, especially SCG, had more opportunity to avoid intersecting the dispersed particles than fast cracks, e.g. lower velocity cracks intersected only about $^1/_2$ of the particles in the crack path compared to fast cracks [82]. Second, they showed that DCB fracture toughnesses measured in samples with gradations of particle volume fraction were determined mainly by the composite character in the area of initial crack propagation, with subsequent crack propagation encountering composite areas with either less or more metal particles than the starting area having limited effect on toughness and crack propagation character [83,84]. Thus while the mechanisms and interrelation of these effects are not clearly established, they show that the initial composite character where fracture initiates has a dominating effect on failure and that crack–particle interactions, and hence composite behavior, change as crack velocity increases.

Becher et al. [85] showed similar *n* values for SCG in both Al_2O_3 and Al_2O_3-SiC whisker composites, but the former occurred at substantially lower *K* and thus had more SCG. Subsequently, dynamic fatigue tests of indented discs of Al_2O_3 with 25 weight% SiC whiskers in biaxial flexure in water at room temperature of Zeng et al. [86] overall showed similar results. Thus though complicated by stress test issues, overall the composite had less susceptibility to SCG than comparable Al_2O_3 with the same porosity and grain size, in part due to *n* values ~ 50–100% greater in the composite. They reported that while the whisker composite was more susceptible to initiation of SCG, it also had greater resistance to low-velocity crack propagation that dominates SCG than did pure Al_2O_3. What the results may be for crack propagation in the direction of the typical alignment texture of the whiskers is apparently unknown (the above tests were with cracks normal to the whisker texture, i.e. the cracks were parallel to the hot pressing direction).

Improved understanding and documentation of both fracture modes, SCG, and of other aspects of composite behavior should be enhanced by studies of crack propagation parallel with phase interfaces. More recently developed tests for such crack propagation make this much more feasible and meaningful. One example of this for ceramic–ceramic (glass–alumina) interfaces is that of Cazzato and Faber [87], and examples of some for metal–ceramic interfaces (also of interest for electronics) are for Al_2O_3-Nb [88,89] and Al_2O_3-Cu [90] interfaces.

While examples of crack bridging by dispersed particles in ceramic composites are common, as was noted earlier, very little study of larger scale behavior and character of cracks in ceramic composites has been made. However, Wu et al. [91,92] showed, using their microradiographic technique, increasing deflection, wandering, branching, and hence diffuseness of cracks resulting from a single crack introduced in ZTA (i.e. Al_2O_3-ZrO_2) composites as the ZrO2 content increased (Fig. 8.15A–C). The resulting diffusiveness and complexity of such large-scale crack character was ≥ that of much coarser grain structure Al_2O_3 (Fig. 2.4). These same investigators also showed that such microradiographic examination of cracks in jade with its natural fibrous structure both fine and substantially oriented and a synthetic uniaxial fiber composite with fine SiC-based fibers were the only ceramics having more widely separated and complex crack branching on such a scale (Fig. 8.15D and E) [93].

The above discussion has focused on crack propagation under static or continuously increasing load. An important area of crack propagation that is being studied more is under cyclical loading, i.e. fatigue crack propagation, where again the issue of crack size can be important; most of the data are for large cracks in typical toughness test specimens such as CT and NB. While some monolithic ceramics exhibit net crack growth under cyclic loading that can be explained by SCG under continuous loading, some cannot. Greater cyclic growth of large cracks may arise from microstructural effects as is discussed in Chap. 2, Sec. III.G. Microstructural effects such as microcracking, crack branching, and especially bridging in the crack wake zone are commonly more prevalent or extensive in composite versus monolithic ceramics, so fatigue crack

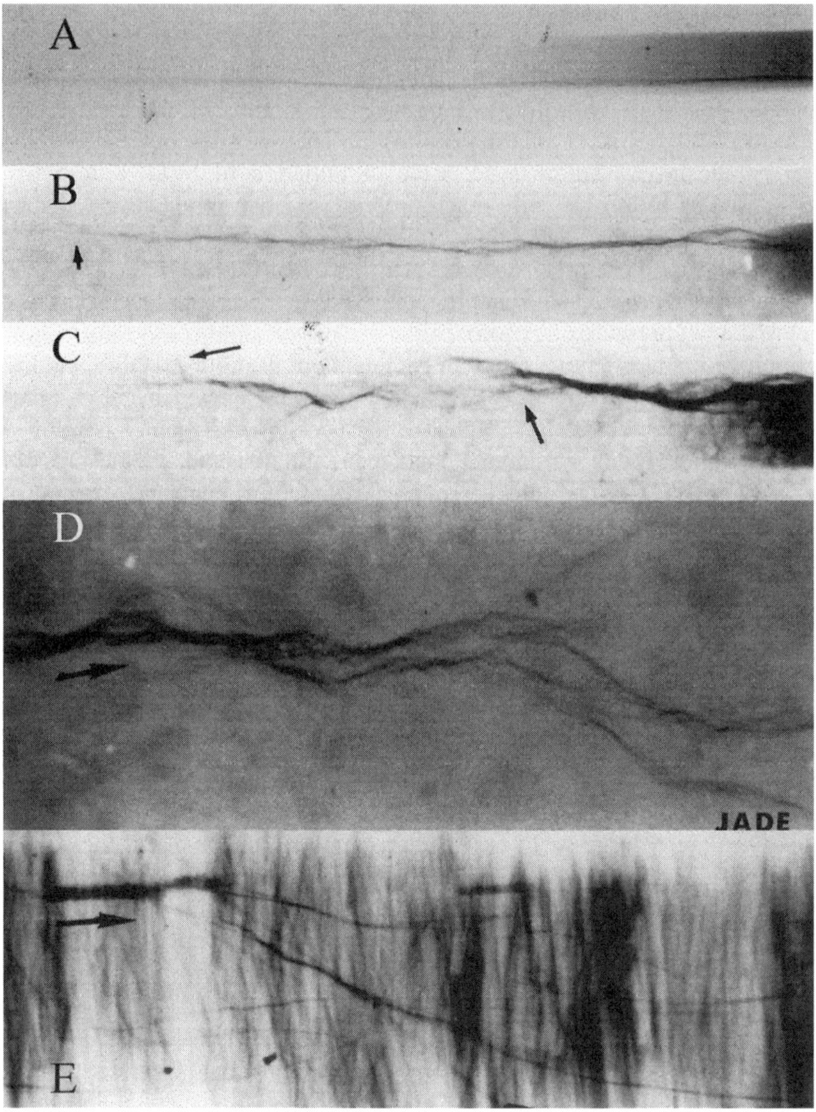

FIGURE 8.15 Microradiographic images of the character of cracks propagating from a notch in DCB specimens of Al_2O_3 with (A)–(C) 0.5, 9, and 19 v/o ZrO_2, (notch, right center, darker area), (D) a fine grain (fiber) jade, and (E) a uniaxial fiber (Nicalon)-ZrO_2 ceramic composite with the fibrous structures aligned ~ normal to the crack (notch, left center in D and E). Note the increasing tortuousness, complexity, width, and diffuseness of the resulting crack character as the content of unstabilized ZrO_2 increases and in the fibrous structures. Compare to Fig. 2.4. ((A) to (C) after Wu et al. [91], published with permission of the ASTM; (D) and (E) from Ref. 93. Published with permission of the American Ceramic Society.)

propagation is frequent and substantial in them. Again, such crack growth can typically be characterized by the Paris relation:

$$dc/dN = C(\Delta K)^{m} \tag{8.6}$$

where dc/dN is the rate of crack growth with the number of cycles (N), C and m are constants for a given body, and ΔK is the increment of stress intensity. Values of m for metals are typically 2–4, but for ceramics they can range to substantially higher values, e.g. ~ 20, though with at least some ceramic composites having lower values, and often higher K values for crack propagation [94]. While much of the fatigue study of composites has been to establish basic aspects of its occurrence rather than details of its microstructural dependence, some information on the latter is available.

Suresh [94] has reviewed fatigue behavior and noted some microstructural effects. Thus Mg-PSZ may have somewhat lower m values in the over-aged condition versus the peak strength condition (e.g. 21 versus 24), and Al_2O_3 with SiC particles lower values than Al_2O_3 by itself (e.g. 7 versus 8–14), and still less with the SiC as whiskers (e.g. m ~ 4). However, few or no details on effects of differing concentrations, sizes, orientations, etc. are available. One partial exception to this is data on WC-Co, e.g. Suresh noted that increased Co content or in the Co mean free path (hence also an increase in G for WC) decreased the rate of fatigue crack growth, and that the fracture path is mainly intergranular along the Co binder, but that transgranular fracture of the WC increases as the WC grain size increases. Again, however it must be noted that documentation and understanding of the fatigue behavior of natural strength controlling flaws in ceramics is limited, especially on specifics of the microstructure, e.g. as noted by Grathwohl and Liu [95].

Consider now some crack propagation studies in ceramic composites; the first and most basic being some model studies of Nadeau and Dickson [96]. They made DT specimens of a commercial soda lime glass (SLG) such that some specimens had grooves of varying width and depths, as well as spacings that were ahead of and normal to the subsequently introduced sharp crack. The grooves were filled with a low-melting sealing glass that had a thermal expansion of 6 versus 8.5 ppm°C^{-1} for the SLG. Their studies of propagating a sharp crack into the region of such artificial microstructure showed that crack arrest and stress intensity for repropagation through the grooves filled with the sealing glass increased as the cross-sectional dimensions of the filled groves increased (e.g. from a fraction to ~ 1 mm), and their spacing increased from a few to tens of mm. The strongest interaction was when the crack bifurcated into two branches nearly parallel with the filled groove, which increased the stress intensity for repropagation to ~ 1.8 MPa·m$^{1/2}$, i.e three times the toughness of the SLG. Crack interaction was velocity dependent; slow moving cracks were not retarded while fast ones were significantly retarded or arrested. These results are clearly consistent with the larger cracks having more interaction, especially with larger particles.

A second set of crack observations are those comparing crack propagation modes as a function of particle–matrix expansion differences. Davidge and Green [97], using ThO_2 spherical particles (50–700 μm) of differing sizes in various glass matrices showed crack propagation basically following the models of Fig. 8.9. Thus in composites with matrices of higher expansion than the dispersed particles, crack propagation tended to be through the particles, while in composites of matrices of lower expansion than the dispersed particles, crack propagation was around the particles. (See also the note at the end of this chapter.) With matrices of lower expansion than the ThO_2 strengths varied inversely with the square root of the ThO_2 particle size for each glass, Chap. 9, Sec. III.A, while those with the opposite expansion difference all fractured spontaneously, so no test results could be obtained. Frey and Mackenzie [31] also showed crack propagation tangentially around ~ spherical ZrO_2 particles with lower expansion than the glass matrix. Composites of ~ spherical Al_2O_3 (125–150 μm) particles in a high expanding glass formed many spontaneous cracks between and ~ radial with the particles, but gross crack propagation tended to be by connection of preexisting cracks to follow an overall path that mostly went around the particles rather than intersecting them. Earlier work by Binns [32] showed that, when there was no expansion difference between the matrix and the dispersed phase, fracture was relatively flat and smooth and became more complex as expansion differences increased. He corroborated that cracks generally propagated around particles (but mainly in the glass) with expansions greater than the glass matrix. With irregularly shaped dispersed particles this tangential propagation occurred mainly at the angular corner extremities of the particles, with fracture further into the glass matrix between the extremities of the particles. When the glass matrix had higher thermal expansion than the dispersed particles, microcracks tended to form, often in an ~ radial fashion, and macrocrack propagation occurred mostly by linking of the smaller cracks. Such microcracking was observed to occur above a threshold particle size at least qualitatively consistent with Eq. (2.4). Faber et al. [98] demonstrated substantially higher (I) toughness in composites of spherical Al_2O_3 particles (~ 30 μm) in a borosilicate (BS) versus an aluminosilicate (AS) glass. The former gave toughnesses up to four times the BS glass at $\phi = 0.3$ (which was apparently a maximum), while the latter gave toughness linearly increasing to 2+ times that of the parent AS glass at the limits of composition tested of $\phi = 0.4$. The higher toughness (and the apparent maximum) with the BS glass were attributed to microcracking (supported by independent observations) due to the greater expansion difference of 5.5 versus 4.4 ppm/°C. They also showed that their toughness versus volume content of dispersed Al_2O_3 with the AS glass was the same as for the composites of ThO_2 particles in glasses of Davidge and Green above.

V. PARTICLE (AND GRAIN) EFFECTS ON FRACTURE TOUGHNESS IN CERAMIC COMPOSITES

A. Composites of Glass Matrices with Dispersed Ceramic Particles

This section addresses composites of ceramic particles dispersed in a (typically silicate) glass made by one of three methods that collectively give a range of composites and the ability to tailor their character. The most versatile method is forming a glass matrix around ceramic particles by dispersing the particles in a molten glass, or much more commonly consolidating a glass matrix from powder via sintering or hot pressing so the glass matrix forms around ceramic particles mixed with glass powder. The second method is via glass systems that undergo considerable crystallization on heat treatment after glass forming, which may give complex microstructures (Chap., 1, Sec. III) but can be a versatile and practical method. The third method is in part a forerunner of the second, namely a number of traditional ceramics such as whitewares, and especially porcelains which are natural composites of glass and crystalline ceramic phases, some of the latter from crystallization, but also frequently as residues from the starting mineral constituents. Though much of the investigation of these materials was done before extensive use of fracture mechanics and fracture toughness, there is some toughness data for these materials. More consideration of these materials will be given in conjunction with the more extensive data on their strengths (Chap. 9, Sec. III).

The dominant and pervasive second phase parameter in particulate composites besides the chemical composition, and hence the physical property differences between the matrix and the dispersed phase, is the volume fraction of second phase particles. There is substantial data to show that, while the quantitative values and their specific trends may vary with the methods and specifics of both fabrication and measurement, there is a broad trend for fracture toughness to increase with increasing amount, i.e. volume fraction, of added phase, ϕ. Many studies do not encompass a sufficient ϕ range to show the limits of such an increase, but those of broader ϕ ranges show toughness increases reaching a maximum, often at $\phi \sim 0.5$, and then decreasing. Such decreases are expected, e.g. since past $\phi = 0.5$ many composites reverse the roles of matrix and dispersed phase. However, the maxima in toughness commonly found may occur at values of $\phi \neq 0.5$ for extrinsic or intrinsic reasons if other mechanisms are involved. Interaction of different dispersed phases, e.g. precipitation of different crystalline phases, or microcracking of the matrix (or of clusters of the original dispersed phase when it is the dominant phase), are examples in some glass matrix composites, and transformation effects are another.

Miyata et al. [99] measured (NB) toughnesses of hot pressed composites of glass matrices having thermal expansions greater (by ~ 2.8 or 5 ppm/°C^{-1})

than the Al_2O_3 (~ 50 μm spherical, or ~ 50 or ~12 μm angular) particles. All combinations increased toughness as φ increased to 0.3, typically by ~ threefold, but with somewhat higher and more varied increases with irregular versus spherical and larger versus smaller angular particles.

Lange [30] measured (DCB) fracture energies for his composites of Al_2O_3 particles in a glass matrix with very similar expansion to that of the Al_2O_3 particles that were of finer sizes averaging ~ 3.5 and 11 μm (and irregular in shape) and larger (~ 44 μm) ~ spherical. Fracture energies increased with increasing φ, and generally with increasing *D*, especially at the largest *D* where increases were nearly fivefold those of the glass itself. He showed that the fracture energies were linear as a function of the mean free particle spacing (λ) per Eq. 9.1 and were generally consistent with his line tension model (Fig. 8.4).

Carlström et al. [77] reported fracture toughnesses of alumina-containing electrical porcelains using three indentation techniques (IF, I, and a modified IF technique) giving respective average values of 1.1, 2.1, and 1.5 $MPa{\cdot}m^{1/2}$ (with coefficients of variation of 10% or less). Such values are ~ 50–200% higher than for similar silicate glasses alone, e.g. ~ 0.7 $MPa{\cdot}m^{1/2}$. Some of this increase may be due to microcracking, indicated by microscopic and acoustic emission techniques, associated mostly with the quartz particles, e.g. Kirchhoff et al. [100] showed that while microcracking is dependent on thermal history as expected, it follows the expected decrease with decreasing quartz particle size. Thus finer particles required greater cooling to cause microcracking, e.g. 1 μm particles had extensive acoustic emission on cooling to ~ 300°C, while ~ 0.6 μm particles required cooling to 100+°C. On the other hand, Banda and Messer [101], who give (NB) toughness in terms of starting quartz particle sizes, showed toughnesses of ~ 2 $MPa{\cdot}m^{1/2}$ at starting particle size (*D*) of < 20–25 μm which then drop rapidly to ~ $^1/_2$ these values by *D* ~ 50 μm, the drop (of both toughness and strength, Fig. 9.4) possibly corresponding to serious microcracking.

Beall et al. [102] obtained different toughness values by different techniques in a crystallized glass, where much of the toughening was attributed to microcracking from the monoclinic canasites crystals ($Ca_5Na_4K_2Si_{12}O_{30}F_4$), as is indicated by the marked toughness decrease as test temperatures increased (Fig. 11.1). NB tests gave ~ 4.4 $MPa{\cdot}m^{1/2}$, while IF values started at ~ 2 $MPa{\cdot}m^{1/2}$ at smaller crack sizes but reached the NB level and were constant for crack sizes ≥ 200 μm, while I values started at ~ 1 and saturated at crack sizes of ≥ 200 μm at ~ 1.8 $MPa{\cdot}m^{1/2}$. Baik et al. [103] reported (I) toughnesses of a glass crystallizing to yield flurophlogopite crystals of 1.2 to 2.2 versus ~ 0.8 $MPa{\cdot}m^{1/2}$ for the parent glass. Anusavicé and Zhang [104] reported toughnesses (IF) in the range of 1.5–2.5 $MPa{\cdot}m^{1/2}$ in a $LiO\text{-}Al_2O_3\text{-}CaO\text{-}SiO_2$ glass that yields a complex crystallized structure and very high (95%) levels of crystallization.

Hing and McMillan [105] measured (NB) fracture energies and other properties of a glass giving $Li_2O{\cdot}2SiO_2$ crystallites giving values from ~ 17 J/m^2

for the starting glass to ~100 J/m^2 generally increasing as ϕ increased from 0 to 0.35–0.67. They showed that these values also increased ~ linearly as a function of λ^{-1}, where λ= the mean spacing between crystallites, noting that this correlation was functionally consistent with the line tension model (Fig. 8.4) but noted that this mechanism of toughening was inadequate to account for the ~ sixfold increase in fracture energies. They did not comment on the translation of these energy values into fracture toughnesses of 1 to ~ 3.5 MPa·$m^{1/2}$ using their E values (measured from deflections during flexure strength testing, which appear to be low by ~ 30–40% across the range of values reported, Chap. 9, Sec. III.A). Cook et al. [80] also indicated substantial increases in toughness due to crystallization of a similar LiO_2-SiO_2 glass, showing that I toughness values increased ~ 20–30% (as did strengths) as the size of the highly elongated crystallites increased (from equivalent radii of 2.8–6.7 μm) and λ nearly doubled from 1.2 to 2.3 μm as ϕ decreased from 0.33 to 0.2. Earlier, Morena et al. [106] also showed ≥ increases in toughness of a crystallized cordierite glass (Pyroceram 9606®) over the parent glass. They also showed their I toughness values increasing with increases in both crystallite size, D, and λ, both from ~ 0.5–1.5 μm, but only at higher indent loads (e.g. > 2–5 N). Govila et al. [107] measured a low fracture energy of ~ 2.2 J/m^2 (i.e. a toughness of ~ 0.6 MPa·$m^{1/2}$ for their crystallized lithium aluminum silicate (LAS) glass (grain size ~ ≤ 1 μm) via indent fracture with flaws ~ 10–100 μm. The low values appear to reflect fracture mainly through the glass phase, and less decrease in indented strengths at larger flaw sizes suggest a trend for increased toughness at larger crack sizes.

Hasselman et al. [108] later followed up on the above indent crack size dependence of toughness in composites of 0–35 v/o of Al_2O_3 spheres (25 ±5 μm) in borosilicate glasses of 75 m/o SiO_2 with B_2O_3/Na_2O molar ratios adjusted to give thermal expansions relative to that of Al_2O_3 of +2.7,+ 0.7, or - 3.7 ppm/°C. They showed all three starting glasses showed crack sizes uniformly increasing as indent load increased to give nearly identical toughness values independent of crack size as expected. However, the composites showed distinct breaks in such crack–load curves indicating a region over which crack sizes being introduced were constrained, which corresponded to toughness being first independent of load and then increasing with load and then saturating (i.e. an R-curve-type effect). They noted that this region of changing crack size–toughness behavior corresponded to the region over which the crack size and the mean free path between the Al_2O_3 particles (λ) were ~ equal and is thus consistent with earlier work indicating strengths being a function of $\lambda^{-1/2}$, due to such constraint of cracks by the spacings between particles (Chap. 9, Sec. III.A). However, note that such λ correlation can reflect other mechanisms, e.g. correlation with the volume fraction dependence of E (see note at the end of this chapter).

Wolf et al. [109] reported that composites of 75 v/o Al_2O_3 particles in glass matrices (e.g. for dental purposes) had (I) toughnesses of ~ 4 MPa·$m^{1/2}$ over a range of glass expansions from ~ 2 ppm/°C less than to ~ that of alumina, with a few percentage decrease as the expansions approached each other. These toughness values were > three times those measured for the glasses themselves (and ~ three times those cited for similar porcelain compositions, which correlated with strengths of the studied bodies being nearly three times those of the similar porcelains cited, consistent with the finer size of the often somewhat tabular Al_2O_3 particles, 0.3 to 10 μm).

B. ZrO_2 Toughened Ceramic Composites

Transformation toughening of ZrO_2 is manifested in ZrO_2 bodies composed entirely of fine grain (e.g. a few μm) tetragonal ZrO_2 (referred to as TZP) or of a cubic stabilized ZrO_2 matrix with fine dispersed tetragonal particles (typically obtained by precipitation heat treatment, resulting in larger, e.g. ~ 50 μm grains, with submicron precipitates isotropically and uniformly distributed within the grains, but often with some excess at grain boundaries). These, especially TZP, bodies were discussed mainly in Chapters 2 and 3. However, a very common use of this mechanism is to incorporate ZrO_2 particles in a matrix of some other composition, with Al_2O_3 being a particularly important and common one due to its high Young's modulus and other attractive properties, uses, and costs, and hence the focus of this section.

The ϕ for a toughness maximum probably varies with the matrix composition, but the dominant factors are the size and composition, i.e. degree of stabilization, of the dispersed ZrO_2 particles. Thus Claussen's original study [16,110] of ZAT showed that both the ϕ value of the maximum and its toughness level increased as the unstabilized ZrO_2 particle size decreased (Fig. 9.5), with the maximum for ZrO_2 particle sizes averaging ~ 1+ μm being at $\phi \sim 0.15$. However, Becher's study [16,111] using sol derived unstabilized ZrO_2 that was finer (≥ 1 μm) in size and much more uniform in both size and spatial distribution showed the toughness (fracture energy) maximum at $\phi \sim 0.1–0.12$ (Fig. 9.6). Lange [112] subsequently corroborated that the toughness maximum for fine unstabilized ZrO_2 was at $\phi \sim 0.1$. He also showed that both the level of the toughness and the ϕ value at which it occurs increased as the degree of partial stabilization was increased, with the maximum toughness and ϕ being those of TZP, i.e. $\phi = 1$(Fig. 9.7).

While the primary toughening mechanism in ZrO_2 toughened bodies is transformation, microcracking can also occur and was proposed as a major source of toughening in ZTA composites [113–116], since larger, less stabilized, and higher volume fractions of ZrO_2 particles can result in substantial microcracking. Ono et al. [37] showed that additions of totally unstabilized ZrO_2 to Al_2O_3 resulted in essentially identical toughness increases as with additions of ZrO_2+ 3 m/o Y_2O_3 as the ZrO_2 additions increased, except that the unstabilized

ZrO_2 additions reached a maximum (I) toughness of ~ 6.5 MPa·m$^{1/2}$ at ~ 50 w/o ZrO_2 (from ~ 2.5–3 MPa·m$^{1/2}$ with no ZrO_2), while ZrO_2+ 3 m/o Y_2O_3 additions continued to increase toughness to a maximum of nearly 8 MPa·m$^{1/2}$ at 70–80 w/o ZrO_2. Lutz and colleagues [117–119] used this to produce bodies of much higher thermal shock resistance (Chap. 11, Sec. III.C). However, the unusual crack propagation in the duplex composites they made, partly by design and partly by chance, in bodies having a dispersion of particles of mixtures of monoclinic ZrO_2 + Al_2O_3 in TZP matrices, depended significantly on limited quantities of porosity left from sintering, which is of interest here. They dispersed, in powder of the selected TZP composition, agglomerates of the selected ZrO_2 + Al_2O_3 composition formed by spray drying and then HIPed or sintered the dual composite with the intent of obtaining a duplex population of microcracks, e.g. like that sketched in Fig. 8.2C. The unexpected result was that HIPing, which gave little or no residual porosity, did not produce any significant amount of microcracking or the resultant desired toughening, but sintering, which left some porosity in the dispersed ZrO_2 + Al_2O_3 (~ 20–60 μm dia.) particles, resulted in substantial microcracking zones along with substantial *R*-curve effects. As revealed by dye penetration, such zones were up to 1–3 mm wide about the mean paths of macrocracks. As microcracking increased, *R*-curve effects and fracture toughness and residual strengths after serious thermal shocks all increased, but initial strengths decreased substantially, e.g. from 1300–1700 MPa to 100–800 MPa [118,119]. The specific mechanisms by which the limited porosity plays such a key role in the microcracking scale is not known, but this indicates that such effects, e.g. possibly similar to those schematically suggested in Fig. 8.1B, deserve investigation. Finally, some bodies processed from melt processed eutectic particles indicate possible unique crack bridging effects [116].

Finally, while much of the toughening in composites with dispersed tetragonal ZrO_2 particles is from transformation of the ZrO_2 particles to the monoclinic phase, and this may be partly replaced, or supplemented, by microcracking, there are other effects to complicate the picture. These include other phases, e.g. possible nontransforming noncubic phases and possible metastable cubic phases, which are topics beyond the scope of this book. However, an important added factor central to this book is effects of dispersed ZrO_2 in addition to or instead of transformation and microcracking, since the presence of such particles can still in principle cause other toughening such as crack deflection, bridging, or branching. Thus Ruf and Evans [41] reported that addition of ~ 20 v/o monoclinic ZrO_2 particles increased (I and NB) toughnesses of ZnO by ~ 70%, with progressively less but still substantial toughening as 4 and 8 w/o Y_2O_3 were added to the ZrO_2. Based on acoustic emission, microcracking was ruled out, while crack deflection and bowing were suggested based on fractography. Further, Langlois and Konaztowicz [120] have subsequently reported ~ 100% increases in toughness (and strength, Chap. 9, Sec. III.B) of Al_2O_3 with 30 v/o cubic ZrO_2 particles. Other important evidence of dispersed ZrO_2 particles causing toughening without transforming or microcracking is the substantial toughness retained in PSZ ZrO_2 crystals with tetragonal precipitates as test temperatures increase to reduce and then eliminate any transformation, yet still have ~ twice the toughness of fully stabilized crystals (Fig 6.6.).

Wang and Stevens [121] investigated Al_2O_3-ZrO_2 composites of various microstructures using I toughness and strength measurements. Bodies fabricated with dispersions of individual, fine (~ 1 μm) ZrO_2 showed toughness increasing rapidly from ~ 3 MPa·m$^{1/2}$ at 0 v/o ZrO_2 and then more slowly, reaching a limit of ~ 7 MPa·m$^{1/2}$ at 12–20 v/o unstabilized ZrO_2. They also noted that increasing addition of ZrO_2 increased the amount of intergranular fracture and decreased the Al_2O_3 grain size from ~ 10 to ~ 3 μm (which, at least in part, accounts for increased intergranular fracture and the marked contrast to their strength trends, though the toughness levels are reasonably consistent with the strength levels). Bodies with addition of 2.5Y-TZP agglomerates (20–50 μm) instead of the unstabilized ZrO_2 particles, i.e. as a step in the direction of composites along the lines of Figure 8.2C, showed a linear increase in toughness starting at 20 v/o and reaching ~ 6+ MPa·m$^{1/2}$ at 40 v/o TZP, i.e. a slower initial increase with somewhat reduced toughness levels. Combinations of both types of ZrO_2 dispersions lead to linear increases of toughnesses, reaching values of 11+ MPa·m$^{1/2}$ at 40 v/o. They attributed toughening of the composites with unstabilized particles alone to transformation and microcrack toughening, that with TZP agglomerated alone to transformation and crack deflection, and all of these with both types of addition. The indicated presence of microcracking with all additions is consistent with the limited strengths (200–500 MPa) and indicates that the combined composite is a probable manifestation of either, or probably both, of those sketched in Figures 8.1B and 8.2C.

French et al. [38] showed IF toughness values decreasing linearly as the content of stabilized, cubic ZrO_2 (+ 8 m/o Y_2O_3) increased in very similar fashions for their *E* values (Sec. III), but the ratio of the end ZrO_2 and Al_2O_3 was higher for toughness, i.e. 0.76 for toughness and 0.6 for *E*. They also noted that there were no *R*-curve effects, which was consistent with the fine grain and particle sizes (about 5 μm for Al_2O_3 alone, somewhat greater for ZrO_2 alone, and somewhat less for mixed compositions), and that fracture of the Al_2O_3 alone was intergranular and that of the ZrO_2 alone was transgranular, with mixed fracture mode for mixed compositions.

Turning to composites with a mullite matrix, which has been the next most used matrix after Al_2O_3 for composites with ZrO_2, toughness increases with modest additions of ZrO_2 have been mixed. Thus Yuan et al. [39] showed NB toughness values increasing by 10–25% as the ZrO_2 addition increased to 25 v/o, i.e. from 2 to ~ 2.5 MPa·m$^{1/2}$ as-measured, increasing faster at lower v/o and possibly saturating at ~ 25 v/o ZrO_2. Probable corrections for the 2.6–8.7% residual porosity reinforce this trend and raise the net increase to ~ 40%. However, Lathabai et el. [122] reported an I toughness of 2.2 MPa·m$^{1/2}$ in their reaction processed ~ 64 w/o mullite + 5 w/o Al_2O_3 + 31 w/o ZrO_2 (about $^1/_2$ tetragonal and $^1/_2$ monoclinic) with 1–2 μm mullite grains and ~ 1 μm (mostly intergranular) ZrO_2 particles. No evidence of transformation toughening or *R*-curve effects was observed, consistent with primarily transgranular fracture and no significant crack bridging or deflection (but high strength of ~ 330 MPa, consistent with the fine grain and particle size). On the other hand, Ishitsuka et al. [40] reported that addition of ~ 5 v/o mullite (mostly as small elongated grains) to 3Y-TZP increased their I toughness from ~ 7.5 to ~ 12+ MPa·m$^{1/2}$, which then decreased with further mullite additions (especially from ~ 30–50 v/o mullite to ~ 5.5 MPa·m$^{1/2}$ at the latter) to ~ 3.6 MPa·m$^{1/2}$ for pure mullite. (Thus while their value for pure mullite is higher, their trend for toughness with ZrO_2 additions on the mullite rich side are not inconsistent with those of Yuan et al. above.)

C. Composites with Nontransforming Particles in Polycrystalline Matrices

This section addresses the fracture toughness of extensively investigated ceramic composites of nontransforming particles in a polycrystalline matrix as a function of the microstructural parameters, especially of the dispersed second phase. Composites are addressed in the order of mixed oxide–nonoxide and then all nonoxide composites (the most important all-oxide composites having been reviewed in the previous section). A detailed review of the many studies of such composites is not attempted here, as it would be too voluminous and ineffective to the goal of this book (see Campbell and El-Rahaiby's [27] compilation). Instead, the focus is on studies that address microstructural parameters, the focus of this book. In particular, having introduced the important parameter of the dispersed particle size, *D*, in ZrO_2 toughened materials above, which is expected from models of both transformation toughening and microcracking, let us consider it more broadly in other nontransforming composites.

Composites of Al_2O_3 with dispersed SiC particles have received considerable attention. Nakahira and colleagues [123] showed that (I) toughness generally rose to a maximum increase of ~ 40% as the SiC content increased, with the maximum toughness value progressively decreasing as the hot pressing temperature increased (from 1600 to 1800°C), and the ϕ for the maximum increased from 0.1 to ≥ 0.4 for SiC with D ~ 2 μm. With 8 μm SiC particles, overall toughness increases were lower, e.g. 25%, at ϕ= 0.1, but again overall toughness values decreased as hot pressing temperatures increased, with the highest temperature (1800°C) giving toughness decreasing, with a modest minimum (~ 25% decrease) at ϕ ~ 0.03. In another study [124] they showed much less effect of hot pressing temperatures, but with similar toughness values, and a distinct maximum for both SiC particle sizes at $\phi \sim 0.05$, the maximum for the larger SiC particle size being ≥25% lower.

Yasuoka et al. [125] also showed that in Al_2O_3-SiC composites the CNB toughness increased, e.g. ≥ 40% at $\phi = 0.3$, but with an inverse dependence on SiC particulate size. Thus SiC particles, $D \sim 4$ µm, gave a lower maximum of ~ 20% increase at $\phi = 0.1$ but an ~ 35% increase with the finest SiC particles (~ 0.4 µm) and a decrease to no increase by $D = 9$ µm. They, however, also showed that the toughness increased ~ 15% as the grain size of the Al_2O_3 matrix increased by ~ 100% from $G < 1$ to > 2 µm, being linear as a function of $D^{1/2}$ (while strength still varied as $D^{-1/2}$). Thompson and Kristic [126] showed CNB toughness only increasing at their upper SiC particle content of 20 v/o, and only by ~ 10%. Similarly, Zhang et al. [127] showed the change in (I) toughness of hot pressed bodies with SiC particles at the maximum addition (24 v/o) was + 42, + 3, and -6% for SiC particles with mean sizes of 2.5, 0.2, and 0.05 µm. Thompson and Kristic, along with Nakahira et al. [123] and Stearns and Harmer [128] have all showed substantial reduction in the Al_2O_3 matrix G as the level of SiC addition increased for a given densification temperature, but with G increasing in the composites at constant ϕ as the processing temperature increased.

Another composite with an oxide matrix and nonoxide particles that has seen considerable investigation is Al_2O_3-TiC, which has been in commercial production for cutting tools and wear applications for a number of years. While much of the development work on this system was done before fracture toughness and mechanics evaluations were commonly available, there is some information on its toughness behavior. Wahi and Ilschner [42] showed that NB toughness using a machined notch increased from 4.1 to 6.7 MPa·m$^{1/2}$ as TiC particle (~ 1 µm) content increased from 0 to 35 v/o, while tests with natural cracks showed a statistically uncertain increase from 4.6 to 5 MPa·m$^{1/2}$. Yasuoka et al. [125] presented some of the limited data on TiC particle size effects showing that for 10 v/o TiC particles, toughness increase was ~ 0 for $D = 0.5$ or 13 µm, with a maximum increase of ~ 20% at $D \sim 5$ µm. Other data also indicates a TiC particle size dependence (Fig. 9.10).

Turning to all nonoxide particulate composites, consider first data for composites of Si_3N_4 with dispersed SiC particles of differing amounts and sizes. Lange's (DCB) toughness [129] clearly showed the level of toughness increasing as D of the dispersed SiC particles increased, whether or not correction is made for the limited amount of porosity in some bodies (Fig. 8.16). Further, only the largest SiC particle size (32 µm) showed any increase in toughness as ϕ increased, being a maximum at $\phi = 0.1$–0.2. Data of Tanaka et al. [130] on the same system, but with finer SiC particle sizes (and use of IF for K), is a logical extension of Lange's data, which is reasonable since they are expected to have similar microstructures of SiC particles dispersed intergranularly between the Si_3N_4 grains. Data of Sawaguchi et al [131] on composites of this system made by

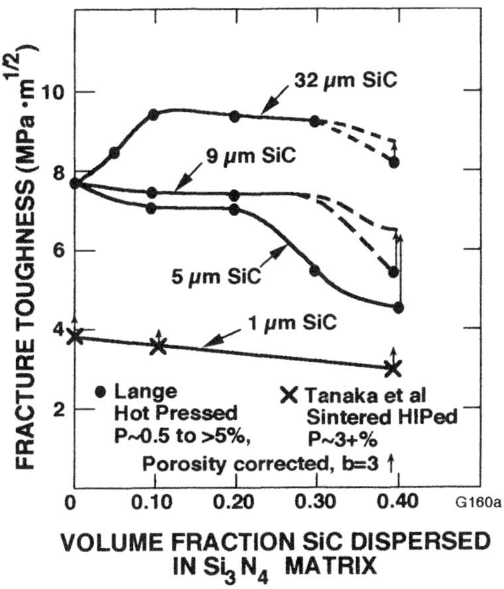

FIGURE 8.16 Fracture toughness of Si_3N_4-SiC particulate composites versus the volume fraction of dispersed SiC particles for different indicated particle sizes of Lange [129] and Tanaka et al. [130] at ~ 22°C. Note vertical arrows indicating corrections for limited porosity in some bodies, and that the SiC particles in all bodies are intergranular, versus those of Sawaguchi et al. [131], which have many of the 50–100 nm SiC particles within the finer (~ $^1/_2$ μm) Si_3N_4 grains. (Original plot (updated). From Ref. 18. Published with permission of the *Journal of Materials Science.*)

CVD production of the mixed powder, which when densified by hot pressing with Al_2O_3 + Y_2O_3 gave Si_3N_4 grains ~ $^1/_2$ μm with SiC particles ~ 50–100 nm, many of which were within the Si_3N_4 grains, with intermediate toughness (I) which was a maximum at ϕ= 0.1. The substantial intrangranular character of the SiC particles is clearly a possible reason why the toughness is not closer to that of Tanaka et al., since for intergranular particles reductions in toughening increases should diminish at small SiC particle sizes. In contrast to the overall substantial effects of dispersed particles on toughness shown, substantial effects of SiC particle size will also be shown on room temperature flexure strengths, but with the particle size effects on strength (Chap. 9, Sec. XII) and toughness being opposite, as is often found for other composites below, and sometimes for monolithic ceramics, especially some self-reinforced Si_3N_4 (Chap. 2, Sec. III.D).

Two other systems that have also received considerable attention are SiC-TiC (Fig. 8.17) and SiC-TiB_2 (Fig. (8.18), including commercially available bodies of

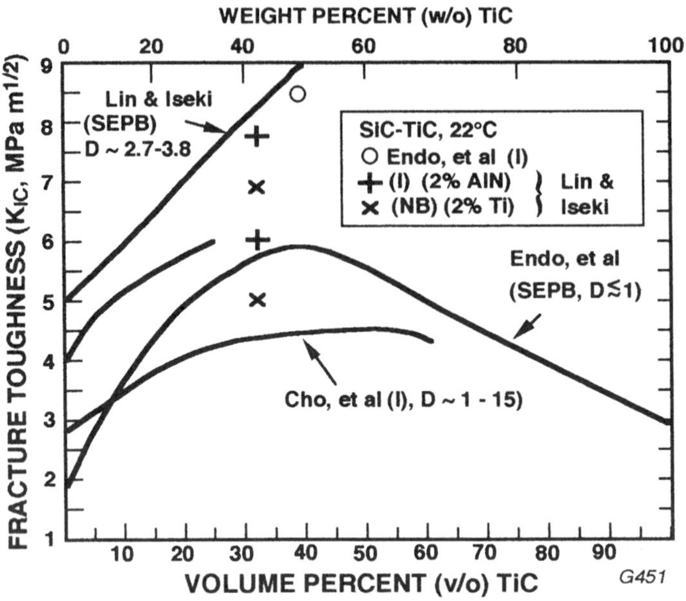

FIGURE 8.17 Fracture toughness versus volume percent TiC particles from Endo et al. [48], Cho et al. [132], Wei and Becher [133], and Lin and Iseki [134,135]. Note (1) designation of toughness tests and different values for testing the same body by different methods (Endo et al.), (2) use of different densification aids (Lin and Iseki), and (3) average *D* values, except the range is shown for Endo et al. (and that of Becher and Wei was similar, i.e. to ~ 11 μm).

sintered α-SiC with 15 w/o TiB_2 and B+C additions (also used in several experimental bodies). Besides either showing or supporting toughness maxima at ~ 40 v/o TiC, the data indicates higher toughnesses with larger TiC particle size over the limited range investigated. A summary of much of the available SiC-TiB_2 data shows similar trends as well as a common range of variation in the level of toughness and the composition for a toughness maximum, e.g. a toughness maximum at ϕ ~ 0.25–0.4 with some indications of increased toughness with increased partucle size (*D*). A complicating factor is SiC grain size variations, better characterized for the SiC-TiB_2 system but also indicated by Lin and Iseki's data [134,135] showing higher toughness with AlN versus Ti additions, with the former giving larger SiC *G* (and TiB_2 *D*) values. Another is different toughness values due to variations in test technique and method, e.g. higher values for NB. Some variations also probably reflect the extent of development (i.e. high for the material of McMurtry et al. [138], which impacts the ranges of dispersed particle and matrix grain sizes. Note that Yoon and

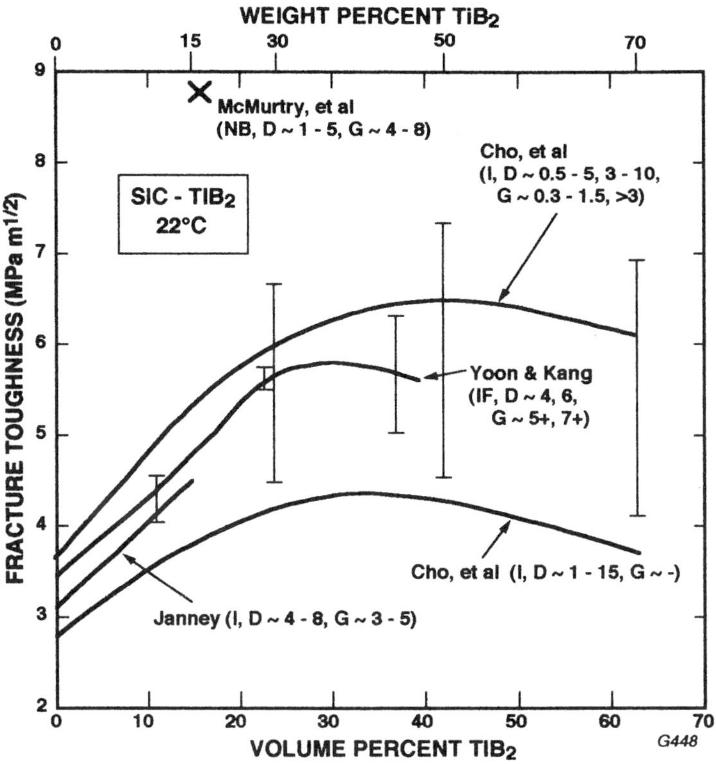

FIGURE 8.18 Fracture toughness of SiC-TiB_2 particle composites versus volume percent TiB_2 from Cho et al. [132], Yoon and Kang [136], Janney [137], and McMurtry et al. [138]. Note toughness measurement methods indicated along with the average particle size (*D*) and average matrix grain size (*G*). Also note (1) that two sets of values shown for Yoon and Kang reflect two different processings, while the extremes of values for three processing variations are shown for data of Cho et al. (the third value was typically substantially closer to the higher one, hence the average trend line being above the average of the two extreme values) and (2) that the materials of Yoon and Kang, Janney, and McMurtry et al. were densified with B–C additions, while the others used Al_2O_3 and Y_2O_3.

Kang [136] showed toughness versus *D* was either more scattered at *D* ~ 2.5–5 μm or passing through a maximum of ~ 5.7 MPa·m$^{1/2}$ at *D* ~ 4 μm, but in either case then constant at ~ 5.4 MPa·m$^{1/2}$ for *D* ~ 6–12 μm (while they showed a continuous ~ 40% decrease in strength over this range). Recently Cho et al. [139] reported that SiC-50 w/o TiB_2 composites hot pressed with Al_2O_3 and Y_2O_3 additions had I toughness of 4.5 MPa·m$^{1/2}$, which increased on annealing and resultant coarsening of both the SiC and TiB_2 phases (and elongation of SiC grains with the α–β transformation to 7.3 MPa·m$^{1/2}$ due to enhanced crack bridging and deflection).

Other composites also show a particle size dependence of toughness. Thus Bellosi et al. [140] showed that addition of coarser (to ~ 7 μm) TiN particles to Si_3N_4 increased K by ~ 60%, while use of finer (<3 μm) TiN increased K by ~ 100% (with similar but smaller trends for strength). Nagaoka et al. [141] reported a K maximum of ~ 7.8 MPa·m$^{1/2}$ in composites of 10 v/o TiN particles in a Si_3N_4 matrix at a TiN D of ~ 4 μm, which was attributed to observed microcracking associated with the TiN particles (and probably assisted by the Si_3N_4 grain boundary phase). Crack deflection or bridging and associated R-curve effects have been shown in many of these composites [133–144]. Petrovic et al. [145] showed (I) toughness increasing to maxima at ~ 40 v/o additions of $MoSi_2$ particles to Si_3N_4 (+ MgO) and then decreasing. Composites with 10 μm $MoSi_2$ particles reached the highest maximum of > 8 versus > 5 MPa·m$^{1/2}$ with 3 μm $MoSi_2$ (from ~ 4.6 MPa·m$^{1/2}$ with no $MoSi_2$ and 1 w/o MgO). Composites with 10 μm $MoSi_2$ particles and 5 w/o MgO reached a maximum of > 6 MPa·m$^{1/2}$ (from 4 at ϕ= 0).

Sigl and Kleebe [146] showed that additional increases in (NB) toughness of B_4C + 20 or 40 v/o TiB_2 particles (~ 3 μm) to ~ 6.0 MPa·m$^{1/2}$ occurred when excess carbon was present, e.g. versus 3–3.5 MPa·m$^{1/2}$ with no excess carbon (and ~ 2.2 MPa·m$^{1/2}$ for B_4C alone). They showed that the excess carbon caused microcracking, mainly at TiB_2-B_4C boundaries, and was the source of the additional toughness.

A potentially interesting system is that of AlN-SiC, which can form either a complete solid solution or two phase bodies, or some mixture of each depending on processing. As discussed earlier, Ruh et al. [45] showed somewhat lower E values for two-phase versus solid solution bodies (Fig. 8.11). Landon and Thevenot [147] reported toughness (I) ~ constant over a range of compositions, i.e. 5.2 ± 0.3 and 4.7 ± 0.2 MPa·m$^{1/2}$ respectively for compositions with 45–90 w/o α-SiC and 45–80 w/o β-SiC despite variations in E (and some of strength, related at least partly to changes in G, Chap. 9, Sec. III.D). Li and Watanabe [148] reported toughness increasing from ~ 3.6 MPa·m$^{1/2}$ to a maximum of ~ 4.6 MPa·m$^{1/2}$ at 10 m/o AlN (i.e. ~ a 30 % increase, about the same as the maximum increase in strength, but this was at 5 m/o AlN, Chap. 9, Sec. III.D).

Even quite low levels of particulate addition to a matrix can have considerable impact in some cases, especially at quite fine particle sizes. Thus more limited additions of fine (e.g. nm scale) SiC particles in Al_2O_3 and Si_3N_4 matrices were noted. Addition of 1–5 v/o of submicron β-SiC particles to $BaTiO_3$ by Hwang and Niihara [43] increased fracture toughness (and Young's modulus and Vickers hardness) by 20–25% along with reducing G to < 1/3 of the value with no additions, i.e. from 1.4 to 0.4 μm and a transition from tetragonal to pseudocubic structure at the higher loadings (Fig. 8.19).

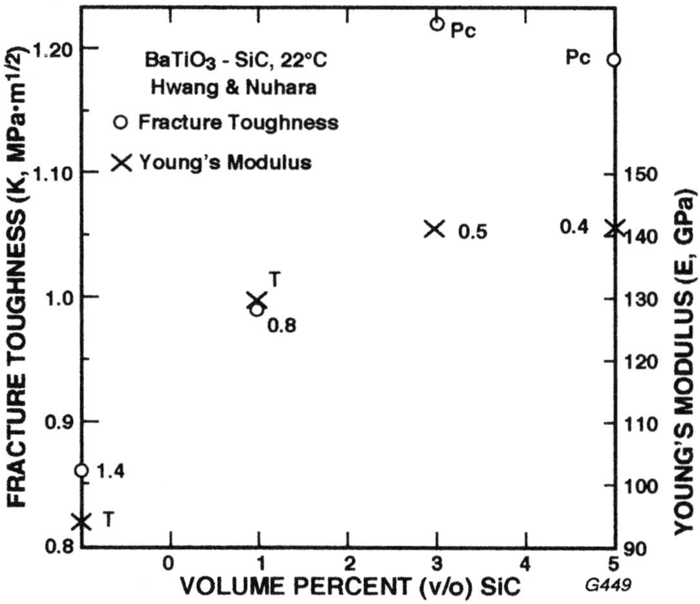

FIGURE 8.19 Young's modulus and fracture toughness versus volume percent (v/o) of submicron β-SiC particles to $BaTiO_3$. Note the initial marked increase in *E* and *K* as well as the grain sizes of the $BaTiO_3$ (in μm) for each composition next to the *K* values, and the phase of the $BaTiO_3$ for each composition next to the Young's modulus value (T= tetragonal and Pc= pseudocubic). (Data after Hwang and Niihara [43].)

(Strengths increased by as much as 100–200%, but with much of this due to reduction of *G* at most sintering temperatures, and some effects of the tetragonal to cubic phase transformation for bodies with 3 and 5 v/o SiC, see Fig. 9.11.) Similarly, limited additions of nanoscale β-SiC particles to Si_3N_4 by Sasaki et al. [44] increased toughness nearly 50% from ~ 4.1 MPa·m$^{1/2}$ to a maximum of ~ 6.1 MPa·m$^{1/2}$ at 5 v/o and then decreasing back to the 0 v/o level at 20 v/o (while strengths increased < $^1/_2$ as much, peaked at 10 v/o, and then rapidly decreased below 0 v/o levels).

Finally note three facts from these observations. First, there is substantial evidence for a particle dependence of toughness in at least several systems, indicating probable broad impact of *D*, e.g. K often increasing with *D* (and *G*), at least to some optimum *D*. Very fine particles may act as an extension of larger particle effects but may have other effects in addition or instead of larger particle effects; more research is needed in this area. Second, there is no clear evidence for the sign of the particle strains from expansion differences with the matrix determining the nature and extent of toughness effects of the dispersed particles. Thus while SiC or TiC particles have thermal expansions < for an Al_2O_3 matrix and these same particles have expansions > for an Si_3N_4 matrix, as do TiB_2 particles in a SiC matrix, all showed toughening with particle additions and an optimum particle size. While microcracking is a probable factor in many and possibly all of these, as shown for the SiC-TiB_2, composites, more extensive and detailed evaluations are needed, e.g. effects of particle size distributions, mixing, agglomeration, and orientation are needed along with statistically significant data. Third, while some composites developed earlier were not tested for *R*-curve effects, most of these composites show such effects with typical large cracks as used for most toughness measurements.

D. Ceramic Platelet and Whisker Composites

A review of the ceramic platelet composites literature shows similar results to particulate composites discussed above, in particular *K* frequently passing through a maximum as ϕ increased. Thus addition of 5–20 vol% Al_2O_3 platelets (e.g. ~ 10 μm dia.) to 2Y- and 3Y-TZP) matrices by Heussner and Claussen [149] using isopressing, sintering, and HIPing increased *K* (CNB) by ~ 15–40% at maxima at ϕ= 0.05, while addition of the platelets to a 12 Ce-TZP matrix gave a minimum in *K* at ϕ= 0.05 (the latter minimum attributed to effects of nontransformable particles suppressing transformation in the Ce system). Huang and Nicholson [150], using a 4.5Y-TZP and tape casting followed by sintering and HIPing showed *K* maxima at ϕ= 0.15–0.3, with IF showing a substantially greater maximum than the CNB (i.e. similar to Heussner and Claussen). They also showed that larger platelets (dia. ~ 12 μm versus ~ 2 and 1 μm) and higher aspect ratios (respectively ~ 12, 5, and 1) gave somewhat greater *K* increases and that orientation of the platelets randomly, parallel, or perpendicular to the stress axis had little effect) (see also Chap. 11, Sec. III.E).

Chen and Chen [47] reported that in situ formation of various hexaluminate platelets with overall dimensions similar to those of the matrix Al_2O_3 grains (e.g. 2–15 μm) resulted in a maximum toughness, i.e. ~ 4.3 versus ~ 3.0 MPa·m$^{1/2}$ for the martix alone, at ϕ= 0.3. Fracture was mixed inter- and transgranular at peak toughness, and higher volume fractions had crack bridging by elongated aluminate grains. Similar but greater increases in (SEPB) toughness from 3.5 to 6 MPa·m$^{1/2}$ were reported by Yasuoka et al. [151] by such in situ growth of hexaluminate platelets combined with inducing some platelet character of the Al_2O_3 grains via doping of 240 ppm SiO_2. The latter addition enhanced intergranular fracture and hence grain bridging and thus was a major factor in this addition being the source of ~ 3/4 the total toughness improvement, but reducing strengths by ~ 6%. Kim et al. [152] reported nearly identical (I) toughness increases, maxima (at 1 m/o), and subsequent modest decreases in Al_2O_3 doped with 0.5 to 3 m/o Na_2O + MgO to grow in situ beta alumina platelets (with greater strength increases to a more pronounced maximum at 0.5 m/o, but with some of this due to *G* reduction). Koyama et al. [153] reported that (CNB) toughness of Al_2O_3 with platelet grains due to low additions of CaO + SiO_2 increased nearly the same as was found by Yasuoka et al. and showed substantial *R*-curve effects. Toughness increased with either diameter (*d*) or thickness (*t*) to the $^1/_2$ power (or $d^{5/6}t^{-1/3}$), which was also true for the lesser increases as *G* increased in bodies with equiaxed grains. However, as noted in Chap. 3, Sec. IV.A, strengths of bodies both with platelet grains and with equiaxed grains increased as toughness deceased, following typical Al_2O_3 σ–$G^{-1/2}$ behavior. An and Chan [154] also demonstrated substantial toughening and *R*-curve effects in Al_2O_3 toughened via in situ formation of 30 v/o $Al_2O_3{\cdot}CaAl_{12}O_{19}$ platelets in Al_2O_3, but noted that this came at the expense of strength, which they acknowledged was inevitable, as is now being increasingly recognized (Chap. 9, Sec. III.E).

Nischik et al. [155] evaluated composites of either Al_2O_3 or SiC platelets (both ~ 15 μm dia. and ~ 1 μm thick) made by variations of powder processing with a mullite matrix via sintering, with and without HF or oxidative pretreatment of the SiC platelets. Toughness (IF, and strengths) were apparently measured for cracks parallel with the pressing or pressure filtration directions. The Al_2O_3 platelet composites (ϕ= 0.1) had either somewhat lower or higher toughnesses than the matrix alone depending on processing, i.e. 1.9–2.8 versus 2.2–2.6 MPa·m$^{1/2}$ for the matrix (but always somewhat to substantially lower strengths than the matrix, i.e. 150–240 versus 260–310 MPa for the matrix). SiC platelet composite toughnesses for ϕ= 0.1 ranged from about as low as with Al_2O_3 platelets to higher values, i.e. 1.9–3.3 MPa·m$^{1/2}$, while strengths ranged from lower to higher values compared to Al_2O_3 platelet composites, i.e. 120–340 MPa, thus higher than for the matrix alone. While HIPing gave among the highest toughnesses (and strengths) it did not give the highest values; varying ϕ from 0.05 to 0.2 did not show clear increases in toughness (or strength), nor did HF etching or preoxidation of the SiC platelets. However, the substantial interfacial, i.e. matrix-SiC platelet, fracture with as-received SiC platelets was reduced with preoxidation that then often resulted in substantial transgranular fracture of the SiC platelets, e.g. often ~ parallel with the plane of the platelets.

Chou and Green [156] investigated the mechanical properties of composites with 10–30 v/o of α-SiC platelets with average diameters and thicknesses (in μm), and aspect ratios of 12, 2, and 6, and 24, 6, 4 in an alumina matrix made by hot pressing. Composites with the larger SiC platelet showed slight decreases of *E* as ϕ increased to 0.15 and then dropped sharply by ϕ= 0.2 and further by ϕ=0.3, while composites with the smaller SiC platelets showed that *E* increased continuously but modestly over this range. The differences were shown to be due to microcracking mainly in the plane of the platelets (~ parallel with the hot pressing surfaces) due to expected size dependence of microcracking, but also a volume fraction dependence. Fracture toughness (IF) measurements for cracks propagating perpendicular to the plane of hot pressing were nearly the same for the smaller platelets, increasing almost linearly from 4.3 to 7.1 MPa·m$^{1/2}$ as ϕ increased from 0 to 0.3. The larger platelet composites' toughnesses paralleled the Young's modulus behavior, i.e. increasing somewhat less to 5.3 MPa·m$^{1/2}$ at .ϕ 0.15 and then decreasing to 3.5 (ϕ= 0.2) and then to ~ 2.9 MPa·m$^{1/2}$ at ϕ= 0.3. They did not measure toughness for the third normal direction (i.e. crack propagation normal to the hot pressing axis, thus ~ parallel with the larger surfaces of the platelets), which they recognized would produce lower toughness values. However, they also measured toughness from indent crack sizes in composites of the smaller platelets with ϕ= 0.3, again showing little difference for the two crack orientations parallel with the hot pressing axis (hence ~ normal to the plane of the platelets), i.e. 6.7–7.0 MPa·m$^{1/2}$ in good agreement both relatively and absolutely with their indent fracture measurements. The indent toughness for cracks normal to the hot pressing axis, i.e. ~ parallel with the platelets, was ~ 3.3 MPa·m$^{1/2}$.

Chaim and Talanker [157] reported substantial (NB) toughness increases with addition of α-SiC platelets (50–250 μm dia., 5-25 μm thickness) in a cordierite glass matrix. Most of the increase occurred from 0 to 10 v/o SiC addition, e.g. from 1.6 to 2.5 MPa·m$^{1/2}$, i.e. ~ 50% increase, followed by only another 10–15% increase at 30 v/o SiC for crack propagation parallel with the hot pressing direction. Crack propagation in the normal direction gave only modestly lower K values at 10 and 20 v/o and no difference at 30 v/o SiC, so preferred orientation had limited effect. Crystallizing the glass matrix increased composite toughnesses by about the same amount it increased the toughness of the matrix alone, which was to ~ 1.9 MPa·m$^{1/2}$. While some of these toughness trends were similar to those for strength, there were also significant differences.

Cooper et al. [158] showed that WOF of alumina-graphite refractories increased linearly from 20 J/m^2 at ~ 5 v/o graphite to ~ 80 J/m^2 at 40 v/o graphite, where the increase appeared to be beginning to saturate.

Mitchel et al. [159] reported toughness (IF) increased to 6.2–7.3 MPa·m$^{1/2}$ for crack propagation normal to the plane of hot pressing (hence the ~ plane of alignment of the platelets) in composites of thin α-SiC platelets (~ 25 μm dia., ϕ=0.2) in a β-SiC matrix. The key to the toughness was coating the SiC platelets with a thin layer of Al_2O_3 powder before mixing into the fine β powder (with B + C sintering additions). For crack propagation parallel with the plane of hot pressing (hence also with the ~ alignment of the planes of the platelets) the toughness was 3.8–4.6 MPam$^{1/2}$, i.e. only slightly above that of the matrix alone (3.5–4.0 MPa·m$^{1/2}$). (However, strengths were the same for both test orientations despite a nearly 100% difference in toughness.)

Baril et al. [25] also showed that uniform addition of up to 30% SiC platelets (11–24 μm dia.) to Si_3N_4 increased K ~ 40% (but left strength unchanged or reduced by up to 20%, while the Weibull modulus was increased by up to 130%). Strength retention was distinctly best with the finest platelets. Weibull modulus results were mixed, while *K* results were generally independent of platelet size. Hanninen et al. [160] reported greater *K* increases respectively with larger vs. smaller platelets and particulates (of similar diameter as the small platelets), but with corresponding greater *S* decreases as the volume fraction of particles or platelets increased. Claar et al. [161] reported high strengths (450–900 MPa), *K* values (11–23 MPa·m$^{1/2}$), and Weibull moduli (21–68) for reaction processing Zr and B_4C to produce in situ formed ZrC grains and ZrB_2 "platelets." However, much and possibly all of the improvement in mechanical properties and specifically toughness is due to the residual Zr content, i.e. properties increased with increased free Zr content, and plastic deformation of Zr ligaments on fracture surfaces was observed. Also, the "platelets" were not disks, as in most of the previous cases above, but were much closer to thicker whiskers (typically several microns in thickness).

More investigation has been conducted on ceramic whisker composites, in part since whiskers were available and showed success in earlier development of Al_2O_3-SiC whisker composites, which are in commercial production and provide a more comprehensive starting point, especially studies of Becher and colleagues [85,162–166]. They showed fracture toughness increasing as ϕ increases, indicating a probable maximum at $\phi \sim 0.3$ (as have others, though some of this maximum may reflect increasing densification difficulty as ϕ increases), as well as that toughness increases were linear as a function of $\phi^{1/2}$ and of $r^{1/2}$ (where *r* is the whisker radius). They also showed that toughness increased as the Al_2O_3 matrix grain size increased, e.g. by ~ 1 MPa·m$^{1/2}$ on going from $G \sim 1$–2 to 4–8 μm, and again on going to ~ 15–20 μm (and that there was considerable transgranular fracture of the Al_2O_3 matrix grains, especially as *G* increased), and that lower surface oxygen on the whisker surfaces gives higher toughnesses.

Other studies are generally consistent with these trends, e.g. Yang and Stevens [59] showed linear increases in fracture energy as ϕ increased (to 0.3), with greater increases with whiskers leached to remove surface oxide versus as-received whiskers, and also showed substantial matrix transgranular fracture. Yasuda et al. [167] showed fracture energy and toughness increasing with both ϕ and *r*, but that they were linear with ϕ^1 not $\phi^{1/2}$ and with r^2/l (where l = whisker length) not $r^{1/2}$ and that the level of the linear increases varied substantially with different sources of whiskers, in part probably reflecting the above whisker dimension effects but also possible effects of composition such as surface oxygen. Krylov et al. [168] reported (NB) toughness (and strength) increasing rapidly from 4.5 MPa·m$^{1/2}$ as whisker diameters increased from 0.1μm, but saturating at 6.4 MPa·m$^{1/2}$ at diameters of ~ 2 μm in composites with 20 v/o whiskers. Baek and Kim [169] reported nearly linear increases in toughness with 20 v/o whiskers as their lengths increased from ~ 9–18 μm, but with differing rates and levels, i.e. 3 to 4.5 and 4.7 to 5.2 MPa·m$^{1/2}$ respectively for NB and CNB tests.

They attributed the increased benefit of longer whiskers to increased whisker pullout and reported that their data compared favorably with a model for effects of fiber (whisker) pullout.

Tiegs's [170] compilation of Al_2O_3-SiC whisker composite data allows some additional observations to be made on this system. Thus the anisotropy of toughness for crack propagation nominally normal versus parallel with the whiskers (i.e. respectively normal and parallel with the plane of hot pressing) varied from 1.1 to 2.8 for mostly 30%, averaging 1.7±0.6. While much of the variation reflects different whisker and fabrication parameters, and the anisotropy is generally substantial, the toughness for crack propagation ~ parallel with the plane of hot pressing and of the ~ whisker alignment is still typically above that of the Al_2O_3 matrix alone, e.g. by as much as 20%.

Consider other oxide matrix whisker composites, where glasses, cordierite, TZP, and mullite matrices have been more common (Table 8.2). These composites also generally show toughness increasing as ϕ increases (e.g. to 0.2–0.3), but with considerable variation in results, much of which probably reflects variations in measurement techniques, matrix values, and processing differences. Wadsworth and Stevens [58] showed some dependence on SiC whisker dimensions similar to those seen in Al_2O_3 but emphasized benefits of increased whisker aspect ratio, but also noted that with similar aspect ratios larger whiskers increased toughness more than smaller ones (but had opposite effects on strength). They cited crack deflection, crack bridging, and load transfer as improving toughness, but noted there was no evidence of whisker pullout as a factor. The importance of whisker aspect ratio was also reported by Okada et al. [176] for their Y-TZP matrix composite with either in situ development, or direct addition, of mullite whiskers, where a pronounced toughness maximum was found at ϕ= 0.15. They showed that an important factor in (I) toughness increasing from ~ 7 to ~ 15 MPa·m$^{1/2}$ was the aspect ratio of the whiskers increasing from ~ 1.3 to ~ 2.5.

Wu et al. [172] showed toughness (and strength) with a mullite matrix reaching maxima (at ϕ= 0.4) for tests with stresses both parallel and perpendicular to the plane of hot pressing. Particular variability was indicated with TZP matrices. Thus Claussen et al. [177] reported toughness doubling from 6 to 12 MPa·m$^{1/2}$ with addition of 30 v/o SiC whiskers (but strengths were reduced by ~ 40%, as was also the case for Yang and Stevens [59], with both possible reactions and matrix grain size reductions due to the whiskers being noted as possible reasons for some of the limitations). However, Becher and Tiegs [178] and Ruh et al. [179] both report substantial and additive increases in toughness of mullite-based composites with combined additions of TZP and SiC whiskers.

Ceramic whisker composites with nonoxide matrices have also been investigated, with Si_3N_4 matrices and SiC whiskers being most common with both similar and different results and thus the focus of review for such nonoxide composites. Buljan et al. [174] were apparently the only investigators to compare directly composites of either SiC particles or whiskers in a Si_3N_4 matrix. They showed a modest toughness decrease of ~ 20% with fine (0.5 μm) SiC particles and a more modest increase with larger (8 μm) SiC particles at ϕ=0.3. These results were in contrast to increases of ~ 40% with SiC whiskers (Table 8.2), but almost all of the increase occurred from ϕ=0.2 to 0.3, indicating greater advantage of the whisker composites (but with some probable resultant anisotropy). (Note however that while both indent and indentation fracture toughness tests of the whisker composites gave similar trends with ϕ, the former gave results ~ 20 % lower.) Singh et al. [180] showed indent toughness increasing by 75% from ϕ=0 to 0.2 (with limited anisotropy found in the matrix itself). Shalek et al.'s [63] CNB tests showed substantial but lesser increases in toughness that were similar for different hot pressing temperatures despite significant changes in the matrix toughness. On the other hand Lundberg et al. [173] reported (I) and (IF) toughness decreasing by 18% from ϕ=0 to 0.2. Kandori et al. [181] showed CNB toughness for ϕ=0.1 decreasing as fabrication temperatures increased, but they were higher than that of the matrix, with this advantage increasing as temperature increased. Finally, Tiegs [170] showed that toughness at ϕ=0.2 increased ~ 10% as the whisker diameter increased from ~ 0.2 to 1.4 μm diameter, but with most of the increase occurring below 1 μm diameter, indicating a saturation of improvements with increased whisker diameter. Also, note that while toughness most commonly increased in these composites, the increases were more limited (even without the one decrease) relative to the toughness and Young's modulus of the matrix, and further that strength changes were at best modest increases and more commonly decreases (Table 2) (Chap. 9, Sec. III.E).

Dusza and Šajgalik [182] reported that Si_3N_4 bodies hot pressed with 10 or 20 w/o Si_3N_4 whiskers gave (IF) toughnesses respectively of 6.4 and 6.3 $MPa{\cdot}m^{1/2}$, similar to the matrix (6.1 $MPa{\cdot}m^{1/2}$) and I values for similar composites. They observed some whisker pullout similar to that for in situ toughened Si_3N_4 and noted that their results were more consistent with the model of Becher et al. [164] than that of Campbell et al. [183].

The first and most basic of two factors that should be noted is that microstructure is important in platelet and whisker composites, with more and clearer results for the latter. Thus improved toughness is indicated with increasing whisker aspect ratio and diameter as well as increased matrix grain size, though there are probable variations and limitations on the impact of these parameters (which may often be different than for strength, as is discussed in Chap. 9). The second factor, which is probably one of the factors changing or limiting microstructural effects, is stresses from thermal expansion differences between the dispersed and matrix phases. Again, while there appear to be variations in such differences, toughening is reported in platelet composites with matrices having greater expansion (TZP-Al_2O_3 and Al_2O_3-SiC) as well as lower expansion (SiC-Si_3N_4) than the platelets. The same is true of whisker composites with respectively greater matrix (Al_2O_3-SiC), ~ neutral (mullite-SiC), and lower (cordierite and other low-expansion glasses, as well as Si_3N_4-SiC) matrix versus whisker expansion. There are also important indications that the bonding between matrices and whiskers can be important, i.e. higher toughness with whiskers with lower surface oxygen content and where glassy phases are found between whiskers and matrix.

E. Ceramic Eutectic Composites

There has been considerable past study of directionally solidified ceramic eutectics, especially on systems giving uniform lamellar or rod structures, with the matrix and the rods or lamellae being both single crystal structures of definite crystal orientations. While much of this was again before fracture mechanics and toughness evaluations were common, so most data on these is presented in Chap. 9, there are a few examples of explicit toughness effects. Thus Stubican and colleagues [184,185] showed that (I) toughness for crack propagation normal to the ZrB_2 rods of the ZrC-ZrB_2 eutectic initially rose very slowly from ~ 3.2 MPa·m$^{1/2}$ at larger spacing (λ) between rods (e.g. > 2.5 μm) and then more rapidly to a maximum of 5 MPa·m$^{1/2}$ at λ= 1.85 μm; it then decreased rapidly to 3 MPa·m$^{1/2}$ at λ= 1.65 μm. Results for crack propagation parallel to the rods were very similar, with both showing pronounced maxima with substantial nonlinear increases and decreases versus $\lambda^{-1/2}$. These toughnesses are in contrast to corresponding values of ~ 1.7 and 1.9 MPa·m$^{1/2}$ respectively for ZrC and ZrB_2 alone. Toughness of ZrC-TiB_2 directionally solidified composites over the range of spacings of 9 to 4+ μm also showed ~ no anisotropy as a function of crack propagation parallel or normal to the solidification direction [185]. However, toughness increased linearly versus $\lambda^{-1/2}$ with no indication of reaching a maximum over this range; it started from a level at or below those for the constituents, and the ease of cracking implied that this system had lower toughness than the ZrC-ZrB_2 system. Note that this use of the spacing of second phase entities, whether particles, rods, or plates, while unfortunately neglected in investigations of most other composites, is an important factor. This is shown in Chap. 9, Sec. III.F, not only for such eutectic composites but also for particulate composites with glass and possibly other matrices, along with the fact that the λ dependence has important connections to the *G* dependence of properties, especially strength, of monolithic ceramics.

Similarly Brumels and Pletka [186] showed that (I) toughness for crack propagation normal to the lamellar structure of directionally solidified NiO-CaO eutectic rose from ~ 1.6 MPa·m$^{1/2}$ at λ ~ 1.3 μm to ~ 2 MPa·m$^{1/2}$ at λ ~ 2 μm. For comparison purposes, the toughness of NiO crystal cleavage is ~ 1.2 MPa·m$^{1/2}$ and that for CaO is expected to be ~ or <that of MgO (~ 1 MPa·m$^{1/2}$) (Table 2.1). Mah et al. [187] reported toughness of directionally solidified eutectic specimens

of alumina-YAG of ~ 4.3 MPa·m$^{1/2}$ [versus ~ 1.4 MPa·m$^{1/2}$ for YAG crystals on the (111) plane]. More recently Yang et al. [188] reported that (I) toughness for crack propagation ~ normal to the lamellar structure of as-directionally solidified $Y_3Al_5O_{12}/Al_2O_3$ eutectic was 2.4 MPa·m$^{1/2}$ versus 2 MPa·m$^{1/2}$ for normal crack propagation. Heat treatment that coarsened the microstructure reduced these values by respectively ~ 30 and 15%.

Though investigated more, detailed data for the Al_2O_3-ZrO_2 system is limited but shows higher toughnesses. Thus Rice et al. [189] obtained (DCB) toughnesses of 6.6 ± 0.6 MPa·m$^{1/2}$ for as fusion cast thin (e.g. < 2 mm thick) plates of ~ eutectic compositions (commercially produced for abrasives that had a colony structure and substantial reduction) (Chap. 9, Sec. III.B.2). Krohn et al. [190] also hot pressed melt-derived eutectic powder that gave a toughness of 15 MPa·m$^{1/2}$, while Homeny and Nick [116] conducted more detailed studies of eutectic Al_2O_3-ZrO_2 powders made via plasma torch melting of particles and subsequent hot pressing of eutectic compositions that contained (a) 0, (b) 4.6, and (c) 9.5 m/o Y_2O_3, giving toughnesses of respectively 6.7, 7.6, and 7.9 MPa·m$^{1/2}$. They attributed the substantial toughness in the first body (i.e with no stabilizer) to extensive microcracking (consistent with there being no Y_2O_3, resultant 90% monoclinic ZrO_2, and resultant lower Young's modulus and especially strength, Chap. 9, Sec. III.B). The source of the substantial toughness in the two bodies with Y_2O_3 additions that had 100% tetragonal ZrO_2 was attributed not to transformation toughening but to crack bridging as well as substantial formation of strings of thin deformed material (much finer than grains or substantial fragments of them) bridging the cracks in the wake region. Mazerolles et al. [191] reported very similar values of 6.8 MPa·m$^{1/2}$ for (I) toughness of oriented eutectics from directional solidification with 3 m/o Y_2O_3. Echigoya et al. [192] reported (I) toughness values for similarly directionally solidified compositions starting at ~ 9 MPa·m$^{1/2}$ with no Y_2O_3 and then dropping to ~ 7, 5, and 4.5 MPa·m$^{1/2}$ at respectively 3, 5, and 13 m/o Y_2O_3 for fracture normal to the growth direction. While the value for crack propagation parallel with the solidification direction was somewhat lower for no Y_2O_3, i.e. at ~ 7.5 MPa·m$^{1/2}$, this limited anisotropy rapidly approached nearly zero as the Y_2O_3 content increased.

Earlier work by Hulse and Batt [193] supports the substantial toughness of directionally solidified Al_2O_3-ZrO_2 eutectics and provides other similar and useful data. Their WOF values for such eutectics of CaO·ZrO_2-ZrO_2, CaO-MgO, and Al_2O_3-ZrO_2(Y_2O_3) eutectics were respectively ~ 40, 90, and 90 J/m^2, i.e. respectively similar to and ~ twice the values they obtained for a commercial alumina. (These values translate to toughnesses of ~ 4, 7, and 7.5 MPa·m$^{1/2}$.) No details on the effects of eutectic microstructure were given. Kennard et al. [194] directionally solidified MgO-$MgAl_2O_4$ eutectic specimens, obtaining WOF values of ~ 25 J/m^2 (giving a toughness of ~ 4 MPa·m$^{1/2}$), noting that this had little relation to the eutectic rod spacing due to the control of mechanical properties by the colony structure. In the above cases the WOF showed little or no decrease and often substantial increase as test temperature increased, as was discussed in Chap. 11, Sec. III.E.

F. Ceramic–Metal Particulate and Wire Composites

Ceramic–metal composites are a subject too large and diverse to address comprehensively here, but there are three sets of such composites that are of value to note here. The first is composites consisting of metal particles dispersed in a ceramic matrix, i.e. analogous to ceramic particulate composites discussed earlier. In such composites, the particles are typically totally isolated from each other and the environment (except for those at the surface), so some of these composites also can compete directly with some all-ceramic materials. The second set are composites in which a minimum amount of metal phase is used for densification and bonding, in particular cermets, especially WC-Co, which besides being an important set of composites and competing with other all-ceramic materials is similar to some of these in terms of microstructure, e.g. of sintered Si_3N_4. The third set of ceramic–metal composites briefly considered here are those with metal wires, filaments, or fibers in a ceramic matrix. Note that many ceramic–metal composites were investigated before fracture toughness and related evaluation was established, but strengths were evaluated. Thus representative examples of such studies are summarized in Chap. 9, Sec. III.G, which further show that different mechanisms may control strength from those that control toughness, e.g. as shown for monolithic ceramics.

A variety of metal particles, e.g. Al [195], Ag [196], Fe [34,81–83], Ni [197], Mo [198,199], NiAl [200], Pb [201], and W [202], have been introduced in pure or alloy form into polycrystalline, and often silicate glass, matrices. Besides the usual dispersed particle and composite parameters, another factor for some of these (and some all-ceramic composites) is the degree of bonding between the particle and the matrix, which for oxide matrices can often be controlled by oxidation of the particle surfaces, as noted in Sec. III. Thus, for example, Krstic et al. [203] showed that hot pressed composites of silicate glass matrices with (100–200 μm) particles of partially oxidized Al or Ni could substantially increase toughness (DCB), e.g. increases > sixfold, from < 1 to > 6 $MPa{\cdot}m^{1/2}$ for $\phi = 0.2$. They noted that such increases, which entailed ductile fracture-toughening in the wake region of the crack, required minimizing the elastic and expansion mismatch stresses between the particles and the matrix and having a good bond between the glass and metal (via oxidation of the metal particle surfaces). Substantial fracture examination showed more extensive crack–particle interactions, e.g. crack bowing between particles when the glass matrix had lower expansion than the particles, since the latter then generally act more as pores due to shrinkage away from the matrix. Similar fracture occurred when the glass expansion was greater than the metal particles and there was not a strong bond between the glass and the particles. Elimination of expansion mismatches, but having elastic mismatches, reduces fracture complexity, i.e. cracks propagated around the particle poles due to stress concentrations there. No significant mismatches of either type resulted in less tortuous fracture in the glass, but random intersections with metal particles allowed more impact of plastic flow with good particle matrix bonding.

Tuan and Brook [197] made composites of Al_2O_3 with Ni by sintering (in a reducing atmosphere) mixtures of fine NiO particles in the Al_2O_3 matrix (achieving densities of 98–99% of theoretical for sintering respectively at 1600 and 1700°C at $\phi=0$ and ~ 95 and 92% at $\phi=0.25$). The resulting Ni particles ranged in size (D) from ~ 0.4 to 2–3.3 μm as ϕ increased from 0.01 to 0.25 and sintering increased from 1600 to 1700°C. Toughness (I) increased with both ϕ and D, with 40–80% of the increases attributed to ductile bridging of the crack in the wake region based on both direct observations of this and their toughness increases being a linear function of $(\phi D)^{1/2}$ as predicted by a model developed by Ashby et al. [201]. Tuan and Brooks also observed that residual oxygen in their Ni particles probably substantially raised its yield stress as sintering temperature increased, adding further to the toughening as predicted by Ashby et al.'s model (the yield stress appears with ϕ and D under the square root). They also showed that the fraction of particles actually intersected by the crack was typically ~ 1/3 the potential interactions, similar to observations of Jessen and colleagues discussed below. Also note that while Krstic et al. [203] did not vary particle size, and their work was done before the model of Ashby et al. was developed, their results appear generally consistent with the model, e.g. as indicated by their high level of toughness increases with larger particles and somewhat higher volume fractions of them.

Breval et al. [204] also made Al_2O_3-Ni composites, but by hot pressing of powder obtained via $Ni(NO_3)_2$ additions to an aluminous sol, resulting in dispersed Ni particles from ~ 20 nm to ~ 50 μm, with the sizes not changing significantly as the Ni content was increased from 10 to 50 w/o, but the number of finer particles increased. Although elastic moduli decreased significantly as the amount of Ni increased due to the lower moduli of Ni (~ $^1/_2$ that of Al_2O_3) as well as increasing porosity as the Ni content was increased, fracture toughness more than doubled on going from 0 to 50 w/o Ni, i.e. from 3–4 to 8.3 $MPa{\cdot}m^{1/2}$.

Flinn et al. [195] concluded that the toughening mechanism of Al_2O_3-Al composites made by the directed oxidation of Al, i.e. Lanxide material, is due to ductile bridging of the crack in the wake region and that the model of Ashby et al. is a reasonable guide to the toughening of this material (see Fig. 8.20). However, they concluded that three factors of metal hardening, debonding from the matrix, and the extent of plastic elongation needed further study and documentation. Rice [15,18] also showed that a factor in this material (and others) that

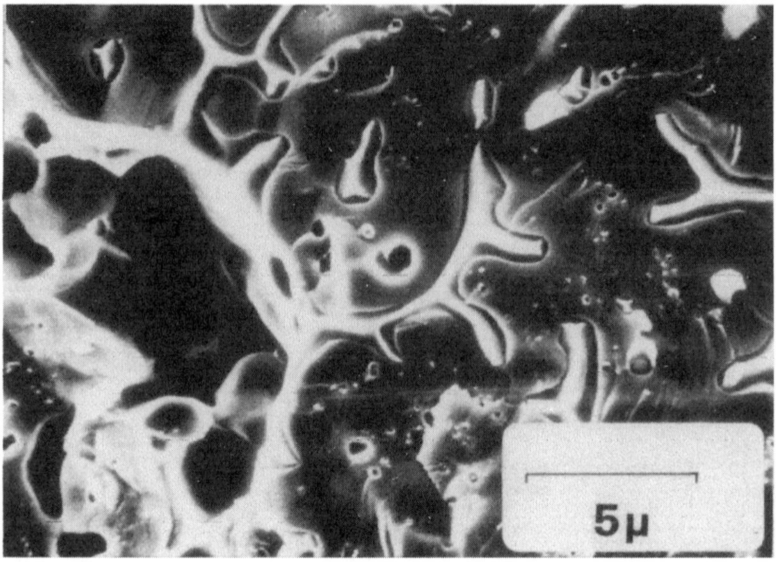

FIGURE 8.20 Fracture of early Lanxide Al_2O_3-Al composites. SEM fracture photos showing general fracture with a substantial and complex pattern of Al (light material) and its extensive necking down and rough transgranular fracture associated with inclusions in the grain. Such effects are factors in the ~ doubling of toughness. (Photo from Ref. 189.)

needs to be considered is the location of the dispersed phase and its relation to fracture mode in composites with polycrystalline matrices. Thus some inclusions may be incorporated within grains and some at grain boundaries, in which case they only affect the resultant fracture process to the extent that it is transgranular (some of which occurs in Lanxide material, Fig. 8.20) or intergranular.

Jessen and Lewis [81,82] specifically studied the degree of interaction of cracks with dispersed spherical Fe-Ni-Co alloy particles in a glass matrix as occurred in DCB toughness tests. Specimens were fabricated with either uniform volume fractions of 0, 0.1, and 0.25 or with sections of differing volume fraction along the specimens and hence of crack propagation. Tests were conducted so toughness would be measured in various sections of the layered specimens, i.e. in some cases after the crack had propagated through some layers. The result was that the layer through which the crack first propagated determined the subsequently measured toughness value. Thus specimens with the initial section of crack propagation having no dispersed particles always gave the toughness for glass without any metal particles (i.e. ~ 0.7 MPa·m$^{1/2}$) regardless of whether the K_{IC} value was actually measured in a subsequent layer of the specimen with metal particles. On the other hand, when the initial section of the specimen had metal particles, the toughness values obtained always reflected that of the initial layer regardless of subsequent layers in which toughness was measured (i.e. where the crack went catastrophic). Thus toughness values were respectively ~ 1.7 and ~ 2.4 MPa·m$^{1/2}$ for specimens having initial sections of crack propagation respectively with ϕ= 0.1 and 0.25 regardless of what compositions followed that in which the actual toughness was measured. In other words, the initial material in which the crack propagated determined its subsequent behavior regardless of the subsequent compositions of these composites encountered.

Yun and Choi [202] showed that (I) toughness of composites sintered from AlN with *W* additions (ϕ 0–0.2) giving *W* particles mostly ≤ 1 μm (and some reaction phases and reduced AlN *G*) increased modestly from ~ 1.8 to ~ 2.4 MPa·m$^{1/2}$ (but strengths decreased from ~ 230 to ~ 200 MPa, which was attributed to some increase in *P* and second phases as ϕ increased).

Consider briefly limited results on ceramics with nano scale dispersed metal particles. Nawa et al. made nanocomposites of Mo particles in 3Y-TZP [198] and Al_2O_3 [199] matrices using Mo powder (0.2 to > 1 μm) mixed with comparable sized Al_2O_3 (α or γ) or TZP powders then hot pressed. Mo (5 v/o) composites using ~ γ-Al_2O_3 powder gave an α-Al_2O_3 matrix with considerable growth of the Al_2O_3 grains (to ~ 8–9 μm, and some elongation, e.g. to 15 μm), but with some MoO_2 and the Mo as elongated layers around parts of the grains and many cracks and branches associated with the grains and Mo resulting in increased (SEPB) toughness of ~ 7.1 MPa·m$^{1/2}$ (IF did not work on these samples, but no increase in strength was found). Use of α-Al_2O_3 powder gave much finer alumina grains (e.g. $^1/_2$–3 μm) with many nanometer Mo particles within the grains (consistent with the Mo acting as a grain growth inhibitor) as well as some Mo particles > 1 μm at grain boundaries, with both the Al_2O_3 grain and the Mo particle sizes increasing as the hot pressing temperatures increased. Fracture toughness (IF) increased as both the volume fraction Mo increased over the range investigated of 20 v/o as well as with hot pressing temperature, hence with increases of both Al_2O_3 grain and the Mo particle sizes (but again with strengths showing the opposite trends). While the fracture mode of their Al_2O_3 was intergranular at finer *G* and transitioning to transgranular fracture at the larger *G*, the fracture mode of the composites was predominantly transgranular, even at the finest *G*, i.e. similar to other composites (Section IV).

Nawa et al.'s 3Y-TZP- Mo composites gave submicron intergranular Mo particles around submicron TZP grains for ϕ < 0.3, but with further increased Mo content fine (e.g. ≤ 10 nm) intragranular particles increasingly appeared, as did larger, more elongated, often interconnected and polycrystalline intergranular Mo particles. IF toughness tests gave toughness values up to ~ 18 MPa·m$^{1/2}$ at ϕ= 0.5, but this was considered uncertain since CNB tests gave a maximum at ϕ= 0.7 of 11.4 MPa·m$^{1/2}$ (consistent with a strength maximum of ~ 2 GPa at ϕ= 0.7).

Sekino et al. [205] made Al_2O_3-Ni composites (ϕ= 0–0.2) by either mixing larger NiO particles with fine (~ 0.2 μm) Al_2O_3 or by addition of NiO via $Ni(NO_3)_2$, with both mixtures reduced to Ni and densified by hot pressing. The two processes producing respectively Ni particles averaging ~ 180 nm (and ~ $^1/_2$ inter- and intragranular) and ~ 100 nm (with much more intragranular particles) both resulted in modest (I) toughness increases, respectively of ~ 25% and ~ 12% for ϕ= 0.2. However, strengths were at a significant maximum of 1 and 1.1 GPa respectively for the two processes, with these values being ~ consistent with the reductions of *G* in the Al_2O_3 from 3.4 μm for no Ni to 2.3 μm for 5 v/o Ni from NiO and 0.9 μm for Ni from the nitrate.

Chou et al. [200] showed an almost linear increase in NB toughness as additions of NiAl particles (*D* ~ 5.5 μm) increased from ~ 3.7 to 9 MPa·$m^{1/2}$ as the addition increased from 0 to 50 v/o. These increases were attributed to crack deflection with some contribution of pullout of elongated NiAl particles and some plastic deformation of them.

Next consider ceramic–metal composites consisting of ceramic grains that are partially or completely bonded with a metallic grain boundary phase, usually of limited volume fraction, i.e. cermets, of which the WC-Co is best known and most extensively studied. While much of the extensive development work on this system was done before the development of fracture mechanics and fracture toughness measurements and studies, a few reasonably comprehensive studies addressing WC particle (grain) size have been conducted. Though there is some scatter as expected, these show toughness increasing with both volume fraction Co and the WC particle size (Fig. 8.21). This correlation might suggest ductile deformation of the Co phase as a toughening mechanism per the above ceramic–metal particle composite behavior, e.g. since increased Co content and WC size should also correlate with increased dimensions of the Co phase. Though the WC grain/particle size range is not sufficient to test this suggestion seriously, it should be noted that some ductile behavior of the Co bonding phase and resultant toughening has long been indicated. More recent study of Marshall et al. [206] has shown that a zone of irreversible strain exists around cracks, mostly in the wake region, which is a major source of higher toughness and is attributed to plasticity and possibly some microcracking in the WC and a TiC cermet studied. This zone was much larger than plastic zones previously thought or indicated by finite element analysis, e.g. over an order of magnitude larger than the WC grain size. However, it was also noted that the zone size diminished with crack velocity, especially for unstable crack growth, indicating that the dynamic toughness may be substantially lower. This may well be an important factor in why increased toughness as the WC particle size increases has limited or no benefit to strength, which decreases as the WC grain/particle size increases (Fig. 9.18). His observations are probably pertinent to the reaction processed ZrC-ZrB_2 (platelet)- Zr composites of Claar et al. [161] noted in the previous section.

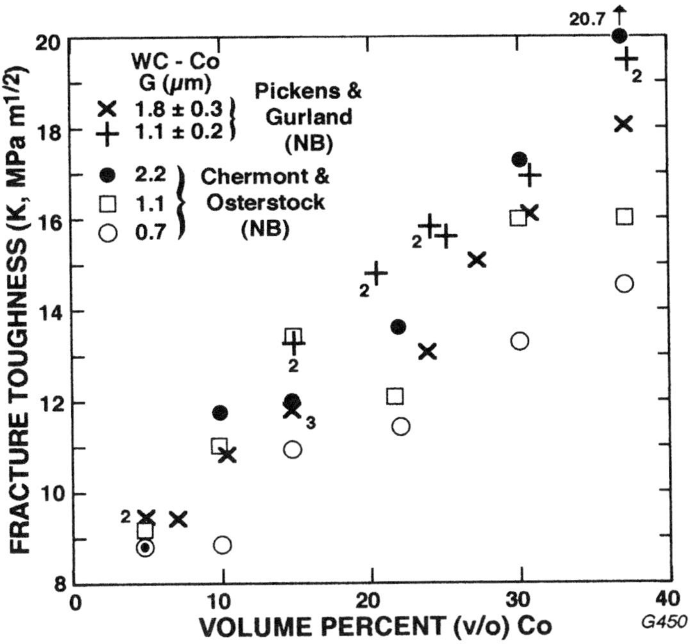

FIGURE 8.21 Fracture toughness of WC-Co versus Co content for various WC grain sizes. (Data from Pickens and Gurland [72,73] and Chermant and Osterstock [76].)

More limited studies of other cermets also provide some evidence for ductile behavior of the bonding phase, e.g. of Cr_3C_2-Ni and increased toughness as the Ni content increases (but limited decrease in toughness as the content of TiC in lieu of Ni increases) [207]. Han and Mecholsky [208] showed that hot pressed composites with WC-10 w/o Co as the matrix with dispersion of 10 w/o of Nb (17 v/o, $D \sim 100$ μm) or Mo (13 v/o, $D \sim 5$ or 55 μm) particles had increased toughness. Specifically they showed that the absolute value of the toughnesses was proportional to the product of the metal particle yield stress and $(D\phi)^{1/2}$, while the increment of toughness was proportional to the latter square root term, hence generally consistent with the model of Ashby et al. [201], suggesting plastic deformation as a mechanism.

The above WC-Co and related composites imply higher toughness with a continuous metal phase, as would seem logical and is supported by other data. Thus Trusty and Yeomans [209] showed that hot pressed Al_2O_3-Fe composites made via wet milling yielding a mainly discontinuous distribution of 20 v/o of Fe doubled the toughness to 6.2 MPa·m$^{1/2}$ (from 3.1 MPa·m$^{1/2}$ for the pure

Al_2O_3), while dry milling that yielded a much more continuous Fe phase tripled toughness to 9.6 MPa·m$^{1/2}$. Pyzik and Beaman's [210] reaction processed composites of B_4C-Al with an isolated B_4C structure generally had higher toughness, e.g. a maximum of ~ 12 MPa·m$^{1/2}$, versus ~ 8.2 MPa·m$^{1/2}$ for bodies with a connected B_4C structure.

A few brief observations are in order for composites of ceramic matrices with metal wires, fibers, or filaments, since these have been of considerable, especially earlier interest. Originally this interest was motivated by the concept of refractory wires providing toughness and related fracture resistance to ceramic matrices, which in turn would provide environmental protection for the metal wires. This was at least part of the motivation for considerable work on such ceramic matrix composites, e.g. made by consolidating powders around wires, investigation of a number of ceramic–metal eutectic systems, or infiltration of porous ceramic preforms with molten metal. However, subsequent showing that cracks commonly formed in the ceramic matrix which allowed environmental access to the wires and development of other systems greatly reduced work on ceramic–metal wire and related composites. Three examples with some pertinence to the microstructural theme of this book are noted here. Zwissler et al. [36] showed NB toughness of composites of 0–15 v/o chopped stainless steel wires (6, 12, or 25 μm dia.) increased linearly by up to 100%, similar to but greater than increases in *E* or strength, apparently due to the absence of cracking (Sec. III; Chap. 9, Sec. III.G). Simpson and Wasylyshyn [211] showed that the work of fracture of Al_2O_3 with 0–12 v/o chopped Mo wires (3.16 mm long and 0.08 mm dia.) increased to ~ 250 times that of Al_2O_3 alone, but strengths were reduced, indicating microcracking. Brennan [212] showed that addition of ~ 25 v/o of Ta wires (0.63 or 1.27 mm dia.) to hot pressed Si_3N_4, while improving impact resistance and hence toughness, still had extensive cracks from the impact (and poor flexure and especially tensile strength). Second, Flinn et al. [195] showed that ductile bridging of the crack wake region was a major factor in the high fracture resistance with increasing crack extension, e.g. to values of 10–25 MPa·m$^{1/2}$ in composites with ~ 20 μm dia. Al alloy fibers introduced by melt infiltration of a dense Al_2O_3 preform with ~ aligned channels. Other aspects of ceramic wire or fiber composites pertinent to the issue of fiber sizes and mismatches in properties are briefly discussed in Chap. 9, Sec. III.G.

VI. GENERAL DISCUSSION

A full discussion of the mechanical properties of composites and their microstructural dependence requires more extensive property and microstructural characterization. An important component is other mechanical property, especially tensile strength, and temperature dependence of properties, which are addressed in Chaps. 9–11, so only a partial discussion focused on toughness, based mostly on large cracks, is presented here. Such cracks often give very different results from smaller cracks, as is noted in Chapters 2, 3, and 9, a key factor being the scale of the crack to that of the microstructural features impacting crack propagation. When the latter are no longer small relative to the crack size, behavior of large cracks may no longer be pertinent to small-scale cracks, i.e. much toughness data with large cracks may not be consistent with strength data reflecting smaller crack behavior, as is shown in Chaps. 3 and 9.

Beginning with macroscopic behavior and progressing to microstructural aspects, consider first the effect of elastic property changes on toughnesses. As noted earlier, since K=($2E\gamma$) and $\gamma \propto E$, $K \propto E$, so increasing E in a composite should increase K. The data for many composites is generally consistent with this, some with quite close correlations, e.g. for $BaTiO_3$ with small amounts of SiC (Fig. 8.19), but there are also important deviations from this, especially for many ceramic–metal composites (Sec. V.F). Except for composites of ceramics with lower E or of metals with higher E, composite moduli are decreased, but toughness is often increased, due to other toughening mechanisms, e.g. plastic deformation, as is discussed below. A specific issue where the composite E and K increase is the extent of K increases versus E values. Becher et al. [164] reported that while composites of two silicate glasses and mullite, or alumina matrices with the same 20 v/o SiC whisker additions, all showed toughness increases, the increases were linear as a function of $(E_C/E_W)^{1/2}$, where the subscripts C and W refer to the composite and the whisker, i.e. greater increases as the modulus of the matrix increased. Data of Table 8.3, while providing some possible agreement, also shows some substantial disagreement with this, poorer results in most Si_3N_4-SiC whisker composites being an important example. This probably reflects other factors such as thermal expansion differences, which Becher et al. also recognized as a source of variation.

Turning to another bulk property correlation, consider ductility of the dispersed phase as in the case of ceramic–metal composites, where both modeling and experiments are reasonably consistent and provide some guidance for improved understanding of mechanisms in other composites. Thus not only do basic models and experiments generally agree, the models predict trends based on material (yield) and microstructure, i.e. on a $(\phi D)^{1/2}$ dependence, that have been corroborated by data. Further, clear microstructural observations of ductile deformation of embedded metal particles have been given. However, we need better understanding of particle–matrix bonding variations, including effects of size and expansion differences, and better definition of the effects of yield stresses and yield extent versus properties of the matrix. Experiments with ceramic fillers of some ductility, e.g. halides, may be of interest scientifically and practically, e.g. to lower density and have better dielectric behavior.

Consider now the effects of thermal expansion differences, which while a bulk property difference, manifests itself primarily as a microstructural factor, moving the discussion to a focus on microstructural issues. A summary of average expansion differences for some key composites is given in Table 8.3, focused on estimates of particle sizes for spontaneous microcrack generation. Though such estimates are uncertain due to probable volume fraction effects and neglect thermal expansion anisotropies of the noncubic constituents as well as elastic anisotropies of all constituents, they indicate that all these composites may exhibit some microcracking. Many composites have not been examined thoroughly for microcracking prior to loading, and especially as a function of loading (e.g. via static modulus decreases and acoustic emission). However, microcracking has been clearly established and studied in SiC-TiB_2 composites [213,214], suggesting that it may be more common in a number of other similar composites. Though studied less, microcracking has also been reported in Al_2O_3-SiC composites by Nakahira et al. [123,124]. In crystallized glasses, differences in thermal expansion (and elastic moduli) may exist between the crystallized particles and the matrix to contribute to microcracking, as may shrinkage strains from the glass–crystal phase transformation strains. Microcracking begins at a minimum particle size and generally increases in effects as D increases (Table 8.3), passing through a maximum and then decreasing in its effects on toughness (e.g. Figs. 2.10, 2.15). However, strength trends often do not reflect toughness trends as measured by large cracks.

Turning to other microstructural aspects of toughening of composites, consider the implications of the microstructural parameters indicated by models summarized in Table 1. Transformation toughening, e.g. in ZTA, is clearly con-

Table 8.3 Summary of Microcracking Parameters for More Investigated Ceramic Composites

Composite	$\Delta\alpha$ (10^{-6}/°C)	ΔT (°C)	E (GPa)	γ (J/m²)	D_S (μm)
Al_2O_3-TiB_2	1.25	1200	500	1	8
Al_2O_3-TiC	1.5	1200	450	1	7
Si_3N_4-SiC	2.0	1200	400	1	4
SiC-TiB_2	2.0	1200	440	1	4
Al_2O_3-SiC	4.5	1200	450	1	0.8
WC-Co	10	800	600	3	0.8

Source: Ref. 15, published with permission of Ceramic Engineering and Science Proceedings.

The matrix or major constituent is given first. $\Delta\alpha$ = the ~ mismatch in polycrystalline thermal expansion coefficients between the composite constituents, ΔT = the approximate temperature below which stress from the expansion mismatch begins to build up, E is the approximate Young's modulus, γ is the approximate fracture energy for microcracking on the particle scale, and D_S is the resulting ~ particle size for spontaneous generation of microcracks (i.e. in the absence of any contribution of external, e.g. applied, stresses).

sistent with the general trends predicted. However, quantitative details are somewhat uncertain, since increased strengths due to reduction of *G* and surface compressive stresses from machining have not been accounted for in transformation toughening theories. Further, the modeling has generally assumed noninteractive particles, which is often at best uncertain.

Crack deflection has no obvious particle size dependence, and so could be a factor in many composites, as has been suggested by a number of investigations (e.g. Ref. 215). Crack pinning should increase with decreasing *D*, but probably with some material-dependent particle size, which has been indicated as a factor in some composites. Crack branching may originate from different sources, e.g. from some pinning as well as microcracking. Similarly, crack bridging may be in part related to both of the preceding mechanisms but is a separate mechanism based on its occurrence and effects in the wake region and related *R*-curve effects, which are typically observed in ceramic composites. Though there are uncertainties in the relation of surface observations of bridging to actual fracture away from the surface (Chap. 2, Sec. II.A), there is no doubt that bridging plays a role in the large crack toughness of many ceramic composites. Again however, there are serious questions of the degree to which such large-scale crack bridging in toughness tests versus effects of such phenomena on strengths that are typically controlled by small cracks with a different propagation history from large cracks for toughness measurements can occur.

Finally consider pullout, which is clearly established as a major factor in the noncatastrophic failure of ceramic fiber composites (where cracks are typically large and probably experience periodic acceleration and deceleration, i.e. similar to large crack toughness tests). Less well established is the extent of pullout mechanisms for whisker and especially platelet composites. While exposure of part of platelet morphology on composite fracture surfaces has led some to suggest pullout as a toughening mechanism in platelet composites, it seems that this partial exposure of platelets may reflect partial microcracking or simple intergranular fracture. Pullout has also been cited as an important toughening mechanism in whisker composites based in part on exposure of parts of whiskers on fracture surfaces. While at least some of the whisker fracture exposure probably again reflects intergranular fracture, there is other evidence to support pullout as a mechanism in whisker composites. A key factor in the latter is the decreased effects of whiskers with increased oxide content on their surface and hence presumably stronger matrix–whisker bonding, since increased matrix–fiber bonding is well known to reduce pullout toughening in fiber composites. While decreasing fiber sizes appear to increase pullout effects (based on surface area), there may be lower limits to this, so the increasing effectiveness of larger whiskers may not be counter to pullout. However, the increase in whisker toughening with increasing Al_2O_3 matrix grain size [83,162,163,170] is opposite to its effects on strength.

VII. SUMMARY

There are reasonable models (including bounding approaches) for predicting elastic properties of many composites, especially those with equiaxed particulate fillers. Uncertainties exist primarily as the particle morphology and contiguity change, and increased preferred orientation occurs, especially with single crystal particles, platelets, and whiskers, that may require attention to their crystal orientation in the body texture and their crystalline anisotropy. These all pose uncertainties for modeling as well as for (generally inadequate) characterization. However, specialized models for the elastic properties of fiber composites may indicate a modeling approach for some composites with substantial orientation of dispersed particles, platelets, and whiskers.

While some intergranular fracture of matrices and dispersed particles, platelets, and whiskers occurs, greater transgranular fracture of one and commonly both phases occurs in most ceramic composites. Though this may reflect better bonding of boundaries due to common higher processing temperatures for composites than corresponding monolithic ceramics, no detailed information is available on possible mechanisms, and better quantification of results is needed. Slow crack growth can clearly occur in composites, especially those with matrices, e.g. silicate glasses, susceptible to SCG, but dispersed particles, especially with suitable bonding to the matrix, can reduce SCG by both increasing the overall stress intensity needed for SCG as well as actually pinning the crack, e.g. as shown in some glass–metal particle composites (Sec. IV).

Toughness almost invariably increases with increasing volume fraction of dispersed phase in composites with any degree of investigation, regardless of physical morphology of the dispersed phase. Some studies have been conducted to high enough dispersed phase contents to show that toughening effects typically reach maxima at $\phi = 0.3$–0.5. Less investigated and thus more uncertain are the effects of dispersed phase dimensions on the extent of toughening, but several studies show that increased dimensions increase toughness. In such cases, a maximum toughness as a function of size is again generally expected and indicated in a few more detailed studies. Such data is particularly important not only for composite design and processing but also for understanding mechanisms, since such dependence on the dimensions of the dispersed phase typically results in opposite trends of toughness and tensile strength, as did grain size dependence in monolithic ceramics. Composite toughness often also increases with matrix grain size, paralleling such effects in monolithic ceramics, again indicating substantial toughness–strength discrepancies in composites. These discrepancies are discussed extensively in Chap. 9 for composites and in Chap. 3 for grain size effects in monolithic ceramics. An important factor in many composites and a critical one in some is orientation of the dispersed phase, and to a lesser extent that of the matrix. Again limited experiments indicating important changes in crack propagation as a function of crack velocity and history indicate that important effects can occur in composites in this widely neglected area, as shown by tests of Jessen and Lewis [81–84] on glass–metal particle composites (Sec. V.F).

Turning to mechanisms, much remains to be established, with it being probable that most composites have more than one active mechanism involved, with their relative importance shifting as the material or microstructural parameters and the mechanical property of interest change. Thus, as noted earlier, mechanisms controlling large scale crack effects may be different from those controlling normal strength. However, while quantitative models are generally only guidelines, some mechanisms appear to be established or probable for certain composites, with fiber pullout being a dominant factor in continuous fiber composites. Similarly, bridging in the crack wake region by ductile metal particles, fibers, or filaments seems established as a major factor in toughness of many ceramic metal composites. Wake bridging appears to be a common factor in increased large crack toughness in many ceramic composites, and whisker pullout may be a factor in at least some whisker composites, e.g. those with Al_2O_3 matrices. However, three issues question such wake toughness effects as the general mechanisms controlling strengths of composites, the first being the frequent basic toughness–strength discrepancies. Second is the increasing number of questions of the accuracy and validity of such large crack effects for strength, including the generally neglected issue of how such phenomena change as a function of extent and velocity of crack propagation. The third is that, as is shown in Chap. 9, much composite strength dependence appears to arise due to microstructural effects on machining flaw character and size, as was shown for monolithic ceramics (e.g. Fig. 3.1).

NOTE

Since completion of this chapter work of Swearengen et al. [216] on composites of up to ~ 40 v/o of spherical polycrystalline Al_2O_3 particles (~ 25 μm dia.) in various borosilicate glasses with thermal expansions ranging from ~ 4 ppm/°C lower to ~ 4 ppm/°C greater than that of the alumina was found. They showed E increasing ~ linearly as ϕ increased, for a total increase of ~ 100%. Using a vibrational technique to mark developing fracture surfaces they showed that cracks interacted with particles as outlined in Fig. 8.9, the marking technique showing significant crack velocity changes as cracks approached particles. They found similar increases in toughness whether the glass had higher or lower expansion than the alumina particles. (The contrast to results of Davidge and Green [97] where bodies with ThO_2 particles in a glass matrix with ~ 2 ppm/°C expansion > that of ThO_2 were fractured may reflect effects of the much larger ThO_2 particles relative to the alumina particles here.) The toughness increases were approximately linear with λ^{-1}, but this is also consistent with toughness increasing with E as ϕ increases, though greater increases in K versus E by ~ 50% may reflect some additional mechanism over that due simply to increased E.

REFERENCES

1. P. G. Karandikar and T.-W. Chou. Structural Properties of Ceramic Matrix Composites. Handbook on Continuous Fiber-Reinforced Ceramic Matrix Composites (R. L. Lehman, S. K. El-Rahaiby, and J. B. Wachtman, Jr., eds.). Ceramics Information Analysis Center, West Lafayette, IN and Am. Cer. Soc., Westerville OH, 1995, pp. 355–429.
2. Z. Hashin and S. Shtrickman. A Variational Approach to the Theory of the Mechanical Behavior of Multiphase Materials. J. Mech. Phys. Solids 11:127–140, 1963.
3. K. S. Ravichandran. Elastic Properties of Two-Phase Composites. J. Am. Cer. Soc. 77(5):1178–1184, 1994.
4. Z. Hashin and B. W. Rosen. The Elastic Moduli of Fiber-Reinforced Materials. J. Appl. Mech., 223–232, 6/1964.
5. T. T. Wu. The Effect of Inclusion Shape on the Elastic Moduli of a Two-Phase Material. Intl. J. Solids Structures 2:1–8, 1966.
6. R. M. Christensen. Asymptotic Modulus Results for Composites Containing Randomly Oriented Fibers. Intl. J. Solids Structures 12:537–544, 1976.
7. R. M. Christensen. A Critical Evaluation for a Class of Micro-Mechanics Models. J. Mech. Phys. Solids 38(3):379-404,1990.
8. Z. Hashin. Elasticity of Ceramic Systems. Ceramic Microstructures '76: With Emphasis on Energy Related Applications (R. M. Fulrath and J. A. Pask, eds.). Westview Press, Boulder, CO, 1977, pp. 313–341.
9. B. Paul. Prediction of Elastic Constants of Multiphase Materials. Trans. Metal. Soc. AIME 218:36–41, 1960.
10. L. J. Cohen and O. Ishai. The Elastic Properties of Three-Phase Composites. J. Comp. Mats. 1:390-403, 1967.
11. E. H. Kerner. The Elastic and Thermo-Elastic Properties of Composite Media. Proc. Roy. Soc. (London), Ser. B 69:803–813, 1956.
12. B. Budiansky. On the Elastic Moduli of Some Heterogeneous Materials. Mech. Phys. Solids 13:223–227, 1965.
13. R. W. Rice. Mechanisms of Toughening in Ceramic Matrix Composites. Cer. Eng. Sci. Proc. 2(7-8):661–701, 1981.
14. R. W. Rice. Ceramic Matrix Composites Toughening Mechanisms: An Update. Cer. Eng. Sci. Proc. 6(7-8):589–607, 1985.
15. R. W. Rice. Toughening in Ceramic Particulate and Whisker Composites. Cer. Eng. Sci. Proc. 11(7-8):667–694, 1990.
16. R. W. Rice. Processing of Ceramic Composites. Advanced Ceramic Processing and Technology, Vol. 1 (J. G. P. Binner, ed.). Noyes, Park Ridge, NJ, 1990, pp. 123–213.

17. R. W. Rice. Ceramic Composites: Future Needs and Opportunities. Fiber Reinforced Ceramic Composites, Materials, Processing and Technology (K. S. Mazdiyasni, ed.). Noyes, Park Ridge, NJ, 1990, pp. 451–495.
18. R. W. Rice. Microstructural Dependence of Fracture Energy and Toughness of Ceramics and Ceramic Composites Versus That of Their Tensile Strengths at 22°C. J. Mat. Sci. 31:4503–4519, 1996.
19. A. G. Evans. Perspective on the Development of High-Toughness Ceramics. J. Am. Cer. Soc. 73(2):187–206, 1990.
20. F. F. Lange. Interaction of a Crack Front with a Second Phase Dispersion. Phil. Mag. 22:983–992, 1970.
21. A. G. Evans. The Strength of Brittle Materials Containing Second Phase Dispersions. Phil. Mag. 26:1327–1344, 1972.
22. A. G. Evans and K. T. Faber. Toughening of Ceramics by Circumferential Microcracking. J. Am. Cer. Soc. 64(7):394–398, 1981.
23. K. T. Faber. Toughening of Ceramic Materials by Crack Deflection Processes. Ph.D. thesis, Univ. of Calif., Berkeley, 6/1982.
24. K. T. Faber and A. G. Evans. Crack Deflection Processes-I Theory. Acta. Metall. 31(4):565–576, 1983.
25. D. Baril, S. P. Tremblay, and M. Fiset. Silicon Carbide Platelet Reinforced Silicon Nitride Composites. J. Mat. Sci. 28:5486–5494, 1993.
26. R. W. Rice and D. Lewis III. Ceramic Fiber Composites Based upon Refractory Polycrystalline Ceramic Matrices. Reference Book for Composites Technology 1 (S. M. Lee, ed.). Technomic Press, Lancaster, PA, 1989, pp. 117–142.
27. C. X. Campbell and S. K. El-Rahaiby. Databook on Mechanical and Thermophysical Properties of Particulate-Reinforced Ceramic Matrix Composites. Ceramics Information Analysis Center, West Lafayette, IN and Am. Cer. Soc., Westerville OH, 1995.
28. C. X. Campbell and S. K. El-Rahaiby. Databook on Mechanical and Thermophysical Properties of Whisker-Reinforced Ceramic Matrix Composites. Ceramics Information Analysis Center, West Lafayette, IN and Am. Cer. Soc., Westerville, Ohio, 1995.
29. D. P. H. Hasselman and R. M. Fulrath. Effect of Alumina Dispersions on Young's Modulus of a Glass. J. Am. Cer. Soc. 48(4):218–219, 1965.
30. F. F. Lange. Fracture Energy and Strength Behavior of a Sodium Borosilicate Glass-Al_2O_3 Composite System. J. Am. Cer. Soc. 54(12):614–620, 1971.
31. W. J. Frye and J. D. Mackenzie. Mechanical Properties of Selected Glass-Crystal Composites. J. Mat. Sci. 2:124–130, 1967.
32. D. B. Binns. Some Properties of Two-phase Crystal-Glass Solids. Science of Ceramics, Vol. 1 (G. H. Stewart, ed.). Academic Press, New York, 1962, pp. 315–334.
33. S. W. Freiman and L. L. Hench. Effect of Crystallization on the Mechanical Properties of Li_2O-SiO_2 Glass-Ceramics. J. Am. Cer. Soc. 55(2):26-90, 1972.
34. T. L. Jessen, J. J. Mecholsky, and R. H. Moore. Fast and Slow Fracture in Glass Composites Reinforced with Fe-Ni-Co- Alloy. Am. Cer. Soc. Bul. 65(2):377–381, 1986.

35. I. W. Donald and P. W. McMillan. The Influence of Internal Stresses on the Mechanical Behavior of Glass-Ceramic Composites. J. Mat. Sci. 12:290–298, 1977.
36. J. G. Zwissler, M. E. Fine, and G. W. Groves. Strength and Toughness of a Ceramic Reinforced with Metal Wires. J. Am. Cer. Soc. 60(9-10):390–396, 1977.
37. T. Ono, K. Nagata, M. Hashiba, E. Miura, and Y. Nurishi. Internal Friction, Crack Length of Fracture Origin and Fracture Surface Energy in Alumina-Zirconia Composites. J. Mat. Sci. 24:1974–1978, 1989.
38. J. D. French, H. M. Chan, M. P. Harmer, and G. A. Miller. Mechanical Properties of Interpenetrating Microstructures: The Al_2O_3/c-ZrO_2 System. J. Am. Cer. Soc. 75(2):418–423, 1992.
39. Q.-M. Yuan, J.-Q. Tan, and Z.-G. Jin. Preparation and Properties of Zirconia-Toughened Mullite Ceramics. J. Am. Cer. Soc. 69(3):265–267, 1986.
40. M. Ishitsuka, T. Sato, T. Endo, and M. Shimada. Sintering and Mechanical Properties of Yttria-Doped Tetragonal ZrO_2 Polycrystal/Mullite Composites. J. Am. Cer. Soc. 70(11):C-342–346, 1987.
41. H. Ruf and A. G. Evans. Toughening by Monoclinic Zirconia. J. Am. Cer. Soc. 66(5):328-332, 1983.
42. R. P. Wahi and B. Ilschner. Fracture Behavior of Composites Based on Al_2O_3-TiC. J. Mat. Sci. 15:875–885, 1980.
43. H. J. Hwang and K. Niihara. Perovskite-Type $BaTiO_3$ Ceramics Containing Particulate SiC. J. Mat. Sci. 33:549–558, 1998.
44. G. Sasaki, H. Nakase, K. Suganuma, T. Fujita, and K. Niihara. Mechanical Properties and Microstructure of Si_3N_4 Matrix Composites with Nano-Meter Scale SiC Particles. J. Cer. Soc. Jpn., Intl. Ed. 100:536–540, 1992.
45. R. Ruh, A. Zangvil, and J. Barlowe. Elastic Properties of SiC, AlN, and Their Solid Solutions and Particulate Composites. Am. Cer. Soc. Bull. 64(10):1368–1373, 1985.
46. T. Mah, M. G. Mendiratta, and H. A. Lipsitt. Fracture Toughness and Strength of Si_3N_4-TiC Composites. Am. Cer. Soc. Bull. 60(11):1229–1240, 1981.
47. P.-L. Chen and I.-W. Chen. In-Situ Alumina/Aluminate Platelet Composites. J. Am. Cer. Soc. 75(9):2610–2612, 1992.
48. H. Endo, M. Ueki, and H. Kubo. Microstructure and Mechanical Properties of Hot Pressed SiC-TiC Composites. J. Mat. Sci. 26:3769–3774, 1991.
49. M.-J. Pan, P. A. Hoffman, D. J. Green, and J. R. Hellmann. Elastic Properties and Microcracking Behavior of Particulate Titanium Diboride-Silicon Carbide Composites. J. Am. Cer. Soc. 80(3):692–698, 1997.
50. M. Ferrari and M. Fillipponi. Appraisal of Current Homogenizing Techniques for the Elastic Response of Porous and Reinforced Glass. J. Am. Cer. Soc. 74(10):229–231, 1991.
51. R. W. Rice. Porosity of Ceramics. Marcel Dekker, New York, 1998.
52. D. Lewis, R. P. Ingel, W. J. McDonough, and R. W. Rice. Microstructure and Thermomechanical Properties in Alumina- and Mullite-Boron-Nitride Particulate Composites. Cer. Eng. Sci. Proc. 2(7–8):719–727, 1981.
53. W. S. Coblenz and D. Lewis III. In Situ Reaction of B_2O_3 with AlN and/or Si_3N_4 to Form BN-toughened Composites. J. Am. Cer. Soc. 71(12):1080–1085, 1988.

54. P. G. Valentine, A. N. Palazotto, R. Ruh, and D. C. Larsen. Thermal Shock Resistance of SiC-BN Composites. Adv. Cer. Mat. 1(1):81–86, 1986.
55. K. S. Mazidyasni, R. Ruh, and E. E. Hermes. Phase Characterization and Properties of AlN-BN Composites. Am. Cer. Soc. Bull. 64(8):1149–1154, 1983.
56. W. Sinclair and H. Simmons. Microstructure and Thermal Shock Behavior of BN Composites. J. Mat. Sci. Lett. 6:627–629, 1987.
57. M. Ashizuka, Y. Aimoto, and M. Watanabe. Mechanical Properties of SiC Whisker Reinforced Glass Ceramic Composites. J. Cer. Soc. Jpn., Intl. Ed. 97:783–789, 1989.
58. I. Wadsworth and R. Stevens. Silicon Carbide Whisker-Reinforced Cordierite. Brit. Cer. Trans. 92:101–103, 1993.
59. M. Yang and R. Stevens. Microstructure and Properties of SiC Whisker Reinforced Ceramic Composites. J. Mat. Sci. 26:726–736, 1991.
60. T. Kumazawa, S. Ohta, H. Tabata, and S. Kanzaki. Mechanical Properties of Mullite-SiC Whisker Composite. J. Cer. Soc. Jpn., Intl. Ed. 97:880–888, 1989.
61. N. Tamari, I. Kondoh, T. Tanaka, and H. Katsuki. Mechanical Properties of Alumina-Mullite Whisker Composites. J. Jpn. Cer. Soc., Intl. Ed. 101:704–707, 1993.
62. E. S. Fisher, M. H. Manghnani, J.-F. Wang, and J. L. Routbort. Elastic Properties of Al_2O_3 and Si_3N_4 Matrix Composites with SiC Whisker Reinforcement. J. Am. Cer. Soc. 75(4):908–914, 1992.
63. P. D. Shalek, J. J. Petrovic, G. F. Hurley, and F. D. Gac. Hot-Pressed SiC Whisker/Si_3N_4 Matrix Composites. Am. Cer. Soc. Bull. 65(2):231–256, 1986.
64. J. M. Desmarres, P. Goursat, J. L. Besson, P. Lespade, and B. Capdepuy. SiC Whiskers Reinforced Si_3N_4 Matrix Composites: Oxidation Behavior and Mechanical Properties. J. Eur. Cer. Soc. 7:101–108, 1991.
65. R. W. Rice. Ceramic Fracture Mode—Intergranular vs. Transgranular Fracture. Cer. Trans. 64: Fractography of Glasses and Ceramics III (J. R. Varner, V. D. Frechette, and G. D. Quinn, eds.). Am. Cer. Soc., Westerville, OH, 1996, pp. 1–53.
66. A. Krell and P. Blank. Inherent Reinforcement of Ceramic Microstructures by Grain Boundary Engineering. J. Eur. Cer. Soc. 9:309–322, 1992.
67. Y. K. Baek and C. H. Kim. The Effect of Whisker Length on the Mechanical Properties of Alumina-SiC Whisker Composites. J. Mat. Sci. 24:1589–1593, 1989.
68. R. H. Dauskardt, B. J. Dalgleish, D. Yao, R. O. Ritchie, and P. F. Becher. Cyclic Fatigue-Crack Propagation in a Silicon Carbide Whisker-Reinforced Alumina Composite: Role of Load Ratio. J. Mat. Sci. 28:3258–3266, 1993.
69. J. Gurland and P. Bardzil. Relation of Strength, Composition, and Grain Size of Sintered WC-Co Alloys. J. Metals 203:311–315, 2/1955.
70. M. J. Murray. Fracture of WC-Co Alloys: An Example of Spatially Constrained Crack Tip Opening Displacements. Proc. Roy. Soc. London A 356:483–508, 1977.
71. A. Hara, T. Nishikawa, and S. Yazu. The Observation of the Fracture Path in WC-Co Cemented Carbides Using a Newly Developed Replica Method. Planseeberichte fur Pullver-Metallurgie 18:28–43, 1970.
72. J. R. Pickens. The Fracture Toughness of Tungsten Carbide-Cobalt Alloys as a Function of Microstructural Parameters. Brown Univ. Report, Ph.D. thesis and Technical Report for NSF Grant GH-42182, 4/1977.

73. J. R. Pickens and J. Gurland. The Fracture Toughness of WC-Co Alloys Measured on Single-Edged Notched Beam Specimens Precracked by Electron Discharge Machining. Mat. Sci. Eng. 33:135–142, 1978.
74. B. Roebuck and E. A. Almond. Deformation and Fracture Processes and the Physical Metallurgy of WC-Co Hardmetals. Intl. Mat. Rev. 33(2):90–110, 1988.
75. J. L. Chermant, A. Deschanvres, and F. Osterstock. Toughness and Fractography of TiC and WC. Fracture Mechanics of Ceramics 4 (R. C. Bradt, D. P. H. Hasselman, and F. F. Lange, eds.). Plenum Press, New York, 1978, pp. 891–901.
76. J. L. Chermant and F. Osterstock. Fracture Toughness and Fracture of WC-Co Composites. J. Mat. Sci. 11:1939–1951, 1976.
77. E. Carlström, R. Carlsson, A.-K. Tjernlund, and B. Johannesson. Some Fracture Properties of Alumina-Containing Electrical Porcelains. Fracture Mechanics of Ceramics 8, Microstructure, Methods, Design, and Fatigue (R. C. Bradt, A. G. Evans, D. P. H. Hasselman, and F. F. Lange, eds.). Plenum Press, New York, 1986, pp. 137–142.
78. T. Soma, M. Matsui, I. Oda, and N. Yamamoto. Applicability of Crack Propagation Data to Failure Prediction in Porcelain. J. Am. Cer. Soc. 63(3–4):166–169, 1980.
79. K. K. Smyth and M. B. Magida. Dynamic Fatigue of a Machinable Glass-Ceramic. J. Am. Cer. Soc. 66(7):500–505, 1983.
80. R. F. Cook, S. W. Freiman, and T. L. Baker. Effect of Microstructure on Reliability Predictions for Glass Ceramics. Mat. Sci. Eng. 77:199–212, 1986.
81. T. L. Jessen and D. Lewis III. Limiting Subcritical Crack Growth in Glass-Matrix Composites. J. Am. Cer. Soc. 75(8):3219–3224, 1992.
82. T. L. Jessen and D. Lewis III. Fracture Toughness of Graded Metal-Particulate/Brittle-Matrix Composites. J. Am. Cer. Soc. 73(5):1405–1408, 1990.
83. T. L. Jessen and D. Lewis III. Effect of Crack Velocity on Crack Resistance in Brittle-Matrix Composites. J. Am. Cer. Soc. 72(5):812–821, 1989.
84. T. L. Jessen and D. Lewis III. Effect of Composite Layering on the Fracture Toughness of Brittle Matrix/Particulate Composites. Composites 26(1):67–71, 1995.
85. P. F. Becher, T. N. Tiegs, J. C. Ogle, and W. H. Warwick. Toughening of Ceramics by Whisker Reinforcement. Fracture Mechanics of Ceramics 7, Composites, Impact, Statistics, and High-Temperature Phenomena (R. C. Bradt, A. G. Evans, D. P. H. Hasselman, and F. F. Lange, eds.). Plenum Press, New York, 1986, pp. 61–73.
86. K. Zeng, K. Breder, and D. Rowcliffe. Comparison of Slow Crack Growth Behavior in Alumina and SiC-Whisker-Reinforced Alumina. J. Am. Cer. Soc. 76(7):1673–1680, 1993.
87. A. Cazzato and K. T. Faber. Fracture Energy of Glass-Alumina Interfaces via the Bimaterial Bend Test. J. Am. Cer. Soc. 80(1):181–188, 1997.
88. N. P. O'Dowd, M. G. Stout, and C.F. Shih. Fracture Toughness of Alumina-Niobium Interfaces: Experiments and Analyses. Phil. Mag. A 66 (6):1037–1064, 1992.
89. V. Gupta, J. Wu, and A. N. Pronin. Effect of Substrate Orientation, Roughness, and Film Deposition Mode on the Tensile Strength and Toughness of Niobium-Sapphire Interfaces. J. Am. Cer. Soc. 80(12):3172–3180, 1997.

90. I. E. Remanis, K. P. Trumble, K. A. Rogers, and B. J. Dalgleish. Influence of Cu_2O and $CuAlO_2$ Interfaces on Crack Propagation at Cu/αAl_2O_3 Interfaces. J. Am. Cer. Soc. 80(2):424–432, 1997.
91. C. Cm. Wu, R. W. Rice, and P. F. Becher. The Character of Cracks in Fracture Toughness Measurements of Ceramics. Fracture Mechanics Methods for Cements, Rocks, and Ceramics (S. W. Freiman and E. R. Fuller, Jr., eds.). Am. Soc. for Testing and Materials, STP 745, 1981, pp. 127–140.
92. C. Cm. Wu, S. W. Freiman, R. W. Rice, and J. J. Mecholsky. Microstructural Aspects of Crack Propagation in Ceramics. J. Mat. Sci. 13:2659–2670, 1978.
93. R. W. Rice. Perspective on Fractography. Advances in Ceramics 22, Fractography of Glasses and Ceramics (V. D. Frechette and J. R. Varner, eds.). Am Cer. Soc., Westerville, OH, 1988, pp. 3–56.
94. S. Suresh. Fatigue Crack Growth in Brittle Materials J. Hard. Mat. 2(1–2):29–54, 1991.
95. G. Grathwohl and T. Liu. Strengthening of Zirconia-Alumina During Cyclical Fatigue Testing. J. Am. Cer. Soc. 72(10):1988–1990. 1989.
96. J. S. Nadeau and J. I. Dickson. Effects of Internal Stresses Due to a Dispersed Phase on the Fracture Toughness of Glass. J. Am. Cer. Soc. 63(9–10):517–523, 1980.
97. R. W. Davidge and T. J. Green. The Strength of Two-Phase Ceramic/Glass Materials. J. Mat. Sci. 3:629–634, 1968.
98. K. T. Faber, T. Iwagoshi, and A. Ghosh. Toughening by Stress-Induced Microcracking in Two-Phase Ceramics. J. Am. Cer. Soc. 71(9):C-399–401, 1988.
99. N. Miyata, S. Ichikawa, H. Monji, and H. Jinno. Fracture Behavior of Brittle Matrix, Particulate Composites with Thermal Expansion Mismatch. Fracture Mechanics of Ceramics 7, Composites. Impact, Statistics, and High-Temperature Phenomena (R. C. Bradt, A. G. Evans, D. P. H. Hasselman, and F. F. Lange, eds.). Plenum Press, New York, 1986, pp. 87–102.
100. G. Kirchhoff, W. Pompe, and H.-A. Bahr. Structure Dependence of Thermally Induced Microcracking in Porcelain Studied by Acoustic Emission. J. Mat. Sci. 17:2809–2816, 1982.
101. J. S. Banda and P. F. Messer. The Effect of Quartz Particles on the Strength and Toughness of Whitewares 17, Ceramic Transactions, Fractography of Glasses and Ceramics II (V. D. Fréchette and J. R. Varner, eds.). Am Cer. Soc., Westerville, OH, 1991, pp. 243–261.
102. G. H. Beall, K. Chyung, R. L. Stewart, K. Y. Donaldson, H. L. Lee, S. Baskaran, and D. P. H. Hasselman. Effect of Test Method and Crack Size on the Fracture Toughness of a Chain-Silicate Glass-Ceramic. J. Mat. Sci. 21:1365–1372, 1973.
103. D. S. Baik, K. S. No, and J. S.-S. Chun. Mechanical Properties of Mica Glass-Ceramics. J. Am. Cer. Soc. 78(5):1217–1222, 1995.
104. K. J. Anusavicé and N.-Z. Zhang. Effect of Crystallinity on Strength and Toughness of LiO-Al_2O_3-CaO-SiO_2 Glass-Ceramics. J. Am. Cer. Soc. 80(6):1353–1358, 1997.
105. P. Hing and P. W. McMillan. The Strength and Fracture Properties of Glass-Ceramics. J. Mat. Sci. 8:1041–1048, 1973.

106. R. Morena, K. Niihara, and D. P. H. Hasselman. Effect of Crystallites on Surface Damage and Fracture Behavior of a Glass-Ceramic. J. Am. Cer. Soc. 66(10):673–682, 1983.
107. R. K. Govila, K. R. Kinsman, and P. Beardmore. Fracture Phenomenology of a Lithium-Aluminum-Silicate Glass-Ceramic. J. Mat. Sci. 13:2081–2091, 1978.
108. D. P. H. Hasselman, D. Hirschfeld, H. Tawil, and E. K. Beauchamp. Effect of Inclusions on Size of Surface Flaws in Glass-Crystal Composites. J. Mat. Sci. 20:4050–4056, 1985.
109. W. D. Wolf, K. J. Vaidya, and L. F. Francis. Mechanical Properties and Failure Analysis of Alumina-Glass Dental Composites. J. Am. Cer. Soc. 79(7):1769–1776, 1996.
110. N. Claussen. Fracture Toughness of Al_2O_3 with an Unstabilized ZrO_2 Dispersed Phase. J. Am. Cer. Soc. 59(1–2):49–54, 1976.
111. P. F. Becher. Transient Thermal Stress Behavior in ZrO_2-Toughened Al_2O_3, J. Am. Cer. Soc. 64(1):37–39, 1978.
112. F. F. Lange. Transformation Toughening, Part 4: Fabrication, Fracture Toughness and Strength of Al_2O_3-ZrO_2 Composites. J. Mat. Sci 17:247–254, 1982.
113. N. Claussen, J. Steeb, and R. F. Pabst. Effect of Induced Microcracking on the Fracture Toughness of Ceramics. Fracture Toughness of Al_2O_3 with an Unstabilized ZrO_2 Dispersed Phase. Am. Cer. Soc. Bull. 56(6):559–562, 1977.
114. N. Claussen, R. L. Cox, and J. S. Wallace. Slow Crack Growth of Microcracks: Evidence for One Type of ZrO_2 Toughening. J. Am. Cer. Soc. 65(11):C-190–191, 1982.
115. M. Rühle. Microcrack and Transformation Toughening of Zirconia-Containing Alumina. Mat. Sci. Eng. A105/106:77–82, 1988.
116. J. Homeny and J. J. Nick. Microstructure-Property Relations of Alumina-Zirconia Eutectic Ceramics. Mat. Sci. Eng. A127:123–133, 1990.
117. E. H. Lutz, N. Claussen, and M. V. Swain. K^R-Curve Behavior of Duplex Ceramics. J. Am. Cer. Soc. 74 (1):11–18, 1991.
118. E. H. Lutz and N. Claussen. Duplex Ceramics: II, Strength and Toughness. J. Eur. Cer. Soc. 7:219–226, 1991.
119. E. H. Lutz, M. V. Swain, and N. Claussen. Thermal Shock Behavior of Duplex Ceramics. J. Am. Cer. Soc. 74(1):19–24, 1991.
120. R. Langlois and K. J. Konaztowicz. Toughening in Zirconia-Toughened Alumina Composites with Non-Transforming Zirconia. J. Mat. Sci. Lett. 11:1454–1456, 1992.
121. J. Wang and R. Stevens. Toughening Mechanisms in Duplex Alumina-Zirconia Ceramics. J. Mat. Sci. 23:804–808, 1988.
122. S. Lathabai, D. G. Hay, F. Wagner, and N. Claussen. Reaction-Bonded Mullite/Zirconia Composites. J. Am. Cer. Soc. 79(1):248–256, 1996.
123. A. Nakahira, K. Niihara, and T. Hirai. Microstructural Properties of Al_2O_3-SiC Composites. Yogyo-Kyokai-Shi 94(8):767–772, 1986.
124. K. Niihara, A. Nakahira, T. Uchiyama, and T. Hirai. High-Temperature Mechanical Properties of Al_2O_3-SiC Composites. Fracture Mechanics of Ceramics 7, Composites, Impact, Statistics, and High-temperature Phenomena (R. C. Bradt, A. G.

Evans, D. P. H. Hasselman, and F. F. Lange, eds.). Plenum Press, New York, 1986, pp. 103–116.
125. M. Yasuoka, M. E. Brito, K. Hirao, and S. Kanzaki. Effect of Dispersed Particle Size on Mechanical Properties of Alumina/Non-Oxide Composites. J. Cer. Soc. Jpn., Intl. Ed. 101:865–870, 1993.
126. I. Thompson and V. D. Kristic. Mechanical Properties of Hot-Pressed SiC Particulate-Reinforced Al_2O_3 Matrix. J. Mat. Sci. 27:5765–5768, 1992.
127. D. Zhang, H. Yang, R. Yu, and W. Weng. Mechanical Properties of Al_2O_3-SiC Composites Containing Various Sizes and Fractions of Fine SiC Particles. J. Mat. Sci. Lett. 16:877–879, 1997.
128. L. C. Stearns and M. P. Harmer. Particle-Inhibited Grain Growth on Al_2O_3-SiC: I, Experimental Results. J. Am. Cer. Soc. 79(12):3013–3019, 1996.
129. F. F. Lange. Effect of Microstructure on Strength of Si_3N_4-SiC Composite System. J. Am. Cer. Soc. 56(9):445–446, 1973.
130. H. Tanaka, P. Greil, and G. Petzow. Sintering and Strength of Silicon Nitride-Silicon Carbide Composite. Intl. J. High Temp. Cer. 1:107–118, 1985.
131. A. Sawaguchi, K. Toda, and K. Niihara. Mechanical and Electrical Properties of Silicon Nitride-Silicon Carbide Nanocomposite Material. J. Am. Cer. Soc. 74(5):1142–1144, 1991.
132. K.-S. Cho, Y.-W. Kim, H.-J. Choi, and J.-G. Lee. SiC-TiC and SiC-TiB_2 Composites Densified by Liquid-Phase Sintering. J. Mat. Sci. 31:6223–6228, 1996.
133. Wei and Becher. Improvements in Mechanical Properties in SiC by the Addition of TiC Particles. J. Am. Cer. Soc. 67(8):571–574, 1984.
134. B.-W. Lin and T. Iseki. Different Toughening Mechanisms in SiC/TiC Composites. Brit. Cer. Trans. 91:147–150, 1992.
135. B.-W. Lin and T. Iseki. Effect of Thermal Residual Stress on Mechanical Properties of SiC/TiC Composites. Brit. Cer. Trans. 91:1–5, 1992.
136. J. D. Yoon and S. G. Kang. Strengthening and Toughening Behavior of SiC with Additions of TiB_2. J. Mat. Sci. Lett. 14:1065–1067, 1995.
137. M. A. Janney. Mechanical Properties and Oxidation Behavior of a Hot-Pressed SiC-15 Vol. %-TiB_2 Composite. Am. Cer. Soc. Bull. 66(2):322–324, 1987.
138. C. H. McMurtry, W. D. G. Boecker, S. G. Seshadri, J. S. Zangil, and J. E. Garnier. Microstructure and Material Properties of SiC-TiB_2 Particulate Composites. Am. Cer. Soc. Bull. 66(2):325–329, 1987.
139. K.-S. Cho, H.-J. Choi, J.-G. Lee, and Y. W. Kim. In Situ-Toughening SiC-TiB_2 Composites, J. Mat. Sci: to be published.
140. A. Bellosi, S. Guicciardi, and A. Tampieri. Development and Characterization of Electroconductive Si_3N_4-TiN Composites. J. Eur. Cer. Soc. 9:83–93, 1992.
141. T. Nagaoka, M. Yasuoka, K. Hirao, and S. Kanzaki. Effects of TiN Particle Size on Mechanical Properties of Si_3N_4/TiN Particulate Composites. J. Cer. Soc. Jpn. 100:617–620, 1992.
142. W.-H. Gu, K. T. Faber, and R.W. Steinbrech. Microcracking and R-Curve Behavior in SiC-TiB_2 Composites. Acta Met. Mat. 40(11):3121–3128, 1992.
143. M. K. Bannister and M. V. Swain. Fracture Behavior of a TiB_2-Based Ceramic Composite Material. J. Mat. Sci. 26:6789–6799, 1991.

144. S. V. Nair, P. Z. Q. Cai, and J. E. Ritter. Mechanical Behavior of Silicon Carbide Particulate Reinforced Reaction Bonded Silicon Nitride Matrix Composites. Cer. Eng. Sci. Proc. 13(7–8):81–89, 1992.
145. J. J. Petrovic, M. I. Pena, I. E. Remanis, M. S. Sandlin, S. D. Conzone, H. H. Kung, and D. P. Butt. Mechanical Behavior of $MoSi_2$ Reinforced-Si_3N_4 Matrix Composites. J. Mat. Sci. 80(12):3070–3076, 1997.
146. L. S. Sigl and H.-J. Kleebe. Microcracking in B_4C-TiB_2 Composites. J. Am. Cer. Soc. 78(9):2374–2380, 1995.
147. M. Landon and F. Thevenot. The SiC-AlN System: Influence of Elaboration Routes on the Solid Solution Formation and Its Mechanical Properties. Cer. Intl. 17:97–110, 1991.
148. J.-F. Li and R. Watanabe. Preparation and Mechanical Properties of SiC-AlN Ceramic Alloys. J. Mat. Sci. 26:4813–4817, 1991.
149. K.-H. Heussner and N. Claussen. Yttria- and Ceria-Stabilized Tetragonal Zirconia Polycrystals (Y-TZP, Ce-TZP) Reinforced with Al_2O_3 Platelets. J. Eur. Cer. Soc. 5:193–200, 1989.
150. X.-N. Huang and P. S. Nicholson. Mechanical Properties and Fracture Toughness of αAl_2O_3-Platelet Reinforced Y-TZP Composites at Room and High Temperature. J. Am. Cer. Soc. 76(5):1294–1301, 1993.
151. M. Yasuoka, K. Hirao, M. E. Brito, and S. Kanzaki. High-Strength and High-Fracture-Toughness Ceramics in the $Al_2O_3/LaAl_{11}O_5$ Systems. J. Am. Cer. Soc. 78(7):1853–1856, 1995.
152. H.-D. Kim, I.-S. Lee, S.-W. Kang, and J.-W. Ko. The Formation of $NaMg_2Al_{15}O_{25}$ in an αAl_2O_3 Matrix and Its Effect on the Mechanical Properties of Alumina. J. Mat. Sci. 29:4119–4124, 1994.
153. T. Koyama, A. Nishiyama, and K. Niihara. Effect of Grain Morphology and Grain Size on the Mechanical Properties of Al_2O_3 Ceramics. J. Mat. Sci. 29:3949–3954, 1994.
154. L. An and H. M. Chan. R-Curve Behavior of In-Situ Toughened Al_2O_3:$CaAl_{12}O_{19}$ Ceramic Composites. J. Am. Cer. Soc. 79(12):3142–3148, 1996.
155. C. Nischik, M. M. Seibold, N. A. Travitzky, and N. Claussen. Effect of Processing on Mechanical Properties of Platelet-Reinforced Mullite Composites. J. Am. Cer. Soc. 74(10):2464–2468, 1991.
156. Y.-S. Chou and D. J. Green. Silicon Carbide Platelet/Alumina Composites: II, Mechanical Properties. J. Am. Cer. Soc. 76(6):1452–1458, 1993.
157. R. Chaim and V. Talanker. Microstructure and Mechanical Properties of SiC Platelet/Cordierite Glass-Ceramic Composites. J. Am. Cer. Soc. 78(1):166–172, 1995.
158. C. F. Cooper, I. C. Alexander, and C. J. Hampson. The Role of Graphite in the Thermal Shock Resistance of Refractories. Brit. Cer. Trans. J. 84:57–62, 1985.
159. T. Mitchel, Jr., L. C. DeJonghe, W. J. MoberlyChan, and R. O. Ritchie. Silicon Carbide Platelet/Silicon Carbide Composites. J. Am. Cer. Soc. 78(1):97–103, 1995.
160. M. Hanninen, R. A. Haber, and D. E. Niesz. Mechanical Properties of Silicon Carbide Platelet and Particulate Reinforced Silicon Nitride. Ceramic Transactions 19,

Advanced Composite Materials (M. D. Sacks, ed.). Am. Cer. Soc., Westerville, OH, 1991, pp. 749–755.

161. T. D. Claar, W. B. Johnson, C. A. Anderson, and G. H. Schiroky. Microstructure and Properties of Platelet-Reinforced Ceramics Formed by the Directed Reaction of Zirconium with Boron Carbide. Cer. Eng. Sci. Proc. 10(7-8):599–609, 1989.
162. T. N. Tiegs and P. F. Becher. Alumina-SiC Whisker Composites. Cer. Eng. Sci. Proc. 7(9–10):1182–1186, 1986.
163. P. F. Becher, T. N. Tiegs, and P. Angelini. Whisker Toughened Ceramic Composites. Fiber Reinforced Ceramic Composites, Materials, Processing and Technology (K. S. Mazdiyasni, ed.). Noyes, Park Ridge, NJ, 1990, pp. 311–327.
164. P. F. Becher, C.-H. Hsueh, P. Angelini, and T. N. Tiegs. Toughening Behavior in Whisker Reinforced Ceramic Matrix Composites. J. Am. Cer. Soc. 71(12): 1050–1061, 1988.
165. P. F. Becher, H.T. Lin, and K. B. Alexander. Development of Toughened Ceramics for Elevated Temperatures. Science of Engineering Ceramics '91 (S. Kimura and K. Niihara, eds.). Cer. Soc. Jpn. 99:307–314, 1991.
166. P. F. Becher, E. R. Fuller, Jr., and P. Angelini. Matrix-Grain-Bridging Contributions to the Toughness of Whisker-Reinforced Ceramics. J. Am. Cer. Soc. 74(9): 2131–2135, 1991.
167. E. Yasuda, T. Akatsu, and Y. Tanabe. Influence of Whiskers' Shape and Size on Mechanical Properties of SiC Whisker-Reinforced Al_2O_3. J. Cer. Soc. Jpn., Intl. Ed. 99:51–57, 1991.
168. A. V. Krylov, S. M. Barinov, D. A. Ivanov, N. A. Mindlina, L. Parilak. J. Dusza, F. Lofaj, and E. Rudnayova. Influence of SiC Whisker Size on Mechanical Properties of Reinforced Alumina. J. Mat. Sci. Lett. 12:904–906, 1993.
169. Y. K. Baek and C. H. Kim. The Effect of Whisker Length on the Mechanical Properties of Alumina-SiC Whisker Composites. J. Mat. Sci. 24:1589–1593, 1989.
170. T. Tiegs. Structural and Physical Properties of Ceramic Matrix Composites. Handbook on Discontinuously Reinforced Ceramic Matrix Composites (R. L. Lehman, S. K. El-Rahaiby, and J. B. Wachtman, Jr., eds.). Ceramics Information Analysis Center, West Lafayette, IN and Am. Cer. Soc., Westerville OH, 1995, pp. 225–273.
171. K. P. Gadkaree. Whisker Reinforcement of Glass-Ceramics. J. Mat. Sci. 26:4845–4854, 1991.
172. M. Wu, G. L. Messing, and M. F. Amateau. Laminate Processing and Properties of Oriented SiC-Whisker-Reinforced Composites. Ceramic Trans. 19, Advanced Composite Materials: Processing, Microstructures, Bulk and Interfacial Properties, Characterization Methods, and Applications (M. D. Sacks, ed.). Am. Cer. Soc., Westerville, OH, 1991, pp. 665–667.
173. R. Lundberg, L. Kahlman, R. Pompe, and R. Carlsson. SiC-Whisker-Reinforced Si_3N_4 Composites. Am. Cer. Soc. Bull. 66(2):330–333, 1987.
174. S. J. Buljan, J. G. Baldoni, and M. L. Huckabee. Si_3N_4-SiC Composites. Am. Cer. Soc. Bull. 66(2):347–352, 1987.
175. S. Iio, M. Watanabe, M. Matsubara, and Y. Matsuo. Mechanical Properties of Alumina/Silicon Carbide Whisker Composites. J. Am. Cer. Soc. 72(10):1880–1884, 1989.

176. K. Okada, N. Ōtsuka, R. J. Brook, and A. J. Moulson. Microstructure and Fracture Toughness of Yttria-Doped Tetragonal Zirconia Polycrystal/Mullite Composites Prepared by an In Situ Method. J. Am. Cer. Soc. 72(12):2369–2372, 1989.
177. N. Claussen, K.-L. Weisskopf, and M. Rühle. Tetragonal Zirconia Polycrystals Reinforced with SiC Whiskers. J. Am. Cer. Soc. 69(3):288–292, 1986.
178. P. F. Becher, and T. N. Tiegs. Toughening Behavior Involving Multiple Mechanism: Whisker Reinforced and Zirconia Toughening. J. Am. Cer. Soc. 70(9):651–654, 1987.
179. R. Ruh, K. S. Mazdiyasni, and M. G. Mendiratta. Mechanical and Microstructural Characterization of Mullite and Mullite-SiC-Whisker and ZrO_2-Toughened-Mullite-SiC-Whisker Composites. J. Am Cer. Soc. 71(60):503–512, 1988.
180. J. P. Singh, K. C. Goretta, D. S. Kupperman, and J. L. Routbort. Fracture Toughness and Strength of SiC-Whisker-Reinforced Si_3N_4 Composites. Adv. Cer. Mat. 3(4):357–360, 1988.
181. T. Kandori, S. Kobayashi, S. Wada, and O. Kamigaito. SiC Whisker Reinforced Si_3N_4 Composites. J. Mat. Sci. Lett. 6:1356–1358, 1987.
182. J. Dusza and D. Šajgalik. Mechanical Properties of Si_3N_4 + βSi_3N_4 Whisker Reinforced Ceramics. J. Eur. Cer. Soc. 9:9–17, 1992.
183. G. H. Campbell, M. Rühle, B. J. Dalgleish, and A. G. Evans. Whisker Toughening: A Comparison Between Aluminum Oxide and Silicon Nitride Toughened with Silicon Carbide. J. Am. Cer. Soc. 73(30):521–530, 1990.
184. V. S. Stubican, R. C. Bradt, F. L. Kennard, W. J. Minford, and C. C. Sorell. Ceramic Eutectic Composites. Tailoring Multiphase and Composite Ceramics, Materials Science Research 20 (R. E. Tressler, G. L. Messing, C. G. Pantano, and R. E. Newnham, eds.). Plenum Press, New York, 1986, pp. 103–113.
185. C. C. Sorell, V. S. Stubican, and R. C. Bradt. Mechanical Properties of ZrC-ZrB_2 and ZrC-TiB_2 Directionally Solidified Eutectics. J. Am. Cer. Soc. 69(4):317–321, 1986.
186. M. D. Brumels and B. J. Pletka. Fracture Initiation in the Directionally Solidified NiO-CaO Eutectics. J. Am. Cer. Soc. 70(5):305–310, 1987.
187. T. Mah, A. Parthasarathy, and L. E. Matson. Processing and Mechanical Properties of $Al_2O_3/Y_5Al_5O_{12}$ (YAG) Eutectic Composites. Cer. Eng. Sci. Proc. 11(9–10): 1617–1627, 1990.
188. J.-M. Yang, S. M. Jeng, and S. Chang. Fracture Behavior of Directionally Solidified $Y_3Al_5O_{12}/Al_2O_3$ Eutectic Fiber. J. Am. Cer. Soc. 79(5):1218–1222, 1996.
189. R. W. Rice, C. Cm. Wu, and K. R. McKinney. Unpublished work at the US Naval Res. Lab, cera 1975–1980.
190. U. Krohn, H. Olapinski, and U. Dworak. US patent 4,595,663, 6/17/1986.
191. L. Mazerolles, D. Michel, and R. Portier. Microstructural and Mechanical Behavior of Al_2O^3-ZrO_2 (Y_2O_3) Oriented Eutectics. J. de Phys. 47, (Suppl.2):C1-335–339, 1986.
192. J. Echigoya, Y. Takabayashi, and H. Suto. Hardness and Fracture Toughness of Directionally Solidified Al_2O_3 - ZrO_2 (Y_2O_3) Eutectics. J. Mat. Sci. Lett. 5:153–154, 1986.
193. C. Hulse and J. Batt. The Effect of Eutectic Microstructures on the Mechanical

Properties of Ceramic Oxides. United Aircraft Res. Lab. Report for ONR Contract N00014-69-C-0073, 5/1974.
194. F. L. Kennard, R. C. Bradt, and V. S. Stubican. Mechanical Properties of the Directionally Solidified $MgO-MgAl_2O_4$ Eutectic. J. Am. Cer. Soc. 59(3-4):160–163, 1976.
195. B. D. Flinn, M. Rühle, and A. G. Evans. Toughening in Composites of Al_2O_3 Reinforced with Al. Acta. Metall. 37(11):3001–3006, 1989.
196. J. Wang, C. B. Ponton, and P. M. Marquis. Silver-Toughened Alumina Ceramics. Trans. J. Brit. Cer. Soc. 92:71–74, 1993.
197. W. H. Tuan and R. J. Brooks. The Toughening of Alumina with Nickel Inclusions. J. Eur. Cer. Soc. 6:31–37, 1990.
198. M. Nawa, K. Yamazaki, T. Sekino, and K. Niihara. Microstructure and Mechanical Behavior of 3Y-TZP/Mo Nanocomposites Processing and Novel Interpenetrated Intragranular Microstructure. J. Mat. Sci. 31:2849–2858, 1996.
199. M. Nawa, T. Sekino, and K. Niihara. Fabrication and Mechanical Behavior of Al_2O_3/Mo Nanocomposites Processing and Novel Interpenetrated Intragranular Microstructure. J. Mat. Sci. 29:3185–3192, 1994.
200. W. B. Chou, W. H. Tuan, and S. T. Chang. Preparation of NIAl Toughened Al_2O_3 by Vacuum Hot Pressing. Brit. Cer. Trans. 95(2):71–74, 1996.
201. M. F. Ashby, F. J. Blunt, and M. Bannister. Flow Characteristics of Highly Constrained Metal Wires. Acta Metall. 37(7):1847–1857, 1989.
202. Y.-H. Yun and S.-C. Choi. The Contributions of Microstructural Characteristics and Residual Stress Distribution to Mechanical Properties of AlN/W Composite System. J. Mat. Sci. 33:707–712, 1998.
203. V. Krstic, P. S. Nicholson, and R. G. Hoagland. Toughening of Glasses by Metallic Particles. J. Am. Cer. Soc. 64(9):499–504, 1981.
204. E. Breval, Z. Deng, S. Chiou, and C. G. Pantano. Sol-Gel Prepared Ni-Alumina Composite Materials. J. Mat. Sci. 27:1464–1468, 1992.
205. T. Sekino, T. Nakajima, S. Ueda, and K. Niihara. Reduction and Sintering of a Nickel-Dispersed-Alumina Composite and Its Properties. J. Am. Cer. Soc. 80(5):1139–1148, 1997.
206. D. B. Marshall, W. L. Morris, B. N. Cox, and M. S. Dadkhan. Toughening Mechanisms in Cemented Carbides. J. Am. Cer. Soc. 73(10):2938–2943, 1990.
207. K. Hirano. Toughening Mechanism for Ceramics by a Ductile Metallic Phase. J. Mat. Sci. Lett. 13:1219–1221, 1996.
208. D. Han and J. J. Mecholsky. Fracture Behavior of Metal Particulate-Reinforced WC-Co Composites. J. Mat. Sci Eng. A170:293–302, 1991.
209. P. A. Trusty and J. A. Yeomans. Crack-Particle Interactions in Alumina-Iron Composites. Cer. Eng. Sci. Proc. 14(9–10):908–913, 1993.
210. A. J. Pyzik and D. R. Beaman. Al-B-C Phase Development and Effects on Mechanical Properties of B_4C/Al- Derived Composites. J. Am. Cer. Soc. 78(20):305–312, 1995.
211. L. A. Simpson and A. Wasylyshyn. Fracture energy of Al_2O_3 Containing Mo Wire. J. Am. Cer. Soc. 54(1):56, 1971.
212. J. Brennan. Development of Fiber Reinforced Ceramic Matrix Composites. United

Technologies Research Center Report No. R911848-4 for US Naval Air Systems Command Contract N62269-74-C-0359, 3/1975.

213. W.-H. Gu, K. T. Faber, and R. W. Steinbrech. Microcracking and R-Curve Behavior in SiC-TiB_2 Composites, Acta Metall Mater. 40(11):3121–3128, 1992.

214. W.-H. Gu and K. T. Faber. Tensile Behavior of Microcracking SiC-TiB_2 Composites. J. Am. Cer. Soc. 78(6):1507–1512, 1995.

215. K. T. Faber and A. G. Evans. Crack Deflection Processes—II Experiment. Acta Metall. 31(4):577–584, 1983.

216. J. C. Swearengen, E. K. Beauchamp, and R. J. Egan. Fracture Toughness of Reinforced Glasses, Fracture Mechanics of Ceramics, Crack Growth and Microstructure 4 (R. C. Bradt, D. P. H. Hasselman, and F. F. Lange, eds.). Plenum Press, New York, 1978, pp. 973–987.

9

Particle Dependence of Tensile Strength of Ceramic Composites at ~ 22°C

I. INTRODUCTION

Chapter 8 addressed crack propagation and fracture toughness in ceramic composites (along with elastic properties). This chapter addresses the tensile (flexure) strength of such composites. If fracture toughness were a clear predictor of tensile strength, there would only be limited information to add in this chapter, since toughness and flaw size and character would determine strength. In such cases the focus would thus be primarily on the degree of strength variations, e.g. as measured by Weibull moduli, and the sources and character of flaw variations, which are addressed (Sec. IV). However, more extensive examination of strength–particle dependence reveals frequent opposite trends of toughness and tensile strength that are even more serious than found for monolithic ceramics in Chapters 2 and 3. Processing defects and microstructural heterogeneities are one factor in these strength–toughness differences. However, the often significant opposite trends of strength and toughness with particle parameters are again more fundamentally attributed to effects of crack size differences between most toughness tests and normal strength behavior, which are an important focus of this chapter. Finally note that the order of the types of composites covered in this chapter is very similar to that of Chapter 8 for cross comparison, but some changes have been made for organizational purposes.

Another aspect of this chapter and a source of added information is much earlier work on composites where strengths were studied but toughness measurements were not developed or in common use. This strength data provides insight into other important factors such as the separation of dispersed particles (λ), consideration of which has been largely neglected in the focus on crack bridging and related *R*-curve effects with large cracks. It will be shown that considerable strength data varies as $\lambda^{-1/2}$, and that this dependence is related to the grain size (G) dependence of strengths of monolithic ceramics. The similarities between monolithic and composite ceramics are also shown by strength variations, e.g. Weibull moduli of composites also often being very similar to those of monolithic ceramics.

II. THEORETICAL BACKGROUND

In principle all the models and concepts considered for crack propagation and toughness are applicable to strength. However, as noted above and in Chapters 2 and 3, there is the important issue of the pertinence of mechanisms whose effectiveness is most significant at large crack sizes, not at the much smaller cracks typically controlling strengths. Serious attention to this issue has been largely absent, so effects can often only be inferred. Thus the implication that crack bridging controls strengths needs further examination in view of its significant dependence on crack size (as well as other uncertainties, Chapters 2, 3, 8). Other mechanisms must also be examined in terms of their crack size dependence, as well as in terms of dependence on volume fraction of dispersed phase. In all cases the central issue is typically the crack size relative to the strength controlling microstructure. Thus the effects of stresses in and around dispersed particles, which has not been found to be a major factor in fracture toughness, may be more important when flaws are on the scale of the dispersed particles and where the volume fraction of dispersed phase is modest. At higher volume fractions crack interactions with the particles may be driven more by their number (and size) than their type or level of stress (unless leading to microcracking, discussed below). Clearly possible line tension effects depend directly on the scale of the crack versus that of the second phase spatial and size distributions, diminishing with increasing second phase content (via their effect on resultant spacing). Similarly, crack deflection presumably becomes less effective beyond some levels of second phase size and content. While this is also likely to be true for fiber pullout, probably at higher fiber volume fractions, even then some systems may perform well at very high levels of fibers (e.g. as shown by ropes on the one hand and implied by natural fibrous materials such as jades, Fig. 8.15D).

Another basic factor that must be considered in strength is that in addition to (or instead of) the mechanisms impacting fracture toughness, there can be effects of composite microstructure on failure causing flaws. A model was proposed by Hasselman and Fulrath [1] for ceramic composites where flaw sizes could become constrained by the dispersed phase (Fig. 9.1), similar to the line tension model (Fig. 8.4), but with the flaw size being constrained between two adjacent particles. Thus when the flaw size in the matrix material from external sources, e.g. machining, was smaller than the spacings between dispersed particles, there would be no effect or limit on the flaw size, and hence no or limited effect of the composite microstructure on strength (e.g. Fig. 8.4B). On the other hand, as the particle spacing decreases below the flaw size expected in the matrix, the flaw size would be limited by the spacing between two adjacent particles, so strength would vary as $\lambda^{-1/2}$, with the slope of strength–$\lambda^{-1/2}$ plots being the composite fracture toughness (when corrected for flaw size being a radius and λ a diameter). Note that this mechanism, while

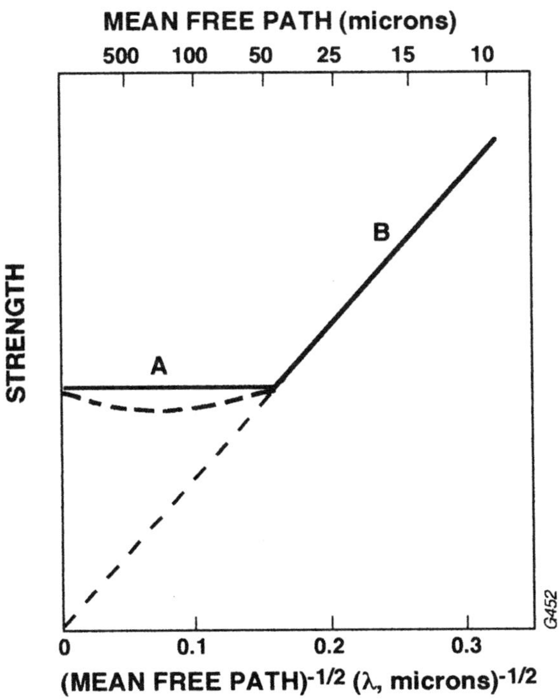

FIGURE 9.1 Schematic plot of tensile (or flexure) strength versus the inverse square root of the mean separation between dispersed particles [$\lambda^{-1/2}$, e.g. with λ obtained from Eq. (9.1)]. Note (1) region A of little or no strength dependence, where the spacings are larger than the flaw size (*c*, e.g. from machining), and (2) that the slope of the line of increasing strength with decreasing λ once $\lambda \leq c$ in region B is the toughness controlling strength (when corrected for flaw size being a radius and λ a diameter, as for monolithic ceramics).

similar to, and possibly impacted by, other effects of composites on machining flaws, is a basic mechanism that can occur independently of other machining flaw effects.

A key flaw factor that needs to be considered is machining flaw generation, where flaw character and size are functions of the local material elastic moduli, toughness and hardness [e.g. via Eq. (3.2)], and machining conditions [2,3]. This is different from such flaws being specifically constrained by the spacing between particles, as was discussed earlier, and can occur independently of, or in addition to, such flaw constraint effects. Subsequent evidence is shown in this chapter that this plays a major role in strengths of composite, as with monolithic, ceramics. More attention to this and other flaw factors is needed, e.g. residual surface compressive stresses generated by machining many ceramics, and possibly somewhat greater in materials with ZrO_2 transformation toughening, which can be important. Another factor is generation of flaws by microcracking and possible linking of microcracks with one another or with machining flaws, again with size and spacing of such cracks relative to the flaw size controlling strength being important.

III. PARTICLE PARAMETER EFFECTS ON TENSILE STRENGTHS OF CERAMIC COMPOSITES

A. Composites with Glass Matrices

1. A Synthetic Glass Matrix Composites

Consider first synthetic composites made by consolidating glass matrices around dispersed crystalline particles, beginning with Binns' [4] earlier, substantial study of such composites with up to 40 v/o of zircon or alumina particles (of ~ 10, 45, or 180 μm dia.) in glass matrices of varying thermal expansions. Both sets of composites with $\phi=0.2$ showed a maximum in E when the thermal expansion of the glass and dispersed particles were matched, progressively decreasing as the expansion difference increased, probably somewhat greater when the glass expansion exceeded that of the dispersed phase versus when it was lower than that of the dispersed phase by the same increment (Chap. 8, Sec. III). Thus the Young's modulus (E) versus glass expansion curves for the two sets of composites were very similar except that the maximum E for the zircon composite was ~ 20% lower than that with Al_2O_3 particles and shifted to lower expansion by ~ the alumina–zircon expansion difference. There was no effect of particle size on E with either type of particle, except for substantially greater decreases for the coarsest Al_2O_3 as the glass expansion increasingly exceeded that of the Al_2O_3.

Biaxial flexure strengths of these composites with the coarsest particles also showed a maximum when the expansions of the glass and the Al_2O_3 or ZrO_2 dispersed phase were matched. The decreases in strengths with increasing expansion mismatch followed those of *E*, again with some probable greater decrease as the glass expansion exceeded that of the particles. All composites had increasing strengths as ϕ increased, but at differing rates, levels, or both. Composites of Al_2O_3 particles with ~ matching matrix expansion and intermediate or finer particles had ~ 20% higher strengths than those with the coarsest particles. However, as the glass expansion progressively exceeded that of the Al_2O_3 particulates, composites with the finest particles reached higher strength levels and a maximum, while composites with intermediate size particles, which had essentially the same strengths as with finer particles when the glass expansion was ≤ that of Al_2O_3, had maximum strengths when the expansion coefficients were ~ the same. At higher glass expansions strengths fell sooner and to a greater degree than with the finer particles, e.g. to ~ 75% that of the finer particle composite. In contrast, composites with 20 v/o of the coarsest Al_2O_3 particles had a maximum strength ~ 2/3 that of composites with intermediate size Al_2O_3 particles when the expansions were ~ the same, and strengths decreased more relative to composites with intermediate and especially finer Al_2O_3 particles as expansion differences increased in either direction, but again with greater decreases with glass expansion exceeding that of Al_2O_3. Composites with zircon particles had much less increase in strengths as ϕ increased with intermediate or coarser particles than with fine particles, so the net result was strengths of the latter ~ 2 times those of the former at ϕ= 0.4 (and ≥or ~ 70% those of composites with respectively intermediate or coarser Al_2O_3 particles). Thus this more comprehensive work shows substantial parallels between Young's modulus and strength behavior with similar effects based on the sign, and especially the magnitude, of the particle–matrix expansion difference, as well as effects of particle size (and composition, and hence probably *E* in the greater decreases in *E* of composites with larger Al_2O_3, but not zircon, particles). Analysis of this data shows a systematic dependence of strength on particle size in conjunction with other key parameters such as ϕ, consistent with behavior of other similar composites, as will be discussed below (Figs. 9.2A, B).

Hasselman and Fulrath's [1] strength data for synthetic composites of Al_2O_3 particles of differing sizes and volume fractions in a glass matrix of nearly the same thermal expansion showed the bilinear behavior of Fig. 1 when plotted versus the mean free spacing between particles, λ, per Fullman's equation,

$$\lambda = 2D(1-\phi)(3\phi)^{-1} \tag{9.1}$$

leading to their model for the strength of composites of discrete ceramic particles in a ceramic matrix. Again, their rationale for this plot and behavior was that at larger λ values machining flaws controlling strengths would be < λ, so that λ would have no effect on strengths as shown, but as λ decreased it would reach a point where λ= the flaw size, beyond which decreasing λ would then constrain the flaw size to ~ λ and hence increase strength in proportion to $\lambda^{-1/2}$. Both the flaw size, which was ~ λ (~ 40 μm) and is consistent with typical machining flaws controlling strengths of machined silicate glasses [2,3], and the slope of the subsequent strength increase at finer λ values, being (~ 0.7 MPa·m$^{1/2}$), i.e. ~ the fracture toughness of the glass, appeared consistent with the model. However, the latter correspondence is somewhat misleading, since λ is taken as the flaw diameter, while the toughness is determined by a slope based on the flaw radius, so the above slope translates into a toughness of ~ 0.5 MP·am$^{1/2}$, i.e. with less good agreement. Hasselman and Fulrath also noted that in the large λ region where strengths were ~ constant with flaws smaller than λ strengths actually had a modest minimum (e.g. ~ 10% lower at higher ϕ), which they attributed to probable stress concentration effects near the Al_2O_3 particles, e.g. due to the large particle–matrix elastic moduli differences. However, while λ is a parameter of some importance, their reliance on it as the controlling parameter is questioned by other evaluations, including a possible more direct role of particle size, as is discussed below (Fig. 9.2B).

Davidge and Green [5] made composites of 10 v/o of ThO_2 spherical particles of different sizes (50 to 700 μm) in three different glass matrices ranging in expansion from ~ 30 to ~ 50, and ~ 75% that of ThO_2. Composites with the lowest expansion glass had no cracks around ThO_2 particles < 60 μm, with at least ~ 25% of particles D > 80 μm having cracks (Chap. 8, Sec. III). In composites with the intermediate expansion glass matrix the threshold for cracking around the spheres increased to D ~ 200 μm, and no preexisting cracks were found in composites with the glass expansion closest to that of ThO_2. Uniaxial flexure strengths of the composites with the glass matrix expansion somewhat below, but approaching that of, the ThO_2 particles were somewhat higher than that of the glass alone at the finest D (~ 150 μm) tested, while those of composites with the two lowest expansion glasses fell somewhat below those of the glasses themselves at the finest ThO_2 particle sizes. Strengths of composites with either of the three glass matrices all decreased linearly as $D^{-1/2}$ decreased, but at two significantly different rates (Fig. 9.2A). Evans presented modeling to support crack sizes being related to the particle sizes, with some variations based on the local stresses [6].

Frye and Mackenzie's [7] biaxial flexure strengths of composites of 125–150 μm spheres of Al_2O_3 or ZrO_2 in glasses tended to ~ scale with the resultant composite Young's moduli with limited glass–particle expansion differences (e.g. ± 2 ppm/°C) from those of the dispersed particles. With larger expansion differences, strengths first decreased, e.g. by 10–20% at ϕ=0.2, and then increased back to or above the glass strength at ϕ=0.4; as noted in Chap. 8, Sec. V.A, the latter bodies had preexisting cracks. The authors noted that the mean free paths between particles in their composites were > the expected flaw sizes, so that the mechanism of Hasselman and Fulrath of particle spacing constraining flaw sizes was not pertinent.

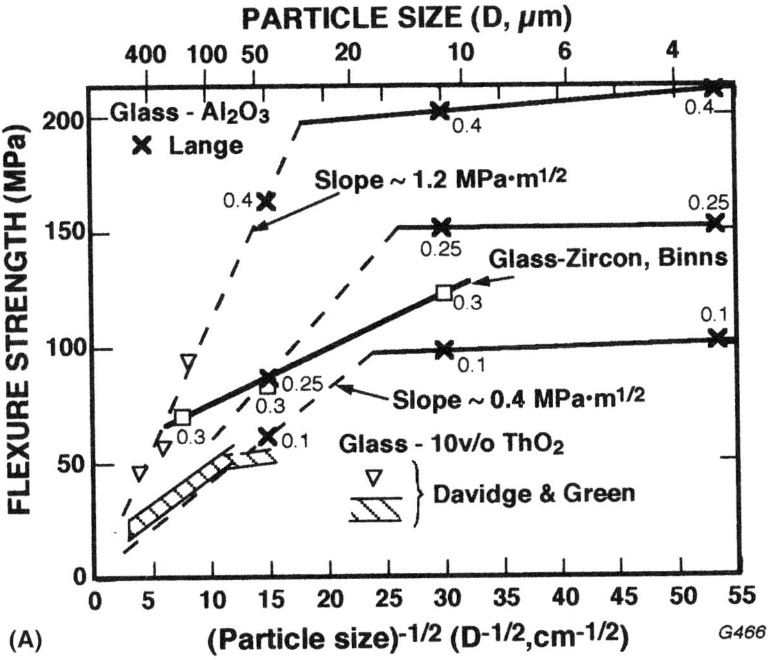

(A)

FIGURE 9.2 Flexure strengths at ~ 22°C of various synthetic glass matrix composites with different crystalline oxide particles and machined or abraded surfaces, as shown (volume fractions indicated in captions, or as decimals next to data points). (A) Data of Binns [4], Davidge and Green [5], and Lange [8], the latter corrected for residual volume fraction porosity (P) of up to 0.07 via e^{-4P} in samples with 40 v/o Al_2O_3 particles. (B) Data of Hasselman and Fulrath [1], Binns [4], and Miyata et al. [9]. Note (1) definite strength dependence on particle size (D), with similar trends, and often similar strength levels, for different composites, with higher strengths at higher volume fraction of particles and less particle–matrix expansion mismatch, and (2) limited effect of irregular versus spherical particles (Miyata et al., half solid circles). See Fig. 9.17 for similar trends for glass matrix–W particle composites.

Lange's [8] uniaxial flexure strengths for his composites of smaller, irregular, or larger spherical Al_2O_3 particles in a glass matrix with very similar expansion to the Al_2O_3 particles tended to follow his fracture energy trends versus ϕ for the finer particle sizes averaging ~ 11 and 3.5 μm, scaling ~ as their increases in E and K with increasing ϕ. However, strengths with the larger (~ 44 μm) particles, which had given much higher toughnesses (Chap. 8, Sec. V.A), gave the lowest strengths, which fell below that of the glass at ϕ= 0.1 and 0.25, giving

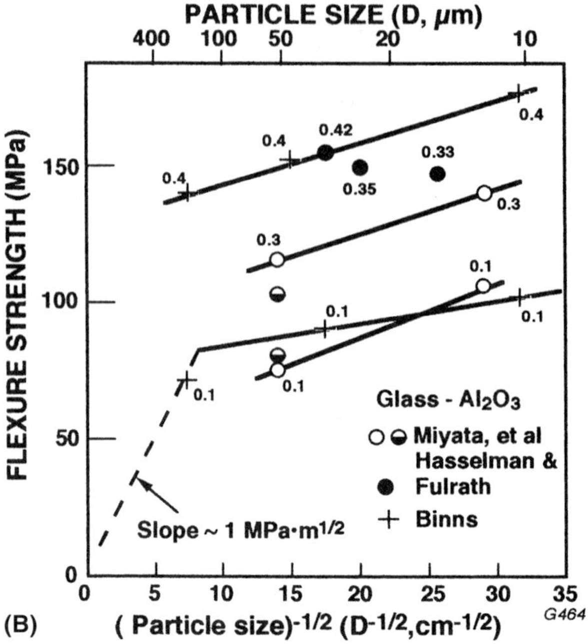

(B)

FIGURE 9.2 Continued

measurable strength increases by ϕ=0.4 but still below those with finer particles. This behavior is inconsistent with his fracture energy results at larger particle size and is thus at least partly inconsistent with his fracture energy results being fairly consistent with his line tension model, since this predicts increasing fracture energy, and hence toughness, as *D* increases (Chap. 8, Sec. II.B). His strength data generally increased ~ linearly with $\lambda^{-1/2}$, suggesting some possible agreement with the line tension model and the flaw size limitation model of Hasselman and Fulrath [1]. However, again the largest particle composite did not appear consistent with this, generally having the lowest strengths and highest fracture energies, and there was no break in the σ–$\lambda^{-1/2}$ line to no strength decreases with further decrease in $\lambda^{-1/2}$ and the σ–$\lambda^{-1/2}$ slope appeared too low, e.g. by ~ 2. Further, there was a distinct particle size dependence to the data that was not recognized (Fig. 9.2A).

Miyata et al.'s [9] strengths of composites of glass matrices having thermal expansions greater (by ~ 2.8 or 5 ppm/°C) than the Al_2O_3 (~ 50 μm spherical, or ~ 50 or ~ 12 μm angular) particles also vary significantly from their toughness trends. While in all cases substantial (e.g. twofold) toughness increases occurred as ϕ increases to 0.3 (Chap. 8, Sec. III.A), strengths always initially decreased below, and frequently barely increased to or slightly above, the glass strengths by ϕ= 0.3. Again, strength levels were lower as the particle–matrix expansion difference increased and the particle size increased (with no great differences between spherical and irregular shaped particles). That microcracking was increased by both the application of external stress and as the matrix–particle expansion differences increased was shown by acoustic emission in conjunction with strength tests. Emission occured at ~ the failure stress of samples with the lowest expansion mismatch and ϕ=0.1 but decreased to ~ 75–95% of failure stress for composites with the larger expansion mismatch and even somewhat lower for all bodies with ϕ=0.3. Surprisingly, composites with 12 μm angular particles had lower thresholds for acoustic emission than did composites with larger spherical particles, but there was little effect of particle shape on strengths, but a substantial effect of particle size (Fig. 9.2B).

Dental composites of 75 v/o Al_2O_3 particles in glass matrices ranging in expansions from ~ 2 ppm/°C < that of alumina to ~ that of alumina made by Wolf et al. [10] via infiltration of molten glass into Al_2O_3 preforms were more consistent with the high toughness values (Chap. 8, Sec. V.A). Thus while toughness values were typically ~ three times those measured for the glasses themselves, strengths were nearly threefold those of the glasses and porcelains cited. These trends are consistent with the finer size of the often somewhat tabular Al_2O_3 particles 0.3 to 10 μm and the above trend for toughness and strength behavior to be closer for finer particles.

Borom [11] in reviewing synthetic glass matrix composite strength behavior noted Miyata and Jinno's [12] reploting of Hasselman and Fulrath's [1] data showing this fell on different curves of strength versus ϕ (again with initial strength decreases from those of the glass at low ϕ and strengths generally increasing as D decreased). He also showed that increases in E of the composites correlated closely with $\lambda^{-1/2}$ and concluded that (1) dispersed particles did not limit flaw sizes (but may produce surface flaws proportional to the particle sizes), and (2) Young's modulus increases were a major source of composite strength increases unless compromised by particle-induced cracking.

However, none of these past evaluations has recognized the central role that composite particle size plays in the strength of these composites as shown here in Figure 9.2, i.e. analogous to the role of grain size for monolithic ceramics (Fig. 3.1). Thus note that Lange's data covering both finer and larger particles relative to expected flaw sizes (mostly 30–50 μm) shows limited but not zero dependence of strength on particle size when this is < the flaw size, but a substantial decrease in strength with decreasing particle size via a $D^{-1/2}$ dependence when D > the expected flaw size. (Note that the larger particle size branch was estimated by assuming a zero intercept for both axes, which may not be true if there are substantial residual stresses contributing to failure, but variations from this slope should not be great.) The mechanism is seen as the same as for the role of grain size in the strength of machined monolithic ceramics (all of the composites were tested as finished by machining or abrasion), i.e. at finer particle sizes machining flaws are < D, so D only has a limited effect on strength via effects on the local E, H, and toughness values affecting the size and character of the surface finishing flaws. However, for $D \geq$ the flaw size, the particles become the flaws. While it might be thought that the effects of D may really reflect a dependence on λ via their close relation, this is unlikely, since at $\phi = 0.1$, $\lambda = 6D$, with $\lambda = D$ only at $\phi = 0.4$, i.e. far too large to be related to normal flaw sizes for most particles. Note also that most of the other data only shows the finer D branch consistent not only with the finer particle sizes but also with the expected flaw sizes being ≥the largest particle size. In the case of the ThO_2 particle composite, the particle sizes are ≥ the flaw sizes, so the data is mostly or completely along the larger D branch, except for a possible transition to the finer D branch at the finer D. Finally note three other key factors about this D dependence and the model for it analogous to that for monolithic ceramics. First, besides the D dependence having the same shape as for the G dependence of strength of monolithic ceramics, and being consistent with the flaws' sizes, the slopes of the larger D branches are < the fracture toughness. Second, this D dependence frequently shows similar strengths for similar v/o of particles from different investigators, as expected from similar composite compositions and flaw sizes. Third, similar results are shown for not only glass–W composites (Fig. 9.17) but also particulate composites of crystalline phases for both the matrix and the dispersed particles (Fig. 9.15).

2. Crystallized Glasses, Porcelains, and Related Composites

Turning to crystallized glasses, Freiman and Hench [13] plotted their strength data for some LiO_2-SiO_2 bodies (normalized by $(E_g/E)^{1/2}$ where E_g and E = the Young's modulus of the parent glass and of the crystallized body) versus $\lambda^{-1/2}$, where λ was the mean spacing between the spherulites (not individual crystallites). Since this plotting resulted in rational, not chaotic, organization of the data, they concluded that strength varied as a function of $\lambda^{-1/2}$. However, they did not associate the differing trends of bodies from different crystallization treatments with any particular mechanism, since there were clearly significantly different trends among the different specimen data sets as well as with models discussed above. Thus their largest data set had strength increasing as $\lambda^{-1/2}$ decreased (and $D^{-1/2}$ increased), while their second largest data set had the opposite trend with $\lambda^{-1/2}$ (and $D^{-1/2}$), i.e. in the first case the trend with $\lambda^{-1/2}$ was not consistent with known or expected mechanisms but was possibly consistent in the second data set, and vice versa for the $D^{-1/2}$ dependence. Since they also showed $E \sim$ doubling as crystallization was completed, this could explain most, but not necessarily all, of their strength increases, but would be inconsistent with significant strength decreases with increased crystallization in one case (but not the other three cases). However, this may reflect stress-induced microcracking, which was not addressed, and illustrates the often serious complexities that can occur in these systems and the need for more characterization and evaluation, as well as consideration of various mechanisms.

Close parallels of Young's modulus and strength were also shown by Atkinson and McMillan [14] in studies of the Li_2O-SiO_2, e.g. showing substantial minima and maxima as a function of heat treatment time at a given temperature. Thus strength and Young's modulus changes were quite similar, but not identical, with maxima and minima patterns as well as changes in numerical values, e.g. of up to ~ threefold for strength and ~ fourfold for *E*. They also noted that strengths showed a $\lambda^{-1/2}$, but not a $D^{-1/2}$, dependence, i.e. strengths varying from ~ 70 to 210 MPa as λ decreased over the range ~ 0.25–0.125 μm and *D* ~ 1.3–1 μm. Freiman and Rice [15] commented that microstructural stresses were probably a factor in their strengths and that this could be a factor in the lack of a correlation with *D*.

Hing and McMillan [16] investigated crystallization of similar LiO_2-SiO_2 glasses to those of Freiman and Hench above but showed strengths following the trend of Fig. 9.3, as well as versus $\lambda^{-1/2}$, with similarity and differences with the results of Freiman and Hench. They noted that the intersection of the two lines at ~ the flaw size and the slope of strength increasing with $\lambda^{-1/2}$ was consistent with the measured glass fracture energy, but this may be fortuitous, e.g. their equation to determine λ appears to be missing a factor of 2/3. They noted that while *E* increased as strength increased (Fig. 9.3), strength increased more than *E*, but the changes in the two are not that much different and there is reasonable correlation between the two (and of λ with *E*). Further, their *E* values are substantially lower than those of Freiman and Hench, e.g. by 30–40% (which is a common occurrence for *E* values from deflections in flexure versus resonance or ultrasonic measurements, as well as from possible stress initiated microcracking in flexure testing). (They noted that microcracking in these materials, though reported by others, was not expected due to similar glass–crystallite expansions, and was not observed.) Other correlations, e.g. of *E* and strength versus λ^{-1}, and that, while the range of *D* values was limited, data was not inconsistent with strength varying as $D^{-1/2}$, indicate the complexity of sorting out mechanisms in many of these complex systems.

Turning to porcelains, residual quartz particles are commonly found to play an important role in their strengths due to either preexisting cracks or stress generated microcracks from them and a general decrease in strengths with higher quartz contents. Plotting Hamano and Lee's [17] strengths versus $D^{-1/2}$ of the quartz particles in fired bodies (based on x-ray analysis) shows a substantial drop in strength (Fig. 9.4A), very similar to that found for bodies in the range of spontaneous microcracking (Figs. 3.1, 3.23). More recent data by Banda and Messer [18] based on starting rather than in situ quartz particle sizes shows a

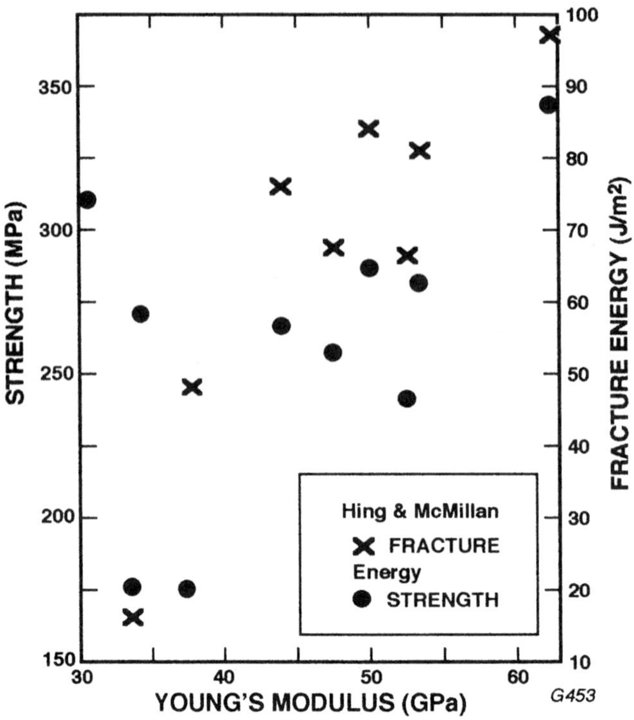

FIGURE 9.3 Strength of LiO_2-SiO_2 specimens versus Young's modulus (measured from deflections in flexure testing) of Hing and McMillan [16]. Note that measured E values are low relative to those reported by Freiman and Hench [13] of ~ 50–90 GPa (as is often the case for such measurements) and that fracture energies showed nearly identical correlation with E.

similar trend, i.e. a rapid decrease in strength over a limited D range and then an apparent leveling out of the strength decrease (Figs. 3.1, 3.23). Thus the differences appear to be mainly or only due to the large differences in D values, much of which stems from the latter work using the starting particle sizes rather than those of the remaining quartz particles. The former measurements of residual quartz particle sizes may not give a good indication of larger remaining sizes that play a larger role in microcracking and resultant mechanical behavior, and thus another source of the difference in quartz particle sizes.

Turning to other related natural ceramic composites, there are indications of strengths correlating with microstructural parameters, especially D (probably because it has been considered more, not necessarily because it is more important or pervasive in its effects). Thus previous plotting [19] of strengths versus $D^{-1/2}$ for flint

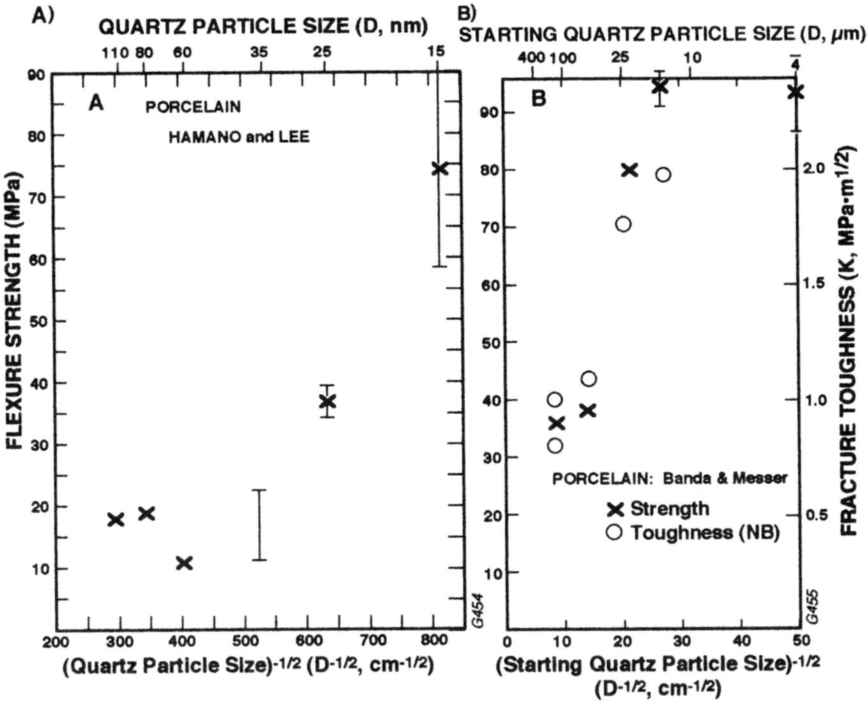

FIGURE 9.4 Strength dependence of porcelains on actual (A) and starting (B) quartz particle sizes. (Respectively after Hamano and Lee [17] and Banda and Messer [18], the latter including data on (NB) toughness.)

particles in earthenware and Al_2O_3 particles in china clay–Nepheline Syenite–Al_2O_3 bodies [20] suggested possible similarities with the $G^{-1/2}$ dependence of monolithic ceramics (Fig. 3.1), e.g. finer and larger D branches intersecting at $D \sim 10$ μm. Similar correlations are seen in some crystallized glasses, e.g. crystallized LiO_2-Al_2O_3-SiO_2 glass data of Utsumi and Sakka [21] also indicate two branches, but data for a crystallized ZnO-Al_2O_3-SiO_2 glass of Stryjak and McMillan [22] appears to be over too small a D range to show a large D branch and thus does not show such behavior. However, again data such as that of Freiman and Hench shows both some similar possible trends as monolithic ceramics and also some clear differences and at least some aspects of the various differing and complex trends that need to be addressed. In the cases where intersections of the two branches are at reasonable flaw sizes, e.g. of the order of 10 μm or so, the branch intersections may reflect the flaw sizes, while at fine, e.g. nm, particle sizes, a $D^{-1/2}$ dependence of strength may reflect microcracking or other correlations, e.g. with λ or ϕ but not due to flaw size–particle relations.

B. ZrO_2 Toughened Composites

1. Powder Processed ZTA Composites

Following the discovery of transformation toughening with metastable tetragonal ZrO_2, especially in fine grain TZP materials (Chap. 2, Sec. III.F, and Chap. 3, Sec. IV.B), many composites of fine metastable tetragonal ZrO_2 particles dispersed in a variety of matrices have been investigated. While the majority of these composites have been with oxide matrices, some have been made with nonoxide matrices despite complications that can arise due to the nonoxidizing atmospheres, and often higher temperatures needed for processing that can result in reaction with, or reduction of, ZrO_2 or both. However, this section focuses on strengths of Al_2O_3-ZrO_2 (ZTA) composites, since these particularly clearly illustrate the microstructural effects. Thus they have been the most extensively developed, used, and understood, since Al_2O_3 is a very desirable and practical matrix. This is because it is compatible with ZrO_2 from both a basic chemical and a processing standpoint and is an excellent matrix because of its properties; its high Young's modulus provides significant constraint of the expansion required by the transformation of the ZrO_2 giving higher tougheners in composites with Al_2O_3 than most, if not all, other matrices (except possibly ZrO_2 itself).

Strength behavior, however, can be significantly different from that of toughness, which as said in Chap. 8, Sec. V.B is a maximum when the ZrO_2 particles have no stabilizer and an average particle size of ~ 1+ μm with $\phi \sim 0.12$ (Figs. 9.5 and 9.6). Thus Claussen's original study [23,24] of ZTA showed that both the ϕ values for toughness maxima and their levels increased as the unstabilized ZrO_2 particle size decreased (Fig.9.5A). However, regardless of the varying levels of toughness increases, the strengths of all his original ZTA bodies progressively decreased as the ZrO_2 content increased, with the strength decreases becoming progressively greater as the ZrO_2 particle size increased. Thus the strength decreases increased as the toughness increases decreased, and hence in this sense they were correlated, i.e. better toughness meant less strength decrease, but there was no increase in strength with increased toughness. The failure of the toughness increases to translate into strength increases was attributed to processing heterogeneities acting as fracture origins, which was reinforced by better toughness and strengths being obtained with finer ZrO_2, which should provide less serious heterogeneities.

This proposed cause of strengths not following the toughness increases due to processing heterogeneities was confirmed by Becher's study [24,25]. Using sol gel processing that produced fine, unstabilized ZrO_2 particles that were much more uniform in both size and spatial distribution in a more uniform dense Al_2O_3 matrix showed toughness (fracture energy) and strength closely following each other, with the maximum for both being at $\phi \sim 0.1$–0.12 (Fig. 9.6). Subsequently conventional powder processing was improved to yield the

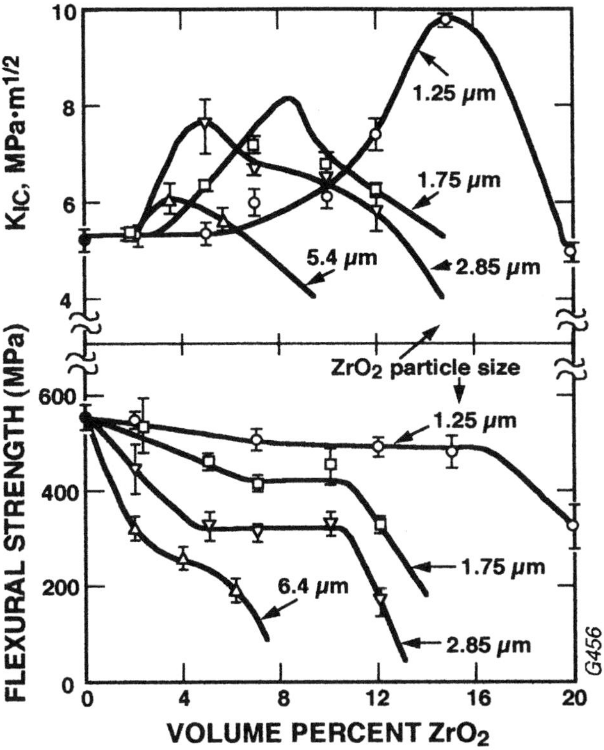

FIGURE 9.5 Plots of (A) fracture toughness and (B) flexure strength versus volume percent of unstabilized ZrO_2 particles in an Al_2O_3 matrix made by conventional powder processing. Note strengths decreasing despite toughness increases, compared to Fig. 9.6. (Data from Ref. 23, plots from Ref. 24. Published with permission of Noyes Publications.)

homogeneity needed to realize at least most of the strength benefits of the ZrO_2 toughening.

Lange [26] subsequently corroborated that the toughness maxima for fine unstabilized ZrO_2 was at $\phi \sim 0.1$, but that progressively higher toughness maxima occurred at higher ϕ values with increasing degrees of partial stabilization of the ZrO_2, with the maximum toughness being that of TZP, i.e. at $\phi = 1$. (This may also allow use of somewhat larger ZrO_2 particles, but this may still be constrained some by possible moisture degradation effects, Fig. 2.9.) Thus Lange's work showed that starting with totally unstabilized ZrO_2 a series of composites with decreasing Al_2O_3 content and increasing toughness and strengths extrapolate to those of TZP bodies (Fig. 9.7).

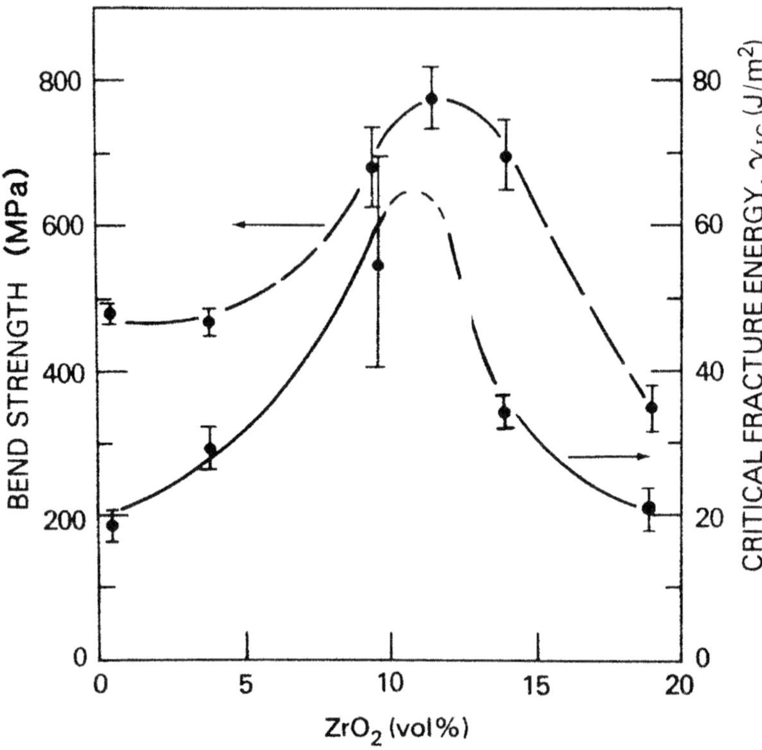

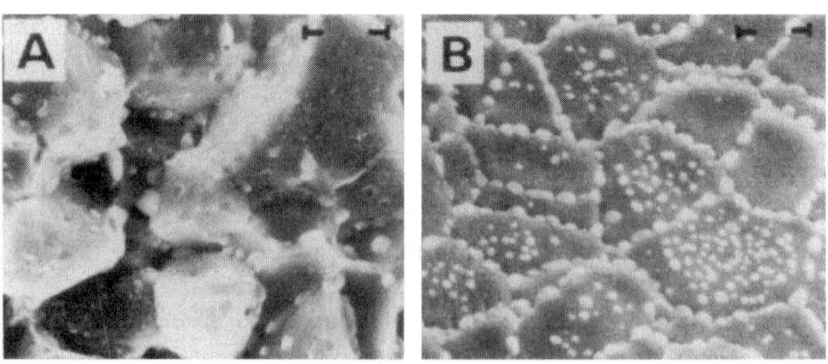

Figure 9.6 Fracture toughness and strength data of Becher [25] for composites of an alumina matrix versus the volume percent of unstabilized ZrO_2 particles. (A) and (B) are SEM photos of respectively a fracture and an as-fired surface showing the uniform microstructure, especially the particle size and distribution of the ZrO_2 as a result of sol–gel composite processing (bars = 5 μm). Note the similar trends of strength and toughness in contrast to Fig. 9.5. (From Ref. 24. Published with permission of Noyes Publications.)

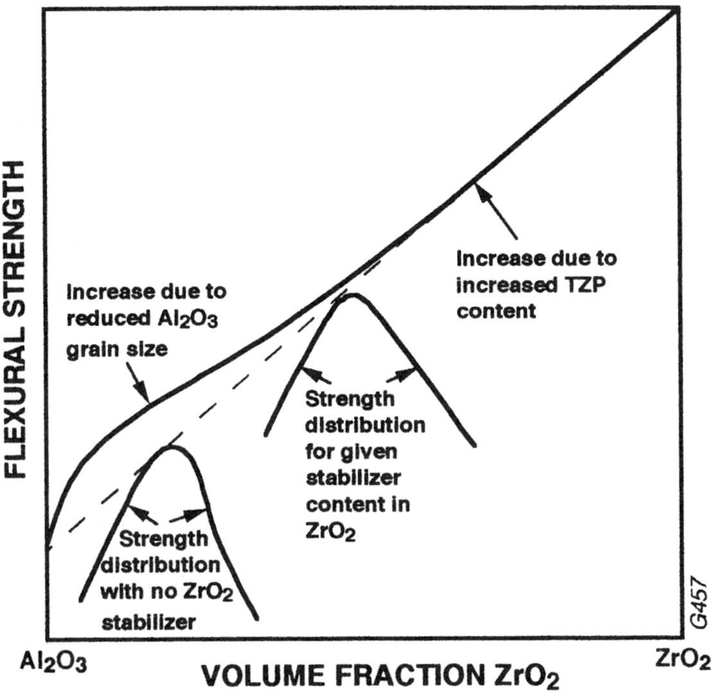

FIGURE 9.7 Schematic summary of the increases in strength of Al_2O_3-ZrO_2 bodies as a function of ZrO_2 and stabilizer content showing combined effects of reduced Al_2O_3 grain size and ZrO_2 stabilized content.

Subsequent work, apparently first by Lange and Hirlinger [27] showed that dispersions of ZrO_2 particles in Al_2O_3 powder compacts can significantly limit Al_2O_3 grain growth, resulting in a finer Al_2O_3 grain size. However, they reported that with uniform mixing at least 2.5 m/o ZrO_2 was needed to control exaggerated grain growth (and more with less uniform mixing). Hori and colleagues [28,29] corroborated such grain growth benefits, even for lower levels of fine ZrO_2 additions, and showed that this made useful increases in strengths of such bodies, despite some lowering of toughness. Thus they showed that the Al_2O_3 grain size dropped from 4 + to ~ 2.5 μm as the ZrO_2 content went from 0 to 0.1 w/o (with a ZrO_2 particle size of 0.2 + μm) and decreased less rapidly as the ZrO_2 content was further increased, reaching G ~ 1.5 μm at 5 w/o ZrO_2 (with its particle size of ~ 0.3 μm). These composition and microstructure changes resulted in composite strengths increasing rapidly at lower additions, and then more slowly, i.e. starting from ~ 310 MPa and reaching ~ 570 MPa as the matrix grain size decreased, consistent with behavior of pure Al_2O_3 (Fig. 3.14), again showing the importance of matrix grain size on mechanical properties of ceramic composites. The strength increases occurred despite the (IF) toughness decreasing from ~ 4.8 MPa·m$^{1/2}$ with no ZrO_2 to a minimum of ~ 3.9 MPa·m$^{1/2}$ at ~ 1.1 w/o ZrO_2, and then rising to ~ 4.6 MPa·m$^{1/2}$ at 5 w/o ZrO_2, again showing basic differences in larger crack toughness behavior from that of strength. The collective mechanisms are schematically summarized in Figure 9.7.

The above effects of ZrO_2 additions in limiting matrix grain growth are expected to impact most composites. Additional explicit evidence of this is provided by Nisida et al.'s [30] study of composites with up to 10 w/o of either fine tetragonal or cubic ZrO_2 particles in MgO. While the ZrO_2 particles coarsened (respectively from ~ 0.3–0.4 to 0.4–1 + μm) as ZrO_2 content and densification temperatures increased, MgO *G* decreased up to ~ three- and fivefold respectively for lower and higher temperature sintering. Resultant strengths were primarily determined by the resultant MgO *G* as shown by comparison to their own MgO without ZrO_2, as well as general agreement with data of Figure 3.5, except for some tendency for deviation to less strength increases at the finer *G* with higher ZrO_2 contents.

While the primary toughening mechanism in ZrO_2 toughened bodies is transformation, microcracking can also occur and was proposed as a major source of toughening in ZTA composites [23,31–33]. Thus duplex designed microstructures using more stabilized, and higher volume fractions of, ZrO_2 particles can result in substantial microcracking with propagation of large cracks and thus can give higher thermal shock resistance (Chap. 8, Sec. V.B, and Chap. 11, Sec. III.C). However, while this produces significant *R*-curve effects and increases fracture toughness and residual strengths after serious thermal shocks, initial strengths decreased substantially, e.g. from 1300–1700 MPa to 100–800 MPa [33].

Wang and Stevens [34] investigated Al_2O_3-ZrO_2 composites of various microstructures using normal ZrO_2 particle dispersions, dispersions of TZP particles, and combinations of these, i.e. duplex structures as above. Bodies sintered with dispersions of individual, fine (~ 1 μm) unstabilized ZrO_2 particles showed modest strengths of ~ 400 MPa decreasing slightly from 0 to 10 v/o ZrO_2, dropping to ~ 200 MPa at 12 v/o and then to 100 MPa at 20 v/o ZrO_2, i.e. essentially a mirror image of their I toughness results (Chap. 8, Sec. V.B). Bodies with addition of both 2.5Y-TZP agglomerates (20–50 μm) and the unstabilized ZrO_2 particles, i.e. composites along the lines of Fig. 2.18 showed a modest linear increase in strength at 10 v/o, reaching ~ 300 MPa at 40 v/o combined additions for sintered bodies and approximately double these values in hot pressed bodies. The increase from ~ 97 to ~ 100% of theoretical density for hot pressing versus sintering appears inadequate to explain the doubling of strength by hot pressing. Reductions of Al_2O_3 and ZrO_2 grain and particle sizes to ~ $^1/_2$ the values in the sintered bodies (respectively ~ 4 and 1 μm) are probably also significant in increased strengths in hot pressed bodies, and some stabilization of the ZrO_2 particles due to some probable reduction in hot pressing is also a possible factor, as is discussed below. While the trend of strengths to increase with v/o additions is consistent with the trends of their I toughnesses, the absolute values of their strengths and the extent of their change with v/o ZrO_2 are clearly not consistent, since their toughness increased ~ threefold. They attributed toughening of the composites with unstabilized particles alone to transformation and microcrack toughening, that with TZP agglomerated alone to transformation and crack deflection, and all of these with both types of addition. The indicated presence of microcracking with all additions is consistent with the limited strengths (especially those from < 100 to 300 MPa) and indicates that the combined composite is a probable manifestation of either, or probably both, of those sketched in Figures 8.1B and 8.2C.

The first of two additional points that should be noted is that while transformation of ZrO_2 particles can be very effective for toughening and strengthening the bodies containing them, either cubic or monoclinic ZrO_2 particles can contribute some toughening, strengthening, or both. Thus monoclinic particles can cause microcracking and resultant toughening or can toughen via crack deflections resulting from the substantial stress fields associated with their presence (without microcracking as in a ZnO matrix) as discussed in Chap. 8, Sec. V.B. The latter is a possibly more extreme example of a more general effect, namely the presence of ZrO_2 particles in a matrix should provide toughening to the extent that they can impede crack propogation due to crack deflection, bowing, or other mechanisms with other nontransforming particles. This is consistent with the recognition that metastable or stable tetragonal ZrO_2 precipitates (i.e. above the tetragonal–monoclinic transformation temperature) in cubic ZrO_2 provide substantial toughening and strengthening, which limits reductions of these as temperature increases to and above the transition temperature (Chap. 6, Sec. III.E). Thus some have reported both toughening and strengthening of Al_2O_3 with nontransforming ZrO_2 additions, e.g. Langlois and Konaztowicz [35] reported (I) toughness increasing more rapidly at lower additions of cubic ZrO_2 particles but increasing by 100+% at the upper addition of 30 v/o, while strength increased by ~ 75% at 10 v/o addition and was then constant at ~ 800 MPa.

The second point is that the processing environment can be important, as is shown by various results such as the use of HIPing. Thus Tsukuma et al. [36] showed strength maxima of ~ 2.5 GPa at 20 w/o Al_2O_3 in 2Y- or 3Y-TZP, which was fairly consistent with CNB, but not I, toughness–composition trends. Hori et al. [37] showed that while sintering of Al_2O_3 with 0–30 w/o unstabilized ZrO_2 gave a narrow maximum of ~ 650 MPa at 15 w/o; HIPing of the bodies after sintering not only increased the maximum strength to 800 MPa but also substantially broadened it to being essentially constant from at least 5 to 15 w/o. IF fracture toughness testing showed no difference between sintered and sintered-HIPed bodies, but it was inconsistent with the strength maximum, as were I toughness measurements, which showed toughness of HIPed bodies progressively falling below that of sintered only bodies as the ZrO_2 content increased, since oxygen loss from ZrO_2 can stabilize it, as is noted below. Tomaszewski [38] found similar but less extreme increases in strength maxima and its breadth (and similar trends for NB toughness) in vacuum versus air sintered bodies of similar composition. He showed that the improved properties of vacuum sintered bodies correlated with increased ZrO_2 stabilization from oxygen deficiency resulting from vacuum sintering. While such environmental effects have received very limited attention, there are sound reasons why they are of probable broad importance, as is discussed further below.

2. Melt Processed ZTA Composites

The above processing and resultant microstructure of the ZTA composites is all via powder processing, which results in ZrO_2 particles at grain boundaries and requires fine Al_2O_3 grain sizes for good strengths. However, melts of the Al_2O_3-ZrO_2 system result in classical, well known eutectic, typically rod (Fig. 9.8), structures that have been used commercially for years for producing tough abrasives and have more recently been investigated as an alternate for producing possibly unique ZTA composites. Since both uses are based on melting to produce powder for making ZTA bodies, they are briefly summarized here; emphasizing strength results.

Commercial abrasives at or near the eutectic composition (70 v/o Al_2O_3-30 v/o ZrO_2) are arc melted, quenched by casting thin sheets (e.g. 1–2 mm thick) and then ground into abrasive particles that are noted for their toughness in grinding wheels. Such materials, despite having no stabilizer added, have predominantly tetragonal ZrO_2 [39] so long as fine microstructures are obtained and maintained. However, both the arc melting with graphite electrodes and their quenching, e.g. by pouring into graphite book molds to form thin sheets for practical crushing, result in a reduced, i.e. black, material. While detailed research has apparently not been conducted on this system, the following mechanism is suggested because the cast material has similar toughnesses to other ZTA bodies (Chap. 8, Sec. V.B) and progressive reduction of ZrO_2 can progressively stabilize it [40,41]. Stabilization results because lattice defects the same as or similar to those from the addition of stabilizing agents are introduced by oxygen deficiencies from reduction. Thus reduced ZrO_2 in ZTA composites may be partially stabilized, allowing transformation toughening or as noted below providing other, e.g. microcracking, toughening. Either is supported by the fact that oxidation of the melt-derived ZTA results in its turning white and developing extensive micro- and larger cracks, mainly at or near Al_2O_3–ZrO_2 interfaces [42], which could also aid in maintaining sharp, tough abrasive edges for effective cutting.

Various investigators have studied polycrystalline bodies made by consolidation of eutectic particles, especially Al_2O_3-ZrO_2. Rice et al. [24,42] hot pressed commercially produced fusion cast and ground 70 v/o Al_2O_3-30 v/o ZrO_2 into fairly dense bodies with strengths of > 500 MPa that were probably limited by residual porosity (due to limitations on fine grinding of the material due to its toughness). Fracture surfaces clearly revealed the eutectic structure and interaction of the crack with that structure (Fig. 9.8). The difficulty of grinding this material, much of the structure not being fine relative to the particle sizes desired for good densification, and the above noted cracking on oxidation, greatly reduced strengths and made this approach impractical for many applications. On the other hand, Krohn et al. [43] hot pressed melt-derived eutectic powder and obtained strengths of > 800 MPa, consistent with their high toughness (15 $MPa{\cdot}m^{1/2}$).

Homeny and Nick [44] conducted more detailed studies of eutectic Al_2O_3-ZrO_2 powders made via plasma torch melting of particles and subsequent hot pressing of eutectic compositions that contained (a) 0, (b) 4.6, and (c) 9.5 m/o Y_2O_3. While these all gave high toughnesses (respectively 6.7, 7.6, and 7.9 $MPa{\cdot}m^{1/2}$), strengths were respectively ~ 200, 750, and 650 MPa. The low strength was attributed to extensive microcracking (which was consistent with there being no Y_2O_3 and resultant 90% monoclinic ZrO_2 content), which was cited as the source of the substantial toughness in this material. However, it was claimed that the two bodies with Y_2O_3 additions that had 100% tetragonal ZrO_2 did not exhibit transformation toughening: toughening was attributed to crack bridging as well as substantial formation of strings of thin deformed material (much finer than grains or substantial fragments of them) bridging the cracks in the wake region.

3. Other ZrO_2 Toughened Composites and Flaw Size Considerations

Briefly consider examples of results for what is probably the next most investigated ZrO_2 toughened system, namely that with a mullite matrix. Data of Yuan et al. [45] showed strength increasing from ~ 100 to ~ 125 MPa, i.e. by up to 25%, as ZrO_2 additions increased from 0 to 25 v/o (Fig. 11.2), with most of the increase occurring by ~ 10 v/o. The levels and trend of the strength increase are consistent with that for toughness (Chap. 8, Sec. V.B) both as measured as well as with probable correction for the 2.6–8.7 residual porosity, which would raise the increase in strength from ~ 110 to ~ 150 MPa. However, the substantial reduction of the mullite grain size (especially of larger grains) generally accompanying increased addition of the ZrO_2 was probably a factor in the strength increases, and the limited increase in ZrO_2 particle size (e.g. 1 to 1.5 μm) as its content was increased may have been a factor in the limitation of strength increases at higher ZrO_2 contents. Strengths of 330 MPa reported by Lathabai et al. [46] for reaction processed mullite + ~ 5 w/o Al_2O_3 and ~ 31 w/o ZrO_2 (~ half

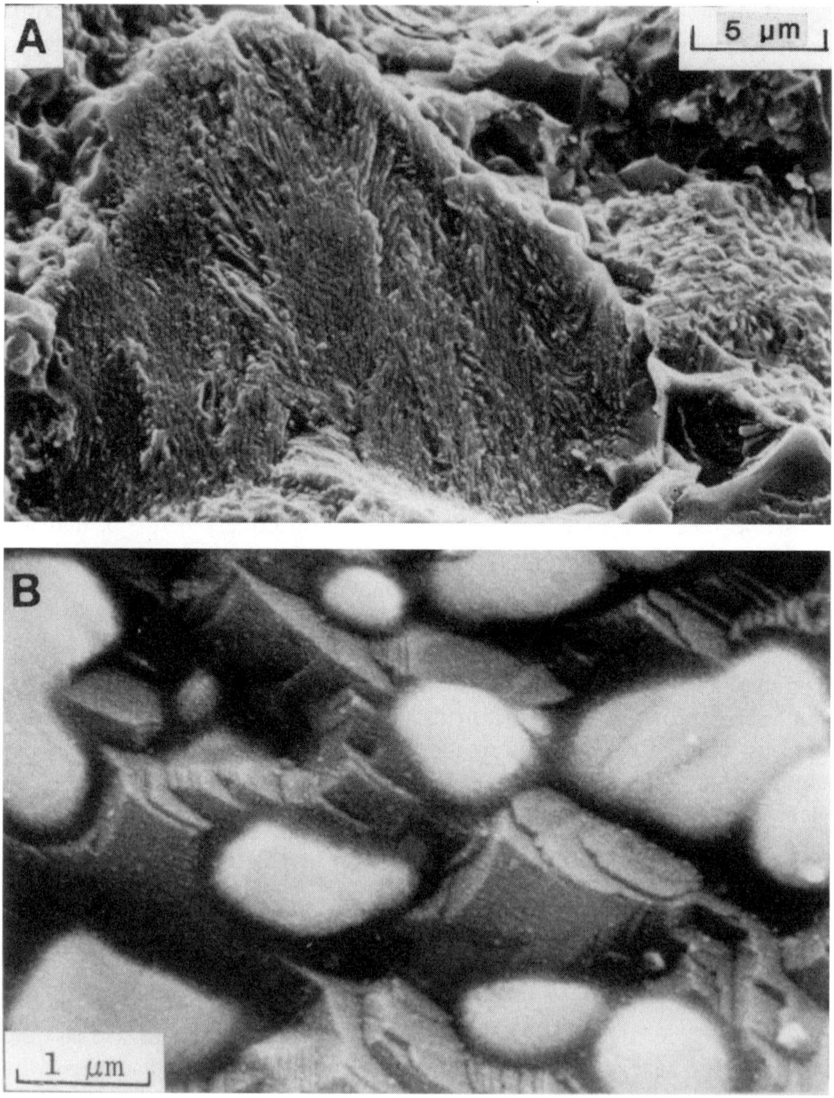

Figure 9.8 Fracture surface of sample hot pressed from Al_2O_3-ZrO_2 eutectic abrasive material. (A) shows a large fractured eutectic particle with the fracture plane ~ parallel with the colony growth direction, and (B) shows a higher magnification of a fracture ~ normal to the eutectic rod structure. The white phase is ZrO_2 and the darker phase is Al_2O_3. (From Ref. 24. Published with permission of Noyes Publications.)

tetragonal and half monoclinic) must be due in substantial part to the finer ZrO_2 particle size and especially the finer mullite grain size (e.g. ~ 1 μm), since no evidence of transformation toughening was found. On the other hand, Ishitsuka et al. [47] showed strengths of TZP steadily decreasing as mullite was added, though again the value of ~ 160 MPa for the pure mullite was probably low due to the larger *G*, since again the addition of ZrO_2 reduced the mullite grain size. However, the only indication of the high toughness values of ~ 12 MPa·m$^{1/2}$ for 10–30 v/o ZrO_2 on strengths was a slight strength decrease at 10 v/o mullite; otherwise there was a smooth continuous strength decrease with increased mullite addition in contrast to substantially greater than proportional toughness decreases from 30 to 50 and 90–100 v/o mullite.

Consider now evaluations of flaw sizes and effects. Failure causing flaws have frequently been found at fracture origins in ZTA samples showing failure from typical, mainly machining, flaws and processing defects, by various investigators. Rice [31] has investigated effects of flaw size on failure of zirconia toughened alumina conventionally produced from mixed powders. He showed that fracture toughnesses calculated from fracture strengths and observed flaw sizes and character agreed with those measured by conventional (DCB) toughness tests with larger crack sizes, but at flaw sizes < ~ 50 μm they showed progressive decreases as flaw size decreased (Fig. 9.9). This was interpreted as being due to mismatch stresses between grains of both materials increasingly contributing to the failure stress as flaw sizes became smaller, ultimately approaching the grain size, i.e. identical to monolithic ceramics, including TZP bodies (see Fig. 3.35) [32].

Finally consider the flaw sizes implied by the strengths and toughnesses, e.g. their ratio as previously discussed by Rice [24,31]. All results for additions of partly stabilized ZrO_2 in his previous survey indicated reasonable flaw sizes of ~ 30–100 μm except those of Claussen [23,32], which gave ~ 100 to > 2000 μm. Ono et al.'s [48] (and Becher's [25]) addition of totally unstabilized ZrO_2 also gave reasonable flaw sizes at lower, especially at ~ 10, v/o, beyond which Ono et al.'s additions gave larger calculated flaw sizes, e.g. to > 2000 μm. On the other hand, flaw size calculations from Ono et al.'s data for additions of partially stabilized ZrO_2 gave reasonable flaw sizes across the complete range of additions from 0 to 100% ZrO_2 [32]. Further, all data indicating reasonable flaw sizes also indicated flaw size minima at intermediate ZrO_2 contents, e.g. ~ 20 w/o, where toughening was at or near its maximum. Calculations from Ono et al.'s data also indicates a flaw size minimum at 80–90 w/o ZrO_2, i.e. again indicating another toughening maximum [32]. More recently reviewed data in this and the previous chapter is generally consistent with these trends from earlier results. Thus Homeny and Nick's data for eutectic compositions with no stabilizer indicated microcracking give lower strengths but substantial toughness,

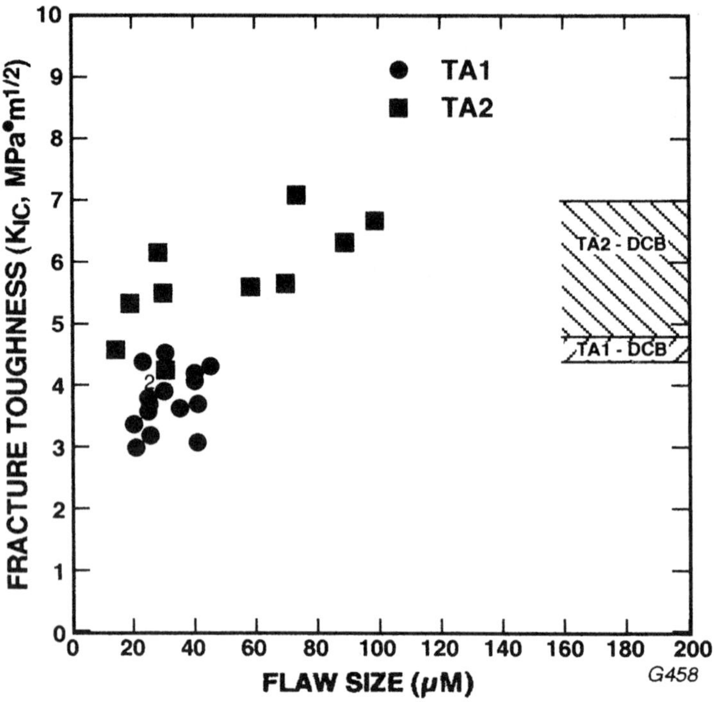

FIGURE 9.9 Fracture toughness calculated from the applied stress at fracture and the flaw size and character at the fracture origin for commercially produced zirconia toughened alumina (Diamonite Products). Note the general agreement with conventional (DCB) toughness tests (cross-hatched areas) at large flaw sizes, but progressively lower values at smaller flaw sizes attributed to increasing contributions of microstructural stresses as for monolithic ceramics (Fig. 3.25). (From Ref. 31. Published with permission of the American Ceramic Society.)

and hence very large calculated flaw sizes, e.g. $G > 1000$ μm [44]. Compositions with some ZrO_2 stabilizer give similar toughnesses but higher strengths, and thus much lower and more reasonable calculated flaw sizes, e.g. of the order of 100 μm. Similarly, data for TZP materials gives very large calculated flaw sizes at low levels of partial stabilization (where strengths are low, quite possibly due to microcracking) and reasonable flaw sizes at higher strengths and levels of stabilization. Konsztowicz and Langois [49] pursued this issue of calculated flaw sizes further for Al_2O_3- ZrO_2 composites showing differences in calculated flaw sizes using I versus NB toughnesses and logical correlations between processing and calculated flaw sizes, and minimum flaw sizes at or near compositions for toughening maxima.

Limited evaluations have been made of flaws in other ZrO_2 toughened bodies. Evaluation of Yuan et al.'s data [45] indicates flaws of a few to several hundred microns, i.e. > normal strength controlling flaws. This is consistent with either *R*-curve effects increasing toughness at large (but not at normal) crack sizes, the occurrence of microcracking, or both. The low strengths of ~ 100 MPa and thermal shock results, including some increase in toughness with decreased Young's modulus and strength, all indicate effects of microcracking.

C. Polycrystalline Oxide Matrix Composites Without Transformation Toughening

The above ZrO_2 toughened composites show that toughness and strength generally follow each other when reasonable homogeneous bodies are made within the range where transformation toughening is dominant. However, many other composites do not show such consistent behavior of toughness and strength and in fact commonly show opposite trends with basic microstructural variables, especially of the dispersed particle size. In particular it is shown that while toughness often increases as particle size increases, the opposite is true for strength, i.e. it typically increases as particle size decreases but may pass through a maximum at finer particle sizes. This section addresses behavior of polycrystalline composites without transformation, showing the noted differences and drawing upon, but also extending, recent reviews of some of these composites [24,32,50,51].

Tiegs and Bowman [51] have shown Al_2O_3-20% SiC particulate composite having a distinct σ maximum, i.e. increasing rapidly from ~ 350 MPa at fine SiC particle sizes to a maximum of ~ 530 MPa at D ~ 2 μm, then decreasing rapidly and then more slowly to ~ 110 MPa at D ~ 17 μm. He attributed most or all of the decrease with increasing SiC particle size to microcracking, which is logical because Eq. (2.4) would predict that spontaneous cracking could start at D ~ 1–2 μm. A maximum of relative strength was also shown by Yasuoka et al. [52] at D < 1 μm, greater than the relative maximum of toughness but then decreasing to lower relative values than relative toughness by D ~ 6 μm for strengths. Both strength and toughness were normalized for the matrix G, since they showed that the matrix G was decreased by the SiC additions and that this added to the strength increases. This effect of the addition of SiC in reducing the matrix G was also shown by Thompson and Kristic [53], as was the benefit of fine SiC particle size. Thus they reported that the addition of ~ 0.5 μm SiC reduced the Al_2O_3 G from 15 μm with no SiC addition to ~ 2+ μm at $\phi = 0.1$ and < 1 μm at $\phi = 0.2$. These changes in G dominated strength increases from ~ 290 to ~ 490 MPa as a function of ϕ (maximum at $\phi = 0.15$), with a limited decrease at $\phi = 0.2$. Strength increased through a maximum in contrast with that of (CNB) toughness, being independent of ϕ at ~ 4.2 MPa·m$^{1/2}$, except for an increase of ~ 15% at ϕ=0.2 (in contrast to a strength decrease there). These differing changes of strength and toughness are indicated by their ratio passing through a maximum at intermediate ϕ The benefits of finer SiC particles was also shown by Nakahira and colleagues [54] via a strength maximum at $\phi \sim 0.1$ of ~ 500 MPa ($D \sim 2$ µm) and ~ 400 MPa ($D \sim 8$ µm), and a decrease in strength with coarsening of the microstructure increasing both D and G.

Higher strengths with finer SiC particle size were also shown by Kim and Kim [55] for Al_2O_3-SiC composites made by the directed oxidation of Al metal (and leaving some residual metal, e.g. Fig. 8.20). Their starting strengths clearly decreased as the size of the added SiC particles increased (Fig. 15; Chap. 11, Sec. III.C).

Considerable research has been conducted on Al_2O_3 composites made with nanoscale SiC particles (despite this presenting some processing challenges), prompted by Niihara's [56] work reporting strength increases from ~ 320 MPa to > 1000 MPa with more limited toughness increases (from 3.2 to 4.7 $MPa{\cdot}m^{1/2}$) with addition of 5 v/o SiC. Thus Zhang et al. [57] showed strengths reaching maxima of ~ 1100, 950, and 900 MPa respectively at SiC particle sizes of 0.05, 0.17, and 2.5 µm, all at ~ 3 v/o SiC starting from ~ 550 MPa without SiC (all with highly polished surfaces). Strengths dropped rapidly and then more slowly as ϕ increased, reaching ~ 800 MPa at 24 v/o, with much less differentiation between the different particle sizes. In contrast, (I) toughnesses for the finer two particle sizes varied only a limited amount from that of the pure Al_2O_3 (~ 3.3 $MPa{\cdot}m^{1/2}$) while decreasing to a minimum of 3 $MPa{\cdot}m^{1/2}$ at 3 v/o, then increasing steadily to ~ 4.7 $MPa{\cdot}m^{1/2}$ at 24 v/o, i.e. clearly different from and in the latter case opposite to the trends for strength. While many obtain less spectacular but still substantial increases in strength, some have also shown considerable improvement after annealing, e.g. Zhao et al. [58] showed increases from 760 to 1000 MPa with $D = 0.15$ µm. It is clear that toughness changes have at best a limited effect on strength increases and that reductions of the Al_2O_3 G, e.g. to ~ 200 nm, is an important factor in these increases (e.g. see Figs. 3.13–3.15). Reduction of flaw sizes [59] or their severity due to surface compressive stresses [60] and less relaxation of these on annealing or crack healing [61] have been indicated as additional factors.

Consider next Al_2O_3-TiC composites, for which there is much less data on this system in the literature, especially on its microstructural dependence, reflecting its earlier, mainly empirical, development a number of (e.g. 40) years ago for use in various wear and cutting tool applications. Rice [32] reviewed the literature and made measurements on some commercial bodies (commonly with ~ 30 v/o TiC and $D \sim < 1$ to 2 + µm). While overall modest increases in toughness often correlated with modest increases in strength as ϕ increases, there was substantial variation, with approximate flaw sizes calculated for strength-to-toughness ratio (Fig. 9.10), ranging from reasonable values of several tens of microns to very questionable values of hundreds of microns. Though finer particle sizes are used, little direct data on effects of TiC particle size is available. Rice showed that significantly increasing the value of the average and especially the average + maximum particle size resulted in lower strengths, especially relative to the toughness (Fig. 9.10), indicating not only better results with finer particles but especially with narrower distributions of finer particles. These trends from earlier data are consistent with specific results of Yasuoka et al. [52], who showed a maximum of their normalized strengths (to account for strength increases due to grain growth inhibition effects on the matrix) at $D \sim 0.5$ μm and decreasing slowly, reaching no increase in strength at $D \sim 12+$ μm (as did toughness, which showed a broad maximum at $D \sim 4$–6 μm). Wahi and Ilschner's [62] study of Al_2O_3-TiC composites, showing strengths of ground specimens increasing from ~ 330 MPa at 0 w/o to a maximum of ~ 540 MPa at 10 w/o TiC and then decreasing back to the 0 w/o TiC

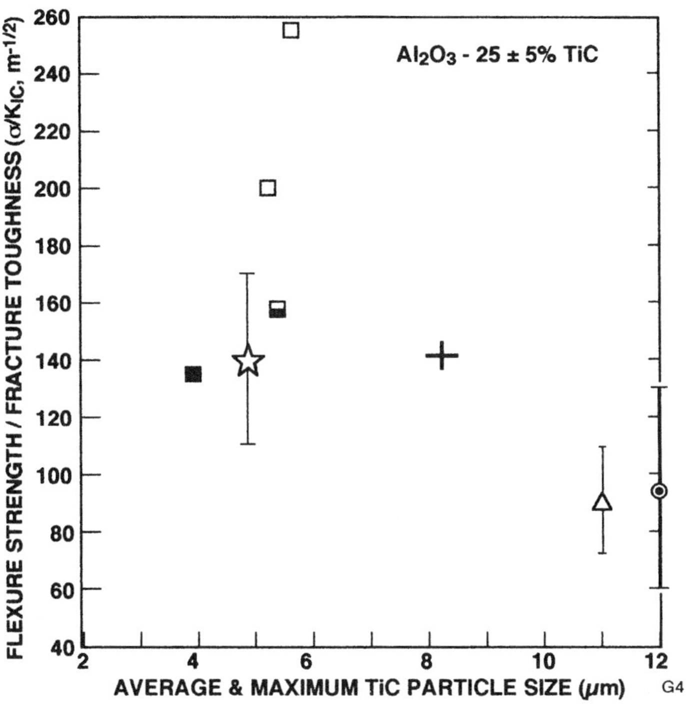

FIGURE 9.10 Strength-to-toughness ratio versus the sum of the average and maximum TiC particle size for Al_2O_3-TiC composites. Note the significant decrease as the particle sizes increase showing negative effects of larger particles and discrepancies between strength and toughness, e.g. by the large flaw sizes implied at larger particle sizes. (From Ref. 32. Published with permission of Ceramic Engineering and Science Proceedings.)

levels as TiC content further increased versus strengths of polished samples continuously increasing to 670 MPa at the highest level of TiC addition (41 w/o, 35 v/o), again indicates important, but widely neglected, effects of machining on composite properties and mechanisms.

Hwang and Niihara's [63] data for $BaTiO_3$ with or without additions of nanosize SiC particles primarily falls into two main groups of data (Fig. 9.11). The first group has $G > 1–2$ μm, often with a significant bimodal grain size distribution, mostly for bodies with no or limited SiC content (and the latter only for more extreme sintering conditions). This group has strengths generally consistent with those of normal $BaTiO_3$ without additives (Fig. 3.3), ranging from the finer to the larger G branch. The other primary group is at substantially finer and more homogeneous G and substantially higher strength (even higher than bodies made with LiF+ MgO additions, Fig. 3.3). There are also a few data

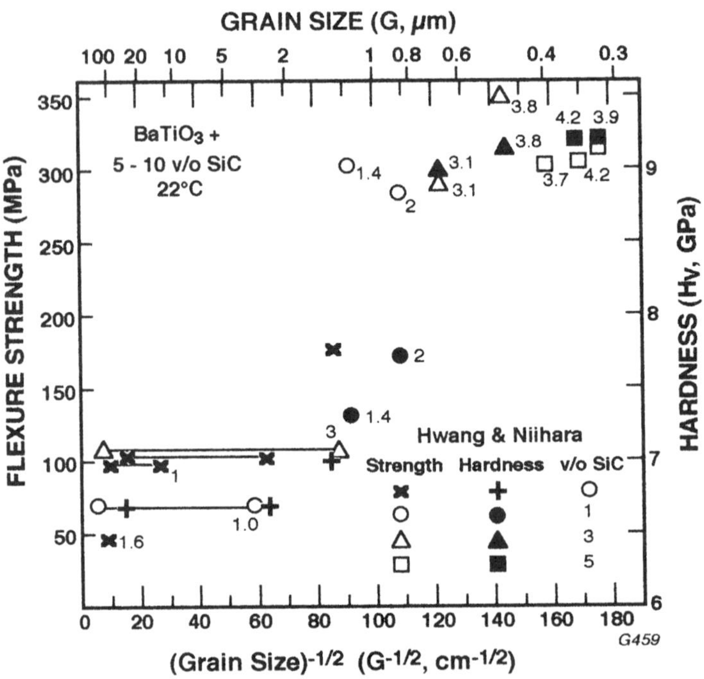

FIGURE 9.11 Strength and hardness (H_V, ~ 9.8 N load) versus the inverse square root of grain size (G) for $BaTiO_3$ with 0–5 v/o of ~ 10 nm SiC particles (compare strengths with Fig. 3.3). Note (1) data points at the ends of horizontal bars reflect the average G for the two grain populations present in bodies processed at higher temperatures resulting in bimodal G distributions, and (2) numbers next to data reflect residual porosity ≥ 1%. (Data from Ref. 63.)

points as transitions between the high and normal strengths. This data appears consistant with plastic deformation control of strength of $BaTiO_3$, i.e. suppressed at very fine *G* resulting from SiC additions without higher temperature processing coarsening the structure, but approaching normal strength behavior with significant grain coarsening, with the larger grains generally playing a dominant role in strength. This intrepretation is supported by the close parallel behavior of hardness. However, some impact of increased *E* values (up to ~ 50%) on strength as the level of SiC addition increases may also be operative, as may both conversion from a tetragonal to a pseudocubic structure as *G* is reduced below $^1/_2$ to 2/3 μm. Modest (~ 30 %) increases in toughness also probably contribute to strength increases but clearly cannot explain strength increases of up to > threefold.

Studies of concrete corroborate both the tradeoffs often seen between toughness and strength and the mechanism causing this tradeoff. Thus Strange and Bryant [64] reported that NB toughness tests of concretes gave increasing toughness as the aggregate (i.e. dispersed stones or stone fragments) sizes increased, but strengths decreased. This difference was attributed to crack nucleation from the aggregate particles thus limiting strengths, since such cracks, whose size is related to that of the aggregate particles, could initiate failure.

D. Composites with Polycrystalline Nonoxide Matrices

Consider next data for composites of all nonoxide constituents, beginning with Si_3N_4-SiC bodies, which have had substantial investigation. Lange [65] showed strengths generally trending in the opposite direction from the SiC particle size dependence of toughness (Fig. 8.16), i.e. strength decreasing as *D* increased while toughness increased across the range of additions (Fig. 9.12). While data of Tanaka et al. [66] appears as a clear extension of the toughness–*D* trends of Lange, it is more uncertain in terms of extending the strength trends, but the differences in strength between the two studies probably reflects differences in finishing and testing and hence flaw populations. This possibility is heightened by considering the strength-to-toughness ratio (~ the reciprocal of the flaw size), which very clearly indicates this data as an extension of Lange's data and the inverse trend of strength and toughness as a function of *D* (Fig. 9.13). Strength decreasing with increasing SiC particle size is clearly shown in plotting strength versus SiC particle size for various volume fractions of SiC from Lange's and other studies (Fig. 9.14). Data of Nakamura et al. [67] and others [68–72] supports this strength–particle size trend, showing strengths of 980 MPa for Si_3N_4 and progressively decreasing for 50 w/o SiC additions of increasing *D* from 0.3 μm (900 MPa) to 1 μm (770 MPa). Also note that Pezzotti and Nishidas' [69] showed strengths that, though decreasing with increasing SiC particle size, were higher than for their Si_3N_4 alone till SiC particle sizes of ~ 35 μm and were progressively lower for all larger SiC particle sizes. They also showed that

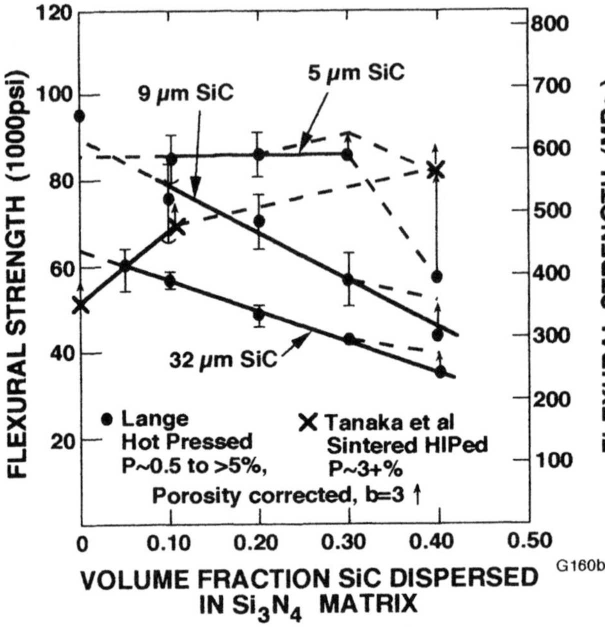

FIGURE 9.12 Strength versus volume fraction SiC particles of various sizes as shown from Lange [65] and Tanaka et al. [66]. Note the limited probable corrections for limited porosity (dashed lines and arrows). Contrast the dependence on *D* with that for toughness (Fig. 8.16). (From Ref. 50. Published with permission of the *Journal of Materials Science*.)

lower strengths with larger particles were associated with fracture initiation from larger SiC particles. Evaluation discussed below showed a $D^{-1/2}$ dependence analogous to the *G* dependence of strengths of monolithic ceramics (Fig. 9.15).

Studies of composites of Si_3N_4 with nanosize SiC particles support the general trends for strengths to increase with finer SiC particle sizes and for there to be better correspondence of strength and toughness. Thus Sawaguchi et al. [70] reported strengths rising from 1 GPa with no SiC to ~ 1.2 across the range of 10–50 v/o additions (and toughness rising from ~ 5 to 7 MPa·m$^{1/2}$ at ϕ= 0.1 and then decreasing to ~ 5.2 MPa·m$^{1/2}$ at ϕ= 0.5) for composites with SiC particles of ~ 50 nm. Sasaki et al. [71] showed that addition of 300 nm SiC particles increased strengths from ~ 780 MPa at ϕ= 0 to a maximum of ~ 920 MPa at ϕ= 0.1, which then decreased substantially. This trend was similar to that of toughness, but the increase of the latter was ~ 46% versus ~ 18% for strength, and both were much greater than the ~ 8% increase in *E*. They noted that the maxima in

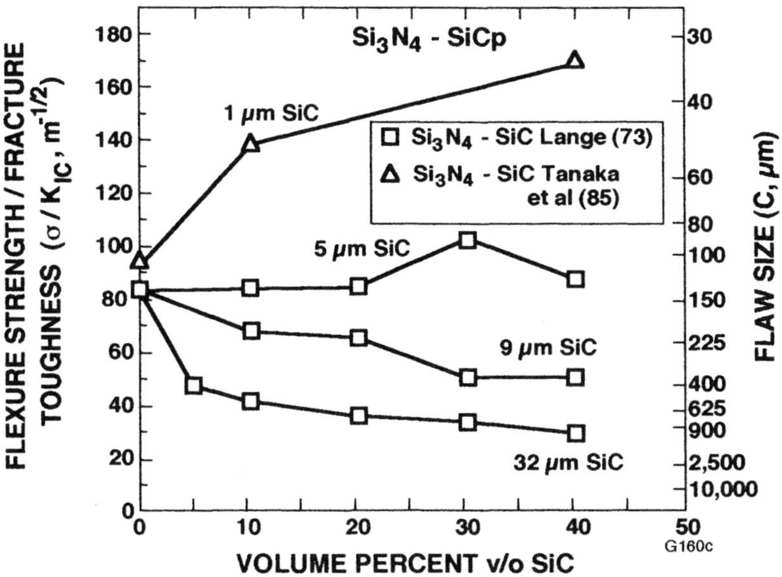

FIGURE 9.13 The ratio of strength to toughness versus volume percent SiC particles for various particle sizes shown from data of Lange [65] and Tanaka et al. [66]. Note the implied flaw sizes (left scale which are ~ the inverse of the ratio indicating reasonable sizes for the finest particle sizes, and clearly excessive for the larger particles) and Tanaka et al.'s data indicated as an extension of Lange's. (From Ref. 50. Published with permission of the *Journal of Materials Science.*)

strength and toughness were associated with the occurrence of rod shaped Si_3N_4 grains in the range of 5–10 v/o SiC and the formation of larger fracture sources beyond such additions. Similarly Tian et al. [72] showed strengths increasing from 750 MPa at 0 v/o to a maximum of ~ 950 MPa at 5 v/o (~ 12% increase) and then decreasing to ~ 650 MPa at 25 v/o, with a similar trend for toughness, but again with a greater (~ 35%) increase to a maximum of ~ 7.7 MPa·$m^{1/2}$. Again, the use of very fine particles commonly resulted in substantial inhibition of matrix grain growth and incorporation of some SiC particles in the Si_3N_4 grains.

Petrovic et al. [73] showed that strengths of Si_3N_4 with additions of 3 μm $MoSi_2$ particles (+ 1 w/o MgO densification aid) were ~ constant at 1 GPa but had a modest maximum of ~ 5% at 30–40 v/o $MoSi_2$ and then a definitive decrease to ~ 900 MPa at 50 v/o, while the addition of 10 μm $MoSi_2$ particles resulted in a decrease in strength generally increasing as ϕ increased, e.g. to ~ 700 MPa. Addition of 10 μm $MoSi_2$ particles (+ 5 w/o MgO densification aid), while

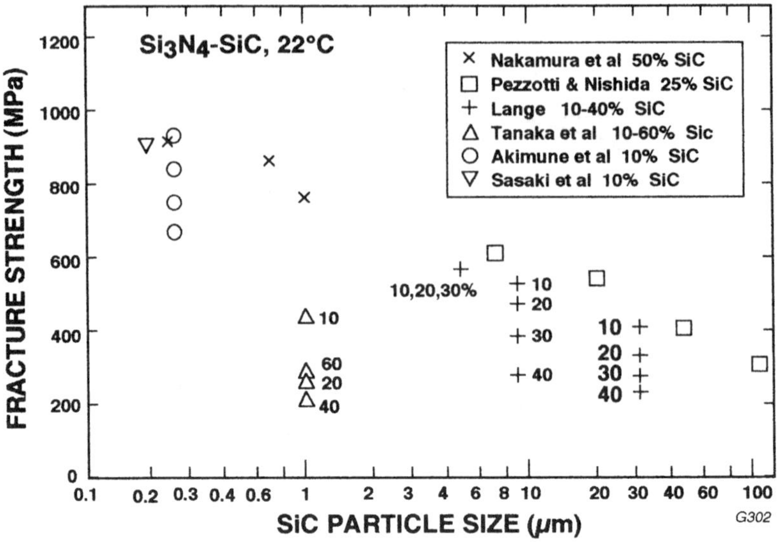

FIGURE 9.14 Strength versus SiC particle size for various indicated volume percent of SiC particles in Si_3N_4 matrix composites from various investigators [65–71]. This shows a clear trend for significant decreases with increasing particle sizes, opposite of toughness trends. (From Ref. 50. Published with permission of the *Journal of Materials Science*.)

giving a modest maximum of strength at 10 v/o of 700 MPa (up from 600 MPa at ϕ=0), then substantially decreased to ~ 300 MPa at 50 v/o. This strength behavior is in marked contrast to that of (I) toughness, which showed all three compositions reaching maxima of > 8 and > 6 MPa·m$^{1/2}$ for composites with the 10 μm $MoSi_2$ particles and respectively for 1 and 5 w/o MgO, and a maximum of ~ 5.5 MPa·m$^{1/2}$ for the composite with 3 μm $MoSi_2$ particles, i.e strength and toughness had essentially opposite dependences on ϕ and D.

Mah et al. [74] reported that strengths of their composites of Si_3N_4 with TiC particles (average and maximum size ~ 8 and 30 μm) decreased from ~ 700 MPa at 0 v/o to ~ 450 MPa at 50 v/o, contrary to the sharp (IF) toughness maximum of ~ 7 MPa·m$^{1/2}$ at 20 v/o (from a baseline of 4–5 MP·am$^{1/2}$), as well as a progressive increase in E with increasing TiC content. Reasonable flaw sizes were generally indicated by strengths and toughnesses, especially for other than the high toughness values, and were consistent with sizes of larger TiC particles or clusters of them found on fractures surfaces, probably as fracture origins. Given the particle size and range and the Si_3N_4-TiC expansion difference, microcracking is the likely cause of strength reduction and probably of the toughness

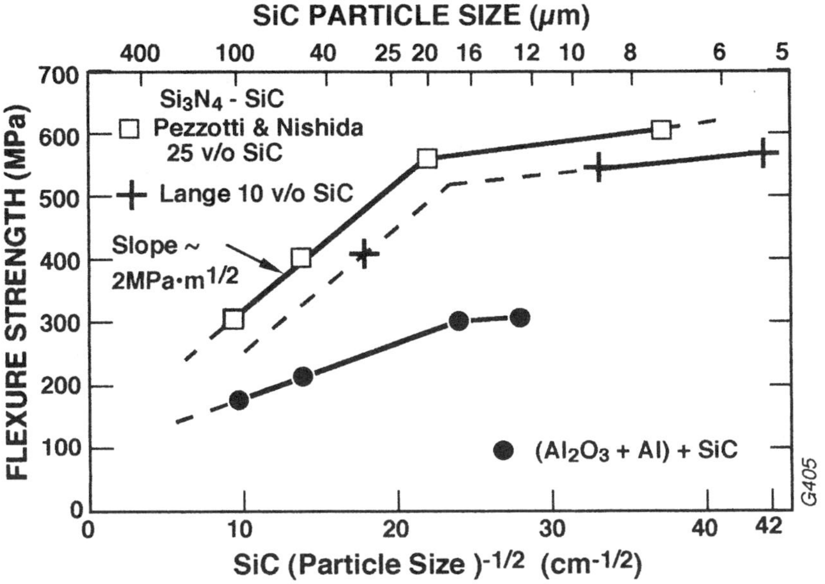

FIGURE 9.15 Strength versus the inverse square root of SiC particle size (*D*) at 22°C. Note the clear bilinear dependence on $D^{-1/2}$ similar to that for glass matrix composites (Figs. 9.2, 9.17) and monolithic ceramics (Fig, 3.1, Chapters 3,11). (From Refs. 55,65,69.)

maxima, hence being another example of opposite trends of strength and toughness due to microcracking.

Turning to SiC-TiB_2 composites, while Yoon and Kang [75] showed (IF) toughness was ~ independent of TiB_2 particle size (at ~ 5.3 MPa·$m^{1/2}$), strength decreased from ~ 580 MPa at D ~ 2 μm to ~ 350 MPa at D = 12 μm. However, strengths of specimens of Cho et al. [76] made with TiB_2 particles 1–15 μm were nearly independent of TiB_2 content to the limit of testing (70 w/o) but with a probable limited minimum of ~ 500 MPa at 50 w/o (in contrast to a clear (I) toughness maximum of 4.2 MPa·$m^{1/2}$ at 50 w/o). While McMurtry et al. [77] did not report the dependence of strength on TiB_2 particle size, they showed that while toughness for ϕ= 0.16 and D ~ 2 μm increased from ~ 6.7 to 9 MPa·$m^{1/2}$ as uniformity of mixing increased, strengths were unchanged. Magley et al. [78] provided evidence of stress-induced microcracking in SiC–15 v/o TiB_2 by measuring residual stress in bodies before and after strength testing by x-ray techniques. After failure, residual stresses were ~ 60% lower, which was attributed to stress-induced microcracking.

Cho et al.'s [76] SiC-TiC composites, similar to their SiC-TiB_2 composites above, showed a more pronounced strength minimum of ~ 400 (from ~ 600) MPa·m$^{1/2}$ versus a toughness maximum of ~ 4.5 (from 2.7) MPa·m$^{1/2}$ at 50 w/o. The strengths of Endo et al. [79] were also 600 MPa at ϕ= 0 but went through a modest maximum of ~ 760 MPa at ~ 30 w/o and then tended to decrease to ~ 400 MPa at 100% TiC, in contrast to a pronounced maximum in (NB) toughness of 6 MPa·m$^{1/2}$ (versus ~ 2 and ~ 3 MPa·m$^{1/2}$ for pure SiC and TiC respectively). Greater differences were shown by Lin and Iseki's [80] SiC-TiC toughnesses, which increased from ~ 4.5 to 9 MPa·m$^{1/2}$ as TiC content increased from 0 to 40 v/o; their strengths decreased from ~ 530 MPa at 0 v/o to ~ 430 MPa at 10 v/o and then remained constant.

In contrast to opposite strength and toughness trends with composite composition, data of Li and Watanabe [81] on SiC with AlN show a maximum in strength at 5 v/o AlN, similar to the maximum in toughness at 10 v/o AlN. Much and possibly all of the rise in strength appears to be due to substantial initial reduction in SiC grain size as AlN was added. The subsequent decrease in strength with larger AlN additions must reflect, at least in part, the decrease in E as AlN content increases. Such effects of composition on E and especially grain size could well differ between strength and toughness and thus be a possible explanation for some differences between them.

Finally, note two aspects of flaw and particle size dependence, starting with strength-to-toughness ratios for the compositions. The above SiC-AlN data shows no clear trend, the ratio averaging ~ 136 m$^{1/2}$, implying a reasonable flaw size of ~ 50 μm, while SiC-TiC data gives values ranging from too low to too high [32,81]. Other composite systems previously evaluated [32] gave either scattered values giving reasonable flaw sizes or (for Si_3N_4-TiC) values decreasing substantially as ϕ increased to either reasonable or larger indicated flaw sizes. Second, the limited data for particle sizes ranging from below to above expected flaw sizes clearly indicate a two-branch behavior (Fig. 9.15) as has been found for glass matrix composites (Figs. 9.2, 9.17) and monolithic ceramics (Fig. 3.1). Thus at finer particle sizes there is limited but generally some increase in strength as D decreases, consistent with substantial other data above showing such behavior in this D range of a few to several microns. However, once the particle size (or particle cluster size) ~ reaches the flaw size, the particles become the flaws, giving greater particle size dependence to the composites. Note that matrix grain size still has an effect on strength, as is commonly observed above, since it impacts the size of machining flaws [e.g. via Eq. (3.2)], as well as effects on local fracture toughness.

E. Platelet and Whisker Composites

Consider first whisker composites, where again much of the emphasis has been on toughness, so there is less data on strength. However, since whisker composites have been extensively investigated, there is reasonable strength data available. Thus Table 8.2 presents a summary of data for whisker composites that shows strength increases commonly being similar to toughness increases, and in the case of composites with lower Young's moduli, strength increases are often similar to increases in *E*. One exception to this is composites of SiC whiskers in Si_3N_4 matrices, which along with composites of SiC whiskers in Al_2O_3 are discussed further below.

The comparisons in Table 8.2 address primarily composition and not specifics such as the effects of whisker dimensions, some of this information is available for Al_2O_3- SiC whisker composites. As noted in Chap. 8, Sec. V.D, toughness commonly increased with diameter, sometimes with length, or both (i.e. with aspect ratio) of the SiC whiskers and was affected by whisker surface character. Baek and Kim [82] showed that doubling whisker lengths from 9 to 18 μm increased strengths with 20 v/o whiskers ~ 20 % (from 380 to 450 MPa), i.e. between the increases in toughness measured by NB and CNB techniques, but closer to the former. They also reported that such increases were consistent with a model for whisker pullout. However, Yasuda et al. [83] reported strengths of composites with the same whisker contents decreasing by ~ 30% over the same range of whisker length increase, i.e. from ~ 700 to ~ 500 MPa (the latter also being the matrix strength without whiskers), with further decreases to ~ 400 MPa at whisker lengths of 50 μm. These observations are consistent with their observed negative effects of increased whisker length on toughness (while increased diameter was reported to improve toughness, its effect on strength were not reported). However, Krylov et al. [84] reported significant (e.g. ~ 140%) increases in strength as whisker diameters increased from 0.1 to 2 μm, with > $^1/_2$ the increase occurring by 0.5 μm, i.e. with the benefits saturating or reaching a maximum at ~ 2 μm (as was also the case for toughness increases, though they were only ~ 1/4 those of strength). A maximum in benefits with increasing whisker dimensions would be expected due to the onset of microcracking, which was indicated by Tiegs and Bowman [51,85] as the source of low strength (~ 180 MPa) in composites with 20 v/o SiC whiskers averaging ~ 5 μm in diameter. The character of whisker surfaces also appears to be a factor in strength, often following that of toughness as reported by Tiegs and Bowman [51] for various whisker treatments. However, Steyer and Faber [86] reported that increasing the thickness of the carbon coating on the whiskers progressively reduced strengths, e.g. from 600 to 400 MPa, as coating thickness went from 0 to 50 Å. On the other hand, a 20 Å carbon coating of the whiskers gave toughnesses that were independent of indent load (versus uncoated whiskers showing toughness starting from ~ 20% lower, increasing to ~ 20% higher as indent load increased).Yang and Stevens [87] showed that the presence of an interfacial silicate film was detrimental to whisker composite properties.

The issue of possible microcracking in composites of SiC whiskers in Al_2O_3 is important, since there is substantial difference in expansion, and composites of SiC platelets in Al_2O_3 clearly exhibit cracking (discussed below); moreover there are indications of cracking in composites with SiC particles, as was discussed earlier. Tiegs' [85] observation above of spontaneous cracking with whiskers ~ 2 μm dia. is an important example. Sato and Kurauchi [88] reported that composites of Al_2O_3 with 25 v/o SiC whiskers ~ 1.5 μm long and 0.5 μm dia. showed substantial acoustic emission on cooling from heating to 600°C, with this increasing substantially as the maximum heating temperature before cooling increased. They indicated that such cracking was indicated by a basic analysis, e.g. reflecting the substantial axial Young's modulus of the whiskers of ~ 700 GPa.

Consider now issues of matrix grain size on strength. Increasing Al_2O_3 grain size was shown to increase toughness across the range of SiC whisker additions, e.g. by ~ 20+% on going from G = 1–2 to 4–8 and then to 15–20 μm [87]. However, Ikeda and Kishi [89] showed that composites of 10–20 w/o SiC whiskers (or platelets or both) in an Al_2O_3 matrix all exhibited strengths decreasing linearly versus $G^{-1/2}$ across the range of resultant G values of ~ 1.5 to 15 μm. The decreases in strength from ~ 400–600 to 100–200 MPa are consistent in both absolute values and relative changes with those of pure alumina (e.g. Figs. 3.13, 3.14). Such grain size effects are clearly a dominant factor in the variations in starting strengths, i.e. with no whiskers ranging from ~ 250 [51] to 550 [83] MPa. These results are in contrast to increased toughness of whisker composites as the Al_2O_3 G increased [90,91].

Strengths of Al_2O_3-SiC whisker composites generally increase as ϕ increases, similar to, but often < the relative increases of toughness, especially at higher whisker contents, due at least in part to processing defects (Section IV). The ratios of strength to toughness generally increase with ϕ and they usually pass through a maximum at various values of ϕ depending on the whiskers and processing, as was previously shown by Rice [32] (Fig. 9.16). Except for some low ratios, mostly at lower or higher relative ϕ values, reasonable flaw sizes are indicated, as are minimum flaw sizes for the maximum ratios.

Composites of SiC whiskers in other oxide matrices also predominantly show strength increasing with increasing SiC whisker content. However, as previously shown by Rice [32], strength-to-toughness ratios generally continuously decrease as ϕ increases, with indicated flaw sizes being reasonable for half or more of these composites. Wadsworth and Stevens' [92] study also corroborates the effect of whisker character on strength as for toughness, e.g. strength increasing with increasing diameter (from 0.7 to 1.2 μm). One exception to this trend for increased strength as ϕ increases in other oxide matrix composites with a TZP matrix, was shown by Zhang et al. [93] and Yang and Stevens [87] who showed respectively strengths decreasing from 1150 to 850 MPa from ϕ = 0 to

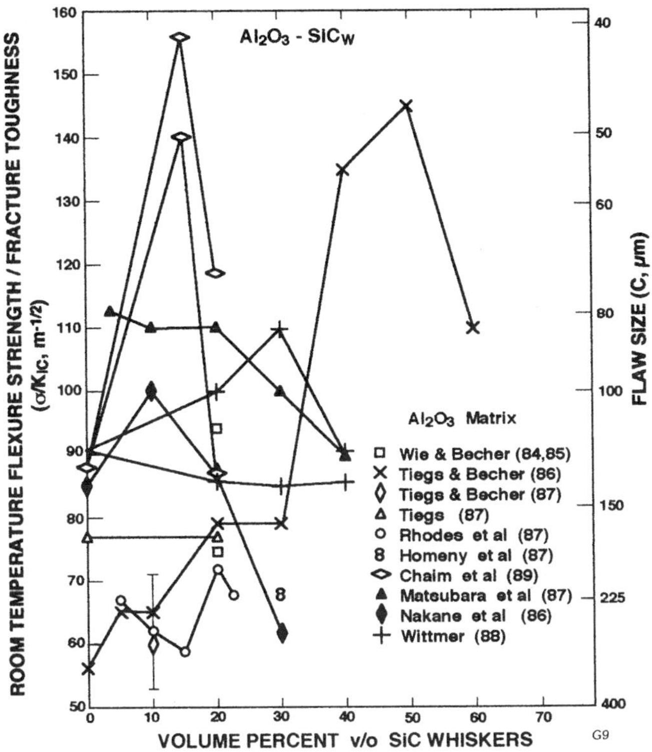

FIGURE 9.16 Strength-to-fracture toughness ratio for various A_2O_3-SiC whisker composite investigations. Note the approximate flaw sizes shown by the right-hand scale. (From Ref. 32. Published with permission of Ceramic Engineering and Science Proceedings.)

0.25 (with most decrease between ϕ= 0.15 and 0.25) and from 1050 to 850 MPa from ϕ= 0 to 0.2 (with the decrease fastest at low ϕ). Zhang et al. showed (NB) toughness increasing ~ linearly over the ϕ range, but the high toughness levels result in somewhat larger than likely flaw sizes. Yang and Stevens cite both larger matrix grain size in the composite and mismatch stresses between the whiskers and the matrix, and implied possible microcracking, as sources of the decreased strengths in the composite.

Turning to composites of Si_3N_4 with SiC whiskers, Rice [32] showed that these had at best limited strength increases, and more often progressive decreases in strengths as ϕ increased. This in turn results in significant decreases in strength-to-toughness ratios, and thus frequent larger than likely calculated flaw sizes as ϕ increases, e.g. to hundreds of microns or more. Neergard and Homeny

[94] showed that BN coated SiC whiskers gave lower strengths than uncoated ones. Olagnon et al. [95] reported a probable but modest maximum of strength of ~ 900 MPa at 15 v/o whiskers, up from 700+ MPa with no or 20 v/o whiskers. This was somewhat inconsistent with I toughness (for cracks perpendicular to the whisker orientation) showing a clear increase only at the highest whisker content of 20 v/o (and apparently no dependence on whisker content for crack propagation parallel with the whisker orientation). Rossignol et al. [96] similarly showed that strength increased from 800 to ~ 950 MPa, with limited or no SENB toughness increases for 30 v/o SiC whiskers. They also showed extensive *R*-curve effects that commenced at smaller crack sizes and increased faster than for the Si_3N_4 alone, e.g. at crack lengths of ~ 1.5 mm versus 2.5 mm, and substantial effects of different whisker sources. Fusakawa and Goto [97] showed that while fracture energy nearly doubled on addition of 30 v/o SiC whiskers, strengths decreased ~ 5%.

Other, more recent studies show modest to negative results with other nonoxide matrices: NbC [98], B_4C [99], TiB_2 [100,101], and TiC [102]. NbC gave strengths increasing ~ 70% from 500+ to ~ 900 MPa (versus ~ 20 % increase in toughness) as ϕ increased from 0 to 0.3 with reasonable calculated flaw sizes, but with slightly lower results as whisker diameter ~ doubled from 0.45 to 1 μm. B_4C gave an ~ 10% increase in toughness and a modest (~ 20%) maximum of strength (ϕ ~ 01–0.2) over the same ϕ range, with some indicated reduction of values with somewhat smaller whiskers. TiB_2 matrices showed mixed and modest changes in toughness and strength over the same ϕ range. TiC gave similar modest (both) effects or substantial strength decreases with limited toughness changes [102]. In all cases except that with B_4C, reduced grain growth due to whisker additions were cited as a factor in strength increases, especially at lower whisker levels.

Dusza et al. [103,104] reported that hot pressing of Si_3N_4 with additions of 10 or 20 w/o Si_3N_4 whiskers (+ Al_2O_3 + Y_2O_3 densification aid) gave respective average strengths of ~ 690 and 510 MPa, though both had essentially the same (IF) toughness of ~ 6.3 MPa·m$^{1/2}$. The discrepancy in strengths was attributed to observed increased fracture initiation from whisker agglomerates or associated pores at the higher whisker loading.

Much less strength data is available for platelet composites, in part because there is much less development and because some investigations do not address strength, even though toughness was measured. Chou and Green [105] reported that composites of Al_2O_3 with 10 v/o of smaller (~ 12 μm dia.) SiC platelets had strengths ~ 1/3 < the Al_2O_3 only, and strength increased only a limited amount with increased platelet additions to 30 v/o, in contrast to ~ 50% increases in toughness. Composites with larger SiC platelets by ~ 2, which resulted in spontaneous cracking and substantial decreases in *E*, as was noted earlier (Chap. 8, Sec. IV.D), suggest that stress-induced cracking may be a factor in the strength loss.

Mitchel et al. [106] reported that composites of Al_2O_3 coated SiC (20 v/o) platelets in a SiC matrix had strengths ~ 10–20% higher that the matrix alone without the nearly 50–100% anisotropy found in toughness. Naschik et al. [107] reported strengths of composites of mullite matrices with 10 v/o SiC platelets (~ 10–25 μm dia.) typically decreasing by up to ~ 50% below that of the matrix alone despite toughness increases of up to ~ 20%. Strength increases of 10–20% were obtained in some cases (with similar toughness increases) when the platelets were oxidized to enhance platelet–matrix bonding.

Strength behavior of composites of aluminas with in situ formed platelets, whose toughnesses were discussed in Chap. 8, Sec. IV.D, are important to consider along with data for alumina bodies with elongated, platelike grains. While Chen and Chen [108] reported that in situ formation of various hexaluminate platelets with overall dimensions similar to those of the matrix Al_2O_3 grains (e.g. 2–15 μm) resulted in a maximum toughness of ~ 4.3 (versus ~ 3.0 MPa·m$^{1/2}$ for the martix alone) at ~ 30 v/o aluminate, no strengths were reported (and E decreased continuously from 420 GPa at 0 v/o to ~ 230 GPa at 100 v/o aluminate). While Yasuoka et al. [109] reported similar but greater increases in (SEPB) toughness from 3.5 to 6 MPa·m$^{1/2}$ by in situ growth of hexaluminate platelets combined with inducing some platelet character of the Al_2O_3 grains via doping of 240 ppm SiO_2, their strengths did not follow toughness trends. Thus small additions reduced G from ~ 5 to ~ 2 μm, increasing strengths from 420 to 660 MPa with no change in toughness, while increased additions increased G and introduced platelets, while increasing toughness to 4.1 and then 6 MPa·m$^{1/2}$ with respectively, no clear increase in strength and a modest (6%) but clearer strength decrease occurred (accompanied by increased intergranular fracture and hence grain bridging). While Kim et al. [110] reported nearly identical (I) toughness increases, maxima (at 1 m/o), and subsequent modest decreases in Al_2O_3 doped with 0.5 to 3 m/o Na_2O + MgO to grow in situ beta alumina platelets, strength increased to a more pronounced maximum at 0.5 m/o, with some and possibly most of this due to G reduction. Koyama et al. [111] reported that (CNB) toughness of Al_2O_3 with platelet grains due to low additions of CaO + SiO_2 increased (~ the same as found by Yasuoka et al.) with either the diameter (d) or the thickness (t) to the $^1/_2$ power (or $d^{5/6}t^{-1/3}$), which was also true for the lesser increases as G increased in bodies with equiaxed grains. However, as noted earlier (Chap. 3, Sec. III.A), strengths of both bodies with platelet grains and those with equiaxed grains both increased as toughness deceased. Similarly, An and Chan [112] showed that in situ formation of 30 v/o Al_2O_3·$CaAl_{12}O_{19}$ platelets in Al_2O_3 increased toughness but at the expense of strength, which they acknowledged was inevitable and is now being increasingly recognized (Sec. III.E). As was discussed in Chap. 3, Sec. III.A, data for the above bodies generally agrees with that for other alumina bodies based on their grain sizes.

Baril et al. [113] reported that their composites of Si_3N_4 with 0–30 v/o SiC of smaller platelets (~ 11 μm) had strengths of 800 MPa independent of the v/o platelets, except for a 5–10% decrease at 30 v/o platelets, while use of larger platelets (17 μm) had strengths of 700 MPa that decreased at 20 and more at 30 v/o platelets, the latter by 10–15%. This was in contrast to the linear increase in CNB toughness from ~ 6.4 to 8.1–8.8 MPa·$m^{1/2}$ for 0–30 v/o, with a trend for slightly higher toughness with larger platelet sizes. This toughness increase was similar to, but substantially greater than, that of elastic moduli (10–15%). Fusakawa and Goto [97] showed that addition of up to 30 v/o SiC (10–20 μm dia., ~ 1 m thick) platelets to Si_3N_4 increased fracture energy by an additional 40% over that produced by such additions of whiskers, but strengths decreased with increased additions, e.g. by ~ 40% versus ~ 5% with whiskers. While simultaneous addition of SiC platelets and whiskers gave even higher fracture energies and increased toughnesses over those achieved with platelets alone, toughness still decreased by up to 10–15%, and strengths decreased as much as, or more likely somewhat more than, with platelets alone. The significant strength reductions with platelet additions were attributed to large surfaces of them frequently acting as fracture origins.

Kellet and Wilkinson [114] fabricated Al_2O_3-graphite flake/platelet composites via filter pressing or tape casting and then hot pressing, showing that flexure strengths of bars with tensile surfaces parallel to the plane of pressing were ~ 2/3 those with tensile surfaces normal to the plane of hot pressing. Coarse (~ 75 μm dia.) platelets gave strengths 30–45% lower than fine platelets (~ 4 μm dia.). Strengths decreased with increasing graphite and pore contents, and the authors noted that both had similar effects on strength (as was also suggested for effects of BN flake/platelets, Chap. 11, Sec. III.E). This similarity is supported by the slope of a semilog plot of their strengths versus the sum of the volume fractions of porosity and graphite being ~ 3.6 for the fine and ~ 3 for coarse graphite, i.e. similar to the dependence of E for composites with BN platelets (Fig. 8.13). Ozgen and Bond [115] studied alumina–graphite refractory bodies made with fine (50–150 μm dia.) or coarse (350–420 μm dia.) graphite flakes made with 3/4 coarse (250–420 μm dia.) and 1/4 finer (< 50 μm dia.) fused alumuna grain and silicate-based binder and ~ 20 v/o porosity. Their strengths were only 3–5% those of Kellet and Wilkinson, reflecting the larger sizes of the graphite and alumina as well as the porosity and the binder, and had a greater rate of decrease with increased graphite content, e.g. a semilog slope of ~ 5.6, reflecting expected less graphite preferred orientation.

Naschik et al. [107] showed that addition of 5–20 (mostly 10) v/o SiC platelets (~ 10–25 μm dia., ~ 1 μm thick with various surface treatments) to mullite did not significantly increase, and most commonly reduced, strength, e.g. with reductions up to ~ 50%, despite more frequent increases in toughness, e.g. by ~ 20%. Fracture surfaces showed substantial fracture through large sections of individual platelets, or even larger fracture areas along the platelet–matrix interfaces.

Chaim and Talanker [116] reported bend strengths of their composites of α-SiC platelets (50–250 μm dia., 5–25 μm thick) in a cordierite glass matrix as generally similar to their NB toughness dependence on composition, i.e. a rapid initial rise and then much less increase. Strengths increased from ~ 120 MPa at 0 v/o to 160 MPa at 10 v/o SiC for crack propagation parallel to the hot pressing direction, i.e. somewhat <the 50% increase in toughness, followed by no increase or a limited decrease for 20 and 30 v/o SiC, in contrast to 10–15% increases for toughness. Strengths for crack propagation normal to the hot pressing direction gave similar trends, i.e. like those for toughness, but again somewhat less. On the other hand, crystallization of the glass matrix, while giving somewhat higher toughnesses versus no crystallization, resulted in strengths decreasing from ~ 130 MPa at 0 v/o SiC to ~ 110 and 90 MPa at respectively 10 and 20 v/o SiC.

F. Ceramic Eutectic Composites

Results for composites made from eutectic particles of Al_2O_3-ZrO_2 were discussed in Sec. III.B.2. Limited results of composites made by directional solidification of eutectics are addressed here. Earlier work by Hulse and Batt [117] on Y_2O_3-ZrO_2, $CaZrO_3$- ZrO_2, MgO-CaO, and Al_2O_3-$ZrO_2(Y_2O_3)$ reported strengths for stresses parallel to the solidification direction of commonly ~ 300–700 MPa, with the highest strengths for the latter system (Fig. 11.11). Limited or no effects of solidification rate and resultant changes in fiber spacing were reported (other than possible effects of high thermal stresses) indicating that colony structure dominated. Mah et al. [118] also reported similar strengths, i.e. ~ 370 MPa, for directionally solidified Al_2O_3-$Y_3Al_5O_{12}$ (YAG) eutectic, but again colony structure (~ 200 μm) probably was a dominant factor limiting strength.

Bradt and colleagues [119–121] also measured strengths of directionally solidified eutectics, mostly parallel with the solidification direction. Strengths of ~ 160 MPa across the range of solidification rates used were reported for ZrO_2-MgO eutectic specimens. The lack of effects of growth rate, which had the normal inverse relation to fiber spacings, was attributed to strengths being controlled by the essentially constant colony structure size (~ 230 μm). Effects of residual stresses were indicated by increases and decreases in strengths on annealing. Similarly, no clear variation of strengths of MgO-$MgAl_2O_4$ with solidification rates and resultant MgO fiber spacings (of ~ 1 to 4 μm) were observed, which was again attributed mainly to the presence of a nearly constant colony size (~ 400 μm), as well as some possible contributions of residual stresses and cracking. However, some anisotropy was observed, e.g. strengths for stressing normal to the solidification direction were ~ 3/4 those for stressing parallel to it, and distinct effects of fiber spacing on hardness were observed (since indent sizes were < the colony size, Chap. 10, Sec. III.A). Significant effects of fiber spacing are also indicated in tests of ZrC-ZrB_2 and ZrC-TiB_2 directionally solidified composites in other mechanical properties, though strengths were not measured [121].

G. Ceramic–Metal Particulate and Wire Composites

Nivas and Fulrath [122] reported that composites of glass matrices with dispersions of 10– ≥ 50 v/o of spherical W particles (20–70 μm) all followed the behavior shown in Fig. 9.1, but which varied significantly as a function of the thermal expansion differences. For a glass with an expansion ~ that of W, the observed intersection of the two σ–$\lambda^{-1/2}$ lines was at flaw sizes of ~ 50–57 μm as respectively observed and calculated. Also the slope of the strength increase as $\lambda^{-1/2}$ increased was again nearly the fracture toughness of the glass (again reduced ~ 30%, since λ is a diameter and c a radius), as was found for similar composites with dispersed Al_2O_3 particles (Sec. III.A). Glasses with expansions ~ 1.3 ppm/°C < or 3.1 ppm/°C > that of W gave similar bilinear trends but with respectively somewhat larger observed and calculated flaw sizes, 64–74 μm and 80–92 μm and substantially and even greater reduced slopes of the portion of increasing strength versus $\lambda^{-1/2}$. This translated to progressive decreases in strengths of respectively 100, 70, and 55 MPa as the absolute difference in expansion increases. However, strengths decreased as the expansion of the dispersed particles exceeded that of the matrix and as the size of the dispersed particles increased due to their increasingly acting as voids due to lack of bonding to the matrix. While the pores caused strength reductions, there can still be some toughening due to the pore–particle combinations that limit strength losses, e.g. see Biswas [123]. However, note that this data clearly shows a definite $D^{-1/2}$ dependence (Fig. 9.17), reflecting effects of v/o W and glass matrix character. The lack of a higher D dependence at larger D is attributed to flaw sizes being at or beyond the larger particle sizes.

Turning to finer dispersed particle sizes, Yun and Choi [124] showed that addition of up to 20 v/o of W particles (up to nearly 1 μm) results in a continuous strength decrease to ~ 30%, i.e. opposite of their (I) toughness values, which increased > 50%. These changes in properties were accompanied by x-ray measured compressive stresses (radial to the W particles) that increased from ~ 70 to ~ 100 MPa as the v/o W increased from 5 to 20%. Sekino et al. [125] dispersed Ni particles in Al_2O_3 by reduction of either fine particles of NiO or of solution dispersed $Ni(NO_3)$ giving Ni average particle sizes of respectively 100–224 and 50–130 nm, where the first value in each set is for intra- and the second for intergranular particles, the latter particles being ~ 66 and 82% of the particles respectively for the larger and the finer particles. Both dispersions resulted in

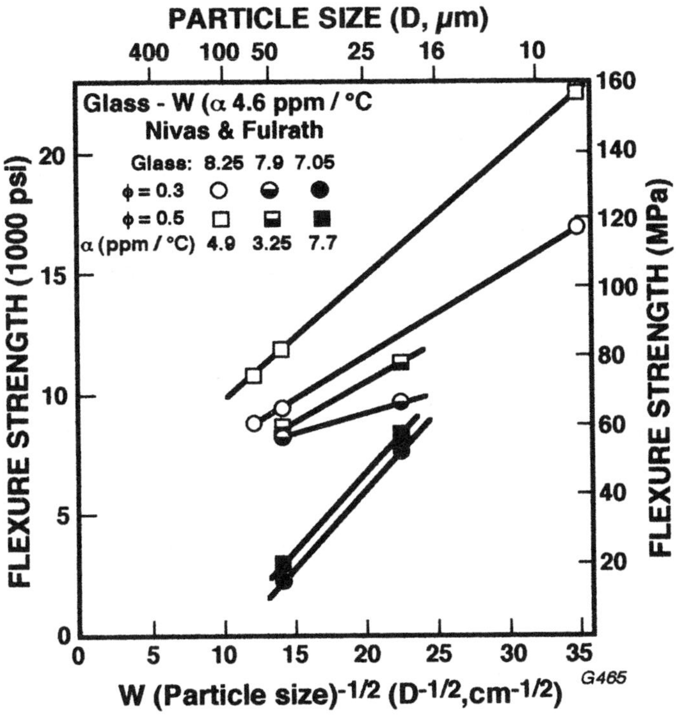

FIGURE 9.17 Strength of spherical W particle–glass matrix composites versus the inverse square root of the W particle size for volume fractions (ϕ) of 0.3 and 0.5 tested at 22°C. Note designations of the three different glass matrices used by their strengths (in 1000 psi) as glasses without any W particles and thermal expansion. (From Ref. 122.)

significant strength maxima of respectively 1.09 and 0.99 GPa for the finer and larger particles, with both decreasing back to the strengths without Ni particles of 700 MPa at 20 v/o Ni. This was in marked contrast to very limited but continuous (I) toughness increases as ϕ increased. Similarly Nawa et al. [126] showed that while addition of 5–20 v/o Mo particles (< 0.2 to > 1 μm) continuously increased (IF) toughness from 3–4.3 to 4.3–7.7 MPa·$m^{1/2}$ as ϕ increased, strengths continuously decreased from 570–880 to 570 to 400 MPa, i.e. opposite to each other. Strengths increased as the grain size was reduced (i.e. consistent with data of Fig. 3.14) as a result of increased Mo addition, decreased hot pressing temperature, or both, while toughness increased as *G* increased, i.e. again opposite to strength, similar to frequent opposite *G* dependence of these properties in monolithic Al_2O_3 (Chaps. 2 and 3).

Pyzik and Beaman [127] formed composites from in situ reaction of B_4C-Al which give complex composites of either discontinuous or continuous B_4C structures, with the latter giving higher strengths, e.g. by ~ 20–50%, while (CNB) toughness showed an opposite trend.

Continuing with composites with continuous metal phases, i.e. mainly cermets, consider WC-Co bodies. Rice [32] reviewed the microstructural dependence of toughness (Chap. 8, Sec. IV.F) and strength, the latter showing strengths generally increasing as the WC grain size decreases, the opposite of toughness. This opposite trend is also seen in the ratio of strength to toughness (Fig. 9.18), which also indicates reasonable calculated flaw sizes and a trend for them to be finer at intermediate Co compositions, i.e. where strength and toughness are both high.

Next, consider results of Donald and McMillan [128] for hot pressed composites of chopped Ni wire ~ 3 mm long and 0.05 or mostly 0.125 mm dia. in

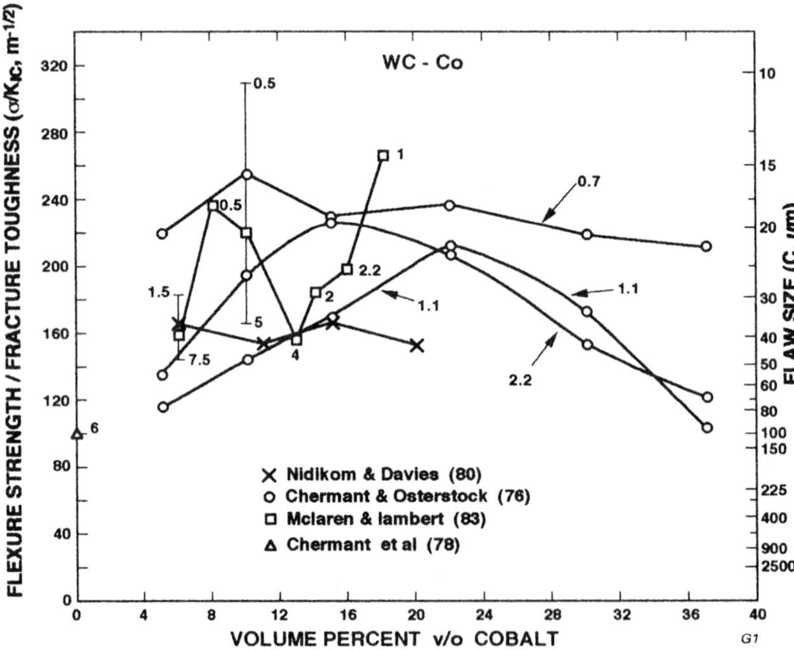

FIGURE 9.18 Strength-to-fracture toughness ratio of WC-Co bodies as a function the volume percent (v/o) Co for various WC grain sizes (numbers next to curves or data points, where there is a bar for the range of data, the grain sizes for the higher and lower strengths are shown at the respective ends of the bars). (From Ref. 32. Published with permission of Ceramic Engineering and Science Proceedings.)

glass matrices having an expansion of ~ 8.3 ppm/°C < that of Ni, ~ the same expansion, or ~ 1.7 ppm/°C > that of Ni. Normal flexural strengths of the composite with the lower expanding matrix dropped precipitously from ~ 220 MPa at 0 v/o to ~ 70 MPa at ~ 1 v/o Ni (with an apparently somewhat slower and less drastic but still substantial decrease, e.g. to ~ 100 MPa with the finer Ni wires segments) and then steadily increased, e.g. to ~ 140 MPa at 30 v/o. Composites with ~ matched expansions started from the same strength, decreased much more slowly to a minimum strength of ~ 110 MPa at 20 v/o Ni, and then increased to ~ 160 MPa at 30 v/o (and to ~ 180 MPa at 40 v/o Ni). Composites with the highest expansion glass behaved very similar to those with ~ matched expansions, except that they showed very little increase in strength beyond the minimum at ~ 20 v/o Ni. Strength behavior of the composite with ~ matched expansions was quite similar to its Young's modulus dependence (Chap. 8, Sec. III), except that the maximum loss of E was ~ 40% versus ~ 70% for strength. There was also similarity between the strength and E dependences for the other two composites, but with greater strength decreases to lower minima relative to those for E. Also, the increased strength for the lower E glass past the minimum was much greater than that for E, while the reverse was indicated for the composite with the higher expansion glass. These strength trends for normal testing (without notches) were similar to results of flexure testing of specimens with notches, except that the strengths of the pure glasses were much lower (~ 70–85 MPa), and strengths increased steadily with increasing v/o Ni, typically more so at higher Ni contents, reaching similar levels at higher loadings as for unnotched specimens. This behavior of notched specimens was quite similar to that for work of fracture (measured from the area under the load-deflection curves) with or without notches, and thus was quite different from the unnotched strength behavior at lower Ni loading.

Zwissler et al. [129] showed that finer (6, 12, or 25 μm dia.) stainless steel chopped wires hot pressed in an FeO matrix increased strengths over the range investigated (0–15 v/o) at a rate of up to ~ twice the rate of increase of Young's modulus and consistent with toughness increases (due mainly to wire bridging in the crack wake). The contrast in these trends with those of Donald and McMillan above are probably due to much less or no cracking in the former composites due to the combination of finer wire size, limited expansion and E differences, e.g. the steel wire expansion was only ~ 14% > that of the FeO. Simpson and Wasylyshyn [130] hot pressed up to 12 v/o Mo wires (3.16 mm dia.) in an Al_2O_3 matrix, which made large (up to 250-fold) increases in the work of fracture, but ultimate strengths were reduced, e.g. by ~ 20–25%, probably due to some cracking. Brennan [131] hot pressed ~ 25 v/o continuous Ta wires (0.63 or 1.27 mm dia.) in a Si_3N_4 matrix, obtaining flexural strengths of 550–700 MPa (and ~ four-fold increases in Charpy impact strengths). However, tensile strengths were only ~ 170 MPa, apparently due to cracking not seen in the flexure tests due to the depth of the wires below the tensile surface and not seeing near as much of the tensile stress in flexure [131,132]. The issue of cracking, especially due to expansion differences, is also indicated in work of Rice and Lewis [132]. They showed that though good strengths (and toughnesses) could be obtained in various composites of ceramics with uniaxial fine SiC-based (Nicalon) fibers in various ceramic matrices, highest values were obtained when the differences in fiber and matrix expansions were < 2–3 ppm/°C.

IV. CERAMIC COMPOSITE RELIABILITY, WEIBULL MODULI, STRENGTH VARIABILITY, AND FRACTOGRAPHY

A major motivation for developing ceramic composites, besides achieving better or unique levels or combinations of properties, has been to increase mechanical reliability. This latter motivation became closely related to study and evaluation of *R*-curve effects, since the increased toughness with increasing crack propagation was seen as reducing the dependence of strength on the starting flaw size and thus reducing the variability of strengths due to varying initial flaw sizes, hence increasing reliability. While issues of the large crack sizes commonly needed for most significant increases in toughness have been noted earlier and will be discussed further in the next section, the issue of reliability, as measured by Weibull moduli, is addressed in this section. Since most investigators have not directly measured Weibull moduli of their composites, values have been estimated by Eq. (3.4) and these estimates compared to other related mechanical properties, not just to strengths.

Rice [32] previously surveyed Weibull moduli of ceramics and ceramic composites, noting that composites of the type that are the focus of much of this book and this chapter typically had a Weibull modulus (m) in the range of 5–15, as did monolithic ceramic materials considered for mechanical applications. (Note that composites of continuous ceramic fibers in a ceramic matrix clearly provide increased reliability, since they result in lack of catastrophic failure and associated notch insensitivity, neither of which has been achieved in the composites addressed here.) Further evaluation of Weibull moduli of ceramic composites considered in this and the previous chapter are generally consistent with this range. This consistency is supported by the limited Weibull moduli reported by the few investigators addressing them. Thus Govila [133] reports $m \sim 7$ for an Al_2O_3 and ~ 10 for a similarly processed Al_2O_3-SiC whisker composite based on ten tests each. Similarly, Akimune et al. [68] report $m = 6–13$ for their Si_3N_4- SiC particulate composites. However, though based on limited tests, Baril et al. [113] report Weibull moduli increasing from ~ 10 to 19–29 as the v/o of SiC platelets in a Si_3N_4 matrix increased from 0 to 30 v/o, with greater increases for smaller SiC platelets, and probably more of the increase occurring by 20 v/o platelets.

Some substantially higher moduli are calculated from some strength data sets, typically for a data point for a fixed composition, but there is substantial uncertainty in their validity given the limited number of tests commonly involved (e.g. 4–10). Thus while some of these higher values may indicate real increases in reliability (which if real may be from other sources, as is noted below), many do not, so that broader trends are of primary interest. Two such trends stand out. The first is that the Weibull moduli for toughness values vary about as widely with similar values as for the corresponding strengths. This seriously questions either the occurrence or the effectiveness of *R*-curve effects, since these would imply decreased variability due to increased effects of the average material-microstructure and less effects of local variations as crack sizes increased. The second, and not surprising, trend is for Weibull moduli for other properties such as *E* or *H* to be far higher, e.g. values of 50 to several hundred. Such high values, though they are few, since adequate data for calculating them is commonly not given, are at least approximately indicative of what the variation of material properties is for properties averaging material behavior over a longer range versus those reflecting a much more local property average. A third but much more uncertain trend is for some possible higher Weibull moduli for strengths and toughnesses for bodies in which the toughening microstructure is developed in situ. This includes crystallized glasses (though changing microstructures and mechanisms are probably an important complication in many of these) and especially in situ development of platelet- or whiskerlike grains [134], with in situ toughened Si_3N_4, as was discussed in Chaps. 2 and 3.

The above issue of whether higher fracture toughness from *R*-curve effects is generally applicable to improved strength and especially reliability is further addressed on the negative side by the results of fractography identifying fracture origins. While limited studies of fracture origins have been made, those that have been conducted all show similar processing defects at fracture origins of composites as for monolithic ceramics, as well as frequent origins from defects of the composite structure. Thus a key example of the former are frequent voids, e.g. as shown by Govila [133] in his study of Al_2O_3- 15 w/o SiC whisker composite, e.g. voids as a result of whisker clustering (i.e. due to whisker "nests" as also frequently seen by other investigators). Rice [32] also showed fracture origins in such whisker composites from matrix rich regions (e.g. Fig. 9.19). Watanabe and Fukuura [135] have shown Al_2O_3-TiC fracture initiation from normal processing defects such as pores, larger grains, etc. Similarly, Cameron et al. [136] showed initiation from heterogeneities, e.g. clusters of larger grains, in reaction hot pressed composites. Other fracture surface observations include frequent exposure, and often failure initiation from clusters of larger particles in particulate composites [54,74] and of larger platelets, with substantial fracture along much or all of one of the large plate faces [116].

Finally, note that other scale factors can be factors in other types of ceramic composites. Thus fibrous monoliths, composites made by introducing an

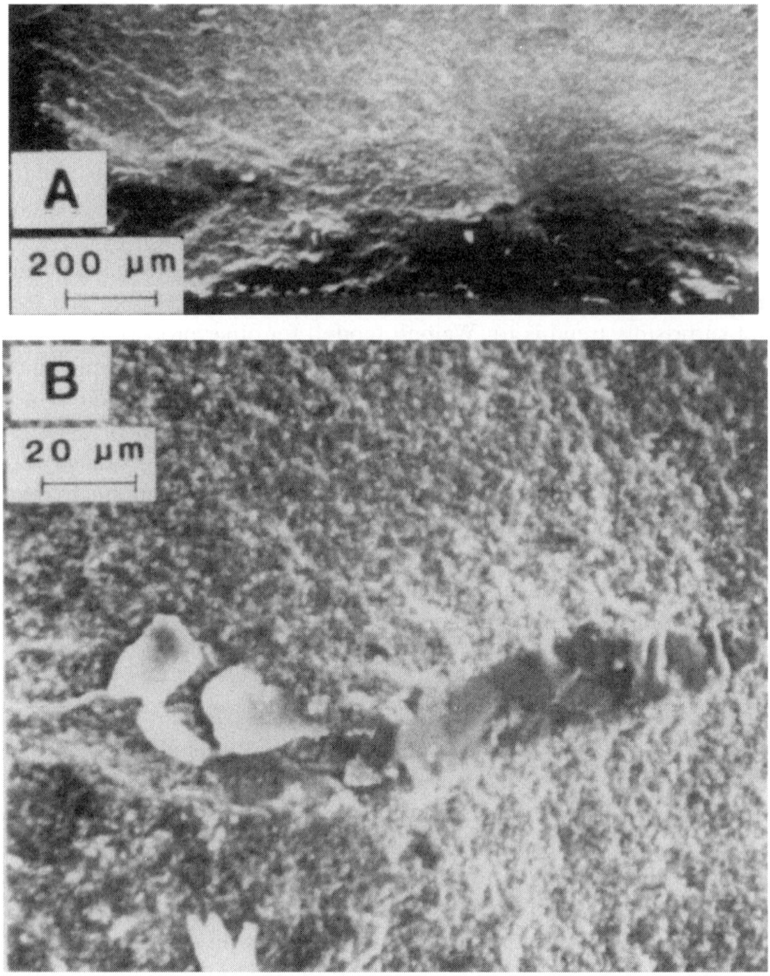

FIGURE 9.19 Fracture initiation from an elongated Al_2O_3 rich region from center right across much of the photo in an Al_2O_3-SiC whisker composite. (A) Lower magnification showing larger fracture area. (B) Higher magnification of specific origin. Note the much larger Al_2O_3 grain size in the matrix agglomerate acting as the fracture origin. (From Ref. 32. Published with permission of Ceramic Engineering and Science Proceedings.)

array of weak interfaces (e.g. of BN in Si_3N_4 via extrusion of an array of BN coated green rods of Si_3N_4), increase strengths as the "fiber" diameter decreases. Simpson [137] showed flexural strengths of ceramic–metal composites he made from particles obtained by consolidating and sintering composite particles made by tape casting multiple, alternating green layers of HfO_2-CeO_2+MgO and Mo increased 50–100% as the particle dimensions decreased from ~ 800 to ~ 70 μm in lateral dimensions (and compressive strengths increased by 15–40%, Chap. 10, Sec. III). This is similar to effects of colony structures (e.g. Figs. 1.9, 1.10, 1.12) on fracture.

V. GENERAL DISCUSSION

A. Toughness–Strength Differences and Strength–Microstructure Mechanisms, Especially via Flaw Sizes

Three sets of issues that need to be discussed are the comparison of toughness and strength results, especially their often significant differences, the mechanisms responsible for toughness and especially strength in composites, and models and improved evaluation of such behavior. With regard to the strength–toughness results, it is important to note that there are broad and significant differences of two types. The first and less common, but still frequently substantial, difference is for opposite dependences on ϕ over at least part of the ϕ range. Thus toughness generally increases, usually to a maximum (that may or may not be in the range investigated), while strengths sometimes show the opposite trend, e.g. Fig. 9.12, or show greater decreases at higher ϕ. While most or all of the latter is due to processing heterogeneities, the former indicates basic differences in mechanisms that are probably related to the source of the other basic difference, namely crack size effects. Thus, as discussed in Chap. 2, Secs. II.A and IV, varying crack–particle interactions in glass–metal composites (Chap. 8, Sec. V.A) and lower toughness in WC-Co (Chap. 8, Sec. IV.F) at higher crack velocities [138] are another indication of basic differences that can occur between toughness and strength tests.

The other, more general and basic difference between strength and toughness behavior are their dependences on microstructure, first and foremost on the sizes of the dispersed second phase and secondarily on the grain size of the matrix. Thus increased toughness with increased matrix grain size often corresponds to decreased strength, i.e. as is also seen in monolithic ceramics. More significant is the same trend for effects of dispersed particle size, i.e. while toughness generally increases as the dimensions of the dispersed phase increase, strengths decrease, e.g. Figures 9.12 and 9.13, directly analogously to the frequent disparities between toughness and strength of monolithic ceramics as a function of *G*. This is attributed to the same cause, namely effects of crack size, since larger dispersed particles can contribute more toughening at larger crack sizes, e.g. due to greater crack deflection or bridging, but strengths are typically determined by smaller flaws that provide fewer opportunities for such toughening effects as crack bridging with larger cracks. Thus toughness tests with arbitrarily introduced cracks will often reflect increased toughening with larger particles, while such larger particles are often fracture origins that control normal strength behavior. Toughness more closely reflects average body behavior due to the use of larger cracks, while strength reflects weak link behavior, generally on a much smaller scale of the microstructure.

Toughness and strength behavior approach consistency with each other when the crack sizes are comparable either on an absolute scale or on a scale relative to the microstructure impacting mechanical properties. Thus large cracks used to determine toughness are pertinent when large cracks also determine strength, e.g. from serious thermal or impact stresses (Chap. 11, Sec. III.C). Alternatively, consistency occurs when smaller cracks are used for measuring toughness, e.g. via fractography from strength tests, or more generally when crack sizes controlling strength are sufficiently large relative to cracks for toughness measurement to reflect the same microstructural effects on both. This typically results with finer microstructutres, e.g. those of ZTA (Fig. 9.6), which have sufficient homogeneity so that there are not large differences in the statistical opportunities for toughening mechanisms to operate over the different crack scales used for much toughness testing versus those controlling strengths. Thus if grain and particle sizes are of the order of 1 μm and flaws are ~ 20 μm in size, then there are ~ 600+ grains in a halfpenny surface flaw and ~ 60 grains along the crack periphery, but only ~ 8 and 5 grains respectively if the grain size is ~ 5 μm. Clearly the former offers more statistical opportunities for toughening to approach the average toughening effects per area of large cracks, while the latter does not.

Turning to the issue of mechanisms, the past focus has been on resistance to crack propagation as typically measured with larger cracks where mechanisms such as crack deflection and especially branching and bridging can have significant effects. However, strength, while sometimes reflecting such toughening effects, commonly does not, as is shown in this chapter for composites and in Chaps. 3 and 6 for monolithic ceramics. The reality is that there are two other general mechanisms impacting strength. The first is the effect of composite character, especially composition, on Young's modulus, which has received some attention, but substantial neglect, and which may or may not be reflected or recognized in toughness tests. The second, often more significant, mechanism is the effects of composite composition and microstructure on flaw character, especially size, typically from machining. These two broader mechanisms can occur in addition to or instead of more specialized mechanisms such as transformation toughening (from tetragonal ZrO_2) or ductile inclusions (in some ceramic–metal composites) as well as more general large crack toughening such as crack deflection, bridging, and branching. Another mechanism that can result in significant strength and toughness differences is residual, especially microstructural, stresses resulting in pre-existing or stressing-induced microcracks or both. While microcracking in composites has been recognized and studied some, it has not been adequately evaluated, especially its effects on flaw size.

Consider the evidence for significant effects of machining flaws on composite strengths. The first and more widely used demonstration of this is flaw sizes calculated from strengths and toughnesses to be commonly a minimum at some intermediate volume fraction of dispersed phase, e.g. as extensively pointed out in Rice's earlier review [32] and addressed in the previous sections. Since machining flaw sizes typically vary as inverse functions of K and H, each raised to different fractional powers [e.g. Eq. (3.2)], and toughness of composites typically passes through a maximum, flaw sizes would commonly pass through minima. Such flaw size minima have been widely indicated [32] (e.g. Figs. 9.10, 9.13, 9.16, 9.18), though not necessarily coincident with the toughness maxima, since the flaw size minima depend on composition and microstructural effects on H and E as well as on K variations with crack size. Thus H has different dependences of composite structure (Chap. 10, Sec. III.A), i.e. directly on ϕ and inversely on D as well as on the matrix G, which is a function of ϕ and dispersed particle powder and matrix powder particle sizes, as well as processing. E also varies with composition. Both H and E variations can shift the c minimum, as can the pertinent toughness controlling flaw formation, which may not be that obtained with large cracks. Thus reductions in matrix grain sizes as a function of composite composition and particle sizes, while possibly having other effects, can impact the sizes of machining-induced flaws due to similar effects. Such effects of limited second phase additions increasing strength via matrix grain size reduction were also shown for monolithic alumina bodies e.g. with limited Mo and W additions to alumina (Fig. 3.14). In addition to flaw size reduction effects, there can be residual surface, especially compressive, stresses from machining that will affect the flaw behavior, e.g. making flaws act as smaller than they really are. Such compressive surface stresses have been reported for machined TZP [139] and for nanocomposites of Al_2O_3-SiC [60].

The second effect of composite character on flaw sizes is via the dependence of composite strength on particle size introduced here in this chapter (Figs. 9.2, 9.15, 9.16). Thus the mechanism is that finer particles are smaller than the flaw sizes induced by machining (or other surface abrasion) and thus have more limited effect on strength via their impact on resultant flaw character, especially size, via their impacts on local Young's modulus, hardness, and toughness. Thus in this finer particle branch there is variable but a generally limited decrease in strength as $D^{-1/2}$ decreases. On the other hand, as the sizes of particles approaches and then exceeds the flaw size, the particles (or larger particle clusters) become the sites for introduction of machining flaws controlling strength with a resultant higher strength–$D^{-1/2}$ dependence in this larger particle size branch. This is directly analogous to the mechanism for strength–grain size effects in monolithic ceramics, as was extensively discussed in Chapter 3, and will be some in Chapter 11. The direct parallel between these mechanisms for monolithic and composite ceramics (including accounting for the effects of matrix grain size in composites having similar effects on flaw generation and hence on strength) provides major corroboration for this mechanism in composites. Additionally, while the specific data base for this machining flaw–strength mechanism is limited, since most composite studies have been with finer particles, the empirical recognition that finer particles give higher strengths than coarser ones is added support for this mechanism, as is data showing limited effects of different finer particles on composite strengths in this chapter. This machining mechanism is also supported by the generally lower strengths of platelet composites, since the platelets often provide a larger, ~ planar interface for preferential machining flaw formation (or microcrack or combined micro- and machining–crack formation).

A simple test of whether machining flaws are significant in controlling strength is to compare strengths for the same set of composite samples with either different machining abrasive grits or the machining direction parallel versus perpendicular to the stress axis in testing. Thus, as noted in Chapter 3, machining typically leaves two sets of flaws, one ~ halfpenny shaped normal to the abrasive motion, hence machining direction, and one of more elongated flaws parallel with the abrasive motion [140]. Thus if machining flaws are controlling strengths, machining test bars parallel to the bar and stress axis thus causes failure from the former flaws, and machining test bars perpendicular to the bar and stress axis causes failure from the second set of flaws. Since both flaw populations are of ~ the same depth [which is a function of the local material toughness, hardness, and elastic moduli, e.g. per Eq. (3.2)], there is strength anisotropy due to elongated flaws controlling failure for machining perpendicular versus machining parallel to the bar axes, the former having lower strengths than the latter. Thus the differences, or the ratios, of the average strengths for parallel and perpendicular machining show whether there is measurable strength anisotropy, and such anisotropy in turn is a clear indicator that machining flaws are controlling strength. In other words, if other sources of failure such as processing defects or microcracks dominate failure, then little or no anisotropy of strength will be found. Similarly, if crack wake toughening mechanisms are significant, so that the effects of the original flaws are limited or zero, there will be limited or no strength anisotropy as a function of machining direction. For reference, the ratios of strengths for typical perpendicular versus parallel diamond grinding of ceramics are ~ 50–60% for grain sizes of ~ 1 μm and increase to ~ 100% (i.e. no anisotropy) for G ~ 50 μm due to the flaw and grain sizes being equal (Fig. 3.1, and decrease again at larger G) [140]. Porosity has little or no effect on such anisotropy unless there is substantial heterogeneity of large pores that control strength.

Rice [140] has measured strength anisotropies of a number of commercial and experimental ceramic composites showing that many have clear machining direction–strength anisotropy. Thus for example commercial crystallized cordierite (Pyroceram 9606), Al_2O_3-20 v/o TiC, and Al_2O_3-7 or 25 w/o SiC whiskers and grain sizes of a few microns had ratios of 70–85%, limited impacts of processing defects on strengths. Another 8 of 15 composites tested fell in the range of 86–96%, and only three did not show the indicated anisotropy. In both cases, especially in the latter group, either processing defects or microcracking (or both) were important sources of failure, thus limiting strength anisotropy as a function of machining direction. Thus, for example, three composites of aluminum- and zirconium-titanate known to have substantial microcracking (e.g. as reflected in strengths of 65–100 MPa) showed anisotropy ratios of 89–96%, i.e. some limited anisotropy despite microcracks and heterogeneous porosity. Such machining evidence, especially on homogeneous composites, shows that machining plays an important role in the strengths of many ceramic composites. Thus the effect of composite character on machining flaws, especially on their depths, needs much more attention as a mechanism for their improved strengths.

Thus an essential perspective for evaluating mechanisms controlling strengths of composites is to recognize that there are a variety of mechanisms that can be operative, and that while in some cases one mechanism may be a major factor, other mechanisms may also impact behavior. Further, there can be shifts in the mechanisms impacting behavior as composite parameters of composition and processing–microstructure change. Thus, for example, increasing the size, quantity, or both of the dispersed phase increases the opportunity and extent of spontaneous, i.e. preexisting, microcracking, e.g. with SiC platelet composites [105]. However, this should also be taken as a sign of possible stress-induced microcracking (which is far too often neglected) and that impacts of other mechanisms should be considered, including possible shifts in them precipitated by changes in microcracking.

Another factor consistent with the above flaw–particle size mechanism is the impact of matrix grain size in the same fashion as for monolithic ceramics, i.e. increasing the local hardness and probably the local toughness controlling machining flaw formation. Thus inhibition of matrix grain growth aids tensile strength, and mutual inhibition of grain and particle growth aids even more. However, the range of matrix grain sizes can vary widely for different types and amounts of additives and processing, e.g. as shown for TiB_2-based bodies by Telle and Petzow [141]. Again, the consistency of data for TiB_2 with different additives (Fig. 3.25) indicates consistency of composite and monolithic strength–microstructure dependence.

B. Probable Mechanisms Controlling Strengths of Specific Composite Types

As key illustrations of the above varying impacts of multiple mechanisms, consider first zirconia toughened materials. While transformation toughening is a dominant factor, this must often compensate for strength decreases due to reductions of E due to ZrO_2 having a lower E than matrices such as Al_2O_3. However, such reductions are often counteracted in part by reductions in matrix grain size as well as of machining flaws, effects of surface compressive stresses from machining, or both. However, there are other mechanisms such as microcracking, e.g. as shown by the work of Claussen and colleagues [31,142] and Homeny and Nick [44], but often at strengths much below those implied by the toughness. Some other mechanism is also indicated by the unusual bridging of cracks by fine fracture or other filaments observed in eutectic Al_2O_3-ZrO_2 bodies of Nick and Homeny. Thus models of transformation toughening are a useful guide but are uncertain in the degree of their quantitative predictions not only because of their idealization (especially assuming dilute, noninteracting transforming particles [143]) but also because of other varied contributions of uncertain quantification due to lack of accurate models and detailed characterization. While the finer microstructural scale of zirconia toughened composites often allows reasonable correspondence between toughness and strength behavior, this is not always so not only due to microstructural heterogeneities as indicated by comparison of Claussen's and Becher's original studies (Figs. 9.5, 9.6) but also due to flaw effects, since these are typically not reflected by most toughness tests.

Metal–ceramic composite toughness is often significantly impacted by ductile elongation of metal particles in the crack wake, but this again raises issues of the extent to which this can apply on the typically much finer scale of normal strength controlling flaws. Higher expansions of some metals relative to common ceramic matrices combined with metal particle size and the extent of metal–ceramic bonding can result in the metal addition often acting more as pores. On the other hand, when this does not occur, stresses may be generated by the particles that influence crack propagation directly, or by microcrack generation in response to a crack stress field or by spontaneous formation, each with some difference in effects on toughness and on strength. As with other composites, there may also be effects on matrix grain size as well as of flaw sizes and local stresses affecting them via effects on machining.

Ceramic particulate composites have much of the possible variations in mechanisms in ceramic–metal composites above, except for the general absence of the ductile toughening. Thus effects on matrix grain size have been noted, and stresses from mismatch in expansions (or altered by elastic property differences) between the dispersed phase and the matrix are clearly a factor. The highest strengths are generally obtained when such stresses are low or zero (see also

Nadeau and Dickson's [144] summary), and generally decrease as they increase, with limited differentiation between stresses from a dispersed phase expansion lower than the matrix versus the reverse, though the latter may be more limiting. With low expansion differences, the mean free separation between particles can be a factor as indicated by work of especially Fulrath and colleagues [1,122] as well as others [123], at least in glass matrix composites. However, as expansion differences increase, this machining flaw effect (which may be similar but not identical to that noted for other composites above) is overcome, probably due to the resulting stresses (and associated strength reductions).

Nanocomposites clearly show at the least a shift in the relative roles of mechanisms, e.g. greater effects of the dispersed particles inhibiting *G* and quite possibly increasing effects on machining flaws. Whether there are unique effects due to the frequent incorporation of nanoparticles within matrix grains, or whether such effects are manifested in machining flaw effects, is unknown.

Pullout has generally been identified as a primary mechanism of toughening in whisker composites based on analogies with fiber composites and exposure of apparently pulled out whiskers on fracture surfaces (e.g. Fig. 8.14), with some modeling along these lines. However, it is uncertain where "intergranular" fracture between whiskers and matrix ends and true pullout begins. Modeling of Campbell et al. [145] indicates that bending failure of random whiskers obviates pullout effects and is an important factor in the limited toughening achieved in whisker composites. The limited and negative effects of whisker coatings to enhance pullout are consistent with this. However, other effects must also be considered in whisker composites, i.e. impacts on *E*, matrix *G* and machining effect and microcracking, spontaneous or stress generated, as well as issues of crack scale to microstructure.

Similar effects and issues are seen for platelet composites, where the equivalence of intergranular fracture around much of many platelets, quite possibly due to interfacial machining flaws, is seen as a more probable cause of platelet exposure on fracture surfaces than any pullout-type effects. Greater sensitivity to spontaneous fracture with larger SiC platelets in Al_2O_3 [146] is consistent with the larger dimensions of platelets relative to whisker and most particle composites and again emphasizes the role of property differences (including probable effects of *E* differences) and is consistent with the generally poorer strength of such composites. Similarly higher strengths with stronger platelet–matrix bonding argues against favorable effects of pullout mechanisms on this scale.

C. Improved Evaluation of Strengths and Related Mechanical Behavior of Ceramic Composites

Briefly consider first the use of interparticle spacing (λ) or particle size (*D*) as parameters in strength and toughness. In composites with glass or single crystal (as in directionally solidified eutectics) matrices that have no equivalence of grain structure themselves, λ defines the same basic physical dimension as G does in a dense monolithic ceramic, i.e. G defines the average separation of nearest neighbor nonadajecent grains, just as λ defines the average separation of particles or eutectic lamellae or rods. Thus the correlation of strength with λ in the latter cases is analogous to the correlation of strength with G in the large G region for monolithic ceramics, which lends support to its use in such cases. However, broader use in more complex structures such as those in crystallized glasses is uncertain, as is similar use of D, since both are related to each other, and both in turn to ϕ, with varying correlations with basic physical properties such as E, as was discussed earlier for results of Hing and McMillan [16].

Turning now to evaluation, the overall need is for broader consideration of possible contributing mechanisms, e.g. as was addressed above, and broader microstructural and especially property and behavior measurement, with much more attention to self-consistency. Thus, at the minimum, besides flexure strength and toughness measurements, Young's modulus should be determined as a function of composite parameters, with preferably both the initial modulus being measured, e.g. by flexural resonance or ultrasonic techniques, as well as of possible changes in E as a function of loading to mechanical failure to detect stress-induced microcracking. Acoustic emission, internal friction, and damping can also be useful in this regard. Different toughness tests should be compared, especially on the basis of crack sizes to critical microstructural scales, e.g. grain and particle sizes as well as particle separations. Hardness testing, especially over a range of loads and with careful examination of indent cracking and compressive strength testing, again possibly accompanied by acoustic emission, can both indicate microcracking or debonding of second phase particles. Flexural strength testing for stressing parallel and then perpendicular to the machining direction can be a valuable tool for probing whether dispersed particles are constraining flaw sizes in the spacing between particles, since in such cases the anisotropy of strength with machining direction should disappear, e.g. as shown as a function of G in monolithic ceramics (Fig. 3.33). Testing with different strain rates, with biaxial loading, and with artificially introduced flaws can also be of use. Often of greater benefit is testing at higher and lower temperatures, for E as well as toughness and strength. Testing at lower temperatures reduces possible environmental effects on microcracking and crack growth and increases expansion mismatch stresses, while testing at higher temperatures reduces the latter stress and environmental effects. Other tests may also be valuable, e.g. electrical and thermal conductivities of the bulk or especially of the tensile surface may be good indicators of microcracking or debonding.

Testing of a broader range of composite microstructures and better characterization of them is also a critical need. Thus tests of a sufficient range of D values to determine the impact of this on strength and related properties is important. Further, since there can be various correlations between microstructural parameters and properties without clear proof of causation, two steps are important. The first is evaluating properties against different microstructural parameters, i.e. G, D, λ, and ϕ to see if there are any unique correlations, and checking all correlations for self-consistency among various properties and possible mechanisms. The second, and especially important, but often neglected step is substantial fracture surface examination, especially, but not exclusively, seeking fracture origins. General examination typically gives a much more accurate picture of the microstructural extremes that may play a role in, or dominate, the composite behavior, and may indicate the relevance of some mechanisms, e.g. bridging is favored more by inter- rather than transgranular fracture. Crack interactions with dispersed particles may also be shown on fracture surfaces (especially but not exclusively) with glass or single crystal matrices (e.g. the latter in eutectic composites). However, identification of fracture origins, and especially quantitative analysis of them, e.g. of flaw sizes and implied fracture toughnesses, can be immensely valuable.

VI. SUMMARY AND CONCLUSIONS

While strengths often increase with the volume fraction second phase, in which case they typically go through a maximum, as toughness generally does, there are a number of cases where strength decreases, i.e. shows opposite dependence from typical toughness tests. Further, many composites show opposite strength dependence on key microstructural parameters such as particle size from that of toughness, which reinforce and more clearly show basic toughness and strength differences attributed to differences in crack size effects.

This review of composite strength behavior reinforces previous evidence for composite composition effects on machining flaw sizes. It was further shown that composites often exhibit the same two-branch strength–$D^{-1/2}$ dependence as the strength–$G^{-1/2}$ dependence for monolithic ceramics as a result of the same underlying impact of composition and microstructure on machining flaw sizes controlling strength. Thus there is a finer particle branch where strengths show variable but limited decreases in strength with increasing particle size due to the flaw sizes being $> D$, but being affected some by the impact on the particles on the local Young's modulus, and especially hardness and toughness controlling flaw formation. On the other hand, there is a larger particle branch where the sizes of the particles (or clusters of them) are about that of the flaw size or greater so that the particle size becomes the flaw size, e.g. due to associated machining flaws, with resultant greater strength dependence on D.

The above flaw mechanisms, given attention because of their lack of recognition, are however only part of the picture, since there are commonly other mechanisms that are also operative to varying degrees, but which vary for different types of composites. Thus transforming toughening clearly commonly, but not necessarily always, plays an important role in zirconia toughened composites, as does ductile elongation of metal particles in ceramic–metal composites, while pullout is often indicated in whisker composites but is more uncertain. In particulate composites, crack deflection, bridging, and branching may play some role, but effects of stress-induced or spontaneous microcracking and effects on flaw sizes are important, often probably more so in such composites, and can also be factors in the above composites. Effects on flaw sizes may occur via dispersed particles constraining flaw sizes (mainly when there are limited particle–matrix mismatch stresses), or much more generally via effects on machining flaws from *E*, *H*, and *K*, or via residual surface compressive stresses. Another important factor is matrix grain size, which is found to be much more pervasive than originally generally thought and can impact all types of composites with polycrystalline matrices and is probably important in nanocomposites.

Finally, a number of recommendations have been made to improve documentation and understanding of the mechanical behavior of ceramic composites. These include a broader range of the type of tests, e.g. more strength, toughness, and Young's modulus testing, complemented by other evaluations, e.g. for acoustic emission and internal friction or damping, as well as over a broader range of temperatures. Broader evaluation of microstructural parameters and evaluation of the consistency of various correlations is also recommended, since correlation does not necessarily mean causation, and microstructural parameters are interrelated. A key factor that needs much more use and correlation with both microstructural aspects of failure and evaluation of mechanisms is fractography, both of the general fracture surfaces and for fracture origin determination and evaluation.

REFERENCES

1. D. P. H. Hasselman and R. M. Fulrath. Proposed Fracture Theory of a Dispersion-Strengthened Glass Matrix. J. Am. Cer. Soc. 49(2):68–72, 1966.
2. R. W. Rice. The Effect of Grinding Direction on the Strength of Ceramics. The Science of Ceramic Machining and Surface Finishing (S. J. Schneider and R. W. Rice, eds.). NBS Special Pub. 348, US Govt. Printing Office, Washington, DC, 1972, pp. 365–376.
3. J. J. Mecholsky, Jr., S. W. Freiman, and R. W. Rice. Effect of Grinding on Flaw Geometry and Fracture of Glass. J. Am. Cer. Soc. 60(3–4):114–117, 1977.
4. D. B. Binns. Some Properties of Two-Phase Crystal-Glass Solids. Science of Ceramics 1 (G. H. Stewart, ed.). Academic Press, New York, 1962, pp. 315–334.
5. R. W. Davidge and T. J. Green. The Strength of Two-Phase Ceramic/Glass Materials. J. Mat. Sci. 3:629–634, 1968.

6. A. G. Evans. The Role of Inclusions in the Fracture of Ceramic Materials. J. Mat. Sci. 9:1145–1152, 1974.
7. W. J. Frye and J. D. Mackenzie. Mechanical Properties of Selected Glass-Crystal Composites. J. Mat. Sci. 2:124–130, 1967.
8. F. F. Lange. Fracture Energy and Strength Behavior of a Sodium Borosilicate Glass-Al_2O_3 Composite System. J. Am. Cer. Soc. 54(12):614–620, 1971.
9. N. Miyata, S. Ichikawa, H. Monji, and H. Jinno. Fracture Behavior of Brittle Matrix, Particulate Composites with Thermal Expansion Mismatch. Fracture Mechanics of Ceramics 7, Composites. Impact, Statistics, and High-Temperature Phenomena (R. C. Bradt, A. G. Evans, D. P. H. Hasselman, and F. F. Lange, eds.). Plenum Press, New York, 1986, pp. 87–102.
10. W. D. Wolf, K. J. Vaidya, and L. F. Francis. Mechanical Properties and Failure Analysis of Alumina-Glass Dental Composites. J. Am. Cer. Soc. 79(7):1769–1776, 1996.
11. M. P. Borom. Dispersion-Strengthened Glass Matrices-Glass-Ceramics, A Case in Point. J. Am. Cer. Soc. 60(1–2):17–21, 1977.
12. N. Miyata and H. Jinno. Theoretical Approach to the Fracture of Two-Phase Glass-Crystal Composites. J. Mat. Sci. 7:973–982, 1972.
13. S. W. Freiman and L. L. Hench. Effect of Crystallization on the Mechanical Properties of Li_2O-SiO_2 Glass-Ceramics. J. Am. Cer. Soc. 55(2):26–90, 1972.
14. D. I. H. Atkinson and P. W. McMillan. Glass-Ceramics with Random and Oriented Microstructures, Part 2. The Physical Properties of a Randomly Oriented Glass-Ceramic. J. Mat. Sci. 11:994–1002, 1978.
15. S. W. Freiman and R. W. Rice. Comment on "Glass-Ceramics with Random and Oriented Microstructures, Part 2. The Physical Properties of a Randomly Oriented Glass-Ceramic." J. Mat. Sci. 13:1130–1134, 1978.
16. P. Hing and P. W. McMillan. The Strength and Fracture Properties of Glass-Ceramics. J. Mat. Sci. 8:1041–1048, 1973.
17. K. Hamano and E. S. Lee. Studies on the Mechanical Properties of Porcelain Bodies. Bull. Tokyo Inst. Tech. 108:95–111, 1972.
18. J. S. Banda and P. F. Messer. The Effect of Quartz Particles on the Strength and Toughness of Whitewares 17. Ceramic Transactions, Fractography of Glasses and Ceramics II (V. D. Fréchette and J. R. Varner, eds.). Am Cer. Soc., Westerville, OH, 1991, pp. 243–261.
19. R. W. Rice. Strength/Grain Size Effects in Ceramics. Proc. Brit. Cer. Soc. 6:119–136, 1966.
20. A. Dinsdale and W. T. Wilkinson. Strength of Whiteware Bodies. Proc. Brit. Cer. Soc. 20:205–257, 1972.
21. Y. Utsumi and S. Sakka. Strength of Glass-Ceramics Relative to Crystal Size. J. Am. Cer. Soc. 53(5):286–287, 1970.
22. A. J. Stryjak and P. W. McMillan. Microstructure and Properties of Transparent Glass-Ceramics. J. Mat. Sci. 13:1794–1804, 1978.
23. N. Claussen. Fracture Toughness of Al_2O_3 with an Unstabilized ZrO_2 Dispersed Phase. Am. Cer. Soc. Bull. 59(1–2):49–54, 1976.
24. R. W. Rice. Processing of Ceramic Composites. Advanced Ceramic Processing

and Technology 1 (J. G. P. Binner, ed.). Noyes, Park Ridge, NJ, 1990, pp. 123–213.
25. P. F. Becher. Transient Thermal Stress Behavior in ZrO_2-Toughened Al_2O_3. J. Am. Cer. Soc. 64(1):37–39, 1978.
26. F. F. Lange. Transformation Toughening, Part 4. Fabrication, Fracture Toughness and Strength of Al_2O_3-ZrO_2 Composites. J. Mat. Sci. 17:247–254, 1982.
27. F. F. Lange and M. M. Hirlinger. Hindrance of Grain Growth in Al_2O_3 by ZrO_2 Inclusions. J. Am. Cer. Soc. 67(3):164–168, 1984.
28. S. Hori, R. Kurita, M. Yoshimura, and S. Somiya. Suppressed Grain Growth in Final-Stage Sintering of Al_2O_3 with Dispersed ZrO_2 Particles. J. Mat. Sci. Lett. 4:1067–1070, 1985.
29. S. Hori, R. Kurita, M. Yoshimura, and S. Somiya. Influence of Small ZrO_2 Additions on the Microstructure and Mechanical Properties of Al_2O_3. Science and Technology of Zirconia III. Adv. in Ceramics 24 (S. Sõmiya, N. Yamamoto, and H. Hanagida, eds.). Am. Cer. Soc., Westerville, OH, 1988, pp. 423–429.
30. A. Nisida, S. Fukuda, Y. Kohtoku, and K. Terai. Grain Size Effect on Mechanical Strength of MgO-ZrO_2 Composite Ceramics. J. Jpn. Cer. Soc., Intl. Ed. 100:203–207, 1992.
31. R. W. Rice. Fractographic Determination of K_{IC} and Effects of Microstructural Stresses in Ceramics. Fractography of Glasses and Ceramics, Ceramic Trans. 17 (J. R. Varner and V. D. Frechette, eds.). Am. Cer. Soc., Westerville, OH, 1991, pp. 509–545.
32. R. W. Rice. Toughening in Ceramic Particulate and Whisker Composites. Cer. Eng. Sci. Proc. 11(7–8):667–694, 1990.
33. E. H. Lutz, M. V. Swain, and N. Claussen. Thermal Shock Behavior of Duplex Ceramics. J. Am. Cer. Soc. 74(1):19–24, 1991.
34. J. Wang and R. Stevens. Toughening Mechanisms in Duplex Alumina-Zirconia Ceramics. J. Mat. Sci. 23:804–808, 1988.
35. R. Langlois and K. J. Konsztowicz. Toughening in Zirconia-Toughened Alumina Composites with Non-Transforming Zirconia. J. Mat. Sci. Lett. 11:1454–1456, 1992.
36. K. Tsukuma, K. Ueda, and M. Shimada. Strength and Fracture Toughness of Isotactically Hot-Pressed Composites of Al_2O_3 and Y_2O_3-Partially-Stabilized ZrO_2. J. Am. Cer. Soc. 68(1):C–4–5, 1985.
37. S. Hori, M. Yoshimura, and S. Sõmiya. Strength-Toughness Relations in Sintered and Isotactically Hot-Pressed ZrO_2-Toughened Al_2O_3. J. Am. Cer. Soc. 69(30):169–172, 1986.
38. H. Tomaszewski. Effect of Sintering Atmosphere on Thermomechanical Properties of Al_2O_3-ZrO_2 Ceramics. Cer. Intl. 15:141–146, 1989.
39. A. K. Kurakose and L. J. Beaudin. Tetragonal Zirconia in Chill-Cast Alumina-Zirconia. J. Canadian Cer. Soc. 46:45–50, 1977.
40. R. Ruh and H. J. Garrett. Nonstoichiometry of ZrO_2 and Its Relation to Tetragonal-Cubic Inversion in ZrO_2. J. Am. Cer. Soc. 50(5):257–261, 1967.
41. R. J. Ackermann, S. P. Garg, and E. G. Rauh. The Lower Phase Boundary of ZrO_{2-x}. J. Am. Cer. Soc. 61(5–6):275–276, 1978.
42. R. W. Rice, B. A. Bender, R. P. Ingel, T. W. Coyle, and J. R. Spann. Tougher Ce-

ramics Using Tetragonal ZrO_2 of HfO_2. Ultrastructure Processing of Ceramics, Glasses, and Composites (L. L. Hench and D. R. Ulrich, eds.). Wiley-Interscience, New York, 1984, pp. 507–523.

43. U. Krohn, H. Olapinski, and U. Dworak. US patent 4,595,663, 6/17/1986.
44. J. Homeny and J. J. Nick. Microstructure-Property Relations of Alumina-Zirconia Eutectic Ceramics. Mat. Sci. Eng. A127:123–133, 1990.
45. Q.-M. Yuan, J.-Q. Tan, and Z.-G. Jin. Preparation and Properties of Zirconia-Toughened Mullite Ceramics. J. Am. Cer. Soc. 69(3):265–267, 1986.
46. S. Lathabai, D. G. Hay, F. Wagner, and N. Claussen. Reaction-Bonded Mullite/Zirconia Composites. J. Am. Cer. Soc. 79(1):248–256, 1996.
47. M. Ishitsuka, T. Sato, T. Endo, and M. Shimada. Sintering and Mechanical Properties of Yttria-Doped Tetragonal ZrO_2 Polycrystal/Mullite Composites. J. Am. Cer. Soc. 70(11):C–342–346, 1987.
48. T. Ono, K. Nagata, M. Hashiba, E. Miura, and Y. Nurishi. Internal Friction, Crack Length of Fracture Origin and Fracture Surface Energy in Alumina-Zirconia Composites. J. Mat. Sci. 24:1974–1978, 1989.
49. K. J. Konsztowicz and R. Langlois. On the Variability of Strength to Toughness Ratio in Zirconia-Alumina Composites. J. Mat. Sci. Lett. 15:1093–1096, 1996.
50. R. W. Rice. Microstructural Dependence of Fracture Energy and Toughness of Ceramics and Ceramic Composites Versus That of Their Tensile Strengths at 22°C. J. Mat. Sci. 31:4503–4519, 1996.
51. T. Tiegs and K. J. Bowman. Fabrication of Particulate-, Platelet-, and Whisker-Reinforced Ceramic Matrix Composites. Handbook on Discontinuously Reinforced Ceramic Matrix Composites (K. J. Bowman, S. K. El-Rahaiby, and J. B. Wachtman, Jr., eds.). Am. Cer. Soc., Westerville, OH, 1995, Ch. 4, pp. 91–138.
52. M. Yasuoka, M. E. Brito, K. Hirao, and S. Kanzaki. Effect of Dispersed Particle Size on Mechanical Properties of Alumina/Non-Oxide Composites. J. Cer. Soc. Jpn., Intl. Ed. 101:865–870, 1993.
53. I. Thompson and V. D. Kristic. Mechanical Properties of Hot-Pressed SiC Particulate-Reinforced Al_2O_3 Matrix. J. Mat. Sci. 27:5765–5768, 1992.
54. A. Nakahira, K. Niihara, and T. Hirai. Microstructural Properties of Al_2O_3-SiC Composites. Yogyo-Kyokai-Shi 94(8):767–772, 1986.
55. I.-S. Kim and I.-G. Kim. Thermal Shock Behavior of Al_2O_3-SiC Composites Made by Directed Melt Oxidation of Al-Alloy. J. Mat. Sci. Lett. 16:772–775, 1997.
56. K. Niihara. New Design Concept of Structural Ceramics—Ceramic Nanocomposites. J. Cer. Soc. Jpn. 99:974–982, 1991.
57. D. Zhang, H. Yang, R. Yu, and W. Weng. Mechanical Properties of Al_2O_3-SiC Composites Containing Various Sizes and Fractions of Fine Particles. J. Mat. Sci. 16:877–879, 1997.
58. J. Zhao, L. C. Stearns, M. P. Harmer, H. M. Chen, and G. A. Miller. Mechanical Behavior of Alumina-Silicon Carbide "Nanocomposites." J. Am. Cer. Soc. 76(2):503–510, 1993.
59. M. Sternitzke, B. Derby, and R. J. Brook. Alumina-Silicon Carbide Nanocomposites by Hybrid Polymer/Powder Processing: Microstructures and Mechanical Properties. J. Am. Cer. Soc. 81(1):41–48, 1998.

60. I. A. Chou, H. M. Chen, and M. P. Harmer. Machining-Induced Surface Residual Stress Behavior in Al_2O_3-SiC Nanocomposites. J. Am. Cer. Soc. 79(9):2403–2409, 1996.
61. A. M. Thompson, H. M. Chen, and M. P. Harmer. Crack Healing and Stress Relaxation in Al_2O_3-SiC "Nanocomposites." J. Am. Cer. Soc. 78(3):567–571, 1995.
62. R. P. Wahi and B. Ilschner. Fracture Behavior of Composites Based on Al_2O_3-TiC. J. Mat. Sci. 15:875–885, 1980.
63. H. J. Hwang and K. Niihara. Perovskite-Type $BaTiO_3$ Ceramics Containing Particulate SiC. J. Mat. Sci. 33:549–558, 1998.
64. P. C. Strange and A. H. Bryant. The Role of Aggregate in the Fracture of Concrete. J. Mat. Sci. 14:1863–1868, 1979.
65. F. F. Lange. Effect of Microstructure on Strength of Si_3N_4-SiC Composite System. J. Am. Cer. Soc. 55(9):445–450, 1973.
66. H. Tanaka, P. Greil, and G. Petzow. Sintering and Strength of Silicon Nitride-Silicon Carbide Composite. Intl. J. High Temp. Cer. 1:107–118, 1985.
67. H. Nakamura, S. Umebayashi, and K. Kishi. Mechanical Properties of Hot Pressed Si_3N_4-SiC Composites. J. Cer. Soc. Jpn., Intl. Ed. 97:1526–1529, 1989.
68. Y. Akimune, T. Ogasawara, and N. Hirosaki. Influence of Starting Powder Characteristics on Mechanical Properties of SiC-Particle/Si_3N_4 Composites. J. Cer. Soc. Jpn., Intl. Ed. 100:468–472, 1992.
69. G. Pezzotti and T. Nishida. Effect of Size and Morphology of Particulate SiC Dispersions on Fracture Behavior of Si_3N_4 Without Sintering Aids. J. Mat. Sci. 29:1765–1772, 1994.
70. A. Sawaguchi, K. Toda, and K. Niihara. Mechanical and Electrical Properties of Silicon Nitride–Silicon Carbide Nanocomposite Material. J. Am. Cer. Soc. 74(5):1142–1144, 1991.
71. G. Sasaki, H. Nakase, K. Suganuma, T. Fujita, and K. Niihara. Mechanical Properties and Microstructure of Si_3N_4 Matrix Composite with Nanometer Scale SiC Particles. J. Cer. Soc. Jpn., Intl. Ed. 100:536-540, 1992.
72. L. Tian, Y. Zhou, and W.-L. Zhou. SiC-Nanoparticle-Reinforced Si_3N_4 Matrix Composites. J. Mat. Sci. 33:797–802, 1998.
73. J. J. Petrovic, M. I. Pena, I. E. Remanis, M. S. Sandlin, S. D. Conzone, H. H. Kung, and D. P. Butt. Mechanical Behavior of $MoSi_2$ Reinforced-Si_3N_4 Matrix Composites. J. Mat. Sci. 80(12):3070–3076, 1997.
74. T. Mah, M. G. Mendiratta, and H. A. Lipsitt. Fracture Toughness and Strength of Si_3N_4-TiC Composites. Am. Cer. Soc. Bull. 60(11):1229–1240, 1981.
75. J. D. Yoon and S. G. Kang. Strengthening and Toughening Behavior of SiC with Additions of TiB_2. J. Mat. Sci. Lett. 14:1065–1067, 1995.
76. K.-S. Cho, Y.-W. Kim, H.-J. Choi, and J.-G. Lee. SiC-TiC and SiC-TiB_2 Composites Densified by Liquid-Phase Sintering. J. Mat. Sci. 31:6223–6228, 1996.
77. C. H. McMurtry, W. D. G. Boecker, S. G. Seshadri, J. S. Zangil, and J. E. Garnier. Microstructure and Material Properties of SiC-TiB_2 Particulate Composites. Am. Cer. Soc. Bull. 66(2):325–329, 1987.
78. D. J. Magley, R. A. Winholz, and K. T. Faber. Residual Stresses in a Two-Phase Microcracking Ceramic. J. Am. Cer. Soc. 73(6):1641–1644, 1990.

79. H. Endo, M. Ueki, and H. Kubo. Microstructure and Mechanical Properties of Hot Pressed SiC-TiC Composites. J. Mat. Sci. 26:3769–3774, 1991.
80. B.-W. Lin and T. Iseki. Effect of Thermal Residual Stress on Mechanical Properties of SiC/TiC Composites. Brit. Cer. Trans. 91:1–5, 1992.
81. J.-F. Li and R. Watanabe. Preparation and Mechanical Properties of SiC-AlN Ceramic Alloys. J. Mat. Sci. 26:4813–4817, 1991.
82. Y. K. Baek and C. H. Kim. The Effect of Whisker Length on the Mechanical Properties of Alumina-SiC Whisker Composites. J. Mat. Sci. 24:1589–1593, 1989.
83. E. Yasuda, T. Akatsu, and Y. Tanabe. Influence of Whiskers' Shape and Size on Mechanical Properties of SiC Whisker-Reinforced Al_2O3. J. Cer. Soc. Jpn., Intl. Ed. 99:51–57, 198 .
84. A. V. Krylov, S. M. Barinov, D. A. Ivanov, N. A. Mindlina, L. Parilak, J. Dusza, F. Lofaj, and E. Rudnayova. Influence of SiC Whisker Size on Mechanical Properties of Reinforced Alumina. J. Mat. Sci. Lett. 12:904–906, 1993.
85. T. Tiegs. Structural and Physical Properties of Ceramic Matrix Composites. Handbook on Discontinuously Reinforced Ceramic Matrix Composites (K. J. Bowman, S. K. El-Rahaiby, and J. B. Wachtman, Jr., eds.). Am. Cer. Soc., Westerville, OH, 1995, Ch. 6, pp. 226–273.
86. T. E. Steyer and K. T. Faber. Fracture Behavior of SiC Whisker-Reinforced Al_2O3 with Modified Surfaces. Cer. Eng. Sci. Proc. 13(9–10):669–677, 1992.
87. M. Yang and R. Stevens. Microstructure and Properties of SiC Whisker Reinforced Ceramic Composites. J. Mat. Sci. 26:726–736, 1991.
88. N. Sato and T. Kurauchi. Microcracking During a Thermal Cycle in Whisker-Reinforced Ceramic Composites Detected by Acoustic Emission Measurements. J. Mat. Sci. Lett. 11:590–591, 1992.
89. K. Ikeda and T. Kishi. Matrix Grain Size Effect and Fracture Behavior on Bending Strength and Fracture Toughness in Multi-Toughened Al_2O_3. Cer. Eng. Sci. Proc. 13(7–8):164–171, 1992.
90. P. F. Becher, C.-H. Hsueh, P. Angelini, and T. N. Tiegs. Toughening Behavior in Whisker Reinforced Ceramic Matrix Composites. J. Am. Cer. Soc. 71(12):1050–1061, 1988.
91. T. N. Tiegs and P. F. Becher. Alumina-SiC Whisker Composites. Cer. Eng. Sci. Proc. 7(9–10):1182–1186, 1986.
92. I. Wadsworth and R. Stevens. The Influence of Whisker Dimensions on the Mechanical Properties of Cordierite/SiC Composites. J. Eur. Cer. Soc. 9:153–163, 1992.
93. Z. Zhang, Y. Huang, L. Zheng, and Z. Jiang. Preparation and Mechanical Properties of SiC Whisker and Nanosized Mullite Particulate Reinforced TZP Composites. J. Am. Cer. Soc. 79(10):2779–2782, 1996.
94. L. J. Neergard and J. Homeny. Mechanical Properties of Beta-Silicon Nitride Whisker/Silicon Nitride Matrix Composite. Cer. Eng. Sci. Proc. 10(9–10): 1049–1062, 1989.
95. C. Olagnon, E. Bullock, and G. Fantozzi. Properties of Sintered SiC Whisker-Reinforced Si_3N_4 Composites. J. Eur. Cer. Soc. 7:265–273, 1991.

96. F. Rossignol, P. Goursat, J. L. Besson, and P. Lespade. Microstructure and Mechanical Behavior of Self-Reinforced Si_3N_4 and Si_3N_4 -SiC Whisker Composites. J. Eur. Cer. Soc. 13:299–312, 1994.
97. T. Fusakawa and Y. Goto. Mechanical Properties of Si_3N_4 Ceramic Reinforced with SiC Whiskers and Platelets. J. Mat. Sci. 33:1647–1651, 1998.
98. A. Kamiya, K. Nakano, and H. Okuda. Fabrication and Properties of SiC Whisker Reinforced NbC Composites. J. Jpn. Cer. Soc., Intl. Ed. 97:934–940, 1989.
99. N. Tamari, H. Kobayashi, T. Tanaka, I. Kondoh, and S. Kose. Mechanical Properties of B_4C-SiC Whisker Composite Ceramics. J. Jpn. Cer. Soc., Intl. Ed. 98:1169–1173, 1990.
100. A. Kamiya and K. Nakano. Mechanical Properties of SiC Whisker-Reinforced TiB_2 Composites Fabricated by Hot-Pressing. J. Jpn. Cer. Soc., Intl. Ed. 101:598–601, 1992.
101. A. Kamiya, K. Nakano, and A. Kondoh. Fabrication and Properties of Hot-Pressed SiC Whisker-Reinforced TiB_2 and TiC Composites. J. Mat. Sci. Lett. 8:566–568, 1989.
102. A. Kamiya, K. Nakano, and H. Okuda. Fabrication and Properties of TiC/SiC Whisker Composites. J. Jpn. Cer. Soc., Intl. Ed. 98:1156–1162, 1990.
103. J. Dusza, D. Sajgalik, and M. Reece. Analysis of Si_3N_4 + β-Si_3N_4 Whisker Ceramics. J. Mat. Sci. 26:6782–6788, 1991.
104. J. Dusza and D. Sajgalik. Mechanical Properties of Si_3N_4 + β-Si_3N_4 Whisker Reinforced Ceramics. J. Eur. Cer. Soc. 9:9–17, 1992.
105. Y.-S. Chou and D. J. Green. Silicon Carbide Platelet/Alumina Composites: II, Mechanical properties. J. Am. Cer. Soc. 76(6):1452–1458, 1995.
106. T. Mitchel, Jr., L. C. DeJonghe, W. J. MoberlyChan, and R. O. Ritchie. Silicon Carbide Platelet/Silicon Carbide Composites. J. Am. Cer. Soc. 78(1):97–103, 1995.
107. C. Naschik, M. M. Seibold, N. A. Travitzky, and N. Claussen. Effect of Processing on Mechanical Properties of Platelet-Reinforced Mullite Composites. J. Am. Cer. Soc. 74(10):2464–2468, 1991.
108. P.-L. Chen and I-W. Chen. In-Situ Alumina/Aluminate Platelet Composites. J. Am. Cer. Soc. 75(9):2610–2612, 1992.
109. M. Yasuoka, K. Hirao, M. E. Brito, and S. Kanzaki. High-Strength and High-Fracture-Toughness Ceramics in the Al_2O_3/$LaAl_{11}O_5$ Systems. J. Am. Cer. Soc. 78(7):1853–1856, 1995.
110. H.-D. Kim, I.-S. Lee, S.-W. Kang, and J.-W. Ko. The Formation of $NaMg_2Al_{15}O_{25}$ in an α-Al_2O_3 Matrix and Its Effect on the Mechanical Properties of Alumina. J. Mat. Sci. 29:4119–4124, 1994.
111. T. Koyama, A. Nishiyama, and K. Niihara. Effect of Grain Morphology and Grain Size on the Mechanical Properties of Al_2O_3 Ceramics. J. Mat. Sci. 29:3949–3954, 1994.
112. L. An and H. M. Chan. R-Curve Behavior of In-Situ Toughened Al_2O_3:$CaAl_{12}O_{19}$ Ceramic Composites. J. Am. Cer. Soc. 79(12):3142–3148, 1996.
113. D. Baril, S. P. Tremblay, and M. Fiset. Silicon Carbide Platelet Reinforced Silicon Nitride Composites. J. Mat. Sci. 28:5486–5494, 1993.

114. B. J. Kellett and D. S. Wilkinson. Processing and Properties of Alumina-Graphite Platelet Composites. J. Am. Cer. Soc. 78(5):1198–1200, 1995.
115. O. S. Ozgen and B. Bond. Effect of Graphite on Mechanical Properties of Alumina/Graphite Materials with Ceramic Bonds. Brit. Cer. Trans. J. 84, 138–142, 1985.
116. R. Chaim and V. Talanker. Microstructure and Mechanical Properties of SiC Platlet/Cordierite Glass-Ceramic Composites. J. Am. Cer. Soc. 78(1):166–172, 1995.
117. C. Hulse and J. Batt. The Effect of Eutectic Microstructures on the Mechanical Properties of Ceramic Oxides. United Aircraft Res. Lab. Report for ONR Contract N00014-69-C-0073, 5/1974.
118. T. Mah, A. Parthasarathy, and L. E. Matson. Processing and Mechanical Properties of $Al_2O_3/Y_5Al_5O_{12}$ (YAG) Eutectic Composites. Cer. Eng. Sci. Proc. 11(9–10): 1617–1627, 1990.
119. F. L. Kennard, R. C. Bradt, and V. S. Stubican. Mechanical Properties of the Directionally Solidified MgO-$MgAl_2O_4$ Eutectic. J. Am. Cer. Soc. 59(3–4):160–163, 1976.
120. V. S. Stubican, R. C. Bradt, F. L. Kennard, W. J. Minford, and C. C. Sorell. Ceramic Eutectic Composites. Tailoring Multiphase and Composite Ceramics, Materials Science Research 20 (R. E. Tressler, G. L. Messing, C. G. Pantano, and R. E. Newnham, eds.). Plenum Press, New York, 1986, pp. 103–113.
121. C. C. Sorell, V. S. Stubican, and R. C. Bradt. Mechanical Properties of ZrC-ZrB_2 and ZrC-TiB_2 Directionally Solidified Eutectics. J. Am. Cer. Soc. 69(4):317–321, 1986.
122. Y. Nivas and R. M. Fulrath. Limitations of Griffith Flaws in Glass–Matrix Composites. J. Am. Cer. Soc. 53(4):188–191, 1970.
123. D. R. Biswas. Strength and Fracture Toughness of Indented Glass-Nickel Compacts. J. Mat. Sci. 15:1696–1700, 1980.
124. Y.-H. Yun and S.-C. Choi. The Contributions of Microstructural Characteristics and Residual Stress Distribution to Mechanical Properties of AlN/W Composite System. J. Mat. Sci. 33:707–712, 1998.
125. T. Sekino, T. Nakajima, S. Ueda, and K. Niihara. Reduction and Sintering of a Nickel-Dispersed-Alumina Composite and Its Properties. J. Am. Cer. Soc. 80(5):1139–1148, 1997.
126. M. Nawa, T. Sekino, and K. Niihara. Fabrication and Mechanical Behavior of Al_2O_3/Mo Nanocomposites Processing and Novel Interpenetrated Intragranular Microstructure. J. Mat. Sci. 29:3185–3192, 1994.
127. A. J. Pyzik and D. R. Beaman. Al-B-C Phase Development and Effects on Mechanical Properties of B_4C/Al-Derived Composites. J. Am. Cer. Soc. 78(20):305–312, 1995.
128. I. W. Donald and P. W. McMillan. The Influence of Internal Stresses on the Mechanical Behavior of Glass–Ceramic Composites. J. Mat. Sci. 12:290–298, 1977.
129. J. G. Zwissler, M. E. Fine, and G. W. Groves. Strength and Toughness of a Ceramic Reinforced with Metal Wires. J. Am. Cer. Soc. 60(9–10):390–396, 1977.
130. L. A. Simpson and A. Wasylyshyn. Fracture Energy of Al_2O_3 Containing Mo Wire. J. Am. Cer. Soc. 54(1):56, 1971.

131. J. J. Brennan. Development of Fiber Reinforced Ceramic Matrix Composites. United Technologies Research Center Report No. R911848-4 for US Naval Air Systems Command Contract N62269-74-C-0359, 3/1975.
132. R. W. Rice and D. Lewis III. Ceramic Fiber Composites Based upon Refractory Polycrystalline Ceramic Matrices. Reference Book for Composites Technology 1 (S. M. Lee, ed.). Technomic Press, Lancaster, PA, 1989, pp. 117–142.
133. R. K. Govila. Fracture of Hot-Pressed Alumina and SiC-Whisker-Reinforced Alumina Composite. J. Mat. Sci. 23:3782–3791, 1988.
134. T. D. Claar, W. B. Johnson, C. A. Anderson, and G. H. Schiroky. Microstructure and Properties of Platelet-Reinforced Ceramics Formed by the Directed Reaction of Zirconium with Boron Carbide. Cer. Eng. Sci. Proc. 10(7–8):599–609, 1989.
135. M. Watanabe and I. Fukuura. The Strength of Al_2O_3 and Al_2O_3-TiC Ceramics in Relation to Their Fracture Sources. Ceramic Sci. Tech. at the Present and in the Future, Japan, 1981, pp. 193–201.
136. C. P. Cameron, J. H. Enloe, L. E. Dolhert, and R. W. Rice. A Comparison of Reaction vs. Conventionally Hot-Pressed Ceramic Composites. Cer. Eng. Sci. Proc. 11(9–10):1190–1202, 1990.
137. F. H. Simpson. Macrolaminate Particle Composite Material Development. Summary Report for Boeing Co. Work on US Navy Bureau of Weapons Contract No. 64-0194-f, 5/1965.
138. D. B. Marshall, W. L. Morris, B. N. Cox, and M. S. Dadkhan. Toughening Mechanisms in Cemented Carbides. J. Am. Cer. Soc. 73(10):2938–2943, 1990.
139. J. S. Reed and A.-M. Lejus. Effect of Grinding and Polishing on Near-Surface Phase Transformation in Zirconia. Mat. Res. Bull. 12:949–954, 1977.
140. R. W. Rice. Effects of Ceramic Microstructural Character on Machining Direction–Strength Anisotropy. Machining of Advanced Materials (S. Johanmir, ed.). NIST Special Pub. 847, US Govt. Printing Office, Washington, DC, 1993, pp. 185–204.
141. R. Telle and G. Petzow. Strengthening and Toughening of Boride and Carbide Hard Material Composites. Mat. Sci. Eng. A 105/106:97–104, 1988.
142. N. Claussen, J. Steeb, and R. F. Pabst. Effect of Induced Microcracking on the Fracture Toughness of Ceramics. J. Am. Cer. Soc. 56(6):559–562, 1977.
143. R. W. Rice. Ceramic Matrix Composites Toughening Mechanisms: An Update. Cer. Eng. Sci. Proc. 6(7–8):589–607, 1985.
144. J. S. Nadeau and J. I. Dickson. Effects of Internal Stresses Due to a Dispersed Phase on the Fracture Toughness of Glass. J. Am. Cer. Soc. 63(9–10):517–523, 1980.
145. G. H. Campbell, M. Rühle, B. J. Dalgleish, and A. G. Evans. Whisker Toughening: A Comparison Between Aluminum Oxide and Silicon Nitride Toughened with Silicon Carbide. J. Am. Cer. Soc. 73(30):521–530, 1990.
146. K. Niihara, A. Nakahira, T. Uchiyama, and T. Hirai. High-Temperature Mechanical Properties of Al_2O_3-SiC Composites. Fracture Mechanics of Ceramics 7, Composites, Impact, Statistics, and High-temperature Phenomena (R. C. Bradt, A. G. Evans, D. P. H. Hasselman, and F. F. Lange, eds.). Plenum Press, New York, 1986, pp. 103–116.

10

Composite Particle and Grain Effects on Hardness, Compressive Strength, Wear, and Related Behavior at ~ 22°C

I. INTRODUCTION

This chapter addresses the effects of particle and (matrix grain) size and related parameters on hardness, compressive strength, and wear and related behavior at nominally 22°C, i.e. it is analogous to Chapters 4 and 5 for monolithic ceramics. There is, however, much less information on these subjects for ceramic composites, especially specifically on particle parameters, where again the term particle is used in the broad sense of any geometry of a dispersed second phase. Thus more of the material in this chapter is on general composite response to such testing, as opposed to specifics of microstructural effects determining that response. After first covering the limited theoretical background, composite systems will be treated under each topic of hardness, compressive strength, etc. in the approximate order of composite compositions considered in Chapters 8 and 9.

II. THEORETICAL BACKGROUND

The theoretical background for ceramic composite behavior of interest in this chapter is about as limited or more so than actual data. Thus much of the guidance must come from general principles and correlations with other composite behavior and behavior of monolithic ceramics.

Consider first hardness. For composites, especially of all crystalline constituents, a first approximation is the rule of mixtures, i.e.

$$H_C = H_M(1-\phi) + \phi H_S \tag{10.1}$$

where the subscripts C, M, and S refer respectively to the composite, matrix, and dispersed second ("particulate") phase, and ϕ is the volume fraction second phase. For multiconstituent composites this Eq. (10.1) can be used repetitively to include each extra phase beyond the second, e.g. by first using the equation to predict the hardness behavior of the two phases present in the greatest amount and then treating that composite as one constituent with the next most prevalent phase, etc. until all phases are included. However, it must again be noted that Eq. (10.1) is an approximation; it is often a useful one, but it may not be feasible to adequately incorporate effects of particle and grain sizes, shapes, orientations, and changing contiguity of one or more of the composite phases as the percolation limits are reached for the various phases as ϕ changes. A modification of this equation was proposed by Lee and Gurland [1] to include the contiguity of each phase as a weighting factor for its volume fraction, which they showed agreed well with WC-Co data.

A special case that has received some consideration is that of partially crystallized glasses, where some glass matrices can undergo deformation required for indentation by densification as well as, or instead of, deformation by volume conserving processes that dominate deformation of crystalline materials. Miyata and Jinno [2] proposed a hardness theory for composites of crystalline or glass spherical particles of isotropic character dispersed in an isotropic glass matrix. They considered two cases where (1) the dispersed phase is harder than the glass matrix and (2) the matrix is harder than the dispersed particles, using Marsh's theory of glass deformation. In case 1 the composite hardness depends on the matrix hardness and flow stress of the matrix, the elastic properties of both phases, and the volume fraction of dispersed phase, while in case 2 hardness and elastic properties vary in the same fashion with the volume fraction dispersed phase. In both cases particle size and spacing do not play a role other than via their variation as the volume fraction dispersed phase (ϕ) changes, i.e. the latter dominates. However, while they showed that the limited data was consistent with their two models, as noted later, whether there is a significant and reliable difference between the rule of mixtures relation Eq. (10.1) and their model is uncertain.

There do not appear to be any models for compressive strength, ballistic performance, wear, or erosion specifically for composites. Correlations of these properties with H and K can be a guide, e.g. per Eqs. (5.1) for erosion and (5.2) for crack sizes introduced by particle impacts and especially abrasive action. However, there are added uncertainties for composites in addition to those for monolithic ceramics (Chaps. 4 and 5). The primary added uncertainty is the possibility of preexisting, or especially stress generated, microcracks due to the two- or multiphase nature of composites, which respectively may not have the same effects on *H* as other properties or not be generated as extensively, or at all, in hardness testing. The latter is a concern in view of the extensive hydrostatic nature of much of the stress during indentation versus generally greater tensile stress generated in compression, wear, etc.

III. EXPERIMENTAL RESULTS

A. Hardness

Miyata and Jinno [2] showed that their H_V (100 gm) data for phase separated PbO-B_2O_3 glasses supported their models for harder and for softer particles relative to the glass matrix, i.e. showing respectively *H* increases of ~ 25% by ϕ= 0.25 and *H* decreases of ~ 30% by ϕ= 0.3. Though the necessary property data was not available to make such specific quantitative comparisons in other systems, they noted that basic trends for phase separated sodium borosilicate glass and for crystallized Li_2O_2-SiO_2 glass were consistent with their model, e.g. in the latter case *H* was independent of crystallite size. However, their plots are not far from linear, as was also noted by French et al. [3], raising the question of how significantly their model deviates from a rule of mixtures.

Some other data supports the dominance of ϕ on *H*, e.g. Roesky and Varner [4] and Stryjak and McMillan [5] for crystallized glasses in respectively the LiO_2-Al_2O_3-SiO_2 and ZnO-Al_2O_3-SiO_2 systems, but they are not clear in differentiating between the rule of mixtures, Miyata and Jinno's, or other mechanisms. Thus in the latter composition, *H* also increased ~ linearly with *D*, but whether this is totally reflected in the ϕ dependence is uncertain. On the other hand, Tashiro and Sakka [6] reported H_V (200 gm) data for a photosensitive LiO_2-SiO_2-based glass varying linearly with ~ D^{-1}, or $D^{-1/2}$ (as also more clearly shown for flexure strength) for higher temperature crystallization (e.g. with *D* ~ 0.8 to 2.3 μm), but a more complex, possibly reverse dependence on *D* from lower heat treatment and finer values (*D* 0.07–1 μm). Further complicating the picture is data of Cook et al. [7] for crystallization of three nearly identical LiO_2-SiO_2-based glasses giving H_V increasing with *D* and λ but not with ϕ Though not presenting detailed quantitative H_v- microstructure data, Donald and McCurrie [8] showed some complexities of crystallized glasses in their MgO-LiO_2-Al_2O_3-SiO_2-TiO_2 systems, i.e., while having an overall trend for *H* to increase with heat treatment and extent of crystallization, they showed wide variations with heat treatment. Thus heat treatments at 800, 900, and 1000°C all gave rapid increases in *H* to a maximum, then a decrease to a minimum, followed by further increases as heat treatment time increased, but with substantial differences in *H*, e.g. of minima and maxima, values.

Besides more detailed study, e.g. of microstructures (including related stresses) and properties, to sort out the above complexities, more study of artificial glass–ceramic composites is needed, but such data is very limited. Wolf et al. [9] reported that H_V values of their composites of 25 v/o of glasses of expansions varying from < to > that of the Al_2O_3 fine grain preforms into which the glasses were infiltrated all gave $H \sim 12$ GPa, which was between the upper and lower bounds [Eqs. (8.1) and (8.2)], i.e. somewhat below the rule of mixtures [Eq. (10.1)].

Turning to all crystalline composites, French et al. [3] reported H_V (10 N) for their Al_2O_3 (~ 5 μm) of ~ 18.4 GPa decreasing linearly as cubic stabilized ZrO_2 (+ 8 m/o Y_2O_3) content was increased, e.g. to 13.9 GPa for only ZrO_2 ($G \sim$ 5+ μm), which was ~ 76% of the Al_2O_3 hardness. Similarly, Ruf and Evans [10] showed that their H_V (200 N) data for ZnO with additions of ZrO_2 of up to 60 v/o was consistent with a rule of mixtures, except for limited deviations to lower values at > 40 v/o, attributed to limited microcracking, i.e. the same behavior as for E (Chap. 8, Sec. 3). More recently Hirano and Inada [11] showed that additions of up to 40 v/o Al_2O_3 to either 4Ce-TZP or 3Y-TZP increased H_V (9.8 N), at or close to a linear function of Al_2O_3 content. However, the trends for each processing temperature shifted downward with increasing temperature and coarsening of the microstructure, with substantially more effect of processing in the Ce-TZP system.

Hwang and Niihara's [12] data for $BaTiO_3$ with 1–5 v/o additions of nanosize SiC particles showed H_V (apparently with 9.8 N) increasing significantly with v/o of SiC similar to flexure strength (Fig. 9.11), fracture toughness and Young's modulus (Fig. 8.19), though complicated by reduced G at lower SiC contents and bimodal G at higher SiC contents and densification temperatures. Nakahira et al. [13] showed that H_V (0.98 N) increased linearly from ~ 19.5 GPa at 0 v/o of ~ 2 μm SiC particles to 25 GPa at 50 v/o in an Al_2O_3 matrix. While some of this increase must be due to the progressive reduction in the Al_2O_3 grain size from ~ 20 μm at 0 v/o SiC to ~ 3 μm at 50 v/o SiC, this contribution is limited, e.g. to ~ 0.5 GPa per Figure 4.2, with which their data is reasonably consistent. Krell and Klaffke [14] showed that H_V (10 kg) of Al_2O_3-35 v/o TiC (fine G) was ~ 21 GPa versus ~ 17–18 GPa for pure Al_2O_3 of the same G (~ 2.5 μm).

Turning to composites in which the matrix and the particulate phase are both nonoxides, the H_V (500 gm) data of Endo et al. [15] for the SiC-TiC system showed more complex behavior (Fig. 10.1) in contrast to their linear trend for E versus composition (Fig. 8.11). This more complex behavior was attributed to probable effects of G and residual stress changes with composition, with support for the latter seen in substantial increases in tensile strengths for the SiC rich portion of the system at elevated versus room temperatures (Chap. 11, Sec. III.E). In contrast to this Sasaki et al. [16] showed H_V (98 N) of Si_3N_4 first increasing slowly from ~ 13 GPa with addition of 0 to 10 v/o of ~ 0.3 μm SiC particles and

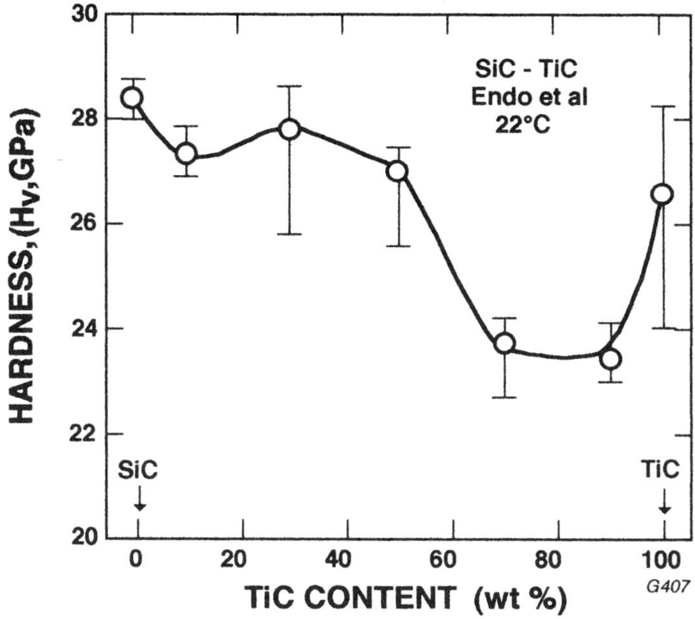

FIGURE 10.1 Vickers hardness (H_V, 500 gm) versus phase content in the SiC-TiC particulate composite system. (From Ref. 15. Published with permission of the *Journal of Materials Science*.)

then rising to ~ 14.3 GPa at 20 v/o, similar to, but more extreme than, the behavior of E (Chap. 8, Sec. III). Bhattacharya and Petrovic [17] showed that H_V (100 N) of $MoSi_2$ increased rapidly from 9.3 to 12.3 GPa with addition of 5 v/o of ~ 0.5 μm SiC particles and then followed an ~ linear trend ~ parallel with, and slowly approaching, the rule of mixtures relation (with H_V SiC ~ 25 GPa), reaching nearly 18 GPa at 40 v/o SiC. The initial rapid rise of H correlates with reductions of the $MoSi_2$ G from ~ 28 to 11 μm for 0 to 5 v/o SiC and the trend toward the rule of mixtures relation to further G reductions to ~ μ5 m at 40 v/0 SiC. Cameron et al. [18] reported H_V (1 kg) of various reaction hot pressed, fine grain/particle (typically a few μm) ceramic composites (Table 10.1) that were typically at least approximately consistent with the rule of mixtures. The H values were also typically close, e.g. within 5–15% of those of composites made by directly hot pressing the composites from mixtures of ceramic powders producing similar microstructures. Landon and Thevenot [19] reported that H_V (1 kg) for AlN with SiC increased from ~ 19 GPa with 30 w/o β-SiC to a possible maximum at 29 GPa at 75 w/o and from ~ 21.5 GPa with 40 w/o α-SiC to a possible maximum or plateau of ~ 28 GPa at 80 w/o. The maximum or plateau at ~ 75–80

TABLE 10.1 Reaction Hot Pressed Ceramic Composite Data

Reaction	Vol% nonoxide	Density (gm/cc)[a]	H_V (1 kg) (GPa)	Ballistic V_{50}[b]	Costs ($/lb)[c]
$4Al + 3SiO_2 + 3C \rightarrow 2Al_2O_3 + 3SiC$	43	3.67 (360)	26	1590	1.11/4.83
$4Al + 3TiO_2 + 3C \rightarrow 2Al_2O_3 + 3TiC$	42	4.29	22	NT	1.58/6.87
$10Al + 3TiO_2 + 3B_2O_3 \rightarrow 5Al_2O_3 + 3TiB_2$	27	4.14	22	NT	1.69/7.97
$8Al + 3SiO_2 + 2B_2O_3 + 4C \rightarrow 4Al_2O_3 + 3SiC + B_4C$	37	3.62 (3.51)	19.9	1580	1.37/6.69
$6Mg_3Si_4O_{10}(OH)_2 + 36Al + 25C + 2B_2O_3 \rightarrow 18MgAl_2O_4 + 24SiC + B_4C + 6H_2O$	31	3.45 (3.31)	15	1540	0.91/4.55
$(La_2O_3 \cdot 6B_2O_3) + 14Al \rightarrow 7Al_2O_3 + 2LaB_6$	35	4.09	21.5	NT	3.55/9.46
$Si_3N_4 + 4Al + 3C \rightarrow (4AlN \cdot 3SiC)$	100	3.24	25.4	1850	3.21/9.41

[a]Top figure is the theoretical density of solid product; bottom figure in ( ) is the value of larger pieces ballistically tested if different from the theoretical density.

[b]Relative velocities at which half the 30 caliber armor piercing (AP) projectiles penetrated through the test ceramic, i.e. higher values are better.

[c]Raw materials costs. Top figure is for the raw materials for reaction hot pressing. Bottom figure is for powders to produce the same product by directly hot pressing of powder mixtures of the same ceramic composite compositions.

w/o SiC may reflect shifts in grain size and the degree of solid solution versus two-phase structure as composition changes.

Baril et al. [20] reported H_V (20 kg) of composites of 0–30 v/o SiC platelets (11 or 17 μm dia.) in a Si_3N_4 matrix modestly increased from ~ 15.5 to 16 GPa over the range of additions for the finest platelets and ~ 5% less for the larger platelets for measurement on planes parallel to the platelet orientation and ~ 5% higher on the normal plane.

Turning to whisker composites, Ashizuka et al. [21] showed H_V (200 gm) of composites of SiC whiskers in a cordierite, anorthite, or diopside matrix increasing nearly linearly as whisker content increased to a maximum followed by significant decreases. This was overall similar to the trends for Young's modulus, fracture toughness, and flexure strength, i.e. with the level of the maximum and the v/o at which it occurs increasing in the listed order of the matrices, e.g. respective maxima at 25, 30, and 40 v/o SiC whiskers. The maximum of H (and other properties) correlated with the maximum whisker content at which near zero porosity was achieved in fabrication. The data thus suggests a rule of mixtures trend, but densification, and especially lack of detailed quantitative information on the whisker orientation and hardness, prevents detailed evaluation.

Tamari et al. [22] showed H_V (4.9 N) was independent of SiC whisker content (0–30 v/o) in B_4C, in part reflecting limited hardness differences in the constituents. Dusza et al. [23] showed that H_V (10 kg) of β-Si_3N_4 added to a S_2N_4 matrix decreased as the whisker content increased, presumably reflecting the lower H of β-Si_3N_4 versus that of α-Si_3N_4, the expected matrix and harder (Fig. 4.13) phase

The most extensive study of ceramic metal composites is of WC-Co bodies, which Lee and Gurland [1] showed agreed well with their modification of Eq. (10.1) to include the contiguity of each phase multiplying its volume fraction. Nawa et al.'s [24] composites of Al_2O_3 with 0 to 20 v/o Mo particles (~ 0.65 μm) showed H_V (9.8 N) decreasing linearly with increasing Mo content, indicating agreement with the rule of mixtures relation [Eq. (10.1)]. Both starting values at 0 v/o Mo of ~ 18 GPa and significant (e.g. 20%) lower values due to increased Al_2O_3 G (from ~ 0.4 to 2.8 μm) are generally consistent with data of Figs. 4.1, 4.2. Chou et al. [25] showed H_V (100 N) of Al_2O_3-NiAl particle (D ~ 5.5 μm) composites decreased linearly from ~ 18.1 to 8 GPa as the volume fraction of NiAl increased from 0 to 50 v/o in fair agreement with Eq (10.1). However, there must be some other factor causing H to decrease more than just due to the much lower H of NiAl (~ 0.3–0.45 GPa) as the v/o NiAl increases, since the Al_2O_3 matrix grain size was reduced with increasing NiAl addition, which would entail some increase in H above the expected rule of mixtures value of nearly 10 GPa. This greater decrease may reflect possible contraction of the NiAl away from the Al_2O_3 matrix or residual stresses in view of the NiAl thermal expansion just over twice that of Al_2O_3.

Turning to ceramic eutectics, the most extensive data is from Bradt and colleagues [26–29], e.g. H_K (300 gm) for directionally solidified MgO-$MgAl_2O_4$ euctectic increased linearly with $\lambda^{-1/2}$, exceeding values for the two constituents over the range of $\lambda^{-1/2}$ investigated (0.3–0.7 $(\mu m)^{-1/2}$). More extensive H_K (500 gm) data for ZrC-ZrB_2 showed H for both transverse and longitudinal sections increasing linearly as a function of $\lambda^{-1/2}$ to a maximum at λ ~ 1.85 μm and then rapidly decreasing linearly with a further increase of $\lambda^{-1/2}$, i.e. at finer spacing, which correlated with breakdown of the well-ordered microstructure. The peak H of ~ 22.5 and 24 GPa respectively for longitudinal and transverse sections were substantially higher than for either constituent (~ 13 and 17 GPa respectively). Stubican and Bradt [27] noted that eutectics of TiC-TiB_2 and SiC-B_4C both also showed H increasing with decreasing λ, but with the highest observed H values not exceeding those of the harder members (TiB_2 and B_4C). Echigoya et al. [30] reported that H_V of longitudinal surfaces of directionally solidified Al_2O_3-ZrO_2(Y_2O_3) eutectics were low (~ 10 GPa) for 0 m/o Y_2O_3 due to cracking, but was ~ 17 GPa at 3 m/o decreasing to 15 GPa at 13 m/o Y_2O_3.

B. Compressive Strength and Ballistic Performance

Very little compressive testing has been done on ceramic composites, presumably in part due to expectations that their strengths will typically be limited by substantial microstructural mismatch stresses, but we thus miss an opportunity to use such testing to help define such mechanisms of failure.

May and Obi [31] measured compressive strengths of crystallized SiO_2-based glasses with 25–35 m/o LiO_2 with 1–3% P_2O_5 for nucleation. While there were broad variations of compressive strengths, e.g. from ~ 10 to 85 MPa, all compositions showed similar trends as a function of heat treatment, namely initial rises in strengths, e.g. to strength maxima a few to 150% that of the glass following early crystallization. These strength increases were followed by strength decreases to minima, e.g. similar to the starting strengths to 60% below them at intermediate heat treatment temperatures and then increases at higher heat treatment temperatures, commonly close to or higher than their initial maximum strengths (Fig. 10.2). Tests as a function of temperature, while all showing overall strength decreases, show substantial variations that should have the potential to allow improved understanding of behavior in these complex systems (Chap. 11, Sec. III.F).

Verma et al. [32] showed marked increases in the compressive strength of composites of SiC platelet (~ 25–50 μm dia., and ~ 1–2 μm thick) in a borosilicate glass matrix chosen for its thermal expansion match with SiC. Strength ~ doubled from ~ 160 MPa for 0 v/o SiC at 10 v/o, rose to 510 MPa at 40 v/o SiC, and then dropped greatly to ~ 200 MPa at 50 v/o. The latter drop is

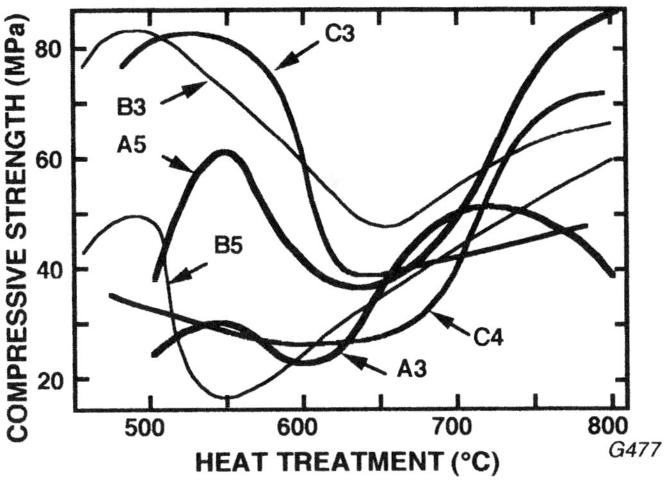

FIGURE 10.2 Compressive strengths versus heat treatment temperature (for 2 hr) for crystallization of Li_2O-SiO_2 glasses. This plot shows nearly half the curves generated and reflects the range and diversity of compressive strengths observed. Compare with Fig. 11.14. (Curves from Ref. 31. Published with permission of the Proceedings of the British Ceramic Society.)

attributed to substantial residual porosity at such high SiC loading. Fracture in such room temperature tests was totally brittle and catastrophic. Tests at elevated temperatures also showed strengths increasing with increasing platelet additions to 40 v/o, but at much reduced strength levels and with substantial plastic deformation.

Simpson [33] compression tested his ceramic–metal composites made from particles obtained by tape casting multiple, alternating green layers of HfO_2-CeO_2+MgO and Mo in his investigation of ceramic–metal composites (based on concepts of Knapp and Shanley [34]). Tapes with individual layers down to 20 μm and total multilayer tape thicknesses of 25–450 μm were cut into rectangular pieces from ~ 800 to ~ 70 μm in lateral dimensions and consolidated and then sintered. Compressive strengths decreased faster initially and then more slowly from ~ 1. 3 GPa with 30 v/o Mo to ~ 650 MPa at 70 v/o Mo (while flexure strength increased from ~ 175 to 400+ MPa). Both strengths increased as the size of laminated particles decreased, e.g. by 50–100% for flexure strengths, and for compressive strengths by 15–40%.

Another useful observation from compressive stressing is that of Suresh et al. [35] on Si_3N_4 with 0, 10, and 20 v/o SiC whiskers. They showed that while the threshold stress range for crack initiation in compressive fatigue testing of

TABLE 10.2 Representative Performance of Commercial Ceramics for 30 Caliber Armor Piercing Bullets

Ceramic	Density (gm/cc)	V_{50}[a]
B_4C	2.50	2090
SiC	3.21	2000
Al_2O_3	3.96	1600
TiB_2	4.50	1280

[a]Relative velocities at which half of the 30 caliber armor piercing (AP) projectiles penetrated through the test ceramic, i.e. higher values are better.

notch beam specimens was higher for 10 and 20 v/o whisker composites, crack extension after 500,000 cycles was respectively 2.2 and 2.4 times that of the matrix alone. On the other hand, the threshold stress range for fatigue crack initiation was 20% lower than for the matrix alone, and crack growth rates were smaller.

Limited ballistic testing of ceramic composites has been conducted, much of which is probably in classified literature. However, preliminary testing of reaction hot pressed composites during exploratory development [18,36] showed intermediate levels of performance (compare Tables 10.1 and 10.2). Also note that comparative testing of a commercial ZTA (~ 30% ZrO_2) gave lower 30 caliber AP V_{50} values of 1375. Further, DU (depleted uranium) penetrator tests of the reaction hot pressed AlN-SiC of Table 1 showed residual penetration the same as for commercial AlN, which approaches, but is poorer than, the performance of other commercial ballistic ceramics.

C. Erosion and Wear

Breder et al. [37] showed that erosion of Al_2O_3-ZrO_2 by sharp airborne abrasive particles followed equations similar to Eq. (5.1) but with greater dependence on toughness. Residual strengths after erosion were an inverse function of particle kinetic energy.

Wada et al. [38,39] reported that H_V (500 gm) of Al_2O_3 showed no increase from ~ 19 GPa at 0 v/o SiC to 10 v/o α-SiC particles (~ 1 μm) or whiskers and then increased respectively to ~ 25 and 28 GPa at 30 v/o, while (I) toughness increased from 4.8 to ~ 5.4 MPa·m$^{1/2}$ over the range 0–30 v/o SiC. Erosion rates for normal impacts of Al_2O_3 and SiC particles (~ 500 μm) entrained in an air stream giving particle velocities of 250–300 m/sec were nearly twice as high with SiC versus Al_2O_3 particles and decreased with increases in both target hardness and toughness, though the correlation with the latter was more scattered. Similarly Wada and Watanabe [40] showed that such particle erosion resistance of Al_2O_3-TiC was higher than that of Al_2O_3, Al_2O_3-SiC, or Si_3N_4.

Yamada et al. [41] showed similar increases in H_V (4.9 N) and somewhat lower increases in (I) toughness on addition of up to 40 v/o SiC particles (~ 0.7 μm) and improved resistance to abrasion with Al_2O_3 or SiC particles, with the resistance of the Al_2O_3-SiC exceeding that of either Al_2O_3 or SiC by themselves when the SiC content was > 20 v/o. The abrasion rate was inversely dependent on *H* but was independent of toughness.

He et al. [42] reported beneficial effects of Al_2O_3-ZrO_2 composites on the transition from a mainly plastic deformation to a mainly fracture dominated mechanism of sliding wear using a test with a Si_3N_4 ball on three flat composite specimens. The load for this sliding wear transition increased from 170 to 350 N as the composition changed from 5 to 20 v/o unstabilized ZrO_2, with much of the effect attributed to reduction of the grain size of the Al_2O_3, especially of the largest grains. However, the lowest wear rate was found with 15 v/o ZrO_2, showing additional benefits of the ZrO_2, which was attributed to reduction of hardness and compressive surface stresses.

The sliding wear behavior of both a 5 a/o Y-TZP + 28 v/o Al_2O_3 and a ZTA composite (Al_2O_3+ 15 w/o, ~ 10 v/o, ZrO_2) were studied by He et al. [43]. They showed that increasing grain size in the former composite resulted in increasing wear due to increasing grain pullout attributed to increased ZrO_2 transformation and resultant grain boundary weakening. At finer grain size where no transformation occurred, the former composite gave a fourfold lower wear rate compared to pure Y-TZP of the same grain size, which was attributed to the higher hardness from the Al_2O_3 addition. Little grain pullout and no microcracking was observed in the finer grain composites, whose wear characteristics were not impacted by sinter-forging of the bodies. The ZTA composite showed opposite dependence on *G*, i.e. increased wear resistance as *G* increased, which was attributed to increased compressive surface stresses from transformation of ZrO_2, and had wear rates reduced two- or threefold by sinter forging with the same resultant grain size.

Wu et al. [44] measured diamond pin on disk (POD), i.e. scratch hardnesses of several ceramic composites versus several monolithic hard ceramics, e.g. constituents of the composites, using a 90° conical diamond pin (0.08 μm radius) with loads of 0.5, 1, or 1.5 kg with the disk specimens rotating so the pin scribed a circle ~ 1 cm radius at a speed of ~ 1 cm/sec. Results showed poorer performance for Al_2O_3 with 30% ZrO_2 or 15 w/o SiC particles, but superior performance for Al_2O_3 with 10% ZrO_2 or 30% SiC whiskers (Fig. 10.3). The poor performance of the Al_2O_3 with 15% SiC was attributed to the large SiC particle size (> 40 μm), while the good performance for the commercial

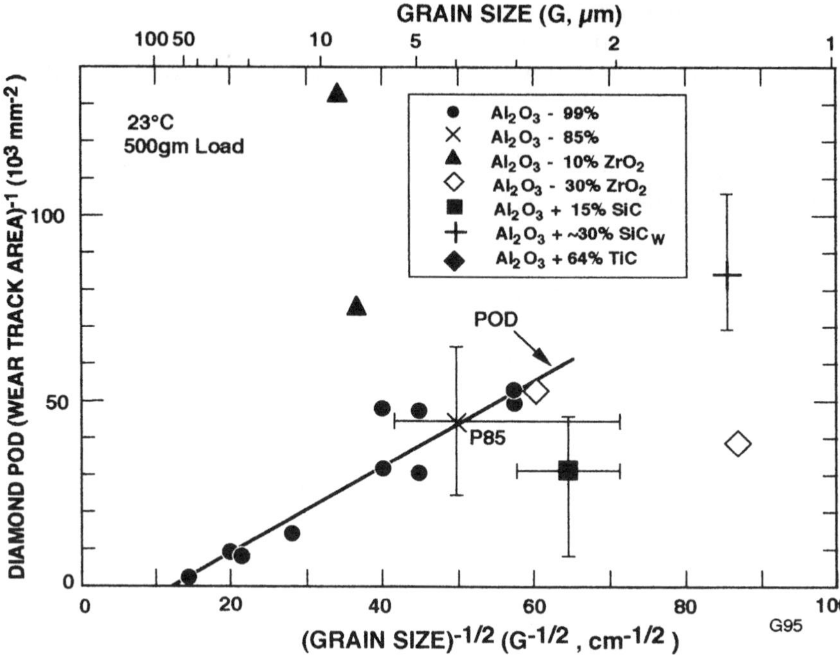

FIGURE 10.3 Wear resistance from diamond pin on disk wear test of some ceramic composites versus the inverse square root of their matrix grain size compared with data for alumina bodies. Note (1) compositions in v/o, (2) composites can have similar, or better, wear resistance than comparable grain size alumina bodies, and (3) poorer performance was associated with bodies having large particles of SiC or microcracking (Al_2O_3-30% ZrO_2). (From Ref. 44. Published with permission of Ceramic Engineering and Science Proceedings.)

Al_2O_3+ 30% SiC whiskers was also approximately consistent with expectations for Al_2O_3 of the same *G* as in the composite matrix (see Fig. 8.14 for a fracture photo of this material). Reaction hot pressed composites of Al_2O_3 with 34 or 47% SiC, 47 or 64 v/o TiC, or 71.4 v/o TiB_2 all with *G* and *D* a few microns, all gave low POD wear rates, e.g. values of 20–90 on the wear scale of Figure 10.3, indicating superior performance relative to pure Al_2O_3 of the same *G*. Fig. 4 shows that these dense fine-grain composites showed very similar appearances of surface plastic deformation in the wear tracks as dense fine-grain monolithic bodies, e.g. of alumina.

Krell and Klaffke [14] reported that their fine *G* (~ 1.5 μm) Al_2O_3-35 v/o TiC had a friction coefficient of 0.3–0.5 that was higher, e.g. by 20–30%, and had more dependence on the H_2O level in the tests than for similar *G* Al_2O_3. The

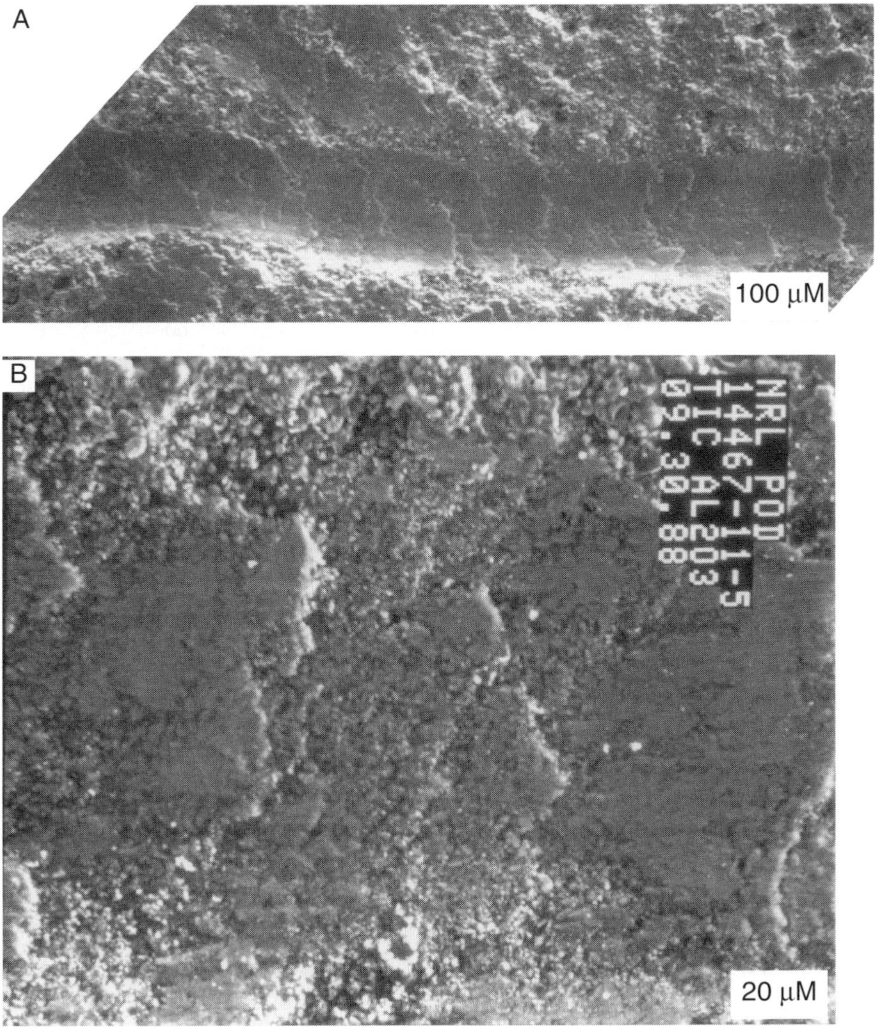

FIGURE 10.4 SEM photos of diamond pin on disk wear tracks on an Al_2O_3-64 v/o TiC reaction processed composite. (A) lower magnification and (B) higher magnification. Note parallel striations indicative of plastic flow and irregular transverse markings indicative of interruptions of this flow (compare to Figs. 5.13, 5.14). (From Ref. 44. Published with permission of Ceramic Engineering and Science Proceedings.)

TABLE 10.3 Relative Abrasive Wear Performance of Various Commercial Ceramics Versus Commercial 85% Alumina

Material[a] (*H*, GPa)	Grit blast[b]	Slurry[c]	Taber[d]	Milling[e]
99% Al_2O_3 (15.2)	1.10	1.36	1.89	10[f]
96% Al_2O_3 (12.6)	1.06	—	1.57	5.6
85% Al_2O_3 (10)	1.0	1.0	1.0	1.0
ZTA (15.7)	0.51	0.37	0.37	2.4
TZP (11.5)	0.38	0.22	0.22	—
Si_3N_4 (16.5)	0.27	—	0.39	—

[a]Commercial materials and data from Diamonite Products. All values normalized to 85% Al_2O_3 = 1.
[b]Measured as the weight loss of plate specimens after a fixed exposure to a continuous grit blasting normal to the surface with 36 grit SiC abrasive.
[c]Measured as the weight loss of plate specimens after a fixed exposure to continuous rotation in a fixed sand slurry composition.
[d]Measured as the weight loss after a fixed number of rotations of plate specimens under two diamond abrasive wheels $^1/_2$ inch wide.
[e]Measured as the weight loss after a fixed exposure of rod specimens to each other with a specific charge of water and rods in a mill (whose size is scaled with rod size). For the tests here 1/4 inch dia. rods were used in a 5 liter mill.
[f]Approximate value, since a somewhat different firing schedule with resultant increased *G* was used for this test.
Source: Courtesy of J. Chakraverty of Diamonite Products.

oscillating wear rate of the composite was also close to but somewhat above that of the comparable alumina.

Tests of commercial ceramics using various industrial screening tests for different wear applications generally show ZTA (Al_2O_3+ 30 v/o ZrO_2) to be better than commercial aluminas but not quite as good as TZP or Si_3N_4 (Table 10.3). However, there are some exceptions to this, e.g. ZTA was not superior to 85% alumina in ball milling tests.

IV. DISCUSSION AND SUMMARY

There is more data for hardness than for other properties of this chapter. This data shows that a linear dependence of hardness (*H*) per Eq. (10.1) is frequently suitable for composites, but that effects of the dispersed phase reducing the matrix grain size [13,17,19] need to be accounted for, and more complex behavior can occur, e.g. Fig. 1, due to reactions, heterogeneities, changes in orientation, etc. Some effect of second phase dimensions is indicated by lower hardness with larger SiC platelets in Si_3N_4 [20]. Definitive effects of interlaminar or rod spacings (λ) in directionally solidified eutectics was shown, i.e. *H* increased as $\lambda^{-1/2}$

increased until the eutectic structure started to break down. This $\lambda^{-1/2}$ dependence of H is analogous to a $G^{-1/2}$ dependence in monolithic ceramics, which is logical since λ simply measures the dimension of the matrix separating the aligned lamelli or rods of the eutectic just as G measures not only the grain dimension but also the separation between next nearest grains. Cracking in eutectics can also occur and reduce H [29] and is probably a factor in the H decreasing as the eutectic structure starts to break down.

Very little data on compressive strength and ballistic performance of composites exists; thus we miss a possible opportunity of using compressive testing as another probe to understand effects of microstructural stresses and microcracking. However, one set of considerable compressive testing [31] clearly shows the diversity and complexity of behavior that can occur in one common crystallized glass system (Fig. 10.2). Also, Simpson's testing of his ceramic–metal composites made of particles of alternating laminated layers of metal and ceramic showing tensile and especially compressive strength increasing as the laminate particle size decreases [32] demonstrates another scale dimension that can impact strengths. Though no detailed studies of particle size effects of composites on their ballistic performance are available (and no definitive information exists on the G dependence of ballistic performance of monolithic ceramics, Chap. 5, Sec. III), preliminary testing of reaction processed ceramic composites showed potential for competitive behavior despite having almost no development (Tables 10.1, 10.2).

Though it is limited, there is more wear data for ceramic composites, which also shows potential for their use. Thus there is similar or greater wear resistance of Al_2O_3-based composites with similar fine matrix grain sizes to those of monolithic alumina bodies (Fig. 10.3), but again there is evidence that even limited microcracking from larger particles (SiC in Al_2O_3, Fig. 10.3) or compositional effect (Al_2O_3-30 v/o ZrO_2) seriously limits wear resistance. Plastic deformation of composites with fine microstructures (Fig. 10.4) is indicated as for fine-grain-size monolithic ceramics. However, data is also needed on friction, which can be higher in some composites [14].

More limited data for abrasion and erosion of ceramic composites indicates performance similar to that for Eq. (5.1) [37], and that some composites, e.g. Al_2O_3 with fine SiC particles [41], can have promising resistance to damage.

REFERENCES

1. H. C. Lee and J. Gurland. Hardness and Deformation of Cemented Tungsten Carbide. Mat. Sci. Eng. 33:125–133, 1978.
2. N. Miyata and H. Jinno. Micromechanics Approach to the Indentation Hardness of Glass Matrix Particulate Composites. J. Mat. Sci. 17:547–557, 1982.

3. J. D. French, H. M. Chan, M. P. Harmer, and G. A. Miller. Mechanical Properties of Interpenetrating Microstructures: The Al_2O_3/c-ZrO_2 System. J. Am. Cer. Soc. 75(2):418–423, 1992.
4. R. Roesky and J. R. Varner. Influence of Thermal History on the Crystallization Behavior and Hardness of a Glass-Ceramic. J. Am. Cer. Soc. 74(5):1129–1130, 1991.
5. A. J. Stryjak and P. W. McMillan. Microstructure and Properties of Transparent Glass-Ceramics. J. Mat. Sci. 13:1794–1804, 1978.
6. M. Tashiro and S. Sakka. Studies on the Mechanical Strength of the Photosensitive Opal Glass. J. Cer. Asn. Jpn. 68(6):72–77, 1960.
7. R. F. Cook, S. W. Freiman, and T. L. Baker. Effect of Microstructure on Reliability Predictions for Glass Ceramics. Mat. Sci. Eng. 77:199–212, 1986.
8. I. W. Donald and R. A. McCurrie. Microstructure and Indentation Hardness of an MgO-LiO_2-Al_2O_3-SiO_2-TiO_2 Glass-Ceramic. J. Am. Cer. Soc. 55(6):289–291, 1972.
9. W. D. Wolf, K. J. Vaidya, and L. F. Francis. Mechanical Properties and Failure Analysis of Alumina-Glass Dental Composites. J. Am. Cer. Soc. 79(7):1769–1776, 1996.
10. H. Ruf and A. G. Evans. Toughening by Monoclinic Zirconia. J. Am. Cer. Soc. 66(5):328–332, 1983.
11. M. Hirano and H. Inada. Fracture Toughness, Strength and Vickers Hardness of Yttria-Ceria-Doped Tetragonal Zirconia/Alumina Composites Fabricated by Hot Isostatic Pressing. J. Mat. Sci. 27:3511–3518, 1992.
12. H. J. Hwang and K. Niihara. Perovskite-Type $BaTiO_3$ Ceramics Containing Particulate SiC. J. Mat. Sci. 33:549–558, 1998.
13. A. Nakahira, K. Niihara, and T. Hirai. Microstructural Properties of Al_2O_3-SiC Composites. Yogyo-Kyokai-Shi 94(8):767–772, 1986.
14. A. Krell and D. Klaffke. Effects of Grain Size and Humidity on Fretting Wear in Fine-Grained Alumina, Al_2O_3/TiC, and Zirconia. J. Am. Cer. Soc. 79(5):1139–1146, 1996.
15. H. Endo, M. Ueki, and H. Kubo. Microstructure and Mechanical Properties of Hot Pressed SiC-TiC Composites. J. Mat. Sci. 26:3769–3774, 1991.
16. G. Sasaki, H. Nakase, K. Suganuma, T. Fujita, and K. Niihara. Mechanical Properties and Microstructure of Si_3N_4 Matrix Composite with Nano-Meter Scale SiC Particles. J. Cer. Soc. Jpn., Intl. Ed. 100:536–540, 1992.
17. A. K. Bhattacharya and J. J. Petrovic. Hardness and Fracture Toughness of SiC-Particle-Reinforced $MoSi_2$ Composites. J. Am. Cer. Soc. 74(10):2700–2703, 1991.
18. C. P. Cameron, J. H. Enloe, L. E. Dolhert, and R. W. Rice. A Comparison of Reaction vs. Conventionally Hot-Pressed Ceramic Composites. Cer. Eng. Sci. Proc. 11(9–10):1190–1202, 1990.
19. M. Landon and F. Thevenot. The SiC-AlN System: Influence of Elaboration Routes on the Solid Solution Formation and Its Mechanical Properties. Cer. Intl. 17:97–110, 1991.

20. D. Baril, S. P. Tremblay, and M. Fiset. Silicon Carbide Platelet Reinforced Silicon Nitride Composites. J. Mat. Sci. 28:5486–5494, 1993.
21. M. Ashizuka, Y. Aimoto, and M. Watanabe. Mechanical Properties of SiC Whisker Reinforced Glass Ceramic Composites. J. Cer. Soc. Jpn., Intl. Ed. 97:783–789, 1989.
22. N. Tamari, H. Kobayashi, T. Tanaka, I. Kondoh, and S. Kose. Mechanical Properties of B_4C-SiC Whisker Composite Ceramics. J. Jpn Cer. Soc., Intl. Ed. 98:1169–1173, 1990.
23. J. Dusza, D. Sajgalik, and M. Reece. Analysis of Si_3N_4+ -Si_3N_4 Whisker Ceramics. J. Mat. Sci. 26:6782–6788, 1991.
24. M. Nawa, T. Sekino, and K. Niihara. Fabrication and Mechanical Behavior of Al_2O_3/Mo Nanocomposites Processing and Novel Interpenetrated Intragranular Microstructure. J. Mat. Sci. 29:3185–3192, 1994.
25. W. B. Chou, W. H. Tuan, and S. T. Chang. Preparation of NIAl Toughened Al_2O_3 by Vacuum Hot Pressing. Brit. Cer. Trans. 95(2):71–74, 1996.
26. F. L. Kennard, R. C. Bradt, and V. S. Stubican. Mechanical Properties of the Directionally Solidified MgO-$MgAl_2O_4$ Eutectic. J. Am. Cer. Soc. 59(3–4):160–163, 1976.
27. V. S. Stubican and R. C. Bradt. Eutectic Solidification in Ceramic Systems. Ann. Rev. Mat. Sci. 11:267–297, 1981.
28. V. S. Stubican, R. C. Bradt, F. L. Kennard, W. J. Minford, and C. C. Sorell. Ceramic Eutectic Composites. Tailoring Multiphase and Composite Ceramics, Materials Science Research 20 (R. E. Tressler, G. L. Messing, C. G. Pantano, and R. E. Newnham, eds.). Plenum Press, New York, 1986, pp. 103–113.
29. C. C. Sorell, V. S. Stubican, and R. C. Bradt. Mechanical Properties of ZrC-ZrB_2 and ZrC-TiB_2 Directionally Solidified Eutectics. J. Am. Cer. Soc. 69(4):317–321, 1986.
30. J. Echigoya, Y. Takabayashi, and H. Suto. Hardness and Fracture Toughness of Directionally Solidified Al_2O_3- $ZrO_2(Y_2O_3)$ Eutectics. J. Mat. Sci. Lett. 5:153–154, 1986.
31. C. A. May and A. K. U. Obi. Compressive Strength of Lithia-Silica Glass Ceramics. Mechanical Properties of Ceramics (2) (R. W. Davidge, ed.). Proc. Brit. Cer. Soc. 25:49–65, 5/1975.
32. A. R. B. Verma, V. S. R. Murthy, and G. S. Murty. Microstructure and Compressive Strength of SiC-Platelet-Reinforced Borosilicate Composites. J. Am. Cer. Soc. 78(10):2732–2736, 1995.
33. F. H. Simpson, Macrolaminate Particle Composite Material Development. Summary Report for Boeing Co. Work on US Navy Bureau of Weapons Contract No. 64-0194-f, 5/1965.
34. W. J. Knapp and F. R. Shanley. Process for Making Laminated Materials. US Patent 3,089,796, 5/1963.
35. S. Suresh, L. X. Han, and J. J. Petrovic. Fracture of Si_3N_4-SiC Whisker Composites Under Cyclic Loads. J. Am. Cer. Soc. 71(3):C–158–161, 1988.
36. R. W. Rice, C. P. Cameron, and P. L. Berneburg. Lower Cost Ceramic Armor Ma-

terials. Proceedings Second Ballistics Symposium on Classified Topics, Johns Hopkins Applied Physics Lab. American Defense Preparedness Assn., 1992.

37. K. Breder, G. DePortu, J. E. Ritter, and D. D. Fabbriche. Erosion Damage and Strength Degradation of Zirconia-Toughened Alumina. J. Am. Cer. Soc. 71(9):770–775, 1988.

38. S. Wada, N. Watanabe, and T. Tani. Solid Particle Erosion of Brittle Materials (Part 6)—The Erosive Wear of Al_2O_3-SiC Composites. J. Jpn. Cer. Soc., Intl. Ed. 96:113–119, 1988.

39. S. Wada, N. Watanabe, and T. Tani. Solid Particle Erosion of Brittle Materials (Part 9)—The Erosive Wear of Al_2O_3-SiC Particle Composites. J. Jpn. Cer. Soc., Intl. Ed. 96:737–740, 1988.

40. S. Wada and N. Watanabe. Solid Particle Erosion of Brittle Materials (Part 7)—The Erosive Wear of Commercial Ceramic Tools. J. Jpn. Cer. Soc., Intl. Ed. 96:323–325, 1988.

41. K. Yamada, N. Kamiya, and S. Wada. Abrasive Wear of Al_2O_3-SiC Composites. J. Cer. Soc. Jpn., Intl. Ed. 99:797, 1991.

42. C. He, Y. S. Wang, J. S. Wallace, and S. M. Hsu. Effect of Microstructure on the Wear Transition of Zirconia-Toughened Alumina. Wear 162–164:314–321, 1993.

43. Y. J. He, A. J. A. Winnubst, A. J. Burggraaf, H. Verweij, P. G. T. Van der Varst, and G. de With. Sliding Wear of ZrO_2-Al_2O_3 Composite Ceramics. J. Eur. Cer. Soc. 17(11):1371–1380, 1997.

44. C. Cm. Wu, R. W. Rice, C. P. Cameron, L. E. Dolhert, J. H. Enloe, and J. Block. Diamond Pin-on-Disk Wear of Al_2O_3 Matrix Composites and Nonoxides. Cer. Eng. Sci. Proc. 12(7–8):1485–1499, 1991.

11

Particle and Grain Effects on Mechanical Properties of Composites at Elevated Temperature

I. INTRODUCTION

This chapter addresses the effects of particle (and matrix grain) parameters on the mechanical properties of ceramic composites as a function of temperature, i.e. analogous to the treatment of grain effects on monolithic ceramics in Chapters 6 and 7. Again, the term particle is used in both a generic sense to include platelet, whisker, and fiber (e.g. in directionally solidified eutectics), and for actual particle composites. The primary particle parameters of interest besides composition are volume fraction (ϕ) and size as well as shape, and orientation. While the amount of desired information is often limited, some useful insight and guidance can be obtained. The focus is again more on changes in mechanical properties at more modest temperature where failure is primarily brittle fracture, in part to aid in understanding fracture at and near room temperature. Thus creep and related deformation are not a focus, but some results are noted, e.g. to emphasize the changes that such behavior often entails. However, thermal shock is addressed, and the limited but important data on thermal shock fatigue and the very limited data on related mechanical fatigue are addressed.

After a brief discussion of the limited theoretical background, the topics treated are elastic and related behavior, crack propagation and fracture energy or toughness, thermal shock and fatigue and related mechanical fatigue, tensile (flexure) strength, and hardness with compressive strength and related behavior.

Again, within each of these topics the data, though limited, is treated in the same general order for composites in earlier chapters, i.e. glass matrix, polycrystalline oxide, all nonoxide, platelet, whisker, eutectic, and then ceramic–metal composites.

II. THEORETICAL BACKGROUND

There is very little specific theoretical background for the focus of this chapter, particle (and matrix grain) dependence of properties, i.e. again similar to the situation for monolithic ceramics in Chapters 6 and 7, but with less certain guidance for composites. Again the most specific guidance is for elastic moduli, but with three added uncertainties and variations over those for monolithic ceramics, namely the specifics of the combination of the moduli of the constituents to yield those of the composite as at room temperature, possible changes due to chemical interactions of the constituents, and variation in moduli of each constituent with temperature. The latter may vary for the constituents, especially when the refractoriness of the constituents varies widely, e.g. for a glass matrix with refractory dispersed phase or a refractory ceramic matrix with less refractory metal dispersed phase.

There is more guidance for the thermal shock behavior of composites, since its dependence on other properties such as tensile strength, Young's modulus, and thermal expansion, as well as thermal conductivity in some cases, is generally accepted (see Chap. 6, Sec. V). However, predicting these composite properties from those of the composite constituents introduces an added degree of uncertainty, which is compounded if microstructural mechanisms such as crack branching and especially significant microcracking occur.

Trends for other mechanical properties may be indicated by their dependence on or correlation with elastic moduli, especially Young's modulus, but this poses the same or more severe uncertainties as for monolithic ceramics. Thus while the trend for fracture toughness would be predicted by that of Young's modulus for simple elastic fracture, the occurrence of behavior such as microcracking and crack branching or bridging, which are composite, microstructure, test, and crack size and character dependent, present complications. Resultant uncertainties are compounded for behavior that only correlates with elastic properties such as hardness and compressive strength, and the relation between the latter two properties in view of the limited and uncertain theoretical dependence of hardness on composite structure and temperature. Wear and related behavior that typically depend on elastic moduli, hardness, and fracture toughness thus reflect the compounded uncertainties of each of these properties and their temperature dependence in predicting trends, thus typically limiting prediction to only rough trends.

III. DATA REVIEW

A. Elastic and Related Behavior

One of the limited studies of the temperature dependence of elastic moduli of ceramic composites is the experimental and summarized literature data for Young's modulus of alumina fibers (FP) and Al_2O_3 +20 w/o Y-ZrO_2 (PRD) versus temperature to 1200–1300°C of Lavaste et al. [1]. The FP fiber had $G \sim 0.5$ μm and the PRD 166 fiber respective Al_2O_3 grain and ZrO_2 particle sizes of 0.34 and 0.15 μm, with both fibers averaging nominally 18 μm diameter and having very limited residual porosity. They reported $E \sim 360$ GPa for the PRD fiber at room temperature, in good agreement with the rule of mixtures expectation, i.e. Eq. (8.1) with $E \sim 400$ and 220 GPa respectively for Al_2O_3 and ZrO_2. Their E value for PRD fibers was very close to that reported by Romie [2] of ~ 380 GPa at room temperature, but both were higher than the value reported from another study [3] of 280 GPa. However, the latter data and that of Lavaste et al. [1] are much closer at higher temperatures, e.g. being respectively 300 and 325 GPa at 800°C and crossing each other twice between 1000 and 1200°C, giving respective values of ~ 190 and 159 GPa at 1200°C. These values are bracketed (except for the one low room temperature value for the PRD fiber) by values for FP fibers of ~ 405 and 300 GPa at room temperature, with the higher FP value being consistent with bulk alumina data. However, the PRD fiber data showed faster E decrease above 1000–1100°C than the alumina fiber (FP) and bulk alumina [4]. This faster decrease of the Al_2O_3-ZrO_2 (PRD) fiber probably reflects at least in part the finer grain and particle sizes and resultant enhanced grain boundary sliding, and probably faster lower temperature decreases in E (Fig. 6.18).

French et al. [5] compiled Young's modulus data for both the components of, as well as for composites of Al_2O_3-50 v/o ZrO_2 and Al_2O_3-50 v/o $Y_3Al_5O_{12}$ (YAG), showing that E for both the composites was very close to the rule of mixtures [Eq. (8.1)] to the limit of the data presented (1200°C). Tsukuma et al. [6] showed an overall nearly linear decrease in E for 2Y-ZrO_2 + 20 w/o Al_2O_3 composite to 800°C, except for a temporarily faster decrease between 200 and 300°C (again consistent with Fig. 6.18). The latter corresponded with a marked internal friction peak just past 200°C.

Gadkaree [7] measured Young's and shear moduli and Poisson's ratio versus temperature for composites of 25 v/o of SiC whiskers hot pressed in either a cordierite or an anorthite matrix. The cordierite matrix composite gave moduli decreases of ~ 11% to 1100°C and the anorthite matrix composite decreases of ~ 5–7% to 1000°C, with indications of accelerating decreases above 1000°C for the former and especially above 1100°C in the latter. The respective changes of Poisson's were ~ + 4% and – 11%, but they are of some statistical uncertainty. Application of the Halpin–Tsai model for fiber composites gave mixed predictions at room temperature and were apparently not applied to the temperature dependence. Lee and Case [8] showed that Young's modulus of Al_2O_3-20 v/o SiC whiskers decreased linearly from 410 GPa at ~ 22°C at a rate of ~ 1.1%/100°C to ~ 370 GPa at 900°C.

Mazdiyasni and Ruh [9] reported Young's modulus of hot pressed composites of Si_3N_4 with 5, 10, or 12.5 w/o additions of BN powder (fine platelets) decreasing, e.g. by ~ 25-33% to 1000°C, ≥ for Si_3N_4 alone (~ 25%), while above 1000°C the rate of decrease accelerated more as the BN level increased.

B. Crack Propagation and Fracture Toughness

Consider first crystallized glasses. Northover and Groves [10] showed that the (NB) fracture toughness of crystallized LiO_2-Al_2O_3-SiO_2 (LAS) decreased from ~ 1.18 MPa·m$^{1/2}$ at ~ 22°C to a minimum of ~ 0.9 MPa·m$^{1/2}$ at ~ 500°C and then increased to ~ 1.2+ MPa·m$^{1/2}$ at the limit of testing at 1000°C (Fig. 11.1); and that high temperature SCG occurred (via intergranular fracture versus transgranular fast fracture). Govila et al. [11] demonstrated in more detail such high-temperature intergranular SCG and corroborated similar strength minima and maxima at ~ 600 and 1000°C (Fig. 11.9), though obtaining a room temperature toughness of only ~ 0.6 MPa·m$^{1/2}$. Majumdar et al. [12] reported (IF) fracture toughness of a similar commercial material but did not measure values between 22 and 850°C, with the value at the latter temperature being < 10% above that of

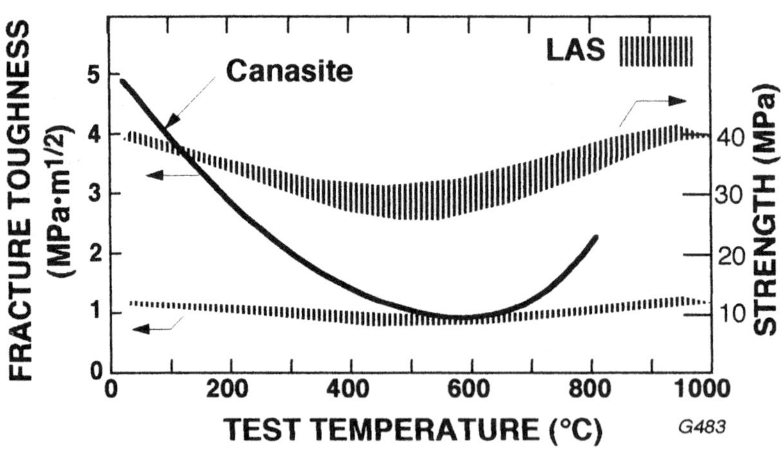

FIGURE 11.1 Fracture toughness and strength of crystallized glasses versus test temperature. Data for LAS [10] and a canasite glass [13]. Note the minimum for both at ~ 500°C and that since mechanical properties normally decrease at higher temperatures there must be a maximum at ≥1000°C, which is corroborated by other tests [11] (Fig. 11.9).

~ 1.4 MPa·m$^{1/2}$ at 22°C and decreasing rapidly above 850°C to ~ 0.25 MPa·m$^{1/2}$ at 1100°C. These studies, especially that of Govila et al., showed that the increase in toughness (and strength) are due respectively to plastic blunting of cracks and high-temperature SCG.

Beall et al. [13] showed the (CNB) fracture toughness of an extensively crystallized canasite glass decreasing from nearly 5 MPa·m$^{1/2}$ at 22°C to a pronounced minimum of ~ 1 MPa·m$^{1/2}$ at ~ 600°C and then increasing to ~ 2 MPa·m$^{1/2}$ at the limit of testing at 800°C (Fig. 11.1). They attributed the marked decrease in toughness with increasing temperature to decreasing microcracking, which they believed was an important factor in the high toughness at ~ 22°C.

Turning to non-glass-based composites, Brandt et al. [14] reported that the NB toughness of a commercial cutting tool composite of Al_2O_3-4 w/o ZrO_2 decreased from ~ 4.3 MPa·m$^{1/2}$ at 22°C by only ~ 10% by 800–1000°C and then by ~ 50% by 1200°C. The latter marked decrease was attributed to the onset of high-temperature slow crack growth based on both the substantial decrease from ~ 50% transgranular fracture at room temperature and especially deviations from linear load deflections at stress intensities > 2 MPa·m$^{1/2}$ at 1200°C.

French et al. [5] measured both CNB and IF toughness for both the constituents and their composites of Al_2O_3-50 v/o ZrO_2 (CZ) and Al_2O_3-50 v/o $Y_3Al_5O_{12}$ (YAG) to 1200°C, all having ~ 2 μm grain and particle sizes. CNB tests showed the Al_2O_3 toughness decreasing from ~ 3.7 to ~ 2.25 MPa·m$^{1/2}$ with most of the decrease by 800–1000°C and a similar ~ 60% decrease for CZ from ~ 2 to ~ 1.25 MPa·m$^{1/2}$, while the composite, though starting lower, decreased less (from ~ 3.1 to 2.5 MPa·m$^{1/2}$), ending up > the Al_2O_3. In contrast to this, CNB tests showed YAG toughness ~ independent of temperature at ~ 1.8 MPa·m$^{1/2}$, but with a modest maximum of ~ 2.2 MPa·m$^{1/2}$ at 1000°C; the YAG composite toughness was constant at ~ 2.8 MPa·m$^{1/2}$, except for a modest decrease between 1000 and 1200°C. On the other hand, IF tests typically showed that toughness values for all constituents and composites at ~ 22°C were ~ 40% < CNB values with similar trends with temperature, except for ZrO_2 and Al_2O_3-50 v/o ZrO_2, which showed less decrease with increasing temperature and increases at higher temperatures to respectively equal and exceeding the toughness of the Al_2O_3. The authors also showed that their Al_2O_3 had essentially all intergranular fracture while the cubic ZrO_2 and especially YAG had mainly transgranular fracture, and the composites had fracture modes intermediate between the constituents at ~ 22°C. However, all bodies showed essentially exclusive intergranular fracture by 800°C.

Tsukuma et al. [6] reported that the NB toughness of their HIPed 2Y-TZP+ 20 w/o Al_2O_3 decreased from 10 MPa·m$^{1/2}$ at 22°C to a minimum of ~ 6.5 MPa·m$^{1/2}$ at ~ 400°C, through a maximum of ~ 8 MPa·m$^{1/2}$ at ~ 700°C and then decreasing to ~ 6 MPa·m$^{1/2}$ at 1000°C. Whether the minimum is related to the more rapid decrease in *E* between 200 and 300°C (Sec. III.A) is not known, but this minimum also appears to correspond to strength minima at 400°C (Fig. 11.10).

Orange et al. [15] showed NB toughness of reaction processed mullite + ~ 20 v/o ZrO_2 decreased to a minimum at ~ 400°C and then rose to a maximum at 800°C (Fig. 11.2), similar to the above 2Y-TZP+ 20 w/o Al_2O_3. Two variations of the composite produced lower toughness values (e.g. to ~ 4.5 MPa·m$^{1/2}$ at 22°C) but with minima and maxima at the same temperatures as for the highest toughness, as was also true for strengths, i.e. they correlated with toughness trends. However, a much more pronounced NB toughness and a lesser strength minimum, and an implied toughness maximum, at ≥ 200°C higher temperature, was reported by Leriche et al. [16,17] for similar reaction processed bodies. These have some similarities with, and differences from, the more limited strength data on such materials.

Niihara et al. [18] showed that I and IF toughness of their hot pressed Al_2O_3+ 5 v/o SiC (2 μm) composite was constant at ~ 4.3 MPa·m$^{1/2}$ (~ 10% above the Al_2O_3 without SiC) till ~ 800°C and then increased to ~ 5 MPa·m$^{1/2}$ at 1200°C (while the Al_2O_3 decreased to ~ 3 MPa·m$^{1/2}$). Fractography showed clear SCG on 1200°C fractures.

Data of Baldoni et al. [19] for an Al_2O_3 -30 v/o TiC composite showed IF toughness decreasing from ~ 3.5 MPa·m$^{1/2}$ at 22°C by 10–15% by 800–1000°C

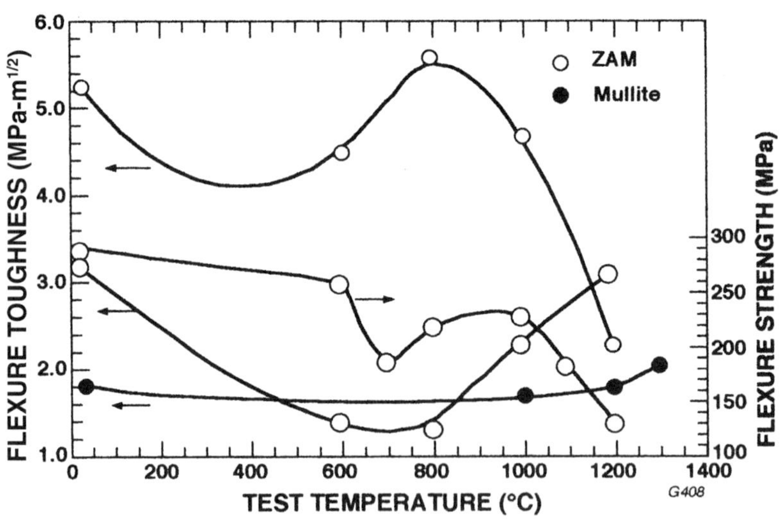

FIGURE 11.2 Fracture toughness versus test temperature of mullite alone and mullite-ZrO_2 composites made by reaction of zircon and alumina [15] and for another similar reaction processed composite (along with strength data) [16,17]. Note the significant minima and subsequent maxima but with different temperatures and property values of occurrence.

and then increasing 10% to ~ 4 MPa·m$^{1/2}$ at 1200°C, the latter associated with SCG. These trends are consistent with those of Brandt et al. [14] for NB toughness of a commercial cutting tool composite of Al_2O_3-30 w/o Ti(C,N) being ~ constant at ~ 3 MPa·m$^{1/2}$ to 1000°C, and with intergranular fracture increasing substantially from ~ 50% at 22°C.

Baldoni et al.'s [19] data for a Si_3N_4 -30 v/o TiC composite showed IF toughness decreasing from ~ 4.7 MPa·m$^{1/2}$ at 22°C by ~15% at 800°C, then increasing to ~ 6.9 MPa·m$^{1/2}$ at 1200°C, the latter associated with SCG. Tai and Watanabe [20] showed that I toughness of their WC+10 w/o Co-20 w/o Al_2O_3 composite was ~ twice that of the Al_2O_3 alone (~ 9 versus ~ 4 MPa·m$^{1/2}$) to ~ 300 °C; it increased to nearly 11 MPa·m$^{1/2}$ at 400°C and then decreased back to ~ 8 MPa·m$^{1/2}$ at their limit of testing of 600°C (while that for Al_2O_3 decreased slightly).

Toughness data for composites of mullite with SiC whiskers showed higher toughness at 1300 versus 22°C along with strength data (Sec. III.E).

C. Thermal Shock

Before considering the effects of temperature on tensile or flexure strength, the subject of effects of transient exposure to higher temperatures, namely thermal shock degradation, is considered. As discussed in Chap. 6, Sec. V, there are two cases to be dealt with, one of avoiding any significant loss of strength as a result of thermal shock exposure, and the other surviving thermal shock with some limited but useful load carrying ability. In the first case it is commonly of interest to start with and maintain substantial strengths, e.g. at least several hundred MPa, while in the latter case starting, and especially postshock, strengths of the order of 10 to 100 MPa are common. Refractories are a common and very important manifestation of the latter.

While the subject of refractories is a large and specialized subject not treatable here, a few comments on the relevance of composite structures are pertinent. High thermal shock resistance is achieved in many refractories by making them quite porous (which also aids their thermal insulation), but there are important cases where extensive porosity is either not sufficient or not suitable, e.g. in steel processing, especially in continuous casting, which places severe thermal shock (and other) demands on refractories. Refractories to meet such demands are commonly made with substantial contents, e.g. 25–40%, of sizeable (e.g. ≥ 100 μm) graphite flakes in matrices of such materials as SiC or Al_2O_3 [21]. Such refractories (often containing 15–30% porosity) have such high thermal shock resistance that use of the thermite reaction of iron oxide and aluminum metal powder to produce molten iron in seconds is needed to test their thermal shock resistance. Such microstructures of coarse graphite flakes in denser refractory carbide matrices or of coarse (e.g. several hundred μm dia.) BN flakes in oxide matrices were also investigated for missile nose tips for military needs [22]. However, more recent work has been heavily focused on much finer composite microstructures in the search for bodies capable of sustaining substantial thermal shock but retaining considerably more strength.

That considerable improvement can be made in thermal shock resistance via composite microstructures was shown by studies of Fairbanks et al. [23] and of Oguma et al. [24] of respectively a cordierite- and a canasite-based glass before and after crystallization. Both showed that crystallization increased the critical quench temperature difference (ΔT_C) for strength loss on quenching into a room temperature water bath ~ twofold (respectively from ~ 130 to ~ 350°C and from ~ 100°C for the parent glass and to 200°C for the crystallized canasite glass), < the ~ threefold increases in strengths, despite some increase in E (and thermal expansion, at least for the canasite system). This doubling of ΔT_C was consistent with the strengths divided by the product of the Young's modulus and the thermal expansion coefficient, especially as shown by more complete characterization of the canasite system. Further, both systems showed ≥ threefold increases in retained strengths for quenching above ΔT_C, i.e. from ~ 5–10 to 25–30 MPa, which correlated with toughness increases of twofold, or substantially more, as crack sizes increased in the crystallized bodies. This crack size dependence of toughness was cited as important in the improved shock resistance of the cordierite system [23]. One uncertainty left in the cordierite system is why annealing the parent glass below the nucleation temperature resulted in not only a doubling of the starting strength but also a nearly 40% increase in ΔT_C.

Becher [25] showed substantial improvements in the thermal shock resistance of Al_2O_3-ZrO_2 composites, with ΔT_C going through a maximum similar to those of strength and fracture energy as a function of the volume fraction of unstabilized ZrO_2 (of high uniformity in size and dispersion, e.g. Fig. 9.6). However, these results were from tests in which the quench medium was boiling water, rather than water at room or lower temperature, and boiling water does not give as severe a thermal shock. Thompson and Rawlings [26] also showed some modestly increased thermal shock resistance, i.e. a ΔT_C of ~ 270°C for Al_2O_3-20 w/o ZrO_2 (+ 3 m/o Y_2O_3) by quenching into an ethylene glycol–water solution at 20°C, which again represents a less severe shock. They also showed that acoustic emission and other acoustic techniques gave additional information, including on precursor events to thermal shock damage and its sources.

Other tests of similar Al_2O_3-ZrO_2 composites quenched into water at ≤ 22°C do not show such improvements. Thus Tomaszewski [27] showed that ΔT_C remained at ~ 200°C with increased addition of unstabilized ZrO_2 to Al_2O_3 despite starting strengths first increasing and then decreasing with increasing ZrO_2 content for quench tests into room temperature water. Beyond 20 w/o ZrO_2 there was essentially no change in the low strengths (~ 100 MPa) with quenching. Sintering the composites in vacuum, which resulted in substantial stabilization (due to ZrO_2 reduction) extended the level of the starting strength and the range of ZrO_2 content over which they occurred but did not change the ΔT_C or the trends at ≥20 w/o ZrO_2, except for some lowering of the low, ~ constant, strengths below and above ΔT_C. Similarly, Sornakumar et al. [28] showed ΔT_C= 200°C for 3Y-TZP-20 w/o Al_2O_3 quenched into room temperature water.

Lutz et al. [29] showed that ΔT_C for their composites consisting of agglomerates of one Al_2O_3-ZrO_2 composition designed to give greater microcracking in a matrix of another Al_2O_3-ZrO_2 composition designed to give less microcracking was also typically ~ 200°C when the bodies had strengths > 200 MPa. Composites giving limited or no loss of strength on quenching into room temperature water all had lower starting strengths, typically 100–180 MPa. The lower strength and good retention of it after thermal shock was attributed to extensive microcracking, as shown by dye penetrant tests. Again note that this microcracking only occurred when some limited porosity remained in the bodies, especially in the agglomerates for greater microcracking, and was not present in HIPed bodies (though whether the primary effect is residual porosity or possible changes in ZrO_2 stabilization due to some reduction in HIPing cannot be unequivocally ascertained).

Yuan et al. [30] thermally shocked mullite-ZrO_2 composite bars by heating them to 1200°C and then removing them from the furnace and exposing them to a stream of room temperature air, giving an estimated cooling rate of 50°C/sec. Postshock measurement showed substantial, ~ 50%, strength loss with no ZrO_2 and progressively less loss as ZrO_2 content increased, with no loss, or possible increase of, strength at higher v/o ZrO_2. Strength losses were generally somewhat greater with three versus one thermal shock (Fig. 11.3). Young's modulus tests showed some decrease on shocking with no ZrO_2, no change with some, and some possible limited increase on shocking at higher ZrO_2 contents, again with some greater effects with three versus one shock. Toughness tests after shocking showed some possible limited, e.g. 10%, increases on shocking. Ishitsuka et al. [31], using quenching into a water bath at 0 (not 22)°C, showed ΔT_C was ~ 250°C for 3Y-TZP alone and ~ 375°C for mullite alone and 3Y TZP-50 v/o mullite composites, with similar low, e.g. 100 MPa, residual strengths. Composites with higher TZP contents showed promise for possible higher ΔT_C and residual strength values, consistent with their higher toughnesses. Though the latter are not fully reflected in strengths (Chap. 9, Sec. III.B), the effect of toughness is consistent with serious thermal shock reflecting larger crack toughnesses values, while strength frequently does not. Orange et al. [15] showed that ΔT_C increased from 250+ to 300+°C and retained strengths increased from ~ 75 to 150 MPa as starting strength decreased for water quench tests of their reaction processed mullite + 20 v/o ZrO_2 composites. The above

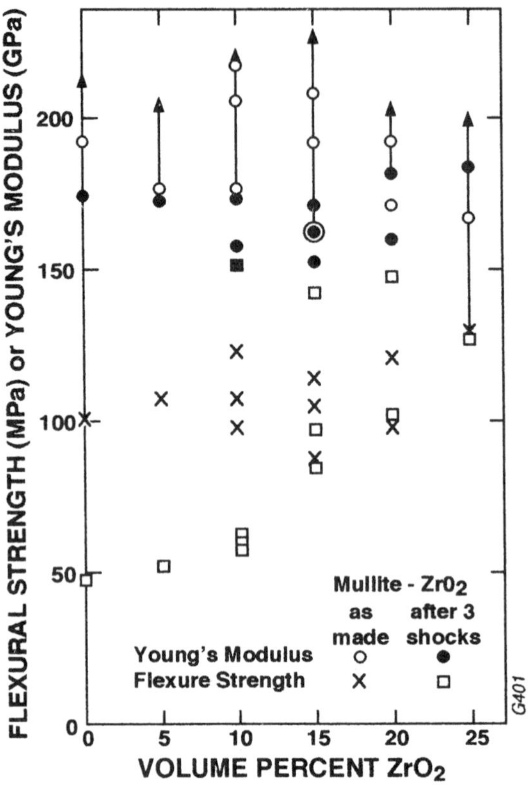

FIGURE 11.3 Strength and Young's modulus at 22°C of mullite-ZrO_2 composites versus volume percent ZrO_2. Data [30] plotted for both as-processed bars and after bars were subjected to three air quench thermal shocks. Arrows with the as-processed E values indicate probable corrections for residual porosities of 2.6–8.7%.

results are consistent with microcracking, which is expected in mullite-ZrO_2 composites, and has been directly shown in directionally solidified mullite-ZrO_2 composites [32].

Kim and Kim [33] showed that the critical ΔT_C for Al_2O_3-SiC composites made by directed oxidation of Al metal was possibly increased by 10–20% over the value of ~ 200°C for Al_2O_3 alone, especially as the SiC particle size decreased from ~ 100 to ~ 13 μm (which increased starting strengths, Fig. 9.15). However, retained strengths after thermal shock damage were increased by addition of SiC particles, especially with coarser SiC particles, e.g. ~ 100% for ~ 100 μm particles, i.e. the opposite trend from that for starting strengths. Residual Al in the composite was seen as not playing a significant role in the composite thermal shock results.

Landon and Thevenot [34] reported that both starting strengths and the critical quench temperature difference for significant strength loss generally decreased in AlN-SiC bodies, e.g. from ~ 800 MPa and ~ 350°C to ~ 450 MPa and ~ 300°C as the SiC content decreased from 90 to 10%, with no clear distinctions between whether α or β-SiC was used.

Tiegs and Becher [35] reported that retained strengths in thermal shock tests of Al_2O_3-20 v/o SiC whiskers progressively increased from ~ 620 MPa with no shock to ~ 700 MPa with a quench of 900°C, but there were slightly reduced retained strengths to ~ 550 MPa with 10 quenches of 900°C. However, these tests were conducted by quenching into boiling instead of room temperature water baths, which as indicated earlier for Al_2O_3-ZrO_2 gives less severe thermal shock results.

Lee and Case [8], while also showing limited effects of additional thermal shocks, showed more changes in thermal shock testing of the same composition whisker composites, with quenching into a room temperature water bath. They evaluated changes in Young's modulus and internal friction showing respectively decreases and increases as the quench temperature differential increased, with both showing substantial rates of change for quenches of > 310°C (Fig. 11.14). These results indicated limited improvement of thermal shock resistance of such composites over that of Al_2O_3 itself, e.g. a ΔT_C of ~ 200°C (Fig. 6.21). Lack of a precipitous drop in E, as commonly shown for strengths, and a rise in internal friction, is more realistic, e.g. since the catastrophic drop in strength at ΔT_C is apparently an artifact of testing limited numbers of specimens and at limited temperature intervals as noted in Chap. 6, Sec. V. The relationship of the above thermal shock behavior to a report of acoustic emissions in similar composites of Al_2O_3-20 v/o SiC whiskers increasing significantly during cooling from thermal cycling to temperatures > 500°C [36] is uncertain.

Consider now composites of either graphite or BN flakes (or platelets) in various matrices, which have been suggested by development of refractories as noted earlier. A study of the underlying mechanical behavior of Al_2O_3–graphite refractories has been reported by Cooper et al. [37] and other aspects of laboratory Al_2O_3–graphite composites were outlined in Chap. 9, Sec. III.C. Composites of various oxides such as BeO, Al_2O_3, MgO, and ThO_2 [22, 38] with additions of 5–30 v/o of large BN flakes (e.g. 75–400 μm) were investigated and showed significantly improved thermal shock resistance over the matrix alone but at substantial loss of initial strengths. It was also noted that addition of as little as 3 v/o W (1–2 μm) particles to MgO substantially increased thermal shock resistance over the matrix alone, but strength data was again not reported [22].

More extensive study has been made of composites of fine BN platelets in various, especially Al_2O_3 or mullite, matrices by Rice and colleagues based on

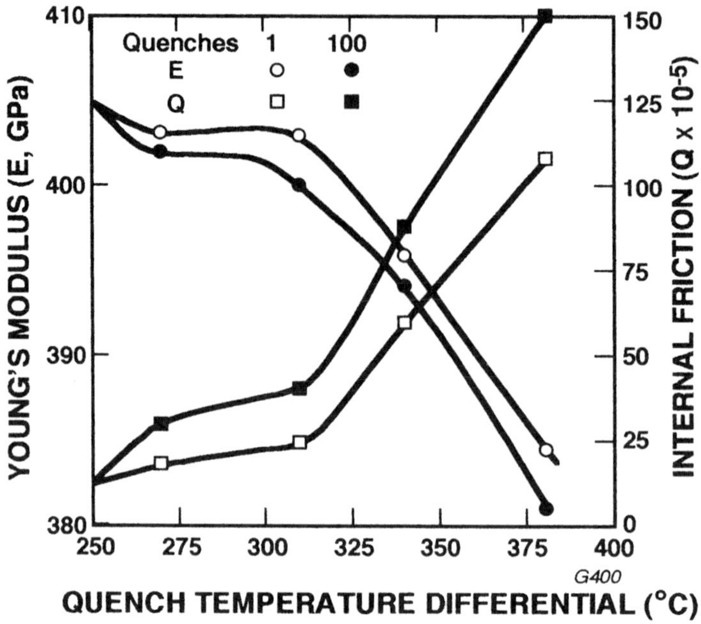

FIGURE 11.4 Young's modulus and internal friction versus quench temperature difference for 10 and 100 quenches of Al_2O_3-20 v/o SiC whisker composites. Note that there is some fatigue effect, i.e. further reduction with repeated quenches at the same temperature difference, but the major effect is from increased quench temperature differences. (From Ref. 8.)

potential needs for improved radar windows and domes for more extreme thermal environments. They first investigated Al_2O_3-BN composites made by hot pressing powders of the ingredients giving Al_2O_3 grain sizes ranging from < 1 to > 5 μm and BN particle thicknesses of 0.1–0.2 μm and diameters of ~ 5 μm [39,40]. Subsequent similar processing with mullite matrices resulted in similar microstructures, but probably more toward a 5 μm matrix grain size (promising results were also demonstrated with similar Si_3N_4-BN composites, but they were not pursued because of expected performance and subsequently demonstrated processing advantages of the oxide matrix, especially mullite-based, composites). Properties of these composites are summarized in Table 11.1, showing substantial anisotropy as expected from substantial orientation of the BN platelets during hot pressing, as well as reasonable strengths attributed to the substantially finer BN size than in earlier composites (Fig. 11.5).

Four additional aspects of the above Al_2O_3- and mullite-BN composite studies should be noted. First, Coblenz [41] conceived of reaction processing

TABLE 11.1 Summary of Properties of Composites With Fine BN Platelet Particles[a]

v/o BN	ρ (gm/cc)	E∣(GPa)	E⊥ (GPa)	K⊥ ($MPam^{1/2}$)	σ⊥ (MPa)	ΔT_C (°C^{-1})
(A) Al_2O_3 + BN[b]						
30	3.3	110	170	2–9	150–500	400–800
40	3.1	81	—	—	75–200	400
50	3.0	77	—	1.3–2.3	150–200	800–1000
(B) Mullite+ BN						
30[c]	2.8	100	130		200–350	400–450
30[d]	2.8	100[e]		2.9–3.7[e]	160–300[e]	350

[a]Density (ρ), Young's modulus (*E*), fracture toughness (*K*), flexure strength (σ), and critical temperature difference on quenching into a room temperature water bath for significant loss of strength (ΔT_C); ∣ and ⊥ refer respectively to properties measured with stress parallel and perpendicular to the hot pressing axis, i.e. normal to the plane of preferred orientation.
[b]Al_2O_3 + 30 v/o BN made by hot pressing powders of mullite and BN with resultant anisotropy. *Source*: Ref. 39.
[c]Mullite + 30 v/o BN made by hot pressing mullite + BN powders and hence resultant anisotropy. *Source*: Ref. 39.
[d]Data for mullite + 30 v/o BN made by reaction hot pressing ingredients to produce mullite + BN in situ and hence resultant approximate isotropy. *Source:* Ref. 41.
[e]Bodies were ~ isotropic (Fig. 11.5C).

that produced more homogeneous and ~ isotropic mullite+ BN composites with lower raw materials costs and similar or easier processing that gave good properties (Table 11.1). Second, better properties and overall performance were found to correlate with more homogeneous, and thus generally finer, microstructures, especially with regard to the matrix grain size (Fig. 11. 5) [39–42].

Third, other compositions have been investigated, e.g. Coblenz and Lewis [41] showed that reaction processing of mullite + 30 v/o BN with ~ 28 v/o residual Si_3N_4 increased initial strengths by ~ 10–15% and ΔT_C by the same or greater percent, e.g. to ≥450°C. Goeuriot-Launay et al. [43] prepared bodies of 70 v/o Al_2O_3 + 30 v/o γ-ALON with increasing volume percents of BN decreasing starting strengths and increasing ΔT_C that are overall consistent with those of Lewis et al. [39] (Fig. 11.6). Mazdiyasni and Ruh [9] showed that Si_3N_4+ 20 w/o BN hot pressed composites had ΔT_C of 700–800°C versus ~ 600°C with no BN and that there were (higher and) more gradual changes in internal friction and less tendency for catastrophic failure in the composite. Valentine et al. [44] showed that SiC-BN composites had increasing ΔT_C with increasing BN content similar to, but greater than, that for Al_2O_3-BN composites (Fig. 11.6).

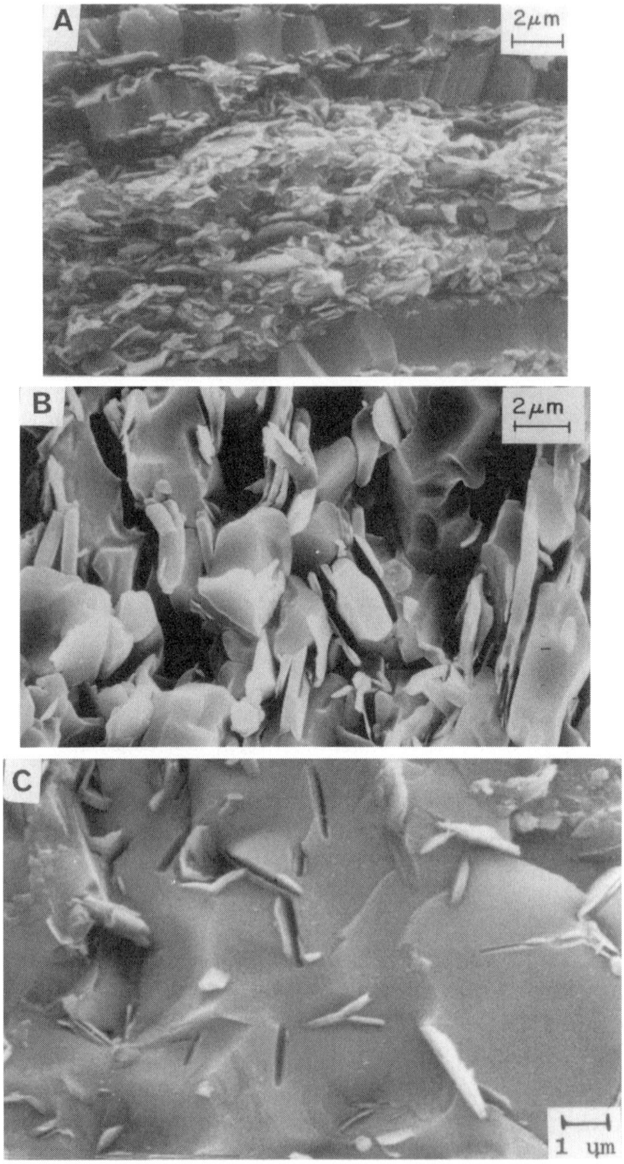

Figure 11.5 Comparison of (A) less homogeneous and (B) more homogeneous Al_2O_3+ 30 v/o BN composites from hot pressing of mixed Al_2O_3 and BN powders. (C) a reaction hot pressed mullite + 30 v/o BN composite. Note the laminar orientation (horizontal, A, and vertical, B, directions) and its coarser character and associated coarser, laminar Al_2O_3 grain structure, especially in (A) versus much more homogeneous and ~ isotropic distribution of BN platelets in (C).

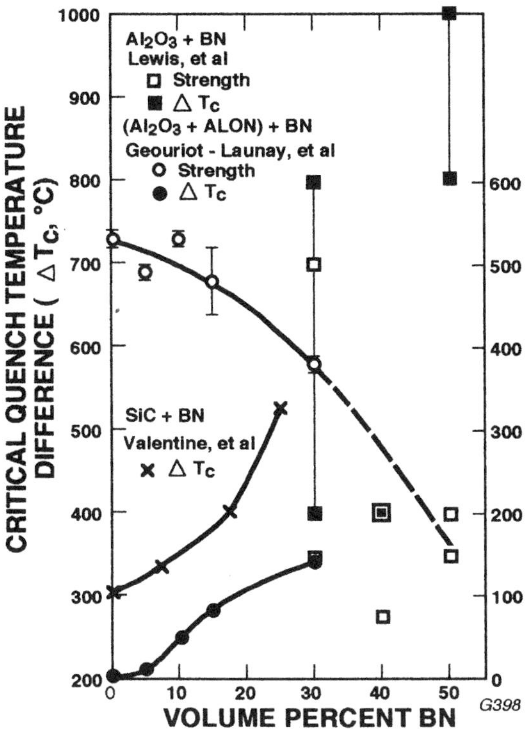

FIGURE 11.6 Critical quench temperature difference for Al_2O_3-BN-based composites of Lewis et al. [39] and Goeuriot-Launay et al. [43] and of SiC-BN composites of Valentine et al. [44], and for starting strengths of the former two composites. Note that (1) the upper and lower points show the range of data [39] and the vertical bars the standard deviations [43], and (2) the trends for both properties are overall consistent between the two alumina-based composites, especially for starting strengths (as are also the trends for SiC-BN), and roughly for ΔT_C, more so for the lower values of Lewis et al. [39].

D. Thermal Shock and Related Mechanical Fatigue

The fourth aspect of Al_2O_3- and mullite-BN composites that should be noted is their thermal shock fatigue, that is, their resistance to degradation and failure under repeated thermal shocking. While this topic has received very little attention, especially for composites designed to give higher strengths than conventional refractories, limited data shows that this is an important subject. Lewis and Rice [45] first investigated this for Al_2O_3- and mullite-BN composites and showed that just a few repeated thermal shock cycles progressively reduced the retained strength beyond that found for a single thermal shock, with the decrease being faster, i.e. in fewer cycles (and possibly somewhat greater), as the quench temperature increased toward ΔT_C, but that there appeared to be a lower limit to the ΔT for such fatigue effects (e.g. ~ 250°C for these BN composites, Fig. 11.7). Less extensive results for mullite-BN composites were similar.

Lewis and Rice [45] tested several ceramics including other composites and monolithic polycrystalline or glass-based ceramics in order to ascertain the mechanisms involved in thermal shock fatigue, with an initial goal being to determine whether the fatigue effect was due to SCG, enhanced by the presence of water in the quench test. They found no quench fatigue effect in repeated quenching at 10°C below ΔT_C (producing ~ 95% of the stress for critical crack growth) of a borosilicate (Code 7740, Pyrex) glass nor in a commercial alumina containing substantial glass phase (AD85). They thus concluded that the quench fatigue observed with the BN composites was not due to environmentally driven SCG, i.e. water, effects, as also shown by other results below. (They noted that

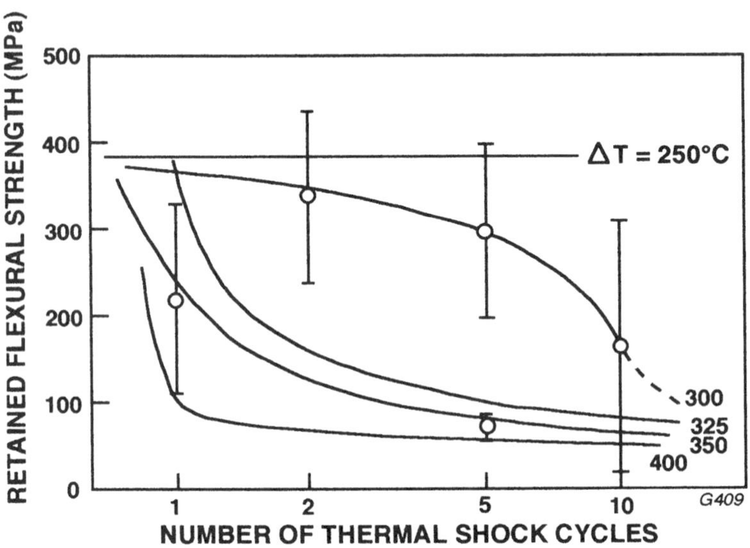

FIGURE 11.7 Plot of the strength retained after increasing numbers of thermal shock cycles at various fixed ΔTs for Al_2O_3-BN composites for quenching into a room temperature water bath. Note the apparent fatigue limit at ΔT ~ 250°C and increasing rates of decrease of strength retained after multiple quenchings as ΔT_C is approached, as well as retained strengths at < ΔT_C approaching (and possibly exceeding) those for a single quench at ΔT_C as the number of cycles increases. (From Ref. 45. Published with permission of Ceramic Engineering and Science Proceedings.)

while water effects were not the cause of the observed fatigue effect, further testing of materials such as soda lime glasses more susceptible to SCG and to substantially more cycles than the ten they used may be useful to explore possible more limited effects of such SCG.)

Lewis and Rice further showed that various ceramics ranged between the above glass and glass-containing aluminas with no observable to substantial multiple quench fatigue effect. Thus a commercial crystallized cordierite–glass body (Pyroceram 9606®) showed a modest effect via increased scatter of retained flexural strengths, but no clear trend for strength reduction for multiple quenches at $<\Delta T_C$ or reductions of retained strength. Tests of a commercial Mg-PSZ (Zircoa 1027) also showed no decrease in ΔT_C or net retained strength but faster rates of strength decrease at and closely past ΔT_C. On the other hand, tests of purer, larger G (~ 30 μm) commercial Al_2O_3 (Lucalox®) showed ΔT_C reduced ~ 10°C in five quench cycles and some possible reduction in net retained strength. More extreme were results for an Al_2O_3-25 v/o TZP composite where five quench cycles reduced ΔT_C from ~ 250 to ≤ 175°C and net retained strengths by ~ $^1/_2$, i.e. similar to effects with the alumina- and mullite-BN composites. The common theme they saw running through all of these tests was the expected increase in microcracking from none in the borosilicate glass to limited amounts in the PSZ, more in the Lucalox, and much more in the composites with TZP or BN. They thus proposed that the quench fatigue effects arose from increasing extent and effect of microcracking, e.g. via the sequence shown schematically in Fig. 11.8.

In order to determine further the causes of the multiple quench fatigue of ceramics, Lewis and Rice [46] conducted some simple mechanical fatigue tests to ascertain how much of this effect was true thermal shock fatigue and how much was basic mechanical fatigue. Having no specific fatigue testing, they conducted some zero-tension tests using repeated flexure with a conventional mechanical test machine at both 20–25 and -196°C (i.e. the latter in liquid N_2) on several ceramics and a granite. Results from two coarser grain materials, Lucalox® alumina and the granite, both of which should have some microcracking, showed particular effects with failures at 2–100 cycles at 80–100% of their nominal flexure, i.e. single loading, flexure strength. Both showed failure in fewer cycles at -196°C versus at 20–25°C and that the number of cycles to failure at the nominal flexure strength increased as the thickness of the specimen increased, especially for the granite. The latter was suggested as possibly being due to stress gradients inherent in flexure (and also thermal shock) tests or to specimen compliance changes leading to increased specimen deflections by the test machine on each cycle. (The former may be a function of grain size relative to specimen dimensions, i.e. with more effect in the granite with larger G than the alumina.) Greater fatigue effects at the lower temperature clearly argue against any effect of SCG, since this is greatly, if not totally, suppressed at

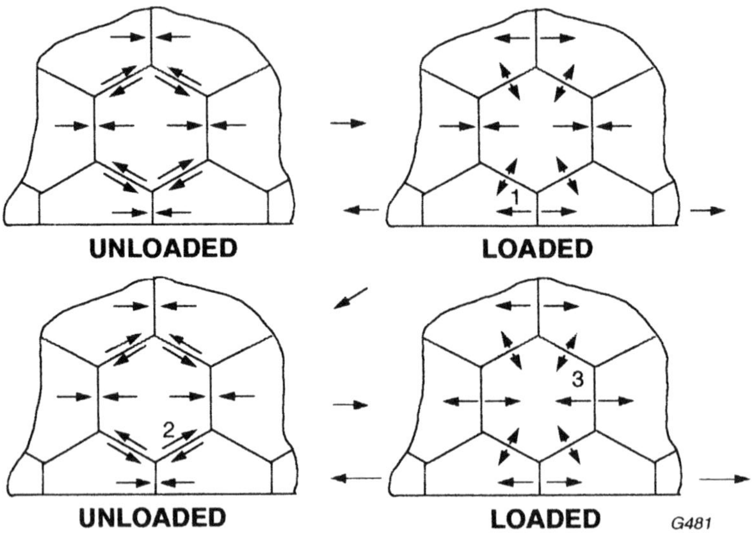

FIGURE 11.8 Schematic of possible progressive grain (or particle) boundary microcracking during both tensile loading and unloading to cause zero-tension mechanical fatigue and thermal shock fatigue in multiple quenching. The concept is based on local grain (particle) stress relaxations due to microcrack formation or extension on loading resulting in incompatibilities inhibiting closure of such microcracks on unloading such that further microcrack extension or forming occurs so the specimen locally does not return to the same local stress state on unloading that it was at before loading. (From Ref. 45,46. Published with permission of Ceramic Engineering and Science Proceedings.)

–196°C, while the lower temperature enhances mismatch stresses between grains due to the thermal expansion anisotropy of Al_2O_3 grains and some of the phases (grains) in granite as well as differential expansion between different phases in the granite.

Based on the above observations, Lewis and Rice [45] proposed that both thermal shock and pure mechanical fatigue resulted from microcracking from mismatch stresses between grains (Fig. 11.8). They recognized that the key to a cyclic fatigue process in a brittle material with macroscopic elastic behavior required a mechanism of progressive damage on a local, e.g. a microstructural, scale with cyclic loading. Some specific scenarios for progressive microcrack extension on each loading cycle were proposed, based on tensile loading forming or extending microcracks such that elastic relaxation of the grains abutting the microcrack result in incompatibilites resisting crack closing on unloading. When such incompatibilities are sufficient to cause further extension or generation of microcracks during unloading, a basic mechanism exists for progressive degradation on subsequent cycles, since the body locally does not return to the same stress state that existed before the loading and unloading. Such microcrack effects should be sensitive not only to the degree and nature of the mismatch stresses between grains, particles, or both, but also on their grain and particle sizes, as well as on local porosity and boundary phases.

Some other observations have been made of repeated thermal shocking, e.g. of mullite-ZrO_2 composites (Fig. 11.3) and of SiC whisker composites. Schneibel et al. [47] investigated the cyclic thermal shock resistance of two commercial Si_3N_4 and Al_2O_3-SiC composites (two with whiskers and one each with particles or fibers) using 10 or 100 shocks by quenching bend bar specimens from 1200°C into a room temperature fluidized bed and then measuring flexure strength at 22°C. This procedure, which is less severe than quenching into room temperature water, showed substantial thermal shock fatigue effects in only the Al_2O_3 whisker and fiber composites.

E. Tensile Strength

Turning now to the temperature dependence of tensile (flexure) strength, consider first glass matrix composites, where limited data is available mainly or only for crystallized glasses. Northover and Groves' [10] study of LiO_2-Al_2O_3-SiO_2 (LAS) bodies showed that fracture toughness and flexural strength both decreased to minima of ~ 2/3 their room temperature values at ~ 500°C and then increased back to, or slightly above, the room temperature values at the limit of their testing of 1000°C (Fig. 11.1). Govila et al. [11] showed similar strength behavior for similar LAS bodies, i.e. a strength minimum at ~ 600°C followed by a sharp maximum at 1000°C, which was similar to that of the parent glass, but with strengths generally ~ twice as high as the minimum, and especially the maximum, extended to higher temperatures (Fig. 11.9). They also demonstrated that the maximum was associated with softening of the residual glass matrix and resultant intergranular SCG in the crystallized body, while subsequent fracture was transgranular.

Borom [48] reported flexure strengths of two Li_2O-SiO_2-based crystallized glasses decreasing substantially with modest temperature increases. One body decreased from ~ 190 MPa at 22°C to a minimum of ~ 140 MPa at ~ 400°C, i.e. a decrease of ~ 24%, while the other had a greater decrease from 235 MPA to ~ 130 MPa at ~ 550°C, which was believed to be at or near a minimum for it. These decreases of ~ 25 and 45–50% are far greater than the ~ 5% decrease in Young's moduli. Tests of both parent glasses and a simulated matrix glass for one crystallized body all showed similar strengths at 22°C of ~ 95 MPa that increased with test temperature by ~ 30% at 400°C, i.e. nearly the same as the crystallized bodies at this temperature. This showed that the real

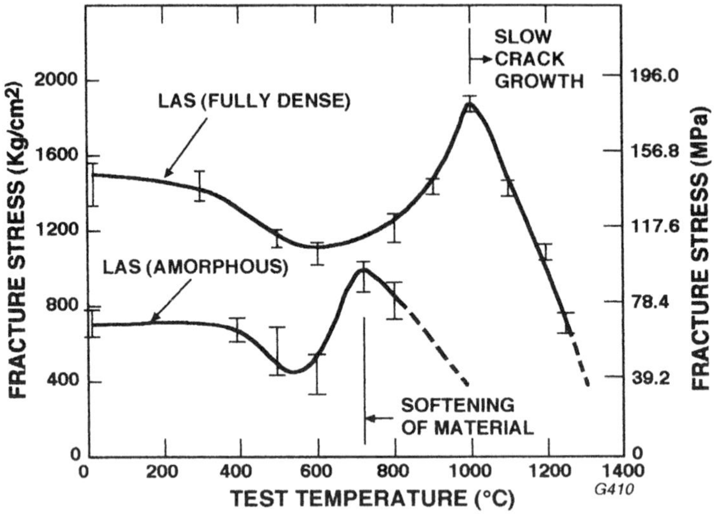

FIGURE 11.9 Strength versus test temperature for the starting Li2O-Al_2O_3-SiO_2 glass and after crystallization. Note the significant increase in strength and reduced dependence on temperature in the crystallized material. (From Ref. 11. Published with permission of the Journal of Materials Science.)

strength decreases of the crystallized bodies was greater due to the increased glass strength, which was attributed to reduced stress corrosion and the onset of plastic flow (which was observed in macroscopic behavior at ~ 450°C).

Consider next Al_2O_3-ZrO_2 composites, starting with sintered-HIPed bodies of 10, 20, or 40 w/o Al_2O_3 with respectively 90, 80, or 60 w/o 2Y-TZP of Tsukuma et al. [6]. These composites showed similar strengths and trends with test temperature (Fig. 11.10), with a trend for strengths to increase with increasing Al_2O_3 content, especially at the highest temperature (1000°C). Thus bodies with no Al_2O_3, i.e. pure 2Y-TZP, had ~ 30% lower strength at ~ 22°C, with the same or greater difference at 1000°C. Greater strengths with Al_2O_3 versus pure 2Y-TZP was also shown in bodies that were only sintered, but with lower strengths and less difference between bodies with and without Al_2O_3, e.g. ~ 20% difference. Note that the strength minimum at ~ 400°C corresponds to a toughness minimum at the same temperature, probably with a more rapid decrease in *E* at 200–300°C, which were noted respectively in Secs. A and B. These in turn are probably related to similar deviations seen in ZrO_2 (Fig. 6.18). Results of Govila [49] for HIPed 3Y-TZP with 20 w/o Al_2O_3, while showing no decrease in strength at 200°C, showed overall similar strength trends with test temperature, as well as a temporarily more rapid strength decrease between 200 and 400°C

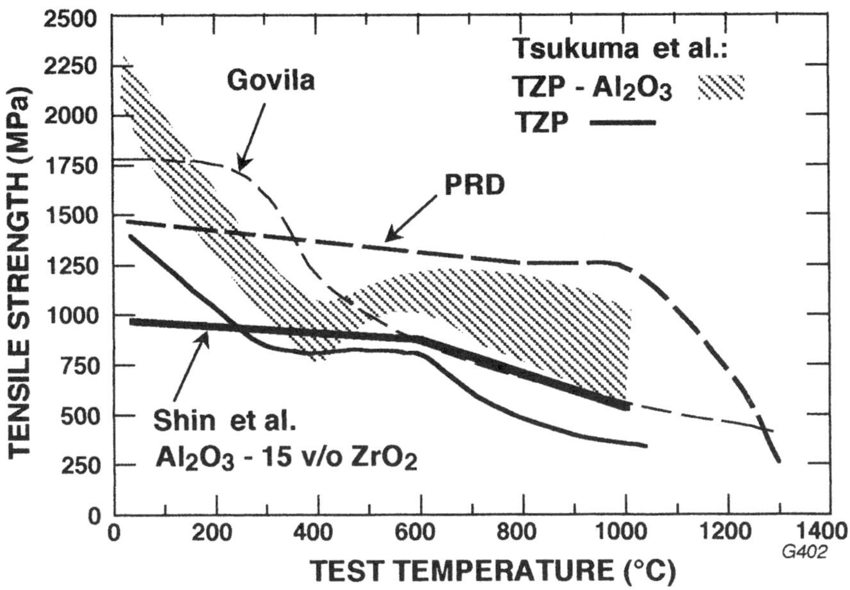

FIGURE 11.10 Strengths of Al_2O_3-ZrO_2 composites versus test temperature [1,6,48,49]. Note the similarity between the behavior of the (PRD) fiber [1] and bulk bodies [6,49, 50].

(Fig. 11.10). Results of Shin et al. [50] for HIPed Al_2O_3-15 v/o ZrO_2 showed similar relative strength decrease with increasing test temperature and lower overall strength as expected, while data of Lavaste et al. [1] for PRD166 fibers of Al_2O_3+ ~ 20 w/o TZP showed similar strength and *E* decreases (Sec. A) till ~ 1000°C, i.e. ~ 1–20%, then a more rapid decrease (Fig. 11.10). Comparison of the latter two results again showed substantial similarities between fiber and bulk ceramic mechanical property trends, e.g. as discussed in Chap. 3, Sec. III.A. Much less decrease, e.g. ~ 20%, in the high 22°C strengths of 330 MPa to 120022°C for reaction processed composites of mullite-6 w/o Al_2O_3 + 31 w/o ZrO_2 is consistent with their strengths not being derived from transformation toughening but probably in part from finer grain and particle sizes maintained at the elevated temperature [51].

Turning to other polycrystalline composites, Niihara et al. [18] showed that their composite of Al_2O_3 with 5 v/o SiC (2 μm) particles having a strength of 500 MPa at ~ 22°C, and the same or slightly less at 600°C, increased slightly to a modest maximum at 1000°C and then rapidly decreased to < 200 MPa at nearly 1300°C. This maximum and the subsequent rapid decrease in strength correspond with an increasing upward swing of fracture toughness from 800 to 1200°C (the latter being the limit of their toughness testing). This data thus also shows some strength variations at more modest temperatures before greater decreases at higher temperatures, though on a much more modest scale.

Baldoni et al. [19] showed that flexure strengths of their Al_2O_3-30 v/o TiC composite decreased gradually from 500 MPa at 22°C to ~ 400 MPa at 1200°C. Their Si_3N_4-30 v/o TiC composite starting from ~ 650 MPa at 22°C decreased modestly to a minimum at ~ 800°C and then went to a sharp but modest maximum of 700 MPa at 1000°C, followed by a sharp drop to <400 MPa at 1200°C. These changes are in contrast to their toughness both showing minima at ~ 800°C and then increasing substantially to the limit of their testing of 1200°C (Sec. B).

Turning to other all nonoxide composites, Endo et al. [52] showed that strengths of SiC-TiC hot pressed from 0.3–0.4 μm particles with B_4C+ C sintering aids had lower strengths for SiC rich bodies but higher strengths of TiC rich bodies with similar but more pronounced complexity at 1500 versus 22°C. Thus strengths with 0 TiC at 1500°C were 600 MPa (versus ~ 1000 MPa at 22°C), increased to a maximum of ~ 750 MPa at 20 v/o TiC, and then decreased (crossing the 22°C strengths at ~ 50 v/o TiC at 600 MPa) to a minimum of ~ 550 MPa at 60 v/o TiC, through a modest maximum at 80 v/o TiC and then to ~ 400 MPa (versus ~ 350 MPa at 22°C) at 100% TiC. Lin and Iseki [53] found that strengths of SiC (+ Al-B-C sintering aids) were constant at ~ 510 MPa and then decreased by ~ 30% at 1400°C, while similarly processed SiC with 10 and 30 v/o TiC (~ 1.5) started at ~ 430 MPa at 22°C and decreased fairly steadily to ~ 300 MPa at 1400°C, with only limited acceleration of decrease from 1200 to 1400°C. (Such SiC+ TiC bodies made without Al-B-C additions had very low strengths of ~ 100 MPa.) Subsequently they [54] showed similar trends for SiC with 2% AlN additions, but with ~ 10+% lower strengths and similar results with 33 v/o TiC and 2% AlN, but higher strengths for 33 v/o TiC with 2% Ti versus AlN additions, i.e. the latter strengths were nearly identical in value and trend as for their SiC. Thus again temperature dependence can be complex and variable with effects of additives often being an important variable. McMurtry et al. [55] showed strengths of commercial αSiC+ ~ 19 v/o TiB_2 (1–5 μm) being ~ constant at ~ 490 MPa to 1200°C, i.e. parallel to strengths of αSiC (both sintered with B-C additions) of ~ 350°C. Note that the higher SiC+ TiB_2 strengths probably reflect, at least in part, finer, more uniform SiC grain size due to grain growth inhibiting effects of the TiB_2 and that there appears to be a limited SiC strength minimum at ≤ 600°C, and then a small increase, while SiC+ TiB_2 indicates a possible limited maximum at ≤ 600°C, and then a small decrease in strength.

Turning to platelet composites, Mazdiyasni et al. [56] reported strengths for AlN with 0 to 10–30 w/o fine BN platelet particles generally decreasing with increasing BN content with such decreases being greater in magnitude at lower versus higher levels of additions. Bodies fabricated without sintering aids (and having 1–6% residual porosity, generally increasing with BN content) all showed lower strengths at 1000 versus 22°C typically by 10–20%, i.e. generally consistent with expectations for decreases in *E*. Between 1000 and 1500°C strength decreases with no BN were pronounced, e.g. ~ 45%, but they progressively decreased as the BN content increased, with the body with the highest (20 w/o) BN content actually showing a strength increase of 19–20% and somewhat higher overall strength than the 15 w/o BN body, despite greater residual porosity with 20 w/o BN. Bodies made with Y_2O_3 or CaH_2 additives and extended to higher BN additions, with similar residual porosity levels, showed more complex trends with strength maxima at 1000–1250°C (often with a minimum at 1000°C with a 1250°C maximum), but generally with lower strengths, more so at the extremes of 22 and 1500°C, especially the former.

Huang and Nicholson [57] introduced up to 40 v/o Al_2O_3 platelets with aspect ratios up to 12 into Y-PSZ composites by die- followed by isopressing or tape casting and lamination, then sinter-HIPing (see also Chap. 8, Sec. V.D). Tape casting gave substantial platelet orientation, which aided densification some and induced some elastic anisotropy and some improvements in strength and toughness, especially at elevated temperatures. Young's and shear moduli at 22°C increased linearly with the v/o platelets, with little effect of aspect ratio; Poisson's ratio also decreased linearly, while hardness H_V (30 kg) increased somewhat more at the 40 v/o than linearly. IF fracture toughness at 22°C increased from ~ 5 to a broad maximum of ~ 7.9 and then decreased to ~ 6 MPa·m$^{1/2}$ at respectively 0, 15–30, and 40 v/o platelets, with similar trends with CNB tests, but with values respectively ~ 20, 60, and 50% those from IF testing. Overall strengths at 22°C showed an opposite trend, i.e. overall decreasing by ~ 30% at 5 and 40 v/o platelets. Opposite toughness and strength trends were also shown for effects of platelet aspect ratio, i.e. toughness increased ~ 10%, while strength decreased ~ 20% (and hardness decreased ~ 5%) as the platelet aspect ratio increased from 1 to 12. Better correlation of CNB and strength results were found in elevated tests at 800 and 1300°C, showing higher values for the composites than the Y-PSZ matrix, especially at 800°C, with somewhat greater effects of platelet alignment at these temperatures than at 22°C.

Becher et al. [58] showed that strengths of Al_2O_3-SiC whisker composites at any given temperature generally increased in proportion to their dependence on whisker loading at 22°C to the limit of their testing (1200°C). All strengths decreased with increasing temperature, e.g. by ~ 15% at 1000°C, consistent with *E* decreases, then more rapidly, e.g. by another 10–15% at 1200°C. Govila [59] obtained flexure strengths for Al_2O_3-15 w/o SiC whisker composites consistent with those of Becher et al. and showed very limited decreases to 600–800°C nominally parallel with those for Al_2O_3 alone and then progressively accelerating in decrease to ~ 1/3 their value at 22°C at 1400°C, i.e greater decrease than for the Al_2O_3 alone. The strength results of Yang and Stevens [60] for Al_2O_3-20 v/o SiC whisker composites were very similar in values and rates of change.

Turning to other oxide matrices, composites of 25 v/o SiC whiskers in a cordierite or an anorthite matrix were shown by Gadkaree [61] to decrease flexure strengths by ~ 10% at 1000°C, i.e. as expected from changes in *E*, and then to accelerate in their decreases, e.g to ~ 54% and ~ 32% respectively of their *E* values at 22°C of ~ 400 and 380 MPa respectively. Kumazawa et al. [62] showed that while strength of their mullite matrix was greater at 1300°C than at 22°C (~ 560 versus ~ 450 MPa), strengths of the composites were ~ 480 MPa with 10–30 v/o SiC whiskers (i.e.~ 15% lower) and then decreased to ~ 420 MPa at 40 v/o SiC at 1300°C. (Strengths at 22°C were higher at ~ 480 MPa from 10 to 30 v/o SiC and then decreased to ~ 410 MPa at 40 v/o SiC.) NB toughnesses were higher than those at 22°C, increasing from ~ 2.7 to ~ 3.6 MPa·$m^{1/2}$ from 0 to 40 v/o SiC whiskers. Resulting strength-to-toughness ratios at 1300°C decreased from ~ 207 to 114 $m^{-1/2}$ over this range of whisker additions, similar to the trend for such values at 22°C (which were ~ 10–20% higher overall, indicating reasonable flaw sizes of ~ 20–80 microns).

Choi and Salem [63] showed that NB and IF toughnesses of Si_3N_4+30 v/o SiC whiskers at 22°C and CNB toughness from 22 to 1200°C for both the composite and the matrix alone were all essentially constant at ~ 5.7 MPa·$m^{1/2}$ with no evidence of *R*-curve effects. This is consistent with both matrix and composite having identical strengths over the range 22–1400°C, decreasing only ~ 5% at 800°C and then progressively more rapidly to ~ 50% at 1400°C, which was attributed to SCG. Fractography showed fracture initiation commonly to occur from processing defects. Zhu et al. [64] reported that stress rupture of Si_3N_4+20 v/o SiC whisker composites resulted in mixed inter- and transgranular fracture and fracture mirrors and mist and hackle boundaries that increased in size as stress decreased, i.e. the same as or similar to normal brittle fracture at lower temperatures. However, fracture at 1200°C was all intergranular with cavity nucleation and growth resulting in a rough fracture region that increased in size as stress decreased, but crack propagation following this creep crack growth was smooth and mirrorlike, indicating catastrophic crack propagation. They reported that SiC whiskers were effective in increasing fracture resistance at 1200°C by crack arrest and crack bridging. Olagnon et al. [65] showed that strengths of their Si_3N_4-SiC whisker composites were lower at 20 v/o than 10 v/o whiskers (~ 700 and 800 MPa respectively at 22~C) but were essentially the same (~ 650 MPa) at 800–1300°C. Thus the 10 v/o composite lost strength faster at lower temperatures, e.g. by ~ 25% to 1000°C, but both decreased faster at higher temperatures, e.g. to ~ 200 MPa at 1300°C. They showed typical brittle fracture features of mirror, mist, and hackle surrounding fracture origins of specimens that failed at 1000°C and increasing areas of rough SCG (followed by smoother fracture) as test temperatures increased to 1200 and 1300°C.

Dusza and Šajgalik [66] also measured both toughness and strength of Si_3N_4 plus 10 or 20 w/o βSi_3N_4 whisker composites from 22 to 1200°C. While the 20% whisker composite showed IF toughness decreasing only a few percent from the room temperature value of 6.3 MPa·m$^{1/2}$ at 800, 1000, and 1200°C, they drastically decreased in strength from 510 MPa at 22°C to ~ 300 MPa at 800°C, followed with much less decrease at higher temperatures. This was in contrast to the higher room temperature strength of 690 MPa (but similar toughness) for the 10% whisker composite decreasing to only 640 MPa at 800°C and to ~ 450 MPa at 1200°C, where fracture at 800°C was brittle but occurred from high-temperature SCG at higher temperatures. The lower properties of the 20% whisker composite were attributed to greater inhomogeneity of the whisker distribution.

The issue of possible effects of high-temperature crack healing was addressed by Moffatt et al. [67] by comparing effects of high temperature annealing and testing on a commercial Al_2O_3-17 v/o SiC whiskers (matrix G ~ 5 μm) and an alumina body similar to the matrix (G ~ 8 μm). They showed that IF toughness of the Al_2O_3 was ~ 3.5 MPa·m$^{1/2}$ at 22 and 800°C and then decreased ~ linearly to ~ 0.4 MPa·m$^{1/2}$at 1400°C, while the whisker composite had only ~ 1.8 MPa·m$^{1/2}$ at 22°C, was modestly higher at 800 and 1000°C, and then accelerated, reaching 5.7 MPa·m$^{1/2}$ at 1400°C. Both the Al_2O_3 and the whisker composite showed large increases in the apparent toughness of cracked specimens after annealing (with no recracking) and testing at room temperature, e.g. maximum values of 28 MPa·m$^{1/2}$ for the Al_2O_3 after annealing at 1400°C and 120 MPa·m$^{1/2}$ for the composite after annealing at 1200°C. However, recracking of specimens before testing returned them to their normal levels. On the other hand, annealing of cracked Al_2O_3 specimens and then testing them at elevated temperatures showed no changes. It was concluded that crack healing occurred in both materials but was more pronounced in the whisker composite, whereas annealing gave maximum results at 1200°C, but the effects of annealing were only significant at lower temperature property measurements.

Carroll and Dharani [68] reviewed literature on the strength dependence of ceramic whisker composites, noting that in general there was less improvement in ultimate strengths at higher temperatures versus those at 22°C. They presented a model for high-temperature failure focusing on thermal expansion coefficients, Poisson's ratios of the constituents along with thermal history, and matrix–whisker friction.

Turning to directionally solidified ceramic eutectics, earlier work to fracture (i.e. area under the stress–strain curves) data of Hulse and Batt [69] showed substantial values for some systems for stressing parallel with the solidification direction (Fig. 11.11). They reported values of ~ 100 J/m^2 at 22 and 1500–1575°C for Al_2O_3-ZrO_2(+Y_2O_3) and values for the CaO-MgO eutectic nearly this high at 22°C and rising over an order of magnitude by ~ 1200°C. Kennard et al. [70,71] showed WOF decreasing some from ~ 20 J/m^2 at 22°C at

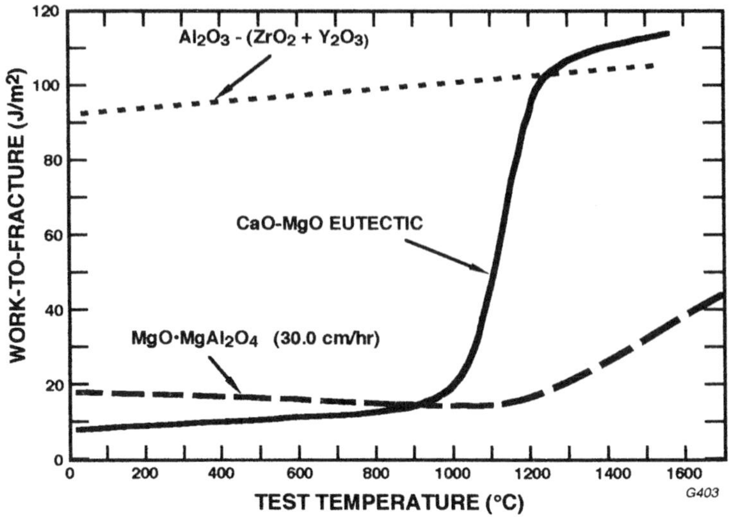

Figure 11.11 Plot of work-to-fracture for Al_2O_3-ZrO_2(+Y_2O_3) and CaO-MgO eutectic composites of Hulse and Batt [67] and work of fracture (WOF) for MgO-$MgAl_2O_4$ composite of Kennard et al. [70,71] versus test temperature. Note that use of these values to calculate toughnesses (an uncertain procedure, especially for the work to fracture values) yields ~ 7, ~ 7, and ~ 4 MPa·$m^{1/2}$ at room temperature. Increases in high temperature values, especially large ones, typically reflect effects of plastic deformation.

1200°C and then increasing to ~ 40 J/m^2 at 1600°C for the MgO-$MgAl_2O_4$ (Fig. 11.11). Mah et al. [72] reported the toughness of alumina-YAG eutectics of ~ 4.4 MPa·$m^{1/2}$ at 22°C, decreasing ~ 10% by 1200°C, and then increasing ~ 10% by 1500°C.

Hulse and Batt [69] reported strengths of their Al_2O_3-ZrO_2(+Y_2O_3) decreasing from room temperature values of ~ 700 MPa only ~ 20% at 1600°C (and lower strengths for slower solidification, hence coarser eutectic structure), a similar trend at lower strength for their CaO·ZrO_2-ZrO_2 eutectic, and a possible increase in strength for their CaO-MgO eutectic (Fig. 11.12). They also showed that considerable ductility occurred in testing some of these composites at elevated temperature. Kennard et al. [70,71] showed strengths of their MgO-$MgAl_2O_4$ eutectic, though modest, also had limited decrease at higher temperature, with somewhat lower, but similar, strength trends for stressing normal to the solidification direction. Mah et al. [72] showed that strengths of their alumina-YAG eutectics decreased from room temperature values of ~ 380 MPa by ~ 25% to ~ 280 MPa at 1500°C. Increases, especially sub-

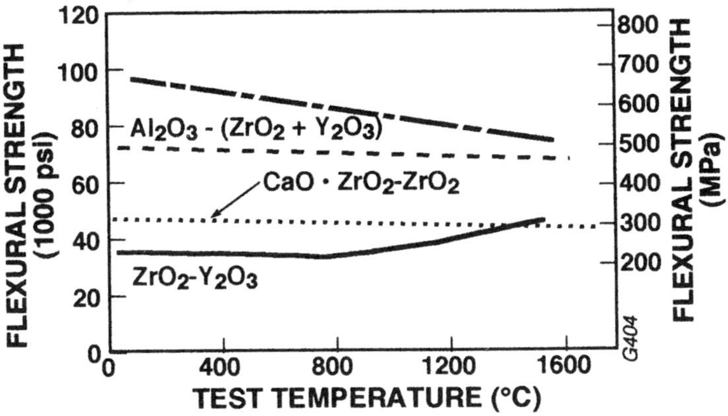

FIGURE 11.12 Flexure strength for directionally solidified eutectics of Al_2O_3-ZrO_2(+Y_2O_3), CaO·ZrO_2-ZrO_2 eutectic, and CaO-MgO of Hulse and Batt [69] versus test temperature. Note (1) tests were with the stress parallel with the solidification direction, except for the case where the stress was normal to this direction (marked perpendicular), and (2) there were lower strengths for slower solidification, hence coarser eutectic structure in the Al_2O_3-ZrO_2(+Y_2O_3), system.

stantial ones, in higher temperature toughnesses are not necessarily reflected in the above strengths, since such toughness increases are typically due to plastic deformation, which does not necessarily increase strength. More recently, Waku et al. [73] reported that directionally solidified specimens of the alumina-YAG eutectic had strengths of 350–400 MPa at both 22 and 1800°C (only ~ 25°C below melting!), with yielding observed at ≥1700°C.

Though the extensive and complex subjects of creep and stress rupture are beyond the scope of this book, a few brief comments are in order, since as strain rates decrease and especially as test temperatures increase, these processes start becoming factors in the temperature dependence of strength. There is often more creep data on some composite ceramics than on strength, e.g. on crystallized glasses, where earlier work by Barry et al. [74] and more extensively by James and Ashbee [75] are of value, but the recent broad survey by Wilkinson [76] is particularly valuable. This shows that rheological models give a reasonable perspective on much creep behavior, especially at lower and higher dispersed phase contents, i.e. below and above the percolation limits for the dispersed phase, with less certainty for the intermediate region transitioning from no to substantial percolation of the dispersed phase. Thus whiskers or needle shaped grains have much lower percolation limits (e.g. ~ 12 v/o) so they can give greater increases in creep resistance, e.g. by of the order of 10 and 60 over respectively platelets and equiaxed particles. Whisker composites also tend to have less dependence on volume fraction addition, processing, and matrix grain size above percolation limits. Thus the work of Lin et al. [77] is cited in noting that creep in Al_2O_3-SiC whisker composites at 1200°C varied as G^{-1} for the matrix versus G^{-2} to G^{-3} for Al_2O_3 and that the G dependence is not present at 1300°C. Desmarres et al. [78] observed that 30 v/o SiC whiskers in a SiAlYON matrix increased strengths by ~ 10% and the temperature for more rapid decrease of strengths from ~ 1000°C for the matrix alone to ~ 1300°C with whiskers. The study of Jou et al. [79] on 50 v/o solid solutions of AlN and SiC also gave microstructural results of interest, namely highest creep rates in inhomogeneous samples (probably due to AlN rich areas), intermediate rates for finer G (~ 3.1 μm), and lower rates for coarser G (~ 5.4 μm) from 1400 to 1525°C and 50 to 120 MPa in bending.

F. Hardness, Compressive Strength, and Other Related Behavior

There is even less data for properties covered in this section than in previous sections, but this limited data is useful. Niihara et al. [18] showed that H_V (4.9 N) of their composite of Al_2O_3 with 5 v/o of 2 μm SiC particles started at the same value for Al_2O_3 alone of 20 GPa at 22°C and at their maximum temperature of 1400°C (~ 1.5 GPa). However, while H_V for the Al_2O_3 alone decreased very nearly linearly on a semilog plot, their composite was clearly bilinear, decreasing more slowly to ~ 1100°C and then more rapidly to 1400°C, with the separation of the two thus being greatest at 1100°C, i.e. ~ 7 and 3 GPa respectively for the composite and the Al_2O_3 alone. This trend for less hardness decrease in the composite at lower temperatures and more at higher temperatures versus their Al_2O_3 is very similar to their flexural strength data, except the strength actually slightly increased to a maximum at ~ 1000°C before beginning its rapid decrease. In contrast to this, Tai and Watanabe [20] reported no clear difference in H_V (196 N) between their Al_2O_3 alone and their composite of (WC+ 10 w/o Co) + 20 w/o Al_2O_3 (~ 0.2 μm) over the range tested (22–600°C).

One of the few studies of compressive strengths of ceramic composites versus temperature is one of May and Obi [80] on various crystallized bodies in the SiO_2 + 25–35 m/o Li_2O system with various heat treatments and amounts (0–3%) of P_2O_5 nucleating agent. As shown in Fig. 11.13, there is a broad diversity of strength levels and trends with temperature, including both some limited maxima at 100–200°C and minima at 500–600°C.

There have also been some compressive creep studies of ceramic composites, e.g. of crystallized glasses, which are also reviewed by Wilkinson [76], often showing substantial differences from tensile creep, e.g. in stress levels. Also, one study of directionally solidified Al_2O_3-YAG composites [81] in comparison with tests of similarly oriented Al_2O_3 and YAG single crystals at 1530°C and lower strain rates (10^{-4}/ sec) showed that the composite creep behavior was close

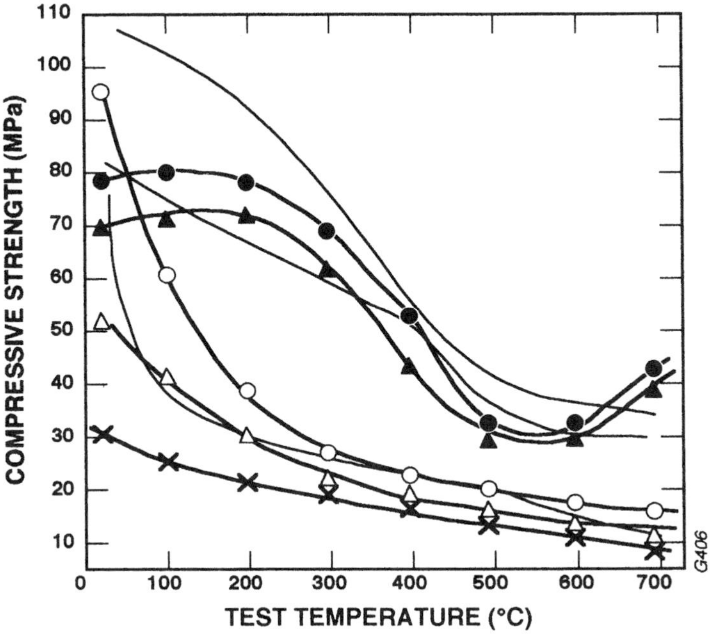

Figure 11.13 Compressive fracture stress for various crystallized SiO_2 + 25–35 m/o Li_2O glasses as a function of temperature. (From Ref. 80. Published with permission of the British Ceramic Society.)

to or slightly less than that of sapphire, while at higher strain rates (10^{-5}/ sec) the YAG reinforced the sapphire phase.

Verma et al. [82] showed that compressive strengths of their composites of SiC platlets (25–50 μm dia., 1–2 μm thick) in a borosilicate glass of matching expansion showed continuous, ~ linear increases in strength at 625 and 700°C to 40 v/o SiC (the limits of compositions that could be fully densified). Substantial plastic flow of the composites occurred, since the glass had very low strength at the test temperatures, with both glass and composites thus exhibiting substantial strain rate sensitivity of strength. Thus while the strengths at elevated temperature were much less than at room temperature, increased viscosity due to the SiC platelets had a significant impact.

IV. DISCUSSION

The first of three major trends of this chapter noted for discussion is the variations that can occur in properties, especially strength, at modest temperatures where testing is often neglected on the frequently incorrect assumption that no significant changes occur at these lower temperatures. Such lower temperature tensile strength behavior, though again given only limited consideration for monolithic ceramics, also revealed important, and sometimes dramatic changes in strength, indicative of underlying mechanisms (Chap. 6). For composites, such lower temperature strength changes are particularly pronounced for crystallized glasses, where substantial variations are seen in toughness and flexure strength (Figs. 11.1, 11.9), as well as compressive strength (Fig. 11.13), the latter being similar to the variety of flexural strength behavior at 22°C for various crystallization schedules in the same compositional system (Chap. 9, Sec. III. A).

Intermediate deviations were observed for toughness and strength of the TZP-Al_2O_3 composite of Tskuma et al. [6], mainly a minimum in both at 400°C (Fig. 11.10). Limited data for other composites often indicates normal strength decreases consistent with decreases in *E* as test temperature increases, but this is typically based on the lowest test temperature > 22°C being > 600–1000°C, i.e. near or at the upper limits where such variations primarily manifest themselves. While some observed variations are substantial, as noted above, some are often more limited, e.g. Al_2O_3-SiC(18), Si_3N_4-TiC(19), and SiC-TiB_2 [55], but these are still potentially significant. For example, all these, especially the latter, must be considered against the typical 10–20% decreases expected from decreases in *E*, which often increases the net deviation.

The above deviations from the normal decrease of strength or toughness with temperature increase must reflect microstructural mismatch stress changes, commonly reductions; but more must be involved. Thus the initial decreases in strength and toughness in Li_2O-SiO_2 crystallized glasses with increasing temperature (Fig. 11.1) has been attributed to decreases in microstructural stress and resultant microcracking and the substantial increase in crack blunting by plastic flow. However, there are issues with both of these mechanisms and with alternative or additional mechanisms. Thus while decreasing microcracking is likely to reduce toughness, at least with some measures of toughness, decreases in strength with reduced microcracking may conceivably occur (mainly with very small microcracks, e.g. < a micron, which is possibly consistent with sizes in many crystallized glasses), though it is contrary to most data, raising the question of what the mechanism of the initial strength decrease is. Reductions of microstructural stresses as temperature increases can also be a factor for both polycrystalline composites and especially for single crystal, i.e. directionally solidified, composites, where plastic deformation at higher temperatures can be an important factor, as shown by observed ductility. However, microstructural stress reduction in these, and to a lesser extent in other bodies, can be complicated by differing changes in elastic anisotropy of crystalline phases as temperature increases, since such anisotropy may increase or decrease, or change from one direction to another as a function of temperature (Chap. 7, Fig. 11.14). Further, while increasing plastic deformation at crack tips can clearly first increase and then significantly decrease strength, e.g. as in glass-based systems, it is more uncertain due to compositional, microstructural, and strain rate sensitivities. Further, the trends for strength and toughness are often opposite as plasticity and especially related SCG increase, with toughness often markedly increasing and strength significantly decreasing.

Three other mechanisms must also be considered. The first and broadest is the general decrease of E with increasing temperature, and resultant decreases in toughness and strength unless overcome or masked by other mechanisms. However, in such cases it still needs to be accounted for, i.e. adding to or subtracting from changes due to other mechanisms. The second mechanism is the often neglected possibility of SCG, e.g. increasing temperatures should progressively accelerate both crack tip reactions of the active species (commonly H_2O) and desorption of the active species, which have opposite effects, resulting in possible toughness, and especially strength, minima. A key test for such effects would be inert atmospheres tests , and possibly tests as a function of increasing strain rate, which would increase strengths with SCG, and possibly decrease them with plasticity, effects. Many unknowns thus remain regarding such mechanisms, and more, and especially more comprehensive, data are needed to resolve these issues, e.g. comparison of toughness and strength behavior for other systems, e.g. for canasite systems (Fig. 11.1) and broader temperature and stressing conditions. The third mechanism, of at least partial crack healing, is often limited due to kinetics but may be a factor for less refractory materials, e.g. glasses, or via oxidation of nonoxide constituents.

Consider next the second trend, namely differing trends of different properties with temperature, especially for the same material, but also between materials. Comparison of Young's modulus, strength, and fracture toughness have been noted above but deserve further attention along with comparison to other properties. Thus there is considerable similarity in the temperature dependences of strength and toughness of the TZP-Al_2O_3 composites of Tsukuma et al. [6], but also some differences. Similarly the deviations noted in E at ~ 200°C may be related to the deviations to minima of toughness and strength at ~ 400°C (and are clearly related to such effects seen in the E–T dependence of ZrO_2 (Fig. 6.18)). Further, while there are similarities in the toughness trends for the two mullite-ZrO_2 composites of Fig. 11.2, there are also significant differences, as is also the case where both strength and toughness were measured. The significant opposite trends of strength and toughness from 1000 to 1200°C most likely reflect opposite effects of deformation processes such as grain boundary sliding on toughness versus strength, i.e. enhancing crack propagation in the latter case and impeding it in the former. However, for this and other ZrO_2 toughened systems, note that an added, simple monotonic decrease of both toughness and strength over and above that due to decreases in E until the ZrO_2 transformation temperature of 800–1200°C does not occur. Note that Al_2O_3-ZrO_2 eutectics also show some opposite trends of work-to-fracture and strength. These strength–toughness differences in ZrO_2 toughened systems again strongly suggest toughening mechanisms associated with the presence of ZrO_2 other than just transformation, as is also shown for TZP and especially PSZ bodies (e.g. (Fig. 6.6). However, more generally, strength–toughness differences in the latter as well as other composite systems show, despite more limited data, both some similarities and some differences similar to those for monolithic ceramics. As with monolithic ceramics, the differences probably reflect significant differences in strain rate–crack velocity effects between various toughness and strength tests.

Both cases of composite hardness versus temperature indicate faster decreases of hardness than for E, i.e. at rates of ~ 2–4% per 100°C for hardness. However, while the Al_2O_3-(WC+ Co) composite showed similar trends of hardness with temperature as the Al_2O_3 alone, this is not the case for the Al_2O_3-SiC composite. Whether the latter reflects differences in the temperature dependences of the two phases and the former similarities in the two is not certain, since data on one of the composite phases was not obtained.

Turning to the third trend, namely property decreases at higher temperatures, typically ≥1000°C, much of this must involve increasing influence of creep processes. Thus the merger of strengths of Al_2O_3-SiC whisker composite with 10 and 20 v/o whiskers at higher temperatures is consistent with creep being less sensitive to v/o whiskers once the percolation limit has been reached. However, much again remains to be more clearly understood and documented.

V. SUMMARY

The broadest messages to take from this chapter are first the overall similarities of the temperature dependence of mechanical properties of ceramic composites and monolithic ceramics and the need for much more documentation and understanding. The latter is even more critical for composites, but the need for both material systems is for more comprehensive study and evaluation. With regard to similarities, it is important to note that these exist at both lower and higher temperatures. Thus tests at moderate temperature are often neglected in both material systems, based on the (often incorrect) assumption that no significant changes occur in this temperature range. This is frequently not the case. Substantial changes were shown in some cases, and other lesser but still significant changes were shown and discussed at temperatures < 600°C. Another broad message is the frequent, and often significant, disparity of toughness due to effects of material, microstructure, temperature, and especially test method and parameters, which is still incompletely documented and understood.

Some other points to note are as follows. Thermal shock resistance can often be increased via composites, with large crack toughnesses generally correlating with retained strengths (but not necessarily with normal strengths). However, significant thermal shock resistance is generally accomplished by mechanisms that limit strength and thermal shock fatigue, which has been demonstrated and can be substantial, is likely to be wide-spread, and probably entails mechanical fatigue (e.g. in zero-tension tests, apparently due to mismatch stresses between grains and particles). Tensile strengths of composites typically decrease increasingly rapidly above 1000–1200°C similar to monolithic ceramics, with rheological creep models generally being good guides. However, composites can have greater resistance to such plastic flow, especially as the percolation limit of the added phase is reached, which occurs much more rapidly as the dispersed phase is elongated in one direction, i.e. whiskers are better. Hardness tests may be useful for probing high-temperature plasticity.

REFERENCES

1. V. Lavaste, J. Besson, M.-H. Berger, and A. R. Bunsell. Elastic and Creep Properties of Alumina-Based Single Fibers. J. Am. Cer. Soc. 78(11):3081–3087, 1995.
2. J. C. Romie. New High-Temperature Ceramic Fiber. Cer. Eng. Sci. Proc. 8(7-8):755–765, 1987.
3. D. J. Pysher, K.C. Goretta, R. S. Hodder, Jr., and R. E. Tressler. Strengths of Ceramics Fibers at Elevated Temperatures. J. Am. Cer. Soc. 72(2):284–288, 1989.
4. S. Sakagushi, N. Murayama, Y. Kodama, and F. Wakai. The Poisson's Ratio of Engineering Ceramics at Elevated Temperature. J. Mat. Sci. Lett. 10:282–284, 1991.
5. J. D. French, H. M. Chan, M. P. Harmer, and Gary A. Miller. High-Temperature Fracture Toughness of Duplex Microstructures. J. Am. Cer. Soc. 79(1):58–64, 1996.
6. K. Tsukuma, K. Ueda, K. Matsushita, and M. Shimada. High-Temperature Strength and Fracture Toughness of Y_2O_3-Partially-Stabilized ZrO_2/Al_2O_3 Composites. J. Am. Cer. Soc. 68(2):C-56–58, 1985.
7. K. P. Gadkaree. Whisker Reinforcement of Glass-Ceramics. J. Mat. Sci. 26:4845–4854, 1991.
8. W. J. Lee and E. D. Case. Cyclic Thermal Shock in SiC-Whisker-Reinforced Alumina Composite. Mat. Sci. Eng. A119:113–126, 1989.
9. K. S. Mazdiyasni and R. Ruh. High/Low Modulus Si_3N_4-BN Composite for Improved Electrical and Thermal Shock Behavior. J. Am. Cer. Soc. 64(7):415–419, 1981.
10. J. P. Northover and G. W. Groves. High-Temperature Mechanical Properties of LiO_2-Al_2O_3-SiO_2 (LAS) Glass Ceramics. J. Mat. Sci. 16:1881–1886, 1981.

11. R. K. Govila, K. R. Kinsman, and P. Beardmore. Fracture Phenomenology of a Lithium-Aluminum-Silicate Glass-Ceramic. J. Mat. Sci. 13:2081–2091, 1978.
12. B. S. Majumdar, T. Mah, and M. G. Mendiratta. Flaw Growth in a Polycrystalline Lithium-Aluminum-Silicate Glass Ceramic. J. Mat. Sci. 17:3129–3139, 1982.
13. G. H. Beall, K. Chyung, R. L. Stewart, K. Y. Donaldson, H. L. Lee, S. Baskaran, and D. P. H. Hasselman. Effect of Test Method and Crack Size on the Fracture Toughness of a Chain-Silicate Glass-Ceramic. J. Mat. Sci. 21:1365–1372, 1973.
14. G. Brandt, B. Johannesson, and R. Warren. Fracture Behavior of Selected Mixed-Ceramic Tool Materials up to 1200°C. Mat. Sci. Eng. A105/106:193–200, 1988.
15. G. Orange, G. Fantozzi, F. Cambier, C. LeBlud, M. R. Anseau, and A. Leriche. High Temperature Mechanical Properties of Reaction-Sintered Mullite/Zirconia and Mullite/Alumina/Zirconia Composites. J. Mat. Sci. 20:2533–2540, 1985.
16. C. H. Henager, Jr., and S. Wada. Processing and Properties of Ceramic Matrix Composites with In Situ Reinforcement. Handbook on Discontinuously Reinforced Ceramic Matrix Composites (K. J. Bowman, S. K. El-Rahaiby, and J. B. Wachtman, Jr., eds.). Am. Cer. Soc., Westerville, OH, 1995, pp. 139–223.
17. A. Leriche, P. Descamps, and F. Cambier. High Temperature Mechanical Behavior of Mullite-Zirconia Composites Obtained by Reaction-Sintered. Zirconia 88:Advances in Zirconia Science and Technology (S. Meriani and C. Palmonari, eds.). Elsevier, New York, 1989, pp. 137–151.
18. K. Niihara, A. Nakahira, T. Uchiyama. and T. Hirai. High-Temperature Mechanical Properties of Al_2O_3-SiC Composites. Fracture Mechanics of Ceramics 7, Composites, Impact, Statistics, and High-Temperature Phenomena (R. C. Bradt, A. G. Evans, D. P. H. Hasselman, and F. F. Lange, eds.). Plenum Press, New York, 1986, pp. 103–116.
19. J. G. Baldoni, S. T. Buljan, and V. K. Sarin. Particulate Titanium Carbide-Ceramic Matrix Composites. Second Intl. Conf. Science Hard Materials (Rhodes, ed.). Inst. Phy. Conf. Ser. No. 75. Adam Hilger Ltd., Bristol, England, 1986, pp. 429–435.
20. W.-P. Tai and T. Watanabe. Elevated-Temperature Toughness and Hardness of a Hot-Pressed Al_2O_3-WC-Co Composite. J. Am. Cer. Soc. 81(1):257–259, 1998.
21. C. F. Cooper. Graphite Containing Refractories. Refractories J. Ref. Assn. Great Brit., No. 6. 1980, pp. 11–21.
22. R. C. Rossi. Thermal-Shock-Resistant Materials. Ceramics in Severe Environments (W. W. Kriegel and H. Palmour III, eds.). Plenum Press, New York, 1971, pp. 123–136.
23. C. J. Fairbanks, H. L. Lee, and D. P. H. Hasselman. Effect of Crystallites on Thermal Shock Resistance of Cordierite Glass-Ceramics. J. Am. Cer. Soc. 67(11):C-236–237, 1984.
24. M. Oguma, K. Chyung, K. Y. Donaldson, and D. P. H. Hasselman. Effect of Crystallization on Thermal Shock Behavior of a Chain-Silicate Canasite Glass-Ceramic. J. Am. Cer. Soc. 70(1):C-2–3, 1987.
25. P. F. Becher. Transient Thermal Stress Behavior in ZrO_2-Toughened Al_2O_3, J. Am. Cer. Soc. 64(1):37–39, 1978.
26. I. Thompson and R. D. Rawlings. Monitoring Thermal Shock of Alumina and Zir-

conia-Toughened Alumina by Acoustic Techniques. J. Mat. Sci. 26:4534–4540, 1991.
27. H. Tomaszewski. Effect of Sintering Atmosphere on Thermomechanical Properties of Al_2O_3-ZrO_2 Ceramics. Cer. Intl. 15:141–146, 1989.
28. T. Sornakumar, V. E. Annamalai, R. Krishnamurthy, and C. V. Gokularthnam. Thermal Shock Resistance of Composites of Alumina and Partially Stabilized Zirconia. J. Mat. Sci. Lett. 12:1253–1254, 1993.
29. E. H. Lutz, M. V. Swain, and N. Claussen. Thermal Shock Behavior of Duplex Ceramics. J. Am. Cer. Soc. 74 (1):19–24, 1991.
30. Q.-M. Yuan, J.-Q. Tan, and Z.-G. Jin. Preparation and Properties of Zirconia-Toughened Mullite Ceramics. J. Am. Cer. Soc. 69(3): 265–267, 1986.
31. M. Ishitsuka, T. Sato, T. Endo, and M. Shimada. Sintering and Mechanical Properties of Yttria-Doped Tetragonal ZrO_2 Polycrystal/Mullite Composites. J. Am. Cer. Soc. 70(11):C-342–346, 1987.
32. V. P. Dravid, M. R. Notis, and C. E. Lyman. Twinning and Microcracking Associated with Monoclinic Zirconia in the Eutectic System Zirconia-Mullite. J. Am. Cer. Soc. 71(4):C-219–221, 1988.
33. I.-S. Kim and I.-G. Kim. Thermal Shock Behavior of Al_2O_3-SiC Composites Made by Directed Melt Oxidation of Al-Alloy. J. Mat. Sci. Lett. 16:772–775, 1997.
34. M. Landon and F. Thevenot. The SiC-AlN System: Influence of Elaboration Routes on the Solid Solution Formation and Its Mechanical Properties. Cer. Intl. 17:97–110, 1991.
35. T. N. Tiegs and P. F. Becher. Thermal Shock Behavior of an Alumina-SiC Whisker Composite. J. Am. Cer. Soc. 70(5):C-109–111, 1987.
36. N. Sato and T. Kurauchi. Microcracking During a Thermal Cycle in Whisker-Reinforced Ceramics Composites Detected by Acoustic Emission Measurement. J. Mat. Sci. Lett. 11:590–591, 1992.
37. C. F. Cooper, I. C. Alexander, and C. J. Hampson. The Role of Graphite in the Thermal Shock Resistance of Refractories. Brit. Cer. Trans. J. 84:57–62, 1985.
38. R. C. Rossi and R. D. Carnahan. Thermal Shock Resistant Ceramic Composite. US Patent 4,007,049, 2/8/1977.
39. D. Lewis, R. P. Ingel, W. J. McDonough, and R. W. Rice. Microstructure and Thermomechanical Properties in Alumina- and Mullite- Boron-Nitride Particulate Composites. Cer. Eng. Sci. Proc. 2(7-8):719–727, 1981.
40. R. W. Rice, W. J. McDonough, S. W. Freiman, and J. J. Mecholsky, Jr. Ablative-Resistant Dielectric Ceramic Articles. US Patent 4,304,870, 12/8/1981.
41. W. S. Coblenz and D. Lewis, III. In Situ Reaction of B_2O_3 with AlN and/or Si_3N_4 to Form BN-Toughened Composites. J. Am. Cer. Soc. 71(12):1080–1085, 1988.
42. R. W. Rice. Processing of Ceramic Composites. Advanced Ceramic Processing and Technology 1 (J. G. P. Binner, ed.). Noyes, Park Ridge, NJ, 1990, pp. 123–213.
43. D. Goeuriot-Launay, G. Brayet, and F. Thevenot. Boron Nitride Effect on the Thermal Shock Resistance of an Alumina-Based Ceramic Composite. J. Mat. Sci. Lett. 5:940–942, 1986.
44. P. G. Valentine, A. N. Palazotto, R. Ruh, and D. C. Larsen. Thermal Shock Resistance of SiC-BN Composites. Adv. Cer. Mat. 1(1):81–86, 1986.

45. D. Lewis and R. W. Rice. Thermal Shock Fatigue of Monolithic Ceramics and Ceramic-Ceramic Particulate Composites. Cer. Eng. Sci. Proc. 2(7-8):712–718, 1981.
46. D. Lewis and R. W. Rice. Comparison of Static, Cyclic, and Thermal Shock Fatigue in Ceramic. Cer. Eng. Sci. Proc. 3(7-8):714–721, 1982.
47. J. H. Schneibel, S. M. Sabol, J. Morrison, E. Ludeman, and C. A. Carmichael. Cyclic Thermal Shock Resistance of Several Advanced Ceramics and Ceramic Composites. J. Am. Cer. Soc. 81(7):1888–1892, 1998.
48. M. P. Borom. Dispersion-Strengthened Glass Matrices-Glass-Ceramics, A Case in Point. J. Am. Cer. Soc. 60(1- 2):17–21, 1977.
49. R. K. Govila. Strength Characterization of Yttria-Partially Stabilized Zirconia/Alumina Composite. J. Mat. Sci. 28:700–718, 1993.
50. D. E.-W. Shin, K. K. Orr, and H. Schubert. Microstructure-Mechanical Property Relations in Hot Isostatically Pressed Alumina and Zirconia-Toughened Alumina. J. Am. Cer. Soc. 73(5):1181–1188, 1990.
51. S. Lathabai, D. G. Hay, F. Wagner, and N. Claussen. Reaction-Bonded Mullite/Zirconia Composites. J. Am. Cer. Soc. 79(1):248–256, 1996
52. H. Endo, M. Ueki, and H. Kubo. Microstructure and Mechanical Properties of Hot Pressed SiC-TiC Composites. J. Mat. Sci. 26:3769–3774, 1991.
53. B.-W. Lin and T. Iseki. Effect of Thermal Residual Stress on Mechanical Properties of SiC/TiC Composites. Brit. Cer. Trans. 91:1–5, 1992.
54. B.-W. Lin and T. Iseki. Different Toughening Mechanisms in SiC/TiC Composites. Brit. Cer. Trans. 91:147–150, 1992.
55. C. H. McMurtry, W. D. G. Boecker, S. G. Seshadri, J. S. Zangil, and J. E. Garnier. Microstructure and Material Properties of SiC-TiB_2 Particulate Composites, Am. Cer. Soc. Bull. 66(2):325–329, 1987.
56. K. S. Mazidyasni, R. Ruh, and E. E. Hermes. Phase Characterization and Properties of AlN-BN Composites. Am. Cer. Soc. Bull. 64(8):1149–1154, 1983.
57. X.-N. Huang and P.S. Nicholson. Mechanical Properties and Fracture Toughness of α-Al_2O_3-Platelet Reinforced Y-TZP Composites at Room and High Temperature. J. Am. Cer. Soc.76(5):1294–1301,1993.
58. P. F. Becher, T. N. Tiegs, J. C. Ogle, and W. H. Warwick. Toughening of Ceramics by Whisker Reinforcement. Fracture Mechanics of Ceramics 7, Composites, Impact, Statistics, and High-Temperature Phenomena (R. C. Bradt, A. G. Evans, D. P. H. Hasselman, and F. F. Lange, eds.). Plenum Press, New York, 1986, pp. 61–73.
59. R. K. Govila. Fracture of Hot-Pressed Alumina and SiC-Whisker-Reinforced Alumina Composite. J. Mat. Sci. 23:3782–3791, 1988.
60. M. Yang and R. Stevens. Microstructure and Properties of SiC Whisker Reinforced Ceramic Composites. J. Mat. Sci. 26:726–736, 1991.
61. K. P. Gadkaree. Whisker Reinforcement of Glass-Ceramics. J. Mat. Sci. 26:4845–4854, 1991.
62. T. Kumazawa, S. Ohta, H. Tabata, and S. Kanzaki. Mechanical Properties of Mullite-SiC Whisker Composite. J. Cer. Soc. Jpn., Intl. Ed. 97:880–888, 1989.
63. S. R. Choi and J. A. Salem. Strength, Toughness and R-Curve Behavior of SiC

Whisker-Reinforced Composite Si_3N_4 with Reference to Monolithic Si_3N_4, J. Mat. Sci. 27:1491–1498, 1992.

64. S. Zhu, M. Mizuno, Y. Nagano, Y. Kagawa, and M. Watanabe. Stress Rupture in Tension and Flexure of a SiC-Whisker-Reinforced Silicon Nitride Composite at Elevated Temperatures. J. Am. Cer. Soc. 79(10):2789–2791, 1996.
65. C. Olagnon, E. Bullock, and G. Fantozzi. Properties of Sintered SiC Whisker-Reinforced Si_3N_4 Composites. J. Eur. Cer. Soc. 7:265–273, 1991.
66. J. Dusza and D. Sajgalik. Mechanical Properties of Si_3N_4 + β-Si_3N_4 Whisker Reinforced Ceramics. J. Eur. Cer. Soc. 9:9–17, 1992.
67. J. E. Moffatt, W. J. Plumbridge, and R. Herman. High temperature Crack Annealing Effects on Fracture Toughness of Alumina and Alumina-SiC Composites. Brit. Cer. Soc. Trans. 95(1):23–29, 1996.
68. D. R. Carroll and L. R. Dharani. Effect of Temperature on the Ultimate Strength and Modulus of Whisker-Reinforced Ceramics. J. Am. Cer. Soc. 75(4):786–794, 1992.
69. C. Hulse and J. Batt, The Effect of Eutectic Microstructures on the Mechanical Properties of Ceramic Oxides. United Aircraft Res. Lab. Report for ONR Contract N00014-69-C-0073, 5/1974.
70. F. L. Kennard, R. C. Bradt, and V. S. Stubican. Mechanical Properties of the Directionally Solidified MgO-$MgAl_2O_4$ Eutectic. J. Am. Cer. Soc. 59(3-4):160–163, 1976.
71. V. S. Stubican, R. C. Bradt, F. L. Kennard, W. J. Minford, and C. C. Sorell. Ceramic Eutectic Composites. Tailoring Multiphase and Composite Ceramics, Materials Science Research 20 (R. E. Tressler, G. L. Messing, C. G. Pantano, and R. E. Newnham, eds.). Plenum Press, New York, 1986, pp. 103–113.
72. T. Mah, A. Parthasarathy, and L. E. Matson. Processing and Mechanical Properties of $Al_2O_3/Y_5A_{15}O_{12}$ (YAG) Eutectic Composites. Cer. Eng. Sci. Proc. 11(9-10): 1617–1627, 1990.
73. Y. Waku, N. Nakagawa, T. Wakamoto, H. Ohtsubo, K. Shimizu, and Y. Kohtoku. High-Temperature Strength and Thermal Stability of a Directionally Solidified Al_2O_3/YAG Eutectic Composite. J. Mat. Sci. 33:1217–1225, 1998.
74. T. I. Barry, L. A. Lay, and R. Morell. High Temperature Mechanical Properties of Cordierite Refractory Glass Ceramics. Proc. Brit. Cer. Soc. 25, Mechanical Properties of Ceramics (2) (R. W. Davidge, ed.):67–84, 1975.
75. K. James and K. H. G. Ashbee. Plasticity of Hot Glass-Ceramics. Prog. Mat. Sci. 21(1):3–59, 1975.
76. D. Wilkinson. Creep Mechanisms in Multiphase Ceramic Materials. J. Am. Cer. Soc. 81(2):275–299, 1998.
77. H.-T. Lin, K. B. Alexander, and P. F. Becher. Grain Size Effect on Creep Deformation of Alumina-Silicon Carbide Composites. J. Am. Cer. Soc. 79(6):1530–1536, 1996.
78. J. M. Desmarres, P. Goursat, J. L. Besson, P. Lespade, and B. Capdepuy. SiC Whiskers Reinforced Si_3N_4 Matrix Composites:Oxidation Behavior and Mechanical Properties. J. Eur. Cer. Soc. 7:101–108, 1991.
79. Z. C. Jou, S.-Y. Kuo, and A. N. Virkar. Elevated-Temperature Creep of Silicon Car-

bide-Aluminum Nitride Ceramics:Role of Grain Size. J. Am. Cer. Soc. 69(11):C-279–81, 1986.
80. C. A. May and A. K. U. Obi. Compressive Strength of Lithia-Silica Glass Ceramics. Proc. Brit. Cer. Soc. 25, Mechanical Properties of Ceramics (2) (R. W. Davidge, ed.):49–65, 1975.
81. T. A. Parthasarathy, T. Mah, and L. E. Matson. Creep Behavior of an Al_2O_3-$Y_3Al_5O_{12}$ Eutectic Composite. Cer. Eng. Sci. Proc. 11(910):1628–1638, 1990.
82. A. R. B. Verma, V. S. R. Murthy, and G. S. Murty. Microstructure and Compressive Strength of SiC-Platelet-Reinforced Borosilicate Composites J. Am. Cer. Soc. 78(10):2732–2736, 1995.

12

Summary and Perspective for the Microstructural Dependence of Mechanical Properties of Dense Monolithic and Composite Ceramics

I. INTRODUCTION

Chapters 2–7 addressed the grain dependence of mechanical properties of nominally dense monolithic ceramics, and Chapters 8–11 addressed the particle (and matrix grain) dependence of mechanical properties of ceramic composites. This chapter presents both a summary and a perspective on both topics, clearly showing the extensive commonality of the dependence of monolithic and composite ceramics (with little or no porosity) on their microstructures. To provide additional perspective beyond this summary, three additional topics are briefly addressed. First, before proceeding to the summary, the grain and particle dependences of some other properties of monolithic and composite ceramics are outlined, e.g. since such dependences must be considered along with those of mechanical properties in applications where both mechanical and nonmechanical properties are important. Then following the summary, needs to improve both the understanding and the performance of ceramic materials and some approaches to doing this are briefly discussed.

II. MICROSTRUCTURAL DEPENDENCE OF OTHER PROPERTIES OF CERAMICS

A. Monolithic Ceramics

Chapters 2–7 clearly show that grain parameters, especially size, and in some cases shape, and orientation play key roles in basic mechanical properties of toughness, tensile and compressive strengths, as well as of hardness, wear, and related behavior. Elastic properties and thermal expansion normally do not depend on grain size unless there is microcracking, but both depend on grain orientation due to crystalline anisotropy of these properties (for essentially all crystalline materials for elastic properties, but only in noncubic materials for expansion) and hence some also on grain shape. Some other properties have varying dependences on grain size, shape, and orientation, and some materials may exhibit special property dependences on these parameters.

Consider first thermal conductivity, where there is no intrinsic grain size (G) dependence in cubic materials. However, there is an intrinsic dependence on grain size, shape, and orientation in noncubic materials, since these parameters determine the tortuosity of the preferred path for heat flow from grain to grain and how much the average conductivity changes from grain to grain due to differing conductivities along different crystal axes. Thermal conductivity anisotropy in noncubic ceramics ranges from very modest to substantial levels, e.g. at 22°C from ~ 10 to 50, and 70% higher along the *c*- versus the *a*-axis in respectively sapphire, rutile, and quartz [1], and ~ 30% estimated for BeO [2], while materials such as micas [3], graphite, and hexagonal BN are more extreme with conductivities in the a versus the *c* directions being respectively nearly 10, ~ 100, and ~ 20-fold (the latter two from manufacturers' literature). The issue of averaging methods to obtain polycrystalline values from single crystal values was addressed by Kumar and Singh [1] showing similar levels of agreement and disagreement to those for other properties addressed in this book. While the issue of grain size effects has received little attention, Williams et al. [4] have shown data for dense, pure Al_2O_3 to increase thermal conductivity at ~ 22°C from ~ 30 to 32, then 33 W/m/K ~ 10% as G increased from ~ 1 to 6, then 16 μm, consistent with theoretical predictions of relative changes. Such G dependence with modest anisotropy indicates greater G dependence in more anisotropic materials.

Electrical conductivity is also isotropic in cubic crystal structures, so there are no intrinsic effects of grain size, shape, or orientation on conductivity in such materials, but there is a basic dependence on these parameters in noncubic materials, i.e. as for thermal conductivity. However, electrical conductivities cover a much broader range of values and thus provide more opportunity for more pronounced anisotropies, as with the converse property, resistivity. For example, sapphire has greater anisotropy of electrical than of thermal conductivity, e.g.

having electrical conductivity ~ 3.3-fold greater parallel versus perpendicular to the *c*-axis [5]. However, there are other materials with greater anisotropy for both ionic and electronic conduction; the ~ 3 orders of magnitude higher conductivity normal versus parallel to the *c*-axis of graphite is an example of the latter. Beta aluminas are examples of the former, having very anisotropic conduction, i.e. about 100 fold greater normal versus parallel to the *c*-axis at 300°C, and thus they essentially show conduction only normal to the *c*-axis via the Na or other alkali metal containing planes [6,7]. However, this anisotropy in conductivity and the converse anisotropy in resistivity can be significantly mitigated in polycrystalline bodies, e.g. Virkar et al. [8] showed that increasing *G* by 50–100-fold from 1–2 to ~ 100 μm reduced resistivity by < 2-fold, from 4.8 to 2.8 Ω ·cm.

Turning to another electrical property, dielectric constant (∈), this is normally independent of grain size, unless some other grain-size-dependent effect such as microcracking or grain boundary concentration of second phases occurs. Thus recent tests showed ∈ for dense, pure alumina independent of grain size over the range tested, *G* ~ 1–6 μm [9]. However, in ferroelectric and related ceramics there can be substantial *G* dependence of ∈, e.g. the maximum ∈ at the Curie point in doped $BaTiO_3$ bodies ranged from ~ 10,000 to 30,000 as *G* increased from ~ 2.5 to 7 μm [10], and ∈ at room temperature for PZT ranged from 700 to 1300 as *G* increased from 2 to ~ 4.4 μm [11]. Dielectric loss, though often entailing other mainly extrinsic mechanisms as a function of grain boundary character, is clearly an important factor in many applications along with ∈. In the above Al_2O_3 study [9], dielectric loss increased with grain size only at the larger grain sizes, i.e. it was ~ constant from *G* ~ 1 to 3 μm, and then rose substantially at *G* = 4–5 μm and still more at *G* ~ 6 μm. Hsueh et al.'s PZT study [11] showed dielectric loss (% tan δ) with a similar behavior, i.e. it was ~ constant at ~ 0.5 from G = 2– to 3+ μm and then increased to ~ 1.3 at *G* = 3.5 μm (for 1100 and 1200°C firings) and ranged from ~ 1.2 at *G* ~ 3.4 μm to between 1.7 and 2.6 at *G*= 4.5 to 5 μm for firing at 1300°C.

At least some of the dielectric constant increase is related to intrinsic increases in Curie temperature that occur as *G* decreases, e.g. from ~ 125 to 140°C as *G* decreased in a PLZT from ~ 4 to 1.5 μm [12]. This and other significant changes in ferroelectric properties are related to domain structure–grain size effects, which can lead to complex behavior, since some properties increase, some decrease, and some are independent of *G*. Since there are similar domain–grain size interactions in ferromagnetic materials, they also show varying effects of grain size. Thus the initial permeability of Ni Zn ferrites was reported to increase ~ 4-fold as *G* increased from < 1 to ~ 50 μm, while the coercive field decreased ≥ 15-fold and the remnant magnetic flux remained unchanged [13].

Scattering of electromagnetic waves intrinsically occurs in noncubic polycrystalline dielectric materials, since the dielectric constant and hence the refractive index vary with crystal direction, resulting in varying changes of these properties across grain boundaries. While the number of scattering events increases as G decreases, the net ray deflection decreases, so the net effect is for more serious scattering as G increases, but this is also a function of wavelength. Scattering decreases as the grain size decreases below the wavelength, e.g. a body with intrinsic transmission and $G \sim 1$ μm will have limited scattering in the infrared and substantial scattering in the visible, and greater scattering in the UV. However, other effects are typically dominant in scattering, since extrinsic effects of impurities and intrinsic effects of residual porosity (often at grain boundaries) are generally much greater than intrinsic effects of refractive index differences across grain boundaries.

Dielectric breakdown, i.e. failure of an electrical insulator at some level of applied electrical field, e.g. in volts per cm, also commonly increases with increasing G [14,15], so breakdown fields are often significantly lower than for finer G, but also lower than single crystal values. However, much if not all of the G dependence is due to extrinsic effects of residual pores and second phases along grain boundaries, often increasing in extent as G increases. The role of larger grain size on dielectric behavior is also reflected in effects of larger grains on the breakdown voltage for TiO_2 doped ZnO varistors. Hennings et al. [16] reported that seeding the varistor body to eliminate large exaggerated grains produced a finer, more uniform grain size giving more consistent onset of nonlinear conduction.

Clearly the significant anisotropy that can occur in the above nonmechanical properties makes grain shape and especially orientation effects important. Also, where there is substantial anisotropy, this can change substantially with temperature due to differing dependences of property values in various crystal directions on temperature.

B. Composite Ceramics

Composite ceramics intrinsically have broader property dependence on microstructure than monolithic ceramics do, as is discussed in Chapters 8–11. For example, while monolithic ceramics normally have no dependence of elastic properties on grain structure unless there is microcracking or preferred orientation (in noncubic ceramics), there is more and broader dependence in composites. Thus the dependence of elastic properties on the nature and volume fraction of the dispersed phase has been addressed in Chapter 8, and the effects of preferred orientation, possible microcracking, and contiguity of the phases (especially for fiber composites) have been discussed to limited extents. Extending the discussion to other properties of composites could be a very large task. Instead, the focus of this section will be to outline the effects of particulate parameters on three important properties that are impacted more in composites than in monoliths. Two of these, namely thermal expansion and conductivity, are particularly pertinent to mechanical properties, especially thermal stresses and thermal shock, and are addressed first.

Consider first thermal expansion, which does not depend on grain size in monolithic ceramics, nor generally on particle size in composites, unless microcracking occurs. However, the thermal expansion of composites clearly depends on the expansions of the matrix and dispersed phase and their volume fractions, and may also be affected by preferred orientations of the matrix grains and dispersed particles if they are noncubic. Composite expansion can also depend on contiguity of the dispersed phase, especially for fibers, but the focus here is on particulate composites. The challenge in calculating composite expansion is to account for the elastic interactions of the matrix grains and the dispersed particles due to differences of thermal strains between them.

A variety of expressions for thermal expansion of various composites have been derived, e.g. as reviewed by Raghava [17] reflecting different results for different dispersed phase geometry and for the same geometry, the latter reflecting the complexity of the problem and the effects of differing assumptions. A simple equation given by Kingery [18] is

$$\alpha_C = [\alpha_M k_M(1-\phi) + \alpha_P k_P\,\phi][k_M(1-\phi) + k_P\,\phi]^{-1} \tag{12.1}$$

where α and k are respectively the linear thermal expansion and the bulk modulus, the subscripts C, M, and P refer to the composite, matrix, and dispersed (particulate) phase, and ϕ is the volume fraction dispersed phase. A simpler equation used with some success for composites of constituents without large differences in elastic properties (e.g. ceramic–metal composites) is [19]

$$\alpha_C = \alpha_M + (\alpha_P - \alpha_M)\,\phi \tag{12.2}$$

However, this clearly does not address the occurrence of factors such as microcracking, e.g. during cooling from fabrication when $\alpha_M > \alpha_P$, or particles acting more as pores due to combinations of $\alpha_P > \alpha_M$, particle size, and limited particle–matrix bonding. Accounting for effects such as varying bonding between Ni particles and an alumina matrix has been reported to have a significant effect on expansion behavior of composites [19].

Turning to thermal conductivity, there is an even greater number of models and more diversity of them than for expansion, reflecting a greater challenge (e.g. Ref. 20 for an earlier survey of models and their fits to data). However, besides the complications of solving the basic problem of thermal conductivity through a densely consolidated (i.e. pore-free) body with dispersed particles of different conductivity than the matrix and having various sizes, shapes, etc., there are three serious complications that can occur. The first, for which there is some theoretical and experimental understanding, is percolation of at least one of the phases, i.e. the onset of that phase forming continuous paths through the body along the axis parallel to which conductivity is of interest. This is important because, as this occurs for a phase of substantially higher conductivity than the other, typically the matrix, phase with which it is mixed, the thermal conductivity increases rapidly. The combined uncertainties in predicting the approach, onset, and extensive percolation combined with incorporating such effects in the prediction of thermal conductivity are a challenging task. The second complication, for which there is less information, is that of having an interfacial phase between many or all of the dispersed particles and the matrix, particularly when such phases have much lower conductivity than the dispersed phase, especially when the latter has high conductivity. The third complication, which is widely overlooked, is that the conductivity of the dispersed phase is not certain, especially for refractory ceramic particulates. This arises because the process of making the particulates can result in factors such as limited impurities or poor crystalline perfection that can substantially reduce their thermal conductivity, as shown by work of Slack on bulk materials (e.g. Refs. 21 and 22). Further, there is no way to ascertain whether the conductivity of the particles is at all consistent with accepted values for the bulk material, except via evaluation of data against models. However, the latter presents significant challenges, since lower than expected composite thermal conductivity, which would result from poorer particulate conductivity, can also result for less particulate percolation or interfacial phases of lower conductivity, or both, either of which can be challenging to ascertain accurately.

Focusing on particle size dependence, note that models not addressing percolation or interfacial layers predict no dependence on dispersed particle size. On the other hand, percolation generally occurs at lower volume fraction particles as the particle size decreases, thus giving increased conductivity with finer particles of good conductivity. However, interfacial phases of lower conductivity tend to reduce body conductivity less with larger particles, since there are fewer larger particles for a given volume fraction of particles and hence fewer interfacial layers and less total area covered by such layers. Thus, for example, the net particulate surface area for a fixed volume fraction of spherical particles varies inversely with the particle diameter, e.g. doubling the particle size gives only half the net particle surface area. Thus such layers can give an opposite dependence on particle size (if layer thickness is independent of G); and if both percolation and interfacial layers are present with similar effects, thermal conductivity behavior similar to that predicted by models neglecting both effects can result.

Consider now some limited experimental data on thermal conductivity of composites. Hasselman et al. [23] showed that thermal conductivity of composites of 15 v/o diamond particles in a cordierite matrix increased as the particle size increased, but that the benefits of larger particle size diminished as temperature increased (Fig. 12.1). (This reduction with temperature is similar to reduced grain size effects on conductivity of monolithic ceramics as temperature increases.) They applied a model including interfacial effects to their data, which

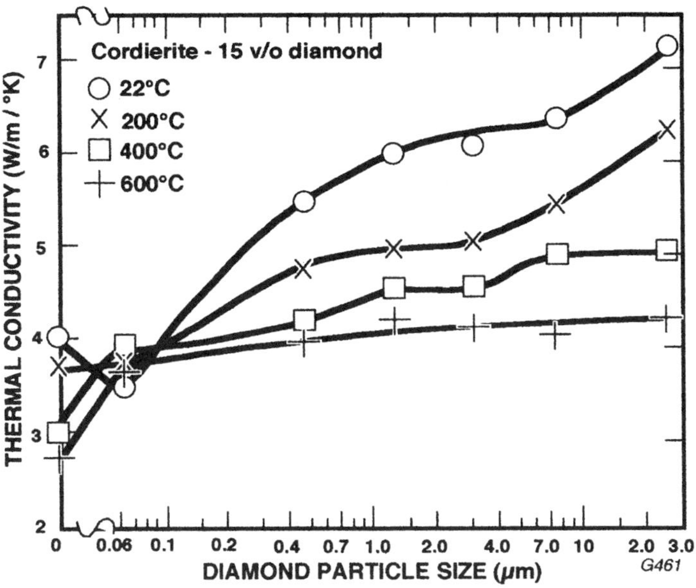

Figure 12.1 Thermal conductivity versus particle size for a composite of 15 v/o diamond particles in a cordierite matrix at various temperatures. Note the decrease in thermal conductivity with decreasing particle size and the decrease in both thermal conductivity and its particle size dependence as temperature increases. Values of the matrix alone are shown at zero particle size. (From Ref. 23.)

was also applied to earlier data for Al_2O_3-SiC particulate composites, showing similar particle size and temperature dependence of composite conductivity (which was typically less than for the matrix alone) [24]. Chen et al. [25] also showed that the thermal conductivity of composites with 10–60 v/o Cu particles in an epoxy matrix showed no significant difference in the increase in conductivity with volume fraction between 11 and 100 μm dia particles. However, composites with 7 μm Al_2O_3 showed modest, but statistically significant, higher conductivity despite Cu having a conductivity ~ 10-fold higher than Al_2O_3, i.e. suggesting that other effects such as interfacial effects (e.g. with the Cu) were probably operative.

Finally, note that Neilsen's model [26] is commonly used, e.g. by those in the important field of developing organic, commonly rubber, matrix composites with dielectric ceramic fillers (e.g. MgO, AlN, and more commonly Al_2O_3 and especially hexagonal BN) for use as thermally conducting, electrical insulating gaskets for electronic components. This model's use is based on both its fitting considerable data (e.g. Ref. 20) and its involving more microstructural dependence, and it reflects both some of the advances and the problems in the field. Thus besides the usual parameters of conductivities and volume fractions of the constituents, it includes particle shape and orientation (but not size) and maximum packing fraction, which reflects essentially a fully percolated volume dependent on particle shape and orientation. However, neither the incorporation nor the determination of these terms is rigorous, and so the important question is raised of whether the frequent fitting of data reflects real correlation with actual composite parameters or just curve fitting by using parameters in the model as adjustable parameters. Note also that conductivities predicted by this equation can become excessively large or infinite, near and above the percolation limit.

Turning now to electrical conductivity, note that many models for thermal conductivity are the same for electrical conductivity and vice versa, e.g. including Nielsen's model [26]. However, the review by McLachlan et al. [27] is recommended as a more current assessment of the field, including more extensive discussion of incorporating percolation effects on electrical conductivity, which unfortunately still is much more phenomenological than rigorous. Electrical conductivity can be even more affected by percolation effects because of the much broader range of electrical versus thermal conductivities, so percolation of particles of a highly conductive phase in a matrix makes large increases in conductivity. For this same reason, electrical conductivity of composites can be more susceptible to corrosion or other chemical actions forming interfacial phases of more limited conductivity, hence exacerbating complications noted above for thermal conductivity. Ota et al. [28] presented data showing the onset of large increases in conductivity (i.e. large decreases in resistivity) at higher volume fraction of conductive particles as particle size increases in composites with organic matrices (often silicone rubbers for switches). While the deformation response of such composites is far greater than for ceramic matrices, the dependence of conductivity on particle volume fraction and size is relevant to many ceramics and ceramic–metal composites.

Data on the changes of either electrical or thermal conductivity as a function of microstructural parameters at elevated temperatures is very limited. Prediction from models is limited not only by their uncertainties but also by the frequently limited data on the conductivity of the constituent materials, with the former exacerbated by interfacial phases and their changes. Thus, for example, the conductivities of crystalline phases typically decrease with increasing temperature, while the thermal conductivity of glassy phases increases, which can be important in ceramic refractories, e.g. combinations of these two effects giving varying conductivities [29,30].

Electronic conductivity typically decreases as temperature increases, while that due to ionic conductivity increases, thus also giving opportunity for more variable and complex behavior of the two combined mechanisms. Again, interfacial phases can be important, as indicated by the anomalous increase in ionic conductivity in composites of CaF_2 or BaF_2 with limited additions of Al_2O_3 at elevated temperatures reported by Fujitsu et al. [31]. Thus at 500°C Al_2O_3 additions in CaF_2 increased conductivities by a maximum of nearly an order of magnitude; then they decreased with further Al_2O_3 addition, with the effect dependent on Al_2O_3 particle sizes. The maximum conductivity occurred at ~ 5, 10, and 15 v/o Al_2O_3 additions with respective particle sizes of 0.06, 0.3, and 8 μm that gave decreases back to the matrix conductivity at respective v/o Al_2O_3 additions of ~ 20, 25, and 30 v/o Al_2O_3. These conductivity effects are attributed to formation of an interfacial phase, i.e. consistent with less addition of finer particles needed because of its greater surface area for such phase formation.

III. SUMMARY OF GRAIN AND PARTICLE SIZE DEPENDENCE OF MECHANICAL PROPERTIES OF MONOLITHIC AND COMPOSITE CERAMICS

A. Grain Dependence of Fracture Mode, Toughness, and Crack Propagation of Monolithic Ceramics

Microcracking occurs in noncubic ceramics spontaneously without an external applied stress, at and above a threshold grain size as a function of material properties, especially the thermal expansion anisotropy, with Eq. 2.4 being a reasonable approximation for spontaneous cracking (Fig. 2.10). While such cracking is typically intergranular and on the scale of the grains, some transgranular cracking has been observed (apparently more at larger *G*), as has variation of the scale of the cracking with *G* (Chap. 2, Secs. II.C and III.C). Besides such general trends, other trends and effects have been identified but leave a great deal to be better documented and understood. Thus such cracking also appears to occur at finer *G* as the applied stress increases, and at least some can also be quite dependent on the occurrence of SCG and may also occur in stressed cubic ceramics of high elastic anisotropy, usually at much larger *G* (e.g. on a cm grain scale, Fig. 1.7B).

While effects of slow crack growth (SCG) on spontaneous microcracking have not been studied in detail, they can clearly be a factor. There has been extensive study of SCG in ceramics due especially to the effects of water, but many microstructural factors are only partially explored. Thus it is known that SCG generally occurs intergranularly, while subsequent fast fracture typically is transgranular (e.g. Fig. 2.5), which can be a substantial aid in identifying and studying it, but there are exceptions (e.g. in some ferrites), for which there is limited understanding. Further, there is some evidence that the rate of SCG significantly decreases as *G* decreases (Fig. 2.8), but documentation and understanding of this factor, which could significantly impact life predictions, is lacking, as is information on the temperature dependence of SCG. While some nonoxides may have intrinsic SCG due to water or other chemicals, some have extrinsic SCG due to oxide-based intergranular phases and fracture. Some oxides such as MgO do not exhibit SCG in single crystals but have SCG in polycrystals via intergranular fracture. Whether this is due to oxide impurities or an intrinsic effect of grain boundaries is unknown. Finally, other degradation due to water and other environmental agents can occur at room and modest temperatures. Thus MgO and especially CaO single- and polycrystals can be degraded by the expansion of hydration products formed in surface cracks and pores, and some TZP bodies can be destroyed by moisture destabilization that is accelerated at temperatures and pressures above the normal ambient (Fig. 2.9). At least some of these latter effects are dependent on *G* as well as chemistry.

Fracture toughness at room temperature, which has been a major focus of research on mechanical properties of ceramics for a number of years, varies from simple to complex dependence on material, microstructure, and test parameters in only partially documented and understood fashions. Roughly there are three interrelated factors that lead from simpler to more complex toughness behavior, the first of which is material character. The simplest material systems are glasses, which generally have less variation in fracture toughness for a given glass and lower values (generally ≤ 1 MPa·m$^{1/2}$). However, even here, trends with other basic physical properties, Young's modulus in this case, are only approximate and show considerable variation (Fig. 12.2). The next simplest type of material is single crystals, which have varying degrees of complexity due to general dependence of toughness on crystal orientation, the extreme of which is highly preferred cleavage, which generally gives lower toughness values (Table 2.1). The prediction of cleavage planes, and hence their multiplicity, toughness values, and interactive effects of these is limited, and the effects of frequently resulting significant mixed mode crack propagation (giving higher apparent toughness values) are often neglected. However, single crystal toughnesses, though not as extensively documented as desired, generally have more limited variations for a given crystal material and crack propagation plane in it than for many polycrystalline bodies. An important but often neglected factor that allows some checking of the self consistency of toughness data is that most fracture toughness values for a given crystal will be substantially less than those of corresponding polycrystals (Table 2.1, Fig. 2.15).

Fine grain polycrystalline materials (e.g $G<$ a few microns) of both cubic and noncubic structures also generally have more limited variations in toughness values for a given material with different tests. Cubic materials tested over a broader range of typical grain sizes often show a tendency for a maximum of toughness, usually a modest one, at some intermediate *G*, with variations in both the overall level of toughness and the *G* of its occurrence (Figs. 2.12, 2.14). The most frequent and extreme variations of toughness occur in noncubic materials,

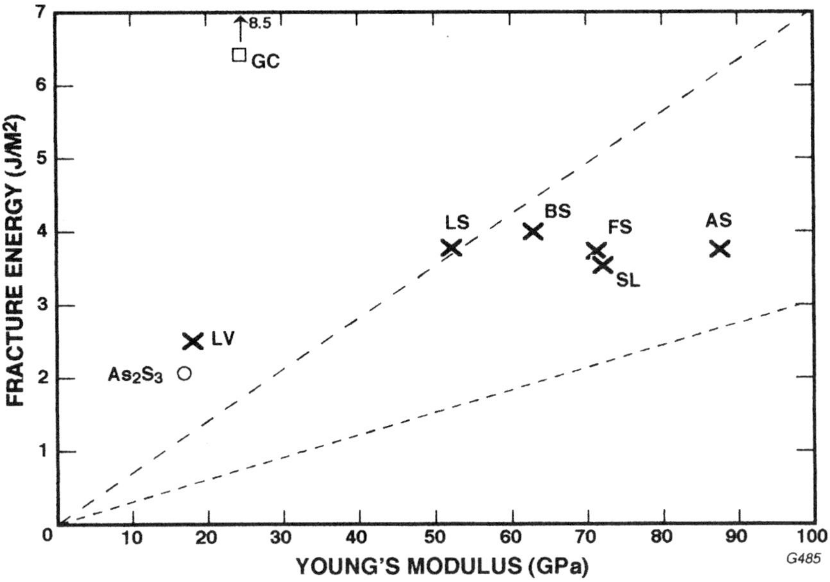

FIGURE 12.2 Fracture energy of glasses versus Young's modulus at ~ 22°C. Note that AS, BS, and LS are respectively alumino-, boro-, and lead-silicate glasses. FS-fused silica, SL = soda lime glass, LV = leached Vycor® (i.e very porous before sintering, ~ 96% silica) and GC = glassy carbon (via polymer pyrolysis) with ~ 20% closed, spherical porosity; while glassy in appearance and behavior (e.g. being isotropic), it is really mostly nanocrystalline graphite. Dashed lines are the upper and lower bounds for the trend of the bulk of the data in the survey of Mechlosky et al. [32].

frequently as a significant maximum as a function of *G* in the commonly investigated *G* range. While there are significant variations in the level and *G* for the maximum, or even its occurrence, depending on material and especially test method and parameters (Figs. 2.16–2.18, most extreme toughnesses are associated with large cracks.

The above significant increases in toughness that occur most extensively in large crack tests of noncubic materials (and ceramic composites, Sec. III.B), can also occur to some extent in (mainly larger grain) cubic materials and are generally related to crack-microstructure interactions that also depend on test method and parameters. The primary mechanism of increased toughness is crack bridging in the crack wake zone commonly resulting in *R*-curve behavior, which develops over a range of crack propagation distances before saturating. This development clearly depends on the extent of crack propagation and hence on the test method and parameters, but this can be complex, since it may also be a function of the initial crack size and character and aspects of the crack propagation that have been incompletely investigated. Crack bridging and its effects are enhanced by intergranular fracture along larger and elongated grains, especially if they continue to have substantial intergranular fracture, which is not commonly the case unless grain boundaries are weaker, e.g. due to boundary phases. Two other mechanisms that also have much of their effect via the crack wake zone are microcracking and transformation toughening, with the latter also sometimes involving microcracking. However, how these, especially microcracking, differ in their origins and character from other bridging effects is not well understood. For example, bridging probably involves some microcracking, but whether this initiates at or near the crack tip or in the wake zone, and the effects of more general and extensive microcracking versus that only due to the crack itself, are not well understood.

Evidence for crack wake effects and bridging as an important component of large-scale crack propagation is well established. Thus *R*-curve effects showing toughness increasing with crack propagation are extensively demonstrated experimentally, and their effect in increasing toughness is demonstrated by reduced toughness when the wake area is removed until further crack propagation develops a new wake zone. Bridging is also seen by microscopic examination of crack wake regions where they intersect specimen surfaces (Figs. 2.4, 8.7, 8.15). However, wake and bridging observations have been restricted almost exclusively to large cracks propagated at limited but unknown velocities and then arrested and examined primarily at or near the intersection of the crack and the, typically machined, specimen surface. Contrary to many earlier and some current assumptions, crack bridging and other wake and *R*-curve effects commonly have limited or no effect on normal strengths of higher strength ceramics. Reasons for this limited relation to strength behavior are discussed below in conjunction with a discussion of the effects of grain parameters on tensile strength; here issues of the nature of wake and bridging observations, themselves pertinent to their applicability to strength, are outlined.

The above observations leave critical questions of crack size effects, low velocities and the arrested aspect of the crack, and possible effects of surface machining flaws and stresses on surface bridging observations. Thus machining flaws commonly in the range of a few tens of microns, with limited increase as G increases, control most strength behavior, raising serious questions of the effects of the orders of magnitude smaller size of such machining flaws from cracks of most toughness tests. The limited observations of crack wake bridging of specimens with as-fired surfaces show much less bridging, indicating that much but not all crack bridging is due to machining effects. The low-velocity-arrested crack aspect of bridging observations raises serious issues of what effect this has on bridging, e.g. whether bridges are sustained as cracks accelerate to, hence assuring significant effects on, catastrophic failure. As noted earlier and in the next section, composite tests raise serious questions of crack velocity on crack propagation, as do tests of $MgAl_2O_4$ crystals. The latter show basic changes in crack propagation with crack velocity, i.e. macroscopic crack propagation shifting from a zigzag pattern on {100} planes at low velocities to flat fracture on {110} planes in (Fig. 2.7) at high velocities.

Another question regarding wake bridging effects on strength is fracture mode behavior. Though it is often not given much scrutiny, it can be an important factor indicative of failure mechanisms. Typically toughness and strength tests at or near room temperature show mostly or exclusively transgranular fracture in dense ceramics (Fig. 2.5), with four exceptions or variations. First, fracture generally transitions to intergranular fracture at finer *G*, e.g. below 1 to a few micron grain sizes, which may often be at least partly extrinsic due to residual grain boundary impurities from use of finer powders and lower processing temperatures to obtain fine *G*. Second, there is, in at least some cases, a transition back to intergranular fracture in cubic materials at very large *G* (e.g. mm to cm scale) and in noncubic materials at much more modest G. Third, residues of some densification aids, e.g. LiF in MgO or $MgAl_2O_4$, often greatly enhance intergranular fracture (as does increased temperature, as is discussed later). Fourth, environmentally driven slow crack growth (SCG), which occurs in many oxides and some nonoxides (some intrinsically and some extrinsically often due to oxide additives) generally occurs by intergranular fracture. However, additive residues have also been observed to enhance grain bridging of cracks while they also lower strength, again raising questions of the impact of bridging on strength, i.e. in this case why a grain boundary phase that weakens the boundaries but enhances grain bridging of cracks still lowers body strength. Similarly, and more broadly, SCG, by mainly intergranular fracture, raises the question of whether this is a significant factor in the surface observations of crack bridging by grains or clusters or fragments of them.

Consider effects of grain shape and orientation and of temperature on fracture toughness. Elongated grains increase bridging effects, mainly with intergranular fracture, and may often correlate with preferred grain orientation because elongated grains often are more prone to orientation in forming precesses. Preferred grain orientation also directly affects fracture toughness, but the extent and nature of the effects are intimately related to material factors, especially preferred cleavage planes, and grain size and shape and their effects, e.g. on fracture mode. Thus oriented grains, especially larger ones and particularly elongated ones, will impact toughness to the extent that crack propagation is parallel or perpendicular to the grain orientation texture and the extent of resultant trans- versus intergranular fracture.

Very little data on toughness at intermediate temperatures exists, so directions of microstructural effects are uncertain (but composite studies in the next section show that significant changes in mismatch stresses occur, impact toughness, and indicate that some parallel effects could occur in monolithic ceramics). However, data at higher temperatures, e.g > 800 to 1000°C, shows three sets of trends: (1) increasing intergranular fracture, which can aid crack bridging but is also commonly a precursor to high temperature SCG and resultant lower strengths, (2) lesser extremes of toughness variations, even before extensive high-temperature SCG or crack tip plasticity occurs, and (3) frequent but strain-rate-dependent increases in toughness at higher temperatures associated with plastic deformation, especially in single crystals. Note that the latter is another example of basic toughness–strength differences as shown by comparison to strength behavior in the following section.

B. Grain Size Dependence of Tensile Strength of Monolithic Ceramics

Turning to tensile or flexure strength at or near room temperature, the extensive focus to explain this has been by crack propagation studies, i.e. of SCG and especially toughness, with limited attention to the nature of flaws controlling strength. Clearly, both are important, but the perspective necessary to understand strength behavior of ceramics, especially its dependence on grain size, must start with the effects of body microstructure and properties impacting flaw populations introduced, which are a major factor in determining strength. The necessity of this perspective can be seen by noting inconsistencies of the toughness-based approach and the consistencies of the flaw-based approach.

Consider first failure from machining flaws, which is most common, and the inconsistencies that an approach based on the microstructural dependence of large crack toughness presents for explaining the microstructural dependence of small crack tensile strength. Strengths of dense machined ceramics show two fundamental aspects of strength–grain size behavior that must be satisfactorily explained in order to understand their strength behavior. The first, which is particularly extensively demonstrated, is the generally modest decrease of strengths as G increases at finer G, followed by a substantially faster strength decrease as G increases at larger G, with strengths at larger G often falling below those for single crystals of the weakest orientation with the same machining (Fig. 3.1). The other aspect of strength behavior of machined samples is the effect of machining variables on resultant strengths, especially increases as the grit size decreases and anisotropy of strength as a function of the machining direction relative to the uniaxial stress axis [33], which is ~ zero when the grain and flaws are about the same actual size, and which increases as the grains decrease or increase in size relative to the flaws (Fig. 3.33). These strength trends, while varying in detail, are the same in overall character for both cubic and noncubic ceramics as well as transformation toughened ceramics (and other toughened ceramic composites, see the next section). Such trends are clearly inconsistent with much data, especially large crack toughness–grain size data, which shows toughness increasing as G increases.

As was discussed extensively in Chapter 3, this strength–grain size behavior is explained by a primary effect and some secondary effects. The primary effect is that machining flaws are larger than finer grains and smaller than larger grains, and to a first approximation they are independent of G and hence are the size of the grains at an intermediate G. The specific flaw size and hence the G at which they are equal are a function of machining, so that the finer G branches shift as a function of machining. (Machining flaws for a given machining operation also do not vary widely for most ceramics, but clearly change some for different ceramics and machining parameters.) However, there can be some secondary dependence of machining flaw size on material parameters, primarily the local Young's modulus, hardness, and toughness controlling formation of the flaws per Eq. (3.2). Note that these values, especially toughness, may be substantially lower than those commonly measured with large-scale cracks because of crack size effects [34], and they reflect local transient higher stress rates and temperatures; toughness and hardness can vary with G. Further, as the flaw and grain sizes approach each other, there can be increasing contributions of mismatch stresses between grains to failure, and the fracture toughness controlling failure is decreasing from polycrystalline to single crystal or grain boundary values (Fig. 2.15). Any or all of these three secondary effects result in strengths decreasing as G increases. Both the calculation of flaw sizes for different G bodies and direct fractographic observations corroborate that the larger and finer G branches intersect when the grain and flaw sizes are ~ equal (recognizing that they are respectively measured by a diameter and a radius), with statistical effects resulting in some variation from absolute equality, as was discussed in Chapter 3.

Also note, first, that the slope of the larger G branch has been shown theoretically and experimentally to be variable and < the polycrystalline toughness due to varying transitions between polycrystalline and single crystal or grain boundary fracture toughnesses (Fig. 2.15). Second, while strengths typically extend below those for single crystals, they must ultimately reverse and approach lower single crystal values at very large G but may follow various paths depending on material, finishing, and especially specimen-grain parameters (Fig. 3.1). Third, as-fired surfaces show similar strength–grain size trends to those for machined surfaces, since the depth of grain boundary grooves increases with G, while the tortuosity of collections of grain boundary grooves to form a single flaw should decrease as G increases. Fourth, there are two variations on the above strength–grain size behavior due to either microplastically induced or grown flaws or for failure from microcracks, each having characteristics different from the above-normal G dependence of flaw failure. For microplastic-induced failure, the larger G strengths do not fall below those for the weakest crystal orientation but instead extrapolate to the lowest single crystal yield stress, which is usually slightly below the crystal failure strength (Figs. 3.1). For microcrack failure, strengths decrease rapidly as *G* increases above the value where microcracking commences, but then the strength decrease begins to saturate as *G* increases further.

Another important source of information on strength behavior is the mirror, mist, hackle, and crack branching behavior widely observed on normal-strength fractures of monolithic ceramics, which, for example, reflects effects of TEA and related stresses [34]. While not quantitatively studied in larger grain ceramics where bridging and *R*-curves are greatest, these fracture patterns do not show deviations from normal patterns that would be expected if large increases in toughness accompanied crack propagation in normal strength failure. Further, data for intermediate *G*, e.g. $< 10\ \mu m$, in materials such as Al_2O_3, where measurable bridging and *R*-curve effects are reported, have fracture patterns consistent with those for finer *G* bodies where bridging and *R*-curve effects are not seen [35].

Though receiving very limited attention, especially for effects of microstructure, mechanical fatigue has been demonstrated under varying tensile loading. This has been shown to occur due to grain mismatch stresses, e.g. in larger grain Al_2O_3, and can be independent of environmental effects, since it can be more severe in liquid N_2, rather than reduced or stopped by low temperatures. A mechanism was proposed based on microcracking during increasing tensile loading leading to local incompatibilities due to elastic relaxations around the microcracks preventing their closure (Fig.11.8). Thus microcracking further extends during unloading so that the local microcrack situation and stress state do not return to the situation at the start of the stress cycle.

Thermal shock generally shows different dependence on grain parameters than on strength. The critical quench temperature difference for serious loss of strength is independent of grain size, but the retained strength after damaging thermal shock generally increases as grain size increases and starting strength decreases. Toughening mechanisms such as crack bridging and *R*-curve effects may be operative in thermal shock to improve retained strengths, as is also suggested by composite effects in the next section. This is logical, because such damage generally involves larger scale cracks that form, but are arrested, thus allowing *R*-curve or other large crack toughening to be operative. Transformation toughening has limited benefit for thermal shock resistance, e.g. higher strengths from such toughening in TZP and PSZ bodies do not necessarily increase ΔT_C, but microcracking in such materials (associated with lower strengths and larger cracks) can significantly aid in retaining reasonable strengths after thermal shock. It is also expected that similar trends will occur from serious impact damage, i.e. impacts of larger or higher velocity particles may leave greater retained strength in bodies in which larger cracks, effected by *R*-curve and related effects, are formed, but are more difficult to propagate.

Consider now the effects of increased temperature on the grain dependence of mechanical properties other than toughness. While data is more limited, there is clearly sufficient data on tensile strength to show that the common assumption that no significant changes occur, other than the decrease expected from *E* decreases (~ 1–3% per 100°C) until temperatures of ≥1000°C, can often be seriously wrong. Thus, for example, Figs. 6.11 ,6.12, 6.14, 6.15, and 6.18 show a variety of significant changes, often varying with *G*, with the last figure also showing a more unusual lower temperature *E* anomaly in ZrO_2. While the more extreme variations are for greater property decreases, some entail temporary property increases. Some of the variations are special cases, such as that for ZrO_2, and especially for Al_2O_3, with the latter apparently due to the onset of twin nucleation of cracks, giving a significant strength minimum for sapphire crystals at ~ 400°C. This also results in similar minima for polycrystalline bodies, though this appears to diminish and then disappear as *G* decreases, as would be expected. However, also note that similar minima are observed in both the hardness and the compressive strength of Al_2O_3 (Fig. 7.6).

Turning to higher temperature behavior, typically at ≥800°C, three effects should be noted. The first, associated with the more limited cases where glassy grain boundary phases are present in sufficient quantity to allow some increasing plastic deformation as temperature increases and strain rates decrease, results in strength commonly having a sharp maximum followed by a precipitous decrease, commonly at ≤ 1000°C. The second and more general occurrence is for strength decreases > the inherent decrease in Young's modulus with increasing temperature when there is limited or no grain boundary phase, or higher strain rates, to typically 1000–1200°C. Above such temperatures strength decreases accelerate, especially at lower strain rates, often with increasing toughness, due to increasing high-temperature slow crack growth and its transitioning to creep and stress rupture processes. Note that before plastic deformation or SCG are prevalent, strength continues to follow an inverse square root of grain size dependence, but transitions to strength increasing with increased *G* at higher temperatures. Note however that while toughness and strength have less severe differences at higher than at lower temperatures, they still can have significant differences that apparently reflect differing effects of crack velocity and strain rates between strength and various toughness tests.

C. Grain Size Dependence of Other Mechanical Properties and Effects of Grain Shape and Orientation on Mechanical Properties of Monolithic Ceramics

Turning to other mechanical properties at room temperature, hardness decreases as *G* increases at finer *G* via a $G^{-1/2}$ dependence. In some few cases this trend may continue at larger *G*, extrapolating toward lower single crystal values, but in most cases it reaches a minimum value below lower single crystal values and then increases to extrapolate to lower single crystal values (Figs. 4.1-4.4,4.7-4.13, 4.15). The minimum is associated with grain cracking and spalling around indents (Figs.4.16-4.20), which reaches a maximum when the indent and grain sizes are similar, apparently due to extrinsic factors such as grain boundary phases and residual porosity and intrinsic factors of grain mismatch stresses due to TEA and EA. As indent load increases, so does the associated cracking, while the *G* location of the minimum and its hardness value decrease, with these trends more pronounced for Vickers versus Knoop indents. However, such cracking is superimposed on the underlying mechanism of indentation, which is now recognized to be plastic flow, e.g. as shown by the crystal orientation dependence of hardness (Figs. 4.21, 7.1).

Compressive strength shows a strong $G^{-1/2}$ dependence, with strengths at larger *G* extrapolating to lower single crystal values. Compressive strengths from well-conducted tests generally approach *H*/3 (~ *E*/10) as an upper limit, but this is not exact, since the *G* dependence of compressive strength is typically > that of *H*. This correlation of compressive strength and *H* indicates that compressive failure typically involves some microplastic processes, e.g. microcrack nucleation and growth. However, compressive failure typically results from cumulative growth and coalescence of many finer cracks under local tensile stresses, but with growth limited by the macro compressive stress. This is clearly supported by the similar *G* dependence of compressive strength with superimposed hydrostatic pressures, but with somewhat increased strength levels and increased evidence of plastic deformation. The cumulative compressive failure from many sources in a body is also supported by higher Weibull moduli for compressive versus tensile failure. Such failure also implies less extreme dependence on larger grains than tensile failure, but no study has been made of this.

The one reasonably comprehensive study of grain size effects on the ballistic performance of Al_2O_3 bodies against lower velocity, smaller (.22) caliber projectiles shows a limited but definite decrease in stopping power as $G^{-1/2}$ decreases (i.e. *G* increases, Fig. 5.7), including sapphire data (i.e. $G = \infty$). Similarly, .30 caliber armor piercing (AP) tests of AlN show a similar *G* dependence of ballistic stopping power, but much less for .50 caliber AP tests (Fig. 5.8). Less comprehensive studies of the *G* dependence of ceramics against higher velocity and caliber projectiles do not indicate any clear dependence of ballistic performance on *G*. It may be a factor, but it is probably in competition with other body factors. Thus limited data indicates a *G* dependence of ballistic performance of ceramics at lower threat levels, but with this decreasing as the threat level increases.

Turning to wear, erosion, and related behavior, there is less data on their grain size dependences, and greater complexity is expected because of variations in the mechanisms resulting from the diversity of conditions for varying aspects of such behavior. However, there are broad trends for increased resistance to wear, etc., as G decreases, commonly as a function of $G^{-1/2}$, i.e consistent with H dependence on G, especially at finer G. Further, there are some results that show marked decreases in wear resistance above certain (apparently) material-dependent grain sizes, e.g. consistent with probable contributions of grain mismatch stresses, wear asperity or eroding particle indentation or both, to local cracking. Limited direct observations show that at least some wear rates are significantly increased by larger grains within the body, but in view of limited overall G effect studies, there has been little study of the effects of grain size distribution on wear.

The effects of grain shape and orientation are much less documented than the effects of grain size. Elongated grain shape can clearly increase crack bridging in the wake region and thus increase large crack toughness values. However this occurs primarily when crack propagation remains intergranular around the elongated grains, i.e. being limited when transgranular fracture of the elongated grains occurs. Grain boundary phases that enhance intergranular fracture can extend the toughening benefits of elongated grains with large cracks, but often at the expense of strength. Elongated grains are larger than equiaxed grains of the same diameter and thus should lower compressive and especially tensile strengths based on size differences, which may be exacerbated by enhanced microcracking from elongated grains. Similarly, it is expected that elongated grains will commonly reduce wear, erosion, etc., resistance due to enhanced fracture and grain pullout. Limited grain orientation data supports the expectation that mechanical properties of polycrystalline bodies will vary with preferred grain orientation in proportion to the orientation dependence of properties in the corresponding single crystals.

D. Particle and Grain Size Dependence of Crack Propagation, Toughness, and Tensile Strength of Ceramic Composites

Elastic moduli of composites do not depend on particle size, D, unless D is large enough to trigger the onset of microcracking, e.g. as estimated by Eq. (2.4). However, elastic moduli clearly depend on the volume fraction added phase and its elastic properties (and the onset, extent, or both of microcracking can be aided by increasing volume fraction second phase). Generally a rule of mixtures estimate of elastic moduli is used, i.e. per Eq. (8.1), since it often gives reasonable results and characterization for selecting and accurately using more sophisticated models, and their parameters are almost universally lacking. However, there can be significant deviations, for which suitable models may or may not be available, e.g. again spontaneous microcracking clearly occurs at finer particle sizes, more extensively, or both as volume fraction of second phase increases, which is not predicted by present models. Further, the extent to which this reflects extrinsic effects due to second phase agglomerates or intrinsic effects due to interaction of mismatch stresses between nearby particles is unknown.

SCG occurs in composites with oxide matrices susceptible to SCG, and also in at least some nonoxide composites with oxide containing grain boundary phases. However, the dispersed phase often inhibits SCG, raises toughness, or both to reduce effects of SCG, but again the issue of crack size effects, i.e. the extent to which this also occurs with normal strength controlling cracks, is generally unknown.

Though not documented as extensively or in detail, it is clear that the microstructural dependence of mechanical properties of ceramic composites closely parallels that of monolithic ceramics, especially for synthetic ceramic particulate composites, with major deviations primarily with ceramic fiber composites. Thus fracture toughness of ceramic composites at and near room temperature also varies substantially with different tests and exhibits substantial *R*-curve effects, especially as the volume fraction of dispersed phase and the coarseness of it and the matrix grain size increase. Again the primary correlation with significant increases of toughness is crack wake bridging by larger dispersed particles. However, this in part reflects general neglect of other possible mechanisms such as those discussed in Chap. 8, Sec. II.B, since the discovery of, and subsequent focus on, crack wake bridging (with the exception of enhanced effects from clustered microcracks, Fig. 8.2, for thermal shock resistance, as is discussed below).

The same issues arise regarding the meaning and applicability of large crack toughness and related wake effects, especially crack bridging, for composites as for monolithic ceramics. Thus observations being based on large cracks intersecting machined surfaces after propagation at generally low, unspecified velocities, followed by arrest, versus typically much smaller strength controlling cracks propagating more into the bulk of the body at accelerating velocity to failure, are again key issues. These concerns are heightened by the two cases where higher crack velocity effects on toughness have been considered in composites; these indicate reduced effectiveness of the composites on fracture toughness of metal particles in a glass matrix [36,37] and WC-Co [38]. Similarly, fracture mode argues against extensive effects of crack wake bridging effects on strength, since the latter is favored by intergranular and interparticle fracture, while most composites tend to enhance transgranular and transparticle fractures. Also, while not quantitatively studied, qualitative observations of fracture mirror, mist, hackle, and crack branching patterns around fracture origins in strength testing of composites question significant increases in toughness during crack propagation to failure. Fracture mirror and related data should show, but also do not support, effects of bridging on strength. Thus while effects of mismatch stresses can decrease mirror sizes in the smaller mirror size range (and in fact control spontaneous failure in a crystallized glass under high energy pulse loading [34,39]), no effects of increased mirror sizes due to *R*-curve effects have been shown. However, as with monolithic ceramics, the basic differences between strength and large crack toughness behavior as a function of microstructure are the most detailed and compelling arguments that crack wake bridging generally has little or no effect on strength.

While many studies have not documented effects of particle or grain size on properties, the considerable number that have all show strengths decreasing as particle or matrix grain size, or both individually, increase, i.e. opposite to toughness dependence (e.g. Figs. 9.13-9.15). Further, limited (and apparently not previously considered) available data covering a sufficient particle size range shows a $D^{-1/2}$ strength dependence for machined samples (Figs. 9.2, 9.15,9.16) the same as for G dependence of monolithic ceramics. Again the mechanisms and specifics of the particle size dependence are also the same as for the G dependence of strength, i.e. as a first approximation flaw sizes do not vary widely with the microstructure of a given composite composition. At finer particle sizes, machining flaws are > the particle (and matrix grain) size, so there is limited effect of particle (or grain) size on strength, resulting in a finer particle size strength branch directly analogous to the finer G branch for monolithic ceramics. As particle size further increases, a size will be reached where machining flaws causing failure are contained in or around individual particles, so at and beyond this particle size the particles become the flaws causing failure (again recalling that flaw sizes are measured by a radius and particle and grain sizes by a diameter). Thus strength decreases more rapidly ~ as $D^{-1/2}$, forming the larger particle branch of the strength–$D^{-1/2}$ behavior.

Similar to the strength–$G^{-1/2}$ behavior of monolithic ceramics, the strength–$D^{-1/2}$ dependence of composites is not necessarily zero along the finer particle strength branch for at least two reasons. First, limited effects of increasing particle (and matrix grain) size can reduce local properties, e.g. hardness and toughness, impacting machining flaw initiation and growth, thus modestly increasing flaw sizes, yielding modest decreases in strength. Second, mismatch stresses from expansion and elastic differences between the particles and matrix could increasingly contribute to failures as flaw sizes increase, approaching larger D sizes, resulting in some strength decreases, which might often be greater than for monolithic ceramics. Besides the direct data showing these trends, their direct correlation with strength–$G^{-1/2}$ effects and mechanisms, this mechanism is also supported by lower strengths generally found for platelet composites, since platelets have larger, ~ flat interfaces along which machining flaws can form. This is further supported by the limited fractography of platelet composites showing failure frequently occurring from such platelet surfaces. However, it is also important to note that other mechanisms can be operative in composites in addition to or instead of machining flaw–particle size interactions. Thus microcracking can clearly occur, as is shown by data with larger or more particles or platelets, or both, as differences in elastic, and especially expansion, properties increase.

Mechanical fatigue under cyclic tensile loading, though examined very little, has been demonstrated in some particulate composites [40]. This is similar to effects shown in monolithic ceramics, again with microstructural mismatch stresses seen as a basic mechanism. Such studies have also shown different ratios of crack propagation at and near machined surfaces from those in the interiors of specimens.

Consider now thermal shock behavior of composites as a transition to higher temperature behavior. Ceramic composites may offer some increases in ΔT_C, but some offer more significant improvements in strength retained on quenching at $\geq\Delta T_C$, e.g. consistent with higher toughnesses at larger crack sizes. However, this typically entails more modest starting strengths, with microcracking being a particularly important mechanism, e.g. with composites based on dual microcrack populations, Figure 8.2C, being a good example. A further limitation is that composites showing improved thermal shock resistance often exhibit thermal fatigue effects, that is, a progressive decrease in thermal shock performance with repeated thermal shock cycling (Fig. 11.7). While composites can still result in some improvements in thermal shock resistance, it is important to note that the improvements may be substantially less than seen from a single thermal shock.

Limited tests of crystallized glasses all show both fracture toughness and tensile strength initially decreasing as temperature increases above room temperature, reaching a minimum, e.g. at ~ 400°C (Fig. 11.1, 11.9). Decreases to minima have been attributed to decreased microcracking as increased temperature reduces mismatch stresses between the grains themselves and the residual glass matrix, which is clearly a candidate mechanism. However, there is no corroboration of this, and the reason for similar trends of toughness and strength, which are often opposite for microcracking, are uncertain, though this may reflect differences in the microcracking in such systems, e.g. their probable finer scale. Beyond the toughness and strength minima, both increase as temperature further increases (Fig. 11.1, 11.9). Strengths must subsequently go through maxima, beyond which they decrease extensively due to increased plastic flow. Whether increasing toughnesses go through similar maxima is not documented, but must occur. Some maxima may be different, e.g. extended to higher temperatures due to plastic deformation, which though varying with strain rate, often is accompanied by decreasing strength.

One other set of intermediate temperature tests is on PSZ single crystals, which show initial decreases in both toughness and strength consistent with expected decreases in transformation toughening; but this decrease reaches a minimum at about twice the toughnesses and strength of CZ crystals. Such remaining higher toughness and strength of PSZ over CZ crystals continues to above the monoclinic–tetragonal transformation temperature, which is attributed to toughening from nontransformation mechanisms from the tetragonal precipitates, such as crack deflection and branching. Such toughening and strengthening arise from mismatches in tetragonal versus cubic ZrO_2 expansion and elastic differences, which exist whether the tetragonal precipitates are metastable or fully stable. At higher temperatures, ceramic composites of all crystalline constituents often show increasing toughness, which is attributed to crack tip blunting due to SCG and creep processes, which are quite strain rate dependent and generally do not reflect higher strengths; in fact they usually correlate with decreasing strengths.

Polycrystalline composites of all refractory crystalline ceramic constituents show accelerating decreases in strength at higher temperatures, similar to refractory monolithic ceramics, but often with less relative decrease. The rates of strength decrease are lower to some extent as the refractoriness of the phases present and their volume fractions increase, and more so with differing particle shape, with whiskers typically giving less rapid decreases. However, some of the ceramic composites retaining strengths to higher temperatures are directionally solidified eutectic (Fig. 11.11,11.12).

E. Particle and Grain Size Dependence of Other Mechanical Properties and Effects of Particle Shape and Orientation on All Mechanical Properties of Composite Ceramics

Hardness data for composites is most extensive for effects of volume fraction of dispersed second phase particles, with much more limited data on effects of particle (or matrix grain) size, and even less on particle (and grain) shape and orientation. Though there are variations, most hardness data is fairly consistent with a rule of mixtures effect of the volume fraction second phase [Eq. (10.1)]. An alternate model proposed for crystallized glasses does not deviate significantly from this, but modeling and data evaluation for WC-Co bodies shows modification of the rule of mixtures via weightings for the contiguity of each phase. Though limited evaluation of grain and particle size effects has been made in only part of the studies, collectively they clearly show hardness increases as either or both sizes decrease, as is expected from the finer G dependence of monolithic ceramics. This reflects a common advantage of composites via their common mutual inhibition of grain and particle growth limiting particle and grain sizes. These trends are consistent with measurements of directionally solidified ceramic composites showing hardness varying as $\lambda^{-1/2}$, where λ is the mean separation between the single crystal rods and is thus also analogous to the $G^{-1/2}$ dependence for monolithic ceramics (Chap. 10, Sec. III.A).

Very few evaluations of effects of particle parameters on compressive strength, ballistic performance, or wear have been made. One set of tests on crystallized glasses at room temperature showed a substantial range of compressive strengths, including some fairly substantial strengths to ~ 100 MPa (up to 150% of the parent glass, Figure 10.2) at 22°C, with varying rates and types of decreases as test temperatures increased (Fig. 11.13). Similarly a limited study showed compressive strengths of up to 500+ MPa in a synthetic composite of SiC platelets in a glass matrix at 22°C, with marked decreases as test temperatures increased, but with composite strengths still well above those of the matrix glass alone. Composites of laminar particles of refractory metal and ceramic layers have shown respectable strengths, e.g. > 1 GPa can be obtained, which is probably due to the finer particle (and grain) size. However such strengths were also a function of the laminar particle dimensions, showing again that other microstructural dimensions can play a role in mechanical properties.

One limited evaluation showed that compressive fatigue crack propagation occurred in a ceramic whisker composite, but not always unfavorably relative to the matrix alone. Limited testing has shown that the ballistic stopping power of a dense, fine particle (and grain) size composite approached performance levels of established monolithic ceramics for armor.

More extensive, but still limited, wear testing of similar composites clearly shows that they can be competitive to established wear resistant ceramics. This is corroborated by some commercial production and use of ceramic composites for wear and related applications, e.g. Al_2O_3-TiC for cutting tools and various wear applications and Al_2O_3-SiC whisker composites for cutting tools and critical wear components for dies for deep drawing of beverage and other cans with an integral bottom.

Even less testing of the above properties has been made at elevated temperature. Testing of a set of Li_2O-SiO_2 crystallized glasses showed net decreases in strength to the limit of testing (700°C, Fig. 11.13) as was expected. However there was a diversity of behavior at more modest temperatures, including widely varying rates of strength decrease with increasing temperature and some modest, temporary strength rises to maxima, some modest, temporary minima, or both. These trends suggest a fairly diverse range of effects, probably due to mismatch stresses.

Little or no data exists on effects of particle or grain shape or orientation on properties of composites addressed in this section. However, while investigation is limited, results clearly show that composites can have good hardness, respectable compressive strengths and ballistic performance, and good wear resistance, all primarily associated with finer microstructures commonly achievable in composites. Thus further investigation and development should be fruitful.

F. Mechanisms Controlling Normal Brittle Tensile Strength Failure of Monolithic and Composite Ceramics

Given the common assumption that crack bridging, related *R*-curve effects, and resultant increased toughness with larger scale crack propagation determined tensile strengths and reliability of materials exhibiting such behavior of monolithic and many composite ceramics, it is useful to review the combined evidence for monolithic and composite ceramics on this and the alternate mechanism. As noted in earlier chapters, many investigators conducted simple, especially indentation, toughness tests and showed or noted observations of crack bridging in their monolithic, and especially composite, ceramics, assuming that these arrested crack observations showed behavior across a range of crack velocities and sizes. However, extensive evidence shows that such effects are far more limited in their control of strengths and reliability than was previously thought, applying primarily to strengths retained after serious damage from thermal stress or shock or mechanical impact, i.e. when cracks on the scale of those showing bridging-*R*-curve effects are developed in components or test specimens. Before proceeding to this summary, it is important to note that two related factors played a major role in the over emphasis of the role of these effects, which are an important component of crack propagation in brittle materials, in determining tensile strength and reliability of some monolithic and many composite ceramics. First was a frequent and often extensive neglect or rejection of the literature, e.g. the rediscovery (and naming) of bridging phenomena, i.e. neglecting earlier observations of bridging in both ceramics and rocks, and the rejection of fractography as a viable tool for corroborating and understanding observations. Second was a narrow range of testing, i.e. limiting or precluding opportunities for testing the self-consistency of data, and interpretation and concepts, with both being important added motivations for this book.

Turning to the summary of evidence seriously questioning, or contrary to, the application of bridging and related, typically large crack phenomena, to normal small-crack strength behavior of monolithic and composite ceramics, major factors are summarized in Table 12.1 for brittle fracture, e.g. at nominally 22°C, under the two headings of test issues and strength/toughness behavior. The most significant and extensive evidence of major discrepancies are those of the microstructural dependences of toughness and strength and of the inconsistencies between strengths and toughness often encountered at high toughness where crack bridging, branching and *R*-curve effects are most substantial. However, the test issues are important, e.g. starting with the observation that toughness values derived from fractography of failed components or test specimens should be, and are generally found to be, most consistent with strength [33,34]. Further, effects of surface finishing and

TABLE 12.1 Summary of Factors Questioning or Contrary to the Application of Large-Scale Crack Bridging and *R*-Curve Effects to Normal Small-Crack Strength Behavior of Dense Monolithic and Composite Ceramics

(A) Test Issues	
Crack/microstructure scale effects	Crack scales in most crack propagation/toughness tests are typically far larger on an absolute scale, and especially relative to that of the microstructure, often allowing effects to occur in such tests that occur much less or not at all with the much smaller cracks normally controlling strength behavior.
Larger grain/particle effects on toughness versus strength	Arbitrary cracks introduced for most crack propagation/ toughness tests will show greatest benefits of large limiting flaws, microstructural features that are often strength limiting flaws, especially of large grains or particles and clusters of these.
Surface finish toughness test effects	Bridging and related observations have been extensively reported without considering effects of the typically machined surfaces, e.g. of machining flaws and surface stresses, despite evidence that as-fired surfaces show less bridging (Fig. 2.4D) and that propagation of cracks along machined surfaces may be quite different from that in the bulk [40].
Crack velocity effects	Bridging and related, e.g. branching, observations are typically made on cracks propagated at low, unmeasured velocities, and then arrested, neglecting possible significant changes of higher crack velocities of strength controlling cracks as they accelerate to failure, e.g. as indicated in single crystals and composites.
(B) Strength/Toughness Behavior	
Microstructural dependences	Toughness, especially for many noncubic monolithic and many composite ceramics, i.e. those that are the main source of bridging and *R*-curve effects, commonly shows significant dependences on grain or particle sizes (or both for composites), e.g. maxima, that are contrary to strengths universally decreasing with increased grain or particle sizes (or both for composites).
Strength versus toughness	In addition to the above basic discrepancies of their microstructural dependences, strengths typically progressively deviate below those expected from toughness of the same bodies as toughness values increase to high levels for a given material.
Fracture mirror behavior	While significant *R*-curve effects impacting strengths should increase fracture mirror sizes (which are related to the toughness controlling failure), there is no evidence showing this, but evidence to the contrary for the one specific data set available for Al_2O_3 bodies.
Reliability/ variation	While *R*-curve and related effects were expected to reduce strength variations, i.e increase reliability by making strengths less dependent on initial flaw variations (and presumably less variation in such large crack toughness values), reliability has generally not increased in bodies with *R*-curve effects over those without such effects, and variation/ scatter in toughness results are often similar to that of strength.

crack velocity on bridging and related effects are a fruitful area for further research. Again, fractography is important, e.g. in this case via effects on mirror and related dimensions.

Another major reason for not accepting large crack toughness results reflecting *R*-curve and related effects for predicting most strength behavior is the evidence and results for the alternative mechanism extensively addressed in this book. This is based on modest size flaws introduced, especially from machining, that control most strengths, as is shown by extensive fractographic studies. Extensive studies also show that machining controls strengths of monolithic ceramics via effects of both the abrasive size and its direction in machining relative to subsequent stressing to measure strength [33]. Though there has been less such study for ceramic composites (presumably because of the assumption that the real control of strengths was via toughness as measured by normal large crack tests), more limited studies clearly show similar machining dependence of composite to monolithic ceramics. These observations are consistent with, and explain, the long established strength–$G^{-1/2}$ behavior of monolithic ceramics, as well as the essentially identical strength–$D^{-1/2}$ behavior shown for ceramic composites in this book. In both cases the thesis is that machining flaws introduced depend not only on the machining conditions (i.e. abrasive, machining depth of cut, speed, etc.) but also on material parameters, specifically, Young's modulus, hardness, and toughness of the material locally around induced flaws, (e.g. per Eqs. (3.2) and (5.2).

Three factors show that such machining-induced flaws are consistent with the microstructural dependence of tensile strength of monolithic and composite ceramics and the generally limited or no effect of large crack *R*-curve effects on tensile strength. First, such flaws are of modest size, e.g. commonly from ~10 μm to a few tens of microns, reflecting the transient indentation of abrasive particles and associated crack generation, as well as probable local elevated temperatures and some residual surface compressive stresses. Small, rather than large, crack toughness values should correlate with the formation of such flaws. Thus propagation of such flaws to determine strength is also likely to be controlled by toughness values for small to modest crack sizes, hence explaining the general lack of *R*-curve effects and thus frequent opposite trends of large crack toughnesses and strengths. Second, the formation of such machining flaws is consistent with the microstructural dependence of tensile strength due to second-order effects of microstructural dependence of flaw size *c* on body properties per Eq. (3.2), especially due to effects of hardness (*H*). Thus since *c* varies inversely with *H*, and *H* varies inversely with *D* and *G*, *c* increases as *D* and *G* increase so that tensile strength decreases with increasing *D* and *G*, as is broadly observed for monolithic and composite ceramics (the former only having a *G* dependence). The size of machining-induced flaws also varies directly with *E* and inversely with *K*, but *E* is not basically dependent on D or G, and small crack toughness values typically have limited microstructural dependence, especially on D and G. Thus even though the dependences of c on E and K are to different powers, their net microstructural dependence is typically limited. Such machining-induced flaw formation also explains the generally poor strengths of platelet composites, since the large platelet–matrix interfaces commonly provide favorable locations for flaw formation and subsequent propagation to failure. Third, there is typically a minimum in c at intermediate volume fractions of dispersed phase that are associated with the typical maximum in toughness (Figs. 9.16 and 9.18, for both smaller and larger cracks, but often more pronounced for the latter). However, this c minimum may be shifted in ϕ value, level, and shape by effects of E and H dependences on ϕ and associated microstructural effects.

Again, when cracks causing failure are larger, e.g. such as those controlling residual strengths after serious thermal stress or impact damage, or from very extensive SCG, R-curve and related effects are probably pertinent to such mechanical behavior. This is the case for continuous ceramic fiber composites where simple, e.g. machining, flaws do not grow continuously, accelerating to failure, thus determining strengths. Behavior of whisker, and probably discontinuous short-fiber, composites may be somewhat in between, though the weight of evidence indicates that whisker composites are much closer in behavior to particulate than continuous fiber composites.

Thus, in summary, there are extensive reasons for discounting most bridging and related R-curve effects in large scale crack toughness measurements of brittle fracture, on normal small crack strengths of monolithic and composite ceramics as summarized in Table 12.1. Though studied much less, there is also evidence that such large scale crack effects often also occur in some porous bodies, mainly at intermediate porosity levels, but that this again has at best limited strength effects, mainly with large cracks, e.g. from serious thermal shock or impact damage, not for normal strength behavior [41,42]. Changes to plastic deformation, e.g. by slip in single crystals and grain boundary sliding, in monolithic ceramics change toughness and strength behavior, reducing differences between large and small crack behavior in some cases, but introducing other differences, probably reflecting effects of crack velocity and strain rate effects. Though data is not as extensive for elevated temperature effects on ceramic composites, they show similar changes and differences as for monolithic ceramics. Thus while there are differences between lower and higher temperature behaviors in different ceramic systems, all raise similar issues of differences between large and small scale crack behavior whether due entirely to brittle fracture processes or to processes involving some plastic deformation, making both areas for further study.

IV. NEEDS AND OPPORTUNITIES TO UNDERSTAND AND BETTER USE THE GRAIN AND PARTICLE DEPENDENCE OF MECHANICAL PROPERTIES OF CERAMICS AND CERAMIC COMPOSITES

A. Testing and Evaluation Needs

There are a variety of needs to improve understanding of the microstructural dependence of properties of both monolithic and composite ceramics, and thus their performance. Clearly a basic one is better microstructural characterization using both direct and indirect methods, the latter via property measurements. More direct documentation of grain and particle character is needed. Even approximate or average values are often not given, nor is the origin of many given values (especially what factor was used to convert linear intercepts to "true" sizes). Almost no data addresses size distributions, less exists for qualitative and none for quantitative characterization of shape, and only limited data exists on grain or particle orientation, reflecting additional needs. Improvements in these measurements should be accompanied by better descriptions of specimen fabrication methods and parameters, since these indicate microstructural characteristics, and the use of such indirect characterization increases as its relation to specific microstructural measurements increases. Modern computerized stereology measurements are an important aid in better characterization.

An even more critical need is for a much wider range of property measurements; determining the extent of microcracking is an important goal. While this is needed better to document properties central to defining mechanisms, it is also an important factor in indirect microstructural characterization. Thus, for example, rather than assuming isotropic bodies and properties in many cases where some anisotropy may exist, determining the degree of isotropy of properties can be valuable. However, the most critical need is for much more comprehensive property measurement. Thus failure to measure elastic moduli along with toughness or strength, and to measure only one of the latter two, are serious constraints on understanding both the properties and the mechanisms. Preferably both static and wave methods should be used to determine at least E, since the former, while often less accurate for absolute values, is an important indicator of microcracking, while the latter methods typically are not but are often better for absolute moduli (before any microcracking). However, use of acoustic emission to detect microcracking is often also valuable, especially in conjunction with other tests such as elastic property changes. Further, given the variability of different toughness tests, it is desirable to measure toughness by at least two or three tests, e.g. indentation, indentation fracture, and possibly a third test. Combinations of other tests, e.g. compressive, hardness, and fatigue tests are also valuable.

There are five other expansions of testing that can be valuable. The first is of other, often nonmechanical, properties, e.g. of thermal or electrical conductivities, especially of composites, since these can often more clearly reveal microstructural factors, especially the onset of percolation, that can also affect mechanical properties, e.g. creep. Second is a broader range of microstructural variables, e.g. of *G*, *D*, or volume fraction second phase. The importance of this is clearly illustrated by the very limited amount of data showing the particle size dependence of tensile strengths (Figs.9.2,9.15,9.16). Third is the closely related aspect of testing composites of the same composition and raw materials, but reflecting different mixing or fabrication methods or both, and hence variations in spatial distributions of microstructures. Fourth, measuring tensile strengths with more than one surface finish and as a function of grinding direction relative to test bar axes can be valuable, e.g. showing the degree of strength dependence on machining flaws. This is particularly important for composites, given the importance of machining flaws in their mechanical behavior and the limited characterization and evaluation of machining effects on their behavior. Fifth is a broader range of temperature testing, especially testing of mechanical properties at modest temperatures, instead of assuming (falsely in some important cases) that properties do not change significantly at modest temperatures. While this is particularly true for tensile strength (Figs. 6.12,6,14,6.15,6.18), it can also apply to other mechanical properties such as elastic moduli (Fig. 6. 18), hardness, and compressive strength (Fig. 7.6).

B. Fabrication and Processing Opportunities to Improve Ceramics

Briefly consider some broader aspects of fabrication on the development and understanding of the mechanical behavior of ceramics, especially composites. While conventional consolidation of powders is the dominant method of fabrication and will continue to be so, variations on this as well as other fabrication technologies offer opportunities for producing novel bodies, especially composite ones, for study as well as possible production. One such step is extension to finer powders and resultant microstructures, especially on a nanoscale. Thus such bodies can extend properties such as strengths and hardnesses to higher levels provided that contaminants on the very high surface area powders can be adequately removed during consolidation while maintaining desired fine microstructures. This has often not been the case, especially for nominally single phase bodies such as MgF_2 (Fig. 3.24) and TiO_2 (Fig.4.5), but it has more commonly been achieved in composites, since higher processing temperatures can better drive off adsorbed species, while the composite structure results in mutual restriction of grain and particle growth. However, the large compaction ratios of most, if not all, very fine powders provide substantial production challenges, which may be reduced or circumvented in some cases by alternant fabrication and processing techniques.

One alternate technique is reaction processing [43,44], such as is used for fabricating mullite-ZrO_2 and various oxide–nonoxide composites. While this is more limited in the compositions to which it is applicable, it is useful for a substantial range of compositions and can offer one to three possible important advantages. The first is that in principle it can produce finer grain and particle sizes for given starting particle sizes, since the resultant body microstructuure is generally formed entirely by nucleation and growth of new phases from the reaction, which if controlled should lead to finer, more uniform microstructures. A key identified need to achieve this is to prevent or limit the extent of transient liquid phase formation, since this greatly increases grain or particle growth, often in a heterogeneous fashion. The second potential advantage is that greater microstructural uniformity can be achieved and that this can translate into greater mechanical reliability . The third potential advantage is that the raw material costs for reactants is often substantially lower than for the resultant product phases (Table 10.1), thus making reaction processed composites somewhat more economical. Some systems may also allow more effective consolidation that can also lower costs.

There are two other alternate fabrication routes, namely melt processing and CVD, that, though they are also more restricted in compositions, can offer similar advantages. Thus PSZ polycrystals and especially single crystals as well as ZTA eutectic bodies are examples of promising composites that have been made with encouraging results via melt processing. While thermal stresses and pores from intrinsic liquid–solid density differences and extrinsically from exsolved gases are a serious challenge, various conventional and novel solidification methods offer important opportunities [43–49]. Conventional directional solidification as for PSZ crystals from skull melting and single crystal eutectics are key examples of important successes, including some of the best high temperature strengths (Figs. 11.11,11.12). Similarly, though probably less recognized, there are important opportunities for preparation of fine uniform microstructures, again especially for composites via CVD [46,47]. Thus codeposition of different compositions have been explored with some promising results, e.g. the formation of TiN precipitates from ~ 3 to 15 nm in dimensions in a Si_3N_4 matrix [47].

V. SUMMARY

This book has extensively reviewed and discussed the dependence of primarily mechanical properties of dense monolithic ceramics and ceramic composites on respectively grain and particle parameters, mainly size, but also to the extent feasible on shape and orientation. Such information, though often not adequately addressed in studying the mechanisms of mechanical behavior, is an essential component of knowledge of the microstructural dependence of properties that is necessary to understanding and improving their performance.

The other key microstructural component is porosity, which has been addressed in a companion book [42].

These books show that microstructural understanding, and ultimately design and control of microstructure, is key to good materials. However, this has to be intimately coupled with fabrication and processing technology, which still holds great promise for further development and invention to produce novel microstructures. This book addresses many of these opportunities, a major component of which is composites, but the reader is also referred to some of the earlier discussions of opportunities [43–46] and more recent ones [48,49].

REFERENCES

1. S. Kumar and R. N. Singh. Thermal Conductivity of Polycrystalline Materials. J. Am. Cer. Soc. 78(3):728–736, 1995.
2. G. A. Slack and S. B. Austerman. Thermal Conductivity of BeO Single Crystals. J. Appl. Phys. 42(12):4713–4717, 1971.
3. A. S. Gray and C. Uher. Thermal Conductivity of Mica at Low Temperature. J. Mat. Sci. 12:959–965, 1977.
4. R. K. Williams, R. S. Graves, M. A. Janney, T. N. Tiegs, and D. W. Yarbrough. The Effects of Cr_2O_3 and Fe_2O_3 Additions on the Thermal Conductivity of Al_2O_3. J. Apl. Phys. 61(10):4894–4901, 1987.
5. F. G. Will, H. G. DeLorenzi, and K. H. Janora. Effect of Crystal Orientation on Conductivity and Electron Mobility in Single-Crystal Alumina. J. Am. Cer. Soc. 75(10):2790–2794, 1992.
6. Prof. B. Dunn, UCLA, private communication, 1998.
7. B. Dunn, G. C. Farrington, and J. O. Thomas. Frontiers in β''-Alumina Research. MRS Bull. 14(9):22–30, 1989.
8. A. V. Virkar, G. R. Miller, and R. S. Gordon. Resistivity-Microstructure Relations in Lithia-Stabilized Polycrystalline β''-Alumina. J. Am. Cer. Soc. 61(5-6):250–252, 1978.
9. S. J. Penn, N. McN. Alford, A. Templeton, X. Wang, M. Reece, and K. Schrapel. Effect of Porosity and Grain Size on the Microwave Dielectric Properties of Sintered Alumina. J. Am. Cer. Soc. 80(7):1885–1888, 1997.
10. P. Hansen, D. Hennings, and H. Schreinemacher. High-K Dielectric Ceramics from Donor/Acceptor-Codoped $(Ba_{1-x}Ca_x)(Ti_{1-y}Zr_y)O_3$ (BCTZ). J. Am. Cer. Soc. 81(5):1269–1273, 1998.
11. C.-C. Hsueh, M. L. Mecartney, W. B. Harrison, M. R. B. Hanson, and B. G. Koepke. Microstructure and Electrical Properties of Fast-Fired Lead Zirconate-Titanate Ceramics. J. Mat. Sci. Lett. 8:1209–1216, 1989.
12. K. Okazaki and K. Nagata. Effects of Grain Size and Porosity on Electrical and Optical Properties of PLZT Ceramics. J. Am. Cer. Soc. 56(2):82–86, 1973.
13. H. Igarashi and K. Okazaki. Effects of Porosity and Grain Size on the Magnetic Properties of NiZn Ferrite. J. Am. Cer. Soc. 60(1-2):51–54, 1977.

14. E. K. Beauchamp. Effect of Microstructure on Pulse Electrical Strength of MgO. J. Am. Cer. Soc. 54(10):484–487, 1971.
15. T. Tunkasiri and G. Rujijanagul. Dielectric Strength of Fine Grained Barium Titanate Ceramics. J. Mat. Sci. Lett. 15:767–769, 1996.
16. D. F. K. Hennings, R. Hartung, and P. J. L. Reijnen. Grain Size Control in Low-Voltage Varistors. J. Am. Cer. Soc. 72(3):645–648, 1990.
17. R. S. Raghava. Thermal Expansion of Organic and Inorganic Matrix Composites:A Review of Theoretical and Experimental Studies. Polymer. Comp. 9(1):1–11, 1988.
18. W. D. Kingery. Introduction to Ceramics. John Wiley, New York, 1960, p. 479.
19. H. A. Bruck and B. H. Rabin. An Evaluation of Rule-of-Mixtures Predictions of Thermal Expansion in Powder Processed Ni-Al_2O_3 Composites. J. Am. Cer. Soc. 82(10):2927–2930, 1999.
20. R. C. Progelhof, J. L. Throne, and R. R. Ruetsch. Methods for Predicting the Thermal Conductivity of Composite Systems:A Review. Polymer Eng. Sci. 16(9):615–625, 1976.
21. G. A. Slack. Thermal Conductivity of Pure and Impure Silicon, Silicon Carbide, and Diamond. J. Appl. Phys. 35(12):3460–3466, 1964.
22. M. P. Borom, G. A. Slack, and J. W. Szymaszer. Thermal Conductivity of Commercial Aluminum Nitride. Am Cer. Soc. Bull. 51(11):852–856, 1972.
23. D. P. H. Hasselman, K. Y. Donaldson, J. Liu, L. J. Gauckler, and P. D. Ownby. Thermal Conductivity of a Particulate-Diamond-Reinforced Cordierite Matrix Composite. J. Am. Cer. Soc. 77(7):1757–1760, 1994.
24. D. P. H. Hasselman and K. Y. Donaldson. Effect of Reinforcement Particle Size on the Thermal Conductivity of a Particulate-Silicon Carbide-Reinforced Alumina Matrix Composite. J. Am. Cer. Soc. 75(11):3137–3140, 1992
25. F. C. Chen, C. L. Choy, and K. Young. A Theory of the Thermal Conductivity of Composite Materials. J. Phys. D, Appl. Phys. 10:571–586, 1977.
26. L. E. Nielsen. The Thermal and Electrical Conductivity of Two-Phase Systems. Ind. Eng. Chem. Fundam. 13(1):17–20, 1974.
27. D. S. McLachlan, M. Blaszkiewicz, and R. E. Newnham. Electrical Resistivity of Composites. J. Am. Cer. Soc. 73(8):2187–2203, 1990.
28. T. Ota, I. Yamai, J. Takahashi, R. E. Newnham, and S. Yoshikawa. Effects of Filler Particle Size on the Electrical Resistance of Conductor-Polymer Composites. Advanced Composite Materials, Cer. Trans. 19 (M. D. Sacks, ed.). Am. Cer. Soc., Westerville,OH, 1991, pp. 381–387.
29. E. Litovsky and M. Shapiro. Gas Pressure and Temperature Dependences of Thermal Conductivity of Porous Ceramic Materials:Part 1, Refractories and Ceramics with Porosity Below 30%. J. Am. Cer. Soc. 75:3425–3429, 1992.
30. E. Litovsky, M. Shapiro, and A. Shavit. Gas Pressure and Temperature Dependences of Thermal Conductivity of Porous Ceramic Materials:Part 2, Refractories and Ceramics with Porosity Exceeding 30%. J. Am. Cer. Soc. 79:1366–1376, 1996.
31. S. Fujitsu, M. Miyama, K. Koumoto, H. Yanagida, and T. Kanazawa. Enhancement of Ionic Conductivity in CaF_2 and BaF_2 by Dispersion of Al_2O_3, J. Mat. Sci. 20:2103–2109, 1985.

32. J. J. Mecholsky Jr., S. W. Freiman, and R. W. Rice. Fracture Surface Analysis of Ceramics. J. Mat. Sci. 11:1310–1319, 1976.
33. R. W. Rice. Machining Flaw Size–Tensile Strength Dependence on Microstructures of Monolithic and Composite Ceramics, to be published.
34. R. W. Rice. Fractographic Determination of K_{IC} and Effects of Microstructural Stresses in Ceramics. Fractography of Glasses and Ceramics. Ceramic Trans. 17 (J. R. Varner and V. D. Frechette, eds.). Am. Cer. Soc., Westerville, OH, 1991, pp. 509–545.
35. H. P. Kirchner and R. M. Gruver. Fracture Mirrors in Alumina Ceramics. Phil. Mag. 27:1433–1446, 1973.
36. T. L. Jessen and D. Lewis III. Effect of Crack Velocity on Crack Resistance in Brittle-Matrix Composites. J. Am. Cer. Soc. 72(5):812–821, 1989.
37. T. L. Jessen and D. Lewis III. Effect of Composite Layering on the Fracture Toughness of Brittle Matrix/Particulate Composites. Composites, 26(1):67–71, 1995.
38. D. B. Marshall, W. L. Morris, B. N. Cox, and M. S. Dadkhan. Toughening Mechanisms in Cemented Carbides. J. Am. Cer. Soc. 73(10):2938–2943, 1990.
39. D. Lewis and J. R. Spann. Fracture Features at Internal Fracture Origins in a Commercial Crystallized Glass. J. Am. Cer. Soc. 65(10):C173–174, 1984.
40. L. Ewart and S. Suresh. Crack Propagation in Ceramics Under Cyclic Loads. J. Mat. Sci. 22:1173–1192, 1987.
41. R. W. Rice. Microstructural Dependence of Fracture Energy and Toughness of Ceramics and Ceramic Composites Versus That of Their Tensile Strengths at 22°C. J. Mat. Sci. 31:4503–4519, 1996.
42. R. W. Rice. Porosity of Ceramics. Marcel Dekker, New York, 1998.
43. R. W. Rice. Processing of Ceramic Composites. Advanced Ceramic Processing and Technology 1 (J. G. P. Binner, ed.). Noyes, Park Ridge, NJ, 1990, pp. 123–213.
44. R. W. Rice. Ceramic Composites: Future Needs and Opportunities. Fiber Reinforced Ceramic Composites, Materials, Processing and Technology (K. S. Mazdiyasni, ed.). Noyes, Park Ridge, 1990, NJ, pp. 451–495.
45. R. W. Rice. Advanced Ceramic Materials and Processes. Design of New Materials (D. L. Cocke and A. Clearfield eds.).Plenum, New York, 1987, pp. 169–194.
46. T. Harai. CVD of Si_3N_4 and Its Codeposites. Emergent Process Methods for High-Technology Ceramics, Materials Science Res. 17. Plenum Press, New York, 1984, pp. 329–345.
47. K. Hiraga, M. Hirabayashi, S. Hiyashi, and T. Hirai. High-Resolution Electron Microscopy of Chemically Vapor-Deposited β-Si_3N_4-TiN Composites. J. Am. Cer. Soc. 66(8):539–542, 1983.
48. A. G. Evans. Perspective on the Development of High-Toughness Ceramics. J. Am. Cer. Soc. 73(2):187–206, 1990.
49. M. P. Harmer, H. M. Chan, and G. A. Miller. Unique Opportunities for Microstructural Engineering with Duplex and Laminar Ceramic Composites. J. Am. Cer. Soc. 75(7):1715–1728, 1992.

Index